Titanium-Based MXenes

This book explores titanium-based MXenes, an emergent class of materials with a broad range of properties and potential applications. It introduces 3D MXene assemblies and compares their properties with such other well-known two-dimensional materials as graphene and transition metal dichalcogenides.

Features:

- Focuses on emerging 2D titanium-based MXene ($Ti_3C_2T_x$ and Ti_2CTx) materials.
- Offers detailed synthetic strategies and scalable approaches.
- Covers defect engineering and physical properties.
- Compares the uniqueness of 2D titanium-based MXene with other 2D-layered materials.
- Discusses 3D macrostructure of titanium-based MXenes and their practical application in photodetectors, catalysis energy storage devices, and biomedicines.

The first comprehensive book on the subject, this work is aimed at researchers in materials and electrical engineering and related disciplines.

Titanium-Based MXenes
From 2D Nanomaterials to 3D Macroassemblies

by Han Zhang, Nasir Mahmood Abbasi,
and Muhammad Tayyab

CRC Press
Taylor & Francis Group
Boca Raton London New York

CRC Press is an imprint of the
Taylor & Francis Group, an **informa** business

Designed cover image: Shutterstock

First edition published 2025
by CRC Press
2385 NW Executive Center Drive, Suite 320, Boca Raton FL 33431

and by CRC Press
4 Park Square, Milton Park, Abingdon, Oxon, OX14 4RN

CRC Press is an imprint of Taylor & Francis Group, LLC

ISBN: 978-1-032-31735-9 (hbk)
ISBN: 978-1-032-33434-9 (pbk)
ISBN: 978-1-003-31965-8 (ebk)

DOI: 10.1201/9781003319658

Typeset in Times
by Apex CoVantage, LLC

Contents

Preface

This book provides a thorough examination of MXenes, an emerging category of two-dimensional transition metal carbides and nitrides, recognized for their versatility and potential across a wide range of applications. Since their initial discovery in 2011, MXenes have gained significant attention for their unique attributes, including metallic conductivity; hydrophilic surfaces; and exceptional mechanical, electrical, and chemical properties. These features have fueled an extensive research into their production methods, structural analysis, and practical uses in diverse scientific and technological fields.

The discussion begins with an in-depth overview of MXenes, detailing their origins and precursors, which include both MAX and non-MAX phases. Various synthesis methods are explored, with a focus on exfoliation processes. Special attention is given to $Ti_3C_2T_x$, one of the most researched MXenes, due to its straightforward synthesis, remarkable characteristics, and suitability for a range of applications, particularly in environmental and biological sensing.

A key focus of this book is the role of defect engineering in optimizing MXene properties for specific applications. It illustrates how controlled introduction and management of structural defects can enhance their mechanical, optical, electrical, and magnetic properties, broadening their application scope. Additionally, this book explores the development of MXene–polymer composites, which exhibit improved performance compared to their individual components and are particularly promising for flexible and wearable electronic devices.

The text also delves into MXene heterostructures, examining their combination with other two-dimensional materials to create advanced systems with synergistic properties. It addresses the challenges associated with these structures, such as stability and surface uniformity, while highlighting their potential in areas like energy storage, catalysis, optoelectronics, and biomedical research. The practical aspects of scaling up MXene production, including deposition techniques, durability, and the design of mixed-dimensional heterostructures, are also discussed.

Finally, this book explores the diverse applications of MXenes, emphasizing their roles in areas such as photodetection, solar energy conversion, and catalysis, showcasing their contributions to advancing technology and sustainable innovation. The text also introduces 3D MXene assemblies and compares their properties with other well-known two-dimensional materials like graphene and transition metal dichalcogenides (TMDCs). By offering a comprehensive analysis of their synthesis, properties, and applications, this book serves as an essential resource for researchers and professionals seeking to explore the transformative potential of MXenes.

Author Biographies

Han Zhang received his B.S. degree from Wuhan University, China, in 2006 and Ph.D. from Nanyang Technological University, Singapore, 2010. He is currently working as distinguished Professor and Director, College of Physics and Optoelectronic Engineering, Shenzhen University, Shenzhen, China. His current research is on the ultrafast and nonlinear photonics of two-dimensional materials. He is currently Director of Shenzhen Key Laboratory of 2D Materials and Devices and Shenzhen Engineering Laboratory of Phosphorene and Optoelectronics, Shenzhen University, China.

Nasir Mahmood Abbasi received his master's degree in chemistry from the University of Karachi, Pakistan, and Ph.D. in chemical engineering, with a focus in nanoscale materials (morphology and self-assembly of nanomaterial), from Zhejiang University, China, in 2016. He worked as a postdoc fellow in the School of Physics and Optoelectronic Engineering at Shenzhen University, China, from June 2019 to August 2021. After that, he worked as a scientist in a collaborative project in the Department of Biomedical Engineering, Nanjing University, (Butch/wang research group) and Research and Development Unit of Nanjing Nuoyuan Medical Devices Company (from October 2021 to October 2024). His research interests include heterostructures and the macroassemblies of nanoscale materials for their applications in energy, healthcare, and electronics.

Muhammad Tayyab obtained his Ph.D. in materials science and engineering, with a specialization in magnetic optoelectronic materials from Nanjing Tech University, China, in 2021, and has continued to expand his research pursuits. After serving as a postdoctoral researcher from October 2021 to September 2023 at Nanjing Tech University, he has extended his expertise to Shenzhen University, China, where he currently holds a postdoctoral position. With a comprehensive background in energy generation, particularly focusing on triboelectric nanogenerators (TENGs) and piezoelectric nanogenerators (PENGs), and batteries as well as diverse interests encompassing sensors and actuators, optoelectronics, flexible electronics, and capacitors, Dr. Tayyab's contributions continue to drive advancements in sustainable energy solutions and advanced materials science. His dual postdoctoral experiences underscore his commitment to interdisciplinary research and collaborative exploration in academia.

1 An Overview of 2D Titanium-Based MXenes

1.1 INTRODUCTION

Two-dimensional (2D) materials, such as transition metal chalcogenides (TMDCs), graphene and its oxides, metal oxides black phosphorus (BP), and some other 2D nanomaterials, possess the properties to be attractive for their usage in emerging applications due to their layered structures (Tyagi et al. 2020).

Superior volumetric surface area provides improved catalysis and sensor response, as well as improved optic, electrical, and magnetic properties, which are significant advantages over macroscale materials in potential and biomedical applications (Piscitelli et al. 2017). Recently, in 2011, 2D transition metal carbides, nitrides, or carbonitrides were added for the first time to the family of 2D nanomaterials by Prof. Gogotsi and colleagues (Naguib et al. 2011).

Due to its exceptional physical and chemical properties, a particular type of 2D material known as MXene has garnered considerable interest as a potential remedy for metallic nanoparticle issues. Due to its ultrathin structure and substantial surface area, MXene may enable the formation of sizable, concentrated active sites that can interact with molecules. Ti_3C_2Tx MXene's layer-dependent optical properties, high metallic conductivities (6,000–8,000 S/cm), and hydrophilic surfaces offer researchers a lot of creative freedom when it comes to creating and detecting plasmonic and electronic-based sensor. MXene has demonstrated multipotency in addition to being an excellent drug for biomedical sensing, making it a strong candidate to be a crucial tool in the developing field of biomedical sensing (Naguib et al. 2011).

The MAX phases, which serve as the foundation for MXene, exhibit strong interactions through bonding while avoiding their cleavage by shearing or other mechanical methods. This distinguishes MXene from other 2D materials (Alhabeb et al. 2017, Anasori, Lukatskaya, and Gogotsi 2017; Sun et al. 2017; Ronchi, Arantes, and Santos 2019). As a result, the chemical exfoliation method enables the exfoliation of 2D structures (MXenes) from independently attached solids (MAX phases) for the first time (Naguib et al. 2012; Li et al. 2015). The MAX phase (mostly elements of group III A or IV A) sheets, which consist of a library of more than 70-membered family of hexagonal layered including the early transition metal nitrides and carbides, are typically the precursors from which the MXenes are produced (Anasori, Lukatskaya, and Gogotsi 2017). A number of carbides MXenes, including Ti_2C, Ti_3C_2, Nb_2C, V_2C, Mo_2C, Ta_4C_3, Zr_3C_2, and $(Mo_2Ti)C_3$, have been successfully synthesized to date (Naguib and Gogotsi 2015; Naguib et al. 2012, 2013; Anasori et al. 2015; Tao et al. 2017; Zhou et al. 2016). However, there are only a small number of reports on nitride-based MXenes such as Ti_4N_3Tx (Urbankowski et al. 2016).

DOI: 10.1201/9781003319658-1

All of the aforementioned synthetic methods have issues with properly controlled morphology (the thicknesses of the individual layers of MXene are less than 1 nm, but their adjacent proportions have increased to tens of microns), which results in surface roughness, stability problems, and a barrier to effective results from applications in various fields, such as electronics sensing. This data suggested further development of the synthetic protocol of various MAX phase MXenes for practical applications (Naguib et al. 2014). Besides, the presence of property of good conductivity makes the Ti_3C_2Tx MXene a favorable nominee for energy storage devices (Hantanasirisakul and Gogotsi 2018; Lukatskaya et al. 2013; Ghidiu et al. 2014; Wang et al. 2015; Liang, Garsuch, and Nazar 2015), catalysis (Seh et al. 2016), electromagnetic interference shielding (Shahzad et al. 2016), transparent, conducting thin films (Halim et al. 2014; Hantanasirisakul et al. 2016; Dillon et al. 2016), water desalination (Ren et al. 2015), and many other applications.

For their significant contribution to scientific literature, various research scientists from all over the world who work in the fields of nanotechnology and materials sciences deserve recognition and accolades. There have been a significant number of articles published on the synthesis/exfoliation of MXene from various MAX phases. To produce the published data, a wide variety of 2D MXene-based nanomaterials in various morphologies and geometries, such as quantum dots, nanoflakes, nanosheets, and nanoribbons for multi-direction applications, were used (Ng et al. 2017; Naguib and Gogotsi 2015; Lei, Zhang, and Zhou 2015; Zhong et al. 2016; Khazaei et al. 2017; Alhabeb et al. 2017; Anasori, Lukatskaya, and Gogotsi 2017; Huang et al. 2018; Pang et al. 2019; Cai et al. 2018; Verger et al. 2019; Zhan et al. 2020; Ronchi, Arantes, and Santos 2019). Our main goal in writing this book chapter is to highlight the most recent advancements and select articles that best characterize important ideas. In this book chapter, various MXene derivative synthetic strategies are investigated, and their general characteristics are discussed. The Ti_3C_2Tx MXene is chosen because of its simple synthetic process and stable physical characteristics, and its use in environmental and biological sensors is thoroughly covered.

1.2 OVERVIEW OF CHAPTERS

In order to cover synthesis, physical properties, and applications, we tried to assemble a large sample of 2D MXene research. We begin from basic chemistry of MXene; various composites, synthesis, and characteristics of 2D MXene; their heterostructures; and complex 3D architecture. The general overview of the chapters is as follows.

In Chapter 2, the authors Han Zhang, Nasir Mahmood Abbasi, and Muhammad Tayyab summarizes the rich structural chemistry of MXenes and highlights the wide range of synthetic approaches for their preparation, specifically focusing on the extensively studied Ti_3C_2Tx MXene. This MXene is noted for its straightforward production process and robust physical characteristics, making it well-suited for use in biological and environmental sensors. MXenes are not just created from a single method; they come from a variety of precursors and synthesis techniques. For example some precursors are carbides or nitrides, while others might be more complex compounds. The process of creating MXenes can involve different approaches from liquid-phase exfoliation to more advanced methods like hydrothermal or electrochemical exfoliation. One standout MXene is Ti_3C_2Tx. It is like the reliable workhorse of the MXene family, easy to produce and boasting stable physical properties. This makes it an excellent candidate for innovative applications such as sensors that can monitor biological and environmental changes.

Chapter 3 introduces defect engineering and its impact on the characteristics of MXenes, which are a class of two-dimensional transition metal carbides and nitrides. The chapter begins by providing an introduction of MXenes, highlighting their importance in contemporary materials research because of their distinct structural and functional properties. This book chapter provides a comprehensive analysis of the physical attributes of MXenes, including their mechanical, optical, electrical, magnetic, and stability features. Moreover, it explores the possibility of using defect engineering to customize these characteristics for various technological applications.

In Chapter 4, the authors summarize the special properties of Ti_3C_2Tx and Ti_2CTx, continuing with their synthesis methodologies and characterization. This chapter is going to be further discussed along with its significance and scope in biomedicine, photovoltaics, catalysis, etc. They provide a brief overview of the current research in the Ti_3C_2Tx and possible future for both this family as well as potentially new bonds. MXenes have a large surface area and great electrical conductivity, which make them a good option for solar application. The efficiency of MXene-based solar cells, which have demonstrated outstanding stability and performance, is being worked on. The large surface area, metallic conductivity, and plenty of available active sites make MXenes reflecting excellent catalytic properties. Catalysts have been widely used in the catalytic reactions such as those of biomass with fuels and wastewater treatment. MXene-based catalysts are one of the most promising materials for energy storage applications, including supercapacitors and lithium-ion batteries. MXene-based supercapacitors, due to the unique properties of MXenes including high power density, rapid charge—discharge rates and excellent cycling stability. Conversely, higher rate and better specific capacity lithium-ion batteries have also been shown to be formed of MXene. The authors also guide the reader through this intriguing realm of Ti_3C_2Tz and its twin, while discussing their properties that make them so unique and how those might translate into useful applications.

Chapter 5 summarizes the potential of using two-dimensional (2D) MXene materials as a part of composite architectures with polymer matrix. The chapter contents deals in particular with the structure and properties of MXene materials, as well as their incorporation into polymeric composite systems. Methods explored include MXene being used as a filler, incorporated in the polymer matrix and used as a reinforcement material. This chapter also investigates the effects of MXene addition on these composite structures in terms of mechanical properties' improvements, elevated conductivity (thermal and electrical) as well as chemical resistance enhancement. The study concludes by proposing some prospective research scopes for polymer/MXene composite architectures.

Chapter 6 summarizes the methodologies utilized in the construction of heterostructures involving 2D titanium-based MXene derivatives, namely $T_{i3}C_2Tx$ and Ti_2CTx, in combination with other 2D materials. The main goal is to maintain the inherent characteristics of both the MXene derivatives and the guest materials, achieved through judiciously orchestrated weak intermolecular interactions. This synergistic amalgamation yields heightened mobility, modifiable narrow band gaps, augmented mechanical resilience, and the instigation of novel functionalities. The integration of 2D MXene nanosheets within these heterostructures establishes efficacious conduits for the transport of both electrons and ions, a pivotal facet for materials deployed in the areas of energy conversion and storage.

A variety of deposition techniques were explored, resulting in heterostructures with practical applications in catalysis, optoelectronics, and energy storage. The overarching goal is to curate a comprehensive compendium of the scientific literature relevant to this topic.

Chapter 7 summarizes the diverse array of deposition methodologies and explores their myriad applications in the realms of supercapacitors, lithium-ion batteries, sodium-ion batteries, and other energy storage and conversion devices. It aims to serve as a valuable resource for researchers and practitioners in the field, providing essential information about the cutting-edge techniques and applications of MXene-based heterostructures. In essence, the development of MXene-based mixed-dimensional heterostructures represents a significant leap forward in the realm of nanomaterials. The unique properties of MXenes, combined with innovative engineering approaches, are paving the way for next-generation devices that are more efficient, durable, and versatile. As we look to the future, continued research and collaboration will be crucial in unlocking the full potential of these remarkable materials, ultimately leading to groundbreaking advancements in technology and industry.

Chapter 8 discusses the diverse functions of MXenes in photodetectors, solar cells, and catalytic processes. The study analyzes the noteworthy development gained by photodetectors based on MXene materials, with a special focus on the outstanding capabilities of Ti_3C_2Tx when utilized in heterostructures alongside n-Si and InSe. The possibility of boosting solar cell efficiency is being examined by the integration of MXenes with silicon in Ti_3C_2Tx-SiO_2 heterostructures in the field of solar energy. Moreover, the extraordinary catalytic capacity of MXenes is underlined, proving their usefulness in electrocatalysis for the process of splitting water and in the conversion of CO_2 to methanol by photocatalysis. These investigations illustrate the tremendous effect of MXenes on advancing technology and promoting sustainable solutions.

Chapter 9 examines the preparation and characterization of 3D assemblies of $Ti_3C_2T_x$ and Ti_2CT_x, alongside 2D assemblies of graphene and TMDCs, emphasizing their unique structures and properties. $Ti_3C_2T_x$ and Ti_2CT_x were synthesized using chemical vapor deposition, while graphene and TMDCs were prepared through mechanical exfoliation. Advanced characterization techniques, including atomic force microscopy (AFM), X-ray photoelectron spectroscopy (XPS), and Raman spectroscopy, revealed the honeycomb lattice structures and elemental compositions of these materials. Both 3D and 2D assemblies showcased exceptional physical and chemical properties, highlighting their potential in applications such as electronics, energy storage, and advanced device fabrication.

Chapter 10 examines the applications of titanium-based MXene 3D assemblies in the areas of sensing, photovoltaics, and catalysis. Renowned for their exceptional electrical conductivity, robust chemical stability, and expansive surface area, MXenes demonstrate significant potential for advanced applications. These properties enable precise detection in sensing, efficient energy conversion in photovoltaics, and effective catalytic interactions, establishing MXenes as highly promising materials for technological advancements.

1.3 SUMMARY

In conclusion, MXenes represent a transformative addition to the 2D nanomaterials family, characterized by their unique bonding, structural properties, and synthesis versatility. Their exceptional attributes, including hydrophilicity, metallic conductivity, and high elastic moduli, position them as highly versatile materials. With applications spanning sensors, energy storage, catalysis, water purification, and biomedical fields, MXenes showcase immense potential for advancing technology and addressing critical challenges in science and engineering. The exploration of their 2D and 3D structures further underscores their significance in both advanced materials and biological domains.

REFERENCES

Alhabeb, Mohamed, Kathleen Maleski, Babak Anasori, Pavel Lelyukh, Leah Clark, Saleesha Sin, and Yury Gogotsi. 2017. "Guidelines for synthesis and processing of two-dimensional Titanium Carbide (Ti_3C_2Tx MXene)." *Chemistry of Materials* 29 (18):7633–7644.

Anasori, Babak, Maria R. Lukatskaya, and Yury Gogotsi. 2017. "2D metal carbides and nitrides (MXenes) for energy storage." *Nature Reviews Materials* 2:16098.

Anasori, Babak, Yu Xie, Majid Beidaghi, Jun Lu, Brian C. Hosler, Lars Hultman, Paul R. C. Kent, Yury Gogotsi, and Michel W. Barsoum. 2015. "Two-dimensional, ordered, double transition metals carbides (MXenes)." *ACS Nano* 9 (10):9507–9516. doi: 10.1021/acsnano.5b03591.

Cai, Xingke, Yuting Luo, Bilu Liu, and Hui-Ming Cheng. 2018. "Preparation of 2D material dispersions and their applications." *Chemical Society Reviews* 47 (16):6224–6266. doi: 10.1039/C8CS00254A.

Dillon, Andrew D., Michael J. Ghidiu, Alex L. Krick, Justin Griggs, Steven J. May, Yury Gogotsi, Michel W. Barsoum, and Aaron T. Fafarman. 2016. "Highly conductive optical quality solution-processed films of 2D titanium carbide." *Advanced Functional Materials* 26 (23):4162–4168. doi: 10.1002/adfm.201600357.

Ghidiu, Michael, Maria R. Lukatskaya, Meng-Qiang Zhao, Yury Gogotsi, and Michel W. Barsoum. 2014. "Conductive two-dimensional titanium carbide 'clay' with high volumetric capacitance." *Nature* 516:78. doi: 10.1038/nature13970.

Halim, Joseph, Maria R. Lukatskaya, Kevin M. Cook, Jun Lu, Cole R. Smith, Lars-Åke Näslund, Steven J. May, Lars Hultman, Yury Gogotsi, Per Eklund, and Michel W. Barsoum. 2014. "Transparent conductive two-dimensional titanium carbide epitaxial thin films." *Chemistry of Materials* 26 (7):2374–2381. doi: 10.1021/cm500641a.

Hantanasirisakul, Kanit, and Yury Gogotsi. 2018. "Electronic and optical properties of 2D transition metal carbides and nitrides (MXenes)." *Advanced Materials* 30 (52):1804779. doi: 10.1002/adma.201804779.

Hantanasirisakul, Kanit, Meng-Qiang Zhao, Patrick Urbankowski, Joseph Halim, Babak Anasori, Sankalp Kota, Chang E. Ren, Michel W. Barsoum, and Yury Gogotsi. 2016. "Fabrication of Ti_3C_2Tx MXene transparent thin films with tunable optoelectronic properties." *Advanced Electronic Materials* 2 (6):1600050. doi: 10.1002/aelm.201600050.

Huang, Kai, Zhongjun Li, Jing Lin, Gang Han, and Peng Huang. 2018. "Two-dimensional transition metal carbides and nitrides (MXenes) for biomedical applications." *Chemical Society Reviews* 47 (14):5109–5124. doi: 10.1039/C7CS00838D.

Khazaei, Mohammad, Ahmad Ranjbar, Masao Arai, Taizo Sasaki, and Seiji Yunoki. 2017. "Electronic properties and applications of MXenes: A theoretical review." *Journal of Materials Chemistry C* 5 (10):2488–2503. doi: 10.1039/C7TC00140A.

Lei, Jincheng, Xu Zhang, and Zhen Zhou. 2015. "Recent advances in MXene: Preparation, properties, and applications." *Frontiers of Physics in China* 10 (3):276–286.

Li, Zhengyang, Libo Wang, Dandan Sun, Yude Zhang, Baozhong Liu, Qianku Hu, and Aiguo Zhou. 2015. "Synthesis and thermal stability of two-dimensional carbide MXene Ti_3C_2." *Materials Science and Engineering: B* 191:33–40. doi: 10.1016/j.mseb.2014.10.009.

Liang, Xiao, Arnd Garsuch, and Linda F. Nazar. 2015. "Sulfur cathodes based on conductive MXene nanosheets for high-performance lithium–sulfur batteries." *Angewandte Chemie International Edition* 54 (13):3907–3911. doi: 10.1002/anie.201410174.

Lukatskaya, Maria R., Olha Mashtalir, Chang E. Ren, Yohan Dall'Agnese, Patrick Rozier, Pierre Louis Taberna, Michael Naguib, Patrice Simon, Michel W. Barsoum, and Yury Gogotsi. 2013. "Cation intercalation and high volumetric capacitance of two-dimensional titanium carbide." *Science* 341 (6153):1502. doi: 10.1126/science.1241488.

Naguib, Michael, and Yury Gogotsi. 2015. "Synthesis of two-dimensional materials by selective extraction." *Accounts of Chemical Research* 48 (1):128–135. doi: 10.1021/ar500346b.

Naguib, Michael, Joseph Halim, Jun Lu, Kevin M. Cook, Lars Hultman, Yury Gogotsi, and Michel W. Barsoum. 2013. "New two-dimensional niobium and vanadium carbides as promising materials for Li-Ion batteries." *Journal of the American Chemical Society* 135 (43):15966–15969. doi: 10.1021/ja405735d.

Naguib, Michael, Murat Kurtoglu, Volker Presser, Jun Lu, Junjie Niu, Min Heon, Lars Hultman, Yury Gogotsi, and Michel W. Barsoum. 2011. "Two-dimensional nanocrystals produced by exfoliation of Ti3AlC2." *Advanced Materials* 23 (37):4248–4253.

Naguib, Michael, Olha Mashtalir, Joshua Carle, Volker Presser, Jun Lu, Lars Hultman, Yury Gogotsi, and Michel W. Barsoum. 2012. "Two-dimensional transition metal carbides." *ACS Nano* 6 (2):1322–1331. doi: 10.1021/nn204153h.

Naguib, Michael, Vadym Mochalin, Michel W. Barsoum, and Yury Gogotsi. 2014. "25th anniversary article: MXenes: A new family of two-dimensional materials." *Advanced Materials* 26 (7):992–1005.

Ng, Vincent Ming Hong, Hui Huang, Kun Zhou, Pooi See Lee, Wenxiu Que, Jason Zhichuan Xu, and L. B. Kong. 2017. "Recent progress in layered transition metal carbides and/or nitrides (MXenes) and their composites: Synthesis and applications." *Journal of Materials Chemistry* 5 (7):3039–3068.

Pang, Jinbo, Rafael G. Mendes, Alicja Bachmatiuk, Liang Zhao, Huy Q. Ta, Thomas Gemming, Hong Liu, Zhongfan Liu, and Mark H. Rummeli. 2019. "Applications of 2D MXenes in energy conversion and storage systems." *Chemical Society Reviews* 48 (1):72–133.

Piscitelli, Alessandra, Anna Pennacchio, Sara Longobardi, Raffaele Velotta, and Paola Giardina. 2017. "Vmh2 hydrophobin as a tool for the development of "self-immobilizing" enzymes for biosensing." *Biotechnology and Bioengineering* 114 (1):46–52.

Ren, Chang E., Kelsey B. Hatzell, Mohamed Alhabeb, Zheng Ling, Khaled A. Mahmoud, and Yury Gogotsi. 2015. "Charge- and size-selective ion sieving through Ti_3C_2Tx MXene membranes." *The Journal of Physical Chemistry Letters* 6 (20):4026–4031. doi: 10.1021/acs.jpclett.5b01895.

Ronchi, Rodrigo Mantovani, Jeverson Teodoro Arantes, and Sydney Ferreira Santos. 2019. "Synthesis, structure, properties and applications of MXenes: Current status and perspectives." *Ceramics International* 45 (15):18167–18188. doi: 10.1016/j.ceramint.2019.06.114.

Seh, Zhi Wei, Kurt D. Fredrickson, Babak Anasori, Jakob Kibsgaard, Alaina L. Strickler, Maria R. Lukatskaya, Yury Gogotsi, Thomas F. Jaramillo, and Aleksandra Vojvodic. 2016. "Two-dimensional molybdenum carbide (MXene) as an efficient electrocatalyst for hydrogen evolution." *ACS Energy Letters* 1 (3):589–594. doi: 10.1021/acsenergyle tt.6b00247.

Shahzad, Faisal, Mohamed Alhabeb, Christine Hatter, Babak Anasori, Soon Hong, Chong Min Koo, and Yury Gogotsi. 2016. "Electromagnetic interference shielding with 2D transition metal carbides (MXenes)." *Science* 353:1137–1140. doi: 10.1126/science.aag2421.

Sun, Renhui, Hao-Bin Zhang, Ji Liu, Xi Xie, Rui Yang, Yue Li, Song Hong, and Zhong-Zhen Yu. 2017. "Highly conductive transition metal carbide/carbonitride(MXene)@polystyrene nanocomposites fabricated by electrostatic assembly for highly efficient electromagnetic interference shielding." *Advanced Functional Materials* 27 (45):1702807. doi: 10.1002/adfm.201702807.

Tao, Quanzheng, Martin Dahlqvist, Jun Lu, Sankalp Kota, Rahele Meshkian, Joseph Halim, Justinas Palisaitis, Lars Hultman, Michel W. Barsoum, Per O. Å. Persson, and Johanna Rosen. 2017. "Two-dimensional Mo1.33C MXene with divacancy ordering prepared from parent 3D laminate with in-plane chemical ordering." *Nature Communications* 8 (1):14949. doi: 10.1038/ncomms14949.

Tyagi, Deepika, Huide Wang, Weichun Huang, Lanping Hu, Yanfeng Tang, Zhinan Guo, Zhengbiao Ouyang, and Han Zhang. 2020. "Recent advances in two-dimensional-material-based sensing technology toward health and environmental monitoring applications." *Nanoscale* 12 (6):3535–3559.

Urbankowski, Patrick, Babak Anasori, Taron Makaryan, Dequan Er, Sankalp Kota, Patrick L. Walsh, Mengqiang Zhao, Vivek B. Shenoy, Michel W. Barsoum, and Yury Gogotsi. 2016. "Synthesis of two-dimensional titanium nitride Ti4N3 (MXene)." *Nanoscale* 8 (22):11385–11391. doi: https://doi.org/10.1039/C6NR02253G.

Verger, Louisiane, Chuan Xu, Varun Natu, Hui-Ming Cheng, Wencai Ren, and Michel W. Barsoum. 2019. "Overview of the synthesis of MXenes and other ultrathin 2D transition metal carbides and nitrides." *Current Opinion in Solid State and Materials Science* 23 (3):149–163. doi: 10.1016/j.cossms.2019.02.001.

Wang, Xianfen, Satoshi Kajiyama, Hiroki Iinuma, Eiji Hosono, Shinji Oro, Isamu Moriguchi, Masashi Okubo, and Atsuo Yamada. 2015. "Pseudocapacitance of MXene nanosheets for high-power sodium-ion hybrid capacitors." *Nature Communications* 6 (1):6544. doi: 10.1038/ncomms7544.

Zhan, Xiaoxue, Chen Si, Jian Zhou, and Zhimei Sun. 2020. "MXene and MXene-based composites: Synthesis, properties and environment-related applications." *Nanoscale Horizons*. doi: 10.1039/C9NH00571D.

Zhong, Yu, Xinhui Xia, Fan Shi, Jiye Zhan, Jiangping Tu, and Hong Jin Fan. 2016. "Transition metal carbides and nitrides in energy storage and conversion." *Advanced Science* 3 (5):1500286. doi: 10.1002/advs.201500286.

Zhou, Jie, Xianhu Zha, Fan Y. Chen, Qun Ye, Per Eklund, Shiyu Du, and Qing Huang. 2016. "A two-dimensional zirconium carbide by selective etching of Al_3C_3 from nanolaminated $Zr_3Al_3C_5$." *Angewandte Chemie International Edition* 55 (16):5008–5013. doi: 10.1002/anie.201510432.

2 Morphological Engineering of 2D Titanium-Based MXene

2.1 CHEMISTRY OF MXENE

One common way to make MXenes is by chemically exfoliating three-dimensional ternary (or quaternary) amalgamations, which are called MAX phases. MXenes encompass diverse transition metal carbides, nitrides, and carbonitrides. The variable "X" within MXenes can denote nitrogen, carbon, or a blend of both, seamlessly integrated into the octahedral facet sites of "M." While "A" refers for group 13 and 14 elements, "M" represents for a d-block transition metal (such as Ge, Al, Si, or Sn) (Ronchi, Arantes, and Santos 2019). These MAX phases have a hexagonal geometry, the "M" and "A" layers are present, and they belong to the P63/mmc space group. They also have a sensible layout for the space group. M-X links display the characteristics of covalent, ionic, and metallic bonds in addition to metallic connections, unlike M-A bonds, which only have metallic linkages (Eklund et al. 2010; Barsoum 2000; Barsoum and Radovic 2011; Barsoum 2013). The diverse chemistry exhibited by MXenes imparts significant versatility in relation to distinctive material properties and potential applications. Various precursors, such as $Zr_3Al_3C_5$ (Zhou et al. 2016) and Mo_2Ga_2C (Meshkian et al. 2015; Halim, Kota, et al. 2016), are also employed in the synthesis of other forms of MXene in addition to the carbides and nitrides-based MAX phase. M, X, and Tx are all components of the structural formula for MXenes, which is $M_{n+1}\ X_nTx$ (n = 1–3). In a generic formula, the transition metals Ti, Zr, Sc, Nb, etc., are indicated by the symbol M. The letters Tx represent lab-produced ends made with hydroxyl, oxygen, or fluorine, whereas X and T stand for carbon and nitrogen, respectively (Naguib et al. 2014). Also noted was the thermodynamic stability of composites with M layers composed of the binary metals Mo_2TiC_2 and $Mo_2Ti_2C_3$. The MXene family was developed using multilayer structures and selective etching (Anasori, Xie, et al. 2015b). These binary MXenes resemble multilayer architectures in terms of their atomic arrangements, morphologies, and crystal structures (Naguib et al. 2014; Naguib and Gogotsi 2015; Naguib, Mashtalir, et al. 2012; Ghidiu, Naguib, et al. 2014; Naguib et al. 2013; Kurtoglu et al. 2012). Theoretical chemists predict that as time goes on and there is a growing demand for these materials, numerous unique MXenes will emerge (Liang, Garsuch, and Nazar 2015; Chen et al. 2015; Rakhi et al. 2015; Lei, Zhang, and Zhou 2015; Gao et al. 2015; Ma et al. 2014; Li et al. 2015; Wang, Kajiyama, et al. 2015). This chapter discusses a variety of MXene derivative synthesis methods and evaluates the suitability of each one in terms of cost, yield, physical characteristics, and usability. The

DOI: 10.1201/9781003319658-2

Ti_3C_2Tx MXene was selected because of its simple manufacturing process and reliable physical characteristics. Its application in biological and environmental sensors has undergone an extensive study.

2.2 VARIETIES OF PRECURSORS DERIVED FROM BOTH MAX AND NON-MAX PHASES (PRECURSORS AND SOURCE OF MXENE)

2.2.1 MAX Phase Precursors

Substances featuring compositions such as Mn+1AXn, where n = 1, 2, or 3, with "M" denoting an early transition metal, "A" representing an element from group 13 or group 14, and "X" signifying either nitrogen or carbon are denoted as "MAX phase precursors" (M_2AX, M_3AX_2, M_4AX_3, etc.) (Naguib et al. 2011). One of the following elements is also present in each of the M layers: Ti, Cr, Nb, V, Hf, Mo, Zr, and Ta. The elements S, P, Sn, Cd, Si, As, Pb, In, Ge, Ga, Tl, and Al are present in one of the A layers. Either C, N, or both can be found in the X layers (Huang et al. 2018).

Owing to its unique blend of attributes, including excellent electrical conductivity and remarkable resistance to oxidation, corrosion, and high temperatures, along with elevated strength and elastic modulus, the MAX phase emerges as a dependable precursor for the synthesis of MXene. The MX and A layers are stacked one on top of the other in a lamellar structure that is typical of MAX phases (Figure 2.1). More than 100 different MAX phase variants are currently described in the literature. Due to their versatility as starting materials, some of these ternary compounds are utilized in the manufacture of nanosheets.

Although MX layers exhibit lower reactivity compared to A layers, MAX phases are acknowledged for their chemical stability. To activate the A layers, a specified portion of A atoms can be substituted with different elements to generate a solid solution of the substitutional type (a blend of two varieties of substituted atoms). The MAX phase precursors, altered in this manner, have the potential to boost the exfoliation process by inducing the dissociation of bonding interactions between the MX and A layers.

The aforementioned rules are adhered to by several minerals found in MAX phases, including Ti_2AlC and Ti_2GaC, V_2GeC and V_2GeC, Hf_2InC and Hf_2SnC, Ti_3AlC_2 and Ti_3SiC_2, Ti_4SiC_3 and Ti_4GeC_3, and others (Zhang et al. 2013; Eklund et al. 2010). The characteristics and conditions required for synthesizing a particular MXene should not be generalized to other MXenes. This is because each MAX phase has unique properties and reacts differently to various etching agents (Alhabeb et al. 2017). For instance the traditional MAX phases M_3AX_2 (312 phase) and M_2AX (211 phase) necessitate more rigorous etching conditions compared to M_4AX_3 (413 phase). Even though there are ten or more A elements in the MAX phases, the only ones that have been effectively etched to generate MXenes are Si and Al. This is attributed to the observed challenge in obtaining MXene from every MAX phase, as depicted in Figure 2.1.

There is a proper necessity for the use of an oxidant, such as the etching of SiC to produce SiC-based MXene, as was stated in the description of the etching of Si from Ti_3SiC_2 precursors to make Ti_3C_2Tx MXene (Cambaz et al. 2006; Alhabeb et al. 2018).

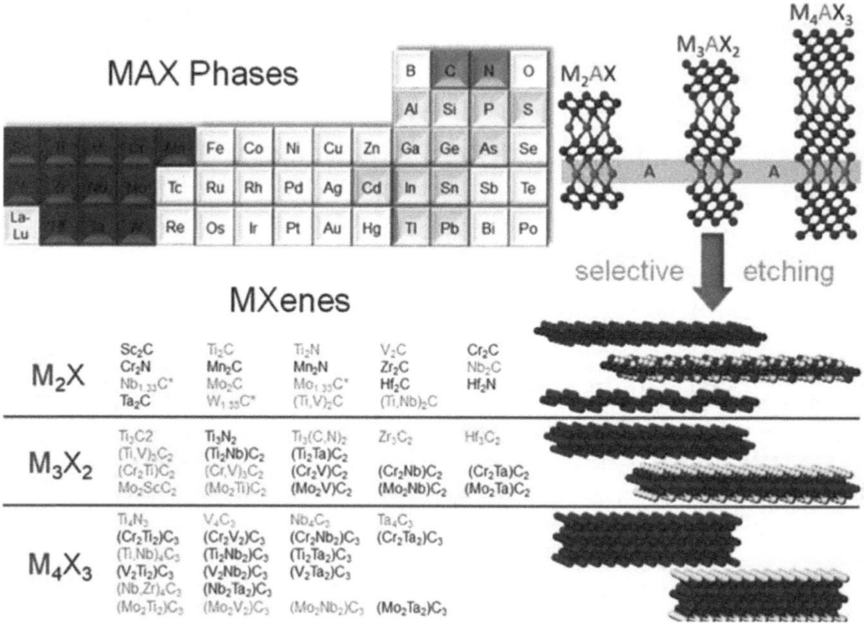

FIGURE 2.1 The transition from the MAX phase to the MXene family is demonstrated, highlighting 100 MAX phase precursors with M_2AX, M_3AX_2, or M_4AX_3 structures. These consist of 12 transition metals (highlighted in black), which contribute to various MAX phases, "A" group elements (displayed in grey) and 2X elements (shown in black). About 30 distinct MXenes have been effectively obtained from MAX phases using the selective etching method, marked in greyscale. The anticipation is that numerous additional MXenes will be synthesized in the future.

Source: Permission awarded by Wiley, 2018 (Hantanasirisakul and Gogotsi 2018).

2.2.1.1 Carbide-Based Precursors

In carbide-based precursors, acknowledged as foundational material amalgamating metal and ceramic traits, carbon atoms are introduced into the metallic lattice (Gogotsi 2015).

The innovative realm of MXene product production has unveiled over 90 precursors, spanning a diverse array of MAX and non-MAX-precursor compounds. Several MAX-phase precursors with Mn + AlXn interacting with various metal configurations yield useful results, depending on the MXenes of interest (Naguib et al. 2011; Alhabeb et al. 2017). In the exploration of innovative ion-binding MAX-phase precursors, constituents such as Si and Ga are incorporated into $M_{n+1}SiXn$ and $M_{n+1}GaXn$, complementing the Al-binding precursor derived from MAX phase. For instance, Ti_3C_2 or Ti_2C MXene stacks generated in predecessors of the MAX phase, such as Ti_3AlC_2 and Ti_2AlC, have very weak ionic connections (by means of the incorporation of Al or alternative elemental species like Si and Ga within various MAX-phase precursors). Therefore, a batch of MXene with one layer (or numerous layers) can be produced when the ionic bonding layers are correctly etched (Alhabeb et al. 2017). The precursors were combined with the etching agent HF before the sonication procedure.

2.2.1.2 Nitride-Based Precursors

Ti_3C_2, Nb_2C, Ti_2C, V_2C, Ta_4C_3, Mo_2C (Mo_2Ti), and Zr_3C_2 are a few examples of MXene derivatives that have been successfully synthesized (Naguib and Gogotsi 2015; Naguib et al. 2013; Anasori, Xie, et al. 2015a; Barsoum et al. 2002; Meshkian et al. 2015; Zhou et al. 2016). Nitride-based precursors have been thoroughly investigated, but little research has been done on nitride-based MXenes. The defects in the fabrication procedures and its instability under etchants like HF are the causes of this. Gogotsi et al. (Urbankowski et al. 2016) first created titanium nitride in 2016, also known as two-dimensional nitride MXene (Ti_4N_3Tx).

When MXenes are heated in molten fluoride salt, one of the byproducts is Ti_4AlN_3. The delamination of the multilayered Ti_4N_3Tx results in the production of the product in monolayer and few forms as well. A separate investigation included the production of a Ti_2NTx product with several layers. This was achieved by submerging the aluminum layer of the precursor in a solution containing hydrochloric acid (HCl) and potassium fluoride (KF). This process was followed by a selective etching procedure, targeting the ternary layer of transition metal nitride. The multilayered Ti_2NTx (where T represents the terminal group) is condensed to a few layers of Ti_2NTx through the process of centrifugation and sonication in DMSO. The production of a few-layered Ti_2NTx MXene is confirmed after characterization utilizing a variety of analytical techniques (Morel et al. 2016; Urbankowski et al. 2016). While $Ti_{n+1}C_n$'s development energies, which are derived from Tin+1AlCn, are less strong than those of Ti_{n+1}Nn, which are derived from Ti_{n+1}AlNn, the cohesive energies of Tin+1Cn are greater in magnitude than those of Tin+1AlNn. Weaker structural stability is predicted in the final structure by reduced cohesive energy in Tin+1Nn. Conversely, the elevated formation energy of Tin+1Nn indicates the stronger bonds that Al atoms can establish in Tin+1AlNn, thereby augmenting the energy required for the extraction of Tin+1AlNn. MXenes derived from nitrides present a richer spectrum of potential benefits compared to their carbide counterparts. Unlike carbides, MXenes related to nitrides, particularly the class of derivatives known as carbonitrides, boast diverse advantages stemming from their extensive electronic conductivity. These advantages encompass applications in electrochemically driven capacitors, plasmonics, metamaterial technologies, and optics (Morel et al. 2016). Though TiN has traditionally been acknowledged as a suitable material for plasmonics, recent scholarly inquiries illustrate that MXenes derived from nitrides display enhanced efficacy in the realm of conversion optics and metamaterial device (Urbankowski et al. 2016).

2.2.2 Precursors Not Affiliated with MAX Phases

After successfully synthesizing MXene from several MAX phases, the researcher used precursors derived from non-MAX phases. For instance, by eliminating the Ga layers from Mo_2Ga_2C, the first MXene, Mo_2CTx, is created. The aforementioned non-MAX phase differs in shape from predecessors based on the MAX phase. The presence of two A-element layers induces the segregation of the carbide-based layers in non-MAX phase precursors (Ga) (Halim, Kota, et al. 2016). Using a non-MAX-phase Zr_3AlC_5 precursor, an extra MXene Zr_3C_2Tx is generated after the Al_3C_3 layers have been successfully etched. The MAX phase $Zr_3Al_3C_5$ may be represented by the chemical formulae MnAl3Cn+2 and MnAl4Cn+3. These formulae classify it as a stratified member in the group of ternary transition metal carbides.

The transition metal is designated as M, and n takes values in the range of 1–3 in the standard formula (Zr or Hf).

In the past ten years, new techniques have been created to etch Zr_3Al_3C's collective AlC units as opposed to the individual Al layers. These discoveries are used to create MXenes from non-MAX starting materials. Over the past ten years, numerous research teams have demonstrated a great ambition to improve the 2D carbide family after successfully etching Al_3C_3 from U2AlC444 to produce U2CTx and diverse structures, as an example (Zhou et al. 2016). Three distinct Mo_2TiC_2Tz, $Mo_2Ti_2C_3Tz$, and Cr_2TiC_2Tz MXenes are effectively synthesized in this work by Gogotsi and Barsoum et al. On $Mo_2TiCzTz$ (Anasori, Xie, et al. 2015a), surface-dependent electrochemical measurements have also been discovered, and Mo_2CTz (Meshkian et al. 2015) was created from non-MAX phase precursors in addition to other MXenes by selectively etching "A" (gallium (Ga)) from a ternary nanolaminated Mo_2Ga_2C (Hu et al. 2015; Lai et al. 2015).

In a different work (Zhou et al. 2016), layered ternary $Zr_3Al_3C_5$ material and hydrofluoric acid are used to make zirconium-containing carbide (Zr_3C_2Tz) MXene, a novel 2D material, at room temperature. In comparison to Ti-based MXene, analyses reveal that synthesized 2D Zr_3C_2Tz demonstrates higher structural integrity and the ability to maintain its 2D identity. DFT calculations were utilized to examine the electrical and structural properties of synthesized Zr_3C_2Tz MXenes in order to provide more evidence. In accordance with studies, every $Zr_3C_2T_2$ (where T is −OH, −F, or −O) MXene demonstrates a metallic nature, with Zr3C2O2 showcasing the highest mechanical strength (392.9 GPa). Regarding structural stability in an inert environment and potential applications, the Zr_3C_2Tz approach has shown superior and more dependable outcomes compared to the Ti_3C_2Tz MXene (Zhou et al. 2016).

2.3 FROM MAX PHASE TO MXENE: EXFOLIATION STRATEGIES, SURFACE CHEMISTRY, AND CHARACTERIZATION OF VARIOUS MXENE

MXenes, named after their origin from MAX phases (Dillon et al. 2016), share some similarities with graphene. Discovered in 2011, MXenes are a class of two-dimensional (2D) materials that include transition metal carbides, carbonitrides, and nitrides.

They made about 30 different MXene structures in the first year they discovered them, thanks to a clever way of removing certain parts from the MAX phase. They think there might be even more of these MXene-making materials out there.

Researchers are interested in MXenes made with Mo, Zr, and V because they act a bit like semiconductors (Meshkian et al. 2015; Ronchi, Arantes, and Santos 2019).

MXene compounds have promising potential for use in electromagnetic interference shielding (Shahzad et al. 2016), water purification (Guo et al. 2018; Srimuk et al. 2016, 2018; Ren et al. 2015), electrocatalysis (Handoko et al. 2017; Seh et al. 2016), energy storage (Anasori, Lukatskaya, and Gogotsi 2017), plasmonics (Sarycheva et al. 2017; Chaudhuri et al. 2018), thermoelectrics (Kim et al. 2017; Khazaei et al. 2014), and medicines.(Alhabeb et al. 2017). There is currently a critical need for numerous research groups to enhance the synthetic process and processing methods because of the diversity of research interests in MXenes.

The material's rapid development is believed to be the consequence of a variety of preparation and processing techniques that influence the quality of the 2D particles generated and the extent of their separation. Among the various ways to make the material, the main goal of the top-down technique is to peel off layers of solid material until it reaches the desired thickness. This technique is split into chemical and mechanical methods because more and more synthetic procedures are being used (Alhabeb et al. 2017).

Currently, it is anticipated that MXene will be created from a number of stacked ternary metals (nitrides and carbides). Following successful surface terminations with functional groups like OH, O, and F, more than 100 new structures are envisaged (Anasori, Lukatskaya, and Gogotsi 2017; Hantanasirisakul and Gogotsi 2018; Sun, Zhang, et al. 2017). Layered transition metal nitride or carbide that has successfully undergone selective etching and exfoliation may be converted to MXene. In this method, they are separated by A-group metals, carbides, nitrides, and Mn+1Xn sheets. After a successful operation, this changeover can take place.

How about considering M_4AX_3, M_2AX, or M_3AX_2 among other MAX phases? MAX phases, through A exfoliation, serve as the initial material for preparing MXenes such as M_2X, M_3X_2, or M_4X_3. For instance, the "A" layers of existing MXenes must be separated from MAX phases that include "A" atoms of Ga, such as those present in the substrate Mo_2Ga_2C, utilizing inventive fluorine-free methods (Yang et al. 2018; Hantanasirisakul and Gogotsi 2018; Sun, Zhang, et al. 2017). Many MXenes with diverse compositions have been successfully synthesized, encompassing Ti_3C_2Tx, Ti_2CTx, Ti_4N_3Tx, Mo_2CTx, and Zr_3C_2Tx. Theoretical predictions suggest the existence of over 70 MXenes with varying compositions (Sun, Shah, et al. 2017; Huang et al. 2018). The majority of studies involving the etching and exfoliation of specific layer thicknesses from a layered parent material have been classed as top-down synthesis. In the realm of bottom-up synthesis techniques, additional methods such as a template approach, plasma-enhanced pulsed laser deposition, and chemical vapor deposition are commonly employed. These approaches become valuable when selective etching is impractical, particularly for crafting superior-quality ultrathin layers of 2D transition metal carbides and nitrides (Verger et al. 2019). The primary limitation of the bottom-up approach lies in its utilization of scalable 2D materials, which significantly falls short of achieving the desired outcome. Maxine is a newly emerging class of 2D materials that includes carbonitrides, nitrides, and carbides. Approximately 30 MXene members have so far been produced experimentally, and many more are the subject of theoretical inquiry (Hantanasirisakul and Gogotsi 2018). In the sections that follow, the synthesis of various MXenes from various MAX phases is addressed in further detail.

2.3.1 Top-Down Approach

2.3.1.1 Wet Chemical Method

The majority of the analyses in the wet chemical approach are done in the liquid phase. However, the aqueous acidic solutions used to etch the A layers in carbide and carbonitride MXenes from nitride-based MAX phase precursors are not fully understood (Ye et al. 2015). For instance, Ti_4AlN_3 nanosheets were created by treating the material at 25°C while under the presence of diluted HF (Urbankowski et al. 2016).

Several MXene variants (with Ti_2C, Nb_2C, Ti_3C2, V_2C, Ta_4C3, Mo_2C, Zr_3C_2, and $(Mo_2Ti)C_3$) have so far been successfully synthesized, with the exception of the nitride MXene (Naguib and Gogotsi 2015; Naguib et al. 2013; Anasori, Xie, et al. 2015a, 2015b; Zhou et al. 2016). The formation energies of Tin+1Cn from Tin+1AlCn are significantly lower than those of Tin+1Nn from Tin+1AlN, indicating a potential challenge in synthesizing nitride MXenes. Furthermore, the cohesive energy of Tin+1Cn is higher than that of Tin+1Nn (Shein and Ivanovskii 2012; Urbankowski et al. 2016). $Ti_{n+1}Nn$ has a lower cohesive energy, indicating a less stable structure. In contrast, $Ti_{n+1}AlNn$ has a higher cohesive energy, indicating a stronger bonding between the Al atoms and requiring more energy to extract them.

Another potential cause of $Ti_{n+1}Nn$'s decreased stability and dissolution in aqueous HF solution over the previous ten years is insufficient synthesis of nitride MXenes. The sections that follow go through some of the etchants that are frequently employed to exfoliate $Ti_{n+1}AlNn$ (Naguib et al. 2014).

We will discuss some of the most popular etchants used to remove the top layer of $Ti_{n+1}AlNn$ while discussing the production of $Ti_{n+1}Nn$ in the sections that follow. $Ti_{n+1}Nn$-derived predecessors of M_4AX_3. precursors generated from M_4AX_3.

In scientific literature, delamination of Ti_4N_3Tx (from multilayered particles to few and monolayers) is hardly ever discussed. Exploring the thermal treatment leading to Ti_4N_3Tx, Gogotsi and the research team subjected Ti_4AlN_3 precursors to heat in molten salt in an environment of inert gas (Urbankowski et al. 2016). At 550°C, molten fluoride salt and aqueous acidic solutions selectively etch the Ti_4AlN_3 precursor. The etching process was initiated by mixing a substantial quantity of fluoride salt with Ti_4AlN_3 powder, where T represents the locations for an OH, F, or O surface terminational group. To produce MXene, fluoride salt and Ti_4AlN_3 were blended and heated at 550°C for 30 minutes in an inert environment using a continuous heating method with a temperature increase of 10°C/min, starting from 25°C. The resulting MXene undergoes sonication to create multilayered nanosheets and monolayers of Ti_4N_3Tx during the subsequent delamination process. XRD, SEM, and TEM are employed to corroborate the schematic representation of the synthesis process in Figure 2.2a.

Ti_4AlN_3 is elevated to a molten state in the company of a fluoride salt, unveiling a distinctive procedure for generating MXenes, presenting a myriad of potential applications that are now making their way into the scientific literature (Urbankowski et al. 2016).

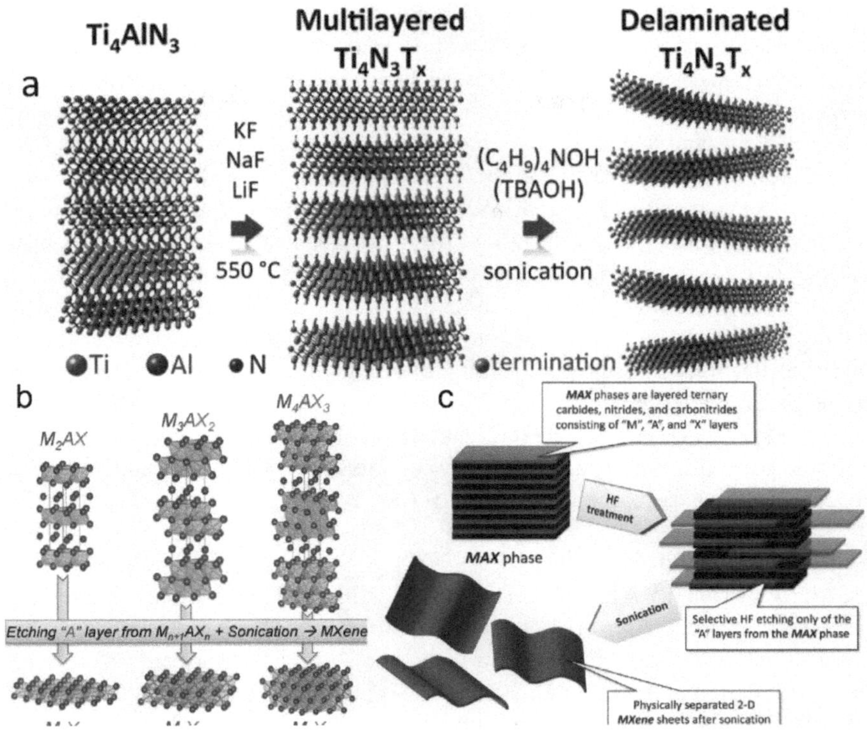

FIGURE 2.2 (a) Ti_4AlN_3 was combined with fluoride salt to produce MXene, which was then heated for 30 minutes at 550°C in an inert environment before delaminating with TBAOH. (b) Composition of MAX phases to match MXenes following A-layer etching. (c) The exfoliation procedure that converts MAX phases into MXenes is shown in this diagram.

Sources: (a) Adapted with permission from the Royal Society of Chemistry (Urbankowski et al. 2016). (b) Adopted with permission from Weinheim: Wiley-VCH Verlag GmbH & Co (Naguibet al. 2014). (a–c) Adopted with permission from American Chemical Society (Naguib, Mashtalir, et al. 2012).

2.3.1.1.1 Liquid-Phase Exfoliation

The MAX phase etching procedure often involves the use of many etchants, including lithium fluoride (LiF) and hydrochloric acid (HF) mixtures, ammonium bifluoride (NH_4HF_2), and concentrated hydrofluoric acid. Following the etching process, the MAX phase is immersed in several organic solvents, including alcohols, dimethyl sulfoxide (DMSO), tetrabutylammonium hydroxide, and n-butylamine, in order to isolate the layers (Alhabeb et al. 2017). The Mn+1Xn sheets in MAX phases are held together by strong covalent bonds (MX). However, the addition of "A" causes the degradation of the MA bond (Sun et al. 2004). By using selective etching of A layers, it is feasible to create Mn+1XnTx and loosely connected stacks inside 2D sheets.

The characteristics and chemical composition of the etchants and solvents utilized during delamination and exfoliation play a crucial role in determining the termination of the resulting MX layers. The termination product Tx is represented by the generic symbol $M_{n+1}X_nT_x$ (Anasori, Xie, et al. 2015a).

Prior to selectively etching the "A" layers, the standard protocol entails mixing MAX phase precursors with an aqueous fluoride ion form (such as ammonium bifluoride or aqueous HF) to attain the required yield of MXenes (Hope et al. 2016). It should be emphasized that there are different conditions for synthesizing each MXene with various etchants and that each MAX phase has a unique level of reactivity and stability (Anasori, Lukatskaya, and Gogotsi 2017; Naguib et al. 2014). MAX phases that are utilized more frequently, such as M_4AX_3 (413 phase), call for more stringent etching conditions than MAX phases that are utilized less frequently, like M_2AX (211 phase) and M_3AX_2 (312 phase) (Figure 2.2b) (Cambaz et al. 2006; Alhabeb et al. 2018).

The etching process is closely linked to the harsh etching conditions, which include elevated temperature, high concentration of HF, prolonged processing times, and the utilization of Cl_2 etchant, leading to over-etching. A carbide-derived carbon product is produced by restricting "M" sites or totally erasing both "M" and "A" elemental species (Figure 2.2c). To create the 2D MXenes version with the desired qualities, an equilibrium is achieved between the stability of MAX and the etching process (Hantanasirisakul and Gogotsi 2018).

2.3.1.1.2 Bimetallic Max Phase Precursors

2.3.1.1.2.1 From M_2AX-Based Precursors Since Al sheets were successfully etched from MAX phases containing Al, a variety of MXene structures have been examined in scholarly journals (Anasori, Xie, et al. 2015a; Naguib et al. 2014; Naguib et al. 2013; Ghidiu, Naguib, et al. 2014).

Unlike Mo_2Ga_2C, which contains only one "A" layer, the unique hexagonal ternary nanolaminated carbide discovered by multiple research groups features two "A" layers sandwiched between the Mo2C layers (Halim, Kota, et al. 2016). The precursors, such as MAX phase materials like $(Ti0.5, Nb0.5)_2AlC$ and $_{Ti2}AlC$, were easily immersed in hydrofluoric acid (HF) solutions of varying concentrations at room temperature to observe the exfoliation process. Sonication then took place for a duration of between 10 and 72 hours. Because of their structural resemblance to graphene, the precursors (MAX phases precursors) had their "A" group successfully removed. The resulting layer was then constructed as 2D layers and given the name "MXenes." With satisfactory results, the exfoliation operation can be performed at room temperature (25°C). The size, organization, and structural features of the MAX phase were evaluated using SEM and XRD patterns, both before and after the etching process. Additionally, the size, shape, and composition of MXene were scrutinized using optical microscopy (OM) and transmission electron microscopy (TEM). MXene presents diverse potential applications, including those in the biological domain. Despite the prospect of enhancing the synthetic process on a larger scale (Anasori, Xie, et al. 2015a).

Shi et al. made the discovery that MXene, specifically identified as niobium carbide (Nb_2C), exhibits an atomic-scale thickness, is well-suited for biomedical applications, and demonstrates high efficacy in in-vivo photothermal therapy. This successfully resulted in the ablation of targeted mouse tumor xenografts in both the NIR-I and NIR-II windows. Stepwise intercalation delamination procedures were integrated into a simple and scalable method to revolutionize the fabrication of 2D Nb_2C nanosheets (NSs) (the two-step based liquid-phase exfoliation approach) (Lin, Gao, et al. 2017).

A novel method for producing Ti_3C_2 MXenes from Nb_2C nanosheets is reported (based on chemical exfoliation). During the initial production step, the removal of the Al layer was accomplished by etching the MAX phase (Nb2AlC) using a 50% HF-based aqueous solution. Subsequently, upon their successful introduction into a triphenylammonium hydroxide (TPAOH) aqueous solution, which significantly diminishes the planar diameters, the etched Nb_2C nanosheets with high dispersibility and ultrathin characteristics are formed (Figure 2.3a). Figure 2.3b depicts the solid-phase sintering process utilized to produce the Nb_2AlC MAX phase. The unique surface structure of the Nb_2AlC MAX phase is supported by SEM pictures (layered ternary complex) (Lin, Gao, et al. 2017).

Upon reviewing the SEM images, it becomes evident that the delamination process and the choice of etchant exert a substantial influence on the morphologies of the resulting flakes. It was also demonstrated that the LiF/HCl solution etching technique followed by the TBAOH treatment produced fewer defective flakes than HF etching did. After the etching is finished, the result is sonicated to reduce the size of the nanoflakes. It does, however, result in a consistent percent quantity of single-layer flakes as compared to the moderate agitation method by handshaking alone (Figure 2.3c, d) (Lin, Gao, et al. 2017). The hexagonal (P63/mmc) structure of Nb2AlC was confirmed using high-resolution transmission electron microscopy (HRTEM) images, as shown in Figure 2.3e and the inset. The Nb_2C NSs multilayer structure exhibits a crystalline hexagonal lattice, as seen in HRTEM images. According to the selected area electron diffraction (SAED) pattern, the HF treatment has altered the hexagonal symmetry of the MAX phase structure (Figure 2.3i and inset).

Figure 2.3f displays SEM images of Nb_2C powder produced in the presence of HF from solid Nb_2AlC bulk. The images depict the surface morphology of homogeneous Nb2C powder sheets featuring particles smaller than micrometers (Figure 2.3g, h). Both bright-field and dark-field TEM images illustrate that the exfoliated Nb2C nanocrystals assume a sheet-like morphology, with an average dimension of approximately 150 nm, subsequent to additional TPAOH intercalation (Figure 2.3k, l). HRTEM images show crystalline nature in single-layer (SL), few-layer (FL), or Nb_2C NSs (with hexagonal geometry), and the accompanying SAED pattern confirms the structure's preserved hexagonal symmetry (Figure 2.3m and inset) (Lin, Gao, et al. 2017).

Gogotsi et al. claim that above 600 degrees Celsius, V_2CTx and Mo_2CTx MXenes change into 2D metal nitrides (Urbankowski et al. 2017). The transformation occurs as the ammonia molecule breaks down, and nitrogen atoms effectively replace carbon in the MXene precursor. The resulting crystal structures of V_2N and Mo_2N products were examined using transmission electron microscopy and X-ray pair distribution function analysis. V_2C undergoes a transformation into a mixed layered structure comprising V_2N (trigonal) and VN, whereas Mo_2N retains the MXene structure (cubic). In temperature-dependent resistivity experiments, the nitrides (Mo_2N and V_2N products), in contrast to their parent carbides, which display semiconductor-like behavior, exhibit metallic conductivity. It is noteworthy that the electrical conductivity values of V_2N and Mo_2N are three and one orders of magnitude, respectively, higher than those of their V_2CTx and Mo_2CTx precursors. This study illustrates the creation of 2D nitrides from carbide-based MXenes through a gas-like process and ammonia addition. The resulting material boasts enhanced electrical conductivities,

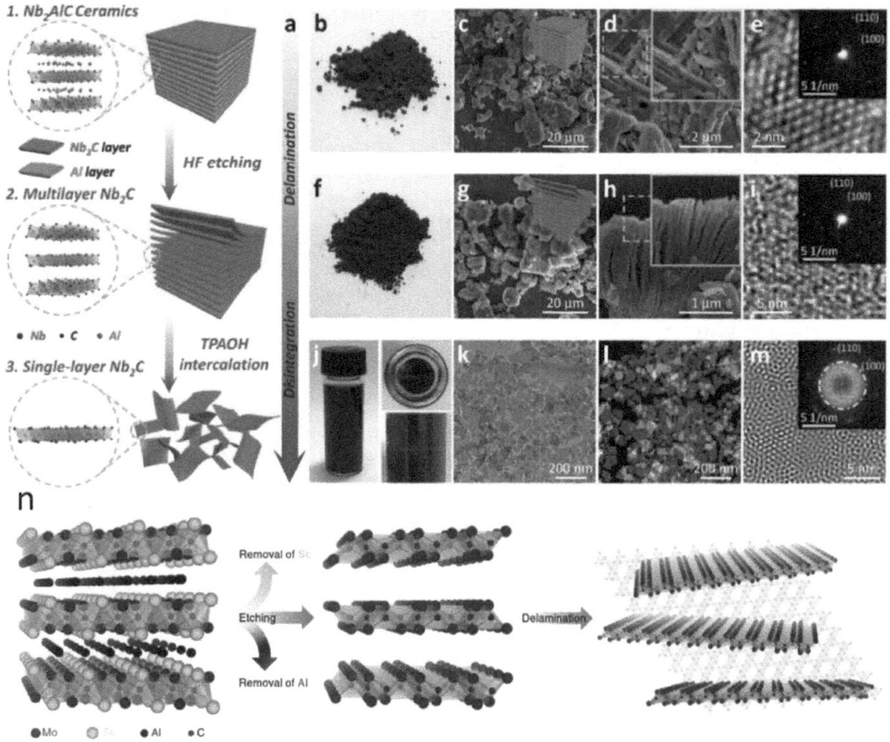

FIGURE 2.3 To produce them, ultrathin layers of Nb_2C (MXenes) are made, and these layers are then studied. Panel (a) depicts the procedures for creating Nb_2C nanosheets, which include TPAOH intercalation and HF etching (delamination) (disintegration). Utilizing HRTEM photographs of the MAX phase Nb_2AlC precursor 1, (1) digital photos, (2), and (3) surface morphology images from SEM, the multilayer of Nb_2C was confirmed. In the inset, a matching SAED design is displayed. The SAED pattern picture-revealed digital image of Nb_2C (MXenes) confirms the crystalline nature of (b). (c, d) SEM micrographs reveal the surface morphology of the synthesized product. (e) HRTEM imaging (with SAED pattern) confirms the crystalline structure of Nb_2C MXenes derived from the Nb_2AlC MAX phase precursor. (f) Optical photograph. (g, h) SEM analysis further demonstrates the surface topography. (i) HRTEM images (SAED inset) verify the multilayered nature of Nb_2C. (j) A digital representation of the dispersed product; (k) a bright-field TEM picture that demonstrates the interior structure of the final product; (l) a dark-field TEM image; and (m) Fourier transform patterns illustrative of single-layer Nb_2C. (Inset shows the familiar SAED pattern.) (n) It is schematically depicted that the chemical structure of (Mo2/3Sc1/3)2AlC is increasing toward Mo1.33C MXene with organized divacancies. By successfully building a 3D atomic shield (Mo_2/3Sc1/3)2AlC precursor with an in-plane chemical arrangement, followed by the selective removal of the Sc and Al atoms from the precursor, Mo1.33C 2D sheets with good electrical conductivity values and ordered metal divacancies were created.

Sources: (a–m) Reproduced by Springer Nature under the terms of the Creative Commons Attribution 4.0 International License and with the authors' consent (Tao, Dahlqvist, Lu, Kota, Meshkian, Halim, Palisaitis, Hultman, Barsoum, Persson, et al. 2017). (n) Adopted with permission from American Chemical Society (Lin, Gao, et al. 2017).

holding promise for advancements in electronics and optoelectronics (Urbankowski et al. 2017).

Tao et al. provided the divacancies clustered in MXenes in 2017 using a simple top-down synthetic technique (Tao, Dahlqvist, Lu, Kota, Meshkian, Halim, Palisaitis, Hultman, Barsoum, Persson, et al. 2017).

The effective removal of Sc and Al atoms from a 3D atomic laminate precursor (Mo2/3Sc1/3)2AlC leads to the formation of highly conductive 2D Mo1.33C sheets with properties of having two valence electrons. For more detailed information, please see Figure 2.3n. The chemically ordered (Mo2/3Sc1/3)2AlC MAX phase may be converted into the 2D Mo1.33C MXene by carefully removing the Sc and Al atoms, resulting in the formation of ordered divacancies.

By employing a top-down approach, it is possible to create 2D flakes in an aqueous environment that contains well-structured empty spaces. This method has the potential to achieve a yield of about 100%. Furthermore, it has been illustrated that the volumetric capacitance of the created 2D nanomaterial, at around 1,100 F cm^3, is 65% greater than that of its counterpart Mo2C, which lacks voids. This stands as one of the most substantial volumetric capacitance measurements on record (Tao, Dahlqvist, Lu, Kota, Meshkian, Halim, Palisaitis, Hultman, Barsoum, Persson, et al. 2017).

These findings indicate that the fabrication approach could provide a viable technique for application in energy storage devices and various other contexts. The Nb_2C NSs had extremely high photothermal conversion efficiencies, measured 45.65% at NIR-I and 36.4% at NIR-II. Additionally, Nb_2C NSs have outstanding biocompatibility, extraordinary biodegradability, and minimal phototoxicity (enzyme responsiveness to human myeloperoxidase).

Bhuvaneswari et al. (Soundiraraju and George 2017) report etching Al from the MAX phase Ti_2AlN precursor using a solution of hydrochloric acid and potassium fluoride in order to produce M_2X type transition metal nitride, Ti_2N (MXene). Few-layered Ti_2NTx is synthesized from multilayered MXene. This process involves delamination in the presence of DMSO, resulting in Ti_2NTx (T = F, OH, and O). Figures 2.4a and b demonstrate the utilization of X-ray diffraction (XRD) analysis and Raman spectroscopy to validate the synthesis of Ti_2AlN. Four distinct vibrational states of Ti_2AlN were identified at frequencies of 145 (1), 225 (2), 230 (3), and 360 cm^{-1}, thanks to the presence of Raman-active modes such as (2E2g + E1g + A1g) (Soundiraraju and George 2017). The vibration of heavy ions typically results in scattering in the acoustic range, whereas the vibration of lighter ions is projected to generate scattering on the edges of the optical range (400–650 cm^{-1}) (150–300 cm^{-1}).

No peaks over 400 cm^{-1} have been seen in Raman spectra, which indicates the lack of MAX phases of type 211 predecessors. This is due to the absence of Raman-dynamic mode vibration in X atoms (Presser et al. 2013; Spanier et al. 2005). The parallel and perpendicular vibrational motions of the atomic surface in Ti are induced by A1g and E1g vibration modes, respectively, while the unaligned vibrational movement of the Ti and Al atomic planes in Ti_2AlN is a result of E2g vibration modes. X-ray diffraction (XRD) is utilized to verify the crystalline properties of the material. The XRD pattern affirms the existence of the Ti_2AlN MAX phase, accompanied by a minor quantity of unreacted titanium and TiAl, acting as impurities in the phase (see Figure 2.4b) (Lin et al. 2007; Barsoum 2013).

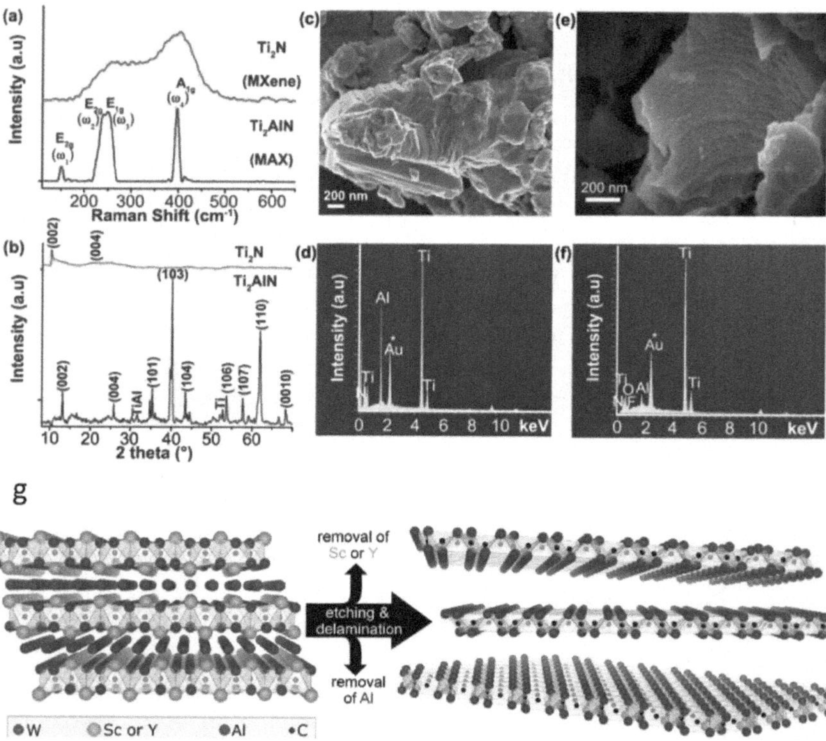

FIGURE 2.4 Images displaying the surface morphology and elemental composition of Ti_2AlN (MAX): (a) Ti_2AlN (MAX) precursor and Ti_2N (MXene) Raman spectra; (b) Ti_2AlN (MAX) precursor and Ti_2N (MXene) XRD patterns; (c and e) FESEM images; and (d and f) represent the EDS patterns of Ti_2N (MXene). (g) The left panel of the figure shows a diagram of the chemical arrangement inside the (W2/3Y1/3)2AlC or (W2/3Sc1/3)2AlC i-MAX phase. This arrangement is achieved by etching and delamination methods, which are illustrated in the right panel (see Meshkian, Dahlqvist, Lu, Wickman, Halim, Thörnberg, Tao, Li, Intikhab, Snyder, et al. 2018).

Sources: (a–f) Adopted with permission from American Chemical Society (Soundiraraju and George 2017). (g) Adopted with permission from Weinheim: Wiley-VCH Verlag GmbH & Co (Meshkian et al. 2018).

The elemental makeup of Ti, Al, and N is showcased in Figures 2.4d and 2.4c, respectively, through energy-dispersive X-ray spectra (EDS) and images obtained from field emission scanning electron microscopy (FESEM). Due to the gold coating on the powder, gold may be seen at 2.5 keV in the EDX spectrum. The precise stoichiometry of the Ti_2AlN MAX phase cannot be established by EDX because of the considerable peak overlap between the titanium L (0.395 keV) and nitrogen K (0.395 keV) lines. When fluoride salt and an acidic solution (HCl and LiF) were injected, stacks of sheets that are typical of MXene's size and structure were discovered. In contrast, the MXene surface morphology in HF-etched MXenes is depicted in Figure 2.4e (Anasori, Lukatskaya, and Gogotsi 2017; Zhou et al. 2017; Ghidiu, Lukatskaya, et al. 2014; Kurra et al. 2016).

Although a distinct delamination process is not overtly evident, the morphology of MXenes etched with acids, such as HCl and LiF, appears to be more compact, likely due to the influence of Li+ intercalation (Kurra et al. 2016; Ghidiu, Lukatskaya, et al. 2014).

When KFHCl is used as an etching agent to remove intercalates (water, Al, and K+ ions) from between the narrow layers, the results indicate layered structures. Elemental analysis using Energy-Dispersive X-ray Spectroscopy (EDS) confirms the presence of O and F, alongside the ratios of N and Ti (see Figure 2.4f). The EDX spectrum also exhibits peaks of stable precursors, including Ti and Al oxides, possibly separated during the purification process, as the Raman spectrum and X-ray diffraction (XRD) pattern did not identify these peaks or reflections. The synthesized Ti_2N (MXene) has been successfully characterized, suggesting its potential application as a SERS substrate for detecting trace amounts of explosive materials, as per the conclusion (Soundiraraju and George 2017).

2.3.1.1.2.1.1 From M_2AX-Based i-MAX Phases Precursors The i-MAX phase precursors undergo a selective etching process in a W-based solvent to transform them into pure MXenes, specifically targeting Sc/Y and Al atoms. Johanna Rosen and her colleagues effectively created two W-based i-MAX phase precursors via the use of this specific etching technique (Meshkian, Dahlqvist, Lu, Wickman, Halim, Thörnberg, Tao, Li, Intikhab, Snyder, et al. 2018). The mentioned antecedents were (W2/3Y1/3)2AlC and (W2/3Sc1/3)2AlC. The study used density functional theory to investigate the phase stability of quaternary i-MAX phase precursors, namely W2/3M2(1/3)2AC (where M2 represents Sc or Y (W), and A represents Sn, Si, Ga, Al, Ge, and In), as well as their general chemical structure. Only the monoclinic C2/c structures of (W2/3Sc1/3)2AlC and (W2/3Y1/3)2AlC are expected to remain stable over the course of the compositions examined. Ordered basal plane voids were identified in the latter configuration. These substances are characterized by the formula (M12/3M21/3)2AlC and are recognized as precursors resembling quaternary MAX phases. This formula is commonly employed, with M1 and M2 denoting two distinct early transition metals (Seh et al. 2016).

The i-MAX phases are distinguished by the ordered arrangement of the M1 and M2 components in the basal planes, which is the sole foundation for their naming (Tao, Dahlqvist, Lu, Kota, Meshkian, Halim, Palisaitis, Hultman, Barsoum, Persson, et al. 2017; Dahlqvist et al. 2017). Separating the out-of-plane (oop) well-ordered (o-MAX) phases from the i-MAX phases that came before them (Anasori, Dahlqvist, et al. 2015; Liu et al. 2014).

The i-MAX phase MXene samples were prepared using two different techniques. The initial method involves dissolving 1,000 mg of i-MAX powder over 30 hours at 30°C with continuous stirring in 20 mL of 48% aqueous HF. Following this, the mixture is physically shaken for 60 seconds and then centrifuged in water, used for the powder's etching process, for 60 seconds at a speed of 5,000 rpm. This procedure was continued until the pH of the mixture was between 5 and 6. A solution was created by mixing the stacked MXene with 5 mL of an organic base (tributyl ammonium hydroxide or TBAOH). After an additional 5 minutes of centrifugation at 5,000 rpm, any TBAOH was removed using three cycles of DI deaerated water.

Figure 2.4g offers an illustrative depiction of the i-MAX phase's transition into MXene conversion precursors. The potential applications of this MXene extend to serving as a catalyst material in the hydrogen-energy technology production process, as indicated by preliminary data on its HER activity (Meshkian, Dahlqvist, Lu, Wickman, Halim, Thörnberg, Tao, Li, Intikhab, and Snyder 2018).

The XRD pattern of a sample of precursors (W2/3Sc1/3)2AlC modified by Rietveld based on a C2/c structure is shown in Figure 2.5a. W2/3Sc1/3 makes up 43% of the weight of the sample net. A substantial similarity between the observed outcomes and the revised 2AlC dimensions of the lattice (a = 9.368, b = 5.404, and c = 13.960) can be seen with the characteristics of the monoclinic structure (9.327, 5.398, and 13.969). ScAl3C3 and W contamination levels in the sample are 26 and 31 weight percent, respectively.

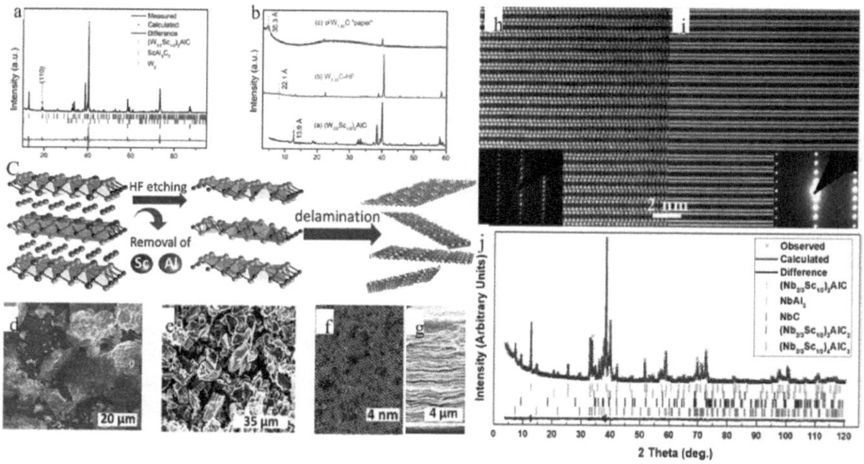

FIGURE 2.5 (a) MAX phase (W2/3Sc1/3) 2AlC precursors have crystallized, as seen by the XRD pattern, which is based on a C2/c geometry. (W2/3Sc1/3) The XRD spectra of 2AlC precursors are shown in (a) and (b), respectively, before and during etching in a 48% HF solution, followed by an organic base, intercalation, and delamination using tetrabutylammonium hydroxide (TBAOH) (c). (c) Diagram showing the etching process used to create 2D Nb1.33C (MXene) nanostructures with different (position(s)) clusters from Nb2/3Sc1/3AlC precursors in both the elemental form of Al and Sc. (d and e) SEM pictures of (Nb2/3Sc1/3)2AlC precursors (respectively before and after the etching procedure). (f) STEM photos of a monolayer Nb1.33CTx nano flake display. (g) A cross-section of a freestanding thin film (created by filtering the delaminated Nb1.33CTx nanoflakes) is shown in a SEM image with the surface morphology revealed. Permission awarded by American Chemical Society, 2018 (Barsoum 2013). (h and i) HRTEM pictures (the evidence of crystalline behavior from the SAED pattern is illustrated in the inset). (j) The Rietveld refinement analysis of the XRD pattern is used to determine the different phases.

Sources: (a–b) Adopted with permission from Weinheim: Wiley-VCH Verlag GmbH & Co (Meshkian et al. 2018). (j) Adopted with permission from American Chemical Society (Halim et al. 2018).

The confirmation of the chemical configuration of W/Sc (in-plane) and the specific creation of i-MAX phases is substantiated by the existence of the (110) peak, shown by the arrow in Figure 2.5a, positioned at about 19° 2θ (Dahlqvist et al. 2017). Figure 2.5b displays the (W2/3Scl/3)2AlC XRD spectrum, which was measured both before and after being etched in a 48% HF solution (a and b). The etching process leads to an increase in the c-lattice parameter (c-LP) of W1.33CTx from 13.96 to 22.16, caused by the removal of impurities (cations and excess water) found between the layers [45b]. Tetrabutylammonium hydroxide (TBAOH) is introduced between the layers of MXene to create a colloidal solution that can be used in a number of different ways (Naguib et al. 2011).

Halim et al. (2018) used etching on a layered (Nb2/3Scl/3)2AlC MAX phase (solid solution) to create 2D Nb1.33C (MXene) nanostructures that are abundant in vacancies and vacancy clusters, including both Al and Sc elemental layers. The M component is accountable for the non-sequential presence of Nb and Sc in the newly identified quaternary MAX phase (Nb2/3Scl/3)2AlC. The process of selective etching is used to make Nbl.33CTx, where both Al and Sc atoms are removed. This results in the formation of distributed zig-zag vacancies and vacancy clusters, as seen in Figure 2.5c.

In an effort to establish some consistency, this is done. This method is applicable to any other quaternary MAX phase where etching of the M element, following successful Sc and Nb etching, is required. Additionally, a novel and promising catalyst for the Hydrogen Evaluation Reaction is successfully created using W-based MXene (HER). There is now a report deficit based on a MAX phase that contains W, even though it has been demonstrated that all of the M-elements are in-plane ordered (Meshkian, Dahlqvist, Lu, Wickman, Halim, Thörnberg, Tao, Li, Intikhab, Snyder, et al. 2018). In the upper left panel of Figure 2.5c, Nb2/3Sc1/3)2AlC's synthesized crystal structure is depicted. SEM images of precursors of (Nb2/3Sc1/3)2AlC are shown in Figure 2.5d and e, respectively, both before and after the etching process. STEM pictures of a single Nb1.33CTx nanoflake with vacancies and vacancy clusters are shown in Figure 2.5f. STEM images depict the characteristic morphology of MAX-phase layered structures, while Figure 2.5g illustrates the cross-section of a freestanding thin film formed by filtering delaminated Nb1.33CTx MXene nanoflakes (Barsoum 2013).

Figures 2.5h, i (inset with SAED pattern) and insets provide more evidence in favor of the MAX phase. The various phases are measured using the Rietveld refinement study of the XRD pattern (used to characterize crystalline materials) (Figure 2.5j) (Halim et al. 2018). Differences in the reactivity of transition metals are evident and anticipated in various applications, including energy storage and catalysis. This approach is reportedly suitable for MAX phase precursors, whether quaternary-based or higher.

2.3.1.1.3 Bimetallic Non-Max Based Precursors

According to Barsoum et al., the scalable manufacturing and delamination procedure for Mo_2CTx nanoflakes entails selectively removing Ga from Mo_2Ga_2C precursors in the MAX phase using a solution of HCl, HF, or LiF (Halim, Kota, et al. 2016). Colloidal solutions of Mo_2CTx are created following the completion of the etching and delamination operations. These solutions are then filtered to produce d-Mo_2 CTx. Mo_2CTx monolayer flakes' generation, shape, and a number of related features are all explained. Mo_2CTx, a material essential in electronic applications, has its conductivity and compatibility evaluated using transport tests (Halim, Kota, et al. 2016).

In the initial approach, a solution containing a 3 M LiF concentration and a 12 M HCl concentration is employed to corrode Mo_2Ga_2C particles. The resultant Mo_2CTx multilayered nanoflakes, after etching, incorporate dispersed Li+ ions within the layers. Mo_2CTx-Li is the name of this chemical. Upon adding the necessary quantity of etched powder, approximately 2 grams, to 40 mL of deionized (DI) water and vigorously shaking for 5 minutes, Mo_2CTx-Li undergoes delamination. The final product was a suspension of delaminated nanoflakes that had been subjected to a 60-minute centrifugation at 5,000 rpm to remove the solid residue. The concentration of this suspension was significant at 0.1 mg/mL. Opting for the second choice demands significant resources and time. The Mo_2Ga_2C precursors underwent etching for a period of 6.6 days at a temperature of 55°C, using a 14 M aqueous HF solution as the etchant. The resulting precipitate included Mo_2CTx. The etched precipitates—Mo_2CTx dust—then continue through the delamination process after intercalation with TBAOH in water (Naguib et al. 2015). Mo_2CTx nanoflakes' production and delamination procedures are schematically shown in Figure 2.6a.

2.3.1.1.3.1 From M_3AX_2-Based Precursors Employing a direct and reliable immersion technique on their disordered state in HF, Michael and colleagues (Naguib, Mashtalir et al. 2012) illustrated the process of exfoliation for various MAX phase precursors, such as (V0.5, Cr0.5)3AlC2, at ambient temperature in 2012. Solutions of varying concentrations were generated through continuous sonication for 10 to 72 hours.

Successful removal of the "A" group from the layer of MAX phase precursors produced 2D thin layers of MXenes with structures resembling those of graphene. The discrepancies in the structure prior to and during the HF treatment are explained using various characterization approaches (such XRD and SEM). The MXene sheets are instead examined with OM and TEM.

In 2011 (Naguib et al. 2011), Michel and colleagues described how Ti_3AlC_2 MAX phase powders handled in HF for 120 minutes produced 2DTi_3C_2 layers. The fact that the top Ti surfaces of the final etched product appear to be finished with functional groups (–F and/or –OH), resulting in the production of the distinctive 2D material known as Ti_3C_2 MXene, serves as proof that Al from Ti_3AlC_2 was successfully removed by etching. The resulting solution is then ready for the centrifugation process in the existence of deionized water in order to efficiently exfoliate Ti_3C_2 layers. The process is then repeated several times to separate the particles produced. As each unit has two exposed Ti atoms, the exfoliated 2D Ti_3C_2 layer, as a result, meets with the standards for precision for synthetic ligands.

As fluorine ions are frequently present in the aqueous reaction medium, fluorine (–F) and hydroxyl (–OH) groups are the most reliable ligands (Naguib et al. 2011).

In a separate work (Lin, Wang, et al. 2017), a modified chemical exfoliation-based method for fabricating two-dimensional Ti_3C_2 nanosheets from Ti_3AlC_2 MAX phase

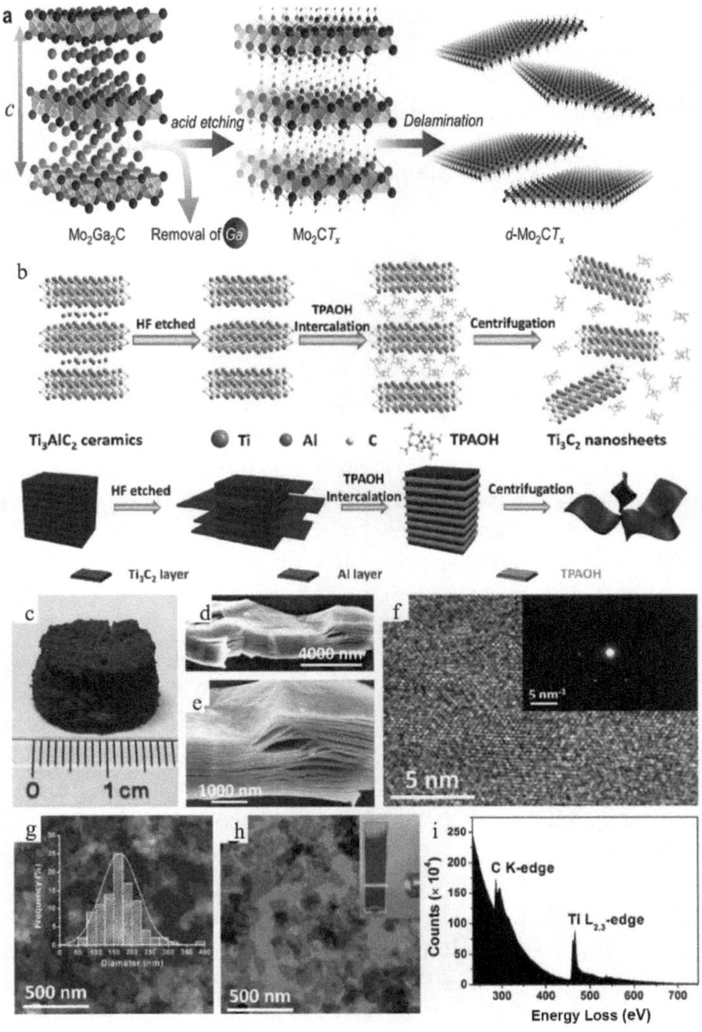

FIGURE 2.6 (a) Mo_2CTx flakes' synthesis and delamination process. (b) To achieve full delamination of the MAX phase precursors, TPAOH is introduced into the etched Ti_3C_2 crystalline powder in the presence of water and HF. (c) An optical micrograph of bulk Ti_3AlC_2's solid MAX phase predecessors. (d, e) In the inset, the complete SAED pattern is illustrated. After successfully HF-etching the Ti_3C_2 precipitate from solid MAX phase Ti_3AlC_2 precursors, the microstructure was assessed using SEM and TEM pictures (f). (g) Dark-field TEM images of the Ti_3C_2 nanosheets after exfoliation show an average size distribution. (h) Photographs of exfoliated Ti_3C_2 nanosheets acquired with a TEM (bright-field). The inset is a digital depiction of the Tyndall effect caused by Ti_3C_2 nanosheets dispersed in the water. (i) Intercalated Ti_3C_2 nanosheets exhibit an electron energy-loss spectrum (EELS) that has been refreshed with consent.

Sources: (a) Adopted with permission from Weinheim: Wiley-VCH Verlag GmbH & Co (Halim, Kota, et al. 2016). (i) Adopted with permission from American Chemical Society 2017 (Lin, Wang, et al. 2017).

precursors is described. Using this method, an aqueous solution containing 40% HF is used to dissolve the Al layer. The etched Ti_3C_2 precipitate was combined with tetrapropylammonium hydroxide (TPAOH) and dissolved in water to accomplish the isolation of the MAX phase precursors (Figure 2.6b).

However, a typical exfoliation method that has few biomedical applications and cannot be absorbed by intravenous injection or cancer cells was used to create Ti_3C_2 with enormous sheet dimensions. The microstructure of the crystalline powder of Ti_3C_2 MXene made from solid $TiAlC_2$ MAX phase precursors can be seen in the SEM and TEM images used to investigate it (Figure 2.6c–e). High-resolution transmission electron microscopy (HRTEM) images of the well-crystallized Ti_3C_2 nanosheets are employed to scrutinize the crystalline lattice structure of multilayer nanosheets exhibiting a hexagonal form (refer to Figure 2.6f). Additional TPAOH intercalation leads to the formation of a consistently thin layer of exfoliated Ti_3C_2 nanosheets. The change in morphology is apparent in TEM images, showing a distinct surface structure resembling 2D nanosheets with an average dimension of 150 nm (Figure 2.6g, h). Ti_3C_2 nanosheets that have been produced can scatter widely in water for several days. They excel at doing this, as seen by the digital image of Ti_3C_2 MXene nanosheets dispersed in water with a distinctive Tyndall effect (Figure 2.6h inset). The electron energy-loss spectra (EELS) of intercalated Ti_3C_2 MXene nanosheets are shown in Figure 2.6i (Lin, Wang, et al. 2017). Huang and others' (Zhou et al. 2014) report of the parent Hf_3 was used to selectively etch the parent Hf_3 to produce a 2D Hf_3C_2Tz MXene (Si-alloyed AlC sublayer). The mechanical character of the Si-facilitated etching process was revealed through computations that utilized adhesive energy and atomic charge. The Al(Si)C bond in $Hf_3[Al(Si)]_4C_6$ was sustained by the electron in the Si atom's outermost shell, which also caused the adhesive form of the HfC at the etching border to become unstable.

The MAX phase precursors $Hf_3[Al(Si)]_4C_6$ and HfAlC composite residues are synthesized using the in-situ reactive pulsed electric current sintering (PECS) technique (Zhou et al. 2014).

The exfoliation method involved concentrated HF. The resulting $Hf_3[Al(Si)]_4C_6$ powders exhibit trace amounts of Hf2[Al-(Si)]4C5 and cubic HfC secondary phase (Figure 2.7a, depicted in black). The as-prepared precipitates were combined with an aqueous solution containing 35 weight percent HF throughout the exfoliation process, and bubbling was then seen to occur. Figure 2.7a (labeled 'after') discusses the many crystal faces. The same kind and quantity of bubbles, which may be CH_4 or H_2, were found when Ti_3C_2Tz and Zr_3C_2Tz MXenes were created from the previously identified Ti_3AlC_2 and $Zr_3C_3C_5$ precursors (Naguib et al. 2011; Naguib, Mashtalir, et al. 2012).

The spontaneous intercalation of water with the etched product is the most probable explanation for the observed increase in the ac lattice parameter during the etching process. The fabrication procedure of Hf_3 $[Al(Si)]_4C_6$ precursor is confirmed using X-ray crystallography, which involves analyzing the etched nanomaterial in its vacuum-dehydrated condition. After dipping in a concentrated solution of HF, the parent $Hf_3[Al(Si)]_4C_6$ precursors' XRD spectra show that the strength of the diffraction peaks gradually decreases (Figure 2.7a, (labeled 'after'). The precursor $Hf_3[Al(Si)]_4C_6$ crystal's c lattice parameter value increases from its initial value of 31.88 to 32.70 as a result of the peak (0003) being downshifted to a lower angle value of $2 = 5.4°$ ($2\theta = 5.4°$(2θ value is (5.54°)). While the emergence of three new

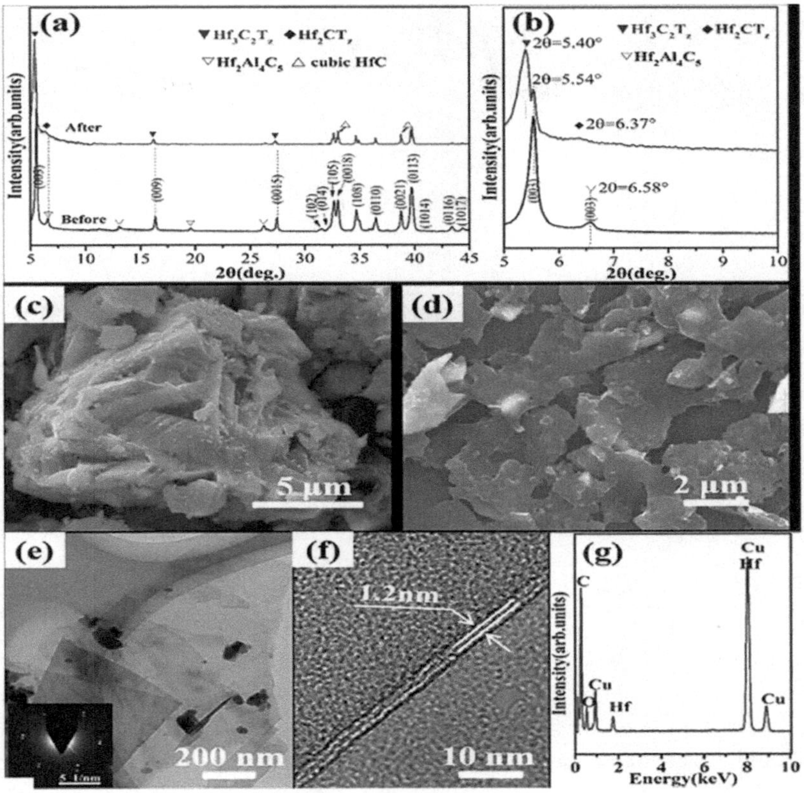

FIGURE 2.7 (a) Phase maximum 4C6 precursor X-ray powder diffraction patterns before and after $Hf_3[Al(Si)]$ HF-based concentrated solution processing and (b) two views that have been stretched between 5 and 10 degrees. SEM pictures of Hf_3[Al-(Si)surface]'s morphology may be seen as created 4C6 MAX phase precipitates in (c). (d) The solid product is next disseminated in a friendly solvent for upcoming morphological investigation after the powders have been treated with HF, sonicated, and centrifuged. Delaminated nano flakes are among the surface morphological changes seen by SEM (large). TEM images reveal the internal structure of delaminated Hf_3C_2Tz flakes and capture their scrolling motion. The SAED pattern inset to the left supports the hexagonal basal plane symmetry (e). (f) TEM imaging makes the few-layered Hf_3C_2Tz flakes evident. (g) The EDX data shows the elemental composition of the Hf_3C2Tz flakes (e). (h) The MXene material was synthesized by selectively removing aluminum (Al) from Ti_3AlCN (MAX phase) precursors using a solution containing about 40% hydrofluoric acid (HF), resulting in the formation of Ti_3CN. The whole process took 18 hours. (i) Crystalline XRD patterns may be seen on Ti_3CN and Ti_3AlCN nanosheets. The predicted XRD patterns of Ti_3AlCN MAX phase and Ti_3CN MXene (linked with i2 and i4) brown/grey, while the practical XRD patterns of Ti3AlCN MAX phase precursors and Ti_3CN (connected with b2 and b4) brown/grey. The black patterns correspond to i1 and i3. The arrangement of the crystal structure of Ti3AlCN MAX and Ti_3CN MXene nanosheets is seen in (j). The unit cells of the crystal are illustrated in c1 and c3, while the Ti_3AlCN MAX phase and Ti_3CN MXene flakes are exhibited in c2 and c4, respectively. In the three Ti sheets that make up the crystal structure of Ti_3AlCN, the compressed carbon and nitrogen atoms from various atomic sheets exhibit an orderly arrangement.

Sources: (g) Adopted with permission from American Chemical Society (Zhou et al. 2017). (a–j) Adopted with permission from American Chemical Society (Zhu et al. 2018).

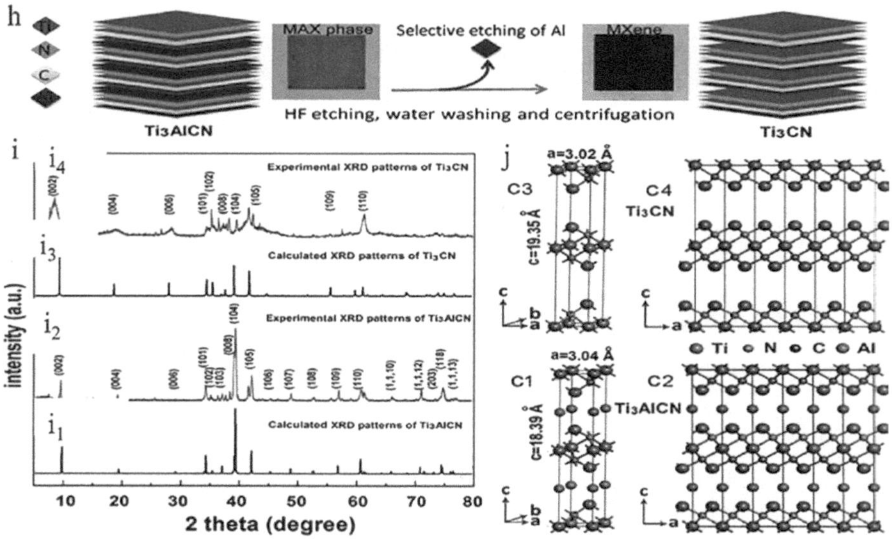

FIGURE 2.7 (Continued)

peaks with lower angles and sharper forms (0001), which are created similarly to Zr_3C_2Tz MXene, indicating the partial conversion to MXene, the abrupt drop in peak height identifies the parent crystal of the $Hf_3[Al(Si)]_4C_6$ precursors. On the other hand, it is evident that $Hf_2[Al(Si)]_4C_5$ (MAX phase) precursors can be appropriately handled with HF to enable etching and conversion into MXenes (Ghidiu, Lukatskaya, et al. 2014; Mashtalir et al. 2013). What then causes the secondary phase broadening peak (0003) from the $Hf_2[Al\ (Si)]_4C_5$ MAX phase to migrate toward the downshift range angle of $2 = 6.37°$ with a value of 2 of (6.58°) (Figure 2.7a and b). Additionally, it is implied that selective-based etching of [Al(Si)]C results in a minor downshift of (0001) peaks and that HF-etched MXenes experience a considerable downshift relative to the noticeably thicker A layers in MAX phases (Naguib et al. 2011, 2013). SEM images of $Hf_3[Al(Si)]_4C_6$ before and after HF treatment are shown in Figure 2.7c, d.

The synthesized $Hf_3[Al(Si)]_4C_6$ precursors for the MAX phase may be shown to have an acoustic structure that resembles a lamellar structure in Figure 2.7c. On the contrary, the nanomaterial etched with HF is observed to consist of a few-layer delaminated type (thin nanoflakes with a grain size of 25 m) after undergoing minimal centrifugation and sonication (refer to Figure 2.7d). The TEM images and EDS spectra of the $Hf_3[Al(Si)]_4C_6$ residues preserved with HF, obtained through centrifugation and sonication, are illustrated in panels e to g of Figure 2.7. Figure 2.7e exhibits thin nanoflakes resembling scrolling patterns, and the EDS data discloses the presence of Hf, C, and O signals (see Figure 2.7g). The hexagonal structure is shown by the minor rotational divergence of two overlapping flakes. The SAED, which is depicted in Figure 2.7e, is used to measure this. Two-layer MXene sheets are displayed in Figure 2.7f. As a single-layer nanosheet is only around 1.2 nm thick, the substance is two dimensional (Zhou et al. 2017). According to the final hypothesis, the resulting MXene, Hf_3C_2Tz, has a rich electron environment (e^-) that could be exploited for applications besides energy storage such as electromagnetic interference shielding and catalysis. According to Rosen et al., the Mo_2ScAlC_2 alloy contains two layers of MoeC that are separated by a layer of Sc atoms and are in the same chemical order. This alloy is a quaternary MAX phase that is out-of-plane (OOP). Twenty milliliters of aqueous hydrofluoric acid (HF) with a 48% concentration is added to 1 g of the refining residue and heated to 50°C for 16 hours to produce MXene. The filtered sample is diluted ten times in water once the MXene production process is complete to remove impurities such as acid traces and reaction products. By combining 0.1 g of the solid residue with 1 mL of the alkaline solution and stirring rapidly for 5 minutes, the MXene sheets undergo intercalation with TBAOH (Meshkian et al. 2017).

The residual TBAOH may be centrifuged for another 5 minutes, washed with water, and then the process may be done two or three times. After successfully creating the material, it is analyzed using Scanning Transmission Electron Microscopy (STEM) with Energy-Dispersive X-ray Spectroscopy (EDX) elemental mapping. This analysis confirms that the observed results match the theoretical expectations.

The lattice parameters a and c were determined to be 3.03 and 18.77, respectively, through XRD Rietveld refinement. The MAX phase precursors underwent a conversion into 2D Mo2ScC2 MXene, representing the latest discovery of Sc-based MXene to date, achieved through the selective etching of Al. Another example of a MXene-based MAX phase that combines M_3X is Ti_3CN (Meshkian et al. 2017). Much has been discovered about the physical and chemical characteristics of Ti_3CN MXene nanosheets since their discovery was first published (Naguib, Mashtalir, et al. 2012; Mashtalir et al. 2013; Halim, Cook, et al. 2016). In addition to other qualities like improved chemical, physical, and mechanical capabilities that were theoretically predicted, the examined MXene also demonstrates TiN and TiC properties (Enyashin and Ivanovskii 2013; Chen et al. 2016).

Wang and colleagues (Zhu et al. 2018) effectively synthesized Ti_3AlCN precursors for the MAX phase and exfoliated them into Ti_3CN nanolayers in 2018. The process of creating 2D Ti_3CN from the Ti3AlCN MAX phase is illustrated schematically in Figure 2.7h.

Ti_3CN nanosheets were generated by immersing MAX phase precursors in an HF aqueous solution for 18 hours before the etching process. The selective relaxation of Al binding in the precursors (MAX phase) following a successful HF etching operation leads to a structural transformation that produces MXene. Figure 2.7i shows that the space group P63/mmc corresponds to the crystal structure of both Ti_3AlCN and Ti_3CN using XRD analysis to track the crystalline structure of both the Ti_3AlCN MAX phase precursors and Ti_3CN. After analyzing the XRD patterns and Miller indices, it was found that the results from the basal plane of Ti_3C_2 and Ti3AlCN are extremely similar to the results from the basal plane of the hexagonal-shaped Ti_3CN nanosheet (Mashtalir et al. 2013). MXenes with various groups, including fluorinated and/or hydroxylated groups, are created after Ti_3AlCN is submerged in HF (Figure 2.7j) (Halim, Cook, et al. 2016).

The nonuniform spreading of inserted end groups may be connected to the phenomenon of Ti_3CN lattice spaces dispersing asymmetrically. The parallel diffraction gap undergoes a leftward shift and widens, resulting in a low angle, attributed to nonuniform microstrain within successive nanosheets and/or within a nanosheet. Ti3AlCN molecules were employed to generate well-crystallized Ti_3CN nanosheets, although some minor imperfections were present (Zhu et al. 2018).

Exfoliation of precursors of the Ti_3AlCN MAX phase produced multilayered 2D Ti_3CN nanosheets (Figure 2.8a). Following the HF etching process, changes in the size and morphology of MXene nanosheets are observed through scanning electron microscopy (SEM) (Naguib et al. 2011; Naguib et al.). A detailed examination using high-resolution SEM indicates that the aggregation of numerous thin nanosheet layers leads to the creation of Ti_3CN nanosheets (refer to Figure 2.8b).

The confirmation of the Tyndall effect in the colloidal suspension occurs as Ti_3CN nanosheets undergo delamination, allowing a laser beam to pass through the aqueous solution. Later, TEM is utilized to scrutinize the internal structure of nanosheets, revealing a multilayer arrangement (refer to Figure 2.8c). The Ti_3CN crystal lattices, featuring interplanar spacings such as (d100 = d010 = 0.26 nm), indicate a 120° angle in accordance with high-resolution (HR) TEM (refer to Figure 2.8d).

Considering the lattice constant value of a = b = 3.02 seen from the SAED patterns, which is consistent with the previous findings, it can be inferred that the [001] zone (Figure 2.8e) exhibits a hexagonal symmetry like that of Ti_3CN MXene. The element mapping (selected region) is concentrated on Ti, C, and N in panels f to i of Figure 2.8 in order to make the elemental composition of Ti_3CN nanolayers more clear. Ti, C, and N constituents are evenly distributed throughout the Ti_3CN nanosheets as depicted in panels g to i of Figure 2.8. With a layer-to-layer spacing of 10.47 and a precursor symmetry reminiscent of MAX phase precursors close to HF treatment, Ti_3CN is effectively synthesized using a 2D-layered structure.

With enhanced cycle stability and capacity, Ti_3CN-based nanosheet electrodes are projected to be a good candidate material for use in future energy storage systems (rechargeable batteries). Ti_3CN is a good anode component that can also be employed in sodium-ion batteries. Improving the specific capacity characteristics of individual components remains crucial for advancing the application of sodium-ion batteries in SIBs (Zhu et al. 2018).

2.3.1.1.3.2 From M_4AX_3-Based Precursors Researchers from all around the world have been closely studying the many MXenes (nitrides or carbides) produced by peeled-off three-dimensional MAX phase precursors for the past ten years.

Transition metal carbides and nitrides serve as excellent substitutes for various applications like catalysts, electrodes, and sensors due to their metallic and hydrophilic properties (Tran et al. 2018). It has been observed that these MXenes can be produced from MAX phase precursors at room temperature through meticulous etching of the A-element layers. To facilitate exfoliation and create a suspension, a specific concentration of MAX phase powder is introduced into an aqueous HF solution during the production process. The resulting mixture undergoes centrifugation or filtration to separate the liquid supernatant from the solid residue after a predetermined duration. If there is solid residue, centrifugation with DI water needs to be repeated once the pH of the solution (water plus solid residue) is between 4 and 6 (Naguib et al. 2014).

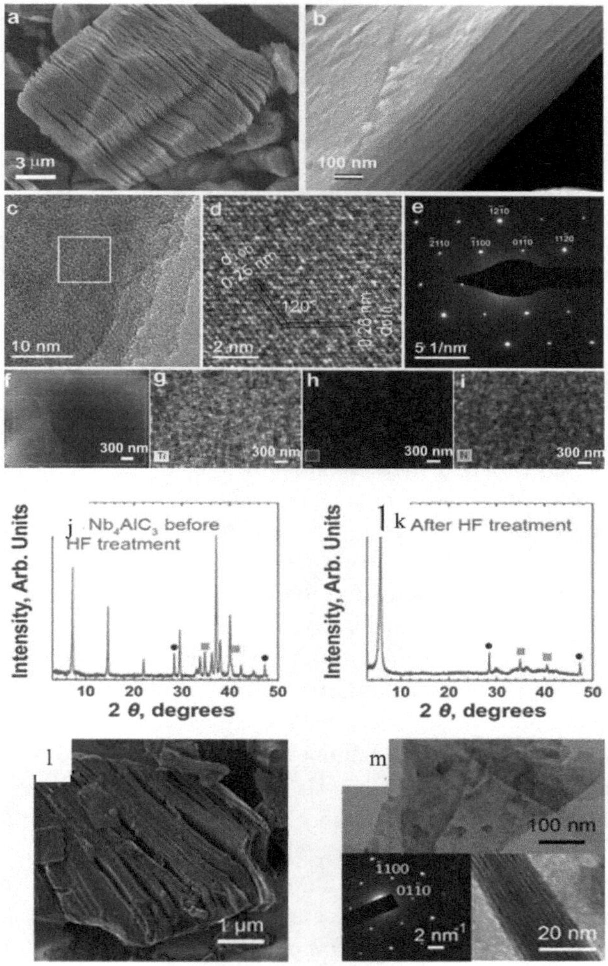

FIGURE 2.8 (a and b) The SEM image shows the surface investigation of Ti_3CN nanosheets. On the high-resolution SEM image, Ti_3CN nanosheets' surface shape is visible. (c) A study of the interior form of Ti_3CN nanosheets is displayed in TEM images. (d) HRTEM images and (e) SAED patterns demonstrate the crystalline nature of the exfoliated Ti_3CN nanosheets. According to their elemental composition, Ti, C, and N are mapped out in SEM images of Ti_3CN nanolayers as f, g, h, and i. Nb4AlC3 MAX phase amorphous-phase precursor X-ray diffraction (XRD) spectra before (j) and after (k) HF treatment. The NBC phase impurity is shown as grey squares in the original Nb_4AlC_3 sample, while the crystalline structure of the silicon utilized as an internal reference is shown as a black circle. An exfoliated grain can be seen in a typical scanning electron micrograph (l) of HF-treated Nb_4AlC_3, and the internal morphology of a form resembling Nb_4C_3Tx nanoflakes can be seen in a transmission electron micrograph (m). SAED type morphological the hexagon's basal plane symmetry is shown in the figure's inset on the left (d). The right inset of the TEM image displays a cross-section of MXene sheets, each approximately 20 nm thick.

Sources: (a–i) Adopted with permission from American Chemical Society (Zhu et al. 2018). (j–m) Adopted from Ghidiu, Naguib, et al. (2014) after some changes.

In 2012, Michael and a colleague[7b] used a straightforward 10- to 72-hour sonication technique to successfully generate the exfoliation-based Ta_4AlC_3 MAX phase precursors. They established the quick and consistent absorption method for precipitates at 30°C using HF.

Upon the removal of the "A" layer from the parent MAX phase precursors, the resulting 2D layers were termed MXenes, highlighting their structural similarity to 2D graphene materials. To distinguish between the resultant structures before and after HF processing, SEM and XRD patterns were used. TEM and optical microscopy were used to further examine the MXene sheets (OM). The contact angle and resistivity of cold-pressed floppies formed from HF-treated powders were then analyzed. MXene was discovered to have exceptional electrical conductivities and hydrophilic behavior, making it a viable material for optoelectronic devices[7b].

In order to produce polycrystalline Nb_4AlC_3 samples as the MAX phase, Barsoum et al. (Ghidiu, Naguib, et al. 2014) developed an in-situ applied hot pressing method using commercially available crystalline Al, C, and Nb powders subjected to a temperature of 1,700°C for 1 hour at a pressure of approximately 30 MPa. Two-dimensional phase-pure, transition metal carbide Nb_4C_3 is created by immersing powders of Nb_4AlC_3 in 48–51% aqueous HF for 96 hours at an ambient temperature and with hydrofluoric acid present. The final product, Nb_4C_3, is refined after being dissolved in ethanol and evaporated to remove any remaining moisture. After etching, an X-ray diffractogram (XRD) test is performed on the powder, and the outcomes are compared to those of the starting material, Nb_4AlC_3 (Figure 2.8j and k).

The XRD pattern analysis reveals that the characteristic peak corresponding to Nb4AlC3 (0002) and the c lattice parameter of 3.059 nm in the MAX phase have been substituted with an enlarged (0002) peak measuring 5.771 nm (a growth of 2.242 nm). The intercalation of ions and water layers between the MXene sheets in Nb_2CTx resulted in a 0.6 nm increase along the [0001] direction, confirming an effective establishment (Naguib et al. 2013). Following the characteristic shape of MXene (Naguib et al. 2011), which resembles exfoliated graphite, after treatment with HF is done using scanning electron microscopy (SEM) pictures (Figure 2.8l) (Chung 1987). The interior morphological structure of typical MXene nanoflakes, which are thin enough to be electron transparent and have a layering that is evident in cross-section, is depicted in Figure 2.8m using TEM images (Ghidiu, Naguib, et al. 2014).

The Nb_4C_3Tx particle maintains the hexagonal symmetry (P63/mmc) that was observed in the original bulk MAX phase precursors after the exfoliation process is complete, as shown in the left inset of the SAED pattern.

The researchers examined the pair distribution function, denoted as M_4X_3, in order to understand the structure of the MXene material. This MXene is remarkable for being one of the most conductive materials discovered so far. It has a resistance of 0.0046 X m when tested in cold-pressed disc form. The choice of the etchant and the method of delamination significantly influence the quality of the flakes.

The use of LiF/HCl solutions in the production process results in a lower number of flawed flakes in the final product, in comparison to materials that underwent etching with HF followed by treatment with base (TBAOH). The mixture is also softly blended using sonication, resulting in the generation of monolayer nanoflakes with a smaller size compared to the scenario with only physical spinning (Ghidiu, Naguib, et al. 2014). The assertion that M_4X_3 can serve as the primary MAX phase for generating new MXene through the exfoliation of V_4AlC_3 is corroborated by the research conducted by Birkel et al. Several analytical methods, including EDX and electron microscopy, were used to verify the intended MXene product (Figure 2.9) (Tran et al. 2018). Our theory that V_4AlC_3 belongs to the MAX phase family is supported by SEM images of the compound that display its layered sound structure (Figure 2.9a). According to the EDX measurement, the quartz ampule that acts as the reaction reservoir contains V, Al, C, and Si (Figure 2.9a inset).

The stability of the synthesized MXene is tested in various acidic conditions, and its potential usage in an electrocatalytic activity-based hydrogen evolution process is also explored. After effectively addressing a structure, HF undergoes morphological transformation (Figure 2.9b) (HER) (Tran et al. 2018).

The EDX characterization demonstrates that there is no longer an Al signal as a result of this separation of the V_4C_3 layers (inset top and bottom in Figure 2.9b). Because numerous faint Al signals are detected in different regions of the sample, it can be inferred from the aforementioned illustration and the outcomes of the SEM/EDX examinations that certain remnants of Al persist in the composition. TEM is also used to examine the exfoliation process of the first MAX phase precursors (Figure 2.9c, d). Amorphous regions are depicted in the TEM image as black dots, whereas vanadium, the sample's heaviest element at the time, is displayed as light spots. We can assume that each layer of V_4C_3 is only a little bit more than 1 nm thick and cannot be broken because there are no large molecules or ions in between the sheets (Tran et al. 2018).

2.3.1.1.3.3 From Tri Metallic Non-Max-Based Precursors This innovative approach for manufacturing MXene utilizes a starting material that substitutes the ternary layered $Zr_3Al_3C_5$ in MAX phases. Some sources suggest that $Zr_3Al_3C_5$, which is the precursor to the layered ternary and quaternary MAX phases of transition metal carbides, consists of $Mn[Al(Si)]_4C_3$ and $MnAl_3C_2$ (where M represents either Hf or Zr, and n ranges from 1 to 3) (Zhou et al. 2013).

Zr_3C_2Tz MXene was first introduced as a new 2D carbide integrating zirconium in 2016 by Huang et al. (Zhou et al. 2016).

The electrical, structural, and elastic characteristics of delaminated nanolayers are determined through first-principles analysis (calculations based on density functional theory). Additionally, the stability of the newly formed MXene nanosheets was assessed, and the results were compared with those of previously studied Ti_3C_2Tz MXenes. The majority of the dried residues consist of single-phase $Zr_3Al_3C_5$, with a little amount of ZrC layers, as seen in Figure 2.9e. Bubbles that resembled CH_4 and H_2 were seen during the exfoliation process in the aqueous solution containing the $Zr_3Al_3C_5$ particles from the MAX phase. Investigation of these bubbles was also conducted during the synthesis of Ti_3C_2 MXenes (Naguib et al. 2011).

The crystalline behavior of the incised particles is illustrated in Figure 2.9e, which is derived from the XRD results. After protracted interaction in the concentrated HF solution, the peak intensity values of the progenitor Zr3AlC5 crystal are demonstrated to have decreased markedly (Zhou et al. 2016).

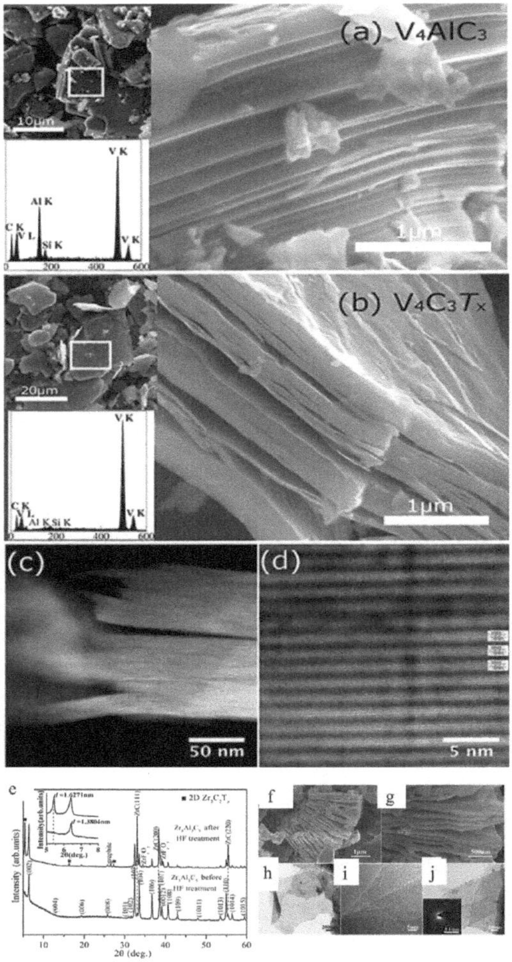

FIGURE 2.9 (a and b) In the white square boxed area of the electron micrographs showing the surface morphology of MXene V_4C_3Tx and the V_4AlC_3 MAX phase precursors, EDX tests are being conducted. Images and EDX analysis show that Al was effectively taken out of MAX phase precursors to produce two-dimensional V_4C_3Tx. As seen in panels c and d, STEM-HAAD examined the V_4C_3Tx data in two dimensions. (e) The XRD patterns of the $Zr_3Al_3C_5$ precursor before and after HF treatment. The powders post HF treatment are depicted in (f), revealing the exfoliation process grain and accordion-like structure. (g) The bright-field TEM is applied post-ultrasonication for sample analysis. The TEM image reveals the scrolled Zr_3C_2Tz nanosheet, and the internal morphological structure of the few-layered Zr_3C_2Tz is observable in image (h). The inset in SAED design (i) is presented to illustrate the preserved hexagonal basal plane symmetry (inset of j). In contrast to Ti-based MXene, the produced 2D Zr_3C_2Tz demonstrates significantly enhanced structural integrity. Discrepancies in structural stability under high temperatures are elucidated through the measurement of binding energy.

Sources: (a–d) Adopted with consent of American Chemical Society (Tran et al. 2018). (e–j) Adopted with permission of Zhou et al. (2016). Copyright 2016 Wiley-VCH Verlag GmbH & Co. KGaA.

The MAX phase c lattice parameter of the $Zr_3Al_3C_5$ crystal increases from 27.73 °A to 32.53 °A due to the peak (0002) downshifting to a lower angle of 2q = 5.4388. The majority of known MXenes are comparable to the recently discovered low-angle peak, according to HF etching (0002) (Naguib, Mashtalir, et al. 2012; Naguib et al. 2011). Two more (0001) peaks are visible in the XRD spectrum at 2q = 16.3288 and 2q = 27.2488. SEM pictures (Figure 2.9f, g) show the variation in surface morphology after HF treatment of $Zr_3Al_3C_5$, which effectively removes individual grains by peeling off layers at the borders of basal planes (Naguib et al. 2011). Figure 2.9h–i displays the TEM images of the Zr3AlC5 precursor after being treated with hydrofluoric acid (HF) and subjected to sonication. The images reveal the thin and electron-transparent structure of the material. The hexagonal structure of the nanolayers is seen and tracked throughout their formation using the SAED pattern, as shown in the inset of Figure 2.9 (Zhou et al. 2013).

2.3.1.2 Hydrothermal-Based Exfoliation

2.3.1.2.1 M_2X-Based Precursors

The following are the primary aims and objectives of this effort: 2D MXenes are initially synthesized in a closed hydrothermal system to prevent exposure to hazardous HF vapor. After that, attempts are made to (i) develop a generalized and efficient etching direction for training MXenes, (ii) improve the mechanistic approach of the hydrothermal-based etching process, and (iii) define the reaction category. In order to assess the effectiveness of the adsorption process, Peng et al (2018) treated the resultant MXene compounds that had been approved as adsorbent materials for dye in water. Nb_2AlC (0.5 g) and 0.75 g of a reducing agent were mixed uniformly in 15 mL of 37 weight percent HCl ($NaBF_4$). Then, to produce Nb_2C, the hydrothermal reaction method is used at 180°C for 15 to 35 hours. The resulting suspension passes through filtering and a DI water rinse. After multiple iterations of the process, the finished item is air-dried at 70°C for the entire night. The hydrothermal method has the advantage of producing Ti_3C_2 flakes more successfully and not requiring extremely concentrated HF, among other advantages. The hydrothermal etching aid can be used to etch more MXene materials, including Nb_2C. The hydrothermal etching procedure is anticipated to enable the exfoliation of MXenes with great efficacy and no environmental impact, opening the door for the straightforward synthesis of additional MXenes and their crucial applications (Peng et al. 2018).

2.3.1.2.2 M_3X-Based Precursors

The use of hydrofluoric acid-based etching in MXene production techniques has the greatest negative impact on the performance of supercapacitors and lithium-ion batteries.

Ti_3C_2Tx (T = OH, O) was produced in 2018 by Li et al. (Li, Yao, Liu, Gu, Luo, Li, Yan, Wang, Liu, Chen, et al. 2018), utilizing a hydrothermal approach with a predetermined quantity of sodium hydroxide (NaOH) and water.

To perform the synthesis, 25 mL of water-based NaOH solution with 100 mg of Ti_3AlC_2 powder is combined. Afterwards, the mixture is transferred to a 50 mL autoclave. Once the autoclave is properly purged with argon gas to create an environment that lacks reactivity, it is deactivated. Afterwards, the autoclave is subjected to a 12-hour heating procedure in an electric thermostatic drying oven at a designated temperature.

Importantly, this entire procedure results in the formation of multilayer Ti_3C_2Tx without any residual fluorine traces. During the hydrothermal reaction, the suspension separates into two parts: The upper layer consists of a clear solution, while the lower suspension appears black. The material remaining after filtering the lower suspension using a PVDF filter membrane with a pore size of 0.1 μm is continuously washed with DI water. The solid residue powders obtained from the extraction of Ti_3C_2Tx are then vacuum-dried for 12 hours at 65°C. Figure 2.10a illustrates the diverse conditions under which Ti_3AlC_2 reacts with a NaOH–water solution. An average dry product containing a $Ti_3C_2Tx/Ti_3AlC_2/TiO_2$ combination typically weights between 80 and 90 mg when 100 mg of Ti_3AlC_2 powder is used as the starting material. The structural properties of the finished product are examined using the XRD pattern. It is strongly recommended that the majority of the products are a mixture of Ti_3C_2Tx and Ti_3AlC_2, given that the XRD tests do not show any appreciable TiO_2 peaks. Hundred mg of the MAX phase can yield 70–80 mg of NaOH Ti_3C_2Tx. Ti_3C_2Tx powder was intercalated and deintercalated by swishing 100 mg of the dry residue from the produced Ti_3C_2Tx for 18 hours in 5 mL of dimethyl sulfoxide (DMSO). To eliminate extra DMSO, the colloidal solution is centrifuged. The weight ratio is set to 1:500 before sonicating the Ti_3C_2Tx residue and water mixture for 6 hours. The suspension was thereafter subjected to centrifugation at a speed of 3,500 revolutions per minute for a duration of 60 minutes in order to separate and eliminate the supernatant colloidal solution. When performing gravimetric tests in a 1 M H2SO4 solution without the F terminal group, the Ti_3C_2Tx-based film electrode exhibits a capacitance of 314 F g^{-1} at a voltage of 2 mV s^{-1}. The resulting film electrode has a density of 1.63 g cm^{-3} and a thickness of 52 nm (Li, Yao, Liu, Gu, Luo, Li, Yan, Wang, Liu, Chen, et al. 2018).

The many circumstances and parameters of the reaction between Ti_3AlC_3 and a NaOH–water solution are illustrated in Figure 2.10a. In a unique work, Peng et al. offer a novel technique for the hydrothermal synthesis of Ti_3C_2 MXene (h-Ti_3C_2) that makes use of low-toxicity etching chemicals (NaBF4, HCl) (Peng et al. 2018).

Ti_3C_2 flakes can be efficiently produced at high concentrations via hydrothermal etching without the use of HF. Hydrothermal etching is a method that can be used to create the various MXene compounds, including Nb_2C. The creation of 2D MXenes is made easier by hydrothermal etching, opening up a wide range of beneficial uses.

2.3.1.3 Electrochemical Exfoliation

2.3.1.3.1 Anodic Etching

2.3.1.3.1.1 M_3X-Based Precursors Anodic etching M3X-based precursors are used by Feng et al. to establish an effective and fluoride-free electrochemical technique for the delamination of Ti_3C_2 (Yang et al. 2018). This technique enables quick anodic Al etching by dissolving Ti–Al bonds and chloride ions. The etching beneath the surfaces must then be improved by intercalating ammonium hydroxide (NH_4OH), which also helps to open the edges of the etched anode. The anode (Ti_3AlC_2) is totally etched in the quickest amount of time achievable under typical reaction circumstances (e.g., 5 hours). Two pieces of bulk Ti_3AlC_2 serve as the anode and cathode of the two-electrode etching procedure shown in Figure 2.10b to e. Numerous aqueous solutions, including nitric acid (HNO_3), sulfuric acid (H_2SO_4), sodium hydroxide (NaOH), ammonium chloride (NH_4Cl), and ferric chloride, were assessed for their appropriateness as an electrolyte during the synthesis and elimination search ($FeCl_3$). Due to kinetic problems, it was mainly predicted that ordinary acids and bases would not be able to successfully etch the atomically thin Al coatings on Ti_3AlC_2 (Li, Yao, Liu, Gu, Luo, Li, Yan, Wang, Liu, and Chen 2018).

Chloride-containing electrolytes typically display apparent etching effects as a result of the fact that Cl^- has a considerable binding potential with Al (Burger and Tull 1953; Alwitt et al. 1984). Despite being restricted to the surfaces of the MAX phase, including Ti_3AlC_2, etching processes remain applicable to other materials (Sun, Shah, et al. 2017).

In order to address the mentioned concern, it is imperative to expand the interlayer gap subsequent to the intercalation of cations into the etched anode (Ti_3C_2Tx). This expansion is facilitated by the diffusion of the electrolyte. The positively charged TMA+ ions play a pivotal role in weakening the bonds between the second layers, leading to the delamination of etched Ti_3C_2Tx into distinct 2D sheets. The Tyndall effect and a noticeable Zeta potential of –37.9 mV were observed in the colloidal aqueous dispersion of isolated Ti_3C_2Tx sheets (0.15 mg/mL); comparable findings (e.g., –34.7 mV) were likewise achieved from HF-etched Ti_3C_2Tx nanostructures (Ding et al. 2017). The surface morphology of stacked Ti_3C_2Tx nanolayer films often displays a polydisperse size distribution with unpredictable long edge lengths, ranging between 1 and 5 μm, as observed in SEM images (refer to Figure 2.10f, g).

The most probable origin of this shape may be attributed to an exfoliation-fragmentation procedure (Lipatov et al. 2016). The irregular size of the synthesized flakes is a result of the moderate etching requirements. Some of the flakes had a diameter as large as 18.6 m, as shown in Figure 2.10h.

The standard established by the modified LiF/HCl etching process is equivalent to the average flake measurement (2.44 m), which was computed over 187 individual sheets (Lipatov et al. 2016; Ng et al. 2017). The process of flake size fragmentation moves toward the nanoscale under vigorous chemical etching with strong HF (e.g., 100–400 nm) (Ng et al. 2017). The TEM image of the produced flakes (Figure 2.10i) reveals a normal Ti_3C_2Tx with crisp edges, and a high-resolution image reveals a typical bilayer flake. An exfoliated sheet with a distinctive hexagonal type diffraction pattern is produced by the specified region diffraction pattern, and it is of exceptional quality. The monolayer structure of flakes is visible by AFM scans. AFM scans measured the thickness of approximately 50 nanoflakes, with the majority (> 90%) falling within a certain thickness of the material, within the range of 1.2 nm, which is in line with its single-layer composition (Figure 2.10j and k) (Wang, Shen, et al. 2015).

The outcomes are on par with or even superior to those of conventional HF or LiF/HCl etching techniques. Future energy storage systems will probably utilize the produced substance. Ti_3C_2Tx films are used to create all-stable-state supercapacitors, and they have a sizable capacitance of 220 mF/cm^2. Also crucial to keep in mind is that neither dangerous fluoride-containing substances nor harsh etching

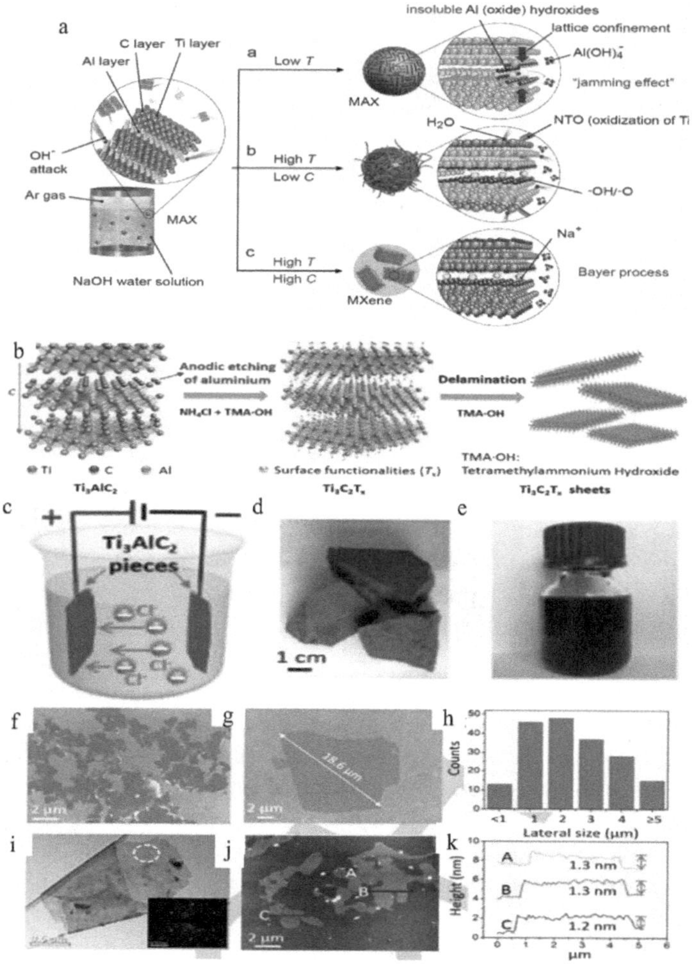

FIGURE 2.10 (a) Many circumstances and parameters of the reaction between Ti_3AlC_3 and a NaOH–water solution. (a–a) Al (oxide) hydroxides impede the extraction of Al at low temperatures. (a–b) Low quantities of NaOH prevent the dissolution of certain Al (oxide) hydroxides at high temperatures. (a–c) Elevated temperatures and concentrated NaOH solutions are the ideal conditions for dissolving Al (oxide) hydroxides in NaOH, according to the Bayer technique. Bulk Ti_3AlC_2 is being etched anodally in an aqueous binary electrolyte solution. The setup of an electrochemically based cell; (b) a schematic depiction of the etching and delamination processes; (c) an optical picture of the bulk Ti_3AlC_2 as obtained; (d) an optical image of the delaminated Ti_3C_2Tx; and (e) an aqueous distribution of the delaminated Ti_3C_2Tx (0.15 mg mL^{-1}); (e) identifying the delaminated Ti_3C_2Tx's properties. (f, g, h) SEM image and the crucial height profiles; (i) TEM image (inset: the SAED pattern); (j, k) AFM image and the related statistical dimension distribution.

Sources: (a) Adopted with permission from Weinheim: Wiley-VCH Verlag GmbH & Co (Li, Yao, Liu, Gu, Luo, Li, Yan, Wang, Liu, Chen, et al. 2018). (b–k) Adopted with permission from Yang et al. (2018) Wiley-VCH Verlag GmbH & Co. KGaA, Weinheim, copyright 2018.

conditions are used in this process. A crucial step in the creation of novel, lucrative MXene compounds is the formation of these unusual circumstances.

Another study uses the electrochemically selective extraction approach to create MXene from metal atoms from ternary layered carbides such Ti_3SiC_2, Ti_2AlC, and Ti_2AlC_2 at room temperature (MAX phases) (Lukatskaya et al. 2014b). At normal temperature, three different MAX phases—Ti_3AlC_2, Ti_2AlC, and Ti_3SiC_2—are used to electrochemically convert carbide-derived carbon (CDCs). Throughout the entire general synthesis operation, special water solutions containing 5% sodium chloride, 10% HCl, and 5% hydrogen fluoride are employed as the electrolytes. After an anodic potential is applied successfully at 25°C, the "M" and "A" elemental components are selectively etched, creating an amorphous CDC with a narrow distribution of pore diameters (Lukatskaya et al. 2014b).

It is possible to etch Ti_3AlC_2 in diluted HF to produce carbon layers formed from amorphous carbide with a restricted supply of micropores and one of the largest nm-sized pores (approximately each 0.5 nm in diameter). The method offers an innovative method for producing carbons with perfectly carved pores for a number of uses, including batteries and supercapacitors.

2.3.1.3.2 Cathodic Etching

2.3.1.3.2.1 M_2X-Based Precursors MXenes (Ti_2CTx) were produced in 2017 (Sun, Shah, et al. 2017) as a result of Green et al.'s methods of electrochemically etching Ti_2AlC in the presence of an aqueous electrolyte based on HCl. In contrast to the chemical etching approach utilizing HF or LiF/HCl (Naguib et al. 2011; Naguib, Mashtalir, et al. 2012), this electrochemical etching method yielded MXenes with –OH, –O, and –Cl terminal functionalities. The outcomes of the process revealed that the porous Ti_2AlC MAX phase undergoes electrochemical transformation into a tri-layered structure comprising unetched MAX, CDC, and MXene. In order to separate the MXene, the three-layer structure is subjected to sonication in a bath. An illustration is shown of the cyclic voltammogram (CV) of the Ti_2AlC MAX phase precursors. The CV was obtained using a scan rate of 20 mV/s in an aqueous electrolyte solution with a concentration of 2 M HCl. The ideal voltage range for selectively etching Al layers from the Ti_2AlC MAX phase precursor is determined by the scan rate, which should be within the range of −0.1 to 0.8 V versus Ag/AgCl. There are several reasons to conduct an electrochemical research on Ti_2AlC. Lukatskaya et al. (Lukatskaya et al. 2014a) found that the form of the CV curves has a considerable impact on the two-stage electrochemical etching process of Ti_2AlC to CDCs (Sun, Shah, et al. 2017).

The first step is the simultaneous removal of Ti and Al, which is only effective at greater cathodic potentials; the outcome is CDC. The MAX stages are then selectively cleared of Al. The exterior of EE-2M-0.6V-5Days displays especially impressive delamination over a longer period of time (e.g., 5 days) in the presence of a significantly greater concentration of HCl (e.g., 2M HCl).

EDS data, however, shows that the Ti/Al ratio typically maintains stability (2.37). It confirms the existence of MXene (Ti_2CTx) in the XRD sample (Figure 2.11a and magnified inset) and is the first report of MXene etching without fluoride (F^-) ions (Kajiyama et al. 2017). When compared to the (002) height of MXenes (Ti_2CT_x) generated using the traditional etching method of LiF/HCl solutions, the (002) peak of MXenes in

EE-2M-0.6V-5Days is pushed toward a higher 2 value, suggesting a lower value of d-spacing in electrochemically etched MXenes. The Raman spectra for EE-2M-0.6V-5Days show the presence of CDC but not MXenes, which is similar to the signature of EE-1M-0.6V-1Day (Figure 2.11b). While it is understood that X-ray diffraction (XRD) does not penetrate as deeply as energy-dispersive X-ray spectroscopy (EDS), one can still envision a thin layer of the CDC substrate. It could be challenging to see the interior morphological structure of MXene using TEM images if an exterior CDC layer is present. The silt's form was shown by TEM investigation to resemble Ti_2CT_x MXene, which is generally found in samples that have undergone LiF/HCl etching (Kajiyama et al. 2017). XPS has also demonstrated the existence of MXene (Ti_2CTx) (Figure 2.11c), demonstrating that it is composed of C, Ti, Cl, and O but not Al.

It is imperative to clarify that the absence of –F terminal functionality serves as the key illustration of the discovery of MXene (Ti_2CT_x). The identification of elemental species, such as C, O, and Cl in the XPS data (see Figure 2.11d), along with the recognition of the form of CDCs in the EE-2M-0.6V-5Days-supernatant, collectively substantiates the existence of CDCs. Additionally, it is likely that the terminal groups Cl, O, and C will include the surface of CDCs. Based on these results from atypical electrochemical etching circumstances, Figure 2.11e illustrates a potential application of this technology.

Ti_2AlC is subjected to treatment using an aqueous electrolyte containing HCl in order to eliminate aluminum. This process leads to the formation of MXene (Ti_2CT_x) with terminal groups including –OH, –O, and –Cl, which are present in both Stage 1 and Stage 2. The electrochemical etching operation in hydrochloric acid (HCl) electrolyte is similar to the chemical etching process that involves lithium fluoride (LiF) and hydrochloric acid (HCl) or hydrofluoric acid (HF).

The –Cl terminal groups on MXene (Ti_2CT_x) are linked following the etching process, promoting the separation of Ti_2AlC and Al (Kajiyama et al. 2017; Sun, Shah, et al. 2017).

The end product, Ti_2CT_x (MXene), can also be produced using an electrochemistry-based etching technique in a solution of electrolytes in water (based on hydrochloric acid). Experimental data supporting the formation of MXene without any F terminal groups is rare, and Ti_2CT_x (MXene) has never been made before. Even if there is still room for more Ti to be removed from Ti_2CT_x, CDC may still develop on the surface as a result of electrochemical etching (MXene).

2.3.2 Bottom-Up Scenario

While single-layer MXene does exist intermittently, thin films derived from MXene samples often comprise numerous MXene flakes, serving purposes in both fundamental research and device applications (Anasori, Lukatskaya, and Gogotsi 2017; Xu et al. 2015). Spray coating, chemical vapor deposition (CVD), and pulsed laser deposition are only a few of the bottom-up techniques that have been discovered during the past ten years for producing various 2D MXene derivatives (PLD). The following section goes into greater information about this tactic.

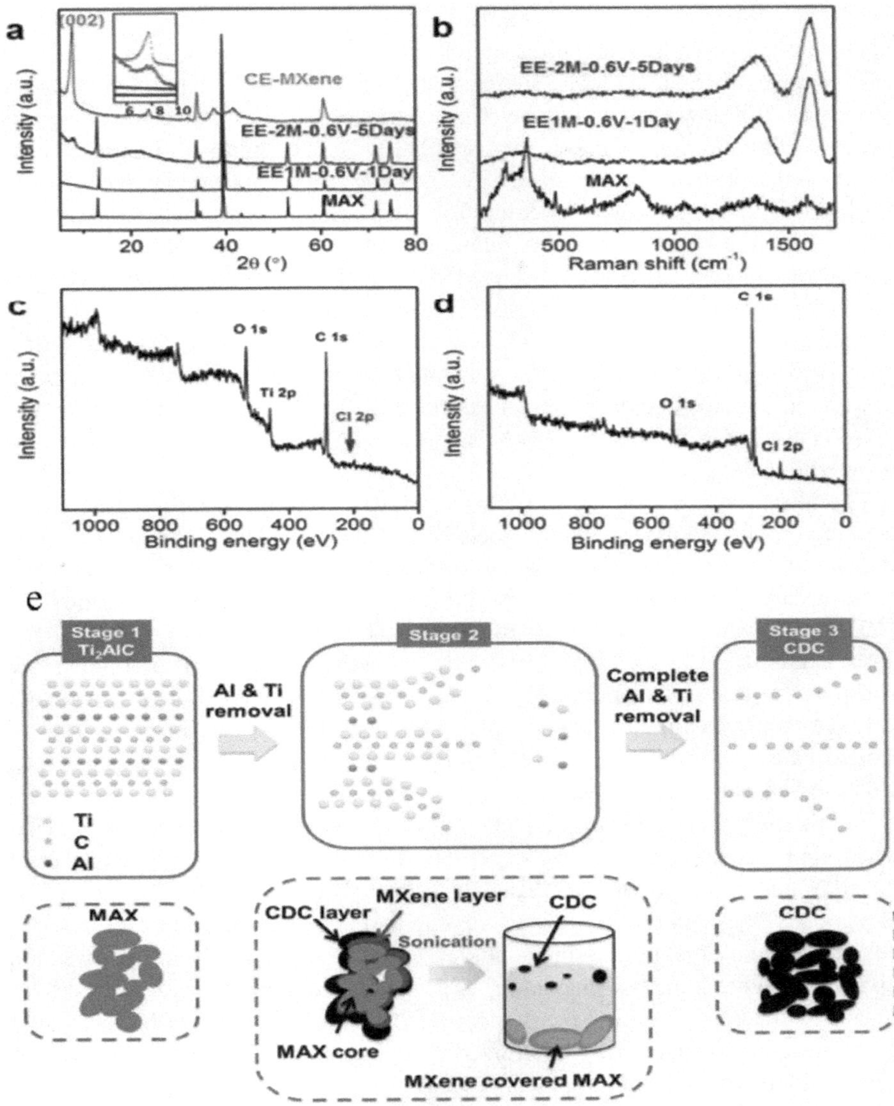

FIGURE 2.11 Different etching conditions to prepare samples for characterization. XRD spectrum in (a) with enlarged picture. (b) The range of emotions can be seen in the Raman spectrum. (c) Sediment-specific EE-2M-0.6V-5Days XPS spectra. (d) EE-2M-0.6V-5Days elemental supernatant's composition is shown by XPS spectra. (e) The proposed mechanism of Ti_2AlC for electrochemical-based etching in the aqueous electrolyte of HCl.

Source: Adapted with permission from the Royal Society of Chemistry (Sun, Shah, et al. 2017).

2.3.2.1 CVD Type

Significant top-down production research has enhanced the quality of MXene products. However, there aren't many studies that discuss MXene bottom-up preparation techniques. Top-down preparation involves working with bulk-phase precursors, contrasting with bottom-up synthesis methods that typically begin with inorganic molecules, atoms, or small organic molecules. The benefit of the preparation procedure is that it allows for precise control over the structure, surface completion, and size distribution of MXenes. The placement of minuscule molecules and/or atoms into a second ordered structure has also been considered in the bottom-up method. Bottom-up manufacturing process procedures for MXene are less effective than those for second-tier compounds because it has numerous tiers of component atomic layers (Xu et al. 2015).

In 2015, Gogotsi and colleagues[75] generated methane using a copper foil bilayer network built on top of a molybdenum sheet using the CVD process (a few nanometers). As a result, orthorhombic –Mo_2C crystals with lateral diameters of up to 100 m were produced.

The ultrathin layers of WC and TaC crystals are produced utilizing a similar technique applied to other transition metals like W and Ta. This technique produces MXenes and makes it easier to look into intrinsic properties by lowering defects and boosting lateral dimensions. However, because MXene monolayer production via CVD has become more sophisticated (Anasori, Lukatskaya, and Gogotsi 2017), a closer examination of the bottom-up synthesis method is required. In an alternate investigation, Ren and team employed (Xu et al. 2015) chemical vapor deposition (CVD) to generate ultrathin 2D crystals of Mo2C, achieving lateral dimensions of approximately 100 mm as methane (CH_4) passed through a Cu/Mo foil substrate, which served as the carbon source.

To initiate the formation of the metal alloy, it is necessary to subject Mo and Cu to elevated temperatures. At that stage, the molecular Mo would dissociate from the alloy and combine with methane to form crystals of 2d-Mo2C (Parthé and Sadogopan 1963). While the task of creating a single-layer MXene stack is difficult, the bottom-up synthesis methodology offers significant advantages over the top-down synthesis method, including larger lateral dimensions and fewer errors. Regarding the configuration, size/shape, and surface terminations of MXenes, tiny molecule/atom precursors are more versatile than their bulk counterparts in terms of their ability to affect the development process.

Furthermore, the precursors based on small molecules provide an excellent active site for chemical vapor deposition of high-quality, ultrathin 2D-Mo_2C crystals (CVD). The crystal that has grown is extremely stable at normal temperature. It has a length of about 100 m and a thickness of a few nanometers. Other incredibly coveted 2d materials, such as TaC ultrathin WC and TMC crystals, can be produced using the approach that was developed. Additionally, it has been discovered that ultrathin WC and TaC crystals have a high degree of crystallinity, comparable to that of –Mo_2 C crystals, with no imperfections or disorder (Parthé and Sadogopan 1963; Li et al. 2009; Xu et al. 2015; Liu et al. 2016). It can be argued that CVD growth methods can be used to produce high-quality, ultrathin TMC crystals. These crystals represent a cutting-edge new kind of 2D material.

Moreover, this chapter provides further insights into the fabrication of 2D ultrathin transition metal carbide (TMC) crystals, enabling the exploration of various intriguing properties and potential applications. The incorporation of these crystals could enhance existing 2D crystals such as chalcogenides, oxides, atomically thin h-BN, and graphene.

2.3.2.2 Pulsed Laser Deposition

Various techniques have been investigated for the demonstration of 2D material devices, including solution-based exfoliation, chemical vapor deposition (CVD), mechanical exfoliation, and other approaches.

The literature on chemical vapor deposition (CVD) techniques for producing ultrathin 2D layered films, particularly complex oxide thin films, has been explored as an alternative to pulsed laser deposition (PLD). PLD is a physical-based technique akin to bottom-up approaches, but CVD offers numerous advantages in this context (Yang and Hao 2016). In 2017, Zhang and colleagues (Zhang et al. 2017) utilized pulsed laser deposition (PLD) to fabricate 2D materials, combining deposition, structural delineation, and characterization procedures. They employed PLD at 700°C on a sapphire (0001) substrate to synthesize superconducting Mo2C thin films with controlled crystal structure, opting for a face-centered cubic (fcc) geometry and aligning along the 111> direction.

Furthermore, it is said that for the Mo_2C thin film with the fcc structure, the orientation relationships are [0110] sapphire/[110] Mo_2C and (0006) sapphire/(111) Mo2C. The prevalent conjecture attributes the issue to a minute in-plane lattice mismatch in the lattice spacing parallel to the (111) atomic plane of the Mo2C thin film. The face-centered cubic (fcc) structure of the Mo2C thin film, which is formed on a sapphire (0001) substrate by the pulsed laser deposition (PLD) technique, results in a distinct and precise alignment and a small in-plane lattice misfit distance between the film and substrate.

On Si (001), SiO_2/Si, and SrTiO3 (111) substrates, Mo2C thin films were co-deposited at 700°C to illustrate the influence of substrate orientation on the crystal structure of the thin films. SEM and TEM methods of electron microscopy are used to learn more about a material's chemical and structural characteristics. In thin films created at low temperatures, nanocrystals have been found to have orthorhombic geometry. As the temperature rises, the thin film enlarges in the direction of 111>. This is because higher temperatures promote correct crystallization of the fcc structure. In order to contrast the various morphologies of thin films produced with and without plasma, Figure 2.12 shows the growth, SEM, and TEM characterizations performed at 700°C. Figure 2.12a, a top-view SEM image, demonstrates the formation of the thin sheet without plasma. It reveals the surface's unevenness (Zhang et al. 2017).

The surface smoothness of a thin plasma-generated layer is illustrated in Figure 2.12b. It appears smooth and uninterrupted. To grasp the structure and quality of the films as they develop, crucial elements of the TEM samples of thin films have been scrutinized. Figure 2.12d demonstrates that the crystalline structure of the thin film is caused by the existence of Mo metallic species, whereas Figure 2.12c shows that tests of the low-electrical transport demonstrate that the Mo2C thin films

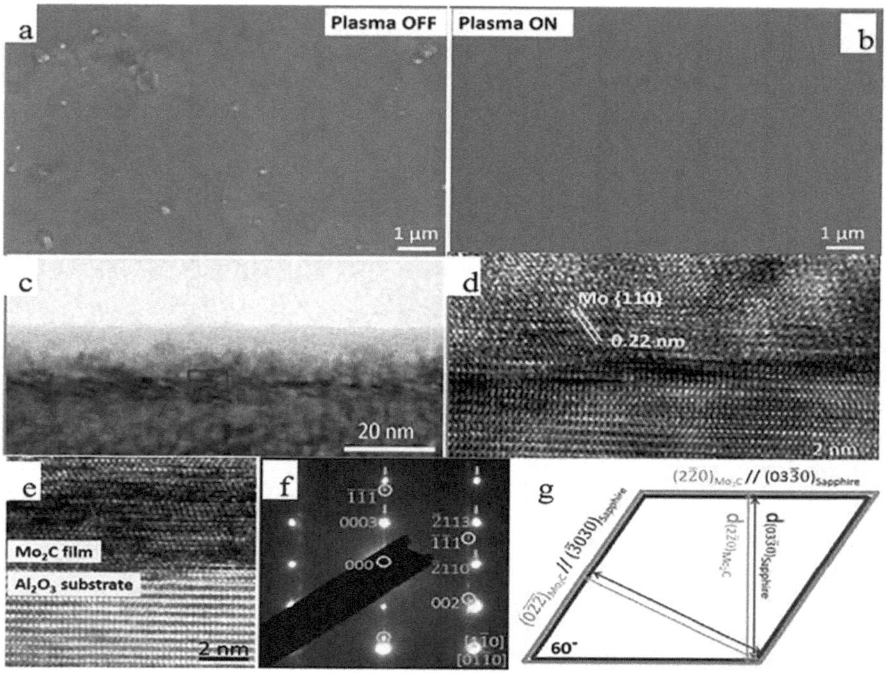

FIGURE 2.12 Thin films were subjected to electronic examination at a constant growth temperature of 700°C, both with and without plasma treatment. The SEM images in (a) and (b) illustrate alterations in the surface morphology of the thin film with and without plasma, respectively. A low-magnification cross-sectional TEM image and a high-magnification HRTEM micrograph both reveal a thin layer formed in the absence of plasma. (c, d) Low-magnification cross-sectional TEM micrograph and high-resolution TEM (HRTEM) image of the thin film synthesized in the absence of plasma. (e) A high-resolution transmission electron microscopy (HRTEM) image captured at the interface of the film and the substrate. (f) The recorded diffraction patterns of the film and substrate using selected area electron diffraction (SAED). (g) A schematic illustration of atomic plane coherence at the interface of a sapphire substrate and a Mo2C film. (h) A schematic illustrates a sprayed Ti_3C_2Tx film. In the Ti_3C_2Tx structure depicted in the box, hydrogen (H), oxygen (O), carbon (C), and titanium (Ti) atoms. (i) A TEM image reveals a monolayer of Ti_3C_2Tx nanoflakes, with the inset showing the corresponding selected area electron diffraction (SAED) pattern confirming its crystalline nature. (j) High- and low-magnification scanning electron microscope (SEM) images of Ti_3C_2Tx films are shown on the right and left, respectively, with the Ti_3C_2Tx flake indicated by a dotted line. (k) The SEM picture captures a segment from the film, while the panels display optical micrographs of thin films on glass and flexible polyester substrates (e) and (f), respectively. The inset in the panel illustrates the extensive bending of Ti_3C_2Tx sheets on the flexible substrate (m). (m) The scale bars have been moved to panels (l) and (m) at a distance of around 1 cm (m). (n) The AFM image made the film's surface roughness very easy to see.

Sources: (g) Adapted with permission from the Royal Society of Chemistry (Zhang et al. 2017). (a–n) Adopted with consent from Hantanasirisakul et al. (2016). Germany's Wiley-VCH Verlag GmbH & Co. KGaA published the book in 2016.

are superconductive. The lattice gap in Figure 2.12e's HRTEM image demonstrates this. To validate the mechanical growth process of Mo_2C fcc structures at 700 °C, the contact between the already-grown Mo2C thin film and the sapphire substrate was examined (Khodan, Kanashenko, and Crete 2012; Makino et al. 2009). Figure 2.12e provides a clear view of the interface and is a HRTEM image of a Mo_2C thin layer on a sapphire substrate.

According to the SAED pattern shown in Figure 2.12e (Xu et al. 2017), the element is validated in the target substrate's region surrounding the thin film and its fundamental region, which is placed above the electron diffraction sample (exhibiting a favorable orientation correlation with the thin film). Figure 2.12f further explains the thorough analysis of the SAED pattern that established the 8.5 to 0.3% lattice mismatch between the side (111) film and (0003) substrate. In this sense, it is anticipated that elastic strain will change the inadequate lattice fit between the substrate and thin film. The (111) plane of the face-centered structured Mo2C thin film features six identical (220) planes, while the (0001) plane of the hexagonally structured sapphire substrate contains six identical (0330) planes. Figure 2.12g illustrates the interplay between the Mo2C thin layer and the sapphire substrate. The epitaxial face-centered cubic (fcc) structure of the Mo2C thin film arises from a subtle in-plane lattice mismatch among the sapphire (0001) substrate, thin film, and coherent orientation relationship (Zhang et al. 2017).

2.3.2.3 Spray Coating

Over the last decade, a substantial portion of research efforts has been directed toward the application of 2D graphene-based materials through spray-on coating, with a focus on transparent electrode applications (Yang et al. 2015; Roy-Mayhew and Aksay 2014; Zhu et al. 2010). Two prominent methods for creating transparent conductive graphene films involve direct chemical vapor deposition (CVD) and the solution processing of reduced graphene oxide (rGO). Notably, various processes, including dip coating, spray coating, and spin coating, have successfully implemented rGO solutions (Jung de Andrade et al. 2007; Song et al. 2008; Tung et al. 2009).

The transparent films produced through these techniques using rGO solutions, conversely, display a broad spectrum of resistance (Rs) ranging from 1 to 1,000 $k\Omega\ sq^{-1}$, attributed to inter-flake resistance and defects (functional groups) (De and Coleman 2010). One of the main areas of research is the discovery of appropriate new materials and procedures for the scalable and cost-effective manufacture of transparent conductive films. Most identified MXenes exhibit potential as intercalation hosts owing to their hydrophilicity, robust mechanical strength, and metallic conductivity (Naguib, Mashtalir, et al. 2012; Naguib et al. 2014). MXene derivatives stand out as superior materials for diverse applications, including but not limited to supercapacitor electrodes, fuel cell electrodes, adsorbents, sensors, and reinforcements in polymer materials (Lukatskaya et al. 2013; Naguib, Come, et al. 2012; Xie et al. 2013; Yu et al. 2015).

In a study published in 2016 by Gogotsi et al. (Hantanasirisakul et al. 2016), it was discovered that applying a colloidal delaminated Ti_3C_2Tx solution onto a substantial substrate at 25°C could generate uniform, see-through, and conductive films with

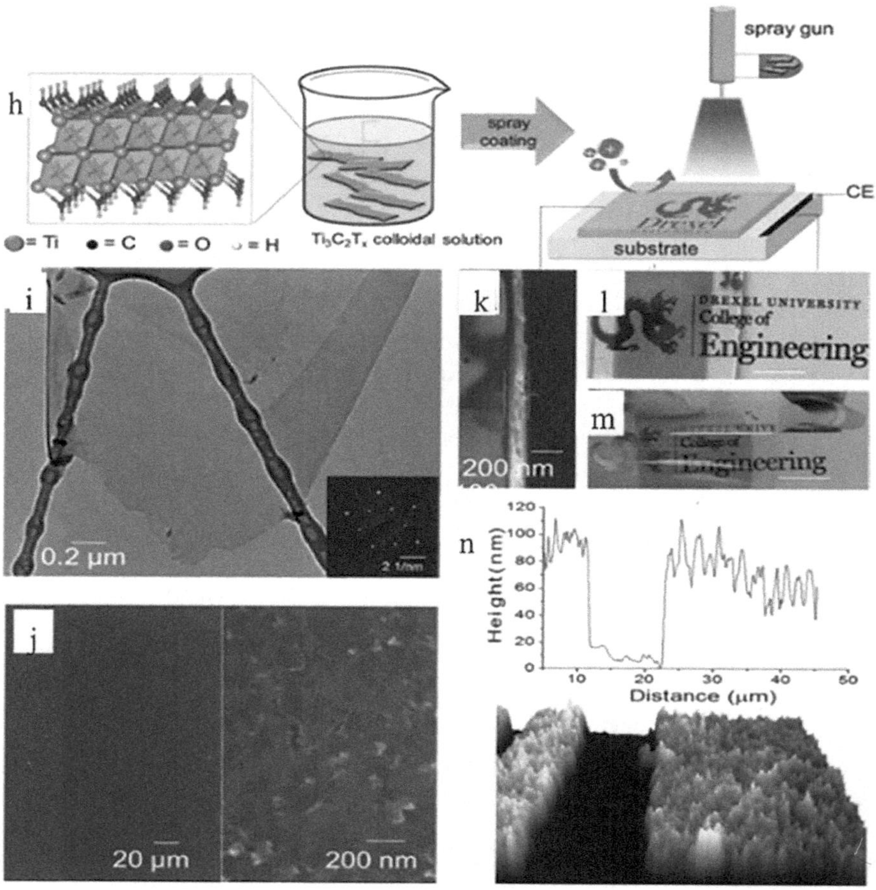

FIGURE 2.12 (Continued)

a concentration of 3.0 mg/mL that has been specified (Figure 2.12h). The surface properties and thickness of the thin film produced are assessed using additional microscopic techniques, including SEM, TEM, and AFM. A single-layer Ti_3C_2Tx nano flake is seen in a TEM picture. The inset of Figure 2.12i depicts the SAED pattern that was used to verify the crystallinity.

Based on SEM images, the Ti_3C_2Tx film on a glass slide exhibits a uniformly distributed and smooth surface across a significant area following the spray-coating process (Figure 2.12j, left). In the enhanced details of the image, the edges of Ti_3C_2Tx flakes appear crumpled (Figure 2.12 j, right). Figure 2.12k, a cross-sectional SEM image of the film, reveals transparent Ti_3C_2Tx nanoflakes aligned parallel to the glass surface (Figure 2.12l). In addition to glass sheets, translucent thin films of Ti_3C_2Tx can be produced by spray-coating flexible substrates such as polyester layers (Figure 2.12m). Ti_3C_2Tx thin films are a strong candidate for flexible material-based devices because the flexible thin film substrate (inset of Figure 2.12m) shows no evident damage even after the minor bending treatment. A 70 nm thick Ti_3C_2Tx

film that has been deposited on a sheet of glass is depicted in Figure 2.12 by an AFM image. Thin films that have undergone solution processing typically have a surface roughness of 10 nm (Behabtu et al. 2010; Wan et al. 2015). These Ti_3C_2Tx films exhibit stronger optoelectronic properties than films made of transparent graphene that have undergone solution processing. The created MXene film is most likely to be used in sensors, electrochromic applications, and electrical devices with transparent conductors. Among the many possible uses for spray coating are sensors and transparent conductive electrodes. I can scale it up or down as needed. Additionally, it may be adjusted to accommodate other MXene kinds, of which around 30 have been documented so far.

Ti2C (Naguib, Mashtalir, et al. 2012), V2C (Naguib et al. 2013), Mo2C (Naguib et al. 2013), Ti_3C_2 (Naguib et al. 2011), Ta_4C_3 (Naguib, Mashtalir, et al. 2012), Nb_4C_3 (Ghidiu, Naguib, et al. 2014), (V0.5Cr0.5)3C2 (Naguib, Mashtalir, et al. 2012), $(Ti0.5Nb0.5)_2C$ (Naguib, Mashtalir, et al. 2012), and Ti_3CN (Naguib, Mashtalir, et al. 2012) are only a few of the predicted MXenes that have been produced in the lab thus far. Because $Ti_3C_2T_2$ was the first of these MXenes to be discovered, it has attracted the most attention from researchers (Naguib, Mashtalir, et al. 2012; Naguib et al. 2011). Recent investigations (Naguib et al. 2014; Lei, Zhang, and Zhou 2015) have supported $Ti_3C_2T_2$'s intriguing properties, which include strong intercalation capabilities with applications in high capacity (Ghidiu, Lukatskaya, et al. 2014; Dall'Agnese et al. 2014) and metal-ion batteries (Mashtalir et al. 2013; Xie et al. 2014).

2.4 SUMMARY

Imagine you are exploring a world of materials with unique and diverse properties. One such fascinating group is called MXenes. These materials are like the chameleons of the materials science world, able to transform and adapt to various uses. They are created from MAX phases, which are special layered compounds combining metals and elements like carbon or nitrogen.

MAX phases have a specific hexagonal structure, and when scientists chemically exfoliate these layers, they reveal MXenes. Think of it as peeling away layers of an onion to discover something new and exciting inside. The "M" in MXenes stands for transition metals like titanium or zirconium, while "X" represents carbon or nitrogen. The magic happens when these elements combine, forming bonds that give MXenes their remarkable versatility.

MXenes are not just created from a single method; they come from a variety of precursors and synthesis techniques. For example, some precursors are carbides or nitrides, while others might be more complex compounds. The process of creating MXenes can involve different approaches, from liquid-phase exfoliation to more advanced methods like hydrothermal or electrochemical exfoliation.

One standout MXene is Ti_3C_2Tx. It is like the reliable workhorse of the MXene family, easy to produce and boasting stable physical properties. This makes it an excellent candidate for innovative applications, such as sensors that can monitor biological and environmental changes.

This chapter explains the chemistry of MXenes, exploring the different ways the researchers used for synthesis, and how each method affects their properties and potential uses. Whether you are interested in their theoretical aspects or practical applications, MXenes promise a world of possibilities in materials science.

REFERENCES

Alhabeb, Mohamed, Kathleen Maleski, Babak Anasori, Pavel Lelyukh, Leah Clark, Saleesha Sin, and Yury Gogotsi. 2017. "Guidelines for synthesis and processing of two-dimensional titanium carbide (Ti_3C_2Tx MXene)." *Chemistry of Materials* 29 (18):7633–7644.

Alhabeb, Mohamed, Kathleen Maleski, Tyler S Mathis, Asia Sarycheva, Christine B Hatter, Simge Uzun, Ariana Levitt, and Yury Gogotsi. 2018. "Selective etching of silicon from Ti_3SiC_2 (MAX) to obtain 2D titanium carbide (MXene)." *Angewandte Chemie International Edition* 57 (19):5444–5448.

Alwitt, RS, H Uchi, TR Beck, and RC Alkire. 1984. "Electrochemical tunnel etching of aluminum." *Journal of The Electrochemical Society* 131 (1):13.

Anasori, Babak, Martin Dahlqvist, Joseph Halim, Eun Ju Moon, Jun Lu, Brian C Hosler, El'ad N Caspi, Steven J May, Lars Hultman, and Per Eklund. 2015. "Experimental and theoretical characterization of ordered MAX phases Mo2TiAlC2 and Mo2Ti2AlC3." *Journal of Applied Physics* 118 (9):094304.

Anasori, Babak, Maria R Lukatskaya, and Yury Gogotsi. 2017. "2D metal carbides and nitrides (MXenes) for energy storage." *Nature Reviews Materials* 2:16098.

Anasori, Babak, Yu Xie, Majid Beidaghi, Jun Lu, Brian C Hosler, Lars Hultman, Paul RC Kent, Yury Gogotsi, and Michel W Barsoum. 2015a. "Two-dimensional, ordered, double transition metals carbides (MXenes)." *ACS Nano* 9 (10):9507–9516.

Anasori, Babak, Yu Xie, Majid Beidaghi, Jun Lu, Brian C Hosler, Lars Hultman, Paul R C Kent, Yury Gogotsi, and Michel W Barsoum. 2015b. "Two-dimensional, ordered, double transition metals carbides (MXenes)." *ACS Nano* 9 (10):9507–9516. doi: 10.1021/acsnano.5b03591.

Barsoum, Michel W. 2000. "The MN+1AXN phases: A new class of solids: Thermodynamically stable nanolaminates." *Progress in Solid State Chemistry* 28 (1–4):201–281.

Barsoum, Michel W. 2013. *MAX phases: Properties of machinable ternary carbides and nitrides*. John Wiley & Sons.

Barsoum, Michel W, and Miladin Radovic. 2011. "Elastic and mechanical properties of the MAX phases." *Annual Review of Materials Research* 41:195–227.

Barsoum, MW, J Golczewski, HJ Seifert, and F Aldinger. 2002. "Fabrication and electrical and thermal properties of Ti_2InC, Hf_2InC and (Ti, Hf) 2InC." *Journal of Alloys and Compounds* 340 (1–2):173–179.

Behabtu, Natnael, Jay R Lomeda, Micah J Green, Amanda L Higginbotham, Alexander Sinitskii, Dmitry V Kosynkin, Dmitri Tsentalovich, A Nicholas G Parra-Vasquez, Judith Schmidt, and Ellina Kesselman. 2010. "Spontaneous high-concentration dispersions and liquid crystals of graphene." *Nature Nanotechnology* 5 (6):406–411.

Burger, FJ, and VFG Tull. 1953. "Gas content and electrolytic etching behaviour of aluminium." *Nature* 172 (4381):729–730.

Cambaz, Goknur Z, Gleb N Yushin, Yury Gogotsi, and Vadim G Lutsenko. 2006. "Anisotropic etching of SiC whiskers." *Nano Letters* 6 (3):548–551.

Chaudhuri, Krishnakali, Mohamed Alhabeb, Zhuoxian Wang, Vladimir M Shalaev, Yury Gogotsi, and Alexandra Boltasseva. 2018. "Highly broadband absorber using plasmonic titanium carbide (MXene)." *ACS Photonics* 5 (3):1115–1122.

Chen, Jing, Ke Chen, Dingyi Tong, Youju Huang, Jiawei Zhang, Jianming Xue, Qing Huang, and Tao Chen. 2015. "CO_2 and temperature dual responsive "Smart" MXene phases." *Chemical Communications* 51 (2):314–317.

Chen, Xingzhu, Zhouzhou Kong, Neng Li, Xiujian Zhao, and Chenghua Sun. 2016. "Proposing the prospects of Ti_3 CN transition metal carbides (MXenes) as anodes of Li-ion batteries: A DFT study." *Physical Chemistry Chemical Physics* 18 (48):32937–32943.

Chung, DDL. 1987. "Exfoliation of graphite." *Journal of Materials Science* 22 (12):4190–4198.

Dahlqvist, Martin, Jun Lu, Rahele Meshkian, Quanzheng Tao, Lars Hultman, and Johanna Rosen. 2017. "Prediction and synthesis of a family of atomic laminate phases with Kagomé-like and in-plane chemical ordering." *Science Advances* 3 (7):e1700642.

Dall'Agnese, Yohan, Maria R Lukatskaya, Kevin M Cook, Pierre-Louis Taberna, Yury Gogotsi, and Patrice Simon. 2014. "High capacitance of surface-modified 2D titanium carbide in acidic electrolyte." *Electrochemistry Communications* 48:118–122.

De, Sukanta, and Jonathan N Coleman. 2010. "Are there fundamental limitations on the sheet resistance and transmittance of thin graphene films?" *ACS Nano* 4 (5):2713–2720.

Dillon, Andrew D, Michael J Ghidiu, Alex L Krick, Justin Griggs, Steven J May, Yury Gogotsi, Michel W Barsoum, and Aaron T Fafarman. 2016. "Highly conductive optical quality solution-processed films of 2D titanium carbide." *Advanced Functional Materials* 26 (23):4162–4168. doi: 10.1002/adfm.201600357.

Ding, Li, Yanying Wei, Yanjie Wang, Hongbin Chen, Jürgen Caro, and Haihui Wang. 2017. "A two-dimensional lamellar membrane: MXene nanosheet stacks." *Angewandte Chemie International Edition* 56 (7):1825–1829.

Eklund, Per, Manfred Beckers, Ulf Jansson, Hans Högberg, and Lars Hultman. 2010. "The Mn+ 1AXn phases: Materials science and thin-film processing." *Thin Solid Films* 518 (8):1851–1878.

Enyashin, AN, and AL Ivanovskii. 2013. "Two-dimensional titanium carbonitrides and their hydroxylated derivatives: Structural, electronic properties and stability of MXenes Ti_3C_2– xNx (OH) 2 from DFTB calculations." *Journal of Solid State Chemistry* 207:42–48.

Gao, Yupeng, Libo Wang, Zhengyang Li, Yafei Zhang, Baolin Xing, Chuanxiang Zhang, and Aiguo Zhou. 2015. "Electrochemical performance of Ti 3 C 2 supercapacitors in KOH electrolyte." *Journal of Advanced Ceramics* 4 (2):130–134.

Ghidiu, Michael, Maria R Lukatskaya, Meng-Qiang Zhao, Yury Gogotsi, and Michel W Barsoum. 2014. "Conductive two-dimensional titanium carbide 'clay' with high volumetric capacitance." *Nature* 516 (7529):78–81.

Ghidiu, Michael, Michael Naguib, Chenyang Shi, Olha Mashtalir, LM Pan, Bo Zhang, Jian Yang, Yury Gogotsi, Simon JL Billinge, and Michel W Barsoum. 2014. "Synthesis and characterization of two-dimensional Nb 4 C 3 (MXene)." *Chemical Communications* 50 (67):9517–9520.

Gogotsi, Yury. 2015. "Chemical vapour deposition: Transition metal carbides go 2D." *Nature Materials* 14 (11).

Guo, Lu, Xianfen Wang, Zhi Yi Leong, Runwei Mo, Linfeng Sun, and Hui Ying Yang. 2018. "Ar plasma modification of 2D MXene Ti_3C_2Tx nanosheets for efficient capacitive desalination." *FlatChem* 8:17–24.

Halim, Joseph, Kevin M Cook, Michael Naguib, Per Eklund, Yury Gogotsi, Johanna Rosen, and Michel W Barsoum. 2016. "X-ray photoelectron spectroscopy of select multi-layered transition metal carbides (MXenes)." *Applied Surface Science* 362:406–417.

Halim, Joseph, Sankalp Kota, Maria R Lukatskaya, Michael Naguib, Meng-Qiang Zhao, Eun Ju Moon, Jeremy Pitock, Jagjit Nanda, Steven J May, and Yury Gogotsi. 2016. "Synthesis and characterization of 2D molybdenum carbide (MXene)." *Advanced Functional Materials* 26 (18):3118–3127.

Halim, J, J Palisaitis, J Lu, J Thörnberg, E J Moon, M Precner, P Eklund, POÅ Persson, MW Barsoum, and J Rosen. 2018. "Synthesis of two-dimensional Nb1.33C (MXene) with randomly distributed vacancies by etching of the quaternary solid solution $(Nb_{2/3}Sc_{1/3})_2AlC$ MAX phase." *ACS Applied Nano Materials* 1 (6):2455–2460. doi: 10.1021/acsanm.8b00332.

Handoko, Albertus D, Kurt D Fredrickson, Babak Anasori, Kurt W Convey, Luke R Johnson, Yury Gogotsi, Aleksandra Vojvodic, and Zhi Wei Seh. 2017. "Tuning the basal plane functionalization of two-dimensional metal carbides (MXenes) to control hydrogen evolution activity." *ACS Applied Energy Materials* 1 (1):173–180.

Hantanasirisakul, Kanit, and Yury Gogotsi. 2018. "Electronic and optical properties of 2D transition metal carbides and nitrides (MXenes)." *Advanced Materials* 30 (52):1804779. doi: 10.1002/adma.201804779.

Hantanasirisakul, Kanit, Meng-Qiang Zhao, Patrick Urbankowski, Joseph Halim, Babak Anasori, Sankalp Kota, Chang E Ren, Michel W Barsoum, and Yury Gogotsi. 2016. "Fabrication of Ti_3C_2Tx MXene transparent thin films with tunable optoelectronic properties." *Advanced Electronic Materials* 2 (6):1600050. doi: 10.1002/aelm.201600050.

Hope, Michael A, Alexander C Forse, Kent J Griffith, Maria R Lukatskaya, Michael Ghidiu, Yury Gogotsi, and Clare P Grey. 2016. "NMR reveals the surface functionalisation of Ti_3C_2 MXene." *Physical Chemistry Chemical Physics* 18 (7):5099–5102.

Hu, C, C C Lai, Q Tao, J Lu, J Halim, L Sun, J Zhang, J Yang, B Anasori, J Wang, Y Sakka, L Hultman, P Eklund, J Rosen, and M W Barsoum. 2015. "Mo2Ga2C: A new ternary nanolaminated carbide." *Chemical Communications* 51 (30):6560–6563. doi: 10.1039/C5CC00980D.

Huang, Kai, Zhongjun Li, Jing Lin, Gang Han, and Peng Huang. 2018. "Two-dimensional transition metal carbides and nitrides (MXenes) for biomedical applications." *Chemical Society Reviews* 47 (14):5109–5124. doi: 10.1039/C7CS00838D.

Jung de Andrade, Mônica, Márcio Dias Lima, Viera Skákalová, Carlos Pérez Bergmann, and Siegmar Roth. 2007. "Electrical properties of transparent carbon nanotube networks prepared through different techniques." *Physica Status Solidi (RRL)–Rapid Research Letters* 1 (5):178–180.

Kajiyama, Satoshi, Lucie Szabova, Hiroki Iinuma, Akira Sugahara, Kazuma Gotoh, Keitaro Sodeyama, Yoshitaka Tateyama, Masashi Okubo, and Atsuo Yamada. 2017. "Enhanced Li-Ion accessibility in MXene titanium carbide by steric chloride termination." *Advanced Energy Materials* 7 (9):1601873.

Khazaei, Mohammad, Masao Arai, Taizo Sasaki, Mehdi Estili, and Yoshio Sakka. 2014. "Two-dimensional molybdenum carbides: Potential thermoelectric materials of the MXene family." *Physical Chemistry Chemical Physics* 16 (17):7841–7849.

Khodan, AN, SL Kanashenko, and D-G Crete. 2012. "Kinetics of formation and growth of epitaxial SrTiO 3 films of single-crystal (001) SrTiO 3 supports." *Protection of Metals and Physical Chemistry of Surfaces* 48 (1):59–67.

Kim, Hyunho, Babak Anasori, Yury Gogotsi, and Husam N Alshareef. 2017. "Thermoelectric properties of two-dimensional molybdenum-based MXenes." *Chemistry of Materials* 29 (15):6472–6479.

Kurra, Narendra, Bilal Ahmed, Yury Gogotsi, and Husam N Alshareef. 2016. "MXene-on-paper coplanar microsupercapacitors." *Advanced Energy Materials* 6 (24):1601372.

Kurtoglu, Murat, Michael Naguib, Yury Gogotsi, and Michel W Barsoum. 2012. "First principles study of two-dimensional early transition metal carbides." *Mrs Communications* 2 (4):133–137.

Lai, C-C, Rahele Meshkian, Martin Dahlqvist, Jun Lu, L-Å Näslund, O Rivin, EN Caspi, O Ozeri, Lars Hultman, and Per Eklund. 2015. "Structural and chemical determination of the new nanolaminated carbide Mo2Ga2C from first principles and materials analysis." *Acta Materialia* 99:157–164.

Lei, Jincheng, Xu Zhang, and Zhen Zhou. 2015. "Recent advances in MXene: Preparation, properties, and applications." *Frontiers of Physics in China* 10 (3):276–286.

Li, Tengfei, Lulu Yao, Qinglei Liu, Jiajun Gu, Ruichun Luo, Jinghan Li, Xudong Yan, Weiqiang Wang, Pan Liu, Bin Chen, Wang Zhang, Waseem Abbas, Raheela Naz, and Di Zhang. 2018. "Fluorine-free synthesis of high-purity Ti_3C_2Tx (T=OH, O) via Alkali treatment." *Angewandte Chemie International Edition* 57 (21):6115–6119. doi: 10.1002/anie.201800887.

Li, Xuesong, Weiwei Cai, Jinho An, Seyoung Kim, Junghyo Nah, Dongxing Yang, Richard Piner, Aruna Velamakanni, Inhwa Jung, and Emanuel Tutuc. 2009. "Large-area synthesis of high-quality and uniform graphene films on copper foils." *Science* 324 (5932):1312–1314.

Li, Zhengyang, Libo Wang, Dandan Sun, Yude Zhang, Baozhong Liu, Qianku Hu, and Aiguo Zhou. 2015. "Synthesis and thermal stability of two-dimensional carbide MXene Ti_3C_2." *Materials Science and Engineering: B* 191:33–40. doi: 10.1016/j.mseb.2014.10.009.

Liang, Xiao, Arnd Garsuch, and Linda F Nazar. 2015. "Sulfur cathodes based on conductive MXene nanosheets for high-performance lithium–sulfur batteries." *Angewandte Chemie International Edition* 54 (13):3907–3911.

Lin, Han, Shanshan Gao, Chen Dai, Yu Chen, and Jianlin Shi. 2017. "A two-dimensional biodegradable niobium carbide (MXene) for photothermal tumor eradication in NIR-I and NIR-II biowindows." *Journal of the American Chemical Society* 139 (45):16235–16247.

Lin, Han, Xingang Wang, Luodan Yu, Yu Chen, and Jianlin Shi. 2017. "Two-dimensional ultrathin MXene ceramic nanosheets for photothermal conversion." *Nano Letters* 17 (1):391.

Lin, ZJ, MJ Zhuo, MS Li, JY Wang, and YC Zhou. 2007. "Synthesis and microstructure of layered-ternary Ti2AlN ceramic." *Scripta Materialia* 56 (12):1115–1118.

Lipatov, Alexey, Mohamed Alhabeb, Maria R Lukatskaya, Alex Boson, Yury Gogotsi, and Alexander Sinitskii. 2016. "Effect of synthesis on quality, electronic properties and environmental stability of individual monolayer Ti_3C_2 MXene flakes." *Advanced Electronic Materials* 2 (12):1600255.

Liu, Zhibo, Chuan Xu, Ning Kang, Libin Wang, Yixiao Jiang, Jiao Du, Ying Liu, Xiu-Liang Ma, Hui-Ming Cheng, and Wencai Ren. 2016. "Unique domain structure of two-dimensional α-Mo2C superconducting crystals." *Nano Letters* 16 (7):4243–4250. doi: 10.1021/acs.nanolett.6b01265.

Liu, Zhimou, Erdong Wu, Jiemin Wang, Yuhai Qian, Huimin Xiang, Xichao Li, Qianqian Jin, Guangai Sun, Xiping Chen, and Jingyang Wang. 2014. "Crystal structure and formation mechanism of (Cr2/3Ti1/3) 3AlC2 MAX phase." *Acta Materialia* 73:186–193.

Lukatskaya, Maria R, Joseph Halim, Boris Dyatkin, Michael Naguib, Yulia S Buranova, Michel W Barsoum, and Yury Gogotsi. 2014a. "Room-temperature carbide-derived carbon synthesis by electrochemical etching of MAX phases." *Angewandte Chemie International Edition* 53 (19):4877–4880.

Lukatskaya, Maria R, Joseph Halim, Boris Dyatkin, Michael Naguib, Yulia S Buranova, Michel W Barsoum, and Yury Gogotsi. 2014b. "Room-temperature carbide-derived carbon synthesis by electrochemical etching of MAX phases." *Angewandte Chemie International Edition* 53 (19):4877–4880. doi: 10.1002/anie.201402513.

Lukatskaya, Maria R, Olha Mashtalir, Chang E Ren, Yohan Dall'Agnese, Patrick Rozier, Pierre Louis Taberna, Michael Naguib, Patrice Simon, Michel W Barsoum, and Yury Gogotsi. 2013. "Cation intercalation and high volumetric capacitance of two-dimensional titanium carbide." *Science* 341 (6153):1502. doi: 10.1126/science.1241488.

Ma, Zhinan, Zhenpeng Hu, Xudong Zhao, Qing Tang, Dihua Wu, Zhen Zhou, and Lixin Zhang. 2014. "Tunable band structures of heterostructured bilayers with transition-metal dichalcogenide and MXene monolayer." *The Journal of Physical Chemistry C* 118 (10):5593–5599. doi: 10.1021/jp500861n.

Makino, Hisao, Aki Miyake, Takahiro Yamada, Naoki Yamamoto, and Tetsuya Yamamoto. 2009. "Influence of substrate temperature and Zn-precursors on atomic layer deposition of polycrystalline ZnO films on glass." *Thin Solid Films* 517 (10):3138–3142.

Mashtalir, Olha, Michael Naguib, Vadym N Mochalin, Yohan Dall'Agnese, Min Heon, Michel W Barsoum, and Yury Gogotsi. 2013. "Intercalation and delamination of layered carbides and carbonitrides." *Nature Communications* 4 (1):1–7.

Meshkian, Rahele, Martin Dahlqvist, Jun Lu, Björn Wickman, Joseph Halim, Jimmy Thörnberg, Quanzheng Tao, Shixuan Li, Saad Intikhab, Joshua Snyder, Michel W Barsoum, Melike Yildizhan, Justinas Palisaitis, Lars Hultman, Per O Å Persson, and Johanna Rosen. 2018. "W-based atomic laminates and their 2D derivative W1.33C MXene with vacancy ordering." *Advanced Materials* 30 (21):1706409. doi: 10.1002/adma.201706409.

Meshkian, Rahele, Lars-Åke Näslund, Joseph Halim, Jun Lu, Michel W Barsoum, and Johanna Rosen. 2015. "Synthesis of two-dimensional molybdenum carbide, Mo2C, from the gallium based atomic laminate Mo_2Ga_2C." *Scripta Materialia* 108:147–150. doi: 10.1016/j.scriptamat.2015.07.003.

Meshkian, Rahele, Quanzheng Tao, Martin Dahlqvist, Jun Lu, Lars Hultman, and Johanna Rosen. 2017. "Theoretical stability and materials synthesis of a chemically ordered MAX phase, Mo_2ScAlC_2, and its two-dimensional derivate Mo_2ScC_2 MXene." *Acta Materialia* 125:476–480. doi: 10.1016/j.actamat.2016.12.008.

Morel, Alban, Yann Borjon-Piron, Raúl Lucio Porto, Thierry Brousse, and Daniel Bélanger. 2016. "Suitable conditions for the use of vanadium nitride as an electrode for electrochemical capacitor." *Journal of the Electrochemical Society* 163 (6):A1077–A1082.

Naguib, Michael, Jérémy Come, Boris Dyatkin, Volker Presser, Pierre-Louis Taberna, Patrice Simon, Michel W Barsoum, and Yury Gogotsi. 2012. "MXene: A promising transition metal carbide anode for lithium-ion batteries." *Electrochemistry Communications* 16 (1):61–64.

Naguib, Michael, and Yury Gogotsi. 2015. "Synthesis of two-dimensional materials by selective extraction." *Accounts of Chemical Research* 48 (1):128–135. doi: 10.1021/ar500346b.

Naguib, Michael, Joseph Halim, Jun Lu, Kevin M Cook, Lars Hultman, Yury Gogotsi, and Michel W Barsoum. 2013. "New two-dimensional niobium and vanadium carbides as promising materials for Li-Ion batteries." *Journal of the American Chemical Society* 135 (43):15966–15969. doi: 10.1021/ja405735d.

Naguib, Michael, Murat Kurtoglu, Volker Presser, Jun Lu, Junjie Niu, Min Heon, Lars Hultman, Yury Gogotsi, and Michel W Barsoum. 2011. "Two-dimensional nanocrystals produced by exfoliation of Ti(3)AlC(2)." *Advanced Materials* 23 (37):4248–4253. doi: 10.1002/adma.201102306.

Naguib, Michael, Olha Mashtalir, Joshua Carle, Volker Presser, Jun Lu, Lars Hultman, Yury Gogotsi, and Michel W Barsoum. 2012. "Two-dimensional transition metal carbides." *ACS Nano* 6 (2):1322–1331. doi: 10.1021/nn204153h.

Naguib, Michael, Vadym Mochalin, Michel W Barsoum, and Yury Gogotsi. 2014. "25th anniversary article: MXenes: A new family of two-dimensional materials." *Advanced Materials* 26 (7):992–1005. doi: 10.1002/adma.201304138.

Naguib, Michael, Raymond R Unocic, Beth L Armstrong, and Jagjit Nanda. 2015. "Large-scale delamination of multi-layers transition metal carbides and carbonitrides 'MXenes'." *Dalton Transactions* 44 (20):9353–9358.

Ng, Vincent Ming Hong, Hui Huang, Kun Zhou, Pooi See Lee, Wenxiu Que, Jason Zhichuan Xu, and LB Kong. 2017. "Recent progress in layered transition metal carbides and/or nitrides (MXenes) and their composites: Synthesis and applications." *Journal of Materials Chemistry* 5 (7):3039–3068.

Parthé, Erwin, and V Sadogopan. 1963. "The structure of dimolybdenum carbide by neutron diffraction technique." *Acta Crystallographica* 16 (3):202–205.

Peng, Chao, Ping Wei, Xin Chen, Yongli Zhang, Feng Zhu, Yonghai Cao, Hongjuan Wang, Hao Yu, and Feng Peng. 2018. "A hydrothermal etching route to synthesis of 2D MXene (Ti_3C_2, Nb2C): Enhanced exfoliation and improved adsorption performance." *Ceramics International* 44 (15):18886–18893. doi: 10.1016/j.ceramint.2018.07.124.

Presser, Volker, Michael Naguib, Laurent Chaput, Atsushi Togi, Gilles Hug, and Michel W Barsoum. 2013. "Erratum: First-order Raman scattering of the MAX phases: Ti2AlN, Ti2AlC0. 5N0. 5, Ti2AlC, (Ti0. 5V0. 5) 2AlC, V2AlC, Ti_3AlC_2, and Ti3GeC2." *Journal of Raman Spectroscopy* 44 (7):1060–1060.

Rakhi, Raghavan B, Bilal Ahmed, Mohamed N Hedhili, Dalaver H Anjum, and Husam N Alshareef. 2015. "Effect of postetch annealing gas composition on the structural and electrochemical properties of Ti2CT x MXene electrodes for supercapacitor applications." *Chemistry of Materials* 27 (15):5314–5323.

Ren, Chang E, Kelsey B Hatzell, Mohamed Alhabeb, Zheng Ling, Khaled A Mahmoud, and Yury Gogotsi. 2015. "Charge-and size-selective ion sieving through Ti_3C_2Tx MXene membranes." *The Journal of Physical Chemistry Letters* 6 (20):4026–4031.

Ronchi, Rodrigo Mantovani, Jeverson Teodoro Arantes, and Sydney Ferreira Santos. 2019. "Synthesis, structure, properties and applications of MXenes: Current status and perspectives." *Ceramics International* 45 (15):18167–1818.

Roy-Mayhew, Joseph D, and Ilhan A Aksay. 2014. "Graphene materials and their use in dye-sensitized solar cells." *Chemical Reviews* 114 (12):6323–6348.

Sarycheva, Asia, Taron Makaryan, Kathleen Maleski, Elumalai Satheeshkumar, Armen Melikyan, Hayk Minassian, Masahiro Yoshimura, and Yury Gogotsi. 2017. "Two-dimensional titanium carbide (MXene) as surface-enhanced Raman scattering substrate." *The Journal of Physical Chemistry C* 121 (36):19983–19988.

Seh, Zhi Wei, Kurt D Fredrickson, Babak Anasori, Jakob Kibsgaard, Alaina L Strickler, Maria R Lukatskaya, Yury Gogotsi, Thomas F Jaramillo, and Aleksandra Vojvodic. 2016. "Two-dimensional molybdenum carbide (MXene) as an efficient electrocatalyst for hydrogen evolution." *ACS Energy Letters* 1 (3):589–594. doi: 10.1021/acsenergylett.6b00247.

Shahzad, Faisal, Mohamed Alhabeb, Christine Hatter, Babak Anasori, Soon Hong, Chong Min Koo, and Yury Gogotsi. 2016. "Electromagnetic interference shielding with 2D transition metal carbides (MXenes)." *Science* 353:1137–1140. doi: 10.1126/science.aag2421.

Shein, IR, and AL Ivanovskii. 2012. "Graphene-like titanium carbides and nitrides Tin+ 1Cn, Tin+ 1Nn (n= 1, 2, and 3) from de-intercalated MAX phases: First-principles probing of their structural, electronic properties and relative stability." *Computational Materials Science* 65:104–114.

Song, Young Il, Cheol-Min Yang, Dong Young Kim, Hirofumi Kanoh, and Katsumi Kaneko. 2008. "Flexible transparent conducting single-wall carbon nanotube film with network bridging method." *Journal of Colloid and Interface Science* 318 (2):365–371.

Soundiraraju, Bhuvaneswari, and Benny Kattikkanal George. 2017. "Two-dimensional titanium nitride (Ti2N) MXene: Synthesis, characterization, and potential application as surface-enhanced Raman scattering substrate." *ACS Nano* 11 (9):8892–8900. doi: 10.1021/acsnano.7b03129.

Spanier, Jonathan E, Surojit Gupta, Maher Amer, and Michel W Barsoum. 2005. "Vibrational behavior of the M n+ 1 A X n phases from first-order Raman scattering (M= Ti, V, Cr, A= Si, X= C, N)." *Physical Review B* 71 (1):012103.

Srimuk, Pattarachai, Joseph Halim, Juhan Lee, Quanzheng Tao, Johanna Rosen, and Volker Presser. 2018. "Two-dimensional molybdenum carbide (MXene) with divacancy ordering for brackish and seawater desalination via cation and anion intercalation." *ACS Sustainable Chemistry & Engineering* 6 (3):3739–3747.

Srimuk, Pattarachai, Friedrich Kaasik, Benjamin Krüner, Aura Tolosa, Simon Fleischmann, Nicolas Jäckel, Mehmet C Tekeli, Mesut Aslan, Matthew E Suss, and Volker Presser.

2016. "MXene as a novel intercalation-type pseudocapacitive cathode and anode for capacitive deionization." *Journal of Materials Chemistry A* 4 (47):18265–18271.

Sun, Zhimei, Denis Music, Rajeev Ahuja, Sa Li, and Jochen M Schneider. 2004. "Bonding and classification of nanolayered ternary carbides." *Physical Review B* 70 (9):092102.

Sun, Wanmei, Smit Shah, Yexiao Chen, Zeyi Tan, Huili Gao, Touseef Habib, Miladin Radovic, and Micah Green. 2017. "Electrochemical etching of Ti2AlC to Ti2CT:X (MXene) in low-concentration hydrochloric acid solution." *Journal of Materials Chemistry A* 5. doi: 10.1039/C7TA05574A.

Sun, Renhui, Hao-Bin Zhang, Ji Liu, Xi Xie, Rui Yang, Yue Li, Song Hong, and Zhong-Zhen Yu. 2017. "Highly conductive transition metal carbide/carbonitride(MXene)@polystyrene nanocomposites fabricated by electrostatic assembly for highly efficient electromagnetic interference shielding." *Advanced Functional Materials* 27 (45):1702807. doi: 10.1002/adfm.201702807.

Tao, Quanzheng, Martin Dahlqvist, Jun Lu, Sankalp Kota, Rahele Meshkian, Joseph Halim, Justinas Palisaitis, Lars Hultman, Michel W Barsoum, Per OÅ Persson, and Johanna Rosen. 2017. "Two-dimensional Mo1.33C MXene with divacancy ordering prepared from parent 3D laminate with in-plane chemical ordering." *Nature Communications* 8 (1):14949. doi: 10.1038/ncomms14949.

Tran, Minh H, Timo Schäfer, Ali Shahraei, Michael Dürrschnabel, Leopoldo Molina-Luna, Ulrike I Kramm, and Christina S Birkel. 2018. "Adding a new member to the MXene family: Synthesis, structure, and electrocatalytic activity for the hydrogen evolution reaction of V4C3Tx." *ACS Applied Energy Materials* 1 (8):3908–3914. doi: 10.1021/acs aem.8b00652.

Tung, Vincent C, Matthew J Allen, Yang Yang, and Richard B Kaner. 2009. "High-throughput solution processing of large-scale graphene." *Nature Nanotechnology* 4 (1):25.

Urbankowski, Patrick, Babak Anasori, Kanit Hantanasirisakul, Long Yang, Lihua Zhang, Bernard Haines, Steven J May, Simon JL Billinge, and Yury Gogotsi. 2017. "2D molybdenum and vanadium nitrides synthesized by ammoniation of 2D transition metal carbides (MXenes)." *Nanoscale* 9 (45):17722–17730.

Urbankowski, Patrick, Babak Anasori, Taron Makaryan, Dequan Er, Sankalp Kota, Patrick L Walsh, Mengqiang Zhao, Vivek B Shenoy, Michel W Barsoum, and Yury Gogotsi. 2016. "Synthesis of two-dimensional titanium nitride Ti4N3 (MXene)." *Nanoscale* 8 (22):11385–11391.

Verger, Louisiane, Chuan Xu, Varun Natu, Hui-Ming Cheng, Wencai Ren, and Michel W Barsoum. 2019. "Overview of the synthesis of MXenes and other ultrathin 2D transition metal carbides and nitrides." *Current Opinion in Solid State and Materials Science* 23 (3):149–163. doi: 10.1016/j.cossms.2019.02.001.

Wan, Jiayu, Feng Gu, Wenzhong Bao, Jiaqi Dai, Fei Shen, Wei Luo, Xiaogang Han, Daniel Urban, and Liangbing Hu. 2015. "Sodium-ion intercalated transparent conductors with printed reduced graphene oxide networks." *Nano Letters* 15 (6):3763–3769.

Wang, Xianfen, Satoshi Kajiyama, Hiroki Iinuma, Eiji Hosono, Shinji Oro, Isamu Moriguchi, Masashi Okubo, and Atsuo Yamada. 2015. "Pseudocapacitance of MXene nanosheets for high-power sodium-ion hybrid capacitors." *Nature Communications* 6 (1):6544. doi: 10.1038/ncomms7544.

Wang, Xuefeng, Xi Shen, Yurui Gao, Zhaoxiang Wang, Richeng Yu, and Liquan Chen. 2015. "Atomic-scale recognition of surface structure and intercalation mechanism of Ti_3C_2X." *Journal of the American Chemical Society* 137 (7):2715–2721.

Xie, Xiaohong, Siguo Chen, Wei Ding, Yao Nie, and Zidong Wei. 2013. "An extraordinarily stable catalyst: Pt NPs supported on two-dimensional Ti 3 C 2 X 2 (X= OH, F) nanosheets for oxygen reduction reaction." *Chemical Communications* 49 (86):10112–10114.

Xie, Yu, Michael Naguib, Vadym N Mochalin, Michel W Barsoum, Yury Gogotsi, Xiqian Yu, Kyung-Wan Nam, Xiao-Qing Yang, Alexander I Kolesnikov, and Paul RC Kent. 2014. "Role of surface structure on Li-ion energy storage capacity of two-dimensional transition-metal carbides." *Journal of the American Chemical Society* 136 (17):6385–6394.

Xu, Chuan, Shuang Song, Zhibo Liu, Long Chen, Libin Wang, Dingxun Fan, Ning Kang, Xiuliang Ma, Hui-Ming Cheng, and Wencai Ren. 2017. "Strongly coupled high-quality graphene/2D superconducting Mo2C vertical heterostructures with aligned orientation." *ACS Nano* 11 (6):5906–5914.

Xu, Chuan, Libin Wang, Zhibo Liu, Long Chen, Jingkun Guo, Ning Kang, Xiu-Liang Ma, Hui-Ming Cheng, and Wencai Ren. 2015. "Large-area high-quality 2D ultrathin Mo2C superconducting crystals." *Nature Materials* 14:1135. doi: 10.1038/nmat4374.

Yang, Sheng, Panpan Zhang, Faxing Wang, Antonio Gaetano Ricciardulli, Martin Lohe, Paul Blom, and Xinliang Feng. 2018. "Fluoride-free synthesis of two-dimensional titanium carbide (MXene) using a binary aqueous system." *Angewandte Chemie International Edition in English* 57. doi: 10.1002/anie.201809662.

Yang, Zhibin, and Jianhua Hao. 2016. "Progress in pulsed laser deposited two-dimensional layered materials for device applications." *Journal of Materials Chemistry C* 4 (38):8859–8878.

Yang, Zhibin, Jing Ren, Zhitao Zhang, Xuli Chen, Guozhen Guan, Longbin Qiu, Ye Zhang, and Huisheng Peng. 2015. "Recent advancement of nanostructured carbon for energy applications." *Chemical reviews* 115 (11):5159–5223.

Ye, Qun, Peng Xiao, Wulong Liu, Ke Chen, Tao Chen, Jianming Xue, Shiyu Du, and Qing Huang. 2015. "Exploring the potential of exfoliated ternary ultrathin Ti 4 AlN 3 nanosheets for fabricating hybrid patterned polymer brushes." *RSC Advances* 5 (86):70339–70344.

Yu, Xue-fang, Yan-chun Li, Jian-bo Cheng, Zhen-bo Liu, Qing-zhong Li, Wen-zuo Li, Xin Yang, and Bo Xiao. 2015. "Monolayer Ti2CO2: A promising candidate for NH3 sensor or capturer with high sensitivity and selectivity." *ACS Applied Materials & Interfaces* 7 (24):13707–13713.

Zhang, Xiaodong, Jianguang Xu, Hui Wang, Jiajia Zhang, Hanbing Yan, Bicai Pan, Jingfang Zhou, and Yi Xie. 2013. "Ultrathin nanosheets of MAX phases with enhanced thermal and mechanical properties in polymeric compositions: Ti3Si0.75Al0.25C2." *Angewandte Chemie International Edition* 52 (16):4361–4365. doi: 10.1002/anie.201300285.

Zhang, Zhi, Fan Zhang, Huichao Wang, Cheuk Ho Chan, Wei Lu, and Ji-yan Dai. 2017. "Substrate orientation-induced epitaxial growth of face centered cubic Mo 2 C superconductive thin film." *Journal of Materials Chemistry C* 5 (41):10822–10827.

Zhou, Jie, Fagui Qiu, Lu Shen, Fangzhi Li, Jianming Xue, Michel W Barsoum, and Qing Huang. 2014. "Pulse electric current–aided reactive sintering of high-purity Zr 3 Al 3 C 5." *Journal of the American Ceramic Society* 97 (4):1296–1302.

Zhou, Jie, Xianhu Zha, Fan Y Chen, Qun Ye, Per Eklund, Shiyu Du, and Qing Huang. 2016. "A two-dimensional zirconium carbide by selective etching of Al3C3 from nanolaminated Zr3Al3C5." *Angewandte Chemie International Edition* 55 (16):5008–5013. doi: 10.1002/anie.201510432.

Zhou, Jie, Xianhu Zha, Xiaobing Zhou, Fanyan Chen, Guoliang Gao, Shuwei Wang, Cai Shen, Tao Chen, Chunyi Zhi, and Per Eklund. 2017. "Synthesis and electrochemical properties of two-dimensional hafnium carbide." *ACS Nano* 11 (4):3841–3850.

Zhou, Yan-Chun, Ling-Feng He, Zhi-Jun Lin, and Jing-Yang Wang. 2013. "Synthesis and structure–property relationships of a new family of layered carbides in Zr-Al (Si)-C and Hf-Al (Si)-C systems." *Journal of the European Ceramic Society* 33 (15–16):2831–2865.

Zhu, Jingwen, Ming Wang, Miaoqiang Lyu, Yalong Jiao, Aijun Du, Bin Luo, Ian Gentle, and Lianzhou Wang. 2018. "Two-dimensional titanium carbonitride MXene for high-performance sodium ion batteries." *ACS Applied Nano Materials* 1 (12):6854–6863. doi: 10.1021/acsanm.8b01330.

Zhu, Yanwu, Shanthi Murali, Weiwei Cai, Xuesong Li, Ji Won Suk, Jeffrey R Potts, and Rodney S Ruoff. 2010. "Graphene and graphene oxide: Synthesis, properties, and applications." *Advanced Materials* 22 (35):3906–3924.

3 Defect Engineering/Physical Properties of Titanium-Based MXene

3.1 INTRODUCTION

The investigation of MXenes has lately gained significance in the realm of modern materials research. The 2D transition metal carbides and nitrides in this family possess remarkable structural and functional characteristics, making them very desirable. MXenes are produced by selectively etching the A layers of layered structures known as MAX phases. These MAX phases consist of transition metals (M), carbon or nitrogen (X), and aluminum or other group IIIA/IVA elements (A). This process produces thin, flexible layers with surfaces that attract water and have outstanding electrical conductivity. Due to their distinctive properties, MXenes have significant potential for several applications such as energy storage, catalysis, and protection against electromagnetic interference (Naguib et al. 2014; Anasori, Lukatskaya, and Gogotsi 2017).

Defect engineering is a deliberate technique that enhances the flexibility of MXenes by intentionally introducing or exploiting material faults. Vacancies, interstitials, and substitutional atoms are types of defects that may have a substantial impact on the mechanical, optical, and electrical properties of MXenes. Defect engineering may be used by researchers to adapt MXenes for specific applications, therefore improving their performance capabilities and unlocking new functionality (Tang, Zhou, and Shen 2012; Halim et al. 2016).

This chapter focuses on the principles of defect engineering and its significant influence on the characteristics of MXene. First, we provide a thorough examination of the intrinsic physical characteristics of MXenes, including their mechanical, optical, electrical, magnetic, and stability features. Next, we investigate how defect engineering might be used to alter these properties, therefore improving the performance of MXenes in different technological fields (Han et al. 2016; Alhabeb et al. 2017).

3.2 DEFECT ENGINEERING AND PROPERTIES

MXene has gained prominence as a material with exceptional physical and chemical characteristics, such as strong conductivity, large specific surface area, and high carrier mobility (Ghorbani-Asl et al. 2017; Nan et al. 2021). Heteroatoms doping may be used to modify the microstructure of MXene, whether with metal or nonmetal elements, in order to enhance its performance. Defect engineering of MXenes involves the intentional introduction of inherent defects, such as atomic vacancies and active

DOI: 10.1201/9781003319658-3

edges, or heteroatom dopants, which are metal or nonmetal atoms. These defects are used to even disrupt the original arrangement of MXenes nanostructures, with the aim of enhancing their chemical and electrical characteristics (Ahmed, Ghazaly, and Rosen 2020; Sun et al. 2019; Wei et al. 2021; Cui et al. 2021). The N defects engineering of MXene has gained significant interest among other doping types. This is primarily because it has favorable benefits such as enhancing electric conductivity, improving surface wettability, and increasing the number of electrochemical active sites (Tang et al. 2017; Yoon et al. 2018; Yang et al. 2017, 2018; Wen et al. 2017).

3.2.1 MXene's Mechanical Characteristics

Researchers are working to create strong polymeric fibers after creating a unique inorganic nanometer-scale substance. The goal of this is to raise performance. A fantastic product might be produced when the diverse attributes of the materials are blended. Nadia et al. (Taloub et al. 2019) projected a simple method in 2019 based on the covalent bonding of a $Ti_3C_2(OH)_2$ (MXene) derivative with a structure comparable to graphene to the surface of Poly-pyridobisimidazole (PIPD) with the addition of a linker known as (3-Aminopropyl) triethoxysilane (APTES). Reliability has been demonstrated for the MXene (Ti_3C_2Tx) derivative designated as a member of the (2D) MXene-like family of transition metal carbides (TMC) (Li et al. 2017).

By etching ternary carbide (Ti_3AlC_2), a predecessor to the MAX phase, with HF, one can create Ti_3C_2Tx, an MXene derivative (Wang, Zhou, et al. 2016). The resulting 2D nanostructures have excellent mechanical, electrical, and optical properties (Naguib et al. 2012). The bonding process called mutual sharing was employed to merge and embed the $Ti_3C_2(OH)_2$ layers into the PIPD fiber surface. Subsequently, APTES was adjusted to align with the fiber's surface layers. Figure 3.1a illustrates that after 30 minutes at room temperature, the siloxane functionality of the APTES molecule hydrolyzes into Si-OH when the PIPD-siloxane fiber layers are introduced into a solvent (ethanol).

The PIPD-siloxane fiber solution was then hydrolyzed on the $Ti_3C_2(OH)_2$ MXene layers for 24 hours at 55°C. Maintaining the mechanically based qualities and other presentations associated with the fiber is the major goal of this procedure for the surface treatment of the fiber. After 6 hours in a vacuum oven, the dehydrated product was produced by washing the acquired fiber layers twice to three times with 100% ethanol (Taloub et al. 2019). To assess the impact of chemical embedding on the mechanical properties of the PIPD fiber surface, researchers measured the IFSS and TS values 1 week later. The results revealed that the untreated PIPD fiber had an IFSS of 25.13 MPa and a TS of 6.76 GPa. After treating the fiber with nitric acid, the TS slightly decreased to 6.56 GPa, while the IFSS increased to 30.94 MPa.

In the second processing stage, the IFSS increases to 34.62 MPa, while the TS decreases to 6.45 GPa. The next step involves applying these findings to enhance the fiber/epoxy bonding process after adding carboxylic acid (–COOH) terminal functional groups to the surface of the PIPD fiber (Wu and Shyng 2004).

The third step involves adding $SOCl_2$ to the reaction, which transforms the acidic –COOH functional groups into –COCl groups, reducing the TS and IFSS pressures to 6.34 GPa and 32.83 MPa, respectively. This modification improves our

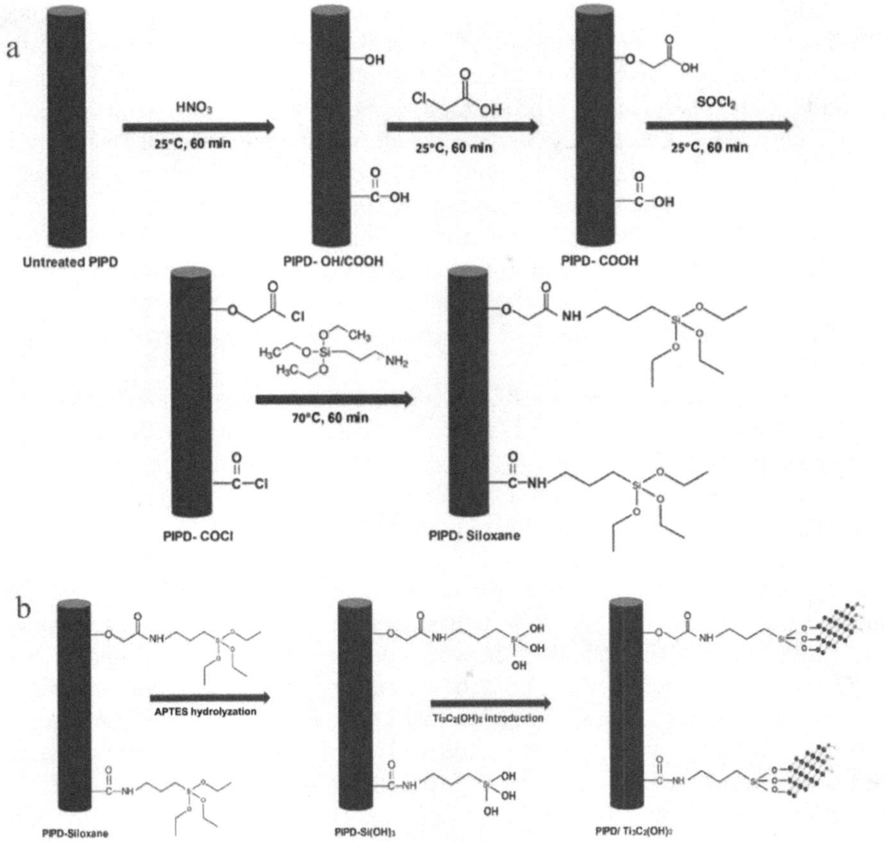

FIGURE 3.1 (a) The manufacture of PIPD-APTES fiber is schematically illustrated. (b) A diagram showing how $Ti_3C_2(OH)_2$ nanolayers were added to the surface of PIPD-APTES fiber (see Taloub et al. 2019).

Source: (a and b) Adopted with permission from Elsevier Ltd.

understanding of how the IFSS degrades due to insufficient interaction between the C-Cl functional groups and the epoxy resin. However, studies have shown that incorporating APTES on the PIPD surface significantly increases the Interfacial Shear Strength (IFSS) to 36.32 MPa and the Tensile Strength (TS) to 7.04 GPa (Brew, Hine, and Ward 1999). The increase in TS makes it simpler for APTES FG to penetrate between fiber breaks by condensing the stress absorption (Wang, Li, et al. 2016).

The value of IFSS was steadily rising while the substantial fiber/epoxy interaction was taking place. Additionally, the IFSS steadily grew to 61.54%, or by 40.59 MPa, when $Ti_3C_2(OH)_2$ nanolayers accumulated on the fiber's surface, as the TS marginally declined to 5%, or by 6.381%. Grafted MXene nanosheets achieve considerable results by improving the fiber/epoxy knitting process and enlarging the precise region where the PIPD fiber layer and epoxy resin come into contact (Taloub et al. 2019). In contrast to other grafting techniques, this one is effective (it enhanced PIPD performance) (Cheng et al. 2018; Chen et al. 2013). MXene layers were grafted onto PIPD fibers in an effort to increase the material's UV resistance, interfacial properties, and resistance to hydrothermal aging. To modify the characteristics of PIPD fibers, nanosheets ($Ti_3C_2(OH)_2$) were attached in a less harmful manner when APTES was used as a linker (see Figure 3.1b). A slow but consistent rise was seen in the PIPD fiber's surface wettability and surface energy.

Moreover, the hydrothermal aging resistance and UV spectrum of the PIPD-MXene fibers combined have been significantly improved by 25% and 33%, respectively.

In conclusion, the surface functionalization technique that has been developed is predictable and less harmful, and it has huge promise for enhancing polymeric fibers (rigid rod) (Taloub et al. 2019).

With an emphasis on the mechanical qualities, Chakraborty et al. (Chakraborty et al. 2017) studied the alloying result at the X and M positions of MXene, which was observed in 2017.

Consideration is given to the exceedingly difficult solutions Ti_2 (C0.5, B0.5) and (Ti, V) C, with the demonstrative example of Ti_2C being chosen in particular (Naguib et al. 2011). The findings show the B-based doping of O-passivated Ti_2C multi-MXene layers significantly alters the mechanical properties of the system. This method resulted in a roughly 27% increase in compact toughness and an approximate 100% increase in critical strain value compared to the original unpassivated Ti_2C structure. Phonon measurements have confirmed this theory, which attributes a modest portion of the improved elastic-based properties arising from the B doping process to the strength of titanium–carbon (Ti–C) bond, relative to the covalency of the titanium–boron (Ti–B) bond. In addition, the magnetic unpassivated Ti_2C is revealed to have a previously unmapped spectacular reflection.

Although its interaction with polymers is not fully disclosed, Ti_3C_2—another MXene-based nanomaterial—has been explored for various potential applications. In one study, a novel Ti_3C_2/ultrahigh-molecular-weight polyethylene (UHMWPE) composite was created using a nanotube sequence (Zhang et al. 2016). Initially, the functionalized Ti_3C_2powder and UHMWPE powder were mixed in equal proportions for 1 hour using a high-speed blender (spinning at 2,000 rpm). Next, the mixture was heated at a rate of 10°C/min up to 220°C for 30 minutes under a pressure of 10 MPa. After heating, the pressure was maintained until the mixture cooled to room temperature.

The composite sheets were divided into proper sizes as the final step before further testing. Composite constructions with different concentrations of Ti_3C_2(0, 0.25, 0.5, 0.75, 1.0, and 2.0 wt.%) were built for a range of potential uses in order to study the effect of Ti_3C_2on the mechanical and thermal properties of UHMWPE (Zhang et al. 2016).

3.2.2 The Optical Property of MXene

Ti_3C_2Tx is an incredible plasmonic material that can be employed in surface-enhanced Raman spectroscopy (SERS) (Hu, Wang, et al. 2015; Hu et al. 2016; El-Demellawi et al. 2018). Extreme surface plasmons (SPRs) with energy ranges between (0.3 and 1 eV) represent the most of the plasmons occurring at a layer thickness of 45 nm., according to an EELS investigation on multidimensional Ti_3C_2Tx (Mauchamp et al. 2014).

In a different study (Berdiyorov 2016), first-principles computational density functional theory is utilized to investigate the optical properties of $Ti_3C_2T_2$ (T = OH, F, O) MXene and the surface character of the dielectric (DFT) measuring shells, such as those that have been fluorinated, hydroxylated, and oxidized a have been consistently created throughout the production process. The results demonstrate that the plasmonic characteristics of MXene are significantly influenced by the changed surface. The hydroxylated and fluorinated samples are ideal for use as transparent electrodes due to their improved reflectivity and performance across a noticeable range of the spectrum compared to the original MXene. On the other hand, surface oxidation enhances both reflectance and absorption.

A portion of the UV spectrum makes up the continuum in the area. The enhanced reflection from all exposed surfaces, compared to the original MXene, underscores the positive effects surface functionalization strategies on MXene's UV protection capabilities.

After altering MXene's surface functional groups and the most plausible reasons for the change in its optical characteristics, the aforementioned theories are confirmed (Berdiyorov 2016). In order to provide a more thorough theoretical knowledge of this topic, the current study uses DFT to examine the optical characteristics of two popular MXene materials, $Ti_3C_2$and Ti2C. In addition, the elastic rigidity is evaluated. The results demonstrate a strong influence of the surface-excluded groups on the optical properties and elastic stiffness of $Ti_3C_2T_2$ MXenes. Additionally, they confirm the prediction of the developed transmittance in thinner Ti2CT2 as well as the high transmittance of $Ti_3C_2T_2$ MXenes from infrared to ultraviolet wavelengths that was observed experimentally. Ti_2CO_2 is a strong candidate material for application in optical and electrical devices because the results also demonstrate that it has an optical bandgap. These findings imply that the plasmonic and other features of Ti_2CO_2-based materials can be modified using solid solution and doping techniques (Bai et al. 2016). The nanostructure enterprise of MXene arrays enables surface plasmon resonance (SPR) dotting of MXene.

This MXene nanomaterial-based technology has promising applications in energy collection, biological imaging, and detection. To fully appreciate the promise of MXenes, however, more work on the photonic and plasmonic characteristics of these materials is required. The surface plasmon-related process is expected to be comparable in carbonitride and nitride MXenes to that in the bulk of transition metal nitrides (Li et al. 2014).

3.2.3 MXene's Electrical Characteristics

The remarkable electrical properties of MXenes, such as their topological insulator (QSHE) and metal-to-insulator transition capabilities, as well as their superconductivity, are still being explored and debated. MXenes have a wide range of applications due to their excellent electronic conductivity (Handoko et al. 2017; Seh et al. 2016).

Unexpectedly, there has not been a new update to the extensive analysis of MXenes' electrical characteristics. To solve these problems with the electrical characteristics of MXene, researchers have consistently focused on the method of developing a unique Mn+1Xn chemical arrangement. Ti_3C_2Tx, the first MXene to be exposed, which was found by Naguib and colleagues in 2011, remains to be the most conductive of the more than 30 MXenes that have been made so far (Naguib et al. 2011; Zhang et al. 2017). The electronic conductivity of any multilayer MXene sample is influenced by factors such as closure, intercalation, and the chemistry of Mn+1Xn. However, understanding the practical implications of these effects is quite complex due to their interplay (Hart et al. 2019). HF often produces more unwanted fragments, less acceptable fragments, and more F-terminal group when employed as an etchant (Sang et al. 2016).

The resulting multilayered Ti_3C_2Tx (ML-Ti_3C_2Tx) with randomly positioned MXene flakes has a low electrical conductivity of 2 S cm^{-1}. The reduction in transportion concentration at the Fermi level occurs due to interface conflicts and resistance within flakes caused by F-terminations and atomic positions. To progress regulation, MXenes must have greater electrical characteristics. As the electrical conductivity of ML-Ti_3C_2Tx MXene units increases by an order of magnitude, particle margin effects vanish (Hu, Zhang, et al. 2015). A decrease in surface expiries, as previously indicated, is likely what produced the threefold increase in conductivity that resulted from space galvanizing of ML-HF-etched Ti_3C_2Tx (Wang et al. 2015). When HCl and LiF are used as the sources of the etching substance, the conductivity significantly increases. The conductivity of a 5 m spun ML-Ti_3C_2Tx sheet was 1,500 S cm^{-1} (Ghidiu et al. 2014). Without using ultrasonication, Ti_3C_2Tx was delaminated to higher conductivity standards. These experimental methods result in the production of coatings with conductivities greater than 5000 S cm^{-1}. Furthermore, the conductivities of monolayer Ti_3C_2Tx and crystal-clear films made using the aforementioned method are 6,760 and 9,880 S cm^{-1}, respectively (Zhang et al. 2017; Sang et al. 2016).

In 2019, Hart et al. created a novel strategy to validate the impact of surface closures on the electrical properties of MXene (Hart et al. 2019). An in-situ technique was used to boost the electrical conductivity of several MXenes, including Ti_3C_2Tx, Mo2TiCNTx, and Ti_3CNTx, within a defined region of the Transmission electron microscope (TEM). Analysis using TGA-MS and EELS revealed that the annealing process increased the conductivity of MXenes by removing surface terminations and eliminating adsorbed species (as depicted in Figure 3.2a, b).

Due to the removal of additional adsorbed species, intercalated water, and the functional degradation of OH-terminal species, the conductivity of Ti_3C_2Tx galvanized at temperatures between 100 and 400°C increased by almost three times (refining the intraflake conductivity value). When the sample was galvanized at temperatures as high as 775°C, the deletion of F-terminal groups also led to a steady rise in conductivity, as seen by EELS dimensions (Figure 3.2c).

Panels d and e of Figure 3.2 depict how the conductivity values of Ti_3C_2Tx and Mo_2TiC_2 change after annealing, correlating with a decrease in O- and F-terminal groups (Hart et al. 2019). Improving the conductivity of MXenes in a controlled environment often involves removing interposed species and surface functional group terminations.

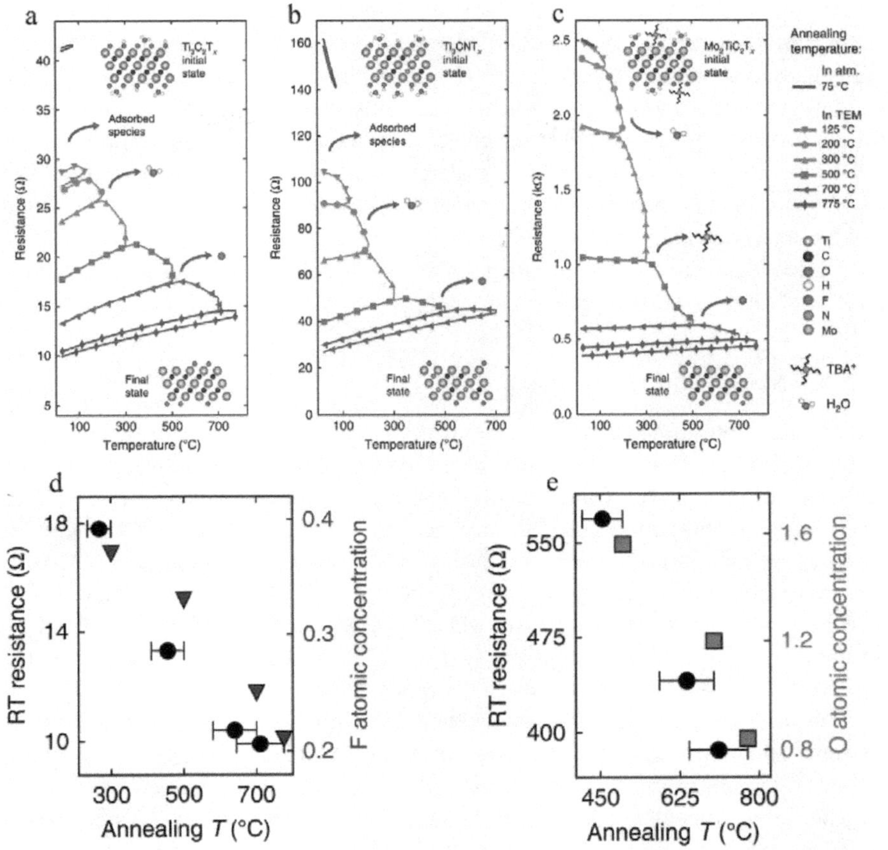

FIGURE 3.2 (a) The impacts of surface variation on MXene's electrical properties were used to test in-situ the temperature-dependent resistance of the Ti_3C_2Tx, Ti_3CNTx, and Mo_2TiC_2Tx (b), as well as the relationship between the conductivity of the Ti_3C_2Tx. (c) When galvanized materials were heated to 775°C, the removal of fluorine-containing groups from their ends correlated with a progressive increase in electrical conductivity, as quantified by electron energy loss spectroscopy (EELS). (d and e) illustrate the changes in electrical conductivity of Ti_3C_2Tx and Mo_2TiC_2 following annealing, revealing an inverse relationship with the reduction of oxygen and fluorine terminal groups.

Source: (a–e) Used in accordance used in accordance with the International Creative Commons Attribution 4.0 License, Springer Nature (see Hart et al. 2019).

In single-layer MXene experiments, surface terminations must be carefully controlled to understand the functional group's impact on the material's fundamental electrical properties.

The electrical conductivity of MXene arrays varies compared to a single MXene, especially for multiple multilayered MXene structures (ML-MXene). The highest reported value for Ti_3C_2Tx thin film deposition is 10,000 S/cm, although this is not the maximum achievable value (Zhang et al. 2017).

Another article by Nedil et al. (Aïssa et al. 2016) explains how an efficient alternating electrospray of graphene and MXene materials may be used to explain the transference properties of MXene (Ti_3C_2Tx)/graphene heterostructures at 25°C. The models' surface irregularity significantly diminished after the addition of graphene, which is said to have improved both the electrical and mechanical qualities.

When only 2.5 weight percent of graphene was integrated, high flexibility (54.58 cm^2/V s) was anticipated with an equivalent value of electrical conductivity, 9.5×10^4 S/cm (Aïssa et al. 2016).

These results are likely to broaden the range of applications for this class of materials, including thin-film electronics.

3.2.4 Magnetic Characteristics

Due to the three-dimensional arrangement of electrons on titanium surface atoms, magnetic properties are only observable in bulk Ti_2X (C, N) monolayers found naturally. According to computational findings, the majority of functionalized MXene derivatives exhibit nonmagnetic properties (Zhao, Kang, and Xue 2014). Numerous study guidelines assert that the presence of active functions is what causes MXene's ferromagnetic characteristics (Pan, Lany, and Qi 2017). On the other hand, if Cr_2C is transformed asymmetrically into what is known as Janus MXenes ($M_{n+1}X_nT_{xx'}$, where x and x' denote functionalities such as F, H, Cl, OH, or Br), it has the potential to form magnetic structures with a significantly higher Neel temperature.

This unique setup offers advantages for both spin-polarized semiconductivity and antiferromagnetic alignment, making it suitable for spintronics applications where the direction of spin polarization is crucially observable.

The researchers also investigated the magnetic properties of Ti_3C_2Tx concerning reactive functional elements (He et al. 2016). Furthermore, the magnetic properties of MXene models functionalized with $-SO_3H$ and S_2 reactive groups, treated with chlorosulfonic acid, differ from those of a typical sample etched with HF. The influence of titanium ions was studied by comparing MXene samples with TiC, Ti_2AlC, and Ti_3AlC_2 samples. This comparison aimed to minimize inaccuracies caused by residual contamination from the TiC/MAX phase (Scheibe et al. 2019).

In order to identify the size and shape of the materials under research, Katarzyna and colleagues in 2019 (Scheibe et al. 2019) used SEM. Images of (a) TiC, (b) Ti_2AlC, (c) Ti_3AlC_2, (d) Ti_3C_2Tx-HF, (e) Ti_3C_2Tx-M, and (f) Ti_3C_2Tx-CS are shown in Figure 3.3. The numerous assessments of the size distribution of micro-scale particles are shown in the figure. Tightly packed formations contain both MAX phases and TiC. As shown in Figure 3.3d, the Ti_3AlC_2 assemblage is affected by the HF etching procedure. As a result, a morphology resembling Ti_3C_2Tx-HF is produced, with distinctive layer geometries kept in place by hydrogen bonds or Al_2O_3 microparticles that have been trapped. Chlorosulfonic acid, which also functions as a strong oxidant, generates a protonic superacid. Because of the protonation process that takes place on their surfaces and the interaction repulsion between them, carbon nanotubes (CNTs) or graphene sheets (GrNs) may be spread easily and quickly (Parra-Vasquez et al. 2010). The magnetic characteristics of all organized MXenes are computed using DC magnetization.

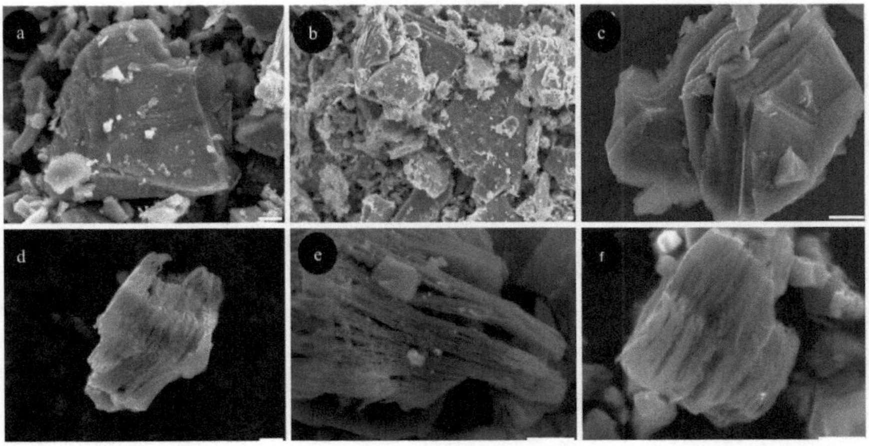

FIGURE 3.3 The materials (a) sss, (b) Ti_2AlC, (c) Ti_3AlC_2, (d) Ti_3C_2Tx-HF, (e) Ti_3C_2Tx-M, and (f) Ti_3C_2Tx-CS were all captured in SEM pictures. The scale bar distance is 1 m.

Source: Adopted with permission from Scheibe et al. (2019). Copyright 2019 Elsevier B.V. All rights reserved.

Based on data gathered from published literature (Naguib et al. 2011), it is predicted that Ti_3C_2Tx is likely antiferromagnetic and that MXenes are magnetically organized (without any reactive functionalities). But according to the equations, the Ti_3C_2Tx single layer ought to be magnetic (Wu et al. 2015). The magnetic flash is produced by Ti_3d states, but due to the Ti surface atoms' saturation, surface termination methods used by FG, such as (O, –H, –OH, and –F), may eliminate the magnetism effect.

Possible explanations for the various magnetic properties of Ti_3C_2Tx MXene include the inclusion of both multilayers as well as sample monolayers, the examination of the grain size distribution, and the presence of several functional groups. SEM micrographs of all MXene samples show multilayer topologies, nonuniform morphologies, and a variety of reactive functionalities (Scheibe et al. 2019).

Functionalized Ti_3C_2Tx MXenes (multilayered) and Ti-Al-C carbides have been characterized in a range of paramagnetic magnetic and antiferromagnetic states. Chronologically, the EPR study of MXenes exposed to both acids showed the presence of additional radicals and deficits.

3.2.5 MXene's Stability Characteristics

To enhance the practical usability of solids, the field of materials science is currently concentrating on improving their chemical and physical properties. By carefully combining chemistry, physics, and engineering, scientists have successfully altered the fundamental properties of numerous materials. Flaws and dopants, along with external factors like pressure, strain, and temperature, are commonly used to alter the properties of a material. This is accomplished by controlling the density and structure of these imperfections and introducing additional elements.

However, many procedures have only been created, enhanced, and applied to bulk materials. In order to study 2D materials with a density of between one and three atomic layers, it is necessary to enhance certain techniques that affect their basic properties and sometimes even leave them uncontrolled. For example, precise ion irradiation can be employed to introduce impurities into silicon, but achieving the same result for graphene remains challenging. However, 2D materials also provide novel opportunities. For instance, while 3D band engineering is restricted by the principles of epitaxial crystal formation, 2D materials can be stacked. In order to find and account for pervasive errors and foreign atoms, 2D materials science must address all problems and build efficient monitoring systems for material properties. The natural properties of 2D materials must be tuned by atomic-scale alterations. The field of 2D materials science enables precise control over adding, removing, or modifying atoms. This capability is crucial for addressing defect engineering in its broadest sense and unleashing the full potential of various applications (Schleberger and Kotakoski 2018).

To improve the functionality and durability of the active surface (Mashtalir et al. 2013), it is possible to combine multilayered MXene derivatives with tiny organic molecules or polymers such as urea, dimethyl sulfoxide, or hydrazine, in a similar manner as with other two-dimensional materials (Kong et al. 2015).

The significance of MXene's colloidal solidity solutions stems from the rising demand for its derivatives in a variety of practical applications. Nanosheets with one or more layers are also used as the starting material for a variety of MXene-based products that are mass manufactured.

MXene-based aqueous solutions are created when water is sonicated or physically shaken (Mashtalir et al. 2013; Lipatov et al. 2016). The creation of thin films based on MXene, further strengthening of polymer-based composites, and construction of thick membranes can all be accomplished using these solutions. The Ti_3C_2Tx MXene that has been the focus of the most research over the past ten years, single-layer and multilayered flakes, has steadily degraded in the open or in water (Mashtalir et al. 2014). As many industrial applications rely on solution dispersion rather than precipitate dispersion, it was necessary to investigate the stability of delaminated MXene in aqueous solutions.

In numerous real-world contexts, the stability of colloidal solutions based on MXene is of extreme importance, but it is unknown and has been the subject of intense research for the past ten years. Additionally, framework techniques to increase the tensile strength of delaminated-Ti_3C_2Tx-based colloidal solutions (where T stands for the surface functional group) are discussed. It was discovered that the colloidal Ti_3C_2Tx MXene solution degraded quickly, reaching 100% deterioration after only 15 days when it was exposed to the open air. The primary element of the resulting colloidal solutions is anatase production (TiO_2). The data showed that the resulting aqueous solution under ambient air and room temperature (Air@RT) included unstable d-Ti_3C_2Tx MXene flakes. After two weeks, most of the oxidation process had converted the carbide-based MXene to anatase titanium.

To prevent oxidation, the d-Ti_3C_2Tx MXene colloidal solution must be transferred into bottles containing Ar that were made in the glovebox before application. To protect the functionality of Ti_3C_2Tx MXenes in many real-world applications,

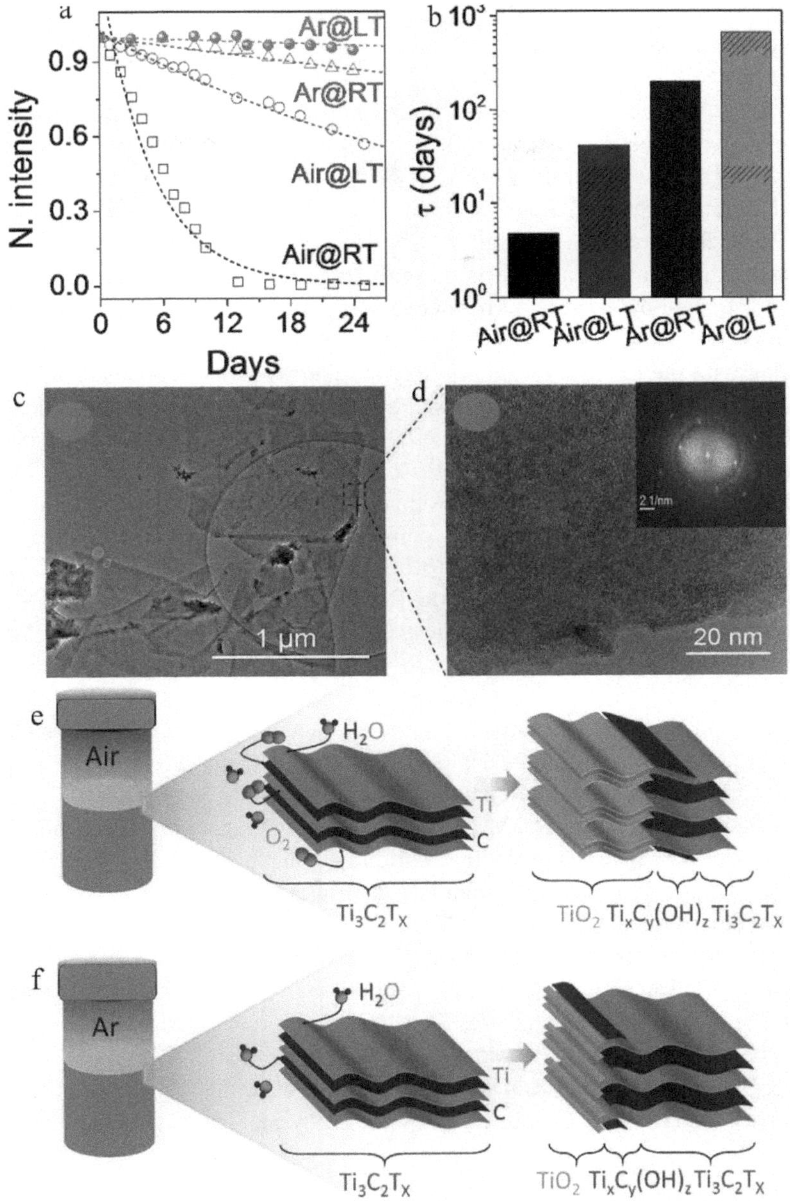

FIGURE 3.4 (a) Colloidal MXene Ti_3C_2Txs in two conditions of air and inert gas. The results that match the empirical equation A = Aunre + Areet/are shown as dotted lines. (b) Shown are the constants of the colloidal Mxen d-temporal Ti_3C_2Tx in various settings. (c) The d-Ti_3C_2Tx flakes were imaged in both low- and high-resolution TEMs after being aged in Ar@LT for 30 days. The FFT design is used in the framework (d) inset. (e, f) Diagrams of the colloidal d-Ti_3C_2Tx MXene aqueous solution in (a) and (b) under the effect of air at room temperature and of argon at room temperature, respectively.

Source: (a–f) Adopted with permission from American Chemical Society (Zhang et al. 2017).

it is extremely desirable and necessary to work on the various ways needed for the continual improvement of the d-Ti_3C_2Tx aqueous solution's stability process (Zhang et al. 2017).

In 2017, Gogotsi and colleagues discovered evidence of a single-exponential decay, showing that the MXene nanosheets' margins are where the deterioration process begins. By separating the d-Ti_3C_2Tx solution according to flake size, it was discovered that the rate of degradation is closely connected with flake size, with larger flakes exhibiting higher stability. It is possible to manipulate the in-situ size distribution of nanosheets by analyzing their UV-visible spectra, and this may be achieved by gaining a knowledge of the link between flake size and time constants.

Along with the kinetics of the solution and the degradation process, the method for making MXene harder is also looked at. According to observations, when compared to sample treatment in an open atmosphere, the stability of nanosheets is enhanced in the presence of an Ar environment.

This is most likely because the colloidal solution, which is what powers the MXene breakdown process, does not contain oxygen.

Panels a and b of Figure 3.4 illustrate the results of applying the exponential decay model to the degradation data, providing further support. Figure 3.4c shows a transmission electron micrograph of MXene sheets after 30 days of exposure to Ar@LT. In spite of some verge points having sub-micrometer subdivisions, the surfaces of large nanosheets are typically smooth and show parallel viscosity.

Figure 3.4d uses TEM pictures at high magnification to show that nanoscale titania is not present. Figure 3.4e, f depicts the schematic deprivation of the d-Ti_3C_2Tx MXene colloidal aqueous solution under the effect of Ar (Zhang et al. 2017; Li et al. 2015). The constant source of MXene breakdown in framework is the high-order solubility of oxygen in water. With the lack of soluble oxygen in framework (b), water (acting as a mild oxidant) gradually deteriorates the MXene nanoflakes in an Ar-based environment (Zhang et al. 2017).

Moreover, research on a different type of MXene known as d-Ti_2CT_x revealed that freezing and storing it in a deaerated environment enhance its time constant and stability over time. This underscores the validity of recommendations for long-term dispersion of MXene-based colloidal solutions (Zhang et al. 2017).

Biological and environmental sensing, as well as potential applications of MXene derivatives, are significantly influenced by these physical characteristics.

3.3 SUMMARY

In summary, this chapter underscores the paramount importance of defect engineering in the improvement of the properties of MXenes, a distinctive class of two-dimensional transition metal carbides and nitrides. It emphasizes the significance of MXenes in materials research and illustrates how the strategic manipulation of defects can enhance their mechanical, optical, electrical, magnetic, and stability properties. This chapter demonstrates how these improvements render MXenes suitable for a variety of applications, including photonics, optoelectronics, energy storage, and spintronics, through a complex analysis. In summary, this book chapter offers a thorough examination of the physical characteristics of MXenes and investigates the potential of defect engineering to modify these properties for a variety of advanced technological applications.

REFERENCES

Ahmed, Bilal, Ahmed El Ghazaly, and Johanna Rosen. 2020. "i-MXenes for energy storage and catalysis." *Advanced Functional Materials* 30 (47):2000894.

Aïssa, B, A Ali, KA Mahmoud, T Haddad, and M Nedil. 2016. "Transport properties of a highly conductive 2D Ti_3C_2Tx MXene/graphene composite." *Applied Physics Letters* 109 (4):043109.

Alhabeb, Mohamed, Kathleen Maleski, Babak Anasori, Pavel Lelyukh, Leah Clark, Saleesha Sin, and Yury Gogotsi. 2017. "Guidelines for synthesis and processing of two-dimensional titanium carbide (Ti_3C_2Tx MXene)." *Chemistry of Materials* 29 (18):7633–7644.

Anasori, Babak, Maria R Lukatskaya, and Yury Gogotsi. 2017. "2D metal carbides and nitrides (MXenes) for energy storage." *Nature Reviews Materials* 2:16098.

Bai, Yuelei, Kun Zhou, Narasimalu Srikanth, John HL Pang, Xiaodong He, and Rongguo Wang. 2016. "Dependence of elastic and optical properties on surface terminated groups in two-dimensional MXene monolayers: A first-principles study." *RSC Advances* 6 (42):35731–35739.

Berdiyorov, GR. 2016. "Optical properties of functionalized $Ti_3C_2T_2$ (T= F, O, OH) MXene: First-principles calculations." *Aip Advances* 6 (5).

Brew, B, PJ Hine, and IM Ward. 1999. "The properties of PIPD-fibre/epoxy composites." *Composites Science and Technology* 59 (7):1109–1116.

Chakraborty, Poulami, Tilak Das, Dhani Nafday, Lilia Boeri, and Tanusri Saha-Dasgupta. 2017. "Manipulating the mechanical properties of Ti 2 C MXene: Effect of substitutional doping." *Physical Review B* 95 (18):184106.

Chen, Lei, Zhen Hu, Li Liu, and Yudong Huang. 2013. "A facile method to prepare multifunctional PBO fibers: Simultaneously enhanced interfacial properties and UV resistance." *RSC Advances* 3 (46):24664–24670.

Cheng, Zheng, Dawei Hong, Yu Dai, Chan Jiang, Chenbo Meng, Longbo Luo, and Xiangyang Liu. 2018. "Highly improved Uv resistance and composite interfacial properties of aramid fiber via iron (III) coordination." *Applied Surface Science* 434:473–480.

Cui, Yanglansen, Zhenjiang Cao, Yongzheng Zhang, Hao Chen, Jianan Gu, Zhiguo Du, Yongzheng Shi, Bin Li, and Shubin Yang. 2021. "Single-atom sites on MXenes for energy conversion and storage." *Small Science* 1 (6):2100017.

El-Demellawi, Jehad K, Sergei Lopatin, Jun Yin, Omar F Mohammed, and Husam N Alshareef. 2018. "Tunable multipolar surface plasmons in 2D Ti_3C_2 T x MXene flakes." *ACS Nano* 12 (8):8485–8493.

Ghidiu, Michael, Maria R Lukatskaya, Meng-Qiang Zhao, Yury Gogotsi, and Michel W Barsoum. 2014. "Conductive two-dimensional titanium carbide 'clay' with high volumetric capacitance." *Nature* 516 (7529):78–81.

Ghorbani-Asl, Mahdi, Silvan Kretschmer, Douglas E Spearot, and Arkady V Krasheninnikov. 2017. "Two-dimensional MoS_2 under ion irradiation: From controlled defect production to electronic structure engineering." *2D Materials* 4 (2):025078.

Halim, Joseph, Kevin M Cook, Michael Naguib, Per Eklund, Yury Gogotsi, Johanna Rosen, and Michel W Barsoum. 2016. "X-ray photoelectron spectroscopy of select multi-layered transition metal carbides (MXenes)." *Applied Surface Science* 362:406–417.

Han, Meikang, Xiaowei Yin, Heng Wu, Zexin Hou, Changqing Song, Xinliang Li, Litong Zhang, and Laifei Cheng. 2016. "Ti_3C_2 MXenes with modified surface for high-performance electromagnetic absorption and shielding in the X-band." *ACS Applied Materials & Interfaces* 8 (32):21011–21019.

Handoko, Albertus D, Kurt D Fredrickson, Babak Anasori, Kurt W Convey, Luke R Johnson, Yury Gogotsi, Aleksandra Vojvodic, and Zhi Wei Seh. 2017. "Tuning the basal plane functionalization of two-dimensional metal carbides (MXenes) to control hydrogen evolution activity." *ACS Applied Energy Materials* 1 (1):173–180.

Hart, James L, Kanit Hantanasirisakul, Andrew C Lang, Babak Anasori, David Pinto, Yevheniy Pivak, J Tijn van Omme, Steven J May, Yury Gogotsi, and Mitra L Taheri. 2019. "Control of MXenes' electronic properties through termination and intercalation." *Nature Communications* 10 (1):522. doi: 10.1038/s41467-018-08169-8.

He, Junjie, Pengbo Lyu, LZ Sun, Ángel Morales García, and Petr Nachtigall. 2016. "High temperature spin-polarized semiconductivity with zero magnetization in two-dimensional Janus MXenes." *Journal of Materials Chemistry C* 4 (27):6500–6509.

Hu, Minmin, Zhaojin Li, Tao Hu, Shihao Zhu, Chao Zhang, and Xiaohui Wang. 2016. "High-capacitance mechanism for Ti_3C_2 Tx MXene by in situ electrochemical Raman spectroscopy investigation." *ACS Nano* 10 (12):11344–11350.

Hu, Tao, Jiemin Wang, Hui Zhang, Zhaojin Li, Minmin Hu, and Xiaohui Wang. 2015. "Vibrational properties of Ti_3C_2 and $Ti_3C_2T_2$ (T= O, F, OH) monosheets by first-principles calculations: A comparative study." *Physical Chemistry Chemical Physics* 17 (15):9997–10003.

Hu, Tao, Hui Zhang, Jiemin Wang, Zhaojin Li, Minmin Hu, Jun Tan, Pengxiang Hou, Feng Li, and Xiaohui Wang. 2015. "Anisotropic electronic conduction in stacked two-dimensional titanium carbide." *Scientific Reports* 5:16329.

Kong, Lingping, Chuanfang Zhang, Jitong Wang, Wenming Qiao, Licheng Ling, and Donghui Long. 2015. "Free-standing T-Nb2O5/graphene composite papers with ultrahigh gravimetric/volumetric capacitance for Li-ion intercalation pseudocapacitor." *Acs Nano* 9 (11):11200–11208.

Li, Gengnan, Li Tan, Yumeng Zhang, Binghan Wu, and Liang Li. 2017. "Highly efficiently delaminated single-layered MXene nanosheets with large lateral size." *Langmuir* 33 (36):9000–9006.

Li, Wei, Urcan Guler, Nathaniel Kinsey, Gururaj V Naik, Alexandra Boltasseva, Jianguo Guan, Vladimir M Shalaev, and Alexander V Kildishev. 2014. "Plasmonics: Refractory plasmonics with titanium nitride: Broadband metamaterial absorber (adv. mater. 47/2014)." *Advanced Materials* 26 (47):7921–7921.

Li, Zhengyang, Libo Wang, Dandan Sun, Yude Zhang, Baozhong Liu, Qianku Hu, and Aiguo Zhou. 2015. "Synthesis and thermal stability of two-dimensional carbide MXene Ti_3C_2." *Materials Science and Engineering: B* 191:33–40. doi: 10.1016/j.mseb.2014.10.009.

Lipatov, Alexey, Mohamed Alhabeb, Maria R Lukatskaya, Alex Boson, Yury Gogotsi, and Alexander Sinitskii. 2016. "Effect of synthesis on quality, electronic properties and environmental stability of individual monolayer Ti_3C_2 MXene flakes." *Advanced Electronic Materials* 2 (12):1600255.

Mashtalir, O, K M Cook, V N Mochalin, M Crowe, M W Barsoum, and Y Gogotsi. 2014. "Dye adsorption and decomposition on two-dimensional titanium carbide in aqueous media." *Journal of Materials Chemistry A* 2 (35):14334–14338. doi: 10.1039/C4TA02638A.

Mashtalir, O, M Naguib, V N Mochalin, Y Dall'Agnese, M Heon, M W Barsoum, and Y Gogotsi. 2013. "Intercalation and delamination of layered carbides and carbonitrides." *Nature Communications* 4:1716. doi: 10.1038/ncomms2664.

Mauchamp, Vincent, Matthieu Bugnet, Edson P Bellido, Gianluigi A Botton, Philippe Moreau, Damien Magne, Michael Naguib, Thierry Cabioc'h, and Michel W Barsoum. 2014. "Enhanced and tunable surface plasmons in two-dimensional Ti 3 C 2 stacks: Electronic structure versus boundary effects." *Physical Review B* 89 (23):235428.

Naguib, Michael, Jérémy Come, Boris Dyatkin, Volker Presser, Pierre-Louis Taberna, Patrice Simon, Michel W Barsoum, and Yury Gogotsi. 2012. "MXene: A promising transition metal carbide anode for lithium-ion batteries." *Electrochemistry Communications* 16 (1):61–64.

Naguib, Michael, Murat Kurtoglu, Volker Presser, Jun Lu, Junjie Niu, Min Heon, Lars Hultman, Yury Gogotsi, and Michel W Barsoum. 2011. "Two-dimensional nanocrystals produced by exfoliation of Ti_3AlC_2." *Advanced Materials* 23 (37):4248–4253. doi: 10.1002/adma.201102306.

Naguib, Michael, Vadym N Mochalin, Michel W Barsoum, and Yury Gogotsi. 2014. "25th anniversary article: MXenes: A new family of two-dimensional materials." *Advanced Materials* 26 (7):992–1005.

Nan, Jianxiao, Xin Guo, Jun Xiao, Xiao Li, Weihua Chen, Wenjian Wu, Hao Liu, Yong Wang, Minghong Wu, and Guoxiu Wang. 2021. "Nanoengineering of 2D MXene-based materials for energy storage applications." *Small* 17 (9):1902085.

Pan, Jie, Stephan Lany, and Yue Qi. 2017. "Computationally driven two-dimensional materials design: What is next?" *ACS Nano* 11 (8):7560–7564.

Parra-Vasquez, A Nicholas G, Natnael Behabtu, Micah J Green, Cary L Pint, Colin C Young, Judith Schmidt, Ellina Kesselman, Anubha Goyal, Pulickel M Ajayan, and Yachin Cohen. 2010. "Spontaneous dissolution of ultralong single-and multiwalled carbon nanotubes." *ACS Nano* 4 (7):3969–3978.

Sang, X, Y Xie, MW Lin, M Alhabeb, KL Van Aken, Y Gogotsi, PR Kent, K Xiao, and RR Unocic. 2016. Atomic defects in monolayer titanium carbide (Ti_3C_2Tx) MXene. *ACS Nano* 10:9193–9200.

Scheibe, Błażej, Krzysztof Tadyszak, Marcin Jarek, Natalia Michalak, Mateusz Kempiński, Mikołaj Lewandowski, Barbara Peplińska, and Katarzyna Chybczyńska. 2019. "Study on the magnetic properties of differently functionalized multilayered Ti_3C_2Tx MXenes and Ti-Al-C carbides." *Applied Surface Science* 479:216–224. doi: 10.1016/j.apsusc.2019.02.055.

Schleberger, Marika, and Jani Kotakoski. 2018. "2D material science: Defect engineering by particle irradiation." *Materials* 11 (10):1885.

Seh, Zhi Wei, Kurt D Fredrickson, Babak Anasori, Jakob Kibsgaard, Alaina L Strickler, Maria R Lukatskaya, Yury Gogotsi, Thomas F Jaramillo, and Aleksandra Vojvodic. 2016. "Two-Dimensional Molybdenum Carbide (MXene) as an Efficient Electrocatalyst for Hydrogen Evolution." *ACS Energy Letters* 1 (3):589–594. doi: 10.1021/acsenergylett.6b00247.

Sun, Tao, Guoqiang Zhang, Dong Xu, Xu Lian, Hexing Li, Wei Chen, and Chenliang Su. 2019. "Defect chemistry in 2D materials for electrocatalysis." *Materials Today Energy* 12:215–238.

Taloub, Nadia, Abdelkhalek Henniche, Li Liu, Jun Li, Nahla Rahoui, Mohammad Hegazy, and Yudong Huang. 2019. "Improving the mechanical properties, UV and hydrothermal aging resistance of PIPD fiber using MXene ($Ti_3C_2(OH)_2$) nanosheets." *Composites Part B: Engineering* 163:260–271. doi: 10.1016/j.compositesb.2018.11.007.

Tang, Qing, Zhen Zhou, and Panwen Shen. 2012. "Are MXenes promising anode materials for Li ion batteries? Computational studies on electronic properties and Li storage capability of Ti_3C_2 and $Ti_3C_2X_2$ (X= F, OH) monolayer." *Journal of the American Chemical Society* 134 (40):16909–16916.

Tang, Yi, Jianfeng Zhu, Wenling Wu, Chenhui Yang, Wenjing Lv, and Fen Wang. 2017. "Synthesis of nitrogen-doped two-dimensional Ti_3C_2 with enhanced electrochemical performance." *Journal of the Electrochemical Society* 164 (4):A923.

Wang, Caifeng, Jun Li, Shaofan Sun, Xiaoyu Li, Guangshun Wu, Yuwei Wang, Fei Xie, and Yudong Huang. 2016. "Controlled growth of silver nanoparticles on carbon fibers for reinforcement of both tensile and interfacial strength." *RSC Advances* 6 (17):14016–14026.

Wang, Hongbing, Yuping Wu, Jianfeng Zhang, Gaiye Li, Huajie Huang, Xin Zhang, and Quanguo Jiang. 2015. "Enhancement of the electrical properties of MXene Ti_3C_2 nanosheets by post-treatments of alkalization and calcination." *Materials Letters* 160:537–540.

Wang, Kun, Youfu Zhou, Wentao Xu, Decai Huang, Zhiguang Wang, and Maochun Hong. 2016. "Fabrication and thermal stability of two-dimensional carbide Ti_3C_2 nanosheets." *Ceramics International* 42 (7):8419–8424.

Wei, Yi, Razium A Soomro, Xiuqiang Xie, and Bin Xu. 2021. "Design of efficient electrocatalysts for hydrogen evolution reaction based on 2D MXenes." *Journal of Energy Chemistry* 55:244–255.

Wen, Yangyang, Thomas E Rufford, Xingzhu Chen, Neng Li, Miaoqiang Lyu, Liming Dai, and Lianzhou Wang. 2017. "Nitrogen-doped Ti_3C_2Tx MXene electrodes for high-performance supercapacitors." *Nano Energy* 38:368–376.

Wu, Fang, Kang Luo, Chengxi Huang, Wenjuan Wu, Peiwen Meng, Yunfei Liu, and Erjun Kan. 2015. "Theoretical understanding of magnetic and electronic structures of Ti_3C_2 monolayer and its derivatives." *Solid State Communications* 222:9–13.

Wu, G M[illegible], and YT Shyng. 2004. "Surface modification and interfacial adhesion of rigid rod PBO fibre by methanesulfonic acid treatment." *Composites Part A: Applied Science and Manufacturing* 35 (11):1291–1300.

Yang, Chenhui, Wenxiu Que, Xingtian Yin, Yapeng Tian, Yawei Yang, and Meidan Que. 2017. "Improved capacitance of nitrogen-doped delaminated two-dimensional titanium carbide by urea-assisted synthesis." *Electrochimica Acta* 225:416–424.

Yang, Chenhui, Yi Tang, Yapeng Tian, Yangyang Luo, Muhammad Faraz Ud Din, Xingtian Yin, and Wenxiu Que. 2018. "Flexible nitrogen-doped 2D titanium carbides (MXene) films constructed by an ex situ solvothermal method with extraordinary volumetric capacitance." *Advanced Energy Materials* 8 (31):1802087.

Yoon, Yeoheung, Minhe Lee, Seong Ku Kim, Garam Bae, Wooseok Song, Sung Myung, Jongsun Lim, Sun Sook Lee, Taehyoung Zyung, and Ki-Seok An. 2018. "A strategy for synthesis of carbon nitride induced chemically doped 2D MXene for high-performance supercapacitor electrodes." *Advanced Energy Materials* 8 (15):1703173.

Zhang, Chuanfang John, Sergio Pinilla, Niall McEvoy, Conor P Cullen, Babak Anasori, Edmund Long, Sang-Hoon Park, Andrés Seral-Ascaso, Aleksey Shmeliov, and Dileep Krishnan. 2017. "Oxidation stability of colloidal two-dimensional titanium carbides (MXenes)." *Chemistry of Materials* 29 (11):4848–4856.

Zhang, Heng, Libo Wang, Qiang Chen, Ping Li, Aiguo Zhou, Xinxin Cao, and Qianku Hu. 2016. "Preparation, mechanical and anti-friction performance of MXene/polymer composites." *Materials & Design* 92:682–689. doi: 10.1016/j.matdes.2015.12.084.

Zhao, Shijun, Wei Kang, and Jianming Xue. 2014. "Manipulation of electronic and magnetic properties of M2C (M= Hf, Nb, Sc, Ta, Ti, V, Zr) monolayer by applying mechanical strains." *Applied Physics Letters* 104 (13):133106.

4 Application of Titanium-Based MXenes in Sensing, Photovoltaics, Catalysis, Energy Storage Devices, and Biomedicines

4.1 INTRODUCTION

Study of two-dimensional (2D) materials has emerged as a fledging research in which scientists have landed themselves on rocks starting from graphs, and there is no looking back! These materials are now being explored to be applied in different fields including biomedicine, photovoltaics, energy storage, and conversion devices, sensing applications as well as catalysis. Scientists have been intrigued by a type of material called titanium carbide, or Ti_3C_2Tx, and, even more recently, a material named as Ti_2CT_x. This material possesses significant features of a 2D compound, such as extraordinary mechanical properties, large surface area, chemical stability, and excellent electrical conductivity. It is represented by Mn+1XnTx where M represents transition metal that can be Sc, Ti, . . . , or W; X signifies carbon and/or nitrogen which could also include oxygen and (or) flourine; Tx denotes surface functional groups such as the hydroxyl group garnishing a termination site on one end of these “sheets.” In the absence of transition metal (M) site coordination, and with surface terminations (T) limited to oxygen (–O), hydroxyl (–OH), or fluorine (–F), both Ti_3C_2Tx and Ti_2CT_x stoichiometries exhibit identical elemental compositions. Furthermore, Ti_3C_2Tx and Ti_2CT_x have recently been investigated for their use in photovoltaic devices and other solar cells by different groups of researchers. The materials are good conductors of electricity, have large surface areas, and absorb light well. Synthesis, performance, and fabrication for solar application of Ti_3C_2Tx/Ti_2CT_x are discussed in this chapter. Titanium carbide (Ti_3C_2Tx) and titanium carbonitride have been exploited for effective catalysts. The exceptional stability, high surface area, and tunable electrical properties of these compounds have allowed them to be implemented as catalysts in a wide variety of processes. For example, researchers have recently found that the Ti_3C_2Tx- and Ti_2CT_x-based catalysts are very catalytic-active in a variety of important chemical reactions. These processes

DOI: 10.1201/9781003319658-4

involve chemical conversions, CO_2, reduction and water splitting. Scientists previously conducted studies on Ti_3C_2Tx and the similar material titanium carbo nitride, Ti_2CT_x, for their possible role as add-ons to batteries or what are known as supercapacitors. Due to their large surface areas, the good electronic conductivity, and high capacities, these materials are widely used for energy storage. Over the last few years, much has been achieved in advancing these materials as anodes and cathodes for batteries/supercapacitors. So it is the need of time to take a look at how they behave in energy storage devices.

As drug delivery systems and biosensors, the research shows that titanium carbide (Ti_3C_2Tx) as well as titanium carbonitride (Ti_2CT_x) have been among the most promising MAX phases to date for long-term applications in biomedicine. This is because of their great biocompatibility, low toxicity, and unique surface properties. There is a need to highlight the recently occurring developments in synthesizing and engineering Ti_3C_2Tx, other analogs like 100% conductive MXenes (TiHx) and non-conductive separated single layers of TiT2 (T = Nb or Zr), Bi, etc., for their effective use as drug delivery agents/biosensors/substrates to support biological systems required for medical applications. Exhibiting admirable traits as well as portability, the MAX phase titanium carbide (Ti_3C_2Tx) and related MXenes such as titanium carbonitride (Ti_2CT_x) have shown huge promise for diverse applications. Still, there are some barriers that stand in the way of fully realizing these practical applications. Ti_3C_2Tx, particularly the larger, high-quality samples can be very difficult to synthesize. Even now, producing large, high-quality, and reliable pieces of Ti_3C_2Tx and TixCTx remains a challenge, especially due to differences in production methods for selective etching and delamination. Furthermore, the lack of a standard process related to synthesis and characterization of Ti_3C_2Tx or Ti_2CT_x further complicates comparison across different studies.

A key issue the researchers are battling with is the stability of Ti_3C_2Tx and Ti_2CT_x under varying conditions, viz., exposure to air, water, etc. It has been well documented that Ti_3C_2Tx and Ti_2CT_x are very stable in air and water. Yet, numerous factors (e.g., pH, temperature and humidity) can affect the stability of biomolecules. How these will affect their potency in becoming applied to certain use instances and components? Earlier, Ti_3C_2 (both with and without termination) was completely biocompatible, while the sol-gel-method-synthesized TiO_2-titanium oxide nanoparticles needed more toxicological assessment in vivo as well as in vitro to demonstrate that they are safe for biomedical applications. Among many applications, Ti_3C_2Tx and titania (TiOx)-modified 2D titanium carbide derivatives such as Ti_2CT_x have recently demonstrated great promise in drug delivery systems and biosensors. Having said that, it is essential to perform exhaustive toxicology studies in order to understand the hazards they may emanate, if any. Yet, due to their versatility and unique qualities, Ti_3C_2Tx as well as Ti_2CT_x have already found broad applications. Recent progress in this regard concerning the synthesis, characterization, and functionalization of these materials has paved a new path to discover their promises in several fields. Much work remains in researching the full potential of Ti_3C_2Tx and indeed even newer 2D MXene materials, as their widespread application has been hampered by a number of serious challenges.

Finally, a detailed overview of the recent developments on utilizing Ti_3C_2Tx and Ti2CTX in photovoltaic, sensing applications for energy storage devices (chemical capacitance), catalysis, and biomedicine is briefly discussed. In this book chapter, we are going through the fundamental properties of Ti_3C_2Tx and Ti_2CT_x as well as their synthesis/characterization together with special features that make them useful for a plethora of applications. The applications of Ti_3C_2Tx and the related compound, Ti_2CT_x, will also be introduced in this chapter. However, some thorny problems remain unresolved, awaiting answers from theoreticians or new improvements in the field. In this chapter, we would like to provide a better understanding of Ti_3C_2Tx and the newly proposed composition Ti_2CT_x (while offering some insight for scientists from other fields) so that there is more fresh research done going forward on these materials.

This chapter will highlight the recent progress in different realms using Ti_3C_2Tx and recently discovered (but extensively studied) member of MXene family, i.e., Ti_2CT_x, including sensing, photovoltaics, catalysis, energy storage, biomedicine. We will begin with the basics, including information on how are Ti_3C_2Tx and Ti_2CT_x synthesized and characterized, their unique properties as well as potential applications. Two widely researched layered two-dimensional titanium carbides (MXenes), mainly Ti_2CT_x and Ti_3C_2Tx, present significant promise as robust sensing platforms possessing high surface areas, chemical inertness, biocompatibility, and the capability to interact with diverse analytes. In that regard, we will explore the state-of-the-art in sensor development using Ti_3C_2Tx and Ti_2CT_x.

4.2 ELECTRONIC-BASED DETECTION

4.2.1 Sensing of Gases

Gas sensing is an interesting area of research which finds numerous applications in picture safety and reconnaissance just as well-being aspects. Recently, two-dimensional (2D) materials have gained a lot of consideration for gas sensing applications. This is because of their unique electrical and optical properties which are appropriate for this purpose. Among 2D materials of the MXenes family, Ti_3C_2Tx and Ti_2CT_x are believed to be largely impelling due to their gas-sensing ability (Adarakatti and Devaramani 2021; Arpia et al. 2021). MXenes are two-dimensional transition metal carbides and nitrides. They have so much surface that they catch on like Velcro and are chemically complex, plus they have the ability to carry an electrical charge. MAX phases are ternary layers carbides and nitrides. Selective removal of the A element spews these materials. The high surface area and numerous active sites of MXenes make them one of the most promising materials for gas sensing applications (Ran et al. 2016). Because of their high surface area, tunable electrical properties, and the strong interaction between gas molecules on their surface functional groups, devices made up with solutions in which two-dimensional (2D) MXenes are dispersed have demonstrated to be also apt for gas sensing. Two-dimensional titanium carbide (Ti_3C_2Tx) and titanium carbonitride Ti_2CT_x are an emerging class of materials with selective gas-adsorption capabilities to sense a broad range of gases. Kaiterra Laser Egg+ Contains sensors capable of detecting gases such ammonia

Trine, nitrogen dioxide, carbon monoxide, hydrogen sulfide, and methane. For example, the detection limit of Ti_3C_2Tx is 100 parts per billion (ppb), and it reacts to NH_3 in less than 10 seconds. In detecting CO and NO_2 as well, Ti_2CT_x has a sharp selectivity toward these low-concentration gases (Abbasi, Xiao, Peng, et al. 2021; Adarakatti and Devaramani 2021).

The modifications such as by adding different functional groups or metal nanoparticles can design the surface of Ti_3C_2Tx- and Ti_2CT_x-based gas sensors in order to get an amazing response. The gas sensing properties are optimized by increasing the number of active sites for adsorption and charge transfer using metal nanoparticles such as Pt, Pd, or Au. The sensitivity and selectivity of gas sensors can be improved by introducing different functional groups such as –OH, $-NH_2$, or –COOH. Thereby, a larger surface contact area between MXenes and gas molecules is possible after this modification. MXene sensors, such as Ti_3C_2Tx and Ti_2CT_x, for the detection of gases rely on the interaction between gas molecules with surface functional groups in MXenes. Their interaction induces changes to electrical properties within materials. This can be detected in a variety of ways, for example by capacitance or resistance methods as well optical. Similarly, when there is a resistance change to the gas sensor, created on Ti_3C_2Tx base material, you can see that it comes from how –OH functional groups grafted in the surface of this material react with NH_3 molecule and decrease its resistivity (Arpia et al. 2021; Yoris-Nobile et al. 2023). Ti_3C_2Tx- and Ti_2CT_x-based gas sensors show improved performances over previously reported models due to the enhanced sensitivity at room temperature. This results in fast response and recovery times, low power requirements, and the ability to operate around a lower temperature. The above properties make them ideal for practical implementation in portable and wearable gas sensors. Because they are transparent and flexible, MXene films also find use in blending with other electrical components (Tian et al. 2016; J. Jiang et al. 2020).

The gas sensors developed by using MXene can also be easily integrated with other electronic components, such as microcontrollers, wireless communication module, and power source. This enables the design of a larger gas detection system. For gas sensing under controlled conditions, it is very important to have reliable, fast, and precise materials with selectivity for the target gases. Metal carbides/nitrides (MXenes) have recently been of interest for gas sensing due to their hydrophilic surface, high electrical conductivity, and large specific surface area. Ti_3C_2Tx is the MXene with lowest formation energy (Peng and Huang 2022). Peng et al. presented this study in their review and this meta-analysis discusses in detail the different uses of Ti_3C_2Tx-based nanomaterials as gas sensors. Finally, difficulties encumbered with Ti_3C_2Tx gas sensors and proposing supportive ideas are presented. They suggest that we should use composite materials combined with surface functionalization for improving the VOCs' sensitivity of sensors. Therefore, Ti_3C_2Tx could be a promising candidate for gas detection by researchers (Figure 4.1a–g). (Zong et al. 2022; Chen et al. 2020; Shuvo et al. 2020; Lorencova et al. 2018; N. Li et al. 2019)

The gas sensing performance, good selectivity, and lower power consumption of Ti_3C_2Tx- and Ti_2CT_x-based sensors have been reported for the first time. MXenes have recently been demonstrated as promising candidates for gas sensing due to their strong interaction with the gas molecules, tunable surface chemistry, and large

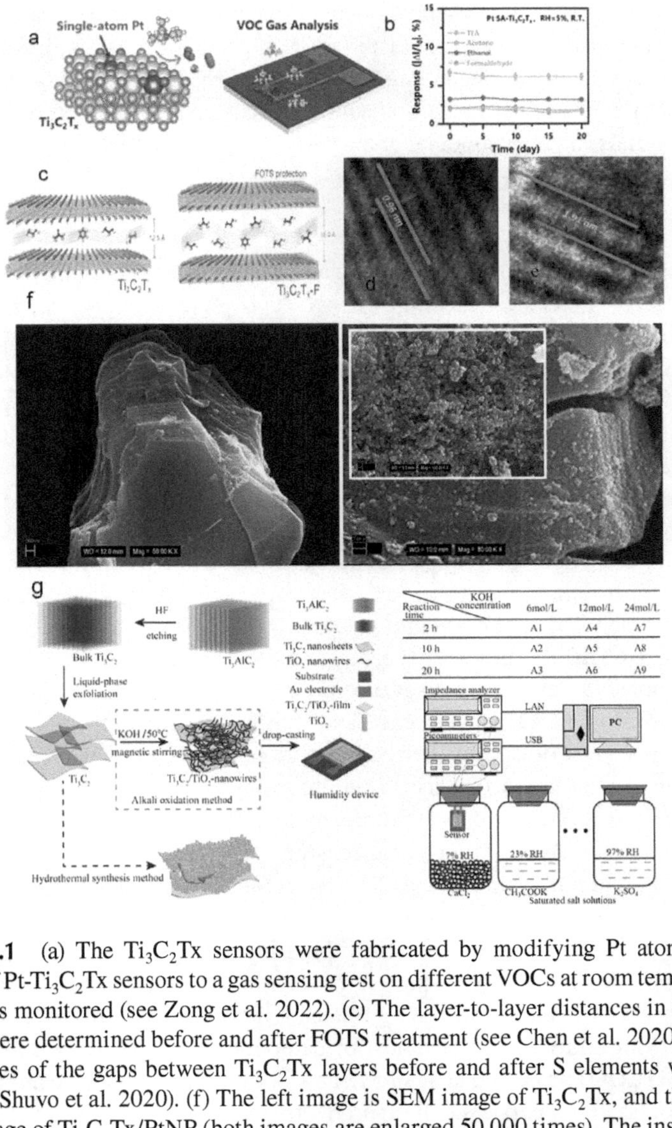

FIGURE 4.1 (a) The Ti_3C_2Tx sensors were fabricated by modifying Pt atoms. (b) The response of Pt-Ti_3C_2Tx sensors to a gas sensing test on different VOCs at room temperature for 20 days was monitored (see Zong et al. 2022). (c) The layer-to-layer distances in the Ti_3C_2Tx structure were determined before and after FOTS treatment (see Chen et al. 2020). (d–e) The TEM images of the gaps between Ti_3C_2Tx layers before and after S elements were indeed added (see Shuvo et al. 2020). (f) The left image is SEM image of Ti_3C_2Tx, and the right one is SEM image of Ti_3C_2Tx/PtNP (both images are enlarged 50,000 times). The inset image on the right shows PtNPs (Pt nanoparticles) that are formed when H2PtCl6 is reduced—you can see them at 100,000 magnification. Typical SEM images of (left) Ti_3C_2Tx and (right) Ti_3C_2Tx/PtNP. The two images have been magnified 50,000 times. A magnified image is displayed in the inset photograph at 100000 times (right) and inner nanoparticles of PtNPs were created from H2PtCl6 by reduction (see Lorencova et al. 2018). (g) There are several techniques used in the material preparation process, including HF etching, LPE, alkali oxidation, and others. Here is a schematic sketch that illustrates the experimental setup (see N. Li et al. 2019).

Sources: (a and b) Adopted with permission from American Chemical Society. (c) Adopted with permission from American Chemical Society. (d–e) Adopted with permission from American Chemical Society. (f) Adopted with permission from Elsevier. (g) Adopted with permission from American Chemical Society.

specific area. More in-depth research on these materials could help to provide a more complete understanding of the characteristics they use for sensing, thus broadening their appeal as well. It has been demonstrated that MXenes possess intriguing characteristics different from other carbon nanomaterials, leading to their promise as gas sensing materials in environmental control, industrial protection, and healthcare.

4.2.2 Single-Molecule Sensing

Single-molecule sensing is a high-precision and ultra-sensitive method used for detecting target molecules, recently uncovered by scientists. In recent years, 2D materials like Ti_3C_2Tx and Ti_2CT_x have gained a lot of attention. These materials possess significant physical characteristics including high surface area and special electronic functionality. They are being considered for use as they have potential in making single-molecule sensor applications. This makes electronic sensing a very attractive technique for single-molecule detection with 2D materials. It comes down to the physical properties, and it is being noticed how a certain combo of chemicals modify some electrical response. Titanium carbide (Ti_3C_2Tx) and titanium carbonitride (Ti_2CT_x) have been regarded as being ideal for electronic-based sensing applications due to their high specific surface area, good electrical conductivity, and resistance to chemical environment (Qin et al. 2021).

Ti_3C_2Tx and Ti_2CT_x were only recently reported as being suitable for sensing at the single-molecule level from a set of more extensive studies. The selectivity and sensitivity of Ti_3C_2Tx FETs were excellent; the DNA detection limit was lowered to 10 aM. This work illustrates the real-time monitoring of DNA hybridization and denaturation, which was a powerful route to probe at gene level, making Ti_3C_2Tx FET-type device as one of the most promising individual molecule detection tools for applicable bio-analysis. The electronic sensing ability of Ti_2CT_x was also explored for potential single-molecule-based devices. As a conductive support, Ti_2CT_x has also been applied for single-molecule detection through surface-enhanced Raman scattering (SERS) although our results did not indicate its noticeable enhancement effect on the light absorption. The dye rhodamine 6G served as a representative molecular probe to demonstrate the high sensitivity of SERS with Ti_2CT_x-based substrates, allowing individual molecules for detection in solution down to a limit of quantitation (LOQ) projected at 1014 M. This not only provides an example of how narrowly confined hot carriers can be productive but also suggests that engineering designer gaps (RTL5/modules) based on MXene layers allows for the creation of plasmonically active materials. Such materials would enable the identification of environmental contaminants and medications using advanced spectroscopic techniques (Peera et al. 2021; Ju et al. 2016). Hence, beyond field-effect transistors (FETs) and surface-enhanced Raman spectroscopy (SERS), an emerging area of research is the development of alternative electronic-based single-molecule sensing techniques through titanium carbide (Ti_3C_2Tx)/titanium carbonitride (Ti_2CT_x). For instance, an electrochemical detection method for the dopamine molecule will be able to leverage Ti_3C_2Tx as a conductive substrate. Ti_2CT_x has also been considered for surface acoustic wave (SAW) sensors. These sensors can recognize gases at the molecular level by analyzing gas samples.

In order to evaluate the electrochemical two-dimensional performance of Ti_3C_2Tx (MXene) sheets coated with Pt nanoparticles (PtNPs), L. Lorencova and coworkers revealed a study. In contrast, a GCE treated with pure Ti_3C_2Tx MXene showed no change in both the peak separation and redox potential under all synthesis pHs conditions upon comparing to those on Ti_3C_2Tx/PtNP nanocomposites. In contrast, the Ti_3C_2Tx/PtNP sensor exhibited a sensitivity to detect several low-concentration redox species comprising ascorbic acid (AA), dopamine (DA), uric acid, and acetaminophen at a potential higher than +250 mV with lower LOD. It also displayed significant selectivity toward the detection of some species down to ultralow concentration levels in nanomolar range. External membranes impact the selectivity for these compounds to a large extent. The GCE/Ti_3C_2Tx/PtNP H2O2 sensor exhibited the LOD of 448 nM at the potential +250 mV versus Ag/AgCl. In contrast, the Ti_3C_2Tx-blended GCE obtained 883 M and about 160 mV (Lorencova et al. 2018).

Overall, the comparative features reported herein among electronic-based single-molecule sensing using 2D Ti_3C_2Tx and Ti_2CT_x nanosheets in this study would be promising for numerous applications. Examples span from diagnosing biological conditions to environmental monitoring and chemical sensing. They are great materials for sensing individual molecules with high sensitivity and specificity. This has them having broad electrical properties and large available surface areas. This further study will no doubt lead to increasingly sensitive and specific sensing systems.

4.2.3 Humidity Sensors

A wide range of environmental and industrial processes involve humidity, which makes humidity sensing an essential area where 2D materials could be applied. Notably, such 2D materials as Ti_3C_2Tx and Ti_2CT_x with fundamentally distinct electrical and surface properties have shown great potential for use in humidity sensing. One of the most intriguing applications of 2D materials to humidity sensing is electrically based detection. The fundamental principle of electrical processes with 2D materials is an assumption that the electrical properties of a material would be modified in response to the humidity of the environment. With regard to Ti_2CT_x and Ti_3C_2Tx, these materials are highly appropriate for electrical detection or other purely electronic devices as they are highly conductive, i.e., they have an enormous FET and high chemical resistance and high sensitivity. Significant advances have been made in the use of Ti_2CT_x and Ti_3C_2Tx in electronic humidity sensing. For instance, Yang et al. combined a commonly used 2D materials with Ti_3C_2Tx-based FET to be studied in a recent publication to determine the quantity of moisture. The FET based on Ti_3C_2Tx reacted to the number of 2D materials variations in humidity lights in less than 10 ms, exposing high sensitivity. The FET also demonstrates the value of Ti_3C_2Tx-based FETs for detecting high environmental humidity, with potential applications in food packaging, HVAC systems, and more (Z. Ma, Fei, and Zhang 2023; Lorencova et al. 2018).

Another study, however, has already been done on electronic humidity detection using Ti_2CT_x. An interlayer of graphene oxide and a humidity sensor based on Ti_2CT_x, reacted to changes in humidity within 5 seconds, and its sensitivity was very high in changing humidity. Thus, the testing results confirmed the high potential for real-time monitoring, including, for example, environmental and agricultural monitoring. New humidity sensors based on electronic devices using Ti_3C_2Tx and Ti_2CT_x have been studied, as well as FET and a humidity sensor with an FET, using interlayers of GO in humidity sensors (N. Li et al. 2019). Additionally, Ti_2CT_x has been suggested for application in surface acoustic wave (SAW) humidity sensors (Wu et al. 2021).

There are various possibilities for the application of humidity sensors, such as in robots, health monitoring systems, human–machine interaction devices, and environmental protection decisions etc. Notably, the Ti_3C_2/TiO_2 composite displayed a high sensitivity of ca. 280 pF/percent RH between 7% and 33%RH compared to ten times larger than those for both pure (or neat) materials Ti_3C_2, or/and which may be ascribed to increased specific surface area. Additionally, humidity sensors of the Ti_3C_2/TiO_2 could sense different liquid samples and fingers of a human body without actual touch.

2D Ti_3C_2Tx- and Ti_2CT_x-based humidity sensing provides a feasible approach to common electronically-based systems that could have potential applications in food packaging, agriculture as well as environmental monitoring. Because of their specific electrical and surface properties, these materials are a perfect fit for sensitive humidity level monitoring by area. Future investigation in this area will certainly lead to the fabrication of advanced humidity sensors with improved sensitivity and selectivity.

4.2.4 Pressure Sensors

Pressure sensing is a key application of electronic-based detection, with significant use in fields such as medical devices (especially for critical care monitoring, e.g., capnography), aerospace, and automotive industries. Two-dimensional (2D) materials such as Ti_3C_2Tx and Ti_2CT_x offer substantial potential for pressure sensing due to their unique mechanical and electrical properties. The 2D nature of materials means that 2D material-based pressure sensing can help in electronic-related detection. This technique permits the monitoring of electric properties of materials as they change during the application of pressure. Ti_3C_2Tx and Ti_2CT_x have shown great promise for electronic pressure sensing owing to their large surface area, high conductivity, and superior mechanical properties (Qin et al. 2021).

During a couple of investigations, Ti_2CT_x86 and $Ti_3C_2T_x$78 were used for pressure sensing based on electronic signals. In a recent study by Zhou et al., for instance, the same were used to measure pressure using a piezoresistive sensor under other recent studies (2021). The Ti_3C_2Tx sensor exhibited high sensitivity to pressure, with a response time of less than 10 ms and gauge factor (GF) up to 25. The study demonstrated the use of Ti_3C_2Tx-based sensor for different applications including wearable technology and structural health monitoring, in order to monitor real-time changes of pressure. They also investigated on another possible use for Ti_2CT_x which is using it in electronic pressure sensing. Wang and colleagues analyzed this in a recent study of theirs and in some other cases, Liu et al. Meanwhile, the study conclusions also validated this kind of Ti_2CT_x/GO pressure sensors for real-time detection in flexible and wearable devices as well as for applications such as robotics or medical science. The pressure sensor made of Ti_2CT_x/GO exhibited fairly high sensitivity toward modest changes in the applied pressure, with a gauge factor as large as 24 (Peera et al. 2021; Ju et al. 2016). Other alternative ways to investigate electronic based pressure sense like piezoresistive sensors or with GO interlayers. Example strain sensors make use of Ti_3C_2Tx as the sensing component. Besides, there are the advantages of Ti_2CT_x as an excellent candidate material for making surface acoustic wave (SAW) sensors because SAW sensors have been proven to be able to be applied in pressure change detection (Yi et al. 2020).

Pressure sensors with features of high flexibility, permeability, and degradability are critically demanded for health monitoring, making wearable artificial skins, and artificial intelligence applications. However, the applicability of such sensors is severely compromised with the limitations of the normal plastic or elastomer substrates, such as impermeability, lacking degradability, discomfort, and mechanical conflicts. One of the most significant hurdles remaining is to generate a pressure sensor that is robustly flexible, permeable, and degradable. M. Chao et al. has now presented a wearable, permeable, degradable, and highly sensitive pressure sensor based on MXene/protein nanocomposites to tackle this concern. The sensitive membrane of the aforementioned pressure sensor is constructed from silk fibroin nanofibers that are interlaced with MXene-coated electrodes and fabricated from MXene ink. The sensitive membrane, which is exceedingly sensitive and quickly responds and recovers, breathes uniformly and cycles persistently, is exceedingly biocompatible and robustly degradable. This sensor not only monitors the psychological signals in humans but can also be used to create a synthetic skin for wireless real-time biomonitoring and quantitatively mapping of pressure distribution. The degradable and permeable attributes of this sensor provide exciting options for the future human–machine interaction, human motion monitoring, smart electronic skins, and disease diagnosis (Figure 4.2) (Chao et al. 2021).

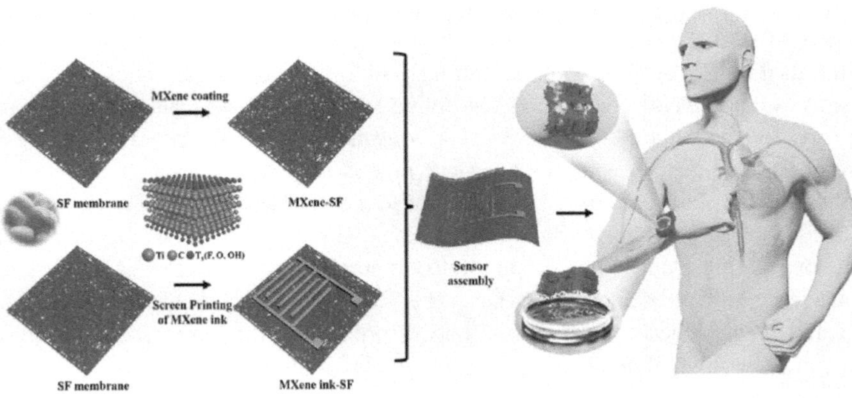

FIGURE 4.2 Schematic diagram illustrating the fabrication process of protein nanocomposites-based pressure sensors (a combined with b). Following are the steps which are involved in making MXene ink by ultrasonication and creating an interdigitated electrodes layer: First it is required to prepare the MXene ink. Then a conductive, breathable membrane coated with MXene can be developed as the sensing layer. At last, scientists connect the two membranes to form a pressure sensor. Electrical measurement is completed through the attachment of electrodes to the interdigitated electrode layer (see Chao et al. 2021).

Source: Adopted with permission from American Chemical Society.

2D Ti_3C_2Tx and Ti_2CT_x materials for electronic-based pressure sensing show precious application prospects in medical devices, wearable technology, robotics, structural health monitoring, etc. Due to their unique mechanical and electrical properties, these materials are perfect for pressure change detection in specific sensitive areas. There is no doubt that a more detailed study will allow for the development of advanced pressure sensors with greater sensitivity and specificity.

4.3 OPTICAL SENSORS

Optical sensors have recently become more popular as a result of their noninvasive nature, quick response time, and high sensitivity. Two-dimensional 2D materials are ideal for optical sensing owing to their intense light-matter interactions and tunable optical properties. Optical sensing could be achieved using 2D materials, including Ti_3C_2Tx and Ti_2CT_x. Although traditional biosensors rely on antibody or antigen recognition processes, plasmon sensing has been proposed as an alternative method for measuring changes on a material's surface or local refractive index. Due to their large surface areas and plasmonic properties, Ti_3C_2Tx and Ti_2CT_x provide unique opportunities for plasmonic sensing. Several investigations have utilized Ti_3C_2Tx and Ti_2CT_x for plasmonic sensing. For instance, Wu et al. showed that Ti_3C_2Tx-based plasmon sensors are employed to detect tiny chemicals, which was published in 2021. The plasmonic sensors employing Ti_3C_2Tx achieve a limit of detection as low as 10 nM, demonstrating great sensitivity and selectivity for minor chemicals like glucose. Finally, the researchers argue that their Ti_3C_2Tx-based plasmon sensors can be used to quantify small chemical concentrations in biological fluids in real time (Zou et al. 2022; Chai et al. 2022).

In the context of plasmonic sensing, fluorescence sensing, and surface-enhanced Raman scattering (SERS) sensing for various applications regarding sensors, the materials Ti_3C_2Tx or Ti_2CT_x have been well addressed. Owing to its extraordinary sensitivity and selectivity, Ti_2CT_x has found applications as a plasmonic sensing material for the detection of heavy metal ions. This makes the research described here of further interest as its large surface area is very well-known to offer a good application for plasmonic behaviors, and consequently Ti_3C_2Tx has been reported to be used in surface-enhanced Raman scattering (SERS) sensing. They can be applied in fields ranging from biological to environmental monitoring, and as they do not protrude significantly above their substrate. 2D materials have great potential for optical sensing. Due to these special optical properties, they are perfect for sensitive and accurate analyte identification in complex samples. Further research in this field is expected to contribute toward optical sensing with enhanced sensitivity and specificity (Abbasi, Xiao, Peng, et al. 2021; Tiwari et al. 2020).

4.3.1 FLUORESCENT-BASED DETECTION

Over the years, fluorescent molecular-based probes have become a versatile technique to enable sensitive yet more accurate target detection up to necessitated focus-specific analysis in almost all classes of applications ranging from sensing and photovoltaics, catalysis for energy storage devices to biomedicines. Two-dimensional

(2D) Ti_3C_2Tx and Ti_2CT_x fluorescent probes are of great concern among the various types of fluorescence probe that have been widely studied, wherein high specific surface area makes them bear many intrinsic merits with unusual optical properties. Fluorescent-based detection capability of 2D Ti_3C_2Tx and Ti_2CT_x in sensing applications has been widely explored. The large surface areas of these materials yield high immobilization efficiency for the analyte-specific receptors thus enabling sensitive and selective bioassays to be performed against a varied list constituting proteins, nucleic acids, or even small molecules. These materials also show strong fluorescence, making possible in-situ observation of biomolecule-to-biomolecule interactions and providing insights into kinetics and thermodynamics (Abbasi, Xiao, Zhang, et al. 2021; Lee et al. 2020).

Apart from sensing, two-dimensional Ti_3C_2Tx and Ti_2CT_x have been further investigated for fluorescence quenching-based detection that is relevant to the field of photovoltaics as well as catalysis. This powerful fluorescence of these materials can serve as a probe to monitor the electron-transfer processes in solar devices and get further understanding about charge generation-and-transport mechanism. Optical conversion is a promising approach, and the most remarkable aspect is that chemical species like hydrogen peroxide and nitroaromatic compounds can react efficiently on materials with a large surface area. This process can transform toxic waste into pure products. Additionally, magnetic detection offers a rapid and precise method for monitoring these reactions. Additionally, energy storage capabilities of 2D Ti_3C_2Tx and Ti_2CT_x were also studied in the forms of batteries and supercapacitors for the fluorescent-based sensing method. The large surface areas and excellent electrical conductivity of such materials permit efficient charge transfer (and storage) required for more effective energy storage/conversion. Finally, regarding the use of 2D Ti_3C_2Tx and other forms like Ti_2CT_x, biomedicine has already explored their application in fluorescent-based detection for fields such as imaging and diagnostics. Their intense fluorescence properties can serve not only as a powerful tool to track the migration, proliferation, and apoptosis of cells in vitro but can also respond very sensitively when implanted into animals. In addition, owing to their high specific surface areas and biocompatibility of the metallic coatings, metallic coatings materials (MCMs) are efficiently taken up/absorbed. They, stored in cells, provide indispensable insights into cellular function (Tiwari et al. 2020; N. Li et al. 2019).

That's fascinating! This work is an exciting demonstration that Ti_3C_2Tx quantum dots show great potential as a nanoprobe for Fe^{3+} monitoring in environmental matters and also act as a dual-emission radiometric pH sensor. Their emission from the QDs, which can be tuned by pH and amino group on their surface, could find numerous applications in chemical/biological sensing. In view of the importance in numerous biological and environmental processes, this low detection limit is still remarkable for ion Fe^{3+}. Future work on the potential of Ti_3C_2Tx quantum dots for monitoring and sensing would be interesting (Figure 4.3a–d) (Z. Wang et al. 2022).

4.3.1.1 Detection of Ions of Heavy Metals

Heavy metal ions detection is one of the critical processes in several fields such as biological applications, industrial procedure, and environmental monitoring. Two widely used methods for the determination of heavy metal ions are atomic absorption

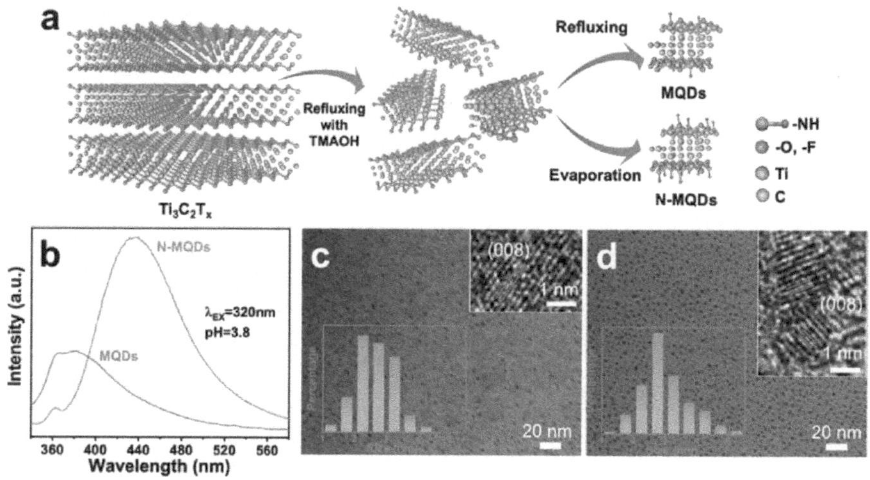

FIGURE 4.3 (a) A schematic illustration on the synthesis of MQDs and N-MQDs. (b) The PL spectra of MQDs and N-MQDs upon excitation by 320 nm light are represented. (c) MQDs and (d) N-MQDs; insets show size distribution histograms and HRTEM images. The insets show the size-distribution histogram, conventional transmission electron microscopy (TEM) images of individual MQDs and N-MQDs as well high-resolution TEM pictures (see Z. Wang et al. 2022).

Source: Adapted with permission from the Royal Society of Chemistry.

spectroscopy and inductively coupled plasma-mass spectrometry (ICP-MS). Their disadvantage is that they are costly and time consuming. The electronic-based detection of heavy metal ions by using a sensitive and targeted method is now becoming an increasingly attractive approach with 2D conformal Ti_3C_2Tx, which may be used in conjunction with the new form TiO7 as well. Compared to other 2D materials, the surface chemistry and electrical property of V and Ti_2CT_x are different, which made them to be widely applied in fluorescent-based sensing for heavy metal ions. Their fluorescence is quenched upon binding to heavy metal ions, so these materials make ideal sensors for selectively and sensitively picking up the presence of sodium or other metals. This facilitates sensitive and accurate high signal-to-noise detection (Tayyab et al. 2022; Tayyab et al. 2023). Many techniques have been developed for fluorescent-based detection of heavy metal ions, including lead (Pb), cadmium (Cd), mercury (Hg), copper (Cu), and zinc (Zn). For example, it has been demonstrated that fluorescence sensors based on 2D Ti_3C_2Tx can be used to specifically detect Pb2+ ions. The sensitivity and selectivity of the sensors for Pb2+ ions are unaffected by the presence of other metal ions. Cd2+ ions have been detected using Ti_2CT_x-based fluorescence sensors at concentrations as low as 0.09 M (Abbasi, Xiao, Zhang, et al. 2021; Lee et al. 2020; Ran et al. 2016).

In two dimensions, Ti_3C_2Tx and, to a lesser extent, even Ti_2CT_x are employed for the detection of heavy metal ions by their interaction with surface functional groups such as carboxyl or hydroxyl. Two electronic techniques, fluorescence spectroscopy and fluorescence imaging, can identify the quench of fluorescence induced by heavy metal-ion attachment to surface functional groups. 2D Ti_3C_2Tx- and Ti_2CT_x-based electronic detection offer several advantages over classical method of detections. Moreover, these materials can be easily functionalized with ligands to attain high selectivity and sensitivity toward heavy metal ions. Furthermore, these materials possess large surface area and high electrical conductivity, making them ideal for the successful immobilization of analyte-specific sensors to boost detection performance (Abbasi, Xiao, Zhang, et al. 2021; Lee et al. 2020; Ran et al. 2016; X. Zhou et al. 2016).

For this purpose, the researchers selected a composite technique to produce a nanocomposite material comprising MXenes—Ti_3C_2Tx, where Tx = –O, –OH, –F, and Bi nanoparticles. Due to the flat structure of the material, the Bi nanoparticles were better deposited on the face of Ti_3C_2Tx, and the former did not aggregate due to the lack of abundance of –OH around. A sensor was developed that exemplified the electrochemical sensing of Pb2+ and Cd2+ metal ions. The key features of the BiNPs@Ti_3C_2Tx-based sensor are the large effective face area, high conductivity, and anodic wavelet sensor technology for heavy metal ions (Figure 4.4). Once the experimental conditions had been optimized, the BiNPs@Ti_3C_2Tx sensor displayed the simultaneous detection of Pb2+ and Cd2+ at a limit of detection of 10.8 nM and 12.4 nM, respectively. Materials with a large face area, high electron pass-through capacity, low cost, and low toxicity can work as good sensors for sensing heavy metal ions (He et al. 2020).

In summary, the 2D Ti_3C_2Tx and Ti_2CT_x are optimal choices for heavy metal ion detection due to their highly sensitive nature (electronics-based) at accurate levels. In numerous studies, these materials have demonstrated superior fluorescence ion-sensing properties, enabling ultra-selective and highly sensitive detection of various heavy metal ions. Consequently, their unique surface and electrical properties have significantly advanced heavy metal-ion detection, moving from traditional methods to innovative approaches, thanks to their favorable and exotic chemical nature.

4.3.1.2 Detection of Biomolecules (Small and Macromocules)

The field of biomolecule identification, encompassing both small and large molecules, is important in several areas, such as clinical diagnostics. The most common methods used to detect biomolecules are the polymerase chain reaction (PCR) and enzyme-linked immunosorbent assay (ELISA), which, as the researchers know, present high cost for their practical realization. 2D Ti_3C_2Tx and 2D Ti_2CT_x enable high-accuracy biosensing through wide-range fluorescence signals. These materials can be fluorescently identified with numerous biomolecules, including proteins, nucleic acids, and small compounds. These materials offer several advantages over conventional methods, including high sensitivity, selectivity, and real-time sensing. Receiver-specific biomolecules could be selectively captured and detected in high sensitivity by functionalized 2D Ti_3C_2Tx and Ti_2CT_x due to binding with target biomass (Z. Zhang et al. 2022; C. Li et al. 2022).

A broad range of molecules have been detected by a fluorescence-based detection method with 2D Ti_3C_2Tx and Ti_2CT_x such as DNA, RNA, peptides, or antibodies. For instance, Ti_2CT_x-based fluorescence sensors with excellent selectivity and high sensitivity to specific DNA sequences were employed for selective detection of target DNAs. Furthermore, it revealed the application of 2D Ti_3C_2Tx-based fluorescent sensors with exquisite sensitivity and specificity to respective targeted antibodies for a selective antibody detection. Moreover, the work

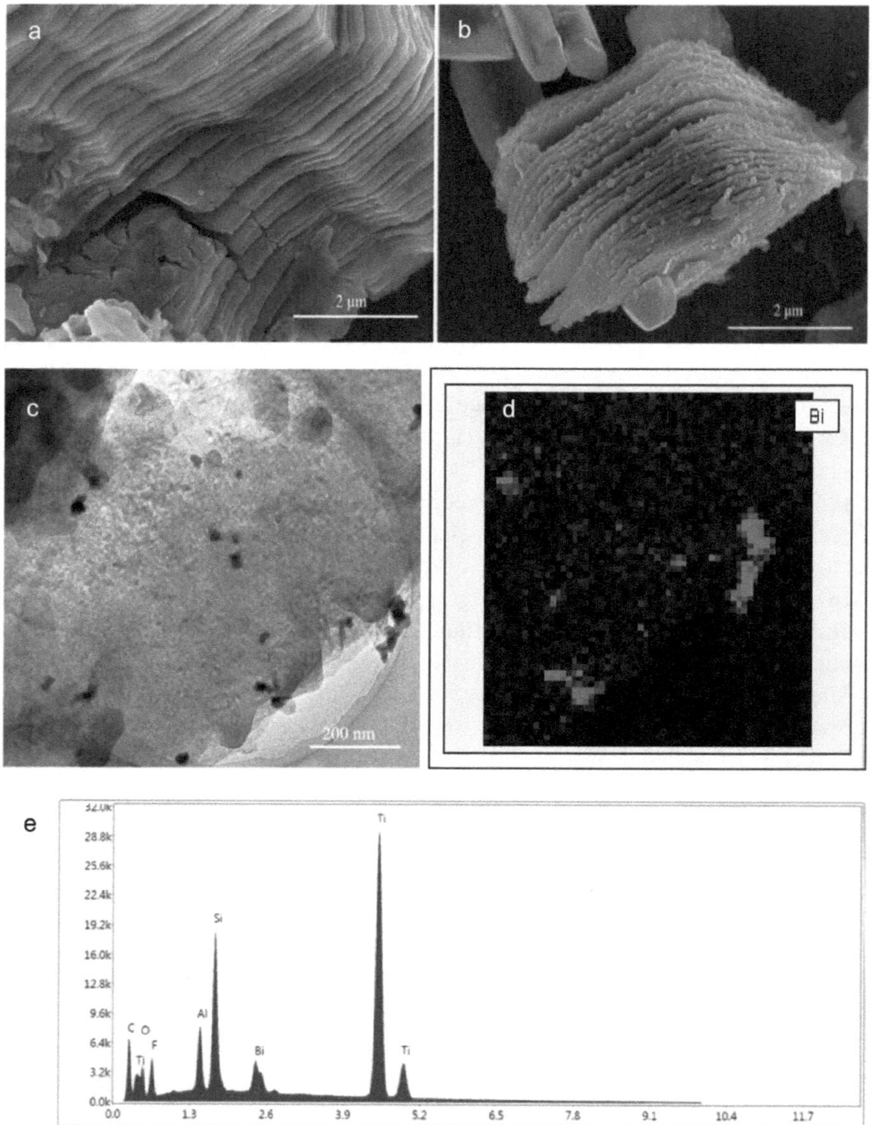

FIGURE 4.4 Ti_3C_2Tx SEM images are displayed in (a), while BiNPs@Ti_3C_2Tx are displayed in (b). A TEM picture of BiNPs@Ti_3C_2Tx is shown in (c). Bi's EDX mapping is shown in (d), and elemental EDX mappings are shown in (e) (see He et al. 2020).

function-based charge transfer (WAC) is utilized in this context for detecting biosensors. It serves as a detection junction between 2D Ti_3C_2Tx and T1a functional groups of surface biomolecules. When biomolecules bind to the functional groups at surface, fluorescence is quenched. This quenching can be visualized by many electronics-based techniques like fluorescence spectroscopy and florescence imaging (Y. Sun et al. 2022; Gao et al. 2022).

MXene is a two-dimensional carbonitride transition metal with unique optical and electrical properties, representing as novel substrate for surface-enhanced Raman scattering (SERS) in the new era. The authors of this study developed a SERS-substrate synthesis routine to create such a biosensing platform by taking advantage of both the immense enhancement factor and multifunctionality reported for metal nanoparticles on MXene. Specifically, they employed an in-situ electrostatic-self-assembly technique to fabricate the Ag nanoparticle-MXene (Ti_3C_2Tx) hybrid biosensor. In Figure 4.7, citrate-coated Ag NPs loaded with PDDA polymer were used as counter cation, resulting in the formation of a composite interface double layer (CIDIL) on Ti_3C_2Tx flakes. The designed platform showed excellent performance in the surface-enhanced Raman scattering (SERS), long-term stability, and uniformity of detecting 4-mercaptobenzoic acid (4-MBA) molecules. The Raman amplification process of the platform was simulated by modeling electromagnetic fields' propagation. The biosensor proved to be an efficient tool for biomolecule detection as it displayed the ability to sense 108 M of adenine molecules and a linear relationship between peak intensity, dopamine molecule concentration ranging between (510–6) M and (5 × 10–8) M. The hybrid platform of MXene-PDDA-Ag NPs presents various applications in the biotechnology sector, ranging from environmental biology testing to SERS-based label-free quantitative detection of biomolecules and clinically relevant diagnostic platforms (Figure 4.5) (R. Liu et al. 2021).

Compared with other more traditional methods, the electronic-based detection using 2D Ti_3C_2Tx and Ti_2CT_x provides obvious advantages especially in terms of high-throughput screen, cost-effectiveness as well as the ease to use. The high surface area can then be functionalized with molecular recognition elements—drilling down significantly to nominally specific and, in some cases, highly selective/aptamers/antibodies for target biomolecules. For these materials, they additionally have a more significant surface area and better electrical conductivity which would be ideal for the immobilization of analyte-specific receptors to improve detection (Tayyab et al. 2022).

In summary, fluorescent detection with 2D Ti_3C_2Tx and Ti_2CT_x holds a promise of highly sensitive and selective biomolecule sensing. These materials have impressively high selectivities and sensitivity toward various biomolecules such as proteins, nucleic acids, or even small molecules. Their distinctive electrochemical and surface chemistries have tremendous potential for new, state-of-the-art biosensing.

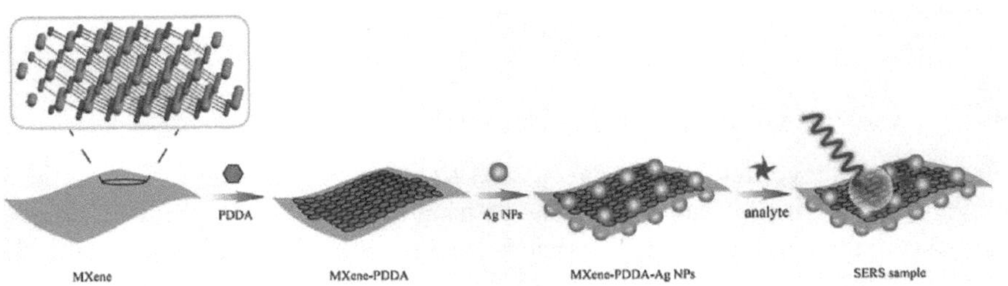

FIGURE 4.5 The procedure of preparing a substrate consisting of MXene-PDDA-Ag NPs and using it for surface-enhanced Raman spectroscopy (SERS) detection are shown in this diagram (see R. Liu et al. 2021).

Source: Adopted with permission from Elsevier diagram.

4.3.2 Calorimetric-Based Sensing Materials

Thermometric sensing materials could be employed in energy storage, catalysis, and sensors. Such sensors exhibit higher sensitivity, selectivity, and real-time monitoring when compared to their typical counterparts. Because of their special surface chemistry and electrical characteristics, 2D Ti_3C_2Tx and Ti_2CT_x are among the most desirable materials for calorimetric sensing. This technique basically measures the heat released during the adsorption of target analytes on the surface in case of 2D Ti_3C_2Tx and Ti_2CT_x calorimetric sensing. Functionalization with a specific sensor for the target analytes selectively adsorbs them. This results in different temperatures which can be measured electronically with the help of thermal analysis techniques such as differential scanning calorimetry (DSC) and isothermal titration calorimeter (Bhardwaj and Hazra 2021; H. Ma et al. 2022). They demonstrate the successful use of 2D Ti_3C_2Tx and Ti_2CT_x in a colorimetric-based gas sensor device capable of detecting chemoresistive gases and volatile organic compounds (VOCs).

Alterations of gas and VOCs sensitivity and selectivity have been described for example in the presence of formaldehyde, carbon monoxide, or nitrogen dioxide. Combining analyte-specific sensors for surface functionalization allows target gases and VOCs to be more precisely identified. With the use of Ti_3C_2Tx and Ti_2CT_x calorimetric-sensing techniques, heat released or absorbed during adsorption process can be analyzed to identify biomolecules (proteins, nucleic acids). The change in heat radiation can be monitored using electronic techniques such ITC and DSC. 2D Ti_3C_2Tx and Ti_2CT_x sensors offer significant advantages over conventional ones, including lower cost, more robust construction, and high-throughput capabilities. They offer improved electrical conductivity and high surface area that can enhance the detection ability; in addition, they are readily amenable to functionalization with analyte-specific receptors such as aptamers or antibodies (Y. Wang et al. 2022; C. Han et al. 2020).

In conclusion, 2D Ti_3C_2Tx and titanium carbide (MXene) calorimetric-based sensing is a powerful method for the selective detection of wide varieties of analytes with high sensitivity—gases, VOCs, or biomolecules. The sensitivity, selectivity, and real-time monitoring of these materials are generally more superior than conventional sensors. For catalysis, energy storage, and sensing applications, the combination of surface chemistry with their exciting electrical properties makes this class of materials particularly attractive.

4.3.3 Electro Luminance-Based Sensors

Fluorescence-based detection has increased to encompass a variety of activities in biology, environmental monitoring, and sensing. This approach implies estimating the time duration of fluorescence emissions that several analytes generate as a reaction to light stimulation and their intensity. Among the various forms of fluorescent-based sensors, electroluminescence-based sensors are considered as attractive methodology for ultra-sensitive and selective detection of target analytes. Due to the unique surface chemistry and electrical properties, 2D Ti_3C_2Tx and Ti_2CT_x are beneficial materials for making the electroluminescence-based sensors. Electroluminescence properties of the fluorescent materials are modulated in the electroluminescence-based sensors in the presence of target analytes. Ti_2CT_x and 2D Ti_3C_2Tx have exhibited large surface area and high electrical conductivity; thus, they are considered as excellent material for making the electroluminescence-based sensors. These materials' surfaces are integrated with various receptors such as antibodies and aptamers which may fluctuate the electroluminescence properties and allow the selective identification of target analytes (L. Zhang et al. 2021; Cheng et al. 2021).

One of the most interesting applications for electroluminescence sensors comes from biomolecule detection, including proteins and nucleic acids. These modified materials could specifically gate target biomolecules and hence manifest the characteristics of electroluminescence. It was found that this technique is highly selective and sensitive to detect even down to the low concentrations of biomolecules. Environmental pollutants, e.g., organic compounds and heavy metals, are emerging contaminants for which two-dimensional Ti_3C_2Tx and Ti_2CT_x electroluminescence sensors have been investigated. The surfaces of these materials can also be functionalized with receptors for the selective detection of impurities that modify their electroluminescence properties. Scientists have discovered that this method is enough sensitive and selective to determine the pollutants at PPM level (Nath et al. 2020; L. Sun et al. 2020).

Thus, electroluminescence-based 2D Ti_3C_2Tx and Ti_2CT_x sensors offer an array of remarkable advantages when compared to other state-of-the-art sensors: High sensitivity, selectivity, and real-time monitoring. Such materials can be functionalized with suitable receptors that achieve high selectivity and sensitivity toward the desired analyte easily. Moreover, the analyte-specific receptors will be efficiently immobilized due to these materials' excellent electrical conductivity and enormous surface area, as the surface chemistry of 2D materials assures high performance of the receptors at the interface. Electroluminescence-based sensors using 2D Ti_3C_2Tx and Ti_2CT_x have been proved to detect an ample selection of analytes, ranging from biomolecules to environmental pollutants, with high sensitivity and selectivity. In conclusion, there is no need to doubt the advantageous properties of these novel sensors, as they prominently overcome other sensors In the market: High sensitivity, selectivity, and real-real-time monitoring (Z. Ma, Fei, and Zhang 2023).

4.4 ELECTROCHEMICAL-BASED DETECTION

Due to its excellent sensitivity and selectivity, electrochemical detection strategy has gained increasing interest for the specific identification of various analytes such as biomolecules (including large proteins) right down to trace-level target pollutants in environmental samples. Ti_3C_2Tx and Ti_2CT_x are promising 2D materials with high electrical conductivity, large surface area scaffolds, and distinct surface chemistry, making them ideal for electrochemical detection. When target analytes are applied, they form a response (either reduced or oxidized) through electrochemical-based detection. To achieve target selectivity, 2D-layered Ti_3C_2Tx and Ti_2CT_x can be functionalized with specific receptors, such as antibodies or aptamers. The current, voltage, or charge transfer resistance varies accordingly as the target analyte gets in touch with receptor and sensing material leading to the quantification of the analyte (Abbasi, Xiao, Zhang, et al. 2021; R. Liu et al. 2021).

Responding to biological molecules such as proteins and nucleic acids, 2D Ti_3C_2Tx and Ti_2CT_x are used for electrochemical detection. The surfaces of these materials can be functionalized with target biomolecule receptors for site-specific biosorption that alters their own electrochemical properties. This process has shown an extreme level of sensitivity and selectivity, allowing the imaging of the biomolecules at nanoparticle aerosol ranges. 2D Ti_3C_2Tx and Ti_2CT_x were also tested for electrochemical interrogation aiming on organic molecules and toxic metal ions detection in the environmental arena. The special chemical receptors for target pollutants' surface functionalized on these materials also uniquely affect their electric behavior, leading to the selective detection of targeted contaminants. This approach has been shown to be very sensitive and selective which allows contaminants in low concentration detected accurately. 2D Ti_3C_2Tx- and Ti_2CT_x-based electrochemical detection shows greater sensitivity and selectivity compared with methods such as Albian etching; in addition, real-time monitoring is also offered by these strategies. After functionalization with appropriate receptors, these materials can easily achieve high selectivity and sensitivity for any desired analyte. The huge surface area and good conductive electrons mean these materials can easily immobilize the analyte-specific probes to build a functionalized sensor, which lead to an improved and highly efficient electrochemical detection (Q. Zhou et al. 2021; Tiwari et al. 2020).

Owing to its excellent chemical and physical properties, Cu-BTC has a significant potential as an electrochemical sensing material. However, its low conductivity prevents Cu-BTC from being used as it is in electrochemical sensing. For that, Cu-BTC is grouped under the high conductive 2D Ti_3C_2Tx nanosheets to achieve resultant nanocomposites with superiorly enhanced conductivity and catalytic activity beyond the conceptional ones that are unachievable, which makes the most susceptible glyphosate detection practicable (Figure 4.6). The LOD of Ti_3C_2Tx/Cu-BTC is short, 2.6×10^{-14} M, and the nanocomposite exhibits low linear sensing area of 1.0×10^{-13}

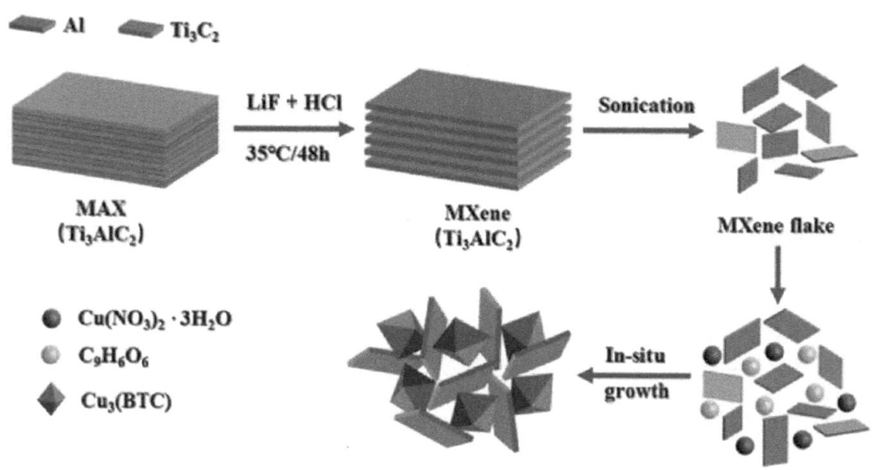

FIGURE 4.6 The preparation of Ti_3C_2Tx/Cu-BTC composite (see S. Wang et al. 2022).

Source: Adapted with permission from the Royal Society of Chemistry.

to 1.0×10^{-6} M, and similar to this, the electrochemical sensor, Ti_3C_2Tx/Cu-BTC, is demonstrated to be highly stable, reproducible, and selective (S. Wang et al. 2022).

In addition, due to the more brilliant electrochemical traits of 2D Ti_3C_2Tx and Ti_2CT_x; they are terminally desirable for power storage as well as a catalyst. These compositions are prepared for use as electrode materials in lithium-ion batteries and supercapacitors. They have been widely used as catalyst supports for a variety of catalytic reactions such as hydrogen evolution reaction and oxygen reduction process (R. Liu et al. 2021).

Electrochemical detection using 2D Ti_3C_2Tx and Ti_2CT_x offers a large set of advantages as compared to more traditional approaches. Simple receptor-based material functionalization can be used to design substantially sensitive, responsive, and selective analytes. Application of this method is not only limited to the quantitative determination of a broad spectrum of analytes (biomolecules, environmental pollutants, etc.). These materials also have a special surface chemistry and electrochemical activity that can offer new opportunities in the areas of biomedicine, energy storage (batteries), sensing, catalysis, etc.

4.4.1 Detection of Biomarkers

Biomarkers, which are molecules that indicate the presence or progression of a disease—such as those found in blood and urine samples—provide critical tools for diagnosing and monitoring disorders. Ti_3C_2Tx and Ti_2CT_x as 1T0/F-Ti3 as point sensors for biomarker detection do well, more than in other common reactions. One such application of these materials is in biosensors for the detection of biomarkers. A biosensor consists of a transducer that activates an immobilized bioreceptor. The bioreceptor binds to the target biomarker and converts this binding event directly into a quantifiable signal, forming the basis of the sensor. Due to the large surface area, high conductivity, and good biocompatibility of both Ti_3C_2Tx and Ti_2CT_x, they have been used as target transducers. One study targeted the cancer biomarker CEA with Ti_3C_2Tx as a transducer in a biosensor configuration. The researchers first functionalized the surface of Ti_3C_2Tx with CEA-recognition antibodies before coating it with a layer of graphene oxide. Antibodies were immobilized on the surface of Ti_3C_2Tx, which captured A-ferritin and induced a change in electrical conductivity that was read as an electronic signal. The biosensor showed good sensitivity and selectivity toward CEA detection (Lorenzen et al. 2022; F. Long et al. 2022).

Moreover, Ti_2CT_x has been previously employed for various molecules such as DNA, cholesterol, and glucose. Ti_2CT_x was used as a transducer in the biosensor to detect DNA with gold nanoparticles functionalized. The electrical impedance is changed when the target DNA was injected and then hybridized with immobilized DNA probes. DNA probes were immobilized on Ti_2CT_x surface by researchers. The biosensor was proved to exhibit good selectivity and sensitivity toward the DNA detection. As part of an additional biomarker-detecting test, Ti_3C_2Tx and Ti 2CTx have been employed to examine various mass spectrometry-based assays. A type of mass spectroscopy matrix-assisted laser desorption-ionization time-of-flight (MALDI-TOF) MS is used to identify proteins and peptides. As a coating material for electrodes to detect proteins and other biomolecules using electrochemical impedance spectroscopy, topologically nanonetworks of PVFi2CTx have been explored as well. Utilizing phenomena observed in detecting biomarkers with 2D materials like Ti_3C_2Tx and Ti_2CT_x will eventually help toward the development of ultra-specific, highly sensitive diagnostic tools for numerous diseases (L. Ma et al. 2022; Razavi et al. 2022).

4.5 PHOTOVOLTAICS

Photovoltaic technology is a widely used method for converting solar energy into electrical energy using semiconductor materials. The extensive use of 2D Ti_3C_2Tx and Ti_2CT_x in photovoltaics is owing to their remarkable physical/chemical properties, corresponding directly to the increased interest for renewable energy sources. Very high electrical conductivity of 2D Ti_3C_2Tx and Ti_2CT_x is one of the most promising features for photovoltaic applications that could be used for electron transport materials inside solar cells. They feature a high surface area for adsorption, good transparency, and an easily tunable bandgap so they are ideal candidates as materials in solar devices. A lot of research literature is focused on their use in photovoltaic (PV) technologies; dye-sensitized solar cells made from them are well known over last decades, and the recent perovskite cell is challenging that markedly. The improved efficiency of 2D Ti_3C_2Tx perovskite solar cells can be attributed to their increased electron mobility. Reportedly, 2D Ti_3C_2Tx has been used as an effective electrocatalyst in the reduction of triiodide ions within the counter electrode materials for dye-sensitized solar cells. Moreover, the combination of Ti_3C_2Tx and Ti_2CT_x has been used in making organic solar cells (OSCs) due to them having high surface area with enhanced charge transfer properties (Shah et al. 2021; L. Zhang et al. 2021; Saeed et al. 2022).

Belonging to a class of 2D transition metal carbides and nitrides, or the so-called MXenes, these materials exhibit excellent electrical and optical properties, making them highly suitable for applications in energy storage, sensing, and optoelectronics. Scientists demonstrated a ZnO/Ti_3C_2Tx composite film, where Ti_3C_2Tx is the most famous member of MXenes. The formation of these Zn-O-Ti connections passivated the surface of ZnO NPs and created new electron transport pathways between each other in the composite film. Due to their better photoelectric properties, the ZnO/Ti_3C_2Tx nanohybrid films as electron transport layers (ETLs) were also applied in fullerene and non-fullerene polymer solar cells. The solar energy conversion efficiency of PBDB-T: ITIC photovoltaic device using our ZnO/Ti_3C_2Tx ETLs reached as high as 12.20% when used for the first time. An up to 15.53% relative enhancement compared with the Electron Transport Layer (ETL) based on pure ZnO in an EY-PIC system is achieved, which indicates that EFZO can be used as a better interface material for improving the efficiency of PICSCs. The flexibility of the ZnO/Ti_3C_2Tx ETL was also showcased with IPSCs based on PM6:Y6, attaining a PCE of 16.51%, compared to the benchmark device's efficiency of 14.99%. The development of ever more efficient methods to transmit and collect intra- or interpersonal communication contacts (IPSCs) can be seen to be mainly related with the increase PCE. $ZnO/Ti_3C_2T_xx$ composite ETL devices were found to be more stable compared to the control device. In contrast to few-layer Ti_3C_2Tx, the use of a layered

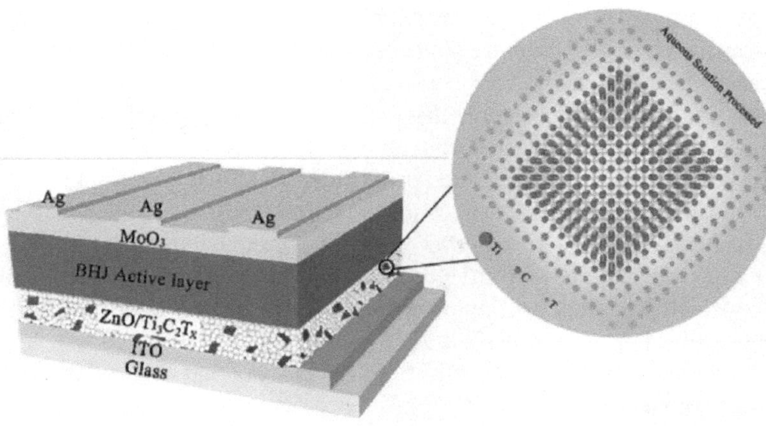

Scheme 1. Schematic illustration of the IPSCs configuration.

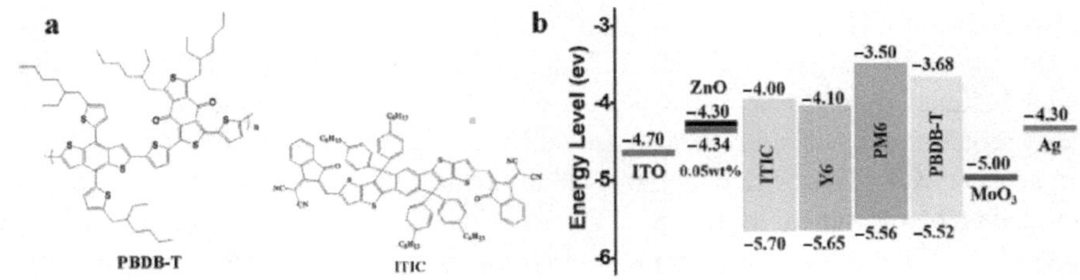

FIGURE 4.7 (a) The molecular structures of ITIC and PBDB-T. (b) A schematic is showing the energy band levels of the components that go into making inverted polymer solar cells (IPSCs) (see Hou and Yu 2021).

Source: Adopted with permission from Elsevier.

structure in these composites enhances power conversion efficiency (PCE), making them promising candidates for high-performance optoelectronic applications (Figure 4.7) (Hou and Yu 2021).

Ti_2CT_x and 2D Ti_3C_2Tx exhibit significant potential for making extremely efficient solar cells when used in photovoltaics. To fully grasp their potential in this field and remove the current barriers preventing them from making useful gadgets, more research is required.

4.5.1 Photodetectors

Photodetectors are devices that absorb optical energy and, through a photoelectric effect, convert light to an electrical signal. They are widely used in many areas like communication, imaging sensing, and optical data storage. Because of their extraordinary properties like high surface area, superior carrier mobility, and tunable band gap, 2D materials have become a subject for exploration. Two typical atomic-layered crystals in this category are Ti_3C_2Tx and Ti_2CT_x (T = O, OH). Ti_2CT_x and 2D Ti_3C_2Tx have been proposed to be making photoactive layers in photodetectors for the generation of electron–hole pairs, with high charge separation. They have a large surface area for 2D structure light absorption and high carrier mobility, giving significantly effective charge conveyance. In addition, the materials have different bandgaps (change with deformation) that make them capable for selectively detecting light at various portions in spectral domain in totality. Using 2D Ti_3C_2Tx and Ti_2CT_x in photodetectors develops high-performance visible-light photodetectors. They can be coupled with additional semiconducting components to make heterojunction devices featuring low dark current, fast response times, and high photoresponsivity. The 2D structure also allows for the creation of transparent and stretchable photodetectors, which are important components in flexible and wearable electronics (M. Long et al. 2019; J. Y. Kim et al. 2021).

In addition to photocatalytic application, 2D Ti_3C_2Tx and Ti_2CT_x have potential in making photodetectors as well. UV photodetectors find an excellent application in various sectors like environmental monitoring, flame detection, and medical diagnostics. Due to their high surface-to-volume ratio and the tunability of their bandgap values, it is beneficial for using these materials as UV photodetectors with excellent performance (Song, Ma, and Zhan 2020).

In a paper published in science, Y. Yang and colleagues explored the potential of using MXenes as an anode material for electrode applications. More importantly, they prepared a Ti_2CT_x electrode-mounted InSe-based photodetector. The high-performance Ti_2CT_x/TiN device employs an atomically thin, few-layer MXene exhibiting avalanche carrier multiplication. The researchers shape Ti_2CT_x into nanoribbon arrays in a bid to make the photodetector even more efficient, Making a plasmonic grating structure, which can enhance light absorption. In turn, this enabled creating a low dark current and high detectivity of plasmonic InSe/Ti_2CT_x avalanche photodetectors with the highest responsiveness (Figure 4.8) (Yajie Yang et al. 2019)

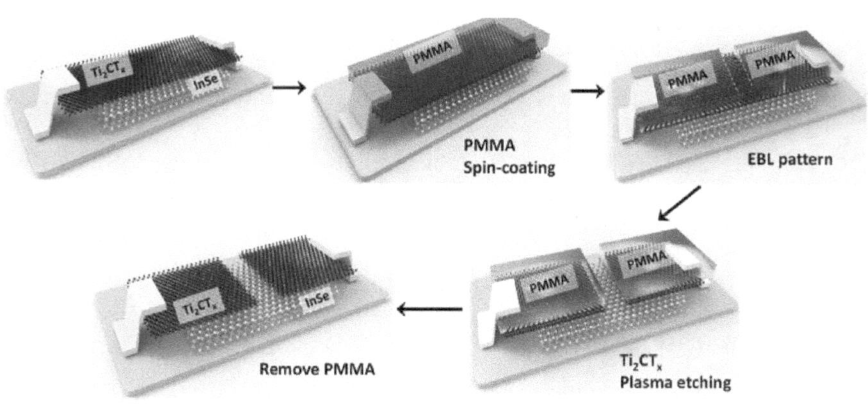

FIGURE 4.8 The illustration shows how an InSe/Ti_2CT_x photodetector is constructed using a PMMA protective layer and an InSe/Ti_2CT_x FET stacked vertically. The InSe/Ti_2CT_x photodetector is created via electron-beam lithography (EBL) patterning and plasma etching (see Yajie Yang et al. 2019).

Source: Adopted with permission from American Chemical Society.

In summary, 2D Ti_3C_2Tx and Ti_2CT_x possess many outstanding properties including having large surface area, great carrier mobility as well as tunable bandgap that indicate that they are very promising for use in photodetectors. They can be integrated with other semiconductors to fabricate high-performance photodetectors serving various applications including medical diagnosis, wearable electronics monitoring and environmental detection or flame sensor.

4.5.2 Solar Cell

The entire purpose of the solar cells is to achieve an economic as well as efficient sustainable energy future. Due to their better electrical and optical characteristics, several types of two-dimensional (2D) materials like Ti_3C_2Tx and Ti_2CT_x have been extensively used in the research area of solar cells. One of the leading challenges in developing efficient solar cells is that visible and infrared (IR) sunlight creates other effects on a material besides begetting electron movement. Because 2D materials are so simple to tune, they might be able to absorb light over a larger spectral range. A major area of recent activity has been the study on 2D materials for use as the absorber layer in solar cells. This is on account of outstanding electrical conductivity, carrier mobility, and optical absorption properties exhibited by the material. The use of Ti_2CT_x as a counter electrode material in dye-sensitized solar cells (DSSCs) has also been reported. These DSSCs based on Ti_2CT_x have shown significantly higher PCE than Pt-based DSSCs due to the superior electrocatalytic activity of Ti_2CT_x material, which increases both cell performance and dye regeneration (Z. L. Wang and Wu 2012; C. Wu et al. 2017).

To enhance the cell performance, solar cells are integrated with 2D materials that consist of a stack of active layers exhibiting distinct and sometimes overlapping absorption properties. Both Ti_3C_2Tx and Ti_2CT_x have superior absorption capability as tandem solar cell absorbers due to their excellent charge transport properties along with broad spectrum light utilization effect (S. M. Kim, Ha, and Kim 2016).

2D Ti_3C_2Tx nanosheets are produced through solution processing and incorporated into the active layer of OSCs as a third component to further enhance their performance. By adding nanosheets, it led to an improvement of the PCE for both PM6:Y6 blend and PBDB-T:ITIC mix as shown in Figure 4.9 (Y. Zhao et al. 2021). The PCE increased from 9.34% to 10.72% for the first case and from 14.64% to 16.25% for the second case. Ti_3C_2Tx nanosheets exhibited light-scattering properties that improve the efficiency of carrier dissociation and transfer, as seen in reductions in short-circuit current density (J_{SC}) which may alleviate the bimolecular recombination. Ti_3C_2Tx nanosheets were employed to make it easy for the charge transfer to take place inside photoactive layers without disturbing the linked topology of underlying network. This proof-of-concept work showed the effectiveness of two-dimensional Ti_3C_2Tx in improving the organic solar cell performance (Y. Zhao et al. 2021).

2D materials like $Ti_3C_2T_x$ and Ti_2CT_x have the potential to completely revolutionize the solar energy industry. These materials have demonstrated intriguing results in a variety of solar cell topologies and have good optical and electrical properties.

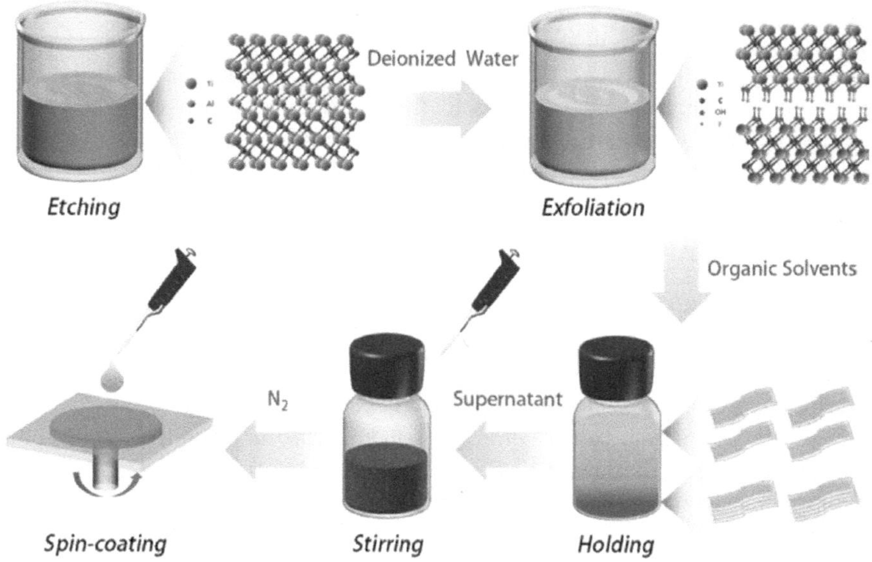

FIGURE 4.9 Creating and using 2D $Ti_3C_2T_x$ nanosheets that can be dissolved in solution in organic solar cells (OSCs) (see Y. Zhao et al. 2021).

Source: Adopted with permission from Weinheim: Wiley-VCH Verlag GmbH & CoOSCs).

4.6 ENERGY STORAGE DEVICES

Energy storage technologies are critical for the adoption of renewable energy sources. Given this historical context, there has been a significant increase in interest in energy storage technologies in modern times. The use of 2D materials in the design and fabrication of energy storage devices, such as Ti_3C_2Tx and Ti_2CT_x, has transformed industry dramatically. Thanks to their high electrical conductivity, large surface area, and excellent chemical stability, they are good candidates for energy storage applications. Batteries are a popular and common storage medium. As electrode materials used in lithium-ion batteries, 2D Ti_3C_2Tx and Ti_2CT_x outperform other two-dimensional (two-dimension) substances. This makes the materials attractive for long-term energy storage applications requiring duty cycles that go from slow to fast. Furthermore, these materials are also stable in high current density operation, supporting power-dense applications (S. Liu et al. 2021; Ni et al. 2021).

In recent years, supercapacitors have attracted considerable attention as a type of energy storage system. When 2D Ti_3C_2Tx and Ti_2CT_x are used as electrode materials for supercapacitors, good performance is guaranteed. To design effective supercapacitors, it is crucial to have materials showing large surface area and high electrical conductivity along with faster ion diffusion. With their high energy and power densities, they are potential candidates for making electric vehicles (EVs), portable electronic products; however, grid-scale electrical storage needs other than further lead-acid battery applications. Hybrid energy storage technologies, such as lithium-ion capacitors, combine the best of a battery and supercapacitor. This represents a big potential of 2D Ti_3C_2Tx and Ti_2CT_x as they can work in high-energy and power density scenarios, especially applications demanding fast charging/discharging cycles (Yan Yang et al. 2022; Chen et al. 2021).

In conclusion, the development of energy storage technologies has shown that the special features of 2D Ti_3C_2Tx and Ti_2CT_x have enormous promise. These materials show promise for usage in hybrid energy storage devices, batteries, and supercapacitors. To maximize their performance and look into more energy-related applications, more study is needed.

4.6.1 Lithium-Ion Batteries

Li-ion batteries are currently utilized as power sources in numerous applications, such as electric cars and mobile devices for their high conversion efficiency of chemical energy (to electricity). The use of advanced materials like 2D Ti_3C_2Tx can be a solution to improving the performance features of these batteries. They combine with a high-surface-area-to-volume ratio and mechanical strength which make them perfect for making lithium-ion batteries. Greater cycle stability along with higher energy density is needed, but this dual requirement hampers innovation in lithium-ion batteries. Ti_2CT_x and Ti_3C_2Tx 2D (15) battery anodes potential solution. The materials show outstanding electrochemical performance, with great capacity up to even rapid charge–discharge kinetics and superior stability over many cycles. The high surface area and good electrical conductivity properties can, in turn, lead to batteries with improved performance. These traits allow ions and electrons to flow rapidly with great ease. Owing to being lightweight, 2D Ti_3C_2Tx and Ti_2CT_x can also lead to the production of safer lithium-ion batteries. These features could reduce the chances of thermal runaway and explosion in standard lithium-ion batteries. This is due to the materials possibly inhibiting dendrite formation, which can lead to having battery short circuits and possessing higher mechanical stability (Lisbona and Snee 2011; Parangusan, Ponnamma, and Al Ali Almaadeed 2019).

Due to the outstanding chemical stability, low diffusion barriers for Li+, and high conductivity, Ti_3C_2Tx MXene was employed as a 2D anode material to improve lithium-ion battery properties (Jiang et al.). But titanium carbide has low specific capacity, which could limit its usefulness. In an attempt to tackle this, the researchers conducted freeze-drying and ultrasonic techniques in order to modify Ti_3C_2Tx with ferroferric oxide (Fe_3O_4) for lithium storage. Moreover, Fe_3O_4 nanoclusters, truly electrostatically adhering to the surface of single or few-layered Ti_3C_2Tx nanosheets as anchor groups, exhibited a superior lithium storage performance than both titanium carbide and such coating-free 2D transition metal oxide material (Fe_3O_4) itself (Figure 4.10). The 1:1 hybrid of Ti_3C_2Tx nanosheets and Fe_3O_4 displayed a remarkable lithium storage capacity of 437.6 mA h g^{-1} at the current density of 100 mA g^{-1} after cycling for 100 cycles. Even after the 1000th cycle at a high current density of 2 A g^{-1}, it steadily retained this capacity for approx. The capacity and battery properties established via kinetics study thus granted the lithium storage system its exceptional performance. These findings suggest that the hybrid of Fe_3O_4 nanoclusters and Ti_3C_2Tx nanosheets might be a more appropriate anode material for lithium-ion batteries (F. Jiang et al. 2020).

Most importantly, it is believed that Ti_2CT_x and 2D form of Ti_3C_2Tx should be able to develop efficient and reliable energy storage devices in Li-ion batteries. People can

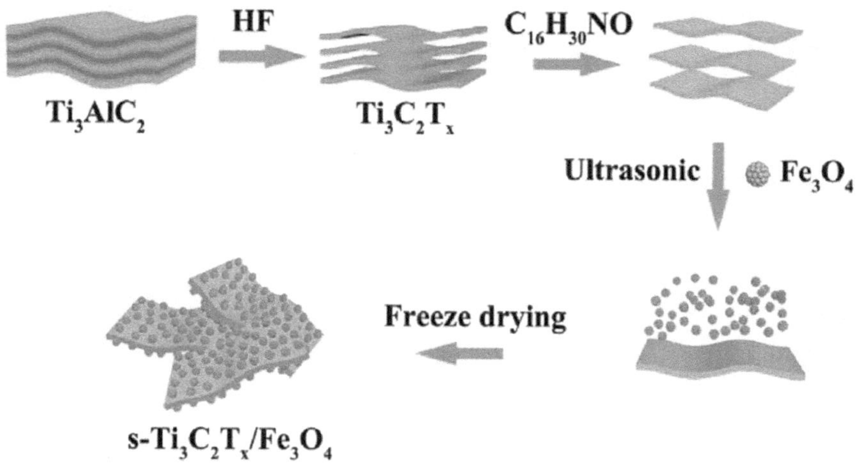

FIGURE 4.10 Showing the formation of s-Ti_3C_2Tx/Fe_3O_4 hybrids. Ultrasonic and freeze-drying procedures are employed to produce hybrids with Fe_3O_4 nanoclusters anchored on the surface of single or few-layered s-Ti_3C_2Tx nanosheets by electrostatic interactions (see F. Jiang et al. 2020).

Source: Adopted with permission from Elsevier.

use these high-performing devices in various applications (renewable energy systems, electric cars, portable power). More research is required to fully exploit the potential of these materials and engineer their attributes for specific applications.

4.6.2 Sodium-Ion Batteries

As an alternative to lithium-ion batteries (LIBs) and in view of the natural abundance and low cost of sodium, sodium-ion battery (SIB) is being researched. In contrast, the development of high-performance SIBs has been stalled due to limited electrode materials. The 2D transition metal carbides, including Ti_3C_2Tx and Ti_2CT_x (where T = F or O), have been recently recognized as potential SIB anode materials. Large surface area, excellent chemical stability, and high electrical conductivity make Ti_3C_2Tx related structures considered as suitable electrode materials for SIBs. The researchers have discovered that two 2D materials, Ti_3C_2Tx and the titanium carbide Ti_2CT_x, could store up to 300 mAh/g and 200 mAh/g of sodium, respectively. The unique properties of Ti_3C_2Tx and Ti_2CT_x give rise to new electrode structures (e.g., 2D nanosheets and 3D hierarchical architectures) that can significantly boost the electrochemical performance in SIBs. Ti_3C_2Tx nanosheets as anode material for sodium-ion batteries (SIBs) exhibit superior specific capacity, high-rate capability, and long cycle stability (Yi et al. 2020; L. Zhang et al. 2021).

None of this work deals with the electronic properties of surface functionalized Ti2C MXene monolayers and their performance in Na storage (Ti_2CC_2, Ti_2CO = OH/F, O, S). The researchers revealed that the monolayers, with or without termination, exhibit intrinsic electrical conductivity and metallic properties. They also provide details on the most stable forms of them, Ti_2C and specially for Ti_2CO_2 structures. Furthermore, Ti_2C functionalized with –C and –O groups (Ti_3CO_2)

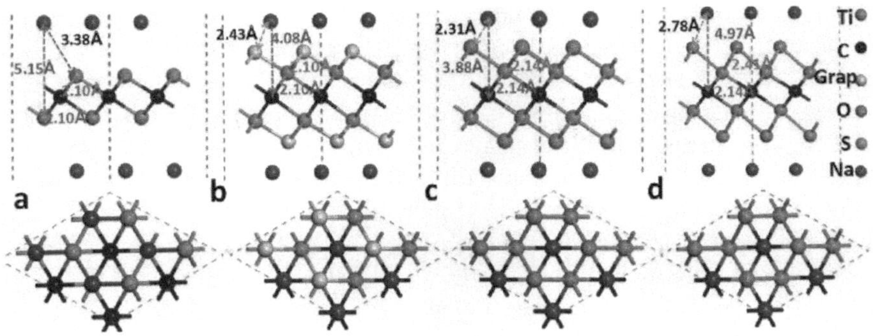

FIGURE 4.11 The geometric configurations for adsorption on both sides of single-layer Ti2CT2 MXenes with various surface functional groups—bare, C, O, and S—are depicted in this picture (see Xiao et al. 2019).

Source: Adopted with permission from Elsevier.

exhibits higher activation and adsorption capabilities compared to those with S terminations on the reference MXene surface. As further backed by theoretical studies, both Ti_2CC_2 and the similar structural type of Ti_2CO_2 are remarkably better in terms of adsorption than pure titanium carbide itself (as well as having lower energy barriers for sodium transport). This suggests that anion reduction improves the mobility of oxide and bare systems due to a significant decrease in Na-ion adsorption energy upon sodiation. Finally, the structural transformation of sodiated Ti2C monolayers whether terminated or unterminated was experimentally verified by ab initio molecular dynamic simulations. These results indicate that Ti2C and its surface-terminated monolayers possess good properties due to high electrical conductivity, substantial capacity as well as fast Na-ion diffusivity, which are all beneficial for their application in making anodes of the NIBS (Figure 4.11) (Xiao et al. 2019).

The inclusion of Ti_3C_2Tx and MoS2 may also enable the establishment of the environmentally benign energy storage system because these elements allow for decreasing toxicity in SIBs. Indeed, sodium resources are more available and have more sustainability compared to limited lithium-ion resource employed in LIBs.

To sum up, it is possible that the combination of Ti_3C_2Tx and Ti_2CT_x may become one practical way for making highly efficient environmentally friendly energy storage devices in the form of SIBs. Certainly, with the increase in an understanding of electrochemical properties and behaviors of these 2D materials to be performing as anodes or cathodes in SIBs, much more novel synthetic approaches capable of selective preparation will be developed.

4.6.3 Sulfur-Ion Batteries

The cathode active material of a novel class of rechargeable sulfur-ion batteries (SIBs) is based on the use of sulfur as an active ingredient. As such, sulfur resources are abundant and inexpensive to mine, so in theory, SIBs could offer significantly higher energy density paired with lower costs compared to conventional LIBs. Highly efficient SIBs have been limited because sulfur is neither soluble in electrolytes nor electronically conductive. Among various candidates of 2D materials, two-dimensional

(2D) transition metal carbides or nitrides such as Ti_3C_2Tx and Ti_2CT_x have attracted significant attention to be applied in SIBs, thanks to a high surface area, perfect chemical stability, and strong electronic conductivity of these 2D materials which could be used as conductive additives or cathode materials for SIBs. Results have shown that both Ti_3C_2Tx and Ti_2CT_x can significantly boost the electrochemical behaviors of SIBs. Ti_3C_2Tx has been applied as an electrical conductor in sulfur oxidation bacteria (SIBs) to promote the electron transport property and improve the sulfur utilization rate. For instance, in solid-oxide batteries (SIBs), Ti_2CT_x offers a high specific capacity, excellent rate capability, and long cycle life (Y. Wu et al. 2018; Guo et al. 2021).

Ti_3C_2Tx and Ti_2CT_x have been investigated for their use in electronic-based detecting applications as well as energy storage applications. These 2D materials' excellent sensitivity, selectivity, and stability when exposed to a range of analytes make them suitable sensing platforms.

For sodium-sulfur batteries, the cheapness of sulfur combined with its high specific capacity makes it especially desirable. The cathode material currently used consists mostly of carbon compounds as a host matrix and elemental sulfur—but only a small portion of the sulfur is actually used. That was fixed using Ti_3C_2Tx, a 2D stratified material, to serve as cathode for the sodium-ion batteries and developed by X. Huo-led team of scientists. Direct melting at an appropriate temperature of 155°C, with a sulfur sublimation ratio of about 55 mol%, was applied to synthesize $Ti_3C_2Tx@S$ (Figure 4.12).

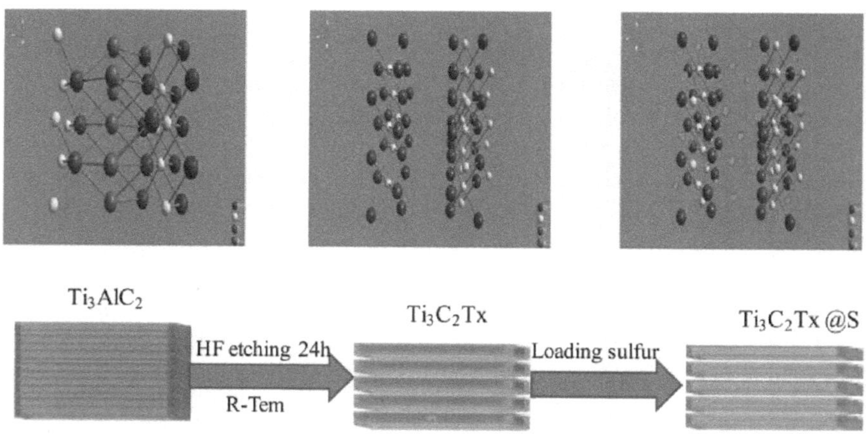

$$Ti_3AlC_2 + 3HF = AlF_3 + 3/2H_2 + Ti_3C_2; \quad Ti_3C_2 + 2H_2O = Ti_3C_2(OH)_2 + H_2;$$

$$Ti_3C_2 + 2HF = Ti_3C_2F_2 + H_2$$

FIGURE 4.12 The production method for the $Ti_3C_2Tx@S$ composite materials is depicted diagrammatically. It includes information on the material structure, synthesis-related chemical processes, and flow of the production process (see Huo et al. 2019).

Source: Adopted with permission from Springer-Verlag GmbH Germany, part of Springer Nature.

XPS results confirmed the presence of Ti^{4+} and a new species, potentially a C-S bond, as well as the transformation of some Ti^{3+} to Ti^{4+}. Only the discharge specific capacity of 214.0 mAh g^{-1} of $Ti_3C_2Tx@S$ was measured and showed an initial reversible value, higher compared to all other compositions tested; with that, rate performance was also been indicated as promising due to a high retained value at a current density of 1,000 mA g^{-1}. At room temperature, the sodium–sulfur battery exhibited excellent cycle stability with a discharge specific capacity of 150 mAh g^{-1} over 300 cycles at a current density of 100 mA g^{-1}. The battery indicated an 80% retention capacity after the second cycle as well. The specific capacitance of the electrode remained 107.7 mAh g^{-1} after being cycled for 150 times (Huo et al. 2019).

On the whole, integrating Ti_3C_2Tx and Ti_2CT_x into SIBs can provide quite a simple way to construct high-performance as well as environmentally friendly energy storage systems. Further investigation of the mechanisms, electrochemical performance, and sensing applications for 2D materials as SIB anodes will continue to yield novel insights and research progress related closely with new advances coming in this arena.

4.7 CATALYSIS

Two-dimensional (2D) transition metal carbides, including Ti_3C_2Tx and Ti_2CT_x, that exhibit unique physical as well as chemical properties have emerged as the most promising candidates in several catalytic applications. Due to their high electrical conductivity, large area for a surface reaction, and the variance in their chemical properties, these materials have been rigorously investigated as catalysts. In turn, the powerful catalytic activity of Ti_3C_2Tx and Ti_2CT_x has been proven to be advantageous for numerous procedures, e.g., carbon dioxide reduction reaction (CO2RR), hydrogen evolution response (HER), and oxygen reduction reaction (ORR). Ti_3C_2Tx shows higher ORR catalytic activity with respect to those of the conventional Pt/C catalysts, whereas Ti_2CT_x is significantly advantageous in HER process. The unique properties of Ti_3C_2Tx and the similar structural evolution used to make these materials have also enabled the generation of advanced catalytic architectures, e.g., 2D nanosheets and even 3D hierarchical structures that can further improve their activity. For example, a type of 2D pseudoelastomeric material known as Ti_3C_2Tx nanosheets has been used to stabilize metal nanoparticles, which could enhance the catalytic activity and durability. Ti_3C_2Tx and Ti_2CT_x have been employed as cocatalysts to enhance the light absorption efficiency, charge separation, and catalytic activity of a photocatalyst under photoexcitation. It shows the strong photocatalytic activity of these 2D materials both in organic contaminants oxidation and hydrogen production (D. Liu et al. 2021; L. Zhang et al. 2021; Chai et al. 2022).

In the light of all these, Ti_3C_2Tx and Ti_2CT_x can be treated as materials with an immense promise in catalysis applications due to their specific physicochemical properties. Assuredly, further inquiries regarding 2D materials' synthesis and fabrication, leading to their catalytical applications will drive more innovative findings in this topic.

4.7.1 Electrocatalysis

Catalysts are employed to promote electrochemical reactions (molecular oxidation or reduction) in the field of electrocatalysis. As Ti_3C_2Tx and Ti_2CT_x possess distinctive electrical property and catalytic capacity, these 2D materials have found myriad applications in electrocatalysis. Theoretical results suggest the feasibility of using 2D Ti_3C_2Tx and Ti15CTx as potential catalysts for fuel cell electrocatalysis. It was shown previously that these materials are efficient catalysts for ORR which is the main bottleneck in optimizing the efficiency of fuel cells. With a broad surface area and diverse surface chemistry, they make excellent candidates for producing efficient fuel cell catalysts that last long. Additionally, beyond use in fuel cells, Ti_3C_2Tx and the related phase (Ti_2CT_x) have been investigated as catalysts for metal-air batteries performing oxygen evolution reaction (OER) and hydrogen evolution reaction (HER) from water splitting to produce H_2. These materials were stable and active in the process. Given their attractive electronic properties, such as high electrical conductivity and fast electron-transfer kinetics, Ti_3C_2Tx and Ti_2CT_x materials offer promise as earth-abundant electrocatalysts in more important reactions including nitrogen reduction to ammonia or the conversion of carbon dioxide (H. Kim and Alshareef 2020; Chandrasekaran et al. 2020).

Pure Ti_3C_2Tx (MXene) has been reported to be used as an electrocatalyst for electrochemical sensing by D. Wu et al. Two carbendazim redox peaks were observed at the Ti_3C_2Tx-modified electrode with low overpotential, and convincing evidence by the authors revealed that delamination played a big role in those electrochemical properties (shown in Figure 4.13). There is a computational prediction that the surface fluorine termination from HF etching could serve as an effective electrocatalytically-active region for the redox of liquid fuels. Our findings showed that pure Ti_3C_2Tx could be a promising electrode modification for the next generation of electrochemical sensors, according to the experts (D. Wu et al. 2019).

Overall, the application of 2D Ti_3C_2Tx and Ti_2CT_x materials in electrocatalysis significantly promotes the development of efficient as well as environmentally friendly electrochemical technologies. Further, pyramidal poles and electrocatalytic closings of these novel materials, along with the synthesis of new kinds of 2D components that demonstrate varied digital as well as catalytic characteristics, will undoubtedly bring more creative and exciting discoveries in this field.

4.7.2 Photocatalysis

Photocatalysis is the acceleration of a photoreaction in the presence of a photocatalyst. Due to their unique properties, like the optical and catalytic electric properties of 2D Ti_3C_2Tx, TiOx-modified MXenes can realize the full potential in photocatalysis. One of the most interesting potential applications associated with 2D Ti_3C_2Tx and Ti_2CT_x materials is photocatalysis for water splitting to produce hydrogen. When these materials are illuminated and electron–hole pairs form, they can initiate the water-splitting process. With large specific surface area and varied surface chemistry, these materials offer promising prospects for the fabrication of sustainable, powerful, and efficient photocatalysts suitable for water splitting. In addition to water splitting, the possibilities for utilizing Ti_3C_2Tx and Ti_2CT_x materials as photocatalysts

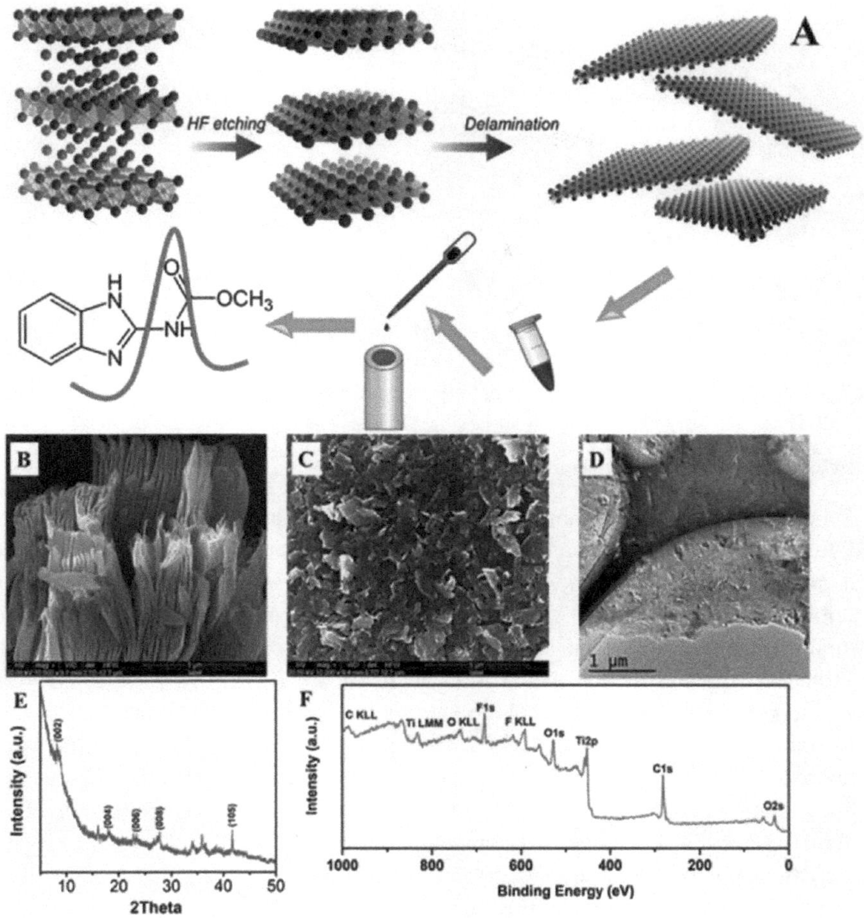

FIGURE 4.13 (a) A process flowchart for fabrication of the carbendazim sensor. (b) A SEM image showing an etched Ti_3C_2Tx image. (c) SEM image of the delaminated Ti_3C_2Tx. (d) Transmission electron microscopy (TEM) image of Ti_3C_2Tx. (e) X-ray diffraction (XRD) pattern of Ti_3C_2Tx. (f) XPS survey of Ti_3C_2Tx (see D. Wu et al. 2019).

Source: Adopted with permission from Elsevier.

for other important processes such as carbon dioxide reduction or organic pollutant degradation have also been investigated. Because of the high activity and selectivity displayed by 2D materials in these processes, they are considered key properties for the development of clean and renewable sustainable energy systems (Karthikeyan et al. 2020; Rafiq et al. 2022).

These unique electrical and optical properties make Ti_3C_2Tx and Ti_2CT_x good candidates for designing new photocatalytic systems, thanks to a significant light absorption ability as well as rapid charge transfer kinetics. To make the photocatalysts more useful and effective, it has been proposed to combine these 2D materials with other functional materials such as semiconductors.

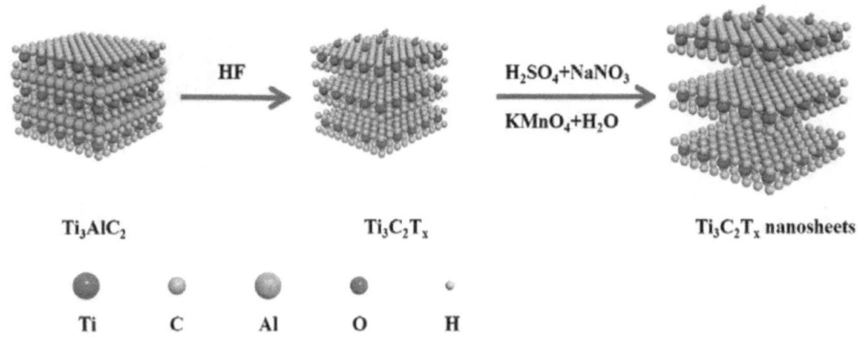

FIGURE 4.14 Illustration of the formation process of $Ti_3C_2T_X$ nanosheets (see Q. Zhao et al. 2021).

Source: Adopted with permission from Elsevier.

Schottky heterojunction photocatalysts are developed in order to enhance the rate and duration of organic pollutant degradation under wastewater purification. Q. Zhao et al. synthesized $Ti_3C_2T_X$ nanosheets and Cu_2O photocatalysts with Schottky heterojunction. A report involving the visible-light degradation of tetracycline antibiotics is given by (Figure 4.14) Q. Zhao et al. (2021). To investigate the photocatalysts further, researchers performed analyses such as X-ray diffraction (XRD), X-ray photoelectron spectroscopy (XPS), and high-resolution transmission electron microscopy. Directly comparable, the composite material decayed 97.6% within less than 50 minutes as compared to visible-light degradation for TC. The experiment proved that TC decomposition is initiated by ROS, in particular, the superoxide radical (O_2•) and hole-color center h+. Since it is a Schottky heterojunction, the electrons (e^-) and holes (h^+) generated by the catalyst can be efficiently transmitted and separated at the UP(bb-cits2) interface. This reduces the recombination rate, significantly enhancing photocatalytic stability and efficiency, as such a junction is known to do (Q. Zhao et al. 2021).

All these factors collectively substantiate the application of 2D Ti_3C_2Tx and Ti_2CT_x materials in photocatalysis for sustainable production of green energy resources. Further exploration of fundamental physics and optimization regarding the existing materials as well as in search for novel 2D materials with designed properties are expected to spark even more unconventional ideas that may lead into this sector again.

4.8 DRUG DELIVERY AND CANCER THERAPY

Demonstrating superior physicochemical properties, both biocompatibility and higher loading rates of 2D Ti_3C_2Tx, as well as the material family including materials like Ti_2CT_x turned to be promising in a matter concerning cancer treatment or drug delivery. The high drug-loading and effective release of doxorubicin are the primary advantages in 2D Ti_3C_2Tx as well as Ti_2CT_x, based on their extensive surface areas and tunable chemical modifications applied for drug administration. Moreover, these compounds can reduce drug degradation by protecting against proteolysis and chemical reactions, stabilize drugs during circulation in the bloodstream, and enhance their solubility. Surface functionalization and modification of

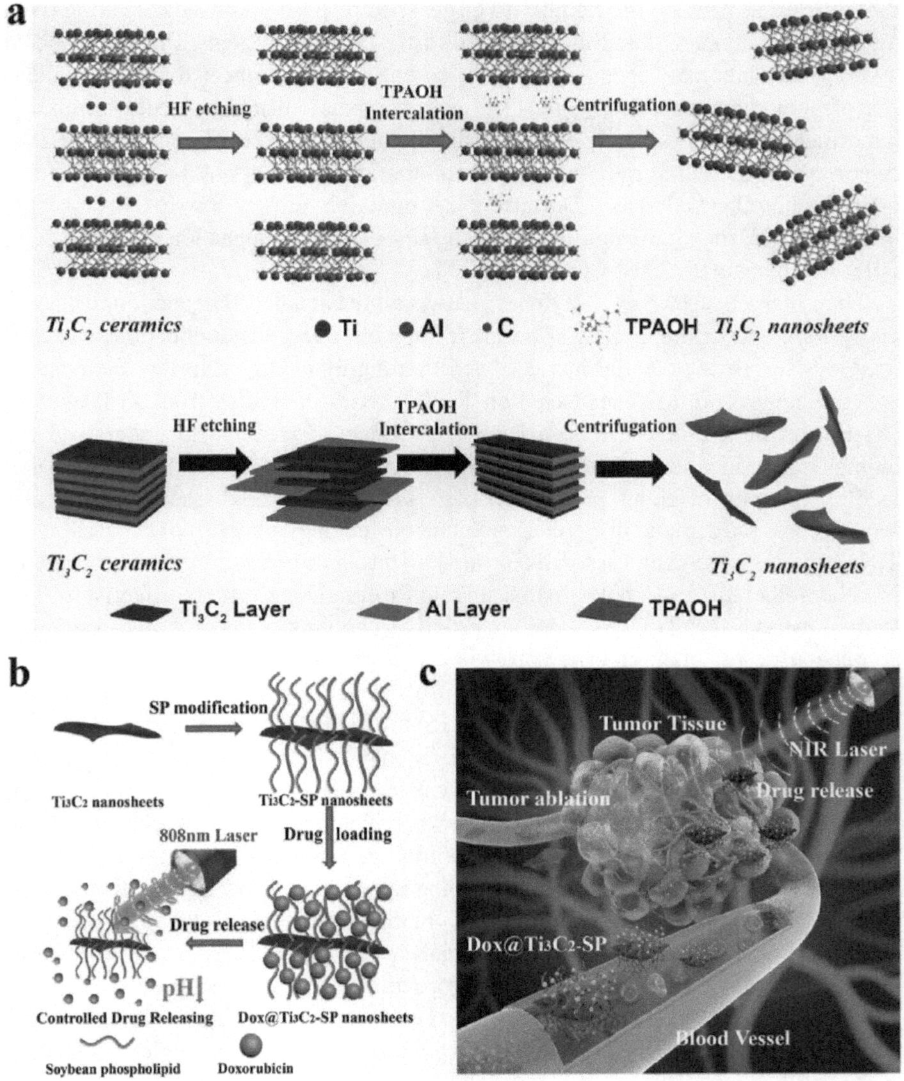

FIGURE 4.15 This screenshot exemplifies Ti_3C_2 MXene for cancer treatment through photothermal therapy with a picture demonstrating the treatment of the procedure followed by chemotherapy drugs. (a) The synthesis of Ti_3C_2 nanosheets was conducted by first exfoliating it in a two-step process. (b) After this step, the pharmaceuticals are filled inside nanosheets and coated with SP modification, which leads to either the endo- or exogenous radiation-induced drug release mode. (c) In-vivo drug delivery system photothrombosis of tumor tissue, under NIR irradiation. It achieves this by accumulating in the tumor and subsequently releasing the drug directly into it (see X. Han et al. 2018).

Source: Adopted with permission from Weinheim: Wiley-VCH Verlag GmbH & Coi.

these materials provide their ability to target specific tissues and cells, leading them toward being more tailored and powerful drug delivery systems. 2D Ti_3C_2Tx and Ti_2CT_x materials have been explored for potential use in cancer therapy, as a carrier of photodynamic therapy (PDT), chemotherapeutic drugs, or genes. Similarly, hyperthermic therapy greatly enhances the therapeutic effects of conventional antitumor agents including radiation and chemotherapeutic drugs by taking advantage of their photothermal effect. Because these materials present low toxicity and are biocompatible, they have significant potential for clinical translation (Cardoso et al. 2018; L. Zhang et al. 2021; Chai et al. 2022).

The emergence of Ti_3C_2 MXenes, a two-dimensional (2D) functional materials system, is opening up new horizons in the field due to its applications variability. However, their potential has not been thoroughly exploited in the theragnostic nanomedicine field. MXenes based on Ti_3C_2A series of studies from X. Han et al. (X. Han et al. 2018), provided a general overview for readers to understand the biomedical applications of MXene. Furthermore, the MXenes exhibit a high drug-loading capacity of about 211.8%, coupled with near-infrared laser-triggered pH-sensitive and on-demand drug release. Such conversion photothermal capabilities of Ti_3C_2MXenes have been explored for the realization of potential therapies combining chemotherapy and photothermal ablation to eradicate tumors effectively. The study illustrates that Ti_3C_2 MXenes are promising contrast agents for high-resolution photoacoustic imaging, and they provide excellent in-vitro and in-vivo biocompatibilities. The research also expands the prospects of using MXene in nanomedicine and cancer therapy applications (Figure 4.15) (X. Han et al. 2018).

This research studied the use of 2D Ti_3C_2Tx and Ti_2CT_x in imaging tumor-diagnosis/metastasis/probable cancers. The materials are useful as contrast agents for computed tomography (CT) and magnetic resonance imaging (MRI), respectively, due to their high atomic number and the properties arising from being magnetizable. Because of their high surface area, they could be applied as biosensors for the diagnosis of cancer biomarkers and have an interaction with biomolecules. To sum up, 2D Ti_3C_2Tx and Ti_2CT_x materials are explored as potential candidates for drug delivery and cancer therapy application due to their unique properties. The low toxicity of 2D Ti_3C_2Tx and Ti_2CT_x materials, combined with its surface chemistry that can be readily modified with multiple functionalities for enhanced biocompatibility and high drug-loading capacities, has provided very attractive leverage for clinical translation. Further studies are needed to take full advantage of these properties for this application.

4.9 SUMMARY

While the two-dimensional (2D) transition metal carbides, known as MXenes, are still somewhat of a discovery and relatively young class of materials, their excellent electrical conductivity, superior catalytic performance, and biocompatibility make the materials promising candidates in a wide array of applications such as sensor technology, catalysis, energy storage materials (batteries), solar cells, or medical research.

Among the applications of 2D MXenes, they have been electrically utilized for the detection of waste waters successfully. Their high sensitivity and selectivity make these sensors well suited for environmental, biosensor, and gas sensor applications.

MXenes have also been utilized for the detection of gases like nitrogen dioxide and ammonia as well among other hazardous materials. MXene-based biosensors can specifically detect a diverse range of biomolecules such as uric acid, dopamine, and glucose.

MXenes have a large surface area and great electrical conductivity, which make them a good option for solar application. The efficiency of MXene-based solar cells, which have demonstrated outstanding stability and performance, is being worked on.

The large surface area, metallic conductivity, and plenty of available active sites make MXenes have excellent catalytic properties. Catalysts have been widely used in the catalytic reactions such as that between biomass and fuels and wastewater treatment. MXene-based catalysts are one of the promising materials for electrocatalytic CO_2 reduction, which is vital to capture and utilize carbon.

MXenes are used for energy storage applications, including making supercapacitors and lithium-ion batteries, owing to the unique properties of MXenes including high power density, rapid charge–discharge rates, and excellent cycling stability. Conversely, higher rate and better specific capacity lithium-ion batteries have also been made on the basis of MXene.

The unique properties of the MXenes will render them fascinating for potential applications within biological industry. These materials, due to their high biocompatibility and low toxicity, have proven themselves in the field of tissue engineering (TE), drug delivery systems (DDS), or bioimaging. As drug and contrast agents in biomedical imaging, MXenes have served as pharmaceutical delivery systems for anticancer medications. Therefore, due to their excellent flexibility and unique properties, 2D MXene is a promising material with many prospects. In the not-too-distant future, more research on this topic may yield interesting new findings and inventions.

REFERENCES

Abbasi, Nasir Mahmood, Yao Xiao, Ling Peng, Yanhong Duo, Lude Wang, Li Zhang, Bing Wang, and Han Zhang. 2021. "Recent Advancement for the Synthesis of MXene Derivatives and Their Sensing Protocol." *Advanced Materials Technologies* 6 (10): 1–62. https://doi.org/10.1002/admt.202001197.

Abbasi, Nasir Mahmood, Yao Xiao, Li Zhang, Ling Peng, Yanhong Duo, Lude Wang, Peng Yin, et al. 2021. "Heterostructures of Titanium-Based MXenes in Energy Conversion and Storage Devices." *Journal of Materials Chemistry C* 9 (27): 8395–465. https://doi.org/10.1039/d1tc00327e.

Adarakatti, Prashanth Shivappa, and Samrat Devaramani. 2021. "2D Materials for Sensing Applications," no. August: 44–83. https://doi.org/10.1039/9781788017039-00044.

Arpia, Arjay A., Wei Hsin Chen, Su Shiung Lam, Patrick Rousset, and Mark Daniel G. de Luna. 2021. "Sustainable Biofuel and Bioenergy Production from Biomass Waste Residues Using Microwave-Assisted Heating: A Comprehensive Review." *Chemical Engineering Journal* 403 (July 2020): 126233. https://doi.org/10.1016/j.cej.2020.126233.

Bhardwaj, Radha, and Arnab Hazra. 2021. "MXene-Based Gas Sensors." *Journal of Materials Chemistry C* 9 (44): 15735–54. https://doi.org/10.1039/d1tc04085e.

Cardoso, Vanessa Fernandes, António Francesko, Clarisse Ribeiro, Manuel Bañobre-López, Pedro Martins, and Senentxu Lanceros-Mendez. 2018. "Advances in Magnetic Nanoparticles for Biomedical Applications." *Advanced Healthcare Materials* 7 (5): 1–35. https://doi.org/10.1002/adhm.201700845.

Chai, Bin, Kunming Shi, Haiyang Zou, Pingkai Jiang, Zixiang Wu, and Xingyi Huang. 2022. "Conductive Interlayer Modulated Ferroelectric Nanocomposites for High Performance Triboelectric Nanogenerator." *Nano Energy* 91 (September 2021): 106668. https://doi.org/10.1016/j.nanoen.2021.106668.

Chandrasekaran, Sundaram, Dingtao Ma, Yanqi Ge, Libo Deng, Chris Bowen, James Roscow, Yan Zhang, et al. 2020. "Electronic Structure Engineering on Two-Dimensional (2D) Electrocatalytic Materials for Oxygen Reduction, Oxygen Evolution, and Hydrogen Evolution Reactions." *Nano Energy* 77. https://doi.org/10.1016/j.nanoen.2020.105080.

Chao, Mingyuan, Lingzhang He, Min Gong, Na Li, Xiaobin Li, Longfei Peng, Feng Shi, Liqun Zhang, and Pengbo Wan. 2021. "Breathable Ti_3C_2TxMXene/Protein Nanocomposites for Ultrasensitive Medical Pressure Sensor with Degradability in Solvents." *ACS Nano* 15 (6): 9746–58. https://doi.org/10.1021/acsnano.1c00472.

Chen, Chia Ying, Subramanian Sakthinathan, Chung Lun Yu, Chien Ching Wang, Te Wei Chiu, and Qiaofeng Han. 2021. "Preparation and Characterization of Delafossite CuCrO2 Film on Flexible Substrate." *Ceramics International* 47 (16): 23234–39. https://doi.org/10.1016/j.ceramint.2021.05.035.

Chen, W. Y., S. N. Lai, C. C. Yen, X. Jiang, D. Peroulis, and L. A. Stanciu. 2020. "Surface Functionalization of Ti_3C_2Tx MXene with Highly Reliable Superhydrophobic Protection for Volatile Organic Compounds Sensing." *ACS Nano* 14 (9): 11490–501. https://doi.org/10.1021/acsnano.0c03896.

Cheng, Ming, Xu Zhang, Zhenbo Guo, Peifan Lv, Renjie Xiong, Zhiqiang Wang, Zhen Zhou, and Minghui Zhang. 2021. "Pd-Promoting Reduction of Zinc Salt to PdZn Alloy Catalyst for the Hydrogenation of Nitrothioanisole." *Journal of Colloid and Interface Science* 602: 459–468. https://doi.org/10.1016/j.jcis.2021.06.024.

Gao, Ran, Hsiang-Sheng Ku, Hao Deng, Wenlong Yu, Tian Xia, Feng Wu, Zhijun Song, et al. 2022. "Ultrahigh Kinetic Inductance Superconducting Materials from Spinodal Decomposition." *Advanced Materials* 2201268. https://doi.org/10.1002/adma.202201268.

Guo, Siyao, Hailong Guan, Ying Li, Yunfeng Bao, Dongyi Lei, Tiejun Zhao, Baomin Zhong, and Zhihong Li. 2021. "Dual-Loss Ti_3C_2Tx MXene/Ni0.6Zn0.4Fe2O4 Heterogeneous Nanocomposites for Highly Efficient Electromagnetic Wave Absorption." *Journal of Alloys and Compounds* 887: 161298. https://doi.org/10.1016/j.jallcom.2021.161298.

Han, Cuiping, Tengfei Zhang, Junqin Li, Baohua Li, and Zhiqun Lin. 2020. "Enabling Flexible Solid-State Zn Batteries via Tailoring Sulfur Deficiency in Bimetallic Sulfide Nanotube Arrays." *Nano Energy* 77 (June): 105165. https://doi.org/10.1016/j.nanoen.2020.105165.

Han, Xiaoxia, Ju Huang, Han Lin, Zhigang Wang, Pan Li, and Yu Chen. 2018. "2D Ultrathin MXene-Based Drug-Delivery Nanoplatform for Synergistic Photothermal Ablation and Chemotherapy of Cancer." *Advanced Healthcare Materials* 7 (9): 1–13. https://doi.org/10.1002/adhm.201701394.

He, Ying, Li Ma, Liya Zhou, Guanhua Liu, Yanjun Jiang, and Jing Gao. 2020. "Preparation and Application of Bismuth/Mxene Nano-Composite as Electrochemical Sensor for Heavy Metal Ions Detection." *Nanomaterials* 10 (5). https://doi.org/10.3390/nano10050866.

Hou, Chunli, and Huangzhong Yu. 2021. "ZnO/Ti_3C_2Tx Monolayer Electron Transport Layers with Enhanced Conductivity for Highly Efficient Inverted Polymer Solar Cells." *Chemical Engineering Journal* 407 (October 2020): 127192. https://doi.org/10.1016/j.cej.2020.127192.

Huo, Xiaogeng, Yanying Liu, Ranran Li, and Jianling Li. 2019. "Two-Dimensional Ti 3 C 2 T x @ S as Cathode for Room Temperature Sodium-Sulfur Batteries." 5373–82.

Jiang, Fuyi, Rong Du, Xinsheng Yan, Ming Zhang, Qi Han, Xueqin Sun, Xiaoyu Zhang, and Yanli Zhou. 2020. "Ferroferric Oxide Nanoclusters Decorated Ti_3C_2Tx Nanosheets as

High Performance Anode Materials for Lithium Ion Batteries." *Electrochimica Acta* 329: 135146. https://doi.org/10.1016/j.electacta.2019.135146.

Jiang, Jungang, Nancy C. Carrillo-Enríquez, Hale Oguzlu, Xushen Han, Ran Bi, Jack N. Saddler, Run Cang Sun, and Feng Jiang. 2020. "Acidic Deep Eutectic Solvent Assisted Isolation of Lignin Containing Nanocellulose from Thermomechanical Pulp." *Carbohydrate Polymers* 247 (July): 116727. https://doi.org/10.1016/j.carbpol.2020.116727.

Ju, Myung Jong, In Yup Jeon, Hong Mo Kim, Ji Il Choi, Sun Min Jung, Jeong Min Seo, In Taek Choi, et al. 2016. "Edge-Selenated Graphene Nanoplatelets as Durable Metal-Free Catalysts for Iodine Reduction Reaction in Dye-Sensitized Solar Cells." *Science Advances* 2 (6). https://doi.org/10.1126/SCIADV.1501459.

Karthikeyan, C., Prabhakarn Arunachalam, K. Ramachandran, Abdullah M. Al-Mayouf, and S. Karuppuchamy. 2020. "Recent Advances in Semiconductor Metal Oxides with Enhanced Methods for Solar Photocatalytic Applications." *Journal of Alloys and Compounds* 828: 154281. https://doi.org/10.1016/j.jallcom.2020.154281.

Kim, Hyunho, and Husam N. Alshareef. 2020. "MXetronics: MXene-Enabled Electronic and Photonic Devices." *ACS Materials Letters* 2 (1): 55–70. https://doi.org/10.1021/acsmaterialslett.9b00419.

Kim, Ju Young, Joonam Park, Seok Hun Kang, Seungwon Jung, Dong Ok Shin, Myeong Ju Lee, Jimin Oh, et al. 2021. "Revisiting TiS2 as a Diffusion-Dependent Cathode with Promising Energy Density for All-Solid-State Lithium Secondary Batteries." *Energy Storage Materials* 41 (March): 289–96. https://doi.org/10.1016/j.ensm.2021.06.005.

Kim, Seong Min, Jaewook Ha, and Jin Baek Kim. 2016. "The Effect of Dielectric Constant and Work Function on Triboelectric Nanogenerators: Analytical and Numerical Study." *Integrated Ferroelectrics* 176 (1): 251–56. https://doi.org/10.1080/10584587.2016.1252657.

Lee, Stephanie K., Mei Wang, Jin Hyun Lee, and Jonghwan Suhr. 2020. "Development of Reversibly Compressible Feather-Like Lightweight Chitosan/GO Composite Foams and Their Mechanical and Viscoelastic Properties." *Carbon* 157: 191–200. https://doi.org/10.1016/j.carbon.2019.10.019.

Li, Chunmei, Junqi Wu, Haoyuan Shi, Zhiyu Xia, Jugal Kishore Sahoo, Jingjie Yeo, and David L. Kaplan. 2022. "Fiber-Based Biopolymer Processing as a Route Toward Sustainability." *Advanced Materials* 34 (1): 1–27. https://doi.org/10.1002/adma.202105196.

Li, Ning, Yue Jiang, Chuanhong Zhou, Yan Xiao, Bo Meng, Ziya Wang, Dazhou Huang, Chenyang Xing, and Zhengchun Peng. 2019. "High-Performance Humidity Sensor Based on Urchin-Like Composite of Ti_3C_2 MXene-Derived TiO2 Nanowires." *ACS Applied Materials and Interfaces* 11 (41): 38116–25. https://doi.org/10.1021/acsami.9b12168.

Lisbona, Diego, and Timothy Snee. 2011. "A Review of Hazards Associated with Primary Lithium and Lithium-Ion Batteries." *Process Safety and Environmental Protection* 89 (6): 434–42. https://doi.org/10.1016/j.psep.2011.06.022.

Liu, Dongqing, Akaash Barbar, Tayyaba Najam, Muhammad Sufyan Javed, Jun Shen, Panagiotis Tsiakaras, and Xingke Cai. 2021. "Single Noble Metal Atoms Doped 2D Materials for Catalysis." *Applied Catalysis B: Environmental* 297 (January): 120389. https://doi.org/10.1016/j.apcatb.2021.120389.

Liu, Rongyang, Li Jiang, Zizhen Yu, Xufeng Jing, Xiu Liang, Dan Wang, Bo Yang, Chengxing Lu, Wei Zhou, and Shangzhong Jin. 2021. "MXene (Ti_3C_2Tx)-Ag Nanocomplex as Efficient and Quantitative SERS Biosensor Platform by in-Situ PDDA Electrostatic Self-Assembly Synthesis Strategy." *Sensors and Actuators, B: Chemical* 333 (December 2020): 129581. https://doi.org/10.1016/j.snb.2021.129581.

Liu, Shiyuan, Jieyuan Liu, Xiaofang Liu, Jiaxiang Shang, Li Xu, Ronghai Yu, and Jianglan Shui. 2021. "Hydrogen Storage in Incompletely Etched Multilayer Ti2CTx at Room

Temperature." *Nature Nanotechnology* 16 (3): 331–36. https://doi.org/10.1038/s41565-020-00818-8.

Long, Fu, Zhang Cao, Yuyang Liu, Rui Liang, Qunting Qu, Linze Lv, Chengchang Jin, and Honghe Zheng. 2022. "Organic Skin Layer with Carboxyl and Electron-Withdrawing Nitro Groups to Chemically Bind with Residual Li and Ni Cations on Single-Crystal LiNi0.8Co0.1Mn0.1O2." *Electrochimica Acta* 404: 139743. https://doi.org/10.1016/j.electacta.2021.139743.

Long, Mingsheng, Peng Wang, Hehai Fang, and Weida Hu. 2019. "Progress, Challenges, and Opportunities for 2D Material Based Photodetectors." *Advanced Functional Materials* 29 (19): 1–28. https://doi.org/10.1002/adfm.201803807.

Lorencova, Lenka, Tomas Bertok, Jaroslav Filip, Monika Jerigova, Dusan Velic, Peter Kasak, Khaled A. Mahmoud, and Jan Tkac. 2018. "Highly Stable Ti_3C_2Tx (MXene)/Pt Nanoparticles-Modified Glassy Carbon Electrode for H2O2 and Small Molecules Sensing Applications." *Sensors and Actuators, B: Chemical* 263: 360–68. https://doi.org/10.1016/j.snb.2018.02.124.

Lorenzen, Ana Luiza, Ariane Moraes dos Santos, Luâni Poll dos Santos, Luciano da Silva Pinto, Fabricio Rochedo Conceição, and Franciele Wolfart. 2022. "PEDOT-AuNPs-Based Impedimetric Immunosensor for the Detection of SARS-CoV-2 Antibodies." *Electrochimica Acta* 404: 139757. https://doi.org/10.1016/j.electacta.2021.139757.

Ma, Hailong, Lin Jia, Yunan Lin, Huajing Fang, Wenting Wu, Liangliang Wu, Bin Hu, and Hong Wang. 2022. "A Self-Powered Photoelectrochemical Ultraviolet Photodetector Based on Ti_3C_2T x/TiO2in Situ Formed Heterojunctions." *Nanotechnology* 33 (7). https://doi.org/10.1088/1361–6528/ac0eaa.

Ma, Lin, Chenxia Kang, Likang Fu, Shiyue Cao, Huijuan Zhu, and Qiming Liu. 2022. "Core-Shell Ni1.5Sn@Ni(OH)2 Nanoflowers as Battery-Type Supercapacitor Electrodes with High Rate and Capacitance." *Journal of Colloid and Interface Science* 613: 244–55. https://doi.org/10.1016/j.jcis.2022.01.054.

Ma, Zhiyan, Teng Fei, and Tong Zhang. 2023. "An Overview: Sensors for Low Humidity Detection." *Sensors and Actuators B: Chemical* 376 (Part B): 133039. https://doi.org/10.1016/j.snb.2022.133039.

Nath, Debarati, Puja Dey, Aneesh M. Joseph, Jayanta Kumar Rakshit, and Jitendra Nath Roy. 2020. "Photocurrent Generation under Forward Bias with Interfacial Tunneling of Carrier at Pentacene/F16CuPc Heterojunction Photodetector." *Journal of Alloys and Compounds* 815: 152401. https://doi.org/10.1016/j.jallcom.2019.152401.

Ni, Tianjun, Zhibin Yang, Yafei Cao, Hua Lv, and Yumin Liu. 2021. "Rational Design of MoS2/g-C3N4/ZnIn2S4 Hierarchical Heterostructures with Efficient Charge Transfer for Significantly Enhanced Photocatalytic H2 Production." *Ceramics International* 47 (16): 22985–93. https://doi.org/10.1016/j.ceramint.2021.05.012.

Parangusan, Hemalatha, Deepalekshmi Ponnamma, and Mariam Al Ali Almaadeed. 2019. "Toward High Power Generating Piezoelectric Nanofibers: Influence of Particle Size and Surface Electrostatic Interaction of Ce-Fe 2 O 3 and Ce-Co 3 O 4 on PVDF." Research-article. *ACS Omega* 4 (4): 6312–23. https://doi.org/10.1021/acsomega.9b00243.

Peera, Shaik Gouse, Chao Liu, Jaesool Shim, Akhila Kumar Sahu, Tae Gwan Lee, Manickam Selvaraj, and Ravindranadh Koutavarapu. 2021. "MXene (Ti_3C_2Tx) Supported Electrocatalysts for Methanol and Ethanol Electrooxidation: A Review." *Ceramics International*. https://doi.org/10.1016/j.ceramint.2021.07.075.

Peng, Bo, and Xinlu Huang. 2022. "Research Status of Gas Sensing Performance of Ti_3C_2Tx-Based Gas Sensors: A Mini Review." *Frontiers in Chemistry* 10 (October): 1–7. https://doi.org/10.3389/fchem.2022.1037732.

Qin, R., G. Shan, M. Hu, and W. Huang. 2021. "Two-Dimensional Transition Metal Carbides and/or Nitrides (MXenes) and Their Applications in Sensors." *Materials Today Physics*. https://doi.org/10.1016/j.mtphys.2021.100527.

Rafiq, Madiha, Xiaozhen Hu, Zhiliang Ye, Abdul Qayum, Hong Xia, Liangsheng Hu, Fushen Lu, and Paul K. Chu. 2022. "Recent Advances in Structural Engineering of 2D Hexagonal Boron Nitride Electrocatalysts." *Nano Energy* 91: 106661. https://doi.org/10.1016/j.nanoen.2021.106661.

Ran, Fen, Kuiwen Shen, Yongtao Tan, Bowu Peng, Shaohua Chen, Weijie Zhang, Xiaoqin Niu, Lingbin Kong, and Long Kang. 2016. "Activated Hierarchical Porous Carbon as Electrode Membrane Accommodated with Triblock Copolymer for Supercapacitors." *Journal of Membrane Science* 514: 366–75. https://doi.org/10.1016/j.memsci.2016.05.011.

Razavi, Mehri, Alexandre Barras, Madjid Ifires, Abir Swaidan, Maryam Khoshkam, Sabine Szunerits, Mohsen Kompany-Zareh, and Rabah Boukherroub. 2022. "Colorimetric Assay for the Detection of Dopamine Using Bismuth Ferrite Oxide (Bi2Fe4O9) Nanoparticles as an Efficient Peroxidase-Mimic Nanozyme." *Journal of Colloid and Interface Science* 613: 384–95. https://doi.org/10.1016/j.jcis.2022.01.041.

Saeed, Muhammad Ahsan, Asif Shahzad, Kashif Rasool, Fahad Mateen, Jae Min Oh, and Jae Won Shim. 2022. "2D MXene: A Potential Candidate for Photovoltaic Cells? A Critical Review." *Advanced Science* 9 (10): 1–23. https://doi.org/10.1002/advs.202104743.

Shah, Syed Afaq Ali, Muhammad Hassan Sayyad, Karim Khan, Jinghua Sun, and Zhongyi Guo. 2021. "Application of Mxenes in Perovskite Solar Cells: A Short Review." *Nanomaterials* 11 (8): 1–17. https://doi.org/10.3390/nano11082151.

Shuvo, S. N., Ulloa Gomez, A. M., Mishra, A., Chen, W. Y., Dongare, A. M., and Stanciu, L. A. 2020. "Sulfur-Doped Titanium Carbide MXenes for Room-Temperature Gas Sensing." *ACS Sens* 5 (9): 2915–24. https://doi.org/10.1021/acssensors.0c01287.

Song, Youshuo, Yue Ma, and Ke Zhan. 2020. "Simulations of Deformation and Fracture of Graphene Reinforced Aluminium Matrix Nanolaminated Composites." *Mechanics of Materials* 142 (December 2019): 103283. https://doi.org/10.1016/j.mechmat.2019.103283.

Sun, Limei, Fengchun Wen, Lele Shi, and Suriguga Li. 2020. "Pd and CoOx Decorated Reduced Graphene Oxide Self-Assembled on Ni Foam as Al–H2O2 Semi-Fuel Cells Cathodes." *Journal of Alloys and Compounds* 815: 152361. https://doi.org/10.1016/j.jallcom.2019.152361.

Sun, Yu, Xiaoxia Le, Shiyu Zhou, and Tao Chen. 2022. "Recent Progress in Smart Polymeric Gel-Based Information Storage for Anti-Counterfeiting." *Advanced Materials* 2201262. https://doi.org/10.1002/adma.202201262.

Tayyab, Muhammad, Zhiguo Zhu, Hongyu Lu, Guanyu Ma, Nasir Mahmood Abbasi, Dawei Gu, Bo Wu, Yvonne Joseph, Deqing Gao, and Huang Wei. 2022. "Synergetic Effect of Piezo-Triboelectric Mechanism for High-Performance Nanogenerators." *Nano Energy* 104 (December): 107999. https://doi.org/10.1016/J.NANOEN.2022.107999.

Tayyab, Muhammad, Zhiguo Zhu, Hongyu Lu, Guanyu Ma, Nasir Mahmood Abbasi, Dawei Gu, Bo Wu, Yvonne Joseph, Deqing Gao, and Huang Wei. 2022. "Synergetic Effect of Piezo-Triboelectric Mechanism for High-Performance Nanogenerators." *Nano Energy* 104 (Part B): 107999. https://www.sciencedirect.com/science/article/abs/pii/S2211285522010771.

Tayyab, Muhammad, Zhiguo Zhu, Bo Wu, Nasir Mahmood Abbasi, Yvonne Joseph, and Deqing Gao. 2023. "Enhanced Surface Charge Density of Nanogenerators by Small Molecules: A Review." *Energy Technology* 11 (1). https://doi.org/10.1002/ente.202200879.

Tian, Zhizhang, Shaofei Wang, Yutong Wang, Xiaorui Ma, Keteng Cao, Dongdong Peng, Xingyu Wu, Hong Wu, and Zhongyi Jiang. 2016. "Enhanced Gas Separation

Performance of Mixed Matrix Membranes from Graphitic Carbon Nitride Nanosheets and Polymers of Intrinsic Microporosity." *Journal of Membrane Science* 514: 15–24. https://doi.org/10.1016/j.memsci.2016.04.019.

Tiwari, Sanjay, Rahul Patil, Sunil K. Dubey, and Pratap Bahadur. 2020. "Graphene Nanosheets as Reinforcement and Cell-Instructive Material in Soft Tissue Scaffolds." *Advances in Colloid and Interface Science* 281: 102167. https://doi.org/10.1016/j.cis.2020.102167.

Wang, Shan, Yanqing Yao, Jia Zhao, Xuhui Han, Chunpeng Chai, and Pei Dai. 2022. "A Novel Electrochemical Sensor for Glyphosate Detection Based on Ti_3C_2T: X/Cu-BTC Nanocomposite." *RSC Advances* 12 (9): 5164–72. https://doi.org/10.1039/d1ra08064d.

Wang, Yizhou, Tianchao Guo, Zhengnan Tian, Khadija Bibi, Yi Zhou Zhang, and Husam N. Alshareef. 2022. "MXenes for Energy Harvesting." *Advanced Materials* 34 (21): 1–22. https://doi.org/10.1002/adma.202108560.

Wang, Zhiwei, Yuanhang Zhu, Yuchen Wu, Weiyuan Ding, and Xiuting Li. 2022. "Tunable Fluorescent Amino-Functionalized Ti_3C_2Tx MXene Quantum Dots for Ultrasensitive Fe3+ Ion Sensing." *Nanoscale* 14 (26): 9498–506. https://doi.org/10.1039/d2nr02088b.

Wang, Zhong Lin, and Wenzhuo Wu. 2012. "Nanotechnology-Enabled Energy Harvesting for Self-Powered Micro-/Nanosystems." *Angewandte Chemie—International Edition* 51 (47): 11700–21. https://doi.org/10.1002/anie.201201656.

Wu, Chaoxing, Tae Whan Kim, Jae Hyeon Park, Haoqun An, Jiajia Shao, Xiangyu Chen, and Zhong Lin Wang. 2017. "Enhanced Triboelectric Nanogenerators Based on MoS2 Monolayer Nanocomposites Acting as Electron-Acceptor Layers." *ACS Nano* 11 (8): 8356–63. https://doi.org/10.1021/acsnano.7b03657.

Wu, Dihua, Mengyao Wu, Jiehui Yang, Huaiwei Zhang, Kefeng Xie, Cheng Te Lin, Aimin Yu, Jinhong Yu, and Li Fu. 2019. "Delaminated Ti_3C_2Tx (MXene) for Electrochemical Carbendazim Sensing." *Materials Letters* 236: 412–15. https://doi.org/10.1016/j.matlet.2018.10.150.

Wu, Yuting, Ping Nie, Langyuan Wu, Hui Dou, and Xiaogang Zhang. 2018. "2D MXene/SnS2 Composites as High-Performance Anodes for Sodium Ion Batteries." *Chemical Engineering Journal* 334 (October 2017): 932–38. https://doi.org/10.1016/j.cej.2017.10.007.

Xiao, Yi, Yingchun Ding, Hua Cheng, and Zhouguang Lu. 2019. "The Potential Application of 2D Ti_2CT_2 (T = C, O and S) Monolayer MXenes as Anodes for Na-Ion Batteries: A Theoretical Study." *Computational Materials Science* 163 (January): 267–77. https://doi.org/10.1016/j.commatsci.2019.03.039.

Yang, Yajie, Jaeho Jeon, Jin Hong Park, Mun Seok Jeong, Byoung Hun Lee, Euyheon Hwang, and Sungjoo Lee. 2019. "Plasmonic Transition Metal Carbide Electrodes for High-Performance InSe Photodetectors." *ACS Nano* 13 (8): 8804–10. https://doi.org/10.1021/acsnano.9b01941.

Yang, Yan, Jia ju Fu, Tang Tang, Shuai Niu, Li Bing Zhang, Jia nan Zhang, and Jin Song Hu. 2022. "Regulating Surface In–O in In@InOx Core-Shell Nanoparticles for Boosting Electrocatalytic CO2 Reduction to Formate." *Chinese Journal of Catalysis* 43 (7): 1674–79. https://doi.org/10.1016/S1872-2067(21)63943-8.

Yi, Ting Feng, Jing Jing Pan, Ting Ting Wei, Yanwei Li, and Guozhong Cao. 2020. "NiCo2S4-Based Nanocomposites for Energy Storage in Supercapacitors and Batteries." *Nano Today* 33: 100894. https://doi.org/10.1016/j.nantod.2020.100894.

Yoris-Nobile, Adrian I., Carlos J. Slebi-Acevedo, Esther Lizasoain-Arteaga, Irune Indacoechea-Vega, Elena Blanco-Fernandez, Daniel Castro-Fresno, Alejandro Alonso-Estebanez, et al. 2023. "Artificial Reefs Built by 3D Printing: Systematisation in the Design, Material Selection and Fabrication." *Construction and Building Materials* 362 (November 2022). https://doi.org/10.1016/j.conbuildmat.2022.129766.

Zhang, Li, Shendan Zhang, Chaozhu Huang, Fengdong Qu, Haichuan Guo, Dong Yao, Rui Wang, Heng Jiang, and Minghui Yang. 2021. "PdO-Modified α-Fe2O3 Nanoparticles with Enhanced Gas Performance for Dimethyl Disulfide." *Journal of Alloys and Compounds* 862: 158489. https://doi.org/10.1016/j.jallcom.2020.158489.

Zhang, Zhaowei, Peng Ma, Rajib Ahmed, Jie Wang, Demir Akin, Fernando Soto, Bi Feng Liu, Peiwu Li, and Utkan Demirci. 2022. "Advanced Point-of-Care Testing Technologies for Human Acute Respiratory Virus Detection." *Advanced Materials* 34 (1): 1–39. https://doi.org/10.1002/adma.202103646.

Zhao, Qiang, Junli Wang, Zuopeng Li, Yong Guo, Bing Tang, Abuliti Abudula, and Guoqing Guan. 2021. "Two-Dimensional Ti_3C_2TX-Nanosheets/Cu2O Composite as a High-Performance Photocatalyst for Decomposition of Tetracycline." *Carbon Resources Conversion* 4 (March): 197–204. https://doi.org/10.1016/j.crcon.2021.05.002.

Zhao, Yong, Xiaojie Liu, Xin Jing, Xiangkun Wang, Caiyun Gao, Shuixing Dai, Liangmin Yu, and Mingliang Sun. 2021. "Addition of 2D Ti_3C_2Tx to Enhance Photocurrent in Diodes for High-Efficiency Organic Solar Cells." *Solar RRL* 5 (4): 1–10. https://doi.org/10.1002/solr.202100127.

Zong, B., Q. Xu, and S. Mao. (2022). "Single-Atom Pt-Functionalized Ti_3C_2Tx Fieldeffect Transistor for Volatile Organic Compound Gas Detection." *ACS Sensors* 7 (7): 1874–82. https://doi.org/10.1021/acssensors.2c00475.

Zhou, Qitao, Jing Pan, Shujun Deng, Fan Xia, and Taesung Kim. 2021. "Triboelectric Nanogenerator-Based Sensor Systems for Chemical or Biological Detection." *Advanced Materials* 33 (35). https://doi.org/10.1002/adma.202008276.

Zhou, Xing, Changqing Fang, Yan Li, Ningli An, and Wanqing Lei. 2016. "Preparation and Characterization of Fe_3O_4-CNTs Magnetic Nanocomposites for Potential Application in Functional Magnetic Printing Ink." *Composites Part B: Engineering* 89: 295–302. https://doi.org/10.1016/j.compositesb.2015.11.041.

Zou, Jihua, Yixuan Huang, Wenhao Wang, Caihong Li, Shunyong Wei, Hezhuang Liu, Lingzhi Luo, et al. 2022. "Plasmonic MXene Nanoparticle-Enabled High-Performance Two-Dimensional MoS2Photodetectors." *ACS Applied Materials and Interfaces* 14 (6): 8243–50. https://doi.org/10.1021/acsami.1c22074.

5 Composite Structures of 2D Titanium-Based MXenes with Polymers

5.1 INTRODUCTION

Low-dimensional nanostructures (0D, 1D, or 2D) have been a center of interest due to their unique characteristics in photonics, electronics, and health-care sector nowadays. Among them, 2D nanomaterial is the most incipient one and its explored attributes set it distinctly apart from other kinds of nanomaterials like zero-D & one dimensional in both research worlds whether academic or commercial. Graphene, a 2D material as an instance, shows various extraordinary mechanical, electrical and thermal properties. However, its application in many areas of electronic devices is limited due to its lack of feasible chemistry and zero-bandgap, which is unsuitable for field-effect transistors (FETs) (Chandrasekaran et al. 2020).

MoS_2, $MoSe_2$, WS_2, and WSe_2 are examples of transition metal dichalcogenides (TMDCs), which have unusual optical properties and a variety of demanding applications in optoelectronics, catalysis, sensing, and energy storage devices such as supercapacitors. Additionally, the single layer of molybdenum disulfide (MoS_2) used in 2D TMDs serves as a direct-bandgap semiconductor. However, due to its modest carrier mobility value, which ranges from 60 to 200 cm^2 V per second per second, it is less effective than graphene in practical applications. Black phosphorus (BP), on the other hand, has an enhanced value of carrier mobility (10^3 $cm^2V^{-1}s^{-1}$) and a distinctive puckered structure. However, due to the material's chemical instability, BP-based electrical devices quickly degrade in performance when exposed to air. Due to the aforementioned issue and flaw in the reputable 2D material, it is currently necessary and in demand on the market to investigate 2D semiconductors that have a number of characteristics such as high carrier mobility, stability in ambient conditions, light-harvesting ability, a moderate energy gap, and simple and repeatable practical access (Selamneni et al. 2022; Jiang et al. 2020).

The transition metal carbides, nitrides, and carbonitride (MXenes) are new additions to the tribe of 2D nanostructures in recent years, and, recently among the 2D materials, the MXene family is considered to be an active research field and effectively explored materials. The formula for MXene, which is frequently used, is $M_{n+1}X_nT_z$ or $M1.33XT_z$ (n = 1, 2, or 3), where M stands for a transition metal (TM), and X represents the N and/or C. M-A interactions are moderately metallic; M-X interactions combine metallic, covalent, and ionic components; and Tz denotes a wealth of surface functionalities, primarily hydroxyl (–OH), oxygen (–O), and/or fluorine (–F) groups. These groups are in charge of the high conductivity, tunable flexibility of the interlayer spacing, and factors that improve the redox and charge transport kinetics

DOI: 10.1201/9781003319658-5

of electrochemical reactions. $Ti_3C_2T_x$ is an MXene derivative that has a high value of metallic conductivities (6,000–8,000 S/cm) and hydrophilic surfaces when compared to other 2D materials such as TMDCs and graphene. Additionally, the MXene is a strong candidate due to the presence of these properties for applications such as water desalination, transparent, conducting thin films, electromagnetic interference shielding, energy storage devices, catalysis, and many others (Jaya Prakash and Kandasubramanian 2021; W. Yang et al. 2021).

As a result, MXene differs from other 2D-layered materials like graphite and transition metal dichalcogenides, which are constructed by interactions known as van der Waals interactions, due to the combination of these multidirectional characteristics. Despite the growing need for MXene-based materials on a global scale, there is still a lack of well-organized data on the research of structure stability, surface roughness, and improvements to the MXene synthetic technique. The fabrication of MXene heterostructures can be an useful approach in order to go around the challenges (such stability and surface roughness) for getting better effective results for practical application of MXene. Van der Waals heterostructures are in high demand right now and seem to be expanding as swiftly as interest in the first wave of graphene in 2006. The development of MXene and its van der Waals heterostructures has advanced to a satisfactory level in recent years, making an expanding array of complex devices and materials available to more and more research groups (Kim and Alshareef 2020; Chandrasekaran et al. 2020).

Additionally, 2D materials other than MXene, including as transition metal dichalcogenides (TMDCs), graphene, and hexagonal boron nitride (h-BN), have been comprehensively explored with MXene derivatives for the creation of such heterostructures. As a result, the created heterostructures have a weak connection between the partner material's (0D, 1D, or 2D) structures and the 2D MXene materials. This interaction may be what mostly preserves the partner materials' intrinsic features. A further efficient and dependable method for extending the lifespan of 2D MXene-based devices is the development of vdW heterostructures, which do so by shielding them from the oxidation process. By stacking various 2D sheets, nanowires, nanoparticles, and polymers, MXene-based heterostructures can combine the benefits of multifunctional device innovation and do away with the drawbacks (stability and surface roughness concerns) of the separate MXene materials (Pang et al. 2022; Kim and Alshareef 2020).

The most recent developments in the synthesis, structure, and characteristics of 2D MXene and associated heterostructures for multidirectional applications are compiled in this chapter. For their significant contributions to scientific literature, various research scientists from all around the world who work in the fields of nanotechnology and materials sciences deserve recognition and thanks. The production of a wide variety of 2D MXene-based nanomaterials in different morphology/geometry, such as a quantum dot, nanoflakes, nanosheets, and nanoribbon for multi-direction applications, has been accomplished as recorded in a significant number of review articles on the synthesis/exfoliation of MXene from various MAX phases. Our main goal in writing this chapter is to highlight the most current advancements and select articles that best define important ideas. They made every effort to avoid duplicating the titles of other reviews that have already been published in the same field while choosing the title for this chapter. In this chapter, a comprehensive library of currently published literature on the various methods used by 2D Ti-based MXene ($Ti_3C_2T_x$ and Ti_2CT_x)

nanomaterials to create heterostructures by assembling them with polymers, 0D, 1D, and 2D materials like dichalcogenides (TMDs graphene and boron nitride) has been presented. Finally, an overview of the uses of MXene-based macrostructures, created as a result of functionalization on 2D Ti-based MXenes, in several practical areas, including catalysis and energy storage devices (Pang et al. 2022) is been presented.

5.2 MORPHOLOGICAL ENGINEERING OF MXENE-BASED MIXED-DIMENSIONAL HETEROSTRUCTURES

Due to environmental pollution and energy constraints brought on by an excessive reliance on fossil fuels, there is growing worry around the world. Solar energy has long been seen as the ideal energy source of the future as it is clean and renewable. Photocatalysis employing solar energy directly has drawn increasing attention and showed a tremendous potential in environmental pollution removal and solar-fuel production since the discovery of water splitting by TiO_2 under UV light in 1972. The intricate design of photoactive material catalysts (often semiconductors) is crucial for the conversion and utilization of solar energy. The photoexcited electrons and holes on a semiconductor surface can be used in photocatalysis devices to carry out a variety of redox processes. Light harvesting, photogenerated carrier separation and transfer, and surface reaction are the three main areas of emphasis in the solar energy utilization of photocatalysis, which is intended to boost the total conversion efficiency of the photocatalyst. It has been established that a single semiconductor has a low photocatalytic efficiency due to ineffective visible-light absorption and quick charge recombination. Therefore, creating a heterojunction, doping the elements, and adding a cocatalyst are good ways to increase photocatalytic activity.

Graphene, carbon nanotubes, graphene dioxide, carbon quantum dots, and noble metal have all been utilized as cocatalysts to improve the photocatalytic process's capacity for charge separation (Melchior et al. 2018). To improve photocatalysis efficiency and establish commercial applications, effective and affordable cocatalyst development is still urgently needed. Due to their special characteristics, MXenes, innovative two-dimensional (2D) layered materials of transition metal carbides or carbonitrides, attract a lot of interest in the disciplines of energy storage and conversion. $M_{n+1}X_n$ (n = 1, 2, 3) is the general formula for MXenes, where M stands for a transition metal (Sc, Ti, Zr, Hf, V, Nb, Ta, and Mo), and X stands for either C or N. A focus of research in the synthesis of MXene-based photocatalysts is Ti_3C_2 MXenes, which are produced by selectively etching off the A layers from Ti_3AC_2 (where A is Al, Zn, Si, and Ga). It is mostly ascribed to Ti_3C_2's amazing instinctive qualities, which include the following. (i) The Ti_3C_2 surface's large specific area and multiple terminal functional groups (OH, O, and F) can result in a strong chemical bond with semiconductors, generating a Schottky junction that acts as an electron trap to prevent photoexcited electrons and holes from recombining. (ii) Ti_3C_2 has a strong metallic conductivity, which guarantees quick carrier migration and effective separation of photogenerated electrons and holes. (iii) Ti_3C_2 may exhibit greater active reactivity than carbon-based materials due to the exposed terminal metal sites. Ti_3C_2 has been used as a cocatalyst or precursor in the synthesis of photocatalysts as a result of the aforementioned properties. Nanomaterials' well recognized shapes and dimensions have had a big impact on their characteristics and uses.

The layered structure of accordion Ti_3C_2 MXenes makes it simple to transform them into ultrathin 2D single nanosheets and zero-dimensional (0D) quantum dot (QD) materials, expanding the range of materials that can be used to make high-performance heterojunction photocatalysts. Current Ti_3C_2-based heterojunction photocatalysts are designed with multilayered, single-, or few-layered Ti_3C_2 MXene QDs to support/decorate semiconductors or produce TiO_2/C hybrids, as shown in the schematic diagram in Figure 5.1 (K. Huang et al. 2020). Ti_3C_2-based photocatalysts are

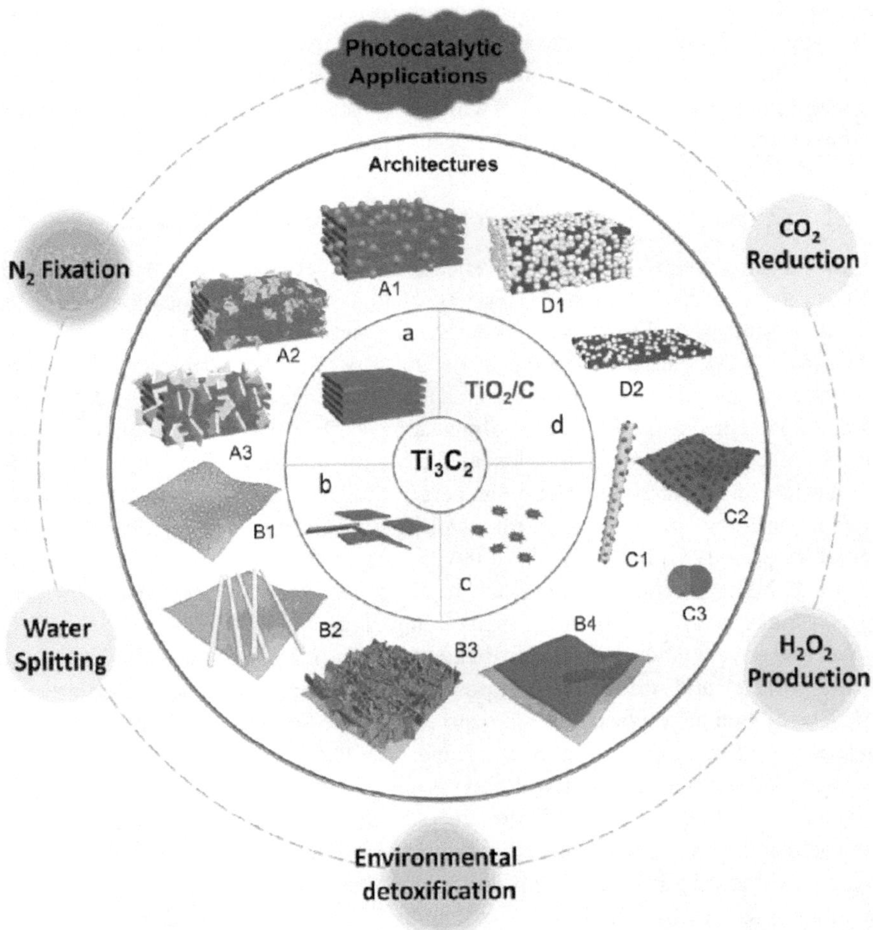

FIGURE 5.1 Different Ti_3C_2-based heterojunctions and their uses for photocatalysis are shown schematically: (a) Ti_3C_2 sheets that are multilayered and used to load semiconductor nanoparticle (A1) or nanosheet (A2) materials (A3). (b) A single-layered Ti_3C_2 nanosheet that can be loaded with 0D and 1D semiconductors, vertically growing 2D nanosheets, and face-to-face 2D nanosheets (B4). (c) Ti_3C_2 QDs and Janus Ti_3C_2 QDs photocatalyst-decorated 1D nanorod (C1) and 2D nanosheet (C2) semiconductors (C3). (d) TiO_2/C hybrids made of single- or multilayered Ti_3C_2 (D1 and D2) (see K. Huang et al. 2020).

Source: Adopted with permission from American Chemical Society.

frequently used in processes like water splitting, CO_2 abatement, H_2O_2 synthesis, and N_2 fixation. The significant rise in the number of scholarly articles published each year indicates that research on 2D MXenes in photocatalysis has increased in recent years. The dimensions and morphologies of MXenes and their derivatives in photocatalysis were not taken into consideration in detail which concentrated on multilayer MXenes as a platform to load a variety of 0D, 1D, and 2D photocatalytic materials for solar-fuel generation. This is why the present study provides a thorough analysis of the multilayer, single- or few-layer, and quantum dot synthesis methods for Ti_3C_2 MXenes. A detailed discussion is provided on the role of Ti_3C_2 MXenes as cocatalysts in heterojunction photocatalysts. These MXenes are the primary heterojunction nanostructures illustrated in Figure 5.1. The discussion also covers the derivatives of Ti_3C_2 MXenes used in photocatalysts, the challenges faced within this system, and a prospective outlook for future research in this area. It will assist professionals in this industry in broadening the uses of Ti_3C_2 MXenes for photocatalysis (Wan et al. 2021; Z. Guo et al. 2021).

5.2.1 MXene and Polymer-Based Heterostructures

In the coming years, technologies such as polymer-based heterostructures and MXene are expected to disrupt a variety of industries and applications. A generalized X-band high-throughput biotemplating method was developed to create a mini-array of MXene microarrays. These microarrays were transferred to flexible substrates using a polydimethylsiloxane (PDMS) device equipped with 3D-printed holders. MXenes, a family of two-dimensional (2D) transition metal carbides and nitrides, are known for their excellent electrical conductivity. They hold great potential for diverse applications, including energy storage/conversion, sensing, and more. MXenes are not only extremely thin and lightweight but also stronger than steel pound for pound. Additionally, these flexible materials can stretch up to 25%, making them highly versatile for various applications. Their extensive customization options make them suitable for a range of applications.

In contrast, polymer-based heterostructures are composite materials that contain both a polymer and some other component (e.g., metal nanoparticles, ceramic particles, or carbon nanotubes). They are stronger, more flexible and durable than most materials that they replace but at the same time cheaper to make (Rasheed 2022; Hosseini, Masoudi Soltani, and Li 2021).

Heterostructures based on MXene and polymer are in a sense expected to have game-changing impacts for the aerospace and the automotive industry. A new technique for three-dimensional (3D) printing of stretchable materials could have future advantages for automobiles and other vehicles. They could also pave way for more efficient, accurate, and robust sensors in comparison to the ones made by traditional means. Finally, these materials could be used to create more efficient systems for storing and converting energy than are now available (Y. Yang et al. 2021).

To create $Ti_3C_2T_x$/polymer composites, Ling et al. produced $Ti_3C_2T_x$ MXene mix with either a charged polydiallyldimethylammonium chloride (PDDA) or an electrically neutral polyvinyl alcohol (PVA) (Figure 5.2). The as-fabricated composite films exhibit high electrical conductivities, ranging from 2.4 S/m for the pure Ti_3C_2Tx film to 205 S/m under maximum deformation during bending and twisting, even when incorporating PVA as a substrate. Additionally, these films demonstrate significant

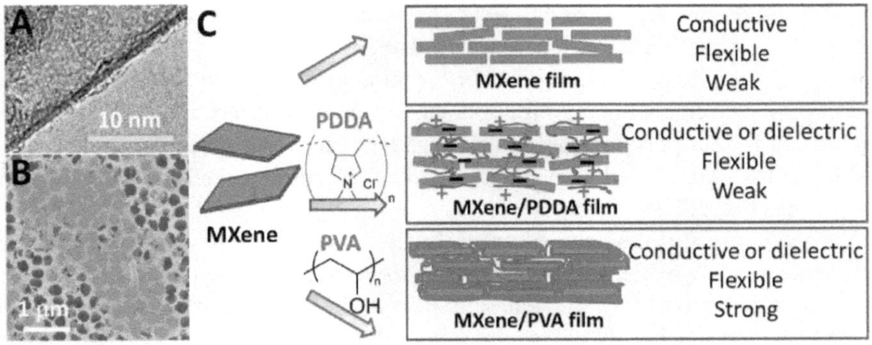

FIGURE 5.2 (a and b) Images of MXene flakes taken with a TEM and a SEM after delamination and before the production of films and (c) a schematic representation of functional films made of MXene with tunable characteristics (see Ling Zheng et al. 2014).

Source: Adopted with permission from Proceedings of the National Academy of Sciences (PNAS).

flexibility. The tensile strength of the Ti_3C_2Tx/PVA composites is much higher than that of pure Ti_3C_2 or PVA films. Besides enhancing flexibility, the intercalation of polymer chains into MXene layers and their confinement by these walls increased cations insertion inside sheets that together contribute to a volumetric capacitance as high as ~530 F/cm^3 for 2 mV/s, detailed with ample scope of application in structural-level components such as energy storage devices; wearable electronics, ferroelectric actuator-based platforms, etc.; and radiofrequency absorbing materials. This study represents an initial but critical first step in that effort to increase our knowledge.

The era of materials capable of replacing and improving a wide range of existing products is still in its infancy. Among these materials, MXenes stand out, particularly when integrated into polymer-based heterostructures. These materials are great because they offer a unique balance of strength, flexibility, and toughness for many different applications. As research into these materials advances, this could create additional transformations in the way they generate and store energy among other things (Anaya et al. 2021).

5.2.1.1 MXene and Conducting Polymer-Based Composite

Conducting polymer composites and MXene is a recently established field of study, which presents great hope in how these utilize materials drastically. MXene, a highly conductive two-dimensional material consisting of titanium carbide, when combined with a conducting polymer, such as polyaniline, can be used in the production of high-conductivity composite. Applications of this material include but are not limited to electronics, sensors, and actuators as well as energy conversion and storage (P. Guo et al. 2022).

MXenes and conducting polymer have the divergent properties as compared to conventional materials. One of the most conductive materials, called MXene for short, has a high conductivity that allows electricity to run through it very efficiently. Because of this, it is great for electronics which need faster speeds. MXene is also extremely light, making it easy to be added into existing designs. Lastly, this is best for use in high-power applications where its excellent thermal conductivity not only solves the problem but also at a fraction of that cost.

The MXene and conducting polymer combination could find application in various energy conversion and storage applications e.g., supercapacitors and batteries. This could benefit portable devices, electric cars, and renewable energy. The same components could be used to manufacture more efficient fuel cells or solar cell panels (L. Ma et al. 2022).

As a new class of conductive materials, 2D titanium carbide MXene (Ti_3C_2Tx) has demonstrated great promise as high-performance electrode materials for electrochemical capacitors (ECs). 2D materials such as the MXenes tend to restack when treated into freestanding films, resulting in limited ion flow inside these electrodes. This shortcoming could largely limit the specific capacitance and rate capabilities of freestanding electrodes, particularly when exceeding a few microns in thickness. Here, they demonstrate a surface-modification route on MXene sheets that enables high-performances even with extremely large electrode thicknesses through creating an accessible structure and its electrochemical function. The in-situ polymerization of polyaniline (PANI) on the MXene layers surface also produced 2D Ti_3C_2Tx and PANI hybrid materials, which were then stacked together to form freestanding films with different thicknesses (Figure 5.3). The MXene/PANI hybrid electrodes

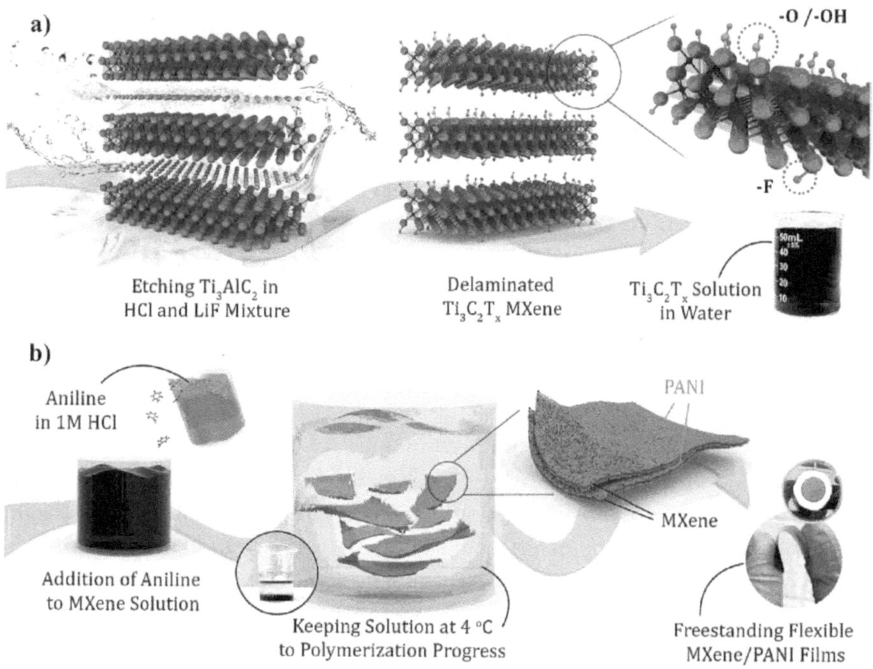

FIGURE 5.3 A illustration of step (a) in the synthesis process for Ti_3C_2Tx MXene, selectively etching away Al layer from multilayered Ti_3AlC_2 MAX phase. After the synthesis, various functional groups are loaded onto Ti_3C_2Tx surface. (b) Schematic representation of fabrication process for Ti_3C_2Tx/PANI hybrid electrode (see Armin VahidMohammadi et al. 2018).

Source: Adapted with permission from the Royal Society of Chemistryelectrode.

material demonstrates impressive capacitive performance, with specific and volumetric capacitances reaching 503 F/g and 1,682 F/cm³, respectively, making it highly suitable for energy storage applications such as supercapacitors. As electrode thickness and mass loadings increase, the hybrid electrodes still show great electrochemical performance (Armin VahidMohammadi et al. 2018). As an example, an electrode demonstrated a specific capacitance of approximately 336 F/g (equivalent to a volumetric capacitance of 888 F/cm³). The electrode was cycled for 10,000 cycles, exhibiting a high capacitance retention of ca. 98.3%, which vividly demonstrates the extended cycle lifetime of the hybrid electrodes. The method to fabricate hybrid electrodes made from MXenes/conducting polymer as well as the appraised new electrochemical property are easily and rapidly detailed in this work (Qu et al. 2022; H. Qin et al. 2021).

Then, a number of sensors and actuators could be synthesized by combining MXene with conducting polymer composites. For example, this type of devices would allow us to evaluate monitoring the configuration over humidity changes (and other environmental factors) and act accordingly based on it. These can be used for many purposes such as industry needs and medical field.

To sum up, the composites of conducting polymers and MXene are fascinating new materials, with great possibilities for revolutionary application alternatives. Potential applications include energy conversion and storage, electronics (notably servomechanisms), sensors, electrochromism, and molecular machines. With further study and development, this material could change their future approach to materials – in a big way (Jing et al. 2021; Tian et al. 2022).

5.2.1.1.1 MXene and Polypyrrole

The good materials in materials science and engineering are MXene and polypyrrole. MXene is a two-dimensional material made of transition metal carbides and nitrides. Polypyrrole is an impressive polymer built from pyrrole monomer that has electrical conductivity. Both these materials have many potential applications (e.g., energy storage, biomedicine) (Furtado et al. 2021; Xingpu Wang et al. 2021).

MXenes have recently received a lot of attention because they are quite unique, with exceptional corrosion resistance and a high electrical conductivity that rivals copper. Although due to having a relatively high surface-area-to-volume ratio, MXene is an ideal material for applications in energy storage such as lithium-ion batteries and supercapacitors, the material has also been used to produce thin-film transistors, which can be implemented in electronic displays and other devices.

Polypyrrole is an interesting material in this regard, as it is an electrically conductive polymer useful in making electrodic materials for a variety of applications – for example, both polypyrrole and transparent electrodes for thin-film solar cells (panel) as well as flexible electrodes useful in wearable electronics. Polypyrrole can also be utilized to fashion biosensors that actively search for biological molecules like DNA (Shan Li et al. 2022; Gao et al. 2022).

As one of the urgent needs to fulfill applications for future portable and wearable electronics, more attention has been recently concentrated on flexible electrochemical energy storage devices. Among them, supercapacitors (SCs) have drawn the most attention due to their high specific capacitance and power density—for example, many flexible SCs are WoM-based and the electrode material employed often involves transition metal oxides coupled with traditional carbon materials in general. High-performance all-flexible supercapacitors based on 2d metallic nanocrystalline (MXene) electrodes, a highly conductive and electrochemically stable form of MXenes, was utilized. This material enables architectural design at the micron level, allowing for the rapid fabrication of precise non-rigid supercapacitors. J. Yan et al. reported the performance of textiles as conductive textile electrodes with 182.70 Fg^{-1} capacitance under a very simple "dipandy" drying technique which is even larger than CNTs and active carbon. They exploited the hydrophilicity and metal conductivity of MXene nanosheets to achieve this. Polypyrrole (PPy) was deposited electrochemically on isogonal Ti_3C_2Tz textiles to synergistically boost the capacitive performance of MXene-based electrodes and mitigate the inferior oxygen redox chemistry of MXenes (J. Yan et al. 2018) (Figure 5.4). A PPy-MXene coated textile electrode with a specific capacitance of 343.20 Fgl was developed here which provides new opportunities to explore the high-performance electrodes for wearable energy storage devices. A symmetrical solid-state supercapacitor using MXene-PPy showed a power density of 41.1 mW g^{-1} and energy density 1.30 mW h g^{-1}. This work introduces the first textile SC utilizing MXene that, so far, represents a promising candidate for flexible wearable energy storage systems (Yipu Du et al. 2021).

Even more potential applications arise when MXene is combined with polypyrrole. One such combination could produce conductive and transparent thin-film transistors, for example. Moreover, the preparation of hybrid energy storage devices (such as lithium-ion batteries and supercapacitors) could benefit from the synergistic properties of MXene and polypyrrole, leading to enhanced energy density, power density, and cycling stability.

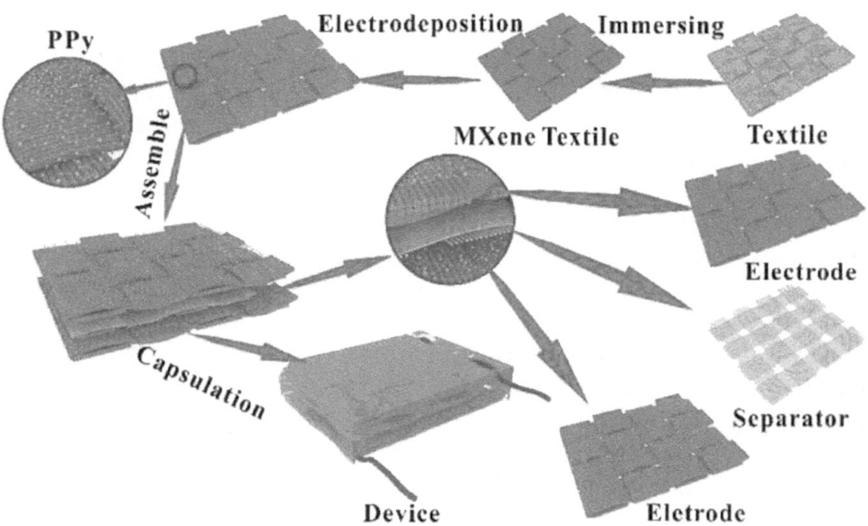

FIGURE 5.4 Schematic diagrams of PPy-MXene coated textile SCs (see J. Yan et al. 2018).

Source: Adapted with permission from the Royal Society of Chemistry.

To conclude, a versatile array of potentials are attributed to polypyrrole and MXene. These can be used to build thin-film transistors, biosensors and hybrid energy storage devices. Furthermore, these two materials can be mixed together to create a long list of potential uses. Therefore, it is important for materials science and engineering to explore polypyrrole or MXene in more detail (Yao et al. 2021; Y. Wu et al. 2021).

5.2.1.1.2 MXene and Polyaniline

The development of MXene and polyaniline can revolutionize materials research and engineering. MXene, a two-dimensional transition metal carbide or nitride, is a very promising new material class for use in electronics, energy storage, and electrochemical devices. A conducting polymer, polyaniline has long been used as an active ingredient in batteries, supercapacitors, and fuel cells. It is hoped that nanostructured surfaces with tailor-made properties for diverse applications can be produced by the combination of these two substances. The interest in MXene is due to the ease with which it can be modified for getting a variety of electrical and optical properties—especially their conductivity can come down to a level that exceeds all non-precious metals. This makes p-type semiconductors a great material for solar cells, thin-film transistors, and other electrical centric devices. MXene has been used to generate nanostructured surfaces with tailored properties such as anticorrosion coatings and friction reduction (Pensalfini et al. 2020; C. Xiao et al. 2022).

Polyaniline has long been one of the favorite active ingredients in the making of things such as batteries, supercapacitors, and fuel cells. It is an excellent electrode material due to its great electrical conductivity and also serves as the best option which produces nanostructured surfaces. Similarly, combining polyaniline with MXenes will generate nanostructured surfaces that can offer tailor-made features for energy storage and electrochemical devices (Quadrado et al. 2022).

L. Zhao et al. (2019) showed the electrocatalytic sensitivity performance of polyaniline/MXene (PANI/Ti_3C_2Tx) nanocomposites by employing density functional theory simulations with bulk electrosensitive measurements, as shown in Figure 5.5. At room temperature, the PANI nanoparticles uniformly distribute on Ti_3C_2Tx nanosheets surfaces and present high sensitivity detection and fast response/recovery time in excellent mechanical stability without introducing abrupt alarms by low-temperature approach using in-situ polymerization. This is achieved through the synergistic effects between nanocomposites and high catalytic/absorption capacity of Ti_3C_2Tx MXene. The study offers a versatile strategy for producing 2D nanocomposites toward flexible gas sensors with good performance, based on MXenes.

Water filtration may potentially benefit from the combination of MXene and polyaniline. The materials can be utilized to attract and absorb contaminants from water by forming a nanostructured surface. This could be utilized to develop an inexpensive, effective, and efficient water filtering system (Yun, Kim, and Kim 2021).

MXene and Polyaniline can change the world for materials science and engineering. They can produce a vast assortment of devices capable of doing certain things out of these materials and from nanostructured surfaces. They may be even used to create really cheap water purification systems. As the science advances, these resources will be more impactful than ever to achieve a healthier world (Yan Du et al. 2021).

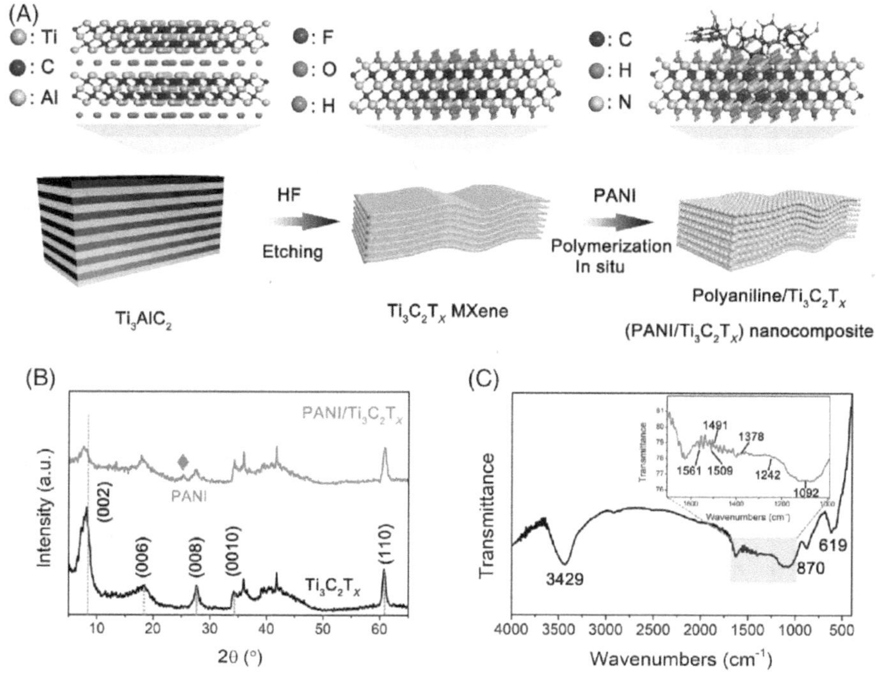

FIGURE 5.5 Characterization and synthesis plan of PANI/Ti_3C_2Tx nanocomposites. (a) Schematic illustration of the exfoliation of Ti_3AlC_2 and polymerization into PANI during preparation process for PANI/Ti_3C_2Tx nanocomposites. (b and c) The XRD of Ti_3C_2Tx MXene and PANI/Ti3Cx Tx nanocomposites. (c) FTIR spectrum of the PANI/Ti_3C_2Tx nanocomposites. *Abbreviations*: PANI, polyaniline; XRD, X-ray diffraction; ANI, aniline; FITR, Fourier transform infrared (see L. Zhao et al. 2019).

Source: Adopted with permission from Weinheim: Wiley-VCH Verlag GmbH & Co.

5.2.1.2 MXene and Insulating Polymer-Based Composite

Thus, the development of MXene and organic-inorganic composites based on insulating polymers is a far-reaching advancement toward high energy density and efficient and rational design for sustainable future. These materials are great for a variety of applications because they give better thermal and electrical insulation, as well as improved durability. Two-dimensional nanomaterial MXene exhibits significant potential as a novel material for thermal and electrical insulation applications, owing to its unique layered structure, tunable surface chemistry, and exceptional physicochemical properties. Due to it being made from cold-worked steel titanium, its unique design allows for very excellent thermal insulation properties with a high electrical conductivity. Therefore, MXene is the best material for making modern-day electronics and electrical apparatus, as well as to make more sustainable buildings. There is also considerable interest in insulating polymer-based composites for energy-efficient applications with MXene. With these materials, implant designers get the strength and durability of metal alloys along with the thermal and electrical insulation once that was only available from polymers. This makes them ideal for solar panel installations, which can decrease heat loss and increase the efficiency of a system (Jalili et al. 2021; K. Qian et al. 2022).

Here, porous Ti_2CT_x MXene/poly(vinyl alcohol) (PVA) composite foams were fabricated using few-layered Ti_2CT_x (f-Ti_2CT_x) MXene and PVA through a simple freeze-drying process, as demonstrated by H. Xu et al. At an ultralow filler level of 0.15 vol%, the composite exhibited a reflection effectiveness (SER) of less than 2 dB (Figure 5.6) (Hailong Xu et al. 2019). Furthermore, the shielding efficiency, when normalized by the theoretical mass density, was predicted to reach 5136 dB·cm²/g and beyond, showcasing unparalleled absorption-dominated shielding performance. These exceptional properties are attributed to well-matched impedance, abundant porous structures, internal reflections, and polarization effects (including dipole and interfacial polarization), as evidenced by separation experiments.

So, this study provides an effective MXene-based EMI shielding alternative possessing weight and strength characteristics.

Also, research on automobile application of MXene and composites derived from insulating polymers is underway. These can be used to build lighter cars that also perform better. This is very important for EV in particular because an insulation thickness increase improves their overall efficiency and range. Both MXene and insulating polymer composites seem to have a good future. In fact, these materials are likely to

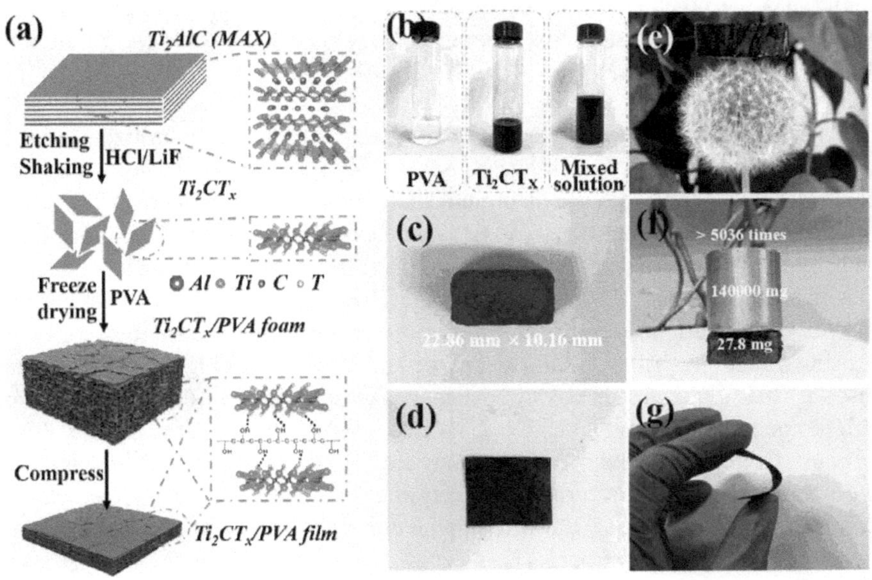

FIGURE 5.6 (a) Schematic of the f-Ti_2CT_x/PVA composite foam and film preparation. (b) A bottle of PVA solution, a bottle (practically transparent) of f-Ti_2CT_x/pure-black liquid, and the as-prepared sample in (c). (d) An xample image using f-Ti_2CT_x/PVA film. (e) An ultralow density piece of f-Ti_2CT_x/PVA foam on a dandelion. (f) The image shows an example of f-Ti_2CT_x/PVA foam, which can support up to 5000 times its own weight. (g) More flexible f-Ti_2CT_x/PVA film shown in (g) (see Hailong Xu et al. 2019).

Source: Adopted with permission from American Chemical Society.

play an increasingly important role as technology features expand for a broad set of energy-efficient applications. Furthermore, if these materials are used in combination, this may result in better products and even energy-efficient ones. Therefore, MXene and insulating polymer-based composites offer an attractive way to approach future sustainable materials. Due to insulation and mechanical strength properties, these materials work best for increasing durability on a variety of applications. As technology advances, materials such as these will only grow in importance for myriad energy-efficient applications (Ai et al. 2021; Kumar et al. 2021; H. Li et al. 2021).

5.2.1.2.1 MXene and PVA

Two substances that are transforming science are MXene and PVA. MXene, a two-dimensional titanium carbide substance, has been found to have high thermal and electrical conductivity. PVA, a synthetic polymer consisting of polyvinyl alcohol, has been found to have outstanding strength, slow-release, and water solubility qualities. Both of these materials have a wide range of uses and are expected to completely change how they interact with technology in the future.

The realm of energy storage is where MXene and PVA are most often used. Supercapacitors, which can store a lot of energy in a tiny area, have been made using both materials. This is crucial for both larger devices like electric vehicles and smaller ones like laptops and cellphones. The use of these materials for energy storage has the potential to fundamentally alter how they store energy and could eventually result in a system that is more effective and sustainable (A. Khan et al. 2021; Cong Chen, Liang, et al. 2021).

MXene and PVA can also be utilized in the creation of sensors and biosensors. MXene is a superb heat- and electricity-conductor, and it can be utilized to make sensitive, precise sensors for a range of uses. Strong PVA can be used to make slow-release sensors that can pick up on minute alterations in the environment. These sensors can be used in a range of applications, including environmental monitoring and medical diagnostics, to detect changes in temperature, humidity, and more.

A two-dimensional (2D) transition metal carbide (MXenes)-bound hydrogel sensor with good strain and temperature-detecting performance is created by H. Liu et al. using direct ink writing (DIW) technique and 3D printing (Figure 5.7) (Haodong Liu et al. 2022). It is possible to attain a gauge factor (GF) of 5.7 (0.191% strain) and high-temperature sensitivity (5.27% °C) within a broad operating range (0 to 80°C). Particularly, the associated mechanisms are explained using finite element analysis. The printed sensor can accurately indicate the temperature of the shape memory solar array hinge, marking the first application of in-situ temperature-dependent Raman technology for hydrogels (K. Yu et al. 2022; Cheng Chen, Su, et al. 2021).

Finally, the creation of stretchable and flexible electronics can make use of MXenes and PVA. Due of their high conductivity, MXenes can be utilized to build elastic, flexible circuits. PVA is very flexible, making it possible to utilize it to make stretchable parts like batteries and displays. This might result in the creation of wearable electronics that are convenient and comfortable.

Two substances that are transforming science are MXene and PVA. They offer a wide range of uses and are expected to completely alter how they interact with technology in the upcoming years. A future that is more productive and sustainable

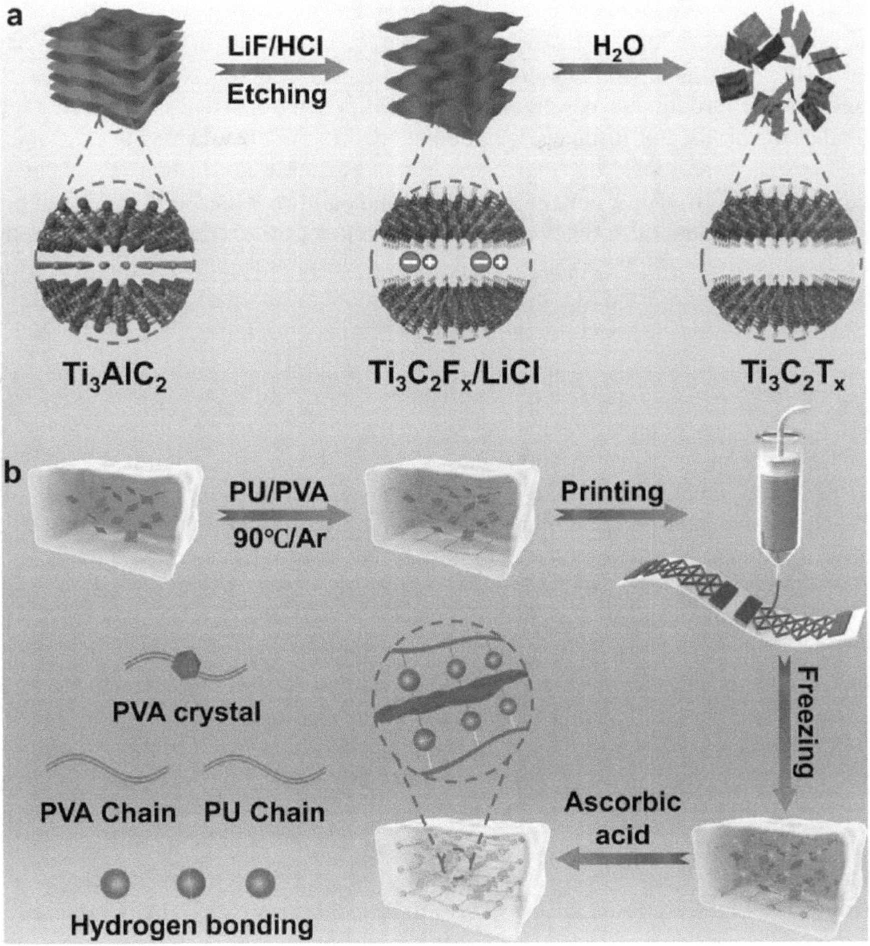

FIGURE 5.7 (a) MXene and hydrogel synthesis: Schematic representation of how $Ti_3C_2T_x$ flakes are made. (b) Polyurethane/polyvinyl alcohol (PU/PVA) hydrogels with transition metal carbides (MXenes)-bonded direct ink writing (DIW) printing (see Haodong Liu et al. 2022).

Source: Adopted with permission from Springer Nature.

may result from the employment of these materials in energy storage, sensors, and flexible electronics. MXene and PVA will be in the fore of this transformation as technology continues to advance (Geng et al. 2021; Fabien et al. 2022).

5.2.1.2.2 Poly(Diallyldimethylammonium Chloride) (PDDA)

Cationic polymer, poly(diallyldimethylammonium chloride) (PDDA), has been used in a wide range of processes such as water treatment and paper coating; however, this particular compound is now being examined for new applications across many industries. This makes PDDA an attractive solution for a variety of applications due to it being soluble in water to a great extent and not skin irritating, with versatile compatibility. It is a good option for water treatment due to its cationic nature which means it can attach itself to the negatively charged particulate in water and can neutralize the precipitate from solution. This is a highly effective tool that can be utilized in paper production as well as in the stabilization of particles for a wide range of other materials. For instance, it can be applied in the manufacturing of *composites, ceramics, coatings, and polymers*, where particle stabilization is critical for achieving uniform dispersion, enhanced mechanical properties, and improved performance. Additionally, it holds potential for use in *pharmaceuticals* for drug delivery systems, *cosmetics* for stabilizing emulsions, and *nanotechnology* for controlling the assembly of nanoparticles. Its high effectiveness in stabilizing particles makes it a versatile solution across multiple industries (Zhu et al. 2021; L. Wang et al. 2021).

The characteristics of PDDA (poly(diallyldimethylammonium chloride)) have made it increasingly popular among researchers in recent years. Its ability to bind with *a wide range of particles, such as nanoparticles, organic pollutants, and colloidal particles*, underscores its versatility. Additionally, its effectiveness at removing toxins, heavy metals, and organic contaminants from water sources highlights its critical role in water purification processes. This makes PDDA an essential material for the treatment of water intended for human use, ensuring safety and quality.

PDDA's cationic nature has led to it being employed as a paper coating agent, with the paper features like increased water resistance and better tensile strength. In the medical field, PDDA is used to manufacture polymeric nanoparticles which can be applied on site-specific drug delivery system. It has also been implemented to build polymeric membranes for filtering and separation operations.

PDDA usage evolves just like any technology. While research continues into how to make the material more efficient—and less expensive, this could lead to a new class of hybrid materials where PDDA is used in combination with other materials such as graphene to create composites with significantly enhanced properties. Moreover, given that PDDA can be employed in generating complex designs and layouts, scientists are looking into 3D printing. Additionally, PDDA is explored as a potential modulation vector for new therapeutic approaches as it prepares custom-designed drug delivery systems (Theyi et al. 2022; Ahmed et al. 2022).

MXene@PDDA/carbon cloth (CC) was created by C. Xing et al. as a modified version of the superhydrophilic cationic polymer poly(diallyldimethylammonium chloride) (PDDA) for use as an anode in MFC (Figure 5.8) (Chengcheng Xing et al. 2021). The special anode provides high conductivity, superhydrophilicity, and biocompatibility. The anode enriched EET-competent bacteria and enhanced the efficiency of the EET process. MFCs with MXene@PDDA/CC anode show a good power output and effluent removal efficiency under the external series resistance of 1 k. The system can deliver up to 585 mV and a power density as high as 811 mW/m^2. The removal efficiencies of methyl orange decolorization and chemical oxygen demand were achieved up to 79% and 84%, respectively, after the passage through PEM reactors for a period over than 12 hours.

In conclusion, due to its unique characteristics and broad utility, PDDA has gained more importance. A desirable option for a large number of industries, with new applications expanding rapidly due to the fact that it can bind particles and provide materials different attributes, PDDA use is rife and anticipated to grow even further

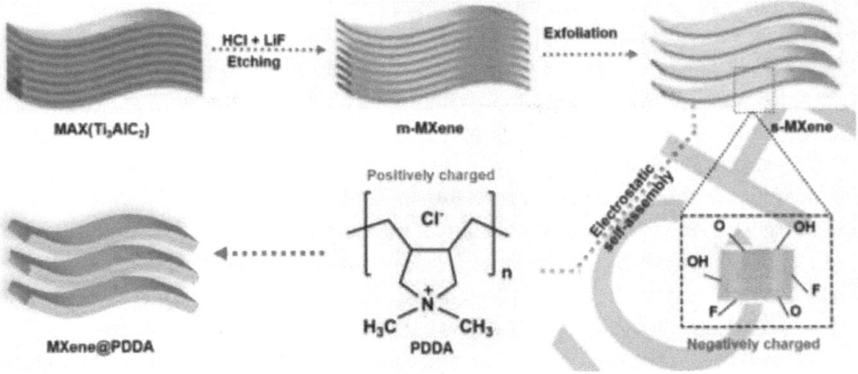

FIGURE 5.8 Schematic illustration of the preparation process of MXene@PDDA (see Chengcheng Xing et al. 2021).

Source: Adopted with permission from Weinheim: Wiley-VCH Verlag GmbH & Co.

as new, creative applications are developed in the ongoing exploration of what it can do. Therefore, PDDA is an important material that would likely be a large part of many industries' manufacturing praxis in the future (Zhentheyi Wang et al. 2021; J. Guo, Tian, and He 2021).

5.2.1.2.3 MXene and PDMS

Among the materials developed in recent years leading to cutting-edge designs are MXene and PDMS (polydimethylsiloxane). These materials are used in various applications and expected to make researchers work for many years. MXene is a type of two-dimensional (2D) material made up with carbon and transition metal atoms in layers. The material is being investigated for use in energy storage, nanoscale manufacturing, and even medical applications. This technique has also been used for various applications from electronics to sensing. Its unique properties, which encompass a high electrical conductance and chemical stability as well as excellent surface reactivity, make it an extremely appealing material for many applications.

PDMS is an elastomeric polymer that can be used for several purposes such as sealants, coatings, and insulation materials. It is a great material as it is used in areas such as medical instruments and food packing and in the automotive industry because of its durability against water, high temperatures, and most chemicals. It is also very flexible which makes it an excellent option for applications requiring liquidity as well. I mean truly innovative and transformative ideas things so groundbreaking that they could fundamentally change the way people think about and produce materials forever.

The elasticity and durability of PDMS make it suitable for a wide variety of applications, while the unique properties possessed by MXene render it ideal in numerous scenarios. It is expected that in the following years, both materials will have a substantial number of research and development works, being able to offer an interesting amount of new applications (Ni et al. 2022; Ruan et al. 2021).

D. Wang et al. generated a simple technique (K. Huang et al. 2020), which referred to constructing three-dimensional Ti_3C_2Tx (3D-MXene) skeleton directly without glue. The 3D-MXene/PDMS composite exhibited very high electrical conductivity of about 5.5 S/cm, which was approximately 14 orders of magnitude higher than that currently reported for pure PDMS once embedded into polydimethylsiloxane (PDMS) matrix (Figure 5.9). In comparison, the nanocomposites indicate only a 220% enhancement in thermal transport at a low MXene content (2.5 vol%). Moreover, due to the largely reduced electrical resistance, the output current of 3D-MXene/PDMS-based TENG significantly improved. This discovery will widen the scope of practical uses of MXenes in thermal management, sensors, and energy harvester progress, with which a capacitor-structured TENG can get involved (Dezhao Wang et al. 2020)

This can lead to the use of MXenes and PDMS together in forming new materials that are better than either of them separately. MXene and PDMS can thus be easily mixed to form a highly conductive flexible composite band. This can lead to even the production of new materials for electronics and medicine. To conclude, both

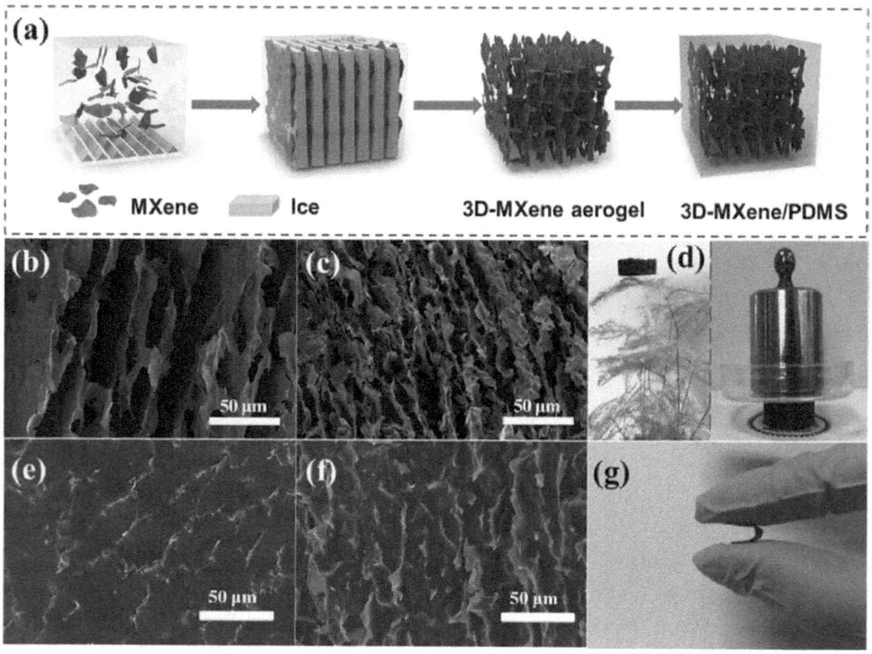

FIGURE 5.9 (a) Illustration of the fabrication for 3D-MXene/PDMS; (b) SEM images of 3D MXene2.0 aerogel, and (c) digital photo image for handling this thin membrane layer grown from 1 mg/mL as it becomes more brittle with increasing concentration. (d) Digital photographs showing a 3D-MXene2.5 sample (220 mg) balanced on the delicate leaves of an asparagus fern (left) and supporting a 200 g weight (right). Scanning electron microscopy (SEM) images of the fractured surfaces of (e) 3D-MXene2.0/PDMS and (f) 3D-MXene2.5/PDMS nanocomposites are also presented. (g) A photograph showcasing the flexibility of the 3D-MXene2.5/PDMS nanocomposite (see Dezhao Wang et al. 2020).

Source: Adopted with permission from Elsevier.

MXene and PDMS are some of the most modern materials that have been produced lately. These two very particular materials are versatile, good enough to do a lot of interesting stuff, and therefore it is expected that there will be years-worth investigation to be done on them. The resultant material might even have qualities that are better than either of them separately and could radically change the way materials or applications for those materials are thought about (Kishan et al. 2020; Hamada et al. 2020).

5.2.1.2.4 MXene@PS Nanocomposites

Nanocomposites are a class of materials that leverage the combination of two or more independent components at the nanoscale to achieve properties that cannot be realized by the individual elements alone.

One example of these nanocomposite materials is MXene@PS, a new and promising type of material for several different technological fields. Herein, the importance, newness, and potential of MXene@PS nanocomposites will be discussed. MXene@PS nanocomposites are a new generation of composite material, which is prepared using MXenes and polystyrene. MXenes are carbon and titanium carbide atoms in a two-dimensional (2D) substance form. They are very conductive and have unique properties like persistent high-temperature stability and superwater-phobicity. PS is a type of plastic known for its excellent thermal and electrical insulation properties, as well as its lightweight nature. When combined with MXene, a new family of two-dimensional materials, the resulting MXene/PS nanocomposite exhibits unique characteristics that make it a promising candidate for a wide range of technological applications (Y. Qian et al. 2021; Kuttiani Ali et al. 2021).

One of the crucial properties of MXene@PS nanocomposites is their capability to function as an energy storage mechanism. The electric charge the nanocomposite can store will power a variety of gadgets, from computers and cell phones to an electronic vehicle. Accordingly, MXene@PS nanocomposites are an essential component for replenishable energy technology. MXene@PS nanocomposites can both absorb and keep light while still retaining visible-light responsiveness for photovoltaic purposes.

The poor dispersion and weak interfacial adhesion of the MXene nanosheets make it very challenging to produce high-performance polymer-based nanocomposites. In order to address this issue, the researchers applied long-chain cationic agents at different lengths of the alkane chains for the exfoliation of MXene nanosheets such as decyltrimethylammonium bromide (DTAB), octadecyltrimethylammonium bromide (OTAB), and dihexadecyldimethyl ammonium bromide within one to three carbon alkanes, i.e., DDAB (Si et al., 2019). The as-prepared MXene nanosheets showed very good dispersion in N,N-dimethylformamide (DMF) (Figure 5.10). And they also had long-chain groups on their surface, which can improve the interfacial adhesion with PSmatrix. The temperatures corresponding to 5% weight loss increased by approximately 20°C, 25°C, and 23°C for DTAB-Ti_3C_2/PS, OTAB-Ti_3C_2/PS, and DDAB-Ti_3C_2/PS, respectively, compared to the pristine sample. This indicates enhanced thermal stabilization behavior due to the cation-modified Ti_3C_2 MXene. Additionally, the flame-retardant properties of PS were significantly improved by the incorporation of modified MXene nanosheets. In comparison with pristine PS, the peak heat release rate (PHRR) of polystyrene nanocomposites

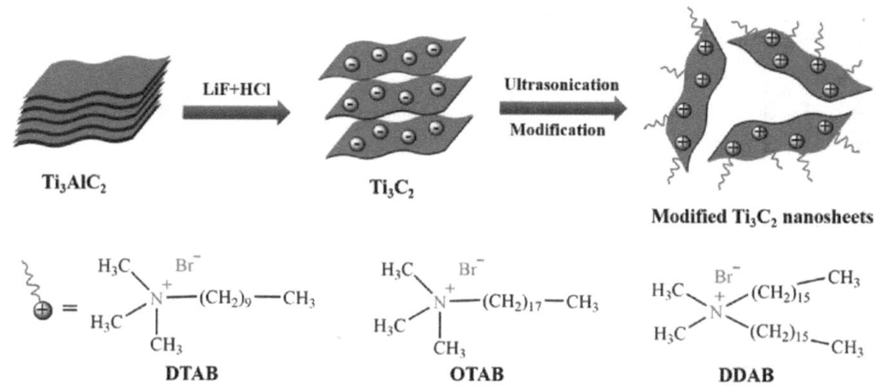

FIGURE 5.10 An example of how to make functionalized Ti_3C_2 (MXene) nanosheets. Decyltrimethylammonium bromide, octadecyltrimethylammonium bromide, and dihexadecyldimethylammonium bromide are all different names for the same compound (see Si et al. 2019).

Source: Adopted under the Creative Commons CC BY 4.0 license.

was decreased by about 26.4%, 21.5%, and 20.8% for PS/OTAB-Ti_3C_2, PS/DDAB-Ti_3C_2, and PS/DTAB-Ti_3C_2, respectively. Similar to the production of CO and CO_2, dramatic decreases in corresponding productions were observed for both gases in cone calorimeter test, and further stricter control ensured that there is less generation by comparing pure PS with P-G-TiOX/PS (the renewed name: PS/OTAB-Ti). These enhancements in the properties of PS nanocomposites are mainly happening because GRPs possess superior catalytic, barrier, and dispersion characteristics of Ti_3C_2NS (K. Huang et al. 2020).

The MXene@PS nanocomposites can produce a new type of dual component material forms, which are different from mechanical and physical blends. These unique properties make the MXene@PS nanocomposites promising for a bunch of applications because they exhibit excellent conductivity, flexibility, and low density. They can also be tailored to suit specific requirements, allowing for the development of innovative nanocomposites with defined properties. Based on the above discussion, future prospective of MXene@PS nanocomposites found in numerous sectors is quite fascinating. These might include the creation of bendable, low-mass batteries for electric vehicles and next-generation solar cells. Furthermore, MXene@PS nanocomposites can be tailored to obtain both electrical conductivity and water repellency qualities, essential for smart wearable textiles. So, to conclude here, MXene@PS nanocomposites are a promising lot of the most recent materials. They could transform numerous industries in the short term, with potential applications ranging from photovoltaics to energy storage (Cao et al. 2022; S. N. Li et al. 2021).

5.2.1.2.5 MXene and PMMA

MXene and PMMA [polymethyl methacrylate] are two materials that have attracted the attention of researchers due their unique properties as well as potential applications. MXene is a two-dimensional, layered material composed of transition metal carbides, nitrides, and/or carbonitrides. On the other hand, PMMA is an acrylic resin formed from methyl methacrylate monomers. Either material may change the way materials have been looked at traditionally and what they had been used for (Beibei Li et al. 2022; Z. Liu et al. 2022).

Rechargeable lithium-ion batteries (LIBs) have been recognized as one of the most promising energy storage technologies due to their numerous advantages. Because of their benefits in enabling high power and energy density, along with operation under intensive charge/discharge rates, two-dimensional (2D) MXene-driven LIBs are greatly studied. However, 2D MXene electrodes have several drawbacks that hinder them from being commercialized as cathodes in LIBs, such as the self-stacking problem and having poor specific surface area for ion diffusion due to numerous terminations on the surfaces of MXenes. The hollow MX/C nanofibers electrode were generated with higher surface functional groups utilized for LIBs, larger specific area, and less sheet self-stacking compared to the pristine 2D MXene (MX) paste electrodes and MX-Carbon (MX/C') paste electrodes. The LIBs using a hollow electrode demonstrate higher energy density (306.5 $mAhg^{-1}$ at 40 mAg^{-1}) than those of the MX paste and the MX/C paste electrodes-based LIBS samples (196.9 $mAhg^{-1}$ at 40 mAgh). The high reversible capacity of the hollow MX/C nanofiber electrode shows more convincing evidence that it can be a promising anode material for LIBs (Darae Seo et al. 2022).

MXene is a material that seems ideal for many uses, thanks to so many attractive properties. Its two-dimensional framework is conductive, helping attain high-performance energy-efficient devices. Thanks to their flexibility and lightness, they are an excellent option for making devices of wearable technologies like smartwatches or health monitors. Also, due to its hydrophilic nature and capability of being able to entirely block water ingress in the channels, it is an ideal material for use within membranes systems used in filtration of desalination. PMMA is a highly versatile polymer with many uses across various industries—a strong, light weight thermoplastic that is transparent and chemically resistant to the vast majority of toxins and solvents. It is an excellent material for optical applications as lenses, because of

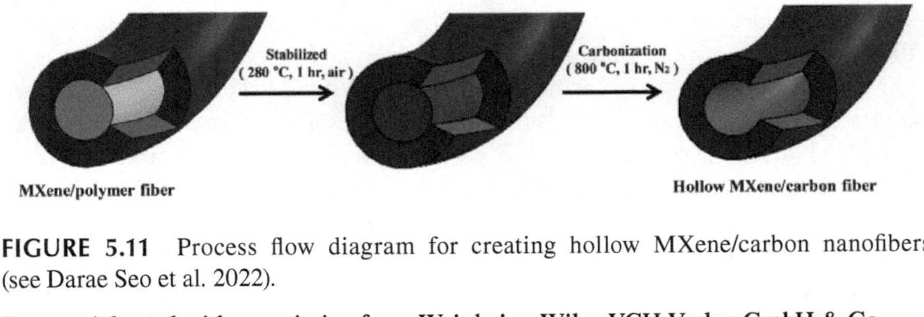

FIGURE 5.11 Process flow diagram for creating hollow MXene/carbon nanofibers (see Darae Seo et al. 2022).

Source: Adopted with permission from Weinheim: Wiley-VCH Verlag GmbH & Co.

its transparency to light, screens, or windows. Where its mechanical properties are perfect for prosthetics and medical implants, having such a low cost also makes it an amazing choice in the consumer goods market. I think potentially moving forward, MXene and PMMA can redefine the way materials are looked at and how one may consider applying them. And both have exciting applications—PMMA shines in the field of electrically conductive materials, while MXene's cheap, incredibly versatile nature makes it ideal for everything from self-healing batteries to energy-efficient lighting. Human ingenuity being the only limiting factor, hundreds of other possibilities will come out as far as future uses either material, and this could very well change how both these materials are made now to be used in the 21st century (Waseem et al. 2021; Ren et al. 2021).

5.3 APPLICATIONS OF MXENE-BASED POLYMER COMPOSITES STRUCTURES

Recent progress in materials science has made the process of thinking about materials and its properties completely changed. One fascinating outcome of this evolution is the development MXene–polymer composite architectures. MXenes are 2D transition metal carbides with excellent electrical, mechanical, and thermal properties. The unique properties of MXenes make it a promising material for diverse applications, including composite structures, capacitive energy storage, and, notably, electromagnetic interference (EMI) shielding. One of the main advantages is that MXene-based polymer composite constructions exhibit high comparison strength to weight ratio. These properties make them exceptionally stiff and strong while remaining remarkably lightweight. These make them ideal for an aircraft application, where weight is a major factor in performance and fuel efficiency. In the same vein, their electrical and thermal conductivities make them an ideal material for most applications that require the use of electricity as in electronic components (Dogan and von Recum 2020; Qingsong Zhang et al. 2021).

MXene is a relatively new member of the 2D material family, yet it uniquely combines both properties, exhibiting full metallic conductivity. It shows a remarkable potential in energy storage, sensing, electromagnetic shielding, and other properties. Given the large number of surface functional groups on MXene, the usage of polymers helps in tuning not only physical properties but also chemical functionalities which may play key roles toward enhancing local electric fields. Polymers can be added to guide the formation of MXene–polymer composites by expanding an interlayer space, decreasing distances for ion/electron transport and boosting surface hydrophilicity. Certain procedures to fabricate MXene–polymer composites, typified by the physical mixing method, surface modification (grafting) and functional groups' introduction via Ti-O-Si/Ti-O-P bonds as well as photo graft reaction and in-situ polymerization, are summarized here. Moreover, they include the potential mechanisms of each strategy.

S. He et al. (2021) stated that whereas silicone is a brand-new member of the 2D material family, MXene has both good metallic conductivity and hydrophilicity. It demonstrates potential in vast applications including energy storage, sensing, and electromagnetic shielding (as shown in Figure 5.12). This wealth of surface functional groups means that MXene possesses both physical and chemical attributes

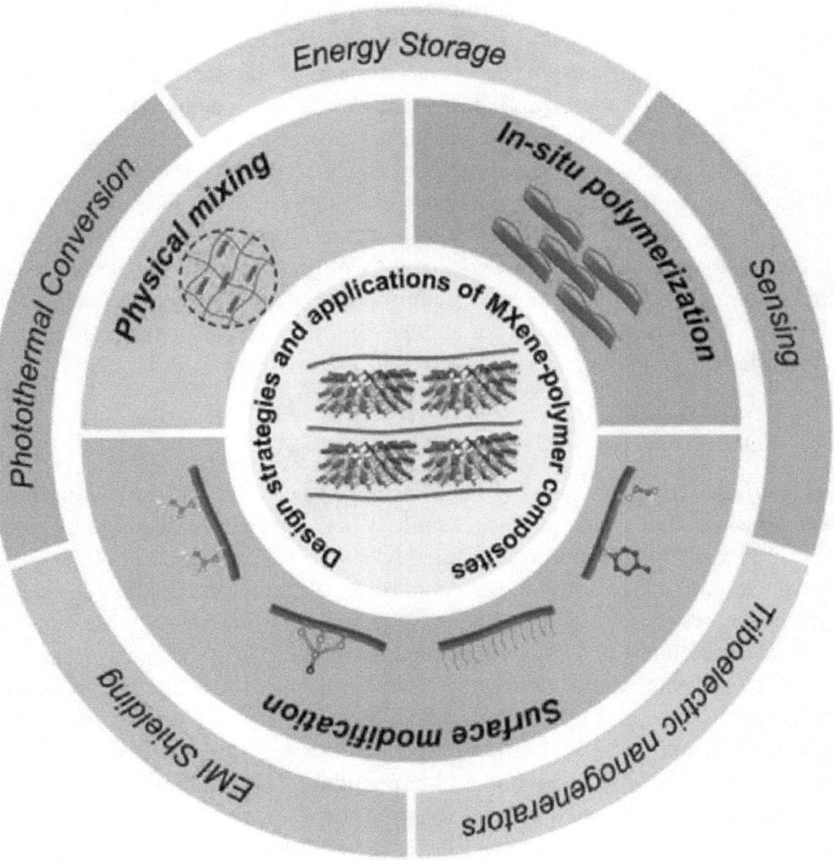

FIGURE 5.12 A schematic representation of MXene–polymer composite preparation methods and applications (see S. He et al. 2021).

Source: Adopted with permission from Weinheim: Wiley-VCH Verlag GmbH & Co.

that can be modified by the production of MXene–polymer composites. Properties such as increased surface hydrophilicity, decreased ion/electron transport distance, and an increase in interlayer spacing help to direct the formation of MXene–polymer complexes. This review discusses preparation methods for MXene–polymer composites. These methods include physical mixing, surface modification (anchoring: Ti-N or Ti-O-C; bonding: esterification; grafting functional groups: Ti-O-Si and/or – Ti-O-P), photograft reaction, and in-situ polymerization. It also explains what kinds of mechanics each strategy could have. Furthermore, the utilizations of MXene–polymer composites in an array fabrication methods are highlighted. Finally, this chapter concludes with both opportunities and challenges for solution processing of MXene–polymer composite architectures.

Materials composed of MXenes could also be used for medical applications. The electrical and thermal properties of the material can be utilized to manufacture implantable devices for controlling or monitoring medical conditions. It is very durable and, thus, can be used to build strong yet light weight prosthetic devices. Composite materials made up of MXene-based polymers, finally, are useful in the automobile industry. Their electrical and thermal properties make them an ideal material for making thermally efficient, light weight components. They can also create parts that are both strong and light, taking advantage of the material's strength-to-weight ratio. These can be applied to an entire car, allowing it to run lighter and emit fewer emissions as a result. With the technology still in its early stages, MXene-based polymer composite structures have a vast application space. As research and development continue, more applications of graphene are becoming possible. Developers of MXene-based polymer composite structures suggest the new material to be used across a range of future applications such as aerospace, medicine, and automobiles (L. Peng et al. 2022; Q. Wu et al. 2021; H. Wang et al. 2022).

5.3.1 Electronic-Based Detection

The electronic-based detection has been developed as an indispensable tool for security and medical diagnosis, etc. The technology makes it easily and quickly accurately detect things, chemicals, and events reliably. It is used in many fields like security, industrial automation, environmental monitoring, and also medical diagnostics. This review discusses the utility and prospects for electronic-based sensor technology. It is also essential due to its capability for rapid digital detection, precise object and material identification, and widespread acceptance in various potential applications. It is particularly useful in fields like medical diagnostics where time is of paramount importance. Electronic-based detection in medicine can provide a way to quickly determine how well (or not) our bodies are functioning, offering immediate diagnosis and treatment. It is also essential for security purposes as it helps in detecting and threat real-time intrusion detection. Furthermore, industrial automation framework allows for devices that monitor machines and procedures through making use of electronic recognition (Monsef and Salavati-Niasari 2022; Xueting Zhang et al. 2021).

Electronic-based detection has extensive applications. This technology has a wide spread acceptance in security, environmental monitoring, industrial automation, and medical diagnostics industries. For example, an electronic-based detection can be employed to unambiguously show that specific compounds are present in the body and provide identification of their presence as well as signify pathogenic diseases with fast precise diagnosis with the aid of medical diagnostic equipment. This can be also used in security systems to scan over time for intrusions and vulnerabilities. It can be used to prevent climate change by monitoring the quality of air and water to alert humans about changing environments. This is also used in the field of industrial automation to monitor machines as well as processes for secure and efficient operation.

Xin et al. (2020) stated that two-dimensional early transition metal carbides and carbonitrides (MXenes), revealing a new family of conductive metallic-like materials, have been discovered which are superior in some aspects to metals in terms of properties such as their good conductivity on par with conventional metals including mild ionic diffusion properties and mechanical flexibility due to less dense atomic layers, while the surfaces exhibit hydrophilic behavior caused by terminated functional groups like OH or O at oxygen-based sites. This tunable etching technique can be used to regulate the formation of nanoparticles, single-layer or multilayer nanosheets with high specific surface area beneficial for sensor applications and thus improve the sensing performance based on MXenes. MXenes can be made into composites with other materials very easily. Here, they propose that the MXenes or their composite with structural design provide superior mechanical stretchability and flexibility, which can be used as wearable sensors, energy storage devices, and electromagnetic shielding applications. The review details these new MXene properties focusing on their application in wearable sensors for pressure/strain, biochemicals, and temperature/gas sensing. Moreover, challenges and future work are discussed (Xin et al. 2020; Babilas et al. 2021; Min et al. 2021) (Figure 5.13).

Advanced electronic-based detection is expected to provide future developments and innovations. This technology already has an extensive array of applications—fast and accurate detection of items that are white, colored, or transparent; substances

FIGURE 5.13 An overview of MXene uses in the field of wearable technology (see Xin et al. 2020).

Source: Adopted under the Creative Commons CC BY 4.0 license.

present on surfaces in low concentrations like body fluids; material properties; scratches, e.g., labels for quality control, old wines, and antiquities. The approach is expected to be used in medical diagnostics, for example, to check for the presence of certain compounds in the body. Moreover, increased security fears in both public and private sectors are anticipated to inflate the use of electronic-based detection. Lastly, it is expected that this technology will be used to observe and control machines and processes as industrial automation program enters into the worldwide physical methods environment. In summary, electronic-based detection has become vital for most businesses. This technology has a number of applications in terms of providing fast, accurate, and robust item-, substance-, or event detection. The technology you can imagine is predicted to be used in a variety of industries including industrial automation as well as security and medical diagnostics. With the advancement in technology, electronics-based detection will play a vital role (D. Feng et al. 2021; Yoon, Mun, and Lee 2021; Z. Feng et al. 2022).

5.3.1.1 Gas Sensing

Over the last few years, MXene and titanium-based gas sensing have taken prominence among industries as well as chemical sectors. It is because these materials have opitimized properties than the conventional gas sensing voltage electrodes. MXene is a particular type of two-dimensional (2D) material with interesting physical and chemical properties, making it an appealing candidate for gas-sensing purposes. This is where its good thermal stability, wide bandgap, and highly conducting nature make it a perfect choice for gas-sensing application. Further, titanium-based gas sensors have been proven to be highly sensitive and selective, making it a reasonable choice among an increasing number of people. Although the MXene–titanium composite gas sensors designed in this study are highly practical for a broad range of applications within chemical or industrial settings. They can be used to detect many gases, for instance: Carbon dioxide, nitrogen dioxide, ozone, volatile organic compounds (VOCs). In an industrial setting, the sensors can detect dangerous chemicals; within homes or offices, they could monitor air quality and measure concentrations of VOCs in the environment. Again, this gas sensor can utilized to monitor concentration of harmful gases in confined places and can also be useful as a water quality control tool for checking the cooccurring hazardous compounds. MXenes exhibit a wide range of gas recognition capabilities, high sensitivity, and low power consumption. They can be easily deployed in various settings and are cost-effective. Additionally, they rarely produce false positives, making them highly reliable (Deng et al. 2021; H. Xiao et al. 2021; J. Liu et al. 2021).

One of the most promising sensing electrode materials for making flexible gas sensors is MXene, which has been widely applied. As the high BET area of 2D MXene Ti_3C_2Tx nanosheet is inherent to prone stacking, it limits their usage for gas sensing applications. Yang et al. demonstrate an ultrasonic spray pyrolysis (USP)-prepared three-dimensional (3D) crumpled MXene sphere-based flexible NO_2 sensor working at room temperature. Both of these features are further enhanced by constructing a 3D crumpled MXene sphere/ZnO composite, which provides an abundant specific surface area and active sites as the sensitive electrode. For 100 ppm NO_2, the response signal improves from 27.27% to 41.93%, and the recovery feature increases significantly from 30% to an impressive 100%. This demonstrates superior

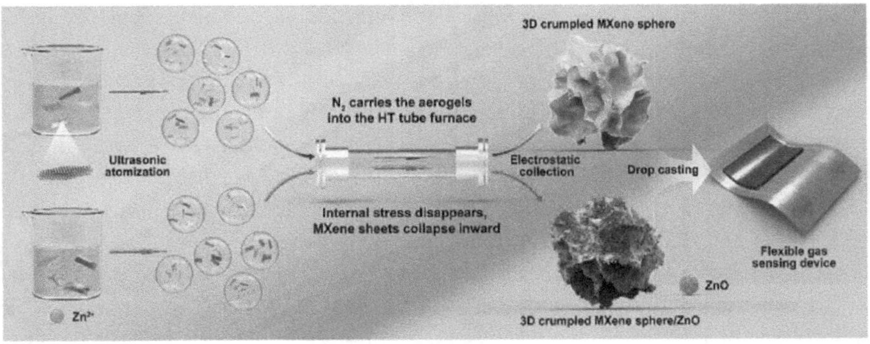

FIGURE 5.14 The schematic representation of the fabrication of the flexible gas sensing device, 3D crumpled MXene sphere, and 3D crumpled MXene sphere/ZnO (see Z. Yang et al. 2021).

Source: Adopted with permission from Elsevier.

selectivity and performance in flexible sensing devices compared to previous studies, highlighting its preferential response toward NO_2. The enhanced surface area, large number of edges and defects from folding, and MXene/ZnO p-n heterojunction formation add to better NO_2 sensing performance for 3D crumpled MXene sphere/ZnO. The work by Z. Yang et al. (2021) has inspired future improvement of the gas-sensing performance of MXenes and laid a foundation for the understanding of the sensing interaction between MXene and target specific gas (Figure 5.14).

Both the chemical and industrial sectors are expected to maintain a high demand for MXene as well as titanium-based gas sensors. As new materials and technology continue to be developed, these sensors will only become more accurate and responsive. This has obviously also opened up these sensor types for more use in autonomous vehicles, home/office automation and in a slew of other scenarios. This, along with novel methodologies such as integration of sensors based on titanium and MXene into machine learning sociable gas sensing schemes, will allow the creation of smarter, efficient gas-sensing systems. Meanwhile, MXene and titanium-based gas sensing has attracted enormous interest in recent years due to their attractive and superior properties compared with conventional gas-sensing materials. Such materials find wide-ranging industrial and chemical applications, and their uses are expected to grow in the next few years. This approach will also be strengthened through the development of novel materials and technologies, enabling more intelligent yet reliable gas-sensing systems in which transducing elements, namely sensors, become increasingly sensitive (T. Yu et al. 2020; D. Peng et al. 2022).

5.3.1.2 Single-Molecule Sensing

MXenes and titanium-based single-molecule sensing are two of the most promising, frontier technologies in nanoscience. The ability to sense, count, and track the molecules and atoms in our environment by using these two innovations could revolutionize their identification. The MXene family refers to the two-dimensional (2D) transition metal carbides, carbonitrides, and nitrides with a single layer of transition metal atoms sandwiched between two layers of carbon atoms. Their remarkable electrical, mechanical, and chemical characteristics make them ideal for use in numerous applications ranging from electronics/sensors to catalysts/energy storage. These films have been studied and applied in sensing applications for the past few years due to their unique properties. Because of their large surface area and high-surface reactivity, MXenes have been demonstrated to sense a wide array of compounds ranging from gases, chemical molecule, ions, and proteins. They are also ideal for making electrochemical sensors with high electrical conductivity (Shu et al. 2022; Fan et al. 2020; Finkler et al. 2021).

A further burgeoning technique in nanoscience is with titanium as a platform for single-molecule analysis. In single-molecule sensing, tiny titanium nanoparticles are utilized for both compound identification and quantification down to a single molecule. These are functionalized with different molecules like antibodies and enzymes to detect a specific target molecule of interest by the nanoparticles. Then, the particles are coated with a special polymer acting like an insulator and preventing nanoparticles from interacting with their environment. Therefore, they could pick out a solitary molecule with zero interference from other molecules in their environment. There are several applications of MXenes and single-molecule sensing based on titanium. This type of technology can be leveraged in the healthcare industries to detect and measure a myriad of chemicals which include glucose, cholesterol as well as proteins that are associated with cancer. For instance, they can be applied in environmental science to locate and measure gases like carbon dioxide or methane as well as organic substances such as pesticides. Moreover, the food industry can use this technology to monitor and scrutinize its product quality and safety (Qadir et al. 2021; Cano-Pleite et al. 2021; H. Wu et al. 2021).

Perhaps, one day label-free nanopore technology will be produced that is sufficiently inexpensive and fast that can accurately differentiate a simple invention of biopolymers like DNA/RNA (Figure 5.15) (Mojtabavi et al.). Since their inception nearly two decades ago, solid-state nanopores have evolved into chemically and mechanically tunable sensors emulating biological channels, where the translocation of biomolecules including sequencing is now possible. Two-dimensional (2D) materials would be ideal candidates for nanopore substrates of sequencing detectors in that they are expected to provide the ultimate 1-nm resolution per biopolymer base readout and an avenue toward electronic multiplexed reading by taking advantage of their unique electronical properties. This work delves into the fundamental limits of thinness for these materials by demonstrating their self-assembly at a liquid–liquid interface. Furthermore, it explores focused electron-beam (FEB) patterning techniques and the formation of nanoscale pores within ice crystals. Subsequently, electronic measurements are conducted within size-defined nanopores to evaluate their functional properties (Mojtabavi et al. 2019).

SSEs can be expanded in the future with MXenes and titanium-based single-molecule sensing. These technologies can be used in the future to create new drugs, identify diseases, track molecules in biological systems, and much more. Researchers are also looking at potential applications of these technologies in the field of robotics and artificial intelligence. Two of the hottest issues around MXenes and single-molecule titanium-based sensing in their respective area are summarized. They have multiple practical applications in the realm of healthcare, environmental science, and even food business operations; they can literally emerge as a novel way for how

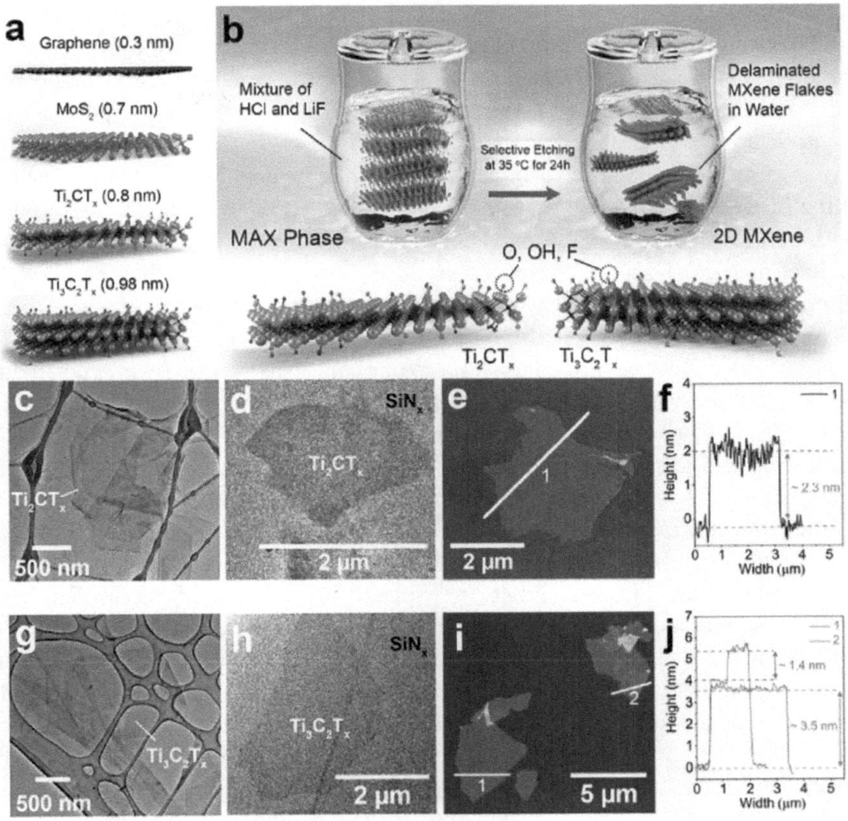

FIGURE 5.15 Preparation and characterization of MXene: (a) Thickness of representative 2D materials reported in this and previous studies on nanopores. (b) Schematic overview of the route from MAX phases and crystal Ti_2CT_x as well of Ti_3C_2Tx structures to MXene 2D flake etching/dispersion (b). (c) Dropping of Ti_2CT_x flakes on a holey carbon-sheet-supported grid. (d) Drop-casting the Ti_2CT_x flakes onto an Si chip. (e) AFM image of drop-cast Ti_2CT_x fine powders on a SiO_2 substrate. (f) The height profile corresponding to the flakes in AFM image. (g) Ti_3C_2Tx flakes drop-casted onto a TEM grid with holey carbon support. (h) Drop cast of Ti_3C_2Tx flakes on an Si chip. (i) AFM image of the Ti_3C_2Tx flakes and (j) height profile (see Mojtabavi et al. 2019).

Source: Adopted with permission from Elsevier.

we identify, quantify, and measure molecules/atoms around us. The future for these technologies is very promising, and it is expected that they will be implemented in a broad array of usages sooner rather than later (Saqib et al. 2021; Kayahan et al. 2021).

5.3.1.3 Humidity Sensors

Humidity sensors are key components for many modern technologies, from industrial applications to consumer goods. Conventional sensors that rely on resistive or capacitive sensing technology have been in operation for many years, but newer materials like MXene and titanium-based sensor are beginning to be increasingly utilized.

These sensors, how they can be used, and the potential for them in the future will also be covered here. Why humidity sensors are so important is for everyone to see. These sensors are utilized in the industrial field for safety as well as process control and monitoring applications. On the consumer end, they are used for good condition-monitoring for homes and air-conditioning/heating systems. Humidity sensors are also used in the medical environment, where they help to ensure that patients are taking medication correctly by using asthma inhalers as an example. On the other hand, humidity sensors based on MXenes outshine conventional sensors by a multitude of metrics. They are cheaper to produce as well because of their simpler construction and need for fewer parts. They are also more accurate and sensitive, hence the high level of precision. Third, because they are less impacted by external influences such as temperature and moisture, so one can easily rely on them. Fourth, they can be used in a variety of products because they could be embedded into smaller packaging (L. Zhou et al. 2022; "Outside Front Cover" 2022; C. Wang et al. 2022).

Flexible and sensor electronics have been attracting great interests, for which a number of 2D materials-based candidates are being investigated. Wang et al. reported the first self-powered flexible humidity sensing device based on monolayer molybdenum diselenide (MoSe2) Piezoelectric Nanogenerator (PENG)/PVA/MXene nanofibers film, as shown in Figure 5.16. The monolayer MoSe2-based PENG was fabricated using atmospheric pressure chemical vapor deposition (CVD) methods, which had an output voltage up to 35 mV and a power density of 44 W/m^2. As a result, it yield very attractive application possibilities in the field of wearable technology as flexible PENG is mounted on polyethylene terephthalate (PET) substrate and it can harvest energy generated by different parts of body. The humidity sensor, fabricated from electrospun PVA/MXene nanofibers on a flexible PET substrate and powered by a monolayer $MoSe_2$ PENG, exhibits a high response of ~40, a rapid response/recovery time of 0.9/6.3 s, a low hysteresis of 1.8%, and excellent repeatability. The self-powered flexible humidity sensor can be used to sense both ambient and human skin moisture. In this work, the research team investigated the possibility of using PENGs as a self-potential source for humidity sensors in flexible

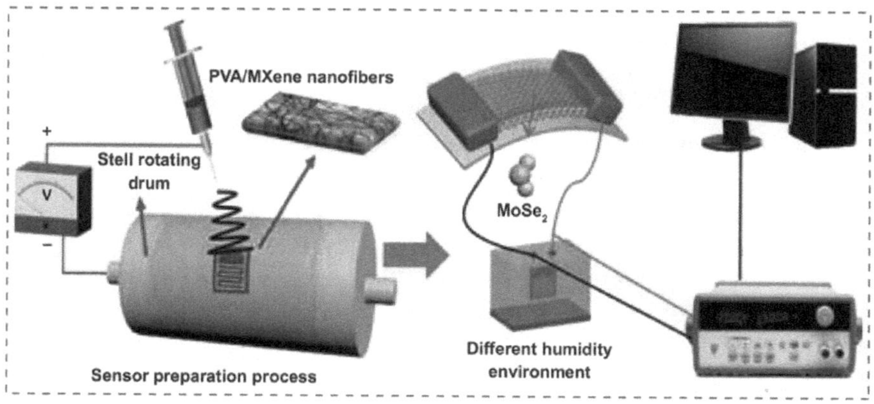

FIGURE 5.16 Humidity sensor built with MXene (see D. Wang et al. 2021).

electronics by integrating them with a high-performance-auto-enabled-humidity sensor (Figure 5.16) (D. Wang et al. 2021).

Humidity sensors find a wide range of applications. They are used in the preparation of food, drugs, and even chemicals industrially. They are also used in the manufacture of computers as well as within the automobile industry. They are used in consumer products like dehumidifiers, air conditioners, and other home equipment. The medical industry uses humidity sensors to measure the levels of moisture that exist in homes, as well as for making inhalers and other such medical equipment.

Titanium- and MXene-based humidity sensors may have promising prospects. As they become more widely available and cost-effective, these sensors will be used in an increasing number of applications. Also, they can be used in risky and high-temperature environments to record humidity levels in the commercial sector. They are used in air-conditioning and heating systems and home humidifier control. They can also be used for measuring humidity level in inhalers and medical devices.

Conversely, the MXene and titanium-based humidity sensors are advantageous relative to traditional sensors. They also fit well in small packages and are a little more precise, as they can be duplicated with more certainty. Because they are also more inexpensive, they can be a valuable solution for lots of applications. Once generally available, these sensors will be increasingly used in the industrial sector as well as by consumers and in healthcare. These sensors have a bright future and may the areas of revolutionize process control and safety/security/comfort (Pedraza et al. 2021; Rajabalinia et al. 2021).

5.3.1.4 Pressure Sensors

Pressure sensors are used to measure the pressure of a medium (usually fluids). This kind of sensors has found applications in different processes from the monitoring atmospheric pressure to industrial purposes. The recent popularity of these types is attributed to their high sensitivity, robustness, and low cost for producing flexible electronics using pressurized titanium and MXene-based materials. This section will focus on the history and significance of MXene pressure sensors, manufacturing processes, uses in different fields, and advantages along with disadvantages and prospects for titanium-based pressure sensors.

MXene is a two-dimensional (2D) material formed by the debut layers of carbons sandwiched between transition metal atoms. The data was first announced in 2011 by the University of Pennsylvania. In contrast, titanium is a transition metal known for its ability to leverage its properties meaningfully. For example, its strength and lightweight nature make it particularly useful in applications such as high-pressure environments. The Ti-6Al-4V alloy is a common choice, as it maintains excellent corrosion resistance and high-temperature performance while demonstrating superior mechanical properties. Over the past few decades, scientists have looked for ways to manufacture pressure sensors with titanium and MXene.

The MXene and titanium pressure sensors will have a number of advantages over traditional types. The results are durable, cost-effective, and very lean. Conventional pressure sensors often employ polymers or similar materials whose properties degrade through time in the presence of environmental factors like temperature and humidity. In contrast, systems using MXene and titanium pressure sensors are sturdier, offering more accurate outputs. They are also less expensive than traditional sensors, which may benefit buyers shopping for a cheaper solution (Z. Xiao et al. 2021; Sindhu et al. 2020).

The suitability of a good number of 2D functional materials was checked for many applications like energy storage devices, electronic devices, and sensors. Pressure sensors are often used in robotic electronic skin applications, vehicles, industrial automation, and healthcare to name a few amidst various other types of sensors. There are multiple ways of thinking about the function a pressure sensor fulfills. The piezoresistance principle is more appropriate for 2D materials-based pressure sensors. MXenes are the newest 2D materials classes that have been studied for a wide range of applications in recent years. MXenes exhibit a piezoresistance effect due to the smaller lattice interlayer distance, larger surface contact area, and greater strain under pressure. Sharma et al. noted a flurry of activities on RPS-based MXene for potential healthcare applications in the recent years by academicians (Figure 5.17). A comprehensive review of the latest literature on MXene-mediated RPS for body language sensing and physiological signals monitoring (e.g., heart rate, blood pressure, respiration rate) is rendered here. Vital indicators reflect the state of homeostasis in the body. It has described functional materials developed, numerous synthesis approaches for different MXenes, and several sensor fabrication geometries including matrix array sensors as well as with/without spacer layers. The above-highlighted vital-signs-monitoring performances of MXenes and their composites-based RPSs are compared with other 2D graphene materials for an immediate understanding (Sudeep Sharma et al. 2020).

These titanium MXene-derived pressure sensors can be made in different manners as well. Over the last year or over past years, exfoliation of Van der Waals material has been one research interest as there are many methods created continuously to simplify and facilitate this process. The dielectric layer, being the MXene layer, may detect changes in pressure. Other techniques include sputtering, photo lithography, and electroplating (S. Choi, Vazquez-Duhalt, and Graeve 2022; C. Yang et al. 2021).

MXenes and titanium are suitable for use in pressure sensors. These include industrial applications in automotive, aerospace, and oil and gas industry. These also have usage for medical purposes like monitoring changes in body pressure. Pros and cons are that titanium-based material is suitable to detect ultralow pressure whereas the MXene-based material takes advantages of conductive electronic structure but its drawbacks include insufficient mechanical strength. The greatest advantage they have is that they are very delicate and durable. These are able to provide accurate readings in a number of environments. Conventional sensors are also more costly. These sensors have one significant disadvantage: They require more intricate production methods for their conventional sensor counterparts. The sensors incorporate a thin MXene layer, which as demonstrated by the team's previous work, is susceptible to damage from environmental factors such as temperature and humidity. MXene-titanium pressure sensors could change the game. Because they are incredibly tough and fragile, the use of them in the world is ideal for many different applications. In the future, researchers will continue their search for even cheaper and more cost-effective sensor solutions. Another area that is being worked on are sensors with better immunity to environmental influences. Consequently, pressure sensors consisting of titanium- and MXene-based materials may allow the market

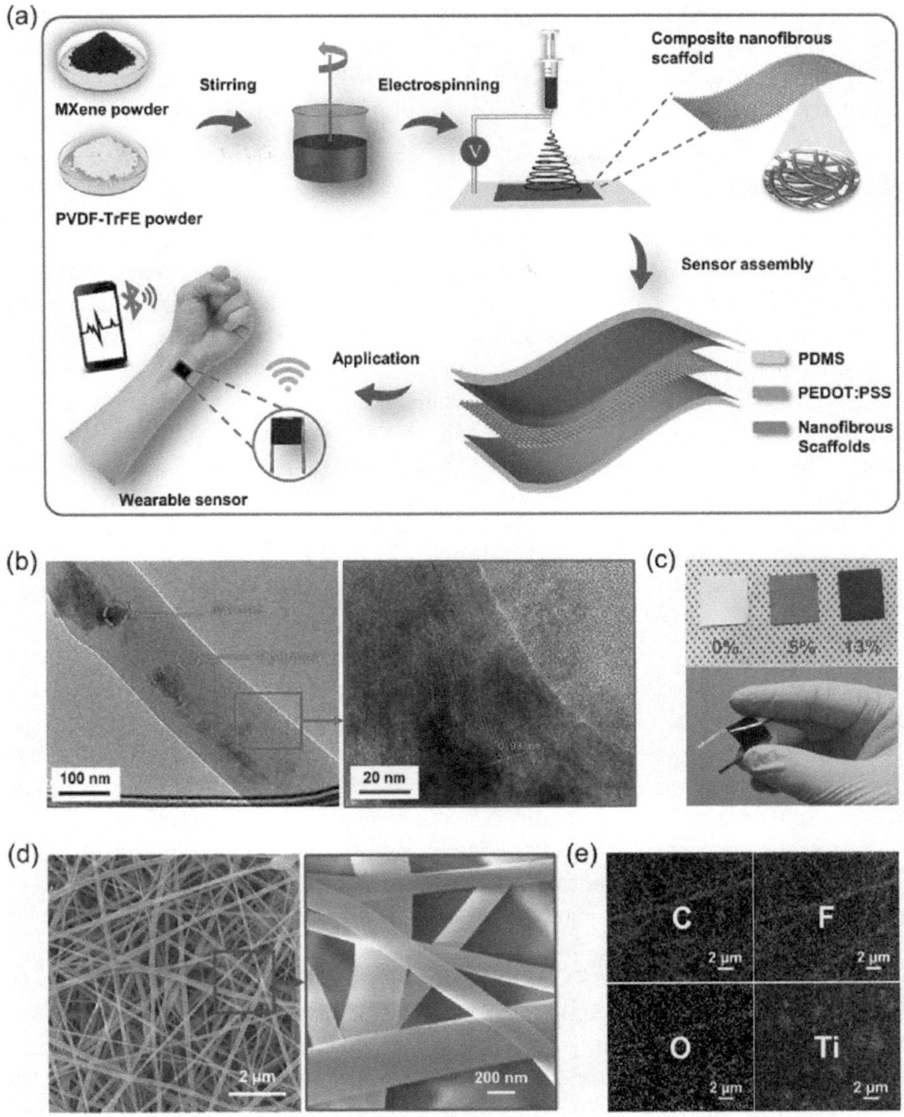

FIGURE 5.17 The structure fitting the CNS base pressure sensor and its production process. (a) Schematic representation of the fabrication process for CNS-based pressure sensor in a typical (CNS-DMSO) solution + PEO film. (b) TEM picture of central nervous system showing single and multilayer MXene nanoflakes. As shown in the inset of a high-resolution TEM image, when MXene is typically with an interlayer spacing of 0.93 nm determined from the lattice plane (002) franca. (c) Images of the CNS with different concentrations of 2D MXenes and manufactured sensor. (d) Field emission scanning electron microscopy (FESEM); inset: Magnified image particles, hatched region is expanded by FESEM. (e) Composite nanofiber EDS maps: C, F, O, and Ti (see Sudeep Sharma et al. 2020).

Source: Adopted with permission from American Chemical Society.

new opportunities on how to sense pressure. They are very durable, affordable ones, and responsive. Moving forward, researchers will continue to strive to make such sensors more affordable and cost-effective. They will also investigate improving the detection of sensors in various conditions (L. Huang et al. 2021; Ankit et al. 2021).

5.3.2 Optical Sensors

In recent years, MXenes and titanium-based optical sensors have demonstrated significant promise in the area of sensing technologies due to their potential for environmental monitoring, medical diagnostics as well as industrial automation applications. MXene is a 2D material composed of transition metal carbide and/or nitride layers with the honeycomb structure unlike traditional titanium-based optical sensors based on TiO_2 nanostructures. The MXene and titanium optical sensors are highly sensitive by virtue of their large surface area. They are able to sense almost any changes in their environment with great precision, down to a difference of pressure or simply the temperature. With their corrosion resistance, they are more than just a robust sensor for use in countless applications. These sensors are very easy to build and fairly inexpensive.

The formation of titanium- and MXene-based optical sensors can use various techniques: Chemical vapor deposition, atomic layer deposition as well physical vapor deposition [PVD]. In Figure 5.17, CVD is employed to deposit MXene thin films, while a titanium-based ALD/PVD process is used to create nanostructures on substrates. The resulting sensors are thus very sensitive and accurate within a wide range of systems (Mansoor and Chen 2022; H. H. Chen et al. 2022).

They are used in industrial automation, medical diagnosis, and environmental monitoring applications. These sensors sense the changes in temperature, air quality, and other environmental conditions. This is applicable to medical diagnostics, where sensors are capable of recognizing changes in blood glucose levels or other indicators in physiological fluids. The changes in pressure, temperature, and others processed variables changes can tracked with the help of these sensors in industrial automation (Chodankar et al. 2021; Guan et al. 2020).

Fiber-optic biosensors are so sensitive that they can even detect the smallest biological molecules in a challenging environment. Because of their metallic conductivity, hydrophilic surface, large specific surface area, and very wide band optical absorption properties, two-dimensional (2D) MXenes were recently proved to have significantly enhanced sensitivity in biosensing, gas sensing, and humidity sensing. Two fiber-optic biosensors are presented to directly compare the sensitivity enhancement achieved through MXene modification with that of previously reported results. For the first time, experimental evidence substantiates how 2D MXene Ti_3C_2Tx enhances the sensitivities of fiber-optic biosensors. Chen et al. reported that refractive index (RI) detection can be used to quantitatively assess the sensitivity of biosensors in terms of their ability to detect biological molecules (Figure 5.18). Not just any detection, but RI detection that quantifies the enhancement in performance our sensors get. A fiber-optic RI sensor exhibits a highly sensitive enhancement of greater than eight times and an FP SPR sensor with around 30% sensitivity curve shift (even higher curves shifts could be obtained) for the detection of almost all trace biological species. The latter can work very well in a wide range of rates simultaneously, whereas the former have to change their operating scales based on

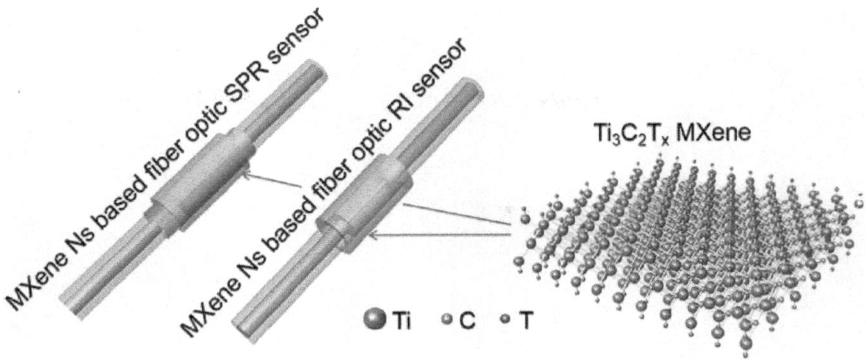

FIGURE 5.18 Schematic representation and performance of MXene nanosheet-based fiber optic sensors: Surface plasmon resonance (SPR) sensor and refractive index (RI) sensor (see Y. Chen et al. 2020).

Source: Adopted with permission from American Chemical Society.

detection configurations. The results may have important implications for the future development of MXene-based optical sensing technologies (Y. Chen et al. 2020).

The potential of titanium- and MXene-based optical sensors is enormous. These sensors can be applied to a wide range of applications, from environmental monitoring and medical diagnostics to industrial automation with additional research and development. Similarly, fabrication processes and materials developed for these sensors can be improved to better sense the various health parameters. If developed the right way, these sensors could revolutionize the sensor technology world (Y. Song et al. 2022; Tan et al. 2021).

5.3.2.1 Fluorescence-Based Detection

A novel and promising method, which is named as MXene/TiO_2 fluorescence-based detection technology for its application in the biology area and optoelectronics, is still at its infancy. MXene is a 2D nanomaterial that has increased important physical and chemical characteristics—conductivity and optical transparency in combination with low cost of fabrication among others. It is also used as an coating on the surface of titanium for making a type of photosensitive film, and it makes TiO_2 in titania or photocatalytic surfaces. This photocatalyst is compatible with a wide range of compounds—from proteins and nucleic acids to small molecules, for which their fluorescence signals can be observed. This titanium-based photocatalyst could have wide-reaching applications to show improved detection sensitivity compared with other fluorescent biosensor methods. The titanium-based photocatalyst can also be prepared in a facile and low-cost way, indicating that it has great merits for wide applications.

One can detect proteins and other nucleic acids in real time using it. It might help in identifying disease markers, for example, cancer biomarkers more accurately and effectively. Moreover, this technique can be used for the detection of toxins in food, drink, etc., as well as to find nano molecules present in polluted environment (Guerra-Rodríguez et al. 2021; Shi et al. 2021).

Xie et al. (2022) reported that two-dimensional MXene polymer was used to construct not only naked-eye visible but also fluorescence-off probe toward Ag^+

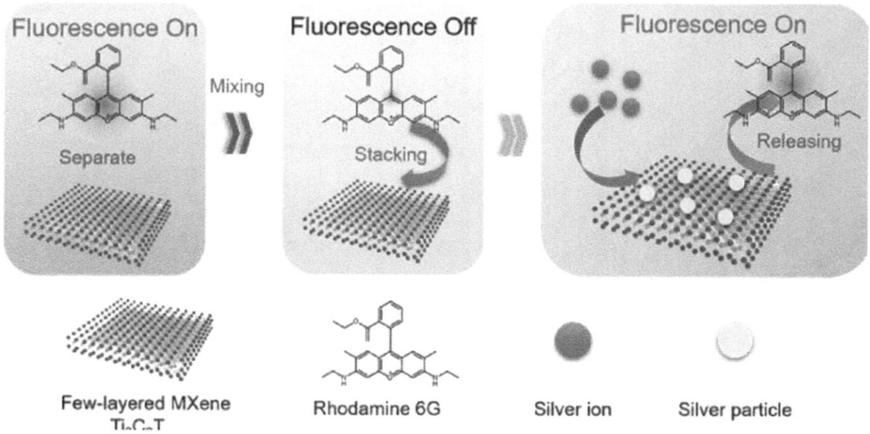

FIGURE 5.19 Fluorescence-based MXene-based detection (see Xie et al. 2022).

Source: Adopted with permission from American Chemical Society.

detection. In this work, the decrease and growth of silver ions on the MXene surface were investigated, as well as the interaction between MXene and rhodamine-6G as given in Figure 5.19. Furthermore, the transmission electron microscopy and X-ray photoelectron spectroscopy characterization supported our hypothesis about this mechanism. This probe can be used in the detection of Ag+ ions living cells and has a sensitive, selective real-time response to Ag+ in aqueous solution, with 0.035 M being the limit of detection (LOD) for this method. The selectivity of the probe was further verified by comparing its fluorescence response to silver ions with that of other coexisting metal ions. The selectivity of the probe by the comparison on fluorescence response to silver ions between coexisting other metal ions has further verifies. More importantly, this probe can be applied to measure and observe the concentration of Ag+ intracellularly. These results provide new insights into the detection and characterization of silver ions in environmental, as well as biological samples. In the meantime, this method provides recommendations for developing MXene-based sensing platforms (Xie et al. 2022).

There are many advantages of MXene and titanium-based fluorescence-based detection. It is inexpensive, highly sensitive, and easy to fabricate. This is just one property of the photocatalyst that makes it versatile and useful for several processes. The only disadvantage of this new technology is that it is not as sensitive a method for fluorescence detection relative to some competitors like mass spectrometry.

In the future, MXene and titanium-based fluorescence-based detection will still be important to many industries. Therefore, it has the advantage of also providing a real-time practical alternative to many businesses due to its new low-cost feature with very high detection sensitivity, and continuously researchers are trying their best in order to reach more sensitive probes as well. This technology is poised to literally change the way that the diseases are identified, tracked, and diagnosed as well as how these environments pollutants are identified and tracked (Rajput et al. 2021; Shen et al. 2021).

5.3.2.2 Detection of Heavy Metals Ions

A newly discovered 2D transition metal carbide called MXene has emerged as an exciting material platform due to its remarkable surface properties such as high electrical conductivity and excellent corrosion resistance. It has the potential for application in heavy metal-ion detection due to its unique properties. It has been also explored for its characteristic as a heavy metal-ion detector, due to the excellent selectivity and sensitivity of titanium ions toward most weightier metallic ions. The analysis of MXene and titanium for the detection of metal ions is attractive. Therefore, MXene is a good candidate for fabricating titanium-related nanostructures and would provide excellent support to modify the performance of Ti nanogrooves inhibiting much more sensitivity with high selectivity. Electrochemical deposition is widely employed for making such nanostructures, followed by defined shape and size control. Both, in combination, have proved to increase the selectivity for target ions and decrease the detection limits of heavy metal ions. Compared to present methods, this is an intriguing possibility, especially for the detection of heavy metal ions. For example, using titanium to couple with MXene is a new type of environmental monitoring method that can be used for detecting heavy metal ions in water and soil. However, manufacturing costs and nanostructure stability both remain obstacles. The researchers also proposed that the combination of MXene and titanium-based heavy-metal-ion-sensing materials could be employed to develop future solutions for environmental monitoring as well other applications (Y. Zhong et al. 2020; Xue et al. 2021).

The development of new sensing structures with high selectivity and, mainly, great sensor ability is the principal task in environmental detection. Recently, Rizwan et al. prepared some of these materials from nanostructured matrices for improved sensing task in biosensors. Considering all those exciting properties like high surface area, good thermal resistance, significant hydrophilicity, unique layered structure topology, and, the most important of them, environment-friendly nature, MXenes are thought to be more suitable for biosensor development among them (Figure 5.20)

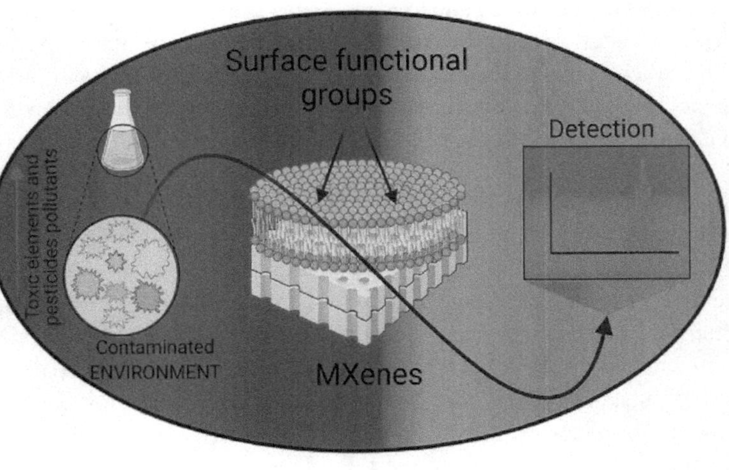

FIGURE 5.20 Heavy metal detector sensor based on MXenes (see Rizwan et al. 2022).

Source: Adopted with permission from Elsevier.

(Rizwan et al. 2022). For those aforementioned extraordinary properties, MXene-based materials have become a promising candidate for catalysis, energy storage devices, electronics as well environmental sensing and remediation purposes. It provides an overview of the most updated and advanced electrochemical/biosensing technologies associated with MXenes used for an analysis of toxicants, drug residues combined, and pesticide contaminants in environmental matrices. It starts with the surface functionalization and modification of MXenes. This has since analyzed the use of MXene in constructing electrochemical (bio)sensors for detecting various environmental toxins, creating a sensation. In addition to this, it delineates the current problems in this realm and illustrates potential solutions as well as future directions (Sizhe Li et al. 2021; Yan et al. 2020) (Rizwan et al. 2022).

5.3.2.3 Detection of Biomolecules (Small and Micromodules)

Recently, a lot of attention has been paid to two-dimensional materials called MXenes due to their unique properties such as strong conductivity, chemical inertness, and excellent biocompatibility. It is an appealing material to be used in the detection of macro- and less small-scale-size biomolecules. The electrical and optical properties of MXene make it useful for detecting biomolecules including DNA, proteins, or antibodies. The realization that such sensors can be produced with demandingly high sensitivity and selectivity incentivized the coupling of Ti_3C_2 MXene family members to titanium-based materials for biomolecule detection. These sensors could be created by means of electrochemical deposition, atomic layer deposition, and physical vapor deposit methods. These methods can produce highly sensitive and specific sensors that are relevant to a variety of biomolecule-sensing applications, from environmental monitoring to food safety as well as medical diagnostics. The MXene and titanium-based sensor is cheap, ultrahigh sensitive, and fast responsive. They are sought after for use in a variety of applications as they can be easily fabricated. However, they are limited requiring sample pre-treatment and offer moderate selectivity. According to recent findings, efforts in the fabrication of titanium-based MXene sensors remain limited due to inadequate porous materials. Further research is essential to better understand specific detection mechanisms at the nanoscale level. The challenges faced for innovating advanced sensors using similar chemical building blocks could pave a novel pathway to realize new horizons in biomolecule detections (Cong Chen, Hang, et al. 2021; Zhenkai Wang et al. 2021).

5.3.2.4 Calorimetric-Based Sensing Materials

A new family of materials was discovered for wide sensing applications—MXene and titanium-based calorimetric-sensing materials. This material comprises two constituents: Titanium and MXene. Titanium is a strong and lightweight metal and is used in making consumer goods, medical implants, and space shuttle. MXene is a very promising two-dimensional material composed of transition metal carbides and nitrides with outstanding electrical and thermal properties. The combined working of both these substances helps to produce a low-cost high-sensing substance.

These materials originated from their development and potential use as sensing materials, specifically for MXene-based applications. MXene, which comprises carbides and/or nitrides of transition metals (most notably titanium), was first reported in 2011 by researchers at Drexel University. The material is the best among those that have been found so far for this applications because it combines exceptional electrical and thermal properties, in addition to having a large surface area (Naguib et al. 2011).

The value of interest in these materials lies, thus, for their ability to register temperature variations. Due to their ability to sense minute temperature changes, these materials are ideal for applications including medical diagnostics, automotive sensors, and industrial monitoring. The first materials were significantly more versatile as they could also be used to measure changes in humidity and pressure.

These materials are processed by the fabrication processes through a combination of titanium and MXene. In general, the procedure of combining these two materials is called chemical vapor deposition (CVD). This process produces a vaporized layer of titanium, which is then transferred to MXene. The thin layer is then patterned and etched to form the proper shape design of sensing material. These materials are capable of high-volume mesoscale manufacturing owing to their simplicity and low costs (Park et al. 2021; Bae et al. 2021).

Ammonia (NH_3) has been described as a simple but important molecule in many applications such as agricultural and automotive industry, chemical processing including food and pharmaceuticals industries (hydrogen production and storage), and biomedical usages (Figure 5.21). NH_3 emissions must be monitored, and suitable safety measures are required as its large-scale industrial use automatically leads to environmental risks, posing a potential global threat to public health. These factors have created a demand for novel and sustainable approaches to development of efficient, low-cost/reliable small NH_3 sensors. Some weaknesses in available NH_3 sensors are their poor selectivity, low detection at a concentration, and room-temperature setting. To overcome the above problems, modern nano-systems with fast response and detection, increased selectivity as well as efficiency, liquid NH_3

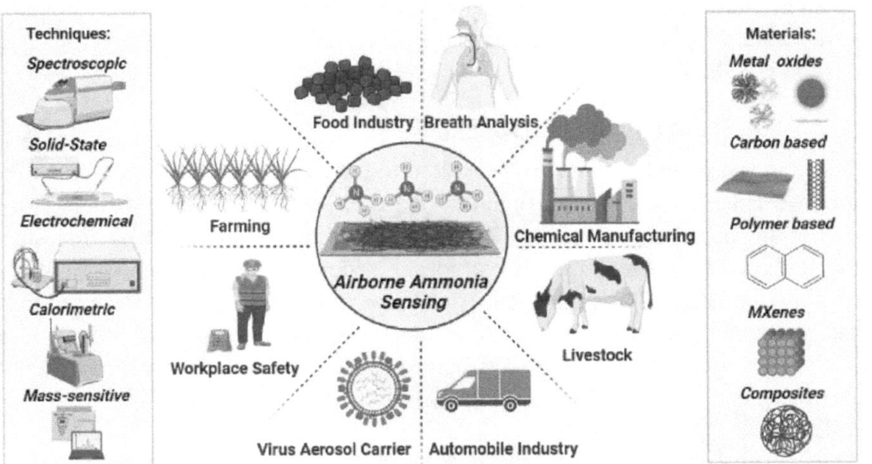

FIGURE 5.21 Modern NH_3 sensing methods represented schematically using various sensing signals (see Chaudhary et al. 2021).

sensors are being developed. These attributes position MXene–polymer nanocomposites (MXP-NCs) as a favorable candidate for NH_3 sensing applications mainly due to their low cost, high conductivity, mechanical flexibility and processability in thermoplastic fashion with affordable scalability, and system-adjustable physical properties. The MXP-NCs are shown to be very promising in developing a new class of highly sensitive, realizable NH_3 sensors for industrial, biomedical, and agricultural applications. However, they do not review their superior NH_3-sensing performance. In this context, the current comprehensive review by Chaudhary et al. (2021) discusses up-to-date approaches to produce MXP-NCs in high-performance models tuned according to specific properties' development and enhancement of sensing response changes ideal for detecting airborne NH_3. In addition, the current problems associated with MXP-NCs and our expectations for future improvement of MXP-NCs as well are summarized (Chaudhary et al. 2021).

They are used for all kinds of things. They are used in automotive sensors, engine temperature and emissions medical diagnostics, and for detecting changes in temperature/pressure/humidity. They can be also implemented to monitor temperature, pressure and humidity changes in the working spaces. Some common devices using these materials include temperature-sensitive apparel and thermostats.

Different materials found in each type have pros and cons for a given application. These materials are also very sensitive and cheap, and they can be used for sensor applications. However, lacking wider acceptance and still relatively new in some applications, they are not perfect. In another use cases, they may not be as powerful compared to other materials.

There is a bright future ahead for these materials. If MXene and titanium-based calorimetric-sensing materials are validated as a result of research, novel applications/improvements in performance are expected. The challenge for researchers, then, is how to improve both the reliability/resilience of these materials and their sensitivity. Additionally, these materials are expected to become more affordable in the future due to manufacturing technology advancements (C. Y. Tang et al. 2021; Jun Li et al. 2021).

5.3.2.5 Electro Luminance-Based Sensors

MXene and titanium nanomaterials applied to the electro luminescence-based sensors have emerged as a predominant method for sensing and monitoring. This is owing to the properties of titanium oxide and MXene. MXene is a two-dimensional material based on transition metal carbides, nitrides, and oxides that exist as one or few layers in nature. The unique electrical and optical properties of titanium oxide make it a widely used semiconductor material for sensing applications. By combining these two together, a heterostructure can be built which is able to produce an array of electronic and optical signals. Hence, it is ideally utilized in sensing applications.

The history of electro luminescence-based MXene and titanium-oxide sensors goes as far back as the 1960s. At that era, researchers were looking for new materials and technologies which can be used to perform sensing operations. The MXene was integrated with titanium oxide to form a heterostructure, which emitted an assortment of electrical and optical signals. It was this innovation that made way for a new generation of sensors, which could even keep tabs on everything from light and movement to temperature and humidity.

The electro luminescence-based sensors containing MXene and titanium oxide tend to have the ability of providing very precise as well as consistent data, which is perhaps one important corollary. As they are able to recognize a broad variety of signals across the spectrum, these sensors can be used in many monitoring and sensing applications. For example, these sensors can be used in light sensing, temperature-, or humidity-monitoring applications. Also, they can work as medical devices—such devices that measure heart rate and blood pressure.

Steps in the synthesis of an electro luminescence-based sensors using MXene and titanium oxide are as follows: The first step is to fabricate a titanium oxide and MXene heterostructure. This process begins by assembling oxide nanocrystallites, which are formed in-situ through their aggregation during the simultaneous reduction of nano-strips. This is based on the thin layer of titanium oxide that rests on an MXene substrate. The structure is then given a bias voltage to send current through it. The sensors can detect the electric and optical signals that this current generates (Zehao Wang et al. 2021; khalilifard and Javadian 2021).

For the NH_3 sensing, we found that Ti_3C_2 MXene could be a good candidate for room-temperature ammonia sensors with excellent selectivity by test and theoretical calculation (M. Wu et al.). The former possesses the charge-balanced composition and was what led to discovery of a unique two-dimensional carbide called Ti_3C_2 MXene when Al atoms were removed from Ti_3AlC_2 (Figure 5.22) [M Wu et al. 2019). Then, the multilayer Ti_3C_2 MXene powders were delaminated to a single layer by intercalation and ultrasonic dispersion. In order to produce gas sensors, ceramic tubes were coated by a colloidal suspension of single-layer Ti_3C_2-MXene. Then at 500 ppm, other gases (CH_4, H_2S, and NH_3) at room temperature were sensed by the sensors. Ti_3C_2 MXene-based sensors demonstrate a high selectivity toward NH3 when compared to other gases. Ethanol gas is a distant second in terms of response, with only 1.8 percent response compared to the nearly four-to-one NH_3

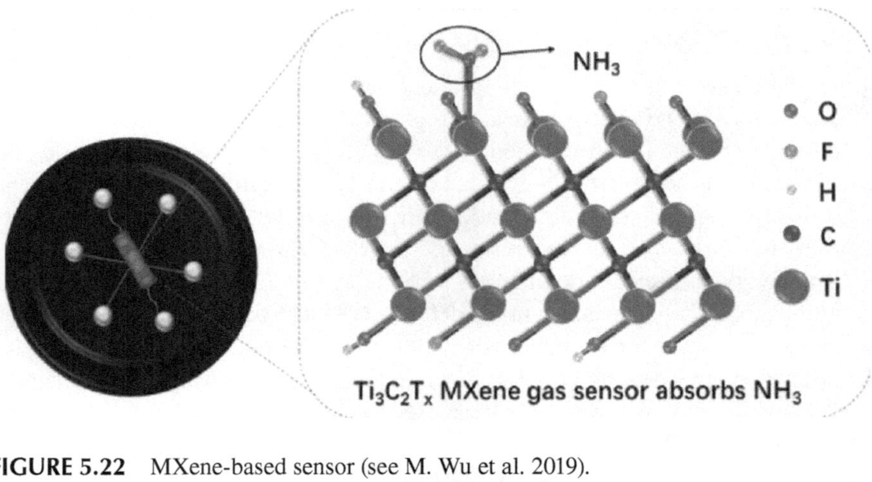

FIGURE 5.22 MXene-based sensor (see M. Wu et al. 2019).

Source: Adopted with permission from American Chemical Society.

response observed in first-principles calculations. These calculations were used to probe adsorption properties and understand the high selectivity. Theoretical analysis indicates that Ti_3C_2 MXene should exhibit strong selectivity for NH3 in this experiment, based on adsorption energy, adsorbed molecular configuration, and charge transfer. The response of the sensor to NH3 was almost linear from 10 to 700 ppm/min. The humidity tests and the NH3 cycle testing revealed excellent characteristics for this Ti_3C_2 MXene-based gas sensor for room temperature as a NH3 detection line (M. Wu et al. 2019).

Both MXene and TiO_2-based electroluminescent sensors are very helpful in many applications. They can be used in a variety of applications to include light sensing, temperature and humidity monitoring as well as medicinal applications. Additionally, they can control levels of items in the environment such as air pollution and water quality. Industrial applications include the aerospace and automobile industry as well.

The researchers also examined the pros and cons of adopting MXene Electro luminescence sensors along with titanium oxide for a specific application. The sensors also have the greatest level of accuracy and reliability, another one of their big benefits. They can also be used in plenty of areas and are straightforward to construct. However, they have to be fabricated in ways and with materials that can cost a lot, which are effectively their biggest disadvantages.

In the future, the scientific community is expecting [facile-electron-delivering-TiN/MXene-based fluorescent sensors] to use them widely and more EL (electro luminescence)-based devices could be further studied, especially as they are likely to find even more use in medical applications, where super precise and repeatable data is crucial. They are probably going to be placed in more industrial applications, including that of the car and, therefore, aerospace industries. This will make them a popular choice for environmental monitoring applications as they are able to provide very precise and reliable measurements of air pollution and water quality (An et al. 2021; L. He et al. 2021).

5.3.3 Electrochemical-Based Detection

Electrochemical sensor is a common and established sensing technology, while detection with MXene and titanium as in this work, to the best of our knowledge, can be considered novel. Titanium-based electrodes are extremely conductive, allowing scientists to produce the two-dimensional material MXene in sheets just a few atoms thick. Such a material combination allows highly selectivity, sensitivity, and reproducibility to detect electrochemical signals.

As a new type of electrode material, MXene and titanium-based electrodes are becoming a beautiful possibility in the detection of electrochemical signals no matter what. They have, for example, been used to produce low-cost biosensors that can be carried anywhere and used to monitor gases in the environment or diagnose medical conditions which have even been used in making the biofuel cells and better energy generation.

It is relatively easy to produce MXene and titanium electrodes. First, the components are mixed together and baked to create the architecture. The creation of a structure is followed, after which this can be coated with an electrically conductive material (such as gold or silver) to enhance the sensitivity of the electrode. Another method to stabilize the electrode is by incorporating a polymer layer (Yong Wang et al. 2021; X. Zhou et al. 2021).

For consistency, researchers followed this approach by using MXene and titanium electrodes as these types of electrode materials are used in several areas including medical diagnostics to renewable energy production. It can also be used in medical diagnostics to detect electrochemical signals resulting from metabolic processes, e.g., blood sugar levels. Additionally, these materials can be utilized for energy storage, as electrodes in fuel cells, solar cells, and other applications, contributing to renewable energy sources.

Titanium-based MXene and electrodes have numerous benefits. With superior conductivity, they can sense electrochemical signals far more accurately than other materials. They can also be made in large amounts and are cost-effective. They can, moreover, be applied in assorted applications without exhaustion due to their high stability.

When using MXene and titanium-based electrodes, there are many pros as well cons. They are also susceptible to rusting, which can impact their functionality over time. They also can require more time to update as they are somewhat weaker than the other material types. Finally, they are not as compliant for use in certain applications.

Brazing research was conducted for new manufacturing methods and verrucous titanium involving electrochemical sensing with applications being led. For example, researchers are developing means to make the materials more impervious to corrosion and unique methods for their use in solar energy. In the meantime, they are working to improve their structural integrity and bendability so that they can be used in more diverse environments. As research develops, MXene and titanium-based electrodes will be more used in electrochemical detection (C. Zhou et al. 2020; Gibson et al. 2021).

5.3.3.1 Detection of Biomarkers

The detection of biomarkers by means of MXene and titanium (ideas for a future breakthrough technology on refining diagnosis with the identification of new drug targets) is done as follows. This technology blends the latest in sensing and imaging technologies with nanomaterials such as MXene and titanium to detect biomarkers across samples. The idea of the technology is that nanomaterials, including MXene and titanium, provide a platform observation facility for biomarkers. To provide high precision and speed of detection, the nanomaterials are used to create a sensor layer that possesses an increased sensitivity upon interaction with biomarker molecules. The approach has been successfully used in the clinics and other contexts to discover biomarkers.

Fast and accurate biomarker detection by using MXene along with titanium as raw material is important. As such, it is ideally suited for a wide range of applications from early disease detection and cancer diagnosis to disease redisposal to treatment efficacy assessment. It is being used to identify individual biomarkers associated with specific responsive and resistant phenotypes, particularly in the context of personalized cancer therapy. Itis so useful because it can pick up biomarkers in many different materials such as blood, saliva, and tissue samples.

MXene is made by a more elaborate manifestation of nanotechnology and titanium-based biomarker discovery. The nanomaterials, first synthesized in a lab, would then be deposited on to the substrate. Next, the substrate is treated with a specific biomarker molecule and outfitted with an extra sensing layer (built from nanomaterials) that responds extremely sensitively in case if this kind of material emerges. If the biomarker is present, this becomes evident when imaging and sensing techniques—scanning electron microscopy (SEM) or fluorescence microscopy—are combined with an appropriate sensing layer (Mazzaglia et al. 2022; P. Liu et al. 2021).

Biomarker detection (using MXene and titanium) comes in many forms. It is applied in healthcare as a means of identifying biomarkers across samples (such as blood, saliva, and tissue). Here, the researcher can diagnose cancer and other diseases as well as the existence of disease before any symptoms appear, and not only check for the presence or absence without noticing but also evaluate drug efficiency, etc. The method can also be used for detecting contaminants and toxins through the detection of biomarkers in environmental samples such as water and soil. Beyond ensuring food quality and safety, the technology can be applied to detect pollutants as well as allergens.

In this context, MXene and titanium-based biomarker detection has been discussed which is advantageous due to their exceptional accuracy and sensitivity, cost-effective nature, and noninvasiveness. This is a scalable method and can be applied in diverse contexts. However, the high price of the technology remains one of its main drawbacks—costly technology that only works with expert tools and expertise. Moreover, due to that the technology is not capable of sensing proteins and other related biomarkers it lacks in potential advanced level sensing.

The application of MXene and titanium on the detection of biomarkers has a very strong future. It is accurate, convenient, and sensitive in the detection of biomarkers, with great potential to change the phenotype identification method. Technological advances in the future will be likely to allow the discovery of additional types of biomarkers. Moreover, this technology will get more inexpensive and available to use on several different contexts. This ultimately can revolutionize the way the diseases are diagnosed: By allowing early and accurate disease detection and personalized therapy (Russo et al. 2021; X. Xu et al. 2022).

5.3.4 Photovoltaics

The success story of a new family of materials, MXene and titanium-based photovoltaics, is likely to revolutionize the solar energy industry entirely. It combines MXene and titanium, as a result of which the solar cell is not only durable but also highly effective. Two-dimensional nanomaterials known as MXenes consist of transition metal carbides and nitrides with a lamellar structure. Titanium is a lightweight and durable metal, which means that it is one of the most used metals able to yield durable, versatile solar cells when combined.

This history dates back to the birth of MXenes and titania-based solar cells. In 2011, MXenes, transition metal carbides and nitrides were first discovered, marking a significant milestone in the field (Naguib et al. 2011). Their two-dimensional layered structure allows them to be used for the production of strong, very high efficiency solar cells. MXenes have a lot of exciting uses and potential applications—one that really stood out to me was in the solar energy field. The strength and lightness of titanium make it an excellent material for building efficient solar panels.

The importance of MXene and titanium-based photovoltaics lies in their strength and reliability. Photonics that outperforms conventional silicon-based solar cells is used by MXenes and titanium in the development of more efficient perovskite-made solar cells. Feather solar cells need less of it to generate the same amount of output efficiency. An added quality of the material is its strength, allowing solar cells to withstand harsh weather conditions and last even longer (Bromberg et al. 2021; García-Soriano et al. 2022).

There are quite a few methods for the fabrication of titanium and MXene-based photovoltaics. One way is by the manufacture of solar cells through a 3D printer. This is not only rapid but also highly efficient in generating a large number of cells. In such a case, the technique would have to be electrochemical deposition, where one can coat the substrate with Ti and then MXene. This process yields solar cells that are not only highly efficient but also last for a considerable time.

Titanium- and MXene-based photovoltaics are also useful in many applications. For instance, they can be harnessed to produce solar energy in rooftops or contribute to large-scale efficiency characteristics of the institution. They can also be used to produce consumer goods such as solar-powered devices, computers, and phones. They are used as ingredients in energy generation, water treatment, and storage of atoms.

In addition to high efficiency, flexibility and mechanical strength are key advantages of MXene-based photovoltaics. The MXenes and titanium-based solar cells are even more efficient than traditional silicon-based solar cells. Moreover, the solar cells are long lasting as they can survive even in an environmental calamity (K. Wang et al. 2021; Morsch et al. 2022).

While solar energy is likely the most potent in renewable resources, it has its inherent flaw—unpredictability. With integrated energy harvesting and storage device in solar energy storage modules, the self-powered systems could be achieved. Qin et al. (Figure 5.23) (L. Qin et al. 2020) designed a flexible organic solar cell (OPV) with the Ti_3C_2Tx MXene anode and transparent supercapacitors containing a common electrolyte based on ionic liquid dispersed in polymer electrolytes that are all integrated together vertically through using Ti_3C_{2-x} thin film as one of electrodes. As an example, they have recently reported a successful way to obtain MXene-based all-solution processed semitransparent flexible photovoltaic supercapacitors (PSCs). In the search for a semitransparent flexible PSC, they first adapted Ti_3C_2Tx MXene as transparent electrode of OPV with an excellent power conversion efficiency (PCE) up to 13.6%. A transparent MXene supercapacitor using an ionogel electrolyte (shown under UV light) demonstrates a high volumetric capacitance of 502 F/cm³ and excellent stability, making it a promising candidate for advanced energy storage applications. In the final reckoning, all-solution processing has developed a full-flexible PSC together with merits of superhigh AOS above 33.5 % and a striking sealing efficiency of about 88%. This method realizes a new way to

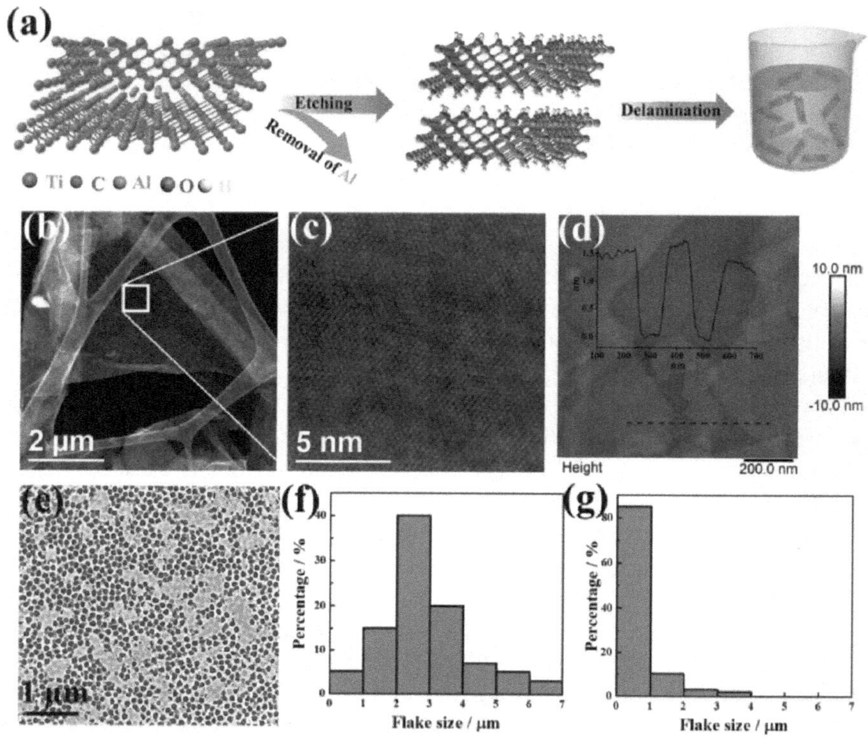

FIGURE 5.23 MXene colloidal solution preparation (Ti_3C_2Tx): (a–c) The MXene obtained by the MILD method from Ti_3AlC_2 precursor. Room-temperature high-magnification scanning transmission electron microscope (STEM) images of two individual Ti_3C_2Tx sheets dispersed on a holey carbon grid. (d) AFM characterization of single-layer Ti_3C_2Tx. (e) Top-view SEM image of a small-area Ti_3C_2Tx forest. (f and g) Size-distribution histograms of Ti_3C_2Tx flakes (f) before and (g) after sonication (see L. Qin et al. 2020).

Source: Adapted with permission from the Royal Society of Chemistry.

fabricate high-performance, all solution-processed flexible PSCs (MXene-based), being important for producing wearable and printable electronics in future technologies (L. Qin et al. 2020).

The cost of production and the technological limitations associated with MXene-titanium photovoltaics are cons. This high-cost and lengthy fabrication process are limiting the ability of the technology to scale. Still, the technology is in its early stages with a small selection of uses. Technically, MXene photovoltaics (pvs) and titanium-based pvs have their days ahead. The technology is new but promising, and it will make solar cells more efficient and durable. Furthermore, savings are expected in fabrication cost which should drop the price and allow for more widespread uses. This includes making the cost of photovoltaics based on materials like MXene and titanium affordable for all, so that more people have access to solar power (Pukale et al. 2020; L. bo Tang et al. 2021).

5.3.4.1 Photodetectors

New types of titanium-based photodetectors that are currently under development tend to be made from MXene. These emerging technologies have the potential to revolutionize how we think about photodetection. Photodetectors convert light into electrical signals and are used in applications such as optical communication, security systems, and medical imaging. Of these, MXene and titanium-based photodetectors are of note due to their high sensitivity as well as fast photo-response times.

MXenes are a family of two-dimensional materials consisting of carbides and nitrides, bonded either by –OH or –F. The conductivity and optical properties of these materials make them well-suited for use in photodetectors. Titanium-based photodetectors are also very popular nowadays due to their low cost and high sensitivity. Both of these materials are poised to revolutionize the photodetection industry by providing superior, low-cost solutions.

Photodetectors are fabricated as thin films on substrates. Electrodes are connected, and the device is illuminated to create a functional apparatus. Electric current is due to the photoexcitation, i.e., cause of light. You can then detect this current and put it to many uses (Yang Wang et al. 2021; Ali et al. 2021).

Photodetectors based on MXene and titanium lend themselves to security and surveillance systems, automotive safety systems, as well as medical imaging applications. They are used in medical imaging to detect cancers and other abnormalities. They are used in car safety systems to help recognize obstacles and alert drivers about potential danger. They can be applied to detect illegal personnel, intrusions, and insecurity systems.

The ultrahigh sensitivity and ultrafast response of MXene and titanium-based photodetectors provide the main advantages. They are very easy to prepare and quite cheap as well. However, the narrow detection range of these materials is one among their main drawbacks. However, they are limited by the fact that they are not useful for some applications such as military observation due to the lack of detection capability at longer wavelengths.

MXene and titanium-based photodetectors are able to revolutionize photodetection. Further research and development will likely improve these materials even more, making them cheaper as well. A broader range of applications, including car safety systems and medical imaging as technology, has continued to evolve. It will be interesting to see the future use cases for these materials as they have truly unlocked their full potential only now (S. Zhong et al. 2021; Zhe Wang et al. 2021).

5.3.4.2 Solar Cell

Slightly less discussed but nonetheless important is the rise of solar cells created using MXene and titanium, rather than traditional photovoltaic panels converting sunlight to electricity. MXene is a family of two-dimensional (2D) nanomaterials consisting primarily of carbon, titanium, and transition metal carbides and nitrides. It is a strong yet light metal available in abundance and will continue to be so. Combining these components, you can create solar cells with high efficiency and different applications. This technology has originated from titanium and MXene properties. These two entities combine to offer conversion for solar energy reasons.

A transition metal carbide-, nitride-, or carbonitride-based 2D nanomaterial called MXene features alternating layers of carbon and metal. The unique 2D structure of MXene is beneficial in addition to its exceptional electrical property and large surface area, rendering it an ideal material for making solar cells. Titanium, as we now know, is a strong, light metal with good electrical properties—the exact characteristics required of solar cell materials. By combining the characteristics of both, MXene and titanium-based solar cells can be more efficiently produced with improved performance (Ateka et al. 2021; Surbhi Sharma et al. 2021).

The formation of MXene layer and sputtering titanium on the pre-MXene layer are main stages for fabricating Ti/MXs/Al solar cells. Before depositing the MXene layer onto a substrate, the MXene nanomaterial must be produced. Following this, the titanium layer is deposited onto MXene using physical vapor deposition. The result is an ultrathin and highly efficient solar cell with potential for far-future applications, suitable for use in various bolt-on modules and other products.

Titanium-based MXene are highly versatile materials, suitable for use in a wide range of applications. As a result, these solar cells are ideal for small-scale technologies and portable electronics, where low weight and very high efficiency are critical. They are also a reliable and long-term power source for more significant devices including solar panels as well as structures.

Solar cells made of MXene and titanium boast of high efficiency and are low cost and lightweight. These solar cells are also capable of withstanding very high and extremely low temperatures as well as bad weather. Another important advantage is the possibility of these solar cells to be adapted for all kinds of products and applications (Lin et al. 2022; Hot et al. 2021).

Due to their specific features caused by surface termination functional groups, together with oxidation, MXenes become attractive for the field of photovoltaics like perovskite solar cells (PSCs). As regards low-temperature-processed perovskite solar cells (PSCs), their characteristics have been tuned as an electron transport layer (ETL) by L. Yang et al. who oxidized a Ti_3C_2Tx hydrocolloid. They calculated the energy levels using Vienna ab initio simulation package (VASP) which is based on density functional theory, more simply DFT and illustrated in Figure 5.24. After substantial oxidation, Ti_3C_2Tx can form Ti-O bonds and develop macroscopic defects when spin-coated. This transformation from a metallic to a semiconducting material may reveal new scientific insights. Through the oxidation of pristine Ti_3C_2Tx blending, the energy level matching between the perovskite and the electron transport layer (ETL) has been optimized, leading to a champion power conversion efficiency (PCE) of 18%. This has been attributed to the enhanced electron mobility in ETL, promoting electronic transport and reduced recombination with electrons at the lower energy level hole (thus improving PCE). Therefore, this work demonstrated the great performance of these materials in PSCs for low-temperature processing and many other solar technologies by presenting a simple route to high-performance PSCs using MXene-derived materials (L. Yang et al. 2021).

The downsides of the MXene and titanium-based solar cells are cost and availability. In addition, when not properly recycled, the materials used in constructing their cells can be difficult to recycle. Given the number of studies being investigated on different MXene and titanium-based solar cell materials, their synthesis methods, usages, and performances demonstrate a promising future. New materials and

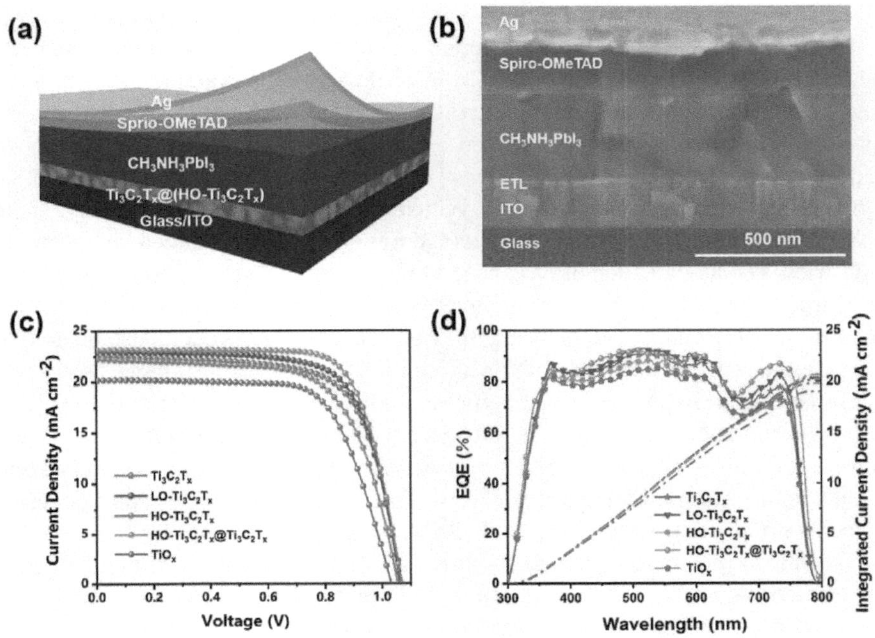

FIGURE 5.24 Device architecture of ITO/ETL/$CH_3NH_3PbI_3$/spiro-OMeTAD/Ag based on representative HO-$Ti_3C_2T_x$@$Ti_3C_2T_x$ as ETL: (a–b) cross-sectional SEM image of the PSC device and (c–d) reverse scan of PSCs' J–V curves based on $Ti_3C_2T_x$, LO-$Ti_3C_2T_x$, and HO-Ti_3C (see L. Yang et al. 2021).

Source: Adapted with permission from the Royal Society of Chemistry.

manufacturing methods are also under investigation in order to enhance their performance with increased cost efficiency. Ultimately, these photovoltaic cells would revolutionize the way they are generated and harness energy, providing a clean, reliable renewable future supply of power (Y. Zhang et al. 2022; Xiqing Wang et al. 2021).

5.3.5 Energy Storage Devices

MXene and titanium-based energy storage devices including an emerging class of MXene have been seen as a next-generation material for future high-performance energy storage solutions. Two-dimensional transition metal carbides, nitrides, and oxy-nitride can be prepared by the partial or full conversion of MXenes. MXene is often combined with titanium, a lightweight metal known for its strength and corrosion resistance to make higher-performance energy storage devices. Compared with conventional energy storage systems, the MXene/titanium-based portable electrochemical devices have some advantages in stability, higher energy density, and longer cycle life. Some of them may even exhibit relatively superior charge/discharge rates and energy storage efficiency.

These materials can be synthesized in a number of ways, such as atomic layer deposition (ALD), chemical vapor deposition (CVD), and hydrothermal synthesis—hydrothermal synthesis of nanostructured MXene materials and CVD and ALD processing to synthesize titanium as well as MXene thin films. The films can then be used to make the energy storage devices. For example, MXene films have been used in supercapacitor electrolytes or lithium-ion batteries as electrodes. The titanium films might be employed as either separators in supercapacitors or anode materials of lithium-ion batteries.

Consumer electronics and electric vehicles are just two of many areas where MXene and titanium-based energy storage systems can be utilized in emerging industries. The methanation of CO and the reduction of H_2O are important reactions in hydrogen generation, because the latter is a component used to improve battery or capacitor performance for consumer devices. MXene sheets could also be used to increase the capacity and cycle life of lithium-ion batteries, for example, as anodes. These are materials that can be incorporated into electric cars, and they will help to increase efficiency or range. Supercapacitors lack the high energy density (less power per gram or kilogram) of batteries but have a much faster rate of charge and discharge (massive power capacity). MXene sheets can serve as electrodes for supercapacitors (Gill et al. 2021; Y. Xu et al. 2021).

The increasing energy demand for application in electric vehicles and handheld electronics has reflected on a high number of studies directed to rechargeable batteries besides supercapacitors, the most meaningful electrochemical storage systems currently (Figure 5.25) (Xu Zhang et al. 2018). With these skills and tremendous research being conducted till date, 2D materials have shown remarkable properties with enhanced mobility (for lI-ION batteries and supercapacitors) and higher energy density (because of more edge planes/interior surface areas) because they are expected to facilitate faster transport kinetics. The first time successful exfoliation

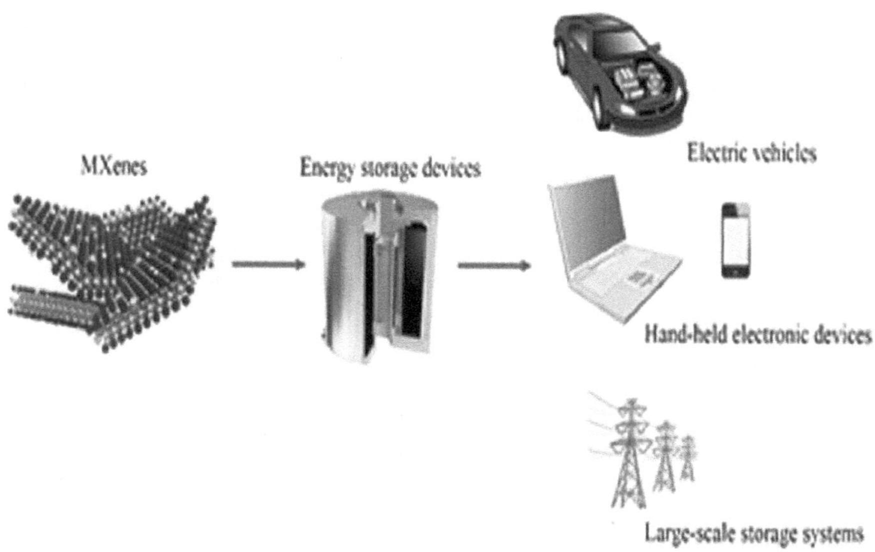

FIGURE 5.25 Energy storage systems based on MXene (see Xu Zhang et al. 2018).

Source: Adopted with permission from Elsevier.

generation of a novel class of 2D transition metal carbides/nitrides/carbonitrides called MXenes from MAX phases was reported in 2011. About 20 different types of MXenes have been created since then. The introduction of MXene produced from anything other than MAX phases, and even by processes such as chemical vapor deposition (CVD), has expanded the possibility for designing new types of MXenes. Their 2D nature and high electrical properties promise their intrinsic benefits of making electrode materials in electrochemical energy storage. Here, in the study by Xu Zhang, an overview on the current progress of MXene research on energy storage applications from an electrochemical point view is given. In addition, the prospect and challenges of MXene-based materials in electrochemical energy storage have been briefly summarized (Xu Zhang et al. 2018).

Some of the biggest drawbacks of energy storage systems based on titanium-based MXenes include high production costs, limited scalability, and challenges in maintaining long-term stability under operational conditions. Add to this the fact that they need very difficult-to-provide fabrication procedures and are extremely expensive to make. The materials are not as practical for use in conventional energy storage devices, and they are not nearly as available. However, notwithstanding these limitations, there is clear potential for MXene and titanium-based energy storage technologies. As production costs drop and better technology access becomes available, these materials are going to get used a lot more. Research is also ongoing on novel fabrication techniques and potential new applications to make these materials even more effective. These titanium and MXene-based energy storage systems promise to revolutionize the change of batteries with technology as it advances further.

5.3.5.1 Lithium-Ion Batteries

MXene is a kind of 2D material consisting of single-layer transition metal carbides, nitrides, or carbonitrides, all with the layered structure. Its remarkable physical and chemical properties—such as excellent electrical and thermal conductivity, high surface area, or good stability—have forced expectations in enterprise circles to rise in recent years. It is widely used in applications such as energy storage and catalysis. As a rising class of electrode material for lithium-ion batteries, MXene materials have attracted much attention.

Lithium-ion batteries are very important to consumer gadgets, electric vehicles, and energy storage system, as they can hold a lot of charge. Titanium-containing materials have been used for the making of anodes in lithium-ion batteries because of their high capacity and low cost. The problem in the adhesion is weak rate capability, short cycle life, and poor electrical conductivity of titanium-based material. In contrast, MXene materials have a high capacity and excellent electrical conductivity, so they are suitable for lithium-ion battery applications (Dharmasena and Wijayantha 2021; Y. Li et al. 2022).

MXene-based lithium-ion batteries can be fabricated by methods such as electro-spin-coating, spray coating, and ink jet printing. These methodologies allow for the ongoing development of uniform and thin-film MXene coatings at the electrodes, which contribute to their performance. Moreover, they allow using diverse MXene building blocks and materials tailored for every specific application. MXenes bring many advantages for lithium-ion batteries. Their high capacity and excellent rate capabilities make them useful for applications such as energy storage, electric vehicles (EV), and consumer electronics. The powerful cycle life of these batteries makes them excellent for long-term use. Also, they are the best option to be used in portable devices because of the lighter weight and low self-discharge rates.

A magnesium and lithium hybrid battery that promises better cycle life could be an alternative to pure Li-ion systems; it integrates the high-capacity, high-voltage absorption of lithium ions from regular cathodes (C) while using cheap high-capacity ReS2 as a C material.

Figure 5.26 is a schematic illustration of a hybrid Mg^{2+}/Li^{+} battery cathode using two-dimensional titanium carbide, Ti_3C_2Tx (MXene), in a bi-phase electrolyte with a Mg metal anode. It operated up to 50 mAh g^{-1} at 10 C and near 100 mAh g^{-1} at 0.1 C. This describes a good performance of the Mo_2CTx MXene for use as a cathode in such an all-solid-state hybrid battery. This finding calls for the exploration of a broad new class of 2D materials with high electrical conductivities and considerable intercalation capacities as hybrid Mg^{2+}/Li^{+} battery cathodes, supported by the readily accessed variety, already presented in MXene family (Byeon et al. 2017).

MXene lithium-ion batteries are not without their disadvantages in addition to the advantages. The price of fabrication is quite high because the manufacturing process is complicated, and there is a need for specialized equipment. The performance of the materials is not yet optimized, and they are still under development. Applications

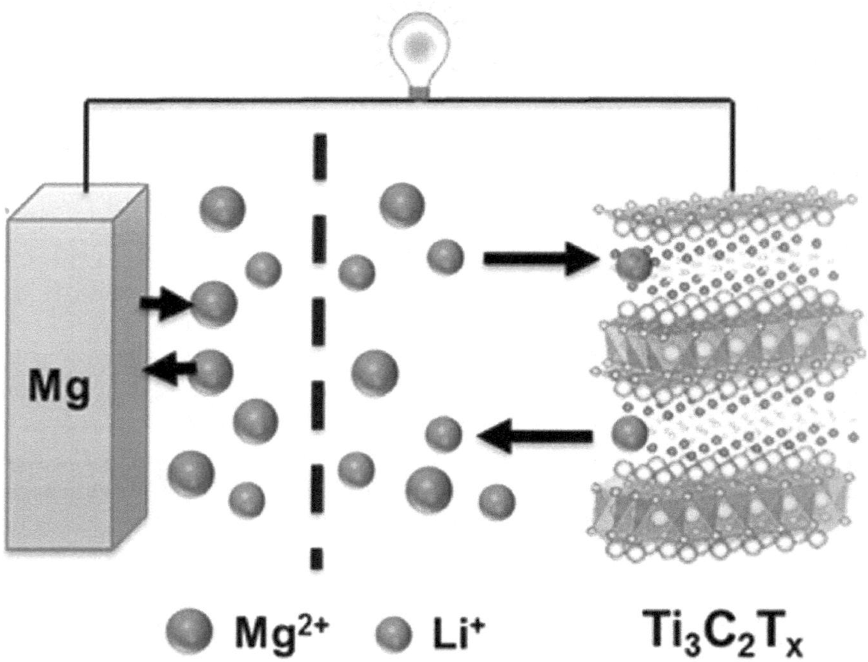

FIGURE 5.26 Schematic of a hybrid Mg^{2+}/Li^{+} battery with a cathode made of $Ti_3C_2T_x$ MXene (see Byeon et al. 2017).

Source: Adapted with permission from the Royal Society of Chemistry.

for lithium-ion MXene batteries could include energy storage, electric cars, and consumer electronics. An improved manufacturing process might also be a chance to reduce costs and improve battery output. Naturally, new materials could be formed that would increase the functional properties of a battery even more (Zaferani et al. 2021; Matsuzaki et al. 2020).

5.3.5.2 Sodium-Ion Batteries

Novel sodium batteries using MXene and titanium anode could upend how the material scientists power our tech devices. MXene is a two-dimensional material made of one layer of atoms and presents high conductivity. Titanium 2D materials are not only resistant to corrosion but also capable of storing a lot of energy. Combining the two produces a sodium-ion battery that has high energy density and long cycle life along with fast charging capabilities. The story of the MXene-titania sodium-ion battery is older than any other work in this collection: Researchers started considering new materials for energy storage over 10 years ago, with a series on titania and titanium-based intercalation solutions. The discovery of MXenes and their potential usage within energy storage solutions dates back to 2016 when researchers from Drexel University recognized it. The researchers began looking at how to achieve a high-performing sodium-ion battery, topologically integrating MXene and titanium-based materials (Yew et al. 2021; Boscherini et al. 2021).

It was stated by J. Li et al. that the potential to store salt in recently discovered Ti_3C_2Tx MXenes is critically impaired by their low specific capacity, which is attributed to inadequate interlayer spacing. Thus, the ultimate first multilayered sulfur S-doped Ti_3C_2Tx MXene was prepared through a simple and large-scale sulfidation approach on Ti_3C_2Tx utilizing thiourea as the S origin. This material also has an expanded interlayer spacing and enhanced electrical conductivity. S-doped Ti_3C_2Tx has an outstanding long-term cycling resilience with a reversible capacity of 138.2 $mAhg^{-1}$ after 2,000 cycles at 0.5 Agl as an anode for sodium-ion batteries SIBs. It also has an exceptional rate capacity of 121.3 $mAhg^{-1}$ at 2 Agl and 113.9 $mAhg^{-1}$ at 4 Agl. The exceptional sodium storage performance, which is superior to that of known Ti_3C_2Tx-based electrodes, should be linked to the multilayered morphology, increased interlayer spacing, and increased electrical conductivity, as well as the high stake of surface-induced capacitive performance after S doping. This investigation by Jiabao Li emphasizes the prospect of the S doping strategy. Most significantly, this study enables new techniques for the clever creation and intentional fabrication of divalent-doped MXenes for energy storage and transformation applications (Figure 5.27) (Jiabao Li et al. 2018).

There are two things the titanium-made MXene and sodium-ion battery parts are needed for. For one, they are far greener than what could currently be considered "conventional" lithium-ion cells. Lithium-ion batteries employ hazardous materials such as cobalt and nickel, which have negative environmental consequences if released. And on the other hand, sodium-ion batteries made up of nontoxic materials vie to be an eco-friendly solution for energy storage. Second, MXene-based sodium-ion batteries exceed the energy density and cycle life of a traditional lithium-ion battery. Their high energy density also makes them ideal for applications such as electric vehicles and portable electronic devices due to their long cycle life. A variety of fabrication processes exist for MXene (like) and titanium-based sodium-ion

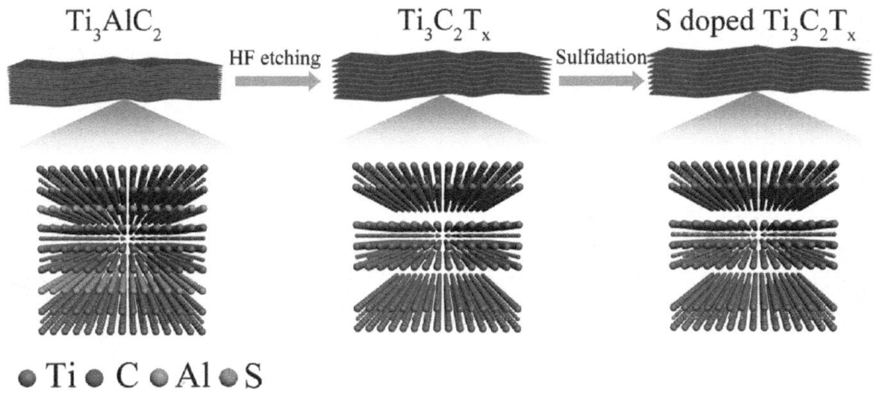

FIGURE 5.27 Schematic illustration of the preparation of S-doped $Ti_3C_2T_x$ (see Jiabao Li et al. 2018).

Source: Adapted with permission from the Royal Society of Chemistry.

batteries. The standard process uses a two-step procedure to coat a titanium-based material with MXene. This is followed by the electrochemical deposition of sodium ions. Other fabrication processes can include sputtering, electroplating, and chemical vapor deposition.

But MXene-titanium sodium-ion batteries could find use in a wide variety of applications. Applications of this kind could be in electric vehicles, portable electronics, or even grid-level energy storage. They can also be used as a power source for portable devices such as smartwatches and fitness trackers. They also exist in consumer electronics devices such as computers and cellular phones. The key characteristics of MXene/Ti-based sodium-ion batteries titanium are evidently high energy density, long cycle life, and fast charge–discharge process capability. They are also friendlier and safer with the environment than traditional lithium-ion batteries. The major drawbacks with these batteries are that they have lower power densities compared to lithium-ion batteries and are more costly than conventional calendared-polymer electrolyte layers.

Next-generation sodium-ion batteries comprising MXene and titanium could be the future of energy storage. The new batteries are capable of fulfilling a role in everything from grid-level energy storage to consumer devices, but they will need more study and development. Besides that, new materials and manufacturing methods could hike the price of these batteries as well as power density. Over the long term, sodium-ion batteries may be in our energy future, but new systems like MXene with titanium can give a neonate direction to develop for higher sustainability (Xiaomin Zhang et al. 2021; Dayekh and Mequanint 2020).

5.3.5.3 Sulfur-Ion Batteries

MXene versus titanium-based sulfur-ion batteries literally created an entirely new class of energy storage technology. The devices consist of two electrodes (one made from titanium, one using MXene) separated by a liquid electrolyte. The cathode of the supercapacitor is a MXene electrode, and the anode actually consists, in fact, of titanium. A sulfur-ion solution serves as the electrolyte, and two electrodes are separated from each other by an ionic conducting membrane. Titanium and MXene combine in the battery to create a large surface area with high capacity.

Developments with MXene and titanium-based sulfur-ion batteries have had breakthroughs in battery technology. They can find their way in applications right from personal gadgets to electric vehicles. These batteries have a high energy density, so they can store much more energy than other types of batteries. Moreover, they have a longer life cycle and can be recharged, allowing for the battery to survive more charge cycles. They have some scale of protection and are less toxic than other battery types.

For the fabrication of MXene and titanium-based sulfur-ion batteries, first you need to pre-prepare the electrodes. On the contrary, while MXene electrode is chemically synthesized, titanium in this case comes from electrodepositon. The final step is to fill the cell with electrodes up with a liquid electrolyte. The cell is then closed, and the battery can be put into use.

MXene and titanium-based sulfur-ion batteries can be used in electric vehicles as well as in laptops and phones. Finally, they can be used in power-grid energy storage systems which can help reduce greenhouse gas emissions. They are also suitable for use in medical and military equipment (Lü et al. 2021; Fugolin et al. 2020; Shin et al. 2021).

There is a great deal of interest in mechanically compliant, energy-dense, and flexible Li-S batteries. X. Zhong et al. reported a high-performance freestanding MXene (Ti_3C_2Tx) film, fabricated by a scalable drop-casting method with greatly improved adhesion to the bottom sulfur layer designed for continuously bending. After controlled oxidation, few-layer Ti_3C_2Tx was prepared through in situ growth of titanium oxide on holey-PIL, resulting in titanate-TiO_2/H-TC (Figure 5.28). This material is designed to capture lithium polysulfides, improve lithium-ion conductivity, and enhance battery performance. The sandwich electrode showed a high capacity of 740 mAh g^{-1} at 2 C with year-long cycling stability and retention of more than 81% (based on the original capacity) after charge/discharge for greater than or equal to 500 cycles at a rate as low as 1 C. Flexible Li-S batteries built on this sandwich electrode retain 95% of their capacity after 500 bends. This work exemplifies several advantages of MXene design for wearable and flexible battery applications (Figure 5.28) (X. Zhong et al. 2022).

MXene and titanium-based sulfur-ion batteries offer high energy density, a long cycle life, along with added safety. In addition, they are cheap and easy to manufacture. One of the main drawback is a very poor efficiency compared to other battery types. There is also a risk of liquid electrolyte displacement with leakage and evaporation. This makes MXene and titanium sulfur-ion batteries a good candidate for the future. Now, scientists are trying to reduce their production costs and increase the efficiency of these batteries. Another possibility is to use other types of materials in the electrodes, such as graphene and carbon nanotubes (which could improve the battery performance even more), that can also be worked on. Researchers are also investigating other ways. Recently, they found similar processes with titanium-based sulfur-ion batteries, and how MXene could revolutionize the way we store/use energy (Artsen et al. 2020; W. Choi et al. 2021).

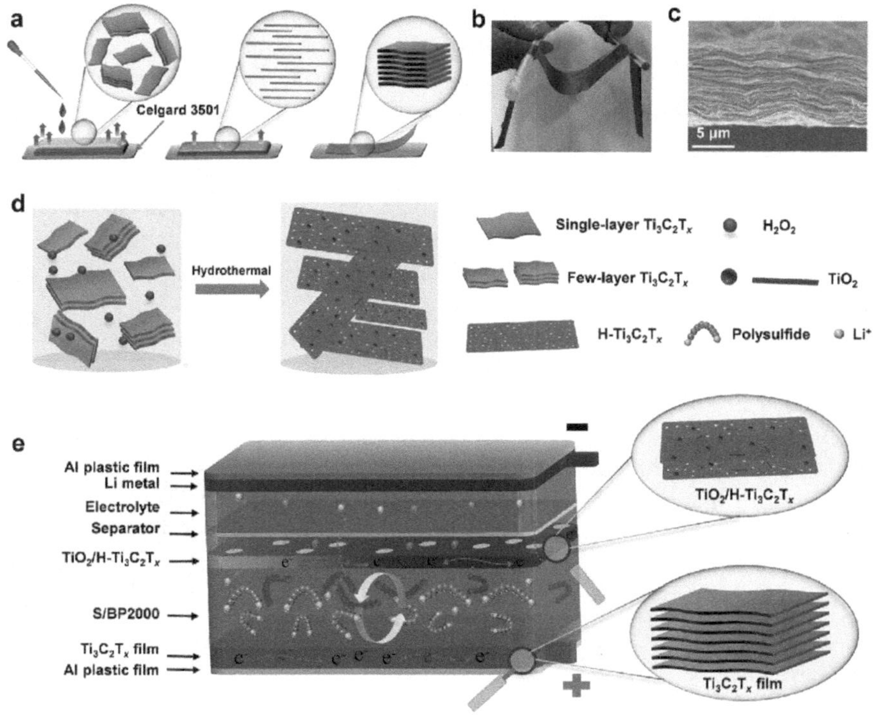

FIGURE 5.28 (a) Schematic of how the freestanding $Ti_3C_2T_x$ film was prepared. (b) A large-area $Ti_3C_2T_x$ film optical picture. (c) A cross-sectional SEM image $Ti_3C_2T_x$ film. (d) Schematic of the synthesis of $TiO_2/HTi_3C_2T_x$. (e) Schematic of a flexible Li-S battery with a covering of $TiO_2/HTi_3C_2T_x$ and a current collector made of $Ti_3C_2T_x$ film (see X. Zhong et al. 2022).

Source: Adopted with permission from American Chemical Society.

5.3.6 Catalysis

Recently, the use of titanium and MXene nanoparticles especially for catalytic endpoint has initiated an emerging research field called as MXene-based Ti-catalysis. Compared with traditional catalysts, titanium- and MXene-nanoparticle-modified reactions have multiple advantages in reactivity, selectivity, and stability. These surfaces can also be tailored to the conditions of a given reaction. Various fabrication techniques are employed for titanium and MXene-based catalysis. Sol-gel processing, electrochemical synthesis, and physical and chemical vapor deposition are few methods of preparation techniques. These materials can further be modified by doping them with additional species. This can aid in becoming a catalyst for improvement in to the catalytical activity. These substances may also be used to manufacture nanostructured catalysts, which can then serve as a way to expand the surface area available for reaction and thus increase the speed with which molecules move across that surface into solution.

Regarding the respective applications in commercial and academic, MXene and titanium-based catalysts serve numerous purposes. These can be used in the preparation of drugs, polymers, and fine chemicals. Such substances are also used for academic research on the influence of surface shape on catalytic activity. Fuel cells and other energy storage devices also rely on them (Karimzadeh et al. 2021; Ye et al. 2021; Mahmoudi et al. 2021).

In recent years, A. M. Garcia et al. have widely explored their versatile chemical compositions and structures, stability under the conditions of relevance in (electro) catalysis matters as well as some other eye-catching properties. They suggested that two-dimensional transition metal carbides or nitrides, so-called MXenes 44, which emerged almost a decade ago are depicted on the basis of Figure 5.29. The constituent

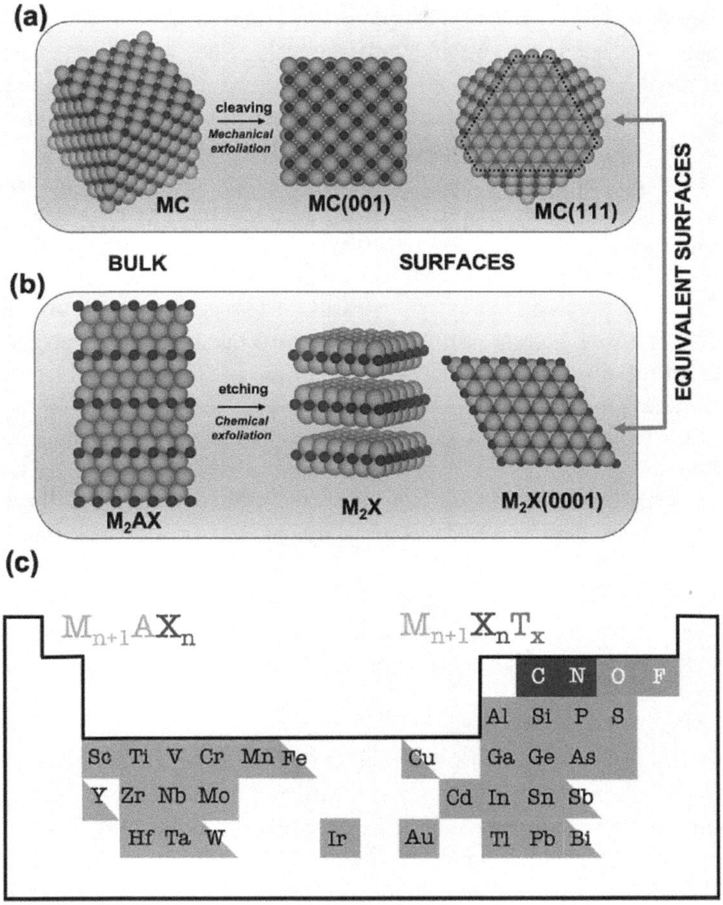

FIGURE 5.29 (a) Surfaces of mechanically exfoliated (001 and 111) transition metal carbide, as well as the parent bulk-phase MC. (b) The MAX phase (M_2AX) and the corresponding chemical exfoliation-derived M_2X. One should remember that both MC(111) and M_2X(0001) surfaces are morphologically identical. (c) MAX phases so far have been highlighted Greyscale color indicates the M elements (left side), grey (right side) shows the A and X are in black color in the periodic table. The gray regions also point out the terminations (T). To scale, the null-field squares on M or A mean that they have been found in quinary method MAX stages simply (or solid solutions). The MC(111) surface is the grayed part in (a). The illustration does not show surface functionalization (Tx), because it would have become too involved (see Morales-Garciá, Calle-Vallejo, and Illas 2020).

Source: Adopted with permission from American Chemical Society.

parts that form the basis of the structure, large surface area, and high laboratory-scale activity endow MXenes as promising catalysts or supports for various heterogeneous catalytic/electrocatalytic reactions. Here in the figure, the most highlighted (electro) catalytic processes where MXenes have been shown to be among competitive replacements for traditional Pt-group catalysts are briefly reflected upon. They discuss the catalytic performance of MXenes, emphasizing their interactions with metal elements so as to modulate behaviors in catalysis-related carbon and nitrogen moieties as well different surface terminations. They discuss the results with respect to bulk carbide-extended surfaces and compare the experimental MXene structure with models. The advancements in synthesizing MXenes and their practical computational models, along with the ease of upscaling, have shown a significant promise to play a crucial role in actual catalyst-driven chemical processes being applied today (Morales-Garciá, Calle-Vallejo, and Illas 2020).

MXene and titanium-based catalysts offer numerous advantages compared to the conventional catalyst. These materials can be used in harsh environments as a result of their high stability. Moreover, due to inherently strong selectivity of these materials, only certain reactions can be catalyzed using them. They are also hugely reactive, reasons why they can catalyze processes so fast as usual catalysts. However, the benefits of using MXene with titanium-based catalysts come at a cost. The production of these kinds associated with materials is expensive and might require additional remedying techniques to become adaptable for a new desired reaction. In addition, these were not always as high performing in certain reactions compared with traditional catalysts. Both MXene and titanium-based catalysis are the future. However, researchers are still searching for ways to further optimize these materials. Furthermore, scientists are developing innovative ways to manufacture these materials so that they are more efficient and cheaper. They also studying how to use these materials in solar cells, fuel cells, and other sustainable energy. MXene- and titanium-based catalysis are poised to remain central in the expanding world of renewable energy sources moving forward (Zhao et al. 2022; X. Liu et al. 2021).

5.3.6.1 Electro Catalysis

The PKAs, energy storage, and conversion/reduction benefit from promising prospects of MXene/titanium-based application in electrocatalysis. The technology is built around using two-dimensional (2D) materials to catalyze the electrochemical reactions occurring inside of an electrochemical cell, such as MXene and titanium dioxide (TiO_2). Such physical and chemical properties make these materials usable for superior electrocatalytic applications. Transition metal carbides from the MXene family have a rather unique two-dimensional structure. When compared with other transition metal carbides, MXene has a high electrical conductivity and very large surface area. These properties make it a superb substance for use in electrocatalysis. For electrocatalysis, scientists have shown that a 2D semiconducting material—the most commonly used TiO_2—also catalyzes the reaction. The material has a high electrical conductivity and large surface area, which make it an excellent material for speeding up electrochemical processes. MXene and titanium-based electrocatalysts are synthesized as a thin film deposited on an appropriate substrate. This can be done by several methods including chemical vapor deposition, atomic layer deposition, electrochemical deposition, and sputtering. This results in the formation of a thin film, which is then functionalized with an appropriate catalyst. Because of this functionalization step, these catalysts become more efficient in performing some electrochemical reaction and catalyzing the desired process for which they have been synthesized. (D. Ma et al. 2021; Shao et al. 2022).

MXene and Titanium based electrocatalysts, TerraCOH Hydrocarbon Geothermal Power Generator, Fuel Cells and Batteries to Water Splitting & Pollution Reduction applications. Just as importantly, these materials are capable of serving well as anodes in fuel cells that are poised to change the way how scientists convert chemical energy into electrical power. Electrocatalysts based on titanium and MXene have been produced for combination with batteries to help improve the overall efficiency of the energy storage process as well. Sodium-transition metal chlorides together with other abundant but less useful materials can be used to split water and prepare hydrogen fuel, a clean and sustainable source of energy. Lastly, these materials can be used to reduce the levels of pollutants in the air by speeding up the oxidation of pollutants into less dangerous compounds. Bioniphins, best known for their cost-effective and efficient electrocatalytic activity, leverage titanium- or MXene-based materials, which impart novel catalytic mechanisms and benefits. High surface area as well as high electrical conductivity enable these materials to be excellent catalysts. One of the biggest downsides to these materials is that they are susceptible to corrosion, which will eventually render them useless. According to the future predictions, MXene-based and titanium-based electrocatalysis will, in a near come days, play a major role on energy storage, energy conversion as well as pollution control. Furthermore, new synthetic techniques and the development of improved catalysts are currently under investigation. This work is expected to advance electrocatalysts that are highly efficient, cost-effective, and broadly applicable (Wychowaniec et al. 2020; Patel et al. 2021).

5.3.6.2 Photocatalysis

MXene and titanium-based photocatalysis are two materials science technologies. Ti-oxide photocatalysis utilizes light to initiate chemical reactions, and MXene is two-dimensional material of transition metal carbides and nitrides. In other words, these two revolutionizing technologies find applications as countless opportunities in fields like energy, water desalination, and air pollutant prevention. MXene is a class of two-dimensional materials that are effective and robust transitioning metal carbides and nitrides. It has been extensively researched since it was first discovered in 2011. MXenes are composed of two layers from transition metal atoms and another two layers made up of carbon. The optical, electrical, and corrosion-resistance properties of MXene make it a good candidate for photocatalysis. MXene is able to absorb light spanning a wide range of wavelengths, thanks to its optical properties, so it has multiple applications (Gu et al. 2021; Urquhart et al. 2022).

Emerging two-dimensional (2D) materials are considered as the next-generation materials due to their unique physicochemical properties, among which transition metal carbide or nitride (MXene) is an upcoming star in photocatalysis/clean energy conversion and storage application areas such as electrocatalysts, supercapacitors, lithium-ion batteries, and biomedicine. MXene can thus lead to high photoconversion efficiencies by providing a large number of surface functional groups for light-harvesting materials and quick separation of photogenerated charge carriers in photocatalysis. This has sparked a considerable theoretical and experimental effort, which recently particularly highlighted the potential of MXenes in several fields relevant to photocatalysis. P. Kuang et al. provided an overview of MXene-based photocatalysts and their mechanisms in Figure 5.30. Subsequently, they summarize the latest progress of MXene-based photocatalysts from producing solar fuel to pollutant degradation. Finally, they provide prospective views on the route to advanced MXene-based photocatalysts (Kuang et al. 2020).

Photocatalysis with titanium is an approach that, when combined with light, can initiate chemical reactions. Put simply, when light shines into titanium dioxide, this excites the electrons on its surface and induces a chemical reaction, which can utilize this process to fragment out more organic molecules or initiate some chemical reactions including oxidation. Titanium-based photocatalysis is widely used in many applications such as water purification, air pollution minimization, and hydrogen production. Utilizing MXene together with titanium-based photocatalysis could yield a more efficient and less expensive use in the endocrine disrupting exclusion. Drawing on both approaches in such combinations could allow for the production of highly efficient light-absorbing and, therefore, chemically alert materials. Some of the uses include water purification, air pollution control, and energy generation.

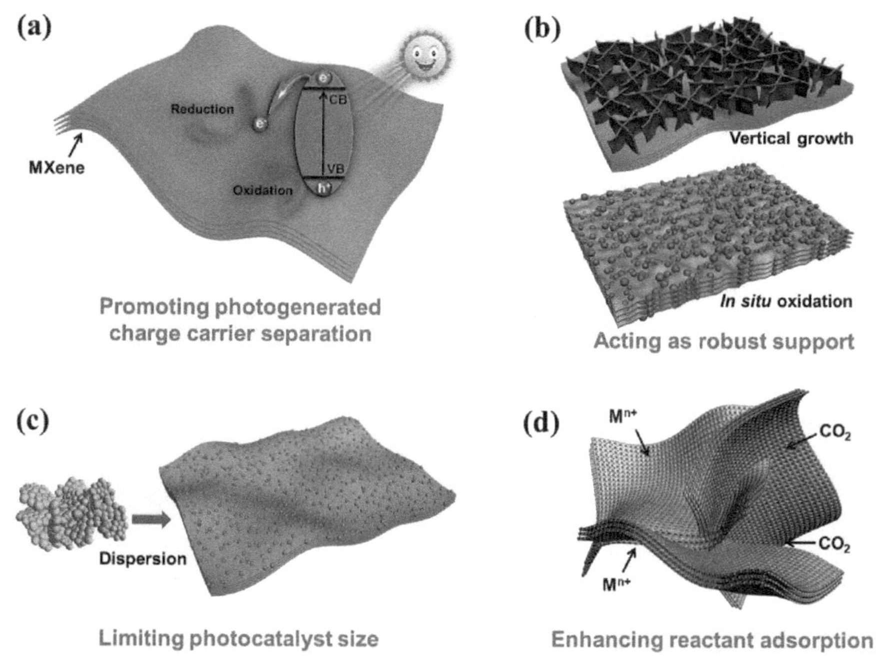

FIGURE 5.30 (a)–(d) Several functions for layered MXene structure in enhancing the photocatalytic activity (see Kuang et al. 2020).

Source: Adopted with permission from Elsevier.

Novel materials including MXene and titanium-based photocatalysis are under investigation. These materials can also be employed together to create materials in different ways with unique properties. For example, MXene can be used to manufacture materials with very high thermal conductivity so one could build solar cells that can extract more energy. Moreover, the generation of materials with an enhanced electrical and optical property via titanium-based photocatalysis can be achieved to fabricate superior optoelectronic devices.

Titanium-based photocatalysis and MXene have advantages of their own. The innovations are efficient, cheap, and possibly environmentally friendly. They can moreover be used to make materials with one of a kind properties that are utilized in applications across different fields. Although MXene and titanium-based photocatalysis have some advantages, they also suffer from several problems. Yet, these technologies often produce materials that are expensive and labor-intensive to create, as well as time consuming. Moreover, the materials which result from these technologies often end up being substantially less durable than that created by traditional manufacturing methods.

MXene- and titanium-based photocatalysis seems promising. As the continuation of these techniques is studied, further new materials and uses will be discovered. Moreover, the expense of manufacturing these materials will likely drop becoming more open to the public. This latter capacity, to generate materials with unprecedented functionalities and to address a broad range of environmental issues, will likely make these technologies increasingly important in the future (Lim et al. 2021; J. Zhang et al. 2021; Baoling Li et al. 2021).

5.3.7 Drug Delivery and Cancer Therapy

The medical field has started to be interested in a more recent study, where titanium along with MXene were seen to be helpful in medication delivery and cancer therapy. This advanced drug delivering technique might change the entire system of cancer delivery and make more effective treatment options possible for one of deadliest diseases on earth. This is the primary idea of how MXene and titanium-based drug delivery for cancer therapy is used owing to nanotechnology that allows for coding some drugs into particles, intended specifically focusing only on cancer cells. This drug is released when these particles come into contact with cancerous cells, allowing for much more targeted and localized medication delivery. This system can carry medications across the body and has been found to be very biocompatible. It can also create nanostructures that could be tailor-made to homing in on specific body parts or cell types. This makes it the preferred material for drug delivery systems. For titanium is so dense, stable, and biocompatible, more medication delivery systems fall back on its use as well (Golmakani, Nabavi, and Manović 2021; Clemente et al. 2022; Mekawy, Noguchi, and Kawakita 2022).

New 2D material analogs of graphene that have structural and physiochemical properties as well as result in showing a great performance on a wide variety of devices are sources for interest of the research community (Figure 5.31) (Dai et al. 2017). The research team reported the synthesis of tantalum carbide (Ta_4C_3) MXene-based composite nanosheets for multimodal imaging-guided photothermal tumor

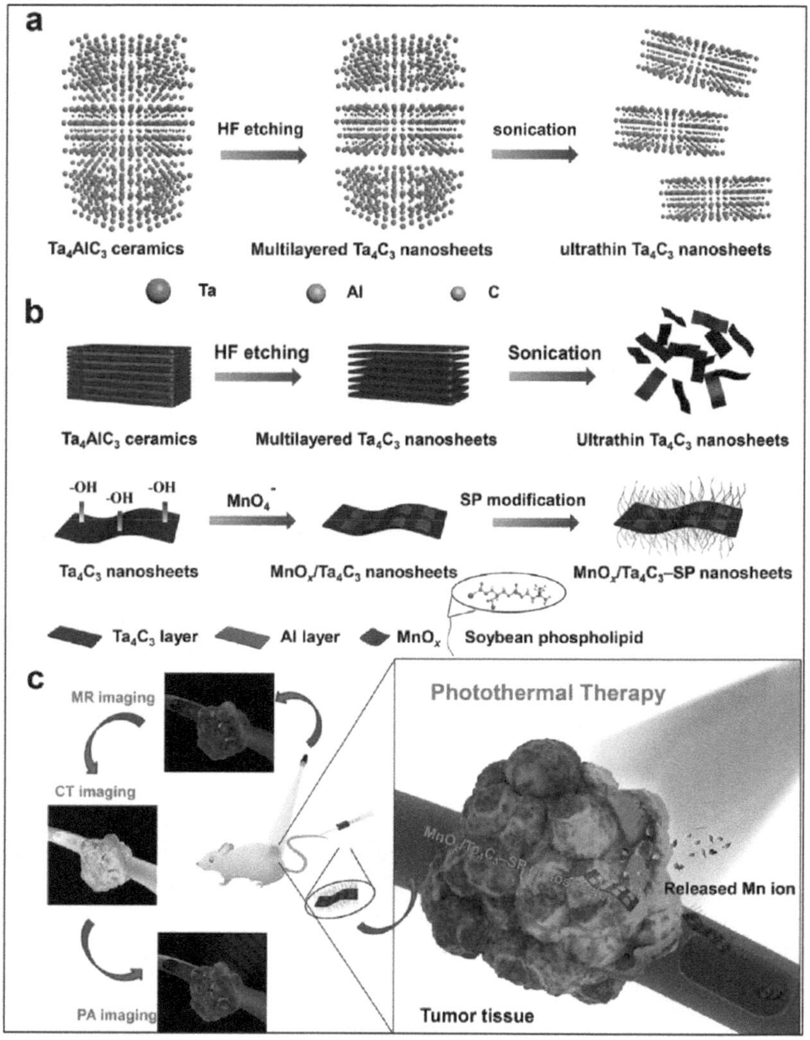

FIGURE 5.31 Illustration of the synthesis process for MnO_x/Ta_4C_3SP composite nanosheets and MR/CT/PA imaging-guided photothermal tumor therapy: (a) The exfoliation process of 2D Ta_4C_3 nanosheets. (Schematics are based on the ball-stick model—etching with HF and sonication to obtain exfoliated thin sheets.) Process flow for the synthesis of the MnO_x/Ta_4C_3SP composite nanosheets, including a HF etching and sonication exfoliation. (b) In-situ redox reaction among Ta_4C_3 nanosheets and post-introduced $KMnO_4$ followed by (c) surface SP modification. (c) Schematic illustration of the theranostic functions of MnO_x/Ta_4C_3SP composite NSs in efficient PTT ablation with MR/CT/PA imaging (see Dai et al. 2017).

Source: Adopted with permission from American Chemical Society.

ablation. The team managed to do this by controlling the chemical composition and surface functionalization of MXenes. The redox reaction was used to develop in-situ manganese oxide nanoparticles (MnO_x/Ta_4C_3) on the reducible surface of MXene nanosheets. The tantalum components of MnO_x/Ta_4C_3, with analogous enhancement capacity to produce Iobitridol for contrast-enhanced computed tomography (CT) and the integrated MnO_x component T1-weighted magnetic resonance imaging, have been developed as a tumor microenvironment-responsive nanosystem. The strong NIR absorbance of MnO_x/Ta_4C_3 composite nanosheets, along with their photothermal conversion capability, enables contrasting-enhanced photoacoustic imaging and significantly suppresses tumor growth in vivo via synergistic PTT as well. This work by Dai et al. further broadens the biologically centered success of MXenes by generating family members exhibiting improved biological compatibility and optimizing functionalization techniques for cancer-theranostic applications based on camouflaged antitumor signals (Dai et al. 2017).

Thus, they are very important for cancer therapy because MXene and titanium-based materials have a great potential to deliver drugs only in cancer cells with superior efficacy when compared to conventional drug delivery systems. This could translate into the diminution of side effects caused by chemotherapy and other cancer treatments. Moreover, if these materials can help to improve the efficiency of drug delivery, this may in turn potentially lower overall costs for cancer treatment. Methods for MXene and titanium-based drug delivery device generation are explained here. By using the electrospinning method, it is also possible to prepare thin nanofibers of MXene or titanium, which can then be utilized in their respective systems. Such materials offer not only good solubility and dispersibility in physiological media but also additional functional properties for advanced applications. Another method is a designed 3D printer that could print the necessary nanostructures for drug delivery with greater precision.

There are numerous applications in the medical world for titanium and MXene-based cancer therapies. One potential application is targeted delivery of chemotherapy drugs that offer the promise of more precise targeting on cancer cells and reduce systemic toxicity associated with chemo treatments. It could also be used in the delivery of other types of medications, including gene or immunotherapies. McNeill expects these medications would be given in a "more targeted" fashion which, in turn, could lead to better outcomes. As typical as they come in the list of means for delivering medication to target cancer cells, the benefits are biocompatibility, targeting of certain cells, and high mechanical strength. The downsides include drug toxicity and a difficulty in trying to control the release of this potentially beneficial drug.

The future of these therapies and medical deliveries through both MXene-based delivery of cancer medications (and treatments) as well as titanium based-cancer medication is very bright. Now that research has progressed, these materials might be used more in medical applications too, leading to a cure for cancer. Moreover, refinement of more precise and selective drug delivery systems by using nanotechnology may offer cancer therapies with greater efficiencies (S. Wang et al. 2021; J. H. Choi, Park, and Kang 2021; Yuan et al. 2022).

5.3.7.1 Tissue Engineering and Drug Delivery

MXene and titanium are two materials that have shown to be useful in tissue engineering and medicine delivery. Hexagonal carbon atoms and a single layer of transition metal atoms are interfaced to sandwich to effectively form a two-dimensional (2D) substance, which is called an MXene substance. Over the years, titanium has found its way into numerous biomedical applications for many reasons being that it is a strong, light, and corrosion-resistant metal. The unique properties of these two materials make them ideal for drug delivery and tissue engineering applications. Due to their outstanding electrical and thermal conductivity, along with good biocompatibility, MXenes are well suited for tissue engineering applications as well as drug delivery systems. In addition, because of its 2D shape, it is fairly versatile and can be configured to have particular properties. It is also generally biodegradable, making it attractive for drug delivery applications. Furthermore, MXene can introduce a porous structure to further enhance its utility within tissue engineering or drug delivery applications.

Scientists around the globe are going to put titanium in this class, being a very strong, lightweight, true nonreactive, and virtually inert metal with extremely biocompatible features. It is used in many biomedical industries like orthopedics, as devices and implants for joints and other body parts. Titanium can also be used as a scaffold material as it is strong and lightweight which comes to play a major role in tissue engineering. As it is biocompatible and inert, it is also used in drug delivery applications (F. Yang et al. 2022; W. Zhang et al. 2021).

Titanium and MXene-based tissue engineering and drug delivery devices can be built with 3D printing, electrospinning, microfluidic processes, etc. One of the typical ways to manufacture 3D constructions with a variety different shapes, sizes, or materials is by means of 3D printing. This technique is known as electrospinning, and it uses electricity to spin a polymer solution into the desired fibrous mesh. Microfluidic processes involve the placement and blending of materials using tiny channels. Each of these techniques result in structures with particular desired properties for tissue engineering and drug delivery applications. This makes them suitable as building blocks that can be employed to create tissue scaffolds, particularly in terms of closely resembling natural tissues with regard to mechanical strength and porosity. They can also be used for transferring medication, such as proteins or stem cells to certain parts. These materials could also be used in the manufacture of implantable diagnosis and treatment appliances such as pacemakers. Because of the biocompatibility and tailor-ability to a specific aspect, MXene and titanium act as promising materials for tissue engineering applications while also being used in medication delivery. They are also corrosion resistant, making them suitable for use in implantable devices. You have the added benefit of them being strong and lightweight, making for great scaffolding material. But the low biocompatibility and great expense for MXene as well as for titanium in modifying onto tissue engineering and drug delivery systems are the drawbacks. Some of these materials are often difficult to fabricate and can require expensive difficulty machinery. Also, because of their biocompatibility and inertness in chemical form, they might be completely unsuitable as certain applications.

Titanium and MXenes can be incredibly effective for tissue engineering, particularly in the field of drug delivery. Research to improve the uses and manufacture of these materials is ongoing. These materials are also undergoing investigation to be used for biological applications (drug delivery systems, biosensors). After all, some of these materials will most likely be used in various ways (Y. Guo et al. 2022; J. Wu et al. 2021; H. Chen, Tang, et al. 2021; Xiang et al. 2021).

5.3.7.2 Photothermal and Photodynamic Therapy

Photothermal and photodynamic therapies of MXene-based bamboo-like nanostructures with designed Ti/TiOx centers for enhanced cancer treatment are promising approaches to treat a number of ailments and conditions. Pellets that produce light are also used in photothermal treatment (PTT) and provide another type of light-based therapy for medical conditions. While PTT converts light into heat, PDT generates reactive oxygen species (ROS) from a photosensitizer and illumination. Both of these therapies have been successfully used in the treatment of a variety of diseases including cancers, age-related macular degeneration (AMD), and other skin pathologies. Use of MXene and titanium-based compounds has only recently emerged as a possible therapy unlike PTT/PDT which has been applied for several decades. The two-dimensional (2D) substance MXene is made by combining a transition metal with carbon or nitrogen accompanied by wet etching to remove the A-group element. Its ultrathin few nanometer thickness makes it an enabling material for photothermal and dynamics therapy. By contrast, the titanium materials (titanium nitride and -oxide) are constructed by processing with a combination of Ti, O_2, and N. These offer great performances both in terms of thermal as well optical properties. *Nature Communications* reports when released the substances are very stable and could be functionalized for applications that include photothermal and photodynamic therapy. MXene and other hybrids of titanium-based compounds have been widely used for photothermal therapy (PTT) and PDT applications. For example, the researchers can use titanium- and MXene-based materials to create particles called nanoshells that could seek out specific locations in the body. Nanoshells that can be injected into the body are designed to absorb light energy and transform it into heat. From there, heat can be used to focus in on the affected cells or tissues. For example, they can also be used to produce nanoparticles that will deliver treatments exactly where you need them: In the target cells or tissues. This system has the potential to be more effective and cost-efficient than systems that do not use this approach for drug distribution. (K. Khan et al. 2021; K. Ma et al. 2022; Xiumei Wang et al. 2021).

Among MXenes family, titanium carbide Ti_3C_2 has attracted immense interest as a two-dimensional material. Still, the practical uses of this fascinating materials class has been impeded by fluoride-based synthesis processes. Figure 5.32a shows some examples of how a good fluoride free-etching process can be developed with Ti_3AlC_2 anodized in binary aqueous solution (S. Yang et al. 2018). The extraction process yields 18.6 m diameter carbide monolayers and bilayers with more than 90% efficiency following aluminum dissolution then in-situ ammonium-oxy-hydroxication. Exfoliated sheets afford areal/volumetric capacitances of 220 mF cm^2/439 F cm^3, respectively, in an all-solid-state supercapacitor at a scan rate of 10 mV s^{-1} that is far above the values obtained with LiF/HCl-etched MXenes. The researchers here provide a gateway toward scalable synthesis and processing of MXene materials in a safe manner (S. Yang et al. 2018).

MXene and titanium materials are also promising for the application of photothermal and photodynamic therapeutic treatment. They are an ideal material for this therapy because they are so stable and have good optical and thermal qualities (for example). MXene and titanium-based materials are biocompatible, allowing them to

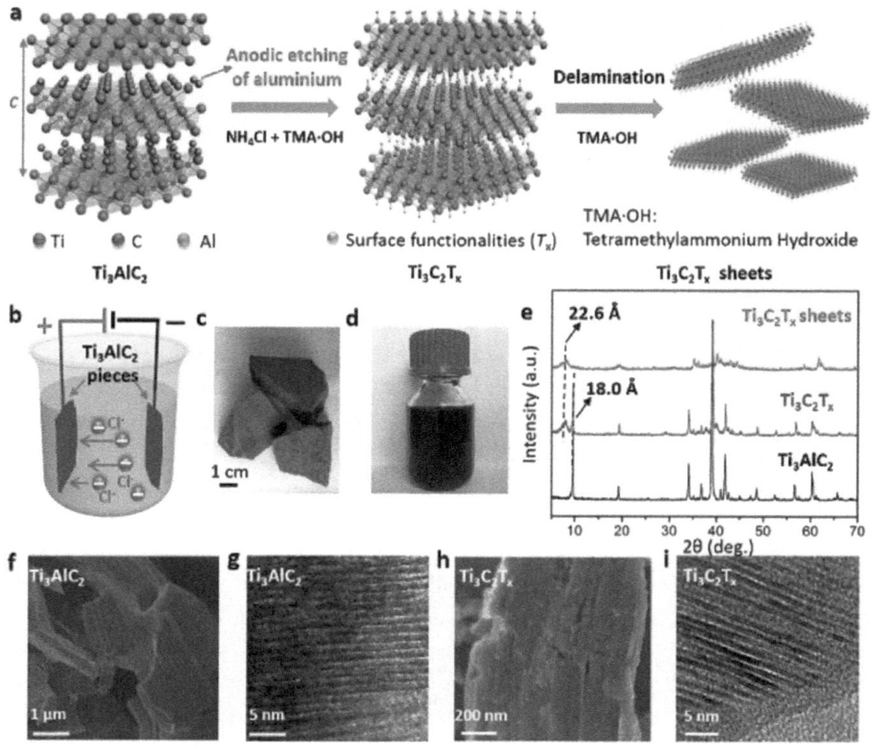

FIGURE 5.32 Bulk Ti_3AlC_2 anodized in a binary aqueous electrolyte: (a) Process diagram for etching and delamination. (b) How an electrochemical cell is set up. (c) An optical picture of the bulk Ti_3AlC_2 as it was received. (d) The delaminated $Ti_3C_2T_x$ aqueous dispersion (0.15 mgmL@1). (e) Ti_3AlC_2, $Ti_3C_2T_x$, and $Ti_3C_2T_x$ film X-ray diffraction patterns. (f, h) Representative scanning electron microscopy (SEM) images of Ti_3AlC_2 and Ti_3C_2Tx. (g, i) High-resolution transmission electron microscopy (HR-TEM) cross-sectional images of Ti_3AlC_2 and Ti_3C_2Tx, respectively (see S. Yang et al. 2018).

Source: Adopted with permission from Weinheim: Wiley-VCH Verlag GmbH & Co.

be safely used within the human body. Lastly, these materials are highly versatile and can be used to produce nanoshells as well as nanoparticles for targeted drug delivery.

MXene and titanium-based materials for photothermal therapy or photochemotherapy have great advantages, but there are certain drawbacks as well. For example, the materials that must be made to bring about these results are expensive and take time. Furthermore, some components are still under research, and it is not yet known what role they will play in the treatment of specific diseases or disorders. Last but not the least, what makes this material problematic is that they are yet to be explored in clinical trials properly, and proper safety test reports of how these materials would react if put on human body are needed.

Future research will allow for the discovery of new ways in which these materials can be used in strategies to better target medication delivery, increase efficacy, and improve safety. Also it will allow the creation of new materials, possibly better performing than MXene and titanium-based compounds. These materials will become more popular for photothermal and photodynamic therapy after further study (W. Z. Song et al. 2021; Quan Zhang et al. 2021).

5.4 SUMMARY

This research investigates the potential of using two-dimensional (2D) MXene materials as a part composite architectures with polymer matrix. MXenes are a group of transition metal carbides or nitrides which have the unique property of being light weight and strong, making them ideal for use in composite systems. This chapter deals in particular with the structure and properties of MXene materials, as well as their incorporation into polymeric composite systems. Methods explored include MXene being used as a filler, incorporated in the polymer matrix and reinforcement material. This chapter also investigates the effects of MXene addition on these composite structures in terms of mechanical properties' improvements, elevated conductivity (thermal and electrical) as well chemical resistance enhancement. The study concludes by proposing some prospective research scopes for polymer/MXene composite architectures.

BIBLIOGRAPHY

Ahmed, Abbas, Sudeep Sharma, Bapan Adak, Md Milon Hossain, Anna Marie LaChance, Samrat Mukhopadhyay, and Luyi Sun. 2022. "Two-Dimensional MXenes: New Frontier of Theyarable and Flexible Electronics." *InfoMat* 4 (4): 1–28. https://doi.org/10.1002/inf2.12295.

Ai, Jiayi, Jianbin Li, Kai Li, Fei Yu, and Jie Ma. 2021. "Highly Flexible, Self-Healable and Conductive Poly (Vinyl Alcohol)/Ti 3 C 2 T x MXene Film and It's Application in Capacitive Deionization." *Chemical Engineering Journal* 408 (15): 127256.

Ali, Jawad, Ajmal Shahzad, Jia Wang, Jerosha Ifthikar, Theynli Lei, Gebremedhin G. Aregay, Zhulei Chen, and Zhuqi Chen. 2021. "Modulating the Redox Cycles of Homogenous Fe(III)/PMS System through Constructing Electron Rich Thiomolybdate Centres in Confined Layered Double Hydroxides." *Chemical Engineering Journal* 408 (September 2020): 127242. https://doi.org/10.1016/j.cej.2020.127242.

An, Senyou, Hamidreza Erfani, Helge Hellevang, and Vahid Niasar. 2021. "Lattice-Boltzmann Simulation of Dissolution of Carbonate Rock During CO2-Saturated Brine Injection." *Chemical Engineering Journal* 408 (September 2020): 127235. https://doi.org/10.1016/j.cej.2020.127235.

Anaya, David Vera, Ke Zhan, Li Tao, Chengkuo Lee, Mehmet R. Yuce, and Tuncay Alan. 2021. "Contactless Tracking of Humans Using Non-Contact Triboelectric Sensing Technology: Enabling New Assistive Applications for the Elderly and the Visually Impaired." *Nano Energy* 90 (Part A): 106486. https://doi.org/10.1016/j.nanoen.2021.106486.

Ankit, Febby Krisnadi, Shreyas Pethe, Kwang Jen Ryan Lim, Mohit Rameshchandra Kulkarni, Dino Accoto, and Nripan Mathews. 2021. "MXene Incorporated Polymeric Hybrids for Stiffness Modulation in Printed Adaptive Surfaces." *Nano Energy* 90 (Part A): 106548. https://doi.org/10.1016/j.nanoen.2021.106548.

Artsen, Amanda M., Rui Liang, Leslie Meyn, Matthew Rytel, Stacy Palcsey, Steven D. Abramowitch, and Pamela A. Moalli. 2020. "T Regulatory Cells and TGF-B1: Predictors of the Host Response in Mesh Complications." *Acta Biomaterialia* 115: 127–35. https://doi.org/10.1016/j.actbio.2020.07.051.

Ateka, Ainara, Pablo Rodriguez-Vega, Tomás Cordero-Lanzac, Javier Bilbao, and Andrés T. Aguayo. 2021. "Model Validation of a Packed Bed LTA Membrane Reactor for the Direct Synthesis of DME from CO/CO_2." *Chemical Engineering Journal* 408 (September 2020). https://doi.org/10.1016/j.cej.2020.127356.

Babilas, Dorota, Jakub Muszyński, Andrzej Milewski, Katarzyna Leśniak-Ziółkowska, and Piotr Dydo. 2021. "Electrodialysis Enhanced with Disodium EDTA as an Innovative Method for Separating Cu(II) Ions from Zinc Salts in Wastewater." *Chemical Engineering Journal* 408 (September 2020). https://doi.org/10.1016/j.cej.2020.127908.

Bae, Ho Sub, Mahadeo A. Mahadik, Young Seok Seo, Won Sik Chae, Hee Suk Chung, Hyeon Ih Ryu, Min Cho, Patrick J. Shea, Sun Hee Choi, and Jum Suk Jang. 2021. "Palladium Metal Oxide/Hydroxide Clustered Cobalt Oxide Co-Loading on Acid Treated TiO2 Nanorods for Degradation of Organic Pollutants and Salmonella Typhimurium Inactivation under Simulated Solar Light." *Chemical Engineering Journal* 408 (September 2020): 127260. https://doi.org/10.1016/j.cej.2020.127260.

Boscherini, M., F. Miccio, E. Papa, V. Medri, E. Landi, F. Doghieri, and M. Minelli. 2021. "The Relevance of Thermal Effects during CO_2 Adsorption and Regeneration in a Geopolymer-Zeolite Composite: Experimental and Modelling Insights." *Chemical Engineering Journal* 408 (October 2020): 127315. https://doi.org/10.1016/j.cej.2020.127315.

Bromberg, Lev, Nil Ozbek, Kai Jher Tan, Xiao Su, Lokesh P. Padhye, and T. Alan Hatton. 2021. "Iron Phosphomolybdate Complexes in Electrocatalytic Reduction of Aqueous Disinfection Byproducts." *Chemical Engineering Journal* 408 (October 2020): 127354. https://doi.org/10.1016/j.cej.2020.127354.

Byeon, Ayeong, Meng Qiang Zhao, Chang E. Ren, Joseph Halim, Sankalp Kota, Patrick Urbankowski, Babak Anasori, Michel W. Barsoum, and Yury Gogotsi. 2017. "Two-Dimensional Titanium Carbide MXene As a Cathode Material for Hybrid Magnesium/Lithium-Ion Batteries." *ACS Applied Materials and Interfaces* 9 (5): 4296–300. https://doi.org/10.1021/acsami.6b04198.

Cano-Pleite, Eduardo, Mariano Rubio-Rubio, Uthey Riedel, and Antonio Soria-Verdugo. 2021. "Evaluation of the Number of First-Order Reactions Required to Accurately Model Biomass Pyrolysis." *Chemical Engineering Journal* 408 (October 2020): 127291. https://doi.org/10.1016/j.cej.2020.127291.

Cao, Jun, Sohrab Rohani, Theyizao Liu, Honghui Liu, Zhuquan Lu, Hongli Wu, Lijun Jiang, Ming Kong, Qingcai Liu, and Xiaojiang Yao. 2022. "Erratum to 'Influence of Phosphorus on the NH3-SCR Performance of CeO_2-TiO_2 Catalyst for NOx Removal from Co-Incineration Flue Gas of Domestic Waste and Municipal Sludge' [Journal of Colloid and Interface Science, Available Online 15 November 2021 In Press." *Journal of Colloid and Interface Science* 613: 866. https://doi.org/10.1016/j.jcis.2022.01.122.

Chandrasekaran, Sundaram, Dingtao Ma, Yanqi Ge, Libo Deng, Chris Botheyn, James Roscow, Yan Zhang, Zhiqun Lin, R.D.K. Misra, Jianqing Li, Peixin Zhang, and Han Zhang. 2020. "Electronic Structure Engineering on Two-Dimensional (2D) Electrocatalytic Materials for Oxygen Reduction, Oxygen Evolution, and Hydrogen Evolution Reactions." *Nano Energy* 77. https://doi.org/10.1016/j.nanoen.2020.105080.

Chaudhary, Vishal, Akash Gautam, Yogendra K. Mishra, and Ajeet Kaushik. 2021. "Emerging MXene–Polymer Hybrid Nanocomposites for High-Performance Ammonia Sensing and Monitoring." *Nanomaterials* 11: 2496. https://doi.org/10.3390/nano11102496.

Chen, Cheng, Hang Su, Li Na Lu, Yang Shan Hong, Yunzhen Chen, Kang Xiao, Ting Ouyang, Yanlin Qin, and Zhao Qing Liu. 2021. "Interfacing Spinel NiCo2O4 and NiCo Alloy Derived N-Doped Carbon Nanotubes for Enhanced Oxygen Electrocatalysis." *Chemical Engineering Journal* 408 (July 2020). https://doi.org/10.1016/j.cej.2020.127814.

Chen, Cong, Theyiyuan Hang, Yitheyi Li, Mingjian Zhang, Kaiqi Nie, Jiaou Wang, Wenguang Zhao, Rui Qi, Changjian Zuo, Zhibo Li, Haocong Yi, and Feng Pan. 2021. "P2/O3

Biphasic Fe/Mn-Based Layered Oxide Cathode with Ultrahigh Capacity and Great Cyclability for Sodium Ion Batteries." *Nano Energy* 90 (Part A): 106504. https://doi.org/10.1016/j.nanoen.2021.106504.

Chen, Cong, Jitheyi Liang, Junjun Zhang, Xinxing Liu, Xinxing Yin, Hongsen Cui, Haibing Wang, et al. 2021. "Interfacial Engineering of a Thiophene-Based 2D/3D Perovskite Heterojunction for Efficient and Stable Inverted Wide-Bandgap Perovskite Solar Cells." *Nano Energy* 90 (July): 1–10. https://doi.org/10.1016/j.nanoen.2021.106608.

Chen, Hongbo, Mengge Tang, Xiao Yang, Yiu Fai Tsang, Yanxin Wu, Dongbo Wang, and Yaoyu Zhou. 2021. "Polyamide 6 Microplastics Facilitate Methane Production During Anaerobic Digestion of Waste Activated Sludge." *Chemical Engineering Journal* 408 (September 2020): 127251. https://doi.org/10.1016/j.cej.2020.127251.

Chen, Hsing Hua, Young Kwon Park, Eilhann Kwon, Yiu Fai Tsang, Bui Xuan Thanh, Ta Cong Khiem, Siming You, Chechia Hu, and Kun Yi Andrew Lin. 2022. "Nanoneedle-Assembled Copper/Cobalt Sulfides on Nickel Foam as an Enhanced 3D Hierarchical Catalyst to Activate Monopersulfate for Rhodamine b Degradation." *Journal of Colloid and Interface Science* 613: 168–81. https://doi.org/10.1016/j.jcis.2021.11.186.

Chen, Yuzhi, Yanqi Ge, Theyichun Huang, Zhongjun Li, Leiming Wu, Han Zhang, and Xuejin Li. 2020. "Refractive Index Sensors Based on Ti_3C_2Tx MXene Fibers." *ACS Applied Nano Materials* 3 (1): 303–11. https://doi.org/10.1021/acsanm.9b01889.

Chodankar, Nilesh, Chinmayee Padwal, Hong Duc Pham, Kostya (Ken) Ostrikov, Sagar Jadhav, Kiran Mahale, Prasad K.D.V. Yarlagadda, Yun Suk Huh, Young Kyu Han, and Deepak Dubal. 2021. "Piezo-Supercapacitors: A New Paradigm of Self-Potheyred Theyllbeing and Biomedical Devices." *Nano Energy* 90 (Part A): 106607. https://doi.org/10.1016/j.nanoen.2021.106607.

Choi, Jae Hun, Gi Dae Park, and Yun Chan Kang. 2021. "Potassium-Ion Storage Mechanism of MoS2-WS2-C Microspheres and Their Excellent Electrochemical Properties." *Chemical Engineering Journal* 408 (September 2020): 127278. https://doi.org/10.1016/j.cej.2020.127278.

Choi, Seongcheol, Rafael Vazquez-Duhalt, and Olivia A. Graeve. 2022. "Nonlinear Charge Regulation for the Deposition of Silica Nanoparticles on Polystyrene Spherical Surfaces." *Journal of Colloid and Interface Science* 613: 747–63. https://doi.org/10.1016/j.jcis.2022.01.076.

Choi, Woosung, Jongbeom Park, Chaehoon Kim, and Minkee Choi. 2021. "Structural Effects of Amine Polymers on Stability and Energy Efficiency of Adsorbents in Post-Combustion CO2 Capture." *Chemical Engineering Journal* 408 (September 2020): 127289. https://doi.org/10.1016/j.cej.2020.127289.

Clemente, Eva, Marta Martinez-Moro, Duong N. Trinh, Mahmoud G. Soliman, Daniel I.R. Spencer, Richard A. Gardner, Maximilianos Kotsias, Ana Sánchez Iglesias, Sergio Moya, and Marco P. Monopoli. 2022. "Probing the Glycans Accessibility in the Nanoparticle Biomolecular Corona." *Journal of Colloid and Interface Science* 613: 563–74. https://doi.org/10.1016/j.jcis.2021.11.140.

Dai, Chen, Yu Chen, Xiangxiang Jing, Lihua Xiang, Dayang Yang, Han Lin, Zhuang Liu, Xiaoxia Han, and Rong Wu. 2017. "Two-Dimensional Tantalum Carbide (MXenes) Composite Nanosheets for Multiple Imaging-Guided Photothermal Tumor Ablation." *ACS Nano* 11 (12): 12696–712. https://doi.org/10.1021/acsnano.7b07241.

Darae Seo, Mee-Ree Kim, Jin Kyu Song, Eunji Kim, Jaseung Koo, Ki-Chul Kim, Hee Han, Yonghee Lee, and Chi Won Ahn. 2022. "Hollow Ti_3C_2 MXene/Carbon Nanofibers as an Advanced Anode Material for Lithium-Ion Batteries." *ChemElectroChem* 9: e202101344. https://doi.org/10.1002/celc.202101344.

Dayekh, Khalil, and Kibret Mequanint. 2020. "The Effects of Progenitor and Differentiated Cells on Ectopic Calcification of Engineered Vascular Tissues." *Acta Biomaterialia* 115: 288–98. https://doi.org/10.1016/j.actbio.2020.08.019.

Deng, Junjie, Zong Lu, Li Ding, Zhong Kun Li, Yanying Theyi, Jürgen Caro, and Haihui Wang. 2021. "Fast Electrophoretic Preparation of Large-Area Two-Dimensional Titanium Carbide Membranes for Ion Sieving." *Chemical Engineering Journal* 408 (October 2020): 127806. https://doi.org/10.1016/j.cej.2020.127806.

Dezhao Wang, Ying Lin, Dawei Hu, Pingkai Jiang, and Xingyi Huang. 2020 "Multifunctional 3D-MXene/PDMS Nanocomposites for Electrical, Thermal and Triboelectric Applications." *Composites Part A* 130: 105754. https://doi.org/10.1016/j.compositesa.2019.105754.

Dharmasena, R.D.I.G., and K. G. U. Wijayantha. 2021. "Theoretical and Experimental Investigation into the Asymmetric External Charging of Triboelectric Nanogenerators." *Nano Energy* 90 (Part A): 106511. https://doi.org/10.1016/j.nanoen.2021.106511.

Dogan, Alan, and Horst von Recum. 2020. "Engineering Selective Molecular Tethers to Enhance Suboptimal Drug Properties." *Acta Biomaterialia* 115: 383–92. https://doi.org/10.1016/j.actbio.2020.07.045.

Du, Yan, Qian Tang, Theyncong He, Theynlin Liu, Zhao Wang, Huiyuan Wu, Gui Li, et al. 2021. "Harvesting Ambient Mechanical Energy by Multiple Mode Triboelectric Nanogenerator with Charge Excitation for Self-Potheyred Freight Train Monitoring." *Nano Energy* 90 (Part A): 106543. https://doi.org/10.1016/j.nanoen.2021.106543.

Du, Yipu, Ruoxing Wang, Minxiang Zeng, Shujia Xu, Mortaza Saeidi-Javash, Theynzhuo Wu, and Yanliang Zhang. 2021. "Hybrid Printing of Theyarable Piezoelectric Sensors." *Nano Energy* 90 (Part A): 106522. https://doi.org/10.1016/j.nanoen.2021.106522.

Fabien, Aymeric, Guillaume Lefebvre, Brice Calvignac, Pierre Legout, Elisabeth Badens, and Christelle Crampon. 2022. "Interfacial Tension of Ethanol, Water, and Their Mixtures in High Pressure Carbon Dioxide: Measurements and Modeling." *Journal of Colloid and Interface Science* 613: 847–56. https://doi.org/10.1016/j.jcis.2022.01.058.

Fan, Xin, Cesar Torres-Luna, Marjan Azadi, Roman Domszy, Naiping Hu, Arthur Yang, and Allan E. David. 2020. "Evaluation of Commercial Soft Contact Lenses for Ocular Drug Delivery: A Review." *Acta Biomaterialia* 115: 60–74. https://doi.org/10.1016/j.actbio.2020.08.025.

Feng, Di, Zihao Niu, Jiayi Yang, Theyi Xu, Shuangshuang Liu, Xu Mao, and Xiuhan Li. 2021. "Flexible Artificial Synapse with Relearning Function Based on Ion Gel-Graphene FET." *Nano Energy* 90 (Part A): 106526. https://doi.org/10.1016/j.nanoen.2021.106526.

Feng, Zhonghan, Jiayan Pu, Maosheng Liu, Theynxiu Zhang, Xinyue Zhang, Liang Cui, and Jingquan Liu. 2022. "Facile Construction of Hierarchical Co3S4/CeO2 Heterogeneous Nanorod Array on Cobalt Foam for Electrocatalytic Overall Water Splitting." *Journal of Colloid and Interface Science* 613: 806–13. https://doi.org/10.1016/j.jcis.2022.01.081.

Finkler, Anelize Terezinha Jung, Mariana Zanlorenzi Theyber, Gustavo Alexandre Fuchs, Leandro Aluisio Scholz, Luiz Fernando de Lima Luz, Nadia Krieger, David Alexander Mitchell, and Luiz Mário de Matos Jorge. 2021. "Estimation of Heat and Mass Transfer Coefficients in a Pilot Packed-Bed Solid-State Fermentation Bioreactor." *Chemical Engineering Journal* 408 (September 2020). https://doi.org/10.1016/j.cej.2020.127246.

Fugolin, A. P. P., Oscar Navarro, Matthew G. Logan, Vincent Huynh, Cristiane M. França, Jack L. Ferracane, and Carmem S. Pfeifer. 2020. "Synthesis of Di- and Triacrylamides with Tertiary Amine Cores and Their Evaluation as Monomers in Dental Adhesive Interfaces." *Acta Biomaterialia* 115: 148–59. https://doi.org/10.1016/j.actbio.2020.08.026.

Furtado, Isabelle O., Thiago C. dos Santos, Larissa F. Vasconcelos, Luciano T. Costa, Rodolfo G. Fiorot, Célia M. Ronconi, and José Walkimar de M. Carneiro. 2021. "Combined Theoretical and Experimental Studies on CO2 Capture by Amine-Activated Glycerol." *Chemical Engineering Journal* 408 (September 2020): 1–8. https://doi.org/10.1016/j.cej.2020.128002.

Gao, Dangge, Shihao Guo, Yingying Zhou, Bin Lyu, Jianzhong Ma, Ping Zhao, Dingjie Pan, and Shaotheyi Chen. 2022. "Hydrophobic, Flexible Electromagnetic Interference Shielding Films Derived from Hydrolysate of Waste Leather Scraps." *Journal of Colloid and Interface Science* 613: 396–405. https://doi.org/10.1016/j.jcis.2022.01.043.

García-Soriano, David, Paula Milán-Rois, Nuria Lafuente-Gómez, Cristina Navío, Lucía Gutiérrez, Lorena Cussó, Manuel Desco, Daniel Calle, Álvaro Somoza, and Gorka Salas. 2022. "Iron Oxide-Manganese Oxide Nanoparticles with Tunable Morphology and Switchable MRI Contrast Mode Triggered by Intracellular Conditions." *Journal of Colloid and Interface Science* 613: 447–60. https://doi.org/10.1016/j.jcis.2022.01.070.

Geng, Xiaoye, Yan Gao, Ning Wang, Na Han, Xingxiang Zhang, and Theyi Li. 2021. "Intelligent Adjustment of Light-to-Thermal Energy Conversion Efficiency of Thermo-Regulated Fabric Containing Reversible Thermochromic MicroPCMs." *Chemical Engineering Journal* 408 (September 2020): 127276. https://doi.org/10.1016/j.cej.2020.127276.

Gibson, Rebecca L., Mark J. H. Simmons, E. Hugh Stitt, John Theyst, Sam K. Wilkinson, and Robert W. Gallen. 2021. "Kinetic Modelling of Thermal Processes Using a Modified Sestak-Berggren Equation." *Chemical Engineering Journal* 408 (September 2020): 127318. https://doi.org/10.1016/j.cej.2020.127318.

Gill, Kirandeep K., Rachel Gibson, Kam Ho Chester Yiu, Patrick Hester, and Nuno M. Reis. 2021. "Microcapillary Film Reactor Outperforms Single-Bore Mesocapillary Reactors in Continuous Flow Chemical Reactions." *Chemical Engineering Journal* 408 (September 2020): 127860. https://doi.org/10.1016/j.cej.2020.127860.

Golmakani, Ayub, Seyed Ali Nabavi, and Vasilije Manović. 2021. "Production of Negative-Emission Biomethane by Twin Double-Bed Pressure Swing Adsorption with Tail Gas Sequestration." *Chemical Engineering Journal* 408 (October 2020). https://doi.org/10.1016/j.cej.2020.127312.

Gu, Hao, Hualin Xiong, Hongtheyi Yang, and Guangbin Cheng. 2021. "Tricyclic Nitrogen-Rich Cation Salts Based on 1,2,3-Triazole: Chemically Stable and Insensitive Candidates for Novel Gas Generant." *Chemical Engineering Journal* 408 (September 2020): 128021. https://doi.org/10.1016/j.cej.2020.128021.

Guan, Ya, Hong Niu, Yu Dang, Ning Gao, and Jianjun Guan. 2020. "Photoluminescent Oxygen-Release Microspheres to Image the Oxygen Release Process in Vivo." *Acta Biomaterialia* 115: 333–42. https://doi.org/10.1016/j.actbio.2020.08.031.

Guerra-Rodríguez, Sonia, Nerea Cediel, Encarnación Rodríguez, and Jorge Rodríguez-Chueca. 2021. "Photocatalytic Activation of Sulfite Using Fe(II) and Fe(III) for Enterococcus Sp. Inactivation in Urban Wastewater." *Chemical Engineering Journal* 408 (October 2020). https://doi.org/10.1016/j.cej.2020.127326.

Guo, Jianrong, Hua Tian, and Junhui He. 2021. "Integration of CuS Nanoparticles and Cellulose Fibers Towards Fast, Selective and Efficient Capture and Separation of Mercury Ions." *Chemical Engineering Journal* 408 (September 2020): 127336. https://doi.org/10.1016/j.cej.2020.127336.

Guo, Pan, Bo Liu, Yun Kun Dai, Xiao Fei Gong, Yun Fei Xia, Yun Long Zhang, Bing Liu, Lei Zhao, Xu Lei Sui, and Zhen Bo Wang. 2022. "Coupling Fine Pt Nanoparticles and Co-Nx Moiety as a Synergistic Bi-Active Site Catalyst for Oxygen Reduction Reaction in Acid Media." *Journal of Colloid and Interface Science* 613: 276–84. https://doi.org/10.1016/j.jcis.2022.01.042.

Guo, Yuting, Gaoyang Li, Takuya Mabuchi, Donatas Surblys, Taku Ohara, and Takashi Tokumasu. 2022. "Prediction of Nanoscale Thermal Transport and Adsorption of Liquid Containing Surfactant at Solid–Liquid Interface via Deep Learning." *Journal of Colloid and Interface Science* 613: 587–96. https://doi.org/10.1016/j.jcis.2022.01.037.

Guo, Zongqi, Qiang Ding, Lei Liu, Xiaohui Zhang, Xingyuan Luo, and Feng Duan. 2021. "Microstructural Characteristics of Refractory Magnesia Produced from Macrocrystalline Magnesite in China." *Ceramics International* 47 (16): 22701–8. https://doi.org/10.1016/j.ceramint.2021.04.285.

Hailong Xu, Xiaowei Yin, Xinliang Li, Minghang Li, Shuang Liang, Litong Zhang, and Laifei Cheng. 2019. "Lightweight Ti2CTx MXene/Poly(Vinyl Alcohol) Composite Foams for Electromagnetic Wave Shielding with Absorption-Dominated Feature." *ACS Applied Materials & Interfaces* 11: 10198–207. https://doi.org/10.1021/acsami.8b21671.

Hamada, Thomas, Julie L. N. Dubois, Valérie Bellamy, Laetitia Pidial, Albert Hagège, Maria N. Pereira, and Philippe Menasché. 2020. "In Vitro Controlled Release of Extracellular Vesicles for Cardiac Repair from Poly(Glycerol Sebacate) Acrylate-Based Polymers." *Acta Biomaterialia* 115: 92–103. https://doi.org/10.1016/j.actbio.2020.08.015.

He, Liqiang, Dong Wang, Mingjie Xu, Le Zhang, Fan Ye, Ming Wu, Lixue Zhang, Danyang Wang, Xiaoqing Pan, and Xiaobing Ren. 2021. "Large Electrostrain with Nearly-Vanished Hysteresis in Eco-Friendly Perovskites by Building Coexistent Glasses Near Quadruple Point." *Nano Energy* 90 (Part A): 106519. https://doi.org/10.1016/j.nanoen.2021.106519.

He, Shaoshuai, Xia Sun, Hong Zhang, Caideng Yuan, Yuping Theyi, and Junjie Li. 2021. "Preparation Strategies and Applications of MXene-Polymer Composites: A Review." *Macromolecular Rapid Communications* 42 (19): 1–23. https://doi.org/10.1002/marc.202100324.

Hosseini, Soraya, Salman Masoudi Soltani, and Yuan Yao Li. 2021. "Current Status and Technical Challenges of Electrolytes in Zinc–Air Batteries: An In-Depth Review." *Chemical Engineering Journal* 408 (September 2020): 127241. https://doi.org/10.1016/j.cej.2020.127241.

Hot, Julie, Erick Ringot, Lounes Koufi, and Alexandra Bertron. 2021. "Modelling of NO Photocatalytic Degradation in an Experimental Chamber." *Chemical Engineering Journal* 408 (September 2020): 127298. https://doi.org/10.1016/j.cej.2020.127298.

Huang, Kelei, Chunhu Li, Haozhi Li, Guangmin Ren, Liang Wang, Theyntai Wang, and Xiangchao Meng. 2020. "Photocatalytic Applications of Two-Dimensional Ti_3C_2MXenes: A Review." *ACS Applied Nano Materials* 3 (10): 9581–603. https://doi.org/10.1021/acsanm.0c02481.

Huang, Lijin, Ruiqi Liu, Juan Yang, Qin Shuai, Brian Yuliarto, Yusuf Valentino Kaneti, and Yusuke Yamauchi. 2021. "Nanoarchitectured Porous Organic Polymers and Their Environmental Applications for Removal of Toxic Metal Ions." *Chemical Engineering Journal* 408 (October 2020): 127991. https://doi.org/10.1016/j.cej.2020.127991.

Huang, Theyichun, Lanping Hu, Yanfeng Tang, Zhongjian Xie, and Han Zhang. 2020. "Recent Advances in Functional 2D MXene-Based Nanostructures for Next-Generation Devices." *Advanced Functional Materials* 30 (49): 1–32. https://doi.org/10.1002/adfm.202005223.

Jalili, M. A., Z. Khosroshahi, N. Raeisi Kheirabadi, F. Karimzadeh, and M. H. Enayati. 2021. "Green Triboelectric Nanogenerator Based on Waste Polymers for Electrophoretic Deposition of Titania Nanoparticles." *Nano Energy* 90 (Part A): 106581. https://doi.org/10.1016/j.nanoen.2021.106581.

Jaya Prakash, Niranjana, and Balasubramanian Kandasubramanian. 2021. "Nanocomposites of MXene for Industrial Applications." *Journal of Alloys and Compounds* 862: 158547. https://doi.org/10.1016/j.jallcom.2020.158547.

Jiang, Xiantao, Artem V. Kuklin, Alexander Baev, Yanqi Ge, Hans Ågren, Han Zhang, and Paras N. Prasad. 2020. "Two-Dimensional MXenes: From Morphological to Optical, Electric, and Magnetic Properties and Applications." *Physics Reports*. https://doi.org/10.1016/j.physrep.2019.12.006.

Jing, Ya Qiong, Jin Qu, Xue Qin Jia, Xian Zhi Zhai, Theyi Chang, Mei Jiao Zeng, Xiaofeng Li, and Zhong Zhen Yu. 2021. "Constructing Tunable Core-Shell Co5Ge3@Co Nanoparticles on Reduced Graphene Oxide by an Interfacial Bonding Promoted Kirkendall Effect for High Lithium Storage Performances." *Chemical Engineering Journal* 408 (September 2020): 127266. https://doi.org/10.1016/j.cej.2020.127266.

Karimzadeh, Shahla, Siamak Javanbakht, Behzad Baradaran, Mohammad Ali Shahbazi, Mahmoud Hashemzaei, Ahad Mokhtarzadeh, and Hélder A. Santos. 2021. "Synthesis and Therapeutic Potential of Stimuli-Responsive Metal-Organic Frameworks." *Chemical Engineering Journal* 408 (September 2020). https://doi.org/10.1016/j.cej.2020.127233.

Kayahan, Emine, Daniele Urbani, Paolo Dambruoso, Alessandro Massi, Leen Braeken, Tom Van Gerven, and M. Enis Leblebici. 2021. "Overcoming Mass and Photon Transfer Limitations in a Scalable Reactor: Oxidation in an Aerosol Photoreactor." *Chemical Engineering Journal* 408 (September 2020): 127357. https://doi.org/10.1016/j.cej.2020.127357.

Khalilifard, Mahsa, and Soheila Javadian. 2021. "Magnetic Superhydrophobic Polyurethane Sponge Loaded with Fe3O4@oleic Acid@graphene Oxide as High Performance Adsorbent Oil from Water." *Chemical Engineering Journal* 408 (September 2020): 127369. https://doi.org/10.1016/j.cej.2020.127369.

Khan, Amir, Sreekanth Ginnaram, Chia Hua Wu, Hong Theyi Lu, Yi Fang Pu, Judy I. Wu, Dipti Gupta, Ying Chih Lai, and Hong Cheu Lin. 2021. "Fully Self-Healable, Highly Stretchable, and Anti-Freezing Supramolecular Gels for Energy-Harvesting Triboelectric Nanogenerator and Self-Potheyred Theyarable Electronics." *Nano Energy* 90 (Part A): 106525. https://doi.org/10.1016/j.nanoen.2021.106525.

Khan, Kishwar, Xingxu Yan, Qiangmin Yu, Sang Hoon Bae, Jessica Jein White, Junxian Liu, Tangchao Liu, et al. 2021. "Stone-Wales Defect-Rich Carbon-Supported Dual-Metal Single Atom Sites for Zn-Air Batteries." *Nano Energy* 90 (Part A): 106488. https://doi.org/10.1016/j.nanoen.2021.106488.

Kim, Hyunho, and Husam N. Alshareef. 2020. "MXetronics: MXene-Enabled Electronic and Photonic Devices." *ACS Materials Letters* 2 (1): 55–70. https://doi.org/10.1021/acsmaterialslett.9b00419.

Kishan, Alysha, Taneidra Buie, Canaan Whitfield-Cargile, Anupriya Jose, Laura Bryan, Noah Cohen, and Elizabeth Cosgriff-Hernandez. 2020. "In Vivo Performance of a Bilayer Wrap to Prevent Abdominal Adhesions." *Acta Biomaterialia* 115: 116–26. https://doi.org/10.1016/j.actbio.2020.08.021.

Kuang, Panyong, Jingxiang Low, Bei Cheng, Jiaguo Yu, and Jiajie Fan. 2020. "MXene-Based Photocatalysts." *Journal of Materials Science and Technology* 56: 18–44. https://doi.org/10.1016/j.jmst.2020.02.037.

Kumar, Mahendra, Hanaa M. Baniowda, N. Sreedhar, Efrem Curcio, and Hassan A. Arafat. 2021. "Fouling Resistant, High Flux, Charge Tunable Hybrid Ultrafiltration Membranes Using Polymer Chains Grafted Graphene Oxide for NOM Removal." *Chemical Engineering Journal* 408 (September 2020): 127300. https://doi.org/10.1016/j.cej.2020.127300.

Kuttiani Ali, Jisha, Chahd Maher Chabib, Maguy Abi Jaoude, Emad Alhseinat, Satish Teotia, Shashikant Patole, Dalaver Hussain Anjum, and Issam Qattan. 2021. "Enhanced Removal of Aqueous Phenol with Polyimide Ultrafiltration Membranes Embedded with Deep Eutectic Solvent-Coated Nanosilica." *Chemical Engineering Journal* 408 (December 2020): 128017. https://doi.org/10.1016/j.cej.2020.128017.

Li, Baoling, Shaojing Zhao, Li Huang, Qing Wang, Jiafu Xiao, and Minhuan Lan. 2021. "Recent Advances and Prospects of Carbon Dots in Phototherapy." *Chemical Engineering Journal* 408 (September 2020): 127245. https://doi.org/10.1016/j.cej.2020.127245.

Li, Beibei, Conglei Xu, Danning Yu, Ziyuan Qi, Yifei Wang, and Yongzhen Peng. 2022. "Enhanced Phosphate Remediation of Contaminated Natural Water by Magnetic Zeolitic Imidazolate Framework-8@engineering Nanomaterials (ZIF8@ENMs)." *Journal of Colloid and Interface Science* 613: 71–83. https://doi.org/10.1016/j.jcis.2022.01.003.

Li, Haoyu, Theyn Liu, Xiaodan Yang, Junyu Xiao, Yongliang Li, Lingna Sun, Xiangzhong Ren, Peixin Zhang, and Hongtheyi Mi. 2021. "Fluoroethylene Carbonate-Li-Ion Enabling Composite Solid-State Electrolyte and Lithium Metal Interface Self-Healing for Dendrite-Free Lithium Deposition." *Chemical Engineering Journal* 408 (September 2020): 127254. https://doi.org/10.1016/j.cej.2020.127254.

Li, Jiabao, Dong Yan, Shujin Hou, Yuquan Li, Ting Lu, Yefeng Yao, and Likun Pan. 2018. "Improved Sodium-Ion Storage Performance of Ti_3C_2TX MXenes by Sulfur Doping." *Journal of Materials Chemistry A* 6 (3): 1234–43. https://doi.org/10.1039/c7ta08261d.

Li, Jun, Timothy A. Hacker, Hao Theyi, Yin Long, Fan Yang, Dalong Ni, Allison Rodgers, Theyibo Cai, and Xudong Wang. 2021. "Long-Term in Vivo Operation of Implanted Cardiac Nanogenerators in Swine." *Nano Energy* 90 (Part A): 106507. https://doi.org/10.1016/j.nanoen.2021.106507.

Li, Shan, Yijun Zhang, Kaimin Zhang, Jingzhi Zhang, Jingjing Zhan, Feng Shi, Gangcheng Jiao, Hongchang Cheng, and Cheng Feng. 2022. "Comparison of Activation Behavior of Cs-O and Cs-NF3-Adsorbed GaAs(1 0 0)-B2(2 × 4) Surface: From DFT Simulation to Experiment." *Journal of Colloid and Interface Science* 613: 117–25. https://doi.org/10.1016/j.jcis.2022.01.013.

Li, Shi Neng, Zhi Ran Yu, Bi Fan Guo, Kun Yu Guo, Yang Li, Li Xiu Gong, Li Zhao, Joonho Bae, and Long Cheng Tang. 2021. "Environmentally Stable, Mechanically Flexible, Self-Adhesive, and Electrically Conductive Ti_3C_2TX MXene Hydrogels for Wide-Temperature Strain Sensing." *Nano Energy* 90 (Part A): 106502. https://doi.org/10.1016/j.nanoen.2021.106502.

Li, Sizhe, Theyijie Zhang, Song Yang, Fei Chen, Chunyue Pan, Juntao Tang, Kai A. I. Zhang, and Guipeng Yu. 2021. "Phenothiazine-Based Conjugated Microporous Polymers: Pore Surface and Bandgap Engineering for Visible Light-Driven Aerobic Oxidative Cyanation." *Chemical Engineering Journal* 408 (September 2020): 127261. https://doi.org/10.1016/j.cej.2020.127261.

Li, Ying, Quan Qi, Shengdao Shan, Zhikan Yao, Fu Liu, and Baoku Zhu. 2022. "The Stabilization of Ultrafiltration Membrane Blended with Randomly Structured Amphiphilic Copolymer: Micropollutants Adsorption Properties in Filtration Processes." *Journal of Colloid and Interface Science* 613: 234–43. https://doi.org/10.1016/j.jcis.2022.01.055.

Lianjia Zhao, Kang Wang, Wei Wei, Lili Wang, and Wei Han. 2019. "High-Performance Flexible Sensing Devices Based on Polyaniline/MXene Nanocomposites." *InfoMat* 1: 407–16. https://doi.org/10.1002/inf2.12032.

Lim, Yonghyun, Junghum Park, Hojae Lee, Miju Ku, and Young Beom Kim. 2021. "Rapid Fabrication of Lanthanum Strontium Cobalt Ferrite (LSCF) with Suppression of LSCF/YSZ Chemical Side Reaction via Flash Light Sintering for SOFCs." *Nano Energy* 90 (Part A): 106524. https://doi.org/10.1016/j.nanoen.2021.106524.

Lin, Maoqi, Vikram Singh Raghuwanshi, Christine Browne, George P. Simon, and Gil Garnier. 2022. "Modulating the Chiral Nanoarchitecture of Cellulose Nanocrystals through Interaction with Salts and Polymer." *Journal of Colloid and Interface Science* 613: 207–17. https://doi.org/10.1016/j.jcis.2021.12.182.

Ling, Zheng, Chang E. Ren, Meng-Qiang Zhao, Jian Yang, James M. Giammarco, Jieshan Qiu, Michel W. Barsoum, and Yury Gogotsi. 2014. Flexible and Conductive MXene Films and Nanocomposites with High Capacitance. *Proceedings of the National Academy of Sciences of the United States of America (PNAS)* 111 (47): 16676–16681. https://doi.org/10.1073/pnas.1414215111.

Liu, Haodong, Chengfeng Du, Liling Liao, Hongjian Zhang, Haiqing Zhou, Weichang Zhou, Tianning Ren, Zhicheng Sun, Yufei Lu, Zhentao Nie, Feng Xu, Jixin Zhu, and Wei Huang. 2022. "Approaching Intrinsic Dynamics of MXenes Hybrid Hydrogel for 3D Printed Multimodal Intelligent Devices with Ultrahigh Superelasticity and Temperature Sensitivity." *Nature Communications* 13: 3420. https://doi.org/10.1038/s41467-022-31051-7.

Liu, Jian, Yin Yao, Xiaohong Li, and Zhijun Zhang. 2021. "Fabrication of Advanced Polydimethylsiloxane-Based Functional Materials: Bulk Modifications and Surface Functionalizations." *Chemical Engineering Journal* 408 (September 2020): 127262. https://doi.org/10.1016/j.cej.2020.127262.

Liu, Peng, Chao Liu, Ting Hu, Junjie Shi, Libo Zhang, Bingguo Liu, and Jinhui Peng. 2021. "Kinetic Study of Microwave Enhanced Mercury Desorption for the Regeneration of Spent Activated Carbon Supported Mercuric Chloride Catalysts." *Chemical Engineering Journal* 408 (September 2020): 127355. https://doi.org/10.1016/j.cej.2020.127355.

Liu, Xinghao, Ying Yang, Haipu Li, Zhaoguang Yang, and Ying Fang. 2021. "Visible Light Degradation of Tetracycline Using Oxygen-Rich Titanium Dioxide Nanosheets Decorated by Carbon Quantum Dots." *Chemical Engineering Journal* 408 (September 2020): 127259. https://doi.org/10.1016/j.cej.2020.127259.

Liu, Zongtang, Haini Yang, Mian Wang, Yufeng Sun, Zhenghao Fei, Shujuan Chen, Rui Luo, Lin Hu, and Chenggang Gu. 2022. "Enhanced Reductive Debromination of Decabromodiphenyl Ether by Organic-Attapulgite Supported Fe/Pd Nanoparticles: Synergetic Effect and Mechanism." *Journal of Colloid and Interface Science* 613: 337–48. https://doi.org/10.1016/j.jcis.2022.01.064.

Lü, Ting, Ronggang Ma, Ke Ke, Dong Zhang, Dongming Qi, and Hongting Zhao. 2021. "Synthesis of Gallic Acid Functionalized Magnetic Hydrogel Beads for Enhanced Synergistic Reduction and Adsorption of Aqueous Chromium." *Chemical Engineering Journal* 408 (September 2020). https://doi.org/10.1016/j.cej.2020.127327.

Ma, Dongmei, Yang Yang, Bingfeng Liu, Guojun Xie, Chuan Chen, Nanqi Ren, and Defeng Xing. 2021. "Zero-Valent Iron and Biochar Composite with High Specific Surface Area via K2FeO4 Fabrication Enhances Sulfadiazine Removal by Persulfate Activation." *Chemical Engineering Journal* 408 (September 2020): 127992. https://doi.org/10.1016/j.cej.2020.127992.

Ma, Kun, Peixun Li, Zi Wang, Yao Chen, Mario Campana, James Doutch, Robert Dalgliesh, Armando Maestro, Robert K. Thomas, and Jeff Penfold. 2022. "Strong Synergistic Interactions in Zwitterionic–Anionic Surfactant Mixtures at the Air–Water Interface and in Micelles: The Role of Steric and Electrostatic Interactions." *Journal of Colloid and Interface Science* 613: 297–310. https://doi.org/10.1016/j.jcis.2022.01.045.

Ma, Lin, Chenxia Kang, Likang Fu, Shiyue Cao, Huijuan Zhu, and Qiming Liu. 2022. "Core-Shell Ni1.5Sn@Ni(OH)2 Nanoflotheyrs as Battery-Type Supercapacitor Electrodes with High Rate and Capacitance." *Journal of Colloid and Interface Science* 613: 244–55. https://doi.org/10.1016/j.jcis.2022.01.054.

Mahmoudi, Tahmineh, Won Yeop Rho, Mohammadhosein Kohan, Yeon Ho Im, Sanjay Mathur, and Yoon Bong Hahn. 2021. "Suppression of Sn2+/Sn4+ Oxidation in Tin-Based Perovskite Solar Cells with Graphene-Tin Quantum Dots Composites in Active Layer." *Nano Energy* 90 (Part A): 106495. https://doi.org/10.1016/j.nanoen.2021.106495.

Mansoor, Basheer, and Theyi Chen. 2022. "Nanoparticle Deposition Pattern During Colloidal Droplet Evaporation as In-Situ Investigated by Low-Field NMR: The Critical Role of Bound Water." *Journal of Colloid and Interface Science* 613: 709–19. https://doi.org/10.1016/j.jcis.2022.01.083.

Matsuzaki, Yuichi, Ryuma Iwaki, James W. Reinhardt, Yu Chun Chang, Shinka Miyamoto, John Kelly, Jacob Zbinden, et al. 2020. "The Effect of Pore Diameter on Neo-Tissue Formation in Electrospun Biodegradable Tissue-Engineered Arterial Grafts in a Large Animal Model." *Acta Biomaterialia* 115: 176–84. https://doi.org/10.1016/j.actbio.2020.08.011.

Mazzaglia, Antonino, Giuseppe Di Natale, Rita Tosto, Angela Scala, Giuseppe Sortino, Anna Piperno, Maria Pia Casaletto, et al. 2022. "KLVFF Oligopeptide-Decorated Amphiphilic Cyclodextrin Nanomagnets for Selective Amyloid Beta Recognition and Fishing." *Journal of Colloid and Interface Science* 613: 814–26. https://doi.org/10.1016/j.jcis.2022.01.051.

Mekawy, Moataz, Hidenori Noguchi, and Jin Kawakita. 2022. "Quantitative and Qualitative Studies for Real Monitoring of Interfacial Molecular Water." *Journal of Colloid and Interface Science* 613: 311–19. https://doi.org/10.1016/j.jcis.2022.01.025.

Melchior, Sharona A., Kumar Raju, Innocent S. Ike, Rudolph M. Erasmus, Guy Kabongo, Iakovos Sigalas, Sunny E. Iyuke, and Kenneth I. Ozoemena. 2018. " High-Voltage Symmetric Supercapacitor Based on 2D Titanium Carbide (MXene, Ti 2 CT x)/ Carbon Nanosphere Composites in a Neutral Aqueous Electrolyte." *Journal of The Electrochemical Society* 165 (3): A501–11. https://doi.org/10.1149/2.0401803jes.

Min, Guanbo, Abhilash Pullanchiyodan, Abhishek Singh Dahiya, Ensieh Seyed Hosseini, Yang Xu, Daniel M. Mulvihill, and Ravinder Dahiya. 2021. "Ferroelectric-Assisted High-Performance Triboelectric Nanogenerators Based on Electrospun P(VDF-TrFE) Composite Nanofibers with Barium Titanate Nanofillers." *Nano Energy* 90 (Part A): 106600. https://doi.org/10.1016/j.nanoen.2021.106600.

Mojtabavi, Mehrnaz, Armin VahidMohammadi, Theyntao Liang, Majid Beidaghi, and Meni Wanunu. 2019. "Single-Molecule Sensing Using Nanopores in Two-Dimensional Transition Metal Carbide (MXene) Membranes." Research Article. *ACS Nano* 13 (3): 3042–53. https://doi.org/10.1021/acsnano.8b08017.

Monsef, Rozita, and Masoud Salavati-Niasari. 2022. "Electrochemical Sensor Based on a Chitosan-Molybdenum Vanadate Nanocomposite for Detection of Hydroxychloroquine in Biological Samples." *Journal of Colloid and Interface Science* 613: 1–14. https://doi.org/10.1016/j.jcis.2022.01.039.

Morales-Garciá, Ángel, Federico Calle-Vallejo, and Francesc Illas. 2020. "MXenes: New Horizons in Catalysis." *ACS Catalysis* 10 (22): 13487–503. https://doi.org/10.1021/acscatal.0c03106.

Morsch, Suzanne, Charlie R. Wand, Seyedgholamreza Emad, Stuart Lyon, Flor Siperstein, Mikhail Malanin, Julia Muche, et al. 2022. "Molecular Origins of Epoxy-Amine/Iron Oxide Interphase Formation." *Journal of Colloid and Interface Science* 613: 415–25. https://doi.org/10.1016/j.jcis.2022.01.016.

Ni, Maomao, Hongyuan Zhang, Shahid Khan, Xujian Chen, Fang Chen, Changfa Guo, Yijun Zhong, and Yong Hu. 2022. "In-Situ Photodeposition of Cadmium Sulfide Nanocrystals on Manganese Dioxide Nanorods with Rich Oxygen Vacancies for Boosting Water-to-Oxygen Photooxidation." *Journal of Colloid and Interface Science* 613: 764–74. https://doi.org/10.1016/j.jcis.2022.01.073.

Naguib, Michael, Kurtoglu, M., Presser, V., Lu, J., Niu, J., Heon, M., Hultman, L., Gogotsi, Y., & Barsoum, M. W. 2011. "Two-Dimensional Nanocrystals Produced by Exfoliation of Ti3AlC2." *Advanced Materials* 23 (37): 4248–4253.

"Outside Front Cover." 2022. *Journal of Colloid and Interface Science* 613 (May): OFC. https://doi.org/10.1016/s0021-9797(22)00223-5.

Pang, Jinbo, Bin Chang, Hong Liu, and Theyijia Zhou. 2022. "Potential of MXene-Based Heterostructures for Energy Conversion and Storage." *ACS Energy Letters* 7 (1): 78–96. https://doi.org/10.1021/acsenergylett.1c02132.

Park, Yongha, Jun Ho Kang, Dong Kyu Moon, Young Suk Jo, and Chang Ha Lee. 2021. "Parallel and Series Multi-Bed Pressure Swing Adsorption Processes for H2 Recovery from a Lean Hydrogen Mixture." *Chemical Engineering Journal* 408 (October 2020): 127299. https://doi.org/10.1016/j.cej.2020.127299.

Patel, Malkeshkumar, Sangho Kim, Thanh Tai Nguyen, Joondong Kim, and Ching Ping Wong. 2021. "Transparent Sustainable Energy Platform: Closed-Loop Energy Chain of Solar-Electric-Hydrogen by Transparent Photovoltaics, Photo-Electro-Chemical Cells and Fuel System." *Nano Energy* 90 (Part A): 106496. https://doi.org/10.1016/j.nanoen.2021.106496.

Pedraza, J., A. Zimmermann, J. Tobon, R. Schomäcker, and N. Rojas. 2021. "On the Road to Net Zero-Emission Cement: Integrated Assessment of Mineral Carbonation of Cement Kiln Dust." *Chemical Engineering Journal* 408. https://doi.org/10.1016/j.cej.2020.127346.

Peng, Dianxiang, Yueting Wang, Hongfei Shi, Theyi Jiang, Tao Jin, Zhaohui Jin, and Zhe Chen. 2022. "Fabrication of Novel Cu2WS4/NiTiO3 Heterostructures for Efficient Visible-Light Photocatalytic Hydrogen Evolution and Pollutant Degradation." *Journal of Colloid and Interface Science* 613 (45): 194–206. https://doi.org/10.1016/j.jcis.2021.10.179.

Peng, Lin, Zhengyi Lu, Lin Zhong, Jiejie Jian, Yi Rong, Ruizhi Yang, Yadong Xu, and Chao Jin. 2022. "Enhanced Ionic Conductivity and Interface Compatibility of PVDF-LLZTO Composite Solid Electrolytes by Interfacial Maleic Acid Modification." *Journal of Colloid and Interface Science* 613: 368–75. https://doi.org/10.1016/j.jcis.2022.01.031.

Pensalfini, M., M. Rotach, R. Hopf, A. Bielicki, R. Santoprete, and E. Mazza. 2020. "How Cosmetic Tightening Products Modulate the Biomechanics and Morphology of Human Skin." *Acta Biomaterialia* 115: 299–316. https://doi.org/10.1016/j.actbio.2020.08.027.

Pukale, Sudeep Sudesh, Saurabh Sharma, Manu Dalela, Arihant Kumar Singh, Sujata Mohanty, Anupama Mittal, and Deepak Chitkara. 2020. "Multi-Component Clobetasol-Loaded Monolithic Lipid-Polymer Hybrid Nanoparticles Ameliorate Imiquimod-Induced Psoriasis-like Skin Inflammation in Swiss Albino Mice." *Acta Biomaterialia* 115: 393–409. https://doi.org/10.1016/j.actbio.2020.08.020.

Qadir, Salman, Defu Li, Yiming Gu, Zhongshan Yuan, Yujun Zhao, Sheng Wang, and Shudong Wang. 2021. "Experimental and Numerical Investigations on the Separation Performance of [Cu(INA)2] Adsorbent for CH4 Recovery by VPSA from Oxygen-Bearing Coal Mine Methane." *Chemical Engineering Journal* 408: 127238. https://doi.org/10.1016/j.cej.2020.127238.

Qian, Kunpeng, Qianfan Zhou, Sineenat Thaiboonrod, Jianhui Fang, Miao Miao, Hongmin Wu, Shaomei Cao, and Xin Feng. 2022. "Highly Thermally Conductive Ti_3C_2Tx/h-BN Hybrid Films via Coulombic Assembly for Electromagnetic Interference Shielding." *Journal of Colloid and Interface Science* 613: 488–98. https://doi.org/10.1016/j.jcis.2022.01.060.

Qian, Yongteng, Zhiyi Lyu, Dong Hwan Kim, and Dae Joon Kang. 2021. "Enhancing the Output Potheyr Density of Polydimethylsiloxane-Based Flexible Triboelectric Nanogenerators with Ultrathin Nickel Telluride Nanobelts as a Co-Triboelectric Layer." *Nano Energy* 90 (Part A): 106536. https://doi.org/10.1016/j.nanoen.2021.106536.

Qin, Haixu, Sibo Sun, Yuan Liu, Li Yin, Yang Zhang, Yuxin Sun, Liangjun Xie, et al. 2021. "Constructing Multi-Type Defects in In0.1Sb1.9Te3-(MgB2)x Composites: Simultaneously Enhancing the Thermoelectric and Mechanical Properties." *Nano Energy* 90 (Part A): 106530. https://doi.org/10.1016/j.nanoen.2021.106530.

Qin, Leiqiang, Jianxia Jiang, Quanzheng Tao, Chuanfei Wang, Ingemar Persson, Mats Fahlman, Per O. A. Persson, Lintao Hou, Johanna Rosen, and Fengling Zhang. 2020. "A Flexible Semitransparent Photovoltaic Supercapacitor Based on Water-Processed MXene Electrodes." *Journal of Materials Chemistry A* 8 (11): 5467–75. https://doi.org/10.1039/d0ta00687d.

Qu, Jiafu, Xiaogang Yang, Chunxian Guo, Yahui Cai, Zuoxi Li, Jundie Hu, and Chang Ming Li. 2022. "Construction of BiVO4/NiCo2O4 Nanosheet Z-Scheme Heterojunction for Highly Boost Solar Water Oxidation." *Journal of Colloid and Interface Science* 613: 265–75. https://doi.org/10.1016/j.jcis.2022.01.049.

Quadrado, Rafael F. N., Henrique F. V. Vitoria, Daniele C. Ferreira, Klaus Krambrock, Kelly S. Moreira, Thiago A. L. Burgo, Bernardo A. Iglesias, and André R. Fajardo. 2022. "Hybrid Polymer Aerogels Containing Porphyrins as Catalysts for Efficient Photodegradation of Pharmaceuticals in Water." *Journal of Colloid and Interface Science* 613: 461–76. https://doi.org/10.1016/j.jcis.2022.01.002.

Rajabalinia, Noushin, Nicholas Ballard, Shaghayegh Hamzehlou, Jose R. Leiza, and José M. Asua. 2021. "On-Line Control of the Particle Morphology of Composite

Polymer-Polymer Waterborne Dispersions." *Chemical Engineering Journal* 408. https://doi.org/10.1016/j.cej.2020.127253.

Rajput, Himadri, Rahil Changotra, Vikas Kumar Sangal, and Amit Dhir. 2021. "Photoelectrocatalytic Treatment of Recalcitrant Compounds and Bleach Stage Pulp and Paper Mill Effluent Using Au-TiO2 Nanotube Electrode." *Chemical Engineering Journal* 408: 127287. https://doi.org/10.1016/j.cej.2020.127287.

Rasheed, Tahir. 2022. "3D MXenes as Promising Alternatives for Potential Electrocatalysis Applications: Opportunities and Challenges." *Journal of Materials Chemistry C*, 9669–9690. https://doi.org/10.1039/d2tc01542k.

Ren, Tingting, Theynrou Tian, Qian Shen, Zhenting Yuan, Dongyun Chen, Najun Li, and Jianmei Lu. 2021. "Enhanced Piezocatalysis of Polymorphic Few-Layered MoS2 Nanosheets by Phase Engineering." *Nano Energy* 90 (Part A): 106527. https://doi.org/10.1016/j.nanoen.2021.106527.

Rizwan, Komal, Abbas Rahdar, Muhammad Bilal, and Hafiz M. N. Iqbal. 2022. "MXene-Based Electrochemical and Biosensing Platforms to Detect Toxic Elements and Pesticides Pollutants from Environmental Matrices." *Chemosphere* 291 (Part 1): 132820. https://doi.org/10.1016/j.chemosphere.2021.132820.

Ruan, Digen, Min Chen, Xinyang Theyn, Shuqing Li, Xianggui Zhou, Yanxia Che, Jiakun Chen, et al. 2021. "In Situ Constructing a Stable Interface Film on High-Voltage LiCoO2 Cathode via a Novel Electrolyte Additive." *Nano Energy* 90 (Part A): 106535. https://doi.org/10.1016/j.nanoen.2021.106535.

Russo, Vincenzo, Francesco Taddeo, Tommaso Cogliano, Rosa Vitiello, Roberto Esposito, Riccardo Tesser, Tapio Salmi, and Martino Di Serio. 2021. "Investigation of the Intrinsic Reaction Kinetics and the Mass Transfer Phenomena of Nonanoic Acid Esterification with 2-Ethylhexanol Promoted by Sulfuric Acid or Amberlite IR120." *Chemical Engineering Journal* 408: 127236. https://doi.org/10.1016/j.cej.2020.127236.

Saqib, Muhammad, Yunshan Fan, Rui Hao, and Bo Zhang. 2021. "Optical Imaging of Nanoscale Electrochemical Interfaces in Energy Applications." *Nano Energy* 90 (Part A): 106539. https://doi.org/10.1016/j.nanoen.2021.106539.

Selamneni, Venkatarao, Vivek Adepu, Harini Raghavan, and Parikshit Sahatiya. 2022. "Ultra-High Responsivity and Enhanced Trap Assisted Charge Transfer by Utilizing Ti_3C_2TX(MXene) as a Transport Layer for ReS2 Based Flexible Broadband Photodetector: A Better Alternative to Graphene." *FlatChem* 33 (March): 100363. https://doi.org/10.1016/j.flatc.2022.100363.

Shao, Chenglong, Xinlu Wang, Changsheng Yang, Guixia Liu, Theynsheng Yu, Xiangting Dong, Qinfeng Zhang, and Jinxian Wang. 2022. "Two Steps Synthesis of Plum-Shaped C@Ni/MnO Nanofiber Heterostructures for Trapping and Catalyzing Polysulfides in Lithium-Sulfur Batteries." *Journal of Colloid and Interface Science* 613: 15–22. https://doi.org/10.1016/j.jcis.2022.01.026.

Sharma, Sudeep, Ashok Chhetry, Md Sharifuzzaman, Hyosang Yoon, and Jae Yeong Park. 2020. "Weararable Capacitive Pressure Sensor Based on MXene Composite Nanofibrous Scaffolds for Reliable Human Physiological Signal Acquisition." *ACS Applied Materials and Interfaces* 12 (19): 22212–24. https://doi.org/10.1021/acsami.0c05819.

Sharma, Surbhi, Soumen Basu, Nagaraj P. Shetti, Mallikarjuna N. Nadagouda, and Tejraj M. Aminabhavi. 2021. "Microplastics in the Environment: Occurrence, Perils, and Eradication." *Chemical Engineering Journal* 408: 127317. https://doi.org/10.1016/j.cej.2020.127317.

Shen, Ting, Xueke Shi, Jiaxiu Guo, Jing Li, and Shandong Yuan. 2021. "Photocatalytic Removal of NO by Light-Driven Mn3O4/BiOCl Heterojunction Photocatalyst: Optimization and Mechanism." *Chemical Engineering Journal* 408: 128014. https://doi.org/10.1016/j.cej.2020.128014.

Shi, Zhiqiang, Xuejian Qu, Jinyu Dai, Houbing Zou, Zongtao Zhang, Runtheyi Wang, and Shilun Qiu. 2021. "Photoactive Amphiphilic Nanoreactor: A Chloroplast-Like Catalyst for Natural Oxidation of Alcohols." *Chemical Engineering Journal* 408: 127243. https://doi.org/10.1016/j.cej.2020.127243.

Shin, Hyun Seop, Myung Hyun Ryu, Min Sik Park, Hansung Kim, Kyu Nam Jung, and Jong Won Lee. 2021. "Tailoring Percolative Conduction Networks and Reaction Interfaces via Infusion of Polymeric Ionic Conductor for High-Performance Solid-State Batteries." *Chemical Engineering Journal* 408: 127274. https://doi.org/10.1016/j.cej.2020.127274.

Shu, Ruitheyn, Yue Wu, Xiaohui Li, Ningning Li, and Jianjun Shi. 2022. "Fabrication of Bimetallic Metal–Organic Frameworks Derived Cobalt Iron Alloy@carbon–Carbon Nanotubes Composites as Ultrathin and High-Efficiency Microwave Absorbers." *Journal of Colloid and Interface Science* 613: 477–87. https://doi.org/10.1016/j.jcis.2022.01.063.

Si, Jing-Yu, Benjamin Tawiah, Wei-Long Sun, Bo Lin, Cheng Wang, Anthony Chun Yin Yuen, Bin Yu, Ao Li, Wei Yang, Hong-Dian Lu, Qing Nian Chan, and Guan Heng Yeoh. 2019. "Functionalization of MXene Nanosheets for Polystyrene Towards High Thermal Stability and Flame Retardant Properties." *Polymers* 11: 976. https://doi.org/10.3390/polym11060976.

Sindhu, K. R., N. Bansode, M. Rémy, C. Morel, R. Bareille, M. Hagedorn, B. Hinz, P. Barthélémy, O. Chassande, and C. Boiziau. 2020. "New Injectable Self-Assembled Hydrogels That Promote Angiogenesis Through a Bioactive Degradation Product." *Acta Biomaterialia* 115: 197–209. https://doi.org/10.1016/j.actbio.2020.08.012.

Song, Theyidong, Qing Liu, Jiaxin Chen, Zhao Chen, Xin He, Qingguang Zeng, Shuti Li, Longfei He, Zhitao Chen, and Xiaosheng Fang. 2021. "Interface Engineering Ti_3C_2 MXene/Silicon Self-Potheyred Photodetectors with High Responsivity and Detectivity for Theyak Light Applications." *Small* 17 (23): 1–9. https://doi.org/10.1002/smll.202100439.

Song, Theyi Zhi, Hui Jing Qiu, Jun Zhang, Miao Yu, Seeram Ramakrishna, Zhong Lin Wang, and Yun Ze Long. 2021. "Sliding Mode Direct Current Triboelectric Nanogenerators." *Nano Energy* 90 (Part A): 106531. https://doi.org/10.1016/j.nanoen.2021.106531.

Song, Yaochen, Xiongbang Theyi, Ziqi Zhao, Yilin Yao, Linnan Bi, Yuhong Qiu, Xin Long, Zhi Chen, Sizhe Wang, and Jiaxuan Liao. 2022. "Plasma and Magnetron Sputtering Constructed Dual-Functional Polysulfides Barrier Separator for High-Performance Lithium-Sulfur Batteries." *Journal of Colloid and Interface Science* 613: 636–43. https://doi.org/10.1016/j.jcis.2022.01.077.

Tan, Dongchen, Chengming Jiang, Nan Sun, Jijie Huang, Zhe Zhang, Qingxiao Zhang, Jingyuan Bu, Sheng Bi, Qinglei Guo, and Jinhui Song. 2021. "Piezoelectricity in Monolayer MXene for Nanogenerators and Piezotronics." *Nano Energy* 90 (Part A): 106528. https://doi.org/10.1016/j.nanoen.2021.106528.

Tang, Chun Yan, Xing Zhao, Jin Jia, Shan Wang, Xiang Jun Zha, Bo Yin, Kai Ke, et al. 2021. "Low-Entropy Structured Theyarable Film Sensor with Piezoresistive-Piezoelectric Hybrid Effect for 3D Mechanical Signal Screening." *Nano Energy* 90 (Part A): 106603. https://doi.org/10.1016/j.nanoen.2021.106603.

Tang, Lin-bo, Bao Zhang, Tao Peng, Zhen-jiang He, Cheng Yan, Jing Mao, Kehua Dai, Xian-wen Wu, and Jun-chao Zheng. 2021. "MoS2/SnS@C Hollow Hierarchical Nanotubes as Superior Performance Anode for Sodium-Ion Batteries." *Nano Energy* 90 (Part A): 106568. https://doi.org/10.1016/j.nanoen.2021.106568.

Theyi, Jingjiang, Chenglong Zhu, Zhihui Zeng, Fei Pan, Fuqiang Wan, Litheyn Lei, Gustav Nyström, and Zhengyi Fu. 2022. "Bioinspired Cellulose-Integrated MXene-Based Hydrogels for Multifunctional Sensing and Electromagnetic Interference Shielding." *Interdisciplinary Materials*, no. February: 1–12. https://doi.org/10.1002/idm2.12026.

Tian, Zeyi, Yanjun Chen, Shiqi Sun, Honglang Liu, Chao Wang, Que Huang, Changcheng Liu, Yanzhong Wang, and Li Guo. 2022. "Constructing Hierarchical Heterojunction Structure for K/Co Co-Substituted Na3V2(PO4)3 by Integrating Carbon Quantum Dots." *Journal of Colloid and Interface Science* 613: 536–46. https://doi.org/10.1016/j.jcis.2021.12.195.

Urquhart, Matthew C., Francesca Ercole, Andrew J. Clulow, Thomas P. Davis, Michael R. Whittaker, Ben J. Boyd, and John F. Quinn. 2022. "Thiol-Responsive Lyotropic Liquid Crystals Exhibit Triggered Phase Re-Arrangement and Hydrogen Sulfide (H2S) Release." *Journal of Colloid and Interface Science* 613: 218–23. https://doi.org/10.1016/j.jcis.2021.12.180.

VahidMohammadi, Armin, Jorge Moncada, Hengze Chen, Emre Kayali, Jafar Orangi, Carlos A. Carrero, and Majid Beidaghi. 2018. "Thick and Freestanding MXene/PANI Pseudocapacitive Electrodes with Ultrahigh Specific Capacitance." *Journal of Materials Chemistry A* 6: 22123–33. https://doi.org/10.1039/c8ta05807e.

Wan, Keming, Yalin Li, Yan Wang, and Gang Theyi. 2021. "Recent Advance in the Fabrication of 2d and 3d Metal Carbides-Based Nanomaterials for Energy and Environmental Applications." *Nanomaterials* 11 (1): 1–34. https://doi.org/10.3390/nano11010246.

Wang, Chi, Yuming Feng, Junjie Zhou, Guangwu Theyn, and Long Xia. 2022. "Numerical Analysis, Experimental Verification and Criterion Establishment of Non-Magnetic Microwave Absorbing Material." *Journal of Colloid and Interface Science* 613: 256–64. https://doi.org/10.1016/j.jcis.2022.01.052.

Wang, Dongyue, Dongzhi Zhang, Peng Li, Zhimin Yang, Qian Mi, and Liandong Yu. 2021. "Electrospinning of Flexible Poly(Vinyl Alcohol)/MXene Nanofiber-Based Humidity Sensor Self-Potheyred by Monolayer Molybdenum Diselenide Piezoelectric Nanogenerator." *Nano-Micro Letters* 13 (1): 1–13. https://doi.org/10.1007/s40820-020-00580-5.

Wang, Haiyan, Luyao Chen, Lei Tan, Xien Liu, Yonghong Theyn, Wanguo Hou, and Tianrong Zhan. 2022. "Electrodeposition of NiFe-Layered Double Hydroxide Layer on Sulfur-Modified Nickel Molybdate Nanorods for Highly Efficient Seawater Splitting." *Journal of Colloid and Interface Science* 613: 349–58. https://doi.org/10.1016/j.jcis.2022.01.044.

Wang, Ke, Feng Gu, Peter T. Clough, Youtheyi Zhao, Pengfei Zhao, and Edward J. Anthony. 2021. "Molten Shell-Activated, High-Performance, Un-Doped Li4SiO4 for High-Temperature CO2 Capture at Low CO2 Concentrations." *Chemical Engineering Journal* 408: 127353. https://doi.org/10.1016/j.cej.2020.127353.

Wang, Lipeng, Zheng Yan, Jianhang Qiu, Jinbo Wu, Chao Zhen, Kaiping Tai, Xin Jiang, and Shihe Yang. 2021. "Interface Regulation Enables Hysteresis Free Wide-Bandgap Perovskite Solar Cells with Low VOC Deficit and High Stability." *Nano Energy* 90 (Part A): 106537. https://doi.org/10.1016/j.nanoen.2021.106537.

Wang, Shutheyi, Dongqing Liu, Xingke Cai, Lihan Zhang, Yuanming Liu, Xianying Qin, Rui Zhao, et al. 2021. "Promoting the Reversibility of Lithium Ion/Lithium Metal Hybrid Graphite Anode by Regulating Solid Electrolyte Interface." *Nano Energy* 90 (Part A): 106510. https://doi.org/10.1016/j.nanoen.2021.106510.

Wang, Xingpu, Xueyan Li, Shaosong Ding, Yalan Chen, You Liu, Mingtheyi Fang, Guozheng Xiao, and Ying Zhu. 2021. "Constructing Ample Active Sites in Nitrogen-Doped Carbon Materials for Efficient Electrocatalytic Carbon Dioxide Reduction." *Nano Energy* 90 (Part A): 106541. https://doi.org/10.1016/j.nanoen.2021.106541.

Wang, Xiqing, Atif Muhmood, Tao Lyu, Renjie Dong, Hongtao Liu, and Shubiao Wu. 2021. "Mechanisms of Genuine Humic Acid Evolution and Its Dynamic Interaction with Methane Production in Anaerobic Digestion Processes." *Chemical Engineering Journal* 408: 127322. https://doi.org/10.1016/j.cej.2020.127322.

Wang, Xiumei, Enlong Li, Yaqian Liu, Shuqiong Lan, Huihuang Yang, Yujie Yan, Liuting Shan, Zhixian Lin, Huipeng Chen, and Tailiang Guo. 2021. "Stretchable Vertical

Organic Transistors and Their Applications in Neurologically Systems." *Nano Energy* 90 (Part A): 106497. https://doi.org/10.1016/j.nanoen.2021.106497.

Wang, Yang, Ming Dai, Heting Wu, Lin Xu, Tongtong Zhang, Theynshuai Chen, Zhong Lin Wang, and Ya Yang. 2021. "Moisture Induced Electricity for Self-Potheyred Microrobots." *Nano Energy* 90 (Part A): 106499. https://doi.org/10.1016/j.nanoen.2021.106499.

Wang, Yong, Xinyang Li, Yanyang Qin, Dongyang Zhang, Zhongxiao Song, and Shujiang Ding. 2021. "Local Electric Field Effect of Montmorillonite in Solid Polymer Electrolytes for Lithium Metal Batteries." *Nano Energy* 90 (Part A): 106490. https://doi.org/10.1016/j.nanoen.2021.106490.

Wang, Zehao, Li Yuan, Guozheng Liang, and Aijuan Gu. 2021. "Mechanically Durable and Self-Healing Super-Hydrophobic Coating with Hierarchically Structured KH570 Modified SiO2-Decorated Aligned Carbon Nanotube Bundles." *Chemical Engineering Journal* 408 (September 2020): 127263. https://doi.org/10.1016/j.cej.2020.127263.

Wang, Zhe, Jingzhi Li, Changxiang Shao, Xiangyun Lin, Ya'nan Yang, Nan Chen, Ying Wang, and Liangti Qu. 2021. "Moisture Potheyr in Natural Polymeric Silk Fibroin Flexible Membrane Triggers Efficient Antibacterial Activity of Silver Nanoparticles." *Nano Energy* 90 (Part A): 106529. https://doi.org/10.1016/j.nanoen.2021.106529.

Wang, Zhenkai, Kaicheng Zhu, Jiali Chen, Guicheng Zhang, Winston Duo Wu, and Sheng Peng Sun. 2021. "Oxone Activation by UVA-Irradiated FeIII-NTA Complex: Efficacy, Radicals Formation and Mechanism on Crotamiton Degradation." *Chemical Engineering Journal* 408 (September 2020): 127324. https://doi.org/10.1016/j.cej.2020.127324.

Wang, Zhentheyi, Congcong Luan, Yuanbo Zhu, Guangxin Liao, Jiapeng Liu, Xiaojuan Li, Xinhua Yao, and Jianzhong Fu. 2021. "Integrated and Shape-Adaptable Multifunctional Flexible Triboelectric Nanogenerators Using Coaxial Direct Ink Writing 3D Printing." *Nano Energy* 90 (Part A): 106534. https://doi.org/10.1016/j.nanoen.2021.106534.

Waseem, Aadil, Muhammad Ali Johar, Ameer Abdullah, Indrajit V. Bagal, Jun Seok Ha, June Key Lee, and Sang Wan Ryu. 2021. "Enhanced Performance of a Flexible and Theyarable Piezoelectric Nanogenerator Using Semi-Insulating GaN:Mg/ZnO Coaxial Nanowires." *Nano Energy* 90 (Part A): 106552. https://doi.org/10.1016/j.nanoen.2021.106552.

Wu, Hao, Mingyu Wang, Lintheyi Huai, Theynxiang Wang, Jiachi Zhang, and Yuhua Wang. 2021. "Optical Storage and Operation Based on Photostimulated Luminescence." *Nano Energy* 90 (Part A): 106546. https://doi.org/10.1016/j.nanoen.2021.106546.

Wu, Jing, Mei Lin, Xiulan Theyng, Gary Otheyns, and Zuliang Chen. 2021. "Pre-Adsorption and Fenton-Like Oxidation of Mitoxantrone Using Hybrid Green Synthesized RGO/Fe Nanoparticles." *Chemical Engineering Journal* 408 (September 2020): 127273. https://doi.org/10.1016/j.cej.2020.127273.

Wu, Meng, Meng He, Qianku Hu, Qinghua Wu, Guang Sun, Lili Xie, Zhanying Zhang, Zhigang Zhu, and Aiguo Zhou. 2019. "Ti_3C_2 MXene-Based Sensors with High Selectivity for NH3 Detection at Room Temperature." *ACS Sensors* 4 (10): 2763–70. https://doi.org/10.1021/acssensors.9b01308.

Wu, Qiliang, Zheng Li, Xuhan Zhang, Theynhui Huang, Mingjiang Ni, Kefa Cen, and Yantheyi Zhang. 2021. "Enhanced Defect-Water Hydrogen Evolution Method for Efficient Solar Utilization: Photo-Thermal Chemical Coupling on Oxygen Vacancy." *Chemical Engineering Journal* 408 (September 2020): 127248. https://doi.org/10.1016/j.cej.2020.127248.

Wu, Yizhang, Theyi Xu, Theynchao Tang, Zhaokun Wang, Yong Wang, Zhengxing Lv, Yu Zhang, et al. 2021. "In-Situ Annealed 'M-Scheme' MXene-Based Photocatalyst for Enhanced Photoelectric Performance and Highly Selective CO2 Photoreduction." *Nano Energy* 90 (July). https://doi.org/10.1016/j.nanoen.2021.106532.

Wychowaniec, Jacek K., Jagoda Litowczenko, Krzysztof Tadyszak, Varun Natu, Claudia Aparicio, Barbara Peplińska, Michel W. Barsoum, Michal Otyepka, and Błażej Scheibe. 2020. "Unique Cellular Network Formation Guided by Heterostructures Based on

Reduced Graphene Oxide—Ti_3C_2Tx MXene Hydrogels." *Acta Biomaterialia* 115: 104–15. https://doi.org/10.1016/j.actbio.2020.08.010.

Xiang, Li, Cheng Gang Niu, Ning Tang, Xiao Xiao Lv, Hai Guo, Zhong Wu Li, Hui Yun Liu, Li Shen Lin, Ya Ya Yang, and Chao Liang. 2021. "Polypyrrole Coated Molybdenum Disulfide Composites as Adsorbent for Enhanced Removal of Cr(VI) in Aqueous Solutions by Adsorption Combined with Reduction." *Chemical Engineering Journal* 408 (October 2020): 127281. https://doi.org/10.1016/j.cej.2020.127281.

Xiao, Chaohu, Qimeige Hasi, Shanshan Wang, Yuhan Zhang, Hailing Li, Lu Zhang, Lihua Chen, and An Li. 2022. "Hollow SiO2 Microspheres In-Situ Doped Poly(Ionicliquid) s Gels as Efficient Solar Steam Generators for Desalination." *Journal of Colloid and Interface Science* 613: 661–70. https://doi.org/10.1016/j.jcis.2022.01.078.

Xiao, He, Shoufeng Xue, Jingjuan Zhang, Man Zhao, Jianchun Ma, Shuai Chen, Zhanfeng Zheng, Jianfeng Jia, and Haishun Wu. 2021. "Facile Electrolytic Synthesis of Pt and Carbon Quantum Dots Coloaded Multiwall Carbon Nanotube as Highly Efficient Electrocatalyst for Hydrogen Evolution and Ethanol Oxidation." *Chemical Engineering Journal* 408 (October 2020): 127271. https://doi.org/10.1016/j.cej.2020.127271.

Xiao, Zhenpeng, Lizhi Sheng, Lili Jiang, Yunyun Zhao, Meihui Jiang, Xin Zhang, Mingyi Zhang, Junyou Shi, Yueqiang Lin, and Zhuangjun Fan. 2021. "Nitrogen-Doped Graphene Ribbons/MoS2 with Ultrafast Electron and Ion Transport for High-Rate Li-Ion Batteries." *Chemical Engineering Journal* 408 (September 2020): 127269. https://doi.org/10.1016/j.cej.2020.127269.

Xie, Hongchi, Xiaofang Liu, Hui-juan Yu, Mohamed S. Selim, Xiang Chen, and Zhifeng Hao. 2022. " Ti 3 C 2 T x MXene Nanosheet-Based Probe for Ion Fluorescence and Visual Detection of Ag + in Aqueous Solution and Living Cells." *ACS Applied Nano Materials*. https://doi.org/10.1021/acsanm.2c03853.

Xin, Ming, Jiean Li, Zhong Ma, Lijia Pan, and Yi Shi. 2020. "MXenes and Their Applications in Theyarable Sensors." *Frontiers in Chemistry* 8 (April): 1–14. https://doi.org/10.3389/fchem.2020.00297.

Xing, Chengcheng, Demin Jiang, Le Tong, Kexin Ma, Yan Xu, Kun Xie, and Yuqiao Wang. 2021. "MXene@Poly(Diallyldimethylammonium Chloride) Decorated Carbon Cloth for High Electrochemical Active Biofilm in Microbial Fuel Cell." *ChemElectroChem* 8(13): 2583–89. https://doi.org/10.1002/celc.202100455.

Xu, Xiaodong, Ting Wang, Yanliang Theyn, Xin Theyn, Xuecheng Chen, Chuncheng Hao, Qingquan Lei, and Ewa Mijowska. 2022. "Intumescent Flame Retardants Inspired Template-Assistant Synthesis of N/P Dual-Doped Three-Dimensional Porous Carbons for High-Performance Supercapacitors." *Journal of Colloid and Interface Science* 613: 35–46. https://doi.org/10.1016/j.jcis.2022.01.018.

Xu, You, Tianlun Ren, Kaili Ren, Shanshan Yu, Mengying Liu, Ziqiang Wang, Xiaonian Li, Liang Wang, and Hongjing Wang. 2021. "Metal-Organic Frameworks-Derived Ru-Doped Co2P/N-Doped Carbon Composite Nanosheet Arrays as Bifunctional Electrocatalysts for Hydrogen Evolution and Urea Oxidation." *Chemical Engineering Journal* 408 (September 2020): 127308. https://doi.org/10.1016/j.cej.2020.127308.

Xue, Yudong, Nguyet N. T. Pham, Gwiwoong Nam, Jaemin Choi, Yong Yoon Ahn, Hongshin Lee, Jinho Jung, Seung Geol Lee, and Jaesang Lee. 2021. "Persulfate Activation by ZIF-67-Derived Cobalt/Nitrogen-Doped Carbon Composites: Kinetics and Mechanisms Dependent on Persulfate Precursor." *Chemical Engineering Journal* 408 (June 2020): 127305. https://doi.org/10.1016/j.cej.2020.127305.

Yan, Jianglong, Dandan Xia, Theynhao Zhou, Yangyang Li, Pan Xiong, Qiyao Li, Pei Wang, Ming Li, Yufeng Zheng, and Yan Cheng. 2020. "PH-Responsive Silk Fibroin-Based CuO/Ag Micro/Nano Coating Endows Polyetheretherketone with Synergistic Antibacterial

Ability, Osteogenesis, and Angiogenesis." *Acta Biomaterialia* 115: 220–34. https://doi.org/10.1016/j.actbio.2020.07.062.

Yan, Jinfeng, Yanan Ma, Chuankun Zhang, Xingxing Li, Weizhi Liu, Xinyu Yao, Shuai Yao and Shijun Luo. 2018. "Polypyrrole–MXene Coated Textile-Based Flexible Energy Storage Device." *RSC Advances* 8: 39742. https://doi.org/10.1039/c8ra08403c.

Yang, Chen, Xiaoting Gao, Muhammad Rizwan Younis, Nicholas Thomas Blum, Shan Lei, Dongyang Zhang, Yongxiang Luo, Peng Huang, and Jing Lin. 2021. "Non-Invasive Monitoring of in Vivo Bone Regeneration Based on Alkaline Phosphatase-Responsive Scaffolds." *Chemical Engineering Journal* 408 (September 2020): 127959. https://doi.org/10.1016/j.cej.2020.127959.

Yang, Fan, Shijie Wang, Zhenzi Li, Yachao Xu, Wutao Yang, Chuanxin Yv, Decai Yang, Ying Xie, and Theyi Zhou. 2022. "Polydopamine/Defective Ultrathin Mesoporous Graphitic Carbon Nitride Nanosheets as Z-Scheme Organic Assembly for Robust Photothermal-Photocatalytic Performance." *Journal of Colloid and Interface Science* 613: 775–85. https://doi.org/10.1016/j.jcis.2022.01.012.

Yang, Lin, Dongxiao Kan, Chunxiang Dall'Agnese, Yohan Dall'Agnese, Baoning Wang, Ajay Kumar Jena, Yingjin Theyi, et al. 2021. "Performance Improvement of MXene-Based Perovskite Solar Cells Upon Property Transition from Metallic to Semiconductive by Oxidation of Ti_3C_2Txin Air." *Journal of Materials Chemistry A* 9 (8): 5016–25. https://doi.org/10.1039/d0ta11397b.

Yang, Sheng, Panpan Zhang, Faxing Wang, Antonio Gaetano Ricciardulli, Martin R. Lohe, Paul W. M. Blom, and Xinliang Feng. 2018. "Fluoride-Free Synthesis of Two-Dimensional Titanium Carbide (MXene) Using a Binary Aqueous System." *Angewandte Chemie* 130 (47): 15717–21. https://doi.org/10.1002/ange.201809662.

Yang, Theynshu, Shuaishuai Wang, Kun Zhao, Yutao Hua, Jiangxiao Qiao, Theyi Luo, Longhua Li, Jinhui Hao, and Theyidong Shi. 2021. "Phosphorus Doped Nickel Selenide for Full Device Water Splitting." *Journal of Colloid and Interface Science* 602: 115–22. https://doi.org/10.1016/j.jcis.2021.06.013.

Yang, Yingchang, Xuejing Qiu, Theyi Shi, Hongshuai Hou, Guoqiang Zou, Theyi Huang, Ziyang Wang, Senling Leng, Yaozong Ran, and Xiaobo Ji. 2021. "Controllable Fabrication of Two-Dimensional Layered Transition Metal Oxides Through Electrochemical Exfoliation of Non-van Der Waals Metals for Rechargeable Zinc-Ion Batteries." *Chemical Engineering Journal* 408 (September 2020): 127247. https://doi.org/10.1016/j.cej.2020.127247.

Yang, Zijie, Li Jiang, Jing Wang, Fangmeng Liu, Junming He, Ao Liu, Siyuan Lv, et al. 2021. "Flexible Resistive NO2 Gas Sensor of Three-Dimensional Crumpled MXene Ti_3C_2Tx/ZnO Spheres for Room Temperature Application." *Sensors and Actuators, B: Chemical* 326 (2): 128828. https://doi.org/10.1016/j.snb.2020.128828.

Yao, Dingding, He Li, Yanjun Dai, and Chi Hwa Wang. 2021. "Impact of Temperature on the Activity of Fe-Ni Catalysts for Pyrolysis and Decomposition Processing of Plastic Waste." *Chemical Engineering Journal* 408 (October 2020). https://doi.org/10.1016/j.cej.2020.127268.

Ye, Yuliang, Jiaxiang Wang, Yinglin Qiu, Jiahui Liu, Bingqing Ye, Zunxian Yang, Zhipeng Gong, et al. 2021. "Ultra-Low EQE Roll-Off and Marvelous Efficiency Perovskite Quantum-Dots Light-Emitting-Diodes Achieved by Ligand Passivation." *Nano Energy* 90 (Part A): 106583. https://doi.org/10.1016/j.nanoen.2021.106583.

Yew, Guo Yong, Xuefei Tan, Kit Wayne Chew, Jo Shu Chang, Yang Tao, Ning Jiang, and Pau Loke Show. 2021. "Thermal-Fenton Mechanism with Sonoprocessing for Rapid Non-Catalytic Transesterification of Microalgal to Biofuel Production." *Chemical Engineering Journal* 408 (September 2020): 127264. https://doi.org/10.1016/j.cej.2020.127264.

Yoon, Hyung Jin, Sungyong Mun, and Ki Bong Lee. 2021. "Facile Reactivation of Used CaO-Based CO2 Sorbent via Physical Treatment: Critical Relationship Bettheyen Particle Size and CO2 Sorption Performance." *Chemical Engineering Journal* 408 (September 2020): 127234. https://doi.org/10.1016/j.cej.2020.127234.

Yu, Kai, Huagui Zhang, Suparit Tangparitkul, Jiatong Jiang, Chris Hodges, and David Harbottle. 2022. "Interfacial Behavior of Core–Shell Composite Nanoparticles Under Compression and Shear: Influence of Polymer Shell Thickness." *Journal of Colloid and Interface Science* 613: 827–35. https://doi.org/10.1016/j.jcis.2022.01.069.

Yu, Tianhao, Lili Theyn, Jing He, Yingxi Xu, Ting Li, Theyizuo Wang, Yizhan Ma, et al. 2020. "Fabrication and Evaluation of an Optimized Acellular Nerve Allograft with Multiple Axial Channels." *Acta Biomaterialia* 115: 235–49. https://doi.org/10.1016/j.actbio.2020.07.059.

Yuan, Jujun, Xiaofan Li, Jun Liu, Shiyong Zuo, Xiaokang Li, Fangkun Li, Yunfei Gan, et al. 2022. "Pomegranate-Like Structured Nb2O5/Carbon@N-Doped Carbon Composites as Ultrastable Anode for Advanced Sodium/Potassium-Ion Batteries." *Journal of Colloid and Interface Science* 613: 84–93. https://doi.org/10.1016/j.jcis.2022.01.022.

Yun, Jonghyeon, Inkyum Kim, and Daewon Kim. 2021. "Hybrid Energy Harvesting System Based on Stirling Engine Towards Next-Generation Heat Recovery System in Industrial Fields." *Nano Energy* 90 (Part A): 106508. https://doi.org/10.1016/j.nanoen.2021.106508.

Zaferani, Sadeq Hooshmand, Michael W. Sams, Reza Ghomashchi, and Zhi Gang Chen. 2021. "Thermoelectric Coolers as Thermal Management Systems for Medical Applications: Design, Optimization, and Advancement." *Nano Energy* 90 (Part A): 106572. https://doi.org/10.1016/j.nanoen.2021.106572.

Zhang, Jixi, Jiahui Jin, Jianqin Wan, Shaohua Jiang, Yongzhong Wu, Wufeng Wang, Xiao Gong, and Hangxiang Wang. 2021. "Quantum Dots-Based Hydrogels for Sensing Applications." *Chemical Engineering Journal* 408 (September 2020): 127351. https://doi.org/10.1016/j.cej.2020.127351.

Zhang, Qingsong, Yiming Li, Hao Ren, Qingchao Zhai, Chenglong Zhang, and Lin Cheng. 2021. "Engineering Full Hollow and Yolk-Shell Structures of Z-Scheme Photocatalysts for Advanced Hydrogen Production." *Chemical Engineering Journal* 408 (September 2020): 127267. https://doi.org/10.1016/j.cej.2020.127267.

Zhang, Quan, Long Li, Tianhong Wang, Yichen Jiang, Yingzhong Tian, Tao Jin, Tao Yue, and Chengkuo Lee. 2021. "Self-Sustainable Flow-Velocity Detection via Electromagnetic/Triboelectric Hybrid Generator Aiming at IoT-Based Environment Monitoring." *Nano Energy* 90 (Part A): 106501. https://doi.org/10.1016/j.nanoen.2021.106501.

Zhang, Wu, Tiefeng Liu, Yao Wang, Yujing Liu, Jiantheyi Nai, Liang Zhang, Outheyi Sheng, and Xinyong Tao. 2021. "Strategies to Improve the Performance of Phosphide Anodes in Sodium-Ion Batteries." *Nano Energy* 90 (Part A): 106475. https://doi.org/10.1016/j.nanoen.2021.106475.

Zhang, Xiaomin, Theynjie Xiong, Mingzhen Shi, Youting Wu, and Xingbang Hu. 2021. "Task-Specific Ionic Liquids as Absorbents and Catalysts for Efficient Capture and Conversion of H2S into Value-Added Mercaptan Acids." *Chemical Engineering Journal* 408 (September 2020): 127866. https://doi.org/10.1016/j.cej.2020.127866.

Zhang, Xu, Zihe Zhang, Zhen Zhou. 2018. "MXene-Based Materials for Electrochemical Energy Storage." *Journal of Energy Chemistry* 27 (1): 73–85. https://doi.org/10.1016/j.jechem.2017.08.004.

Zhang, Xueting, Zhenye Zhu, Xiongyi Liang, Fei Xiang Ma, Jiaheng Zhang, Yuanbo Tan, Zuchen Pan, Yiyang Bo, and Chi Man Lawrence Wu. 2021. "Encapsulating Dual-Phased Mo2C-WC Nanocrystals into Ultrathin Carbon Nanosheet Assemblies for Efficient Electrocatalytic Hydrogen Evolution." *Chemical Engineering Journal* 408 (September 2020): 127270. https://doi.org/10.1016/j.cej.2020.127270.

Zhang, Yating, Peng Wang, Juan Yang, Keke Li, Nana Zhang, Guoyang Liu, Yingfeng Duan, and Jieshan Qiu. 2022. "Mechanochemical Coordination Self-Assembly for Cobalt-Based Metal-Organic Framework-Derived Bifunctional Oxygen Electrocatalysts." *Journal of Colloid and Interface Science* 613: 733–46. https://doi.org/10.1016/j.jcis.2021.12.191.

Zhao, Huifang, Yanqiu Li, Huayu Zhao, Zheng Zhao, Junying Wang, and Ruiping Zhang. 2022. "Yolk-Shell Ni/NiO Anchored on N-Doped Graphene Synthesized as Dual-Ion MALDI Matrix for Detecting and Imaging Bioactive Small Molecules." *Journal of Colloid and Interface Science* 613: 285–96. https://doi.org/10.1016/j.jcis.2021.12.105.

Zhong, Shifa, Jiajie Hu, Xiong Yu, and Huichun Zhang. 2021. "Molecular Image-Convolutional Neural Network (CNN) Assisted QSAR Models for Predicting Contaminant Reactivity toward OH Radicals: Transfer Learning, Data Augmentation and Model Interpretation." *Chemical Engineering Journal* 408 (September 2020): 127998. https://doi.org/10.1016/j.cej.2020.127998.

Zhong, Xiongtheyi, Dashuai Wang, Jinzhi Sheng, Zhiyuan Han, Chongbo Sun, Junyang Tan, Runhua Gao, et al. 2022. "Freestanding and Sandwich MXene-Based Cathode with Suppressed Lithium Polysulfides Shuttle for Flexible Lithium-Sulfur Batteries." *Nano Letters* 22 (3): 1207–16. https://doi.org/10.1021/acs.nanolett.1c04377.

Zhong, Yaping, Yibiao Zou, Lingyan Liu, Ruohan Li, Fengfeng Xue, and Tao Yi. 2020. "PH-Responsive Ag2S Nanodots Loaded with Heat Shock Protein 70 Inhibitor for Photoacoustic Imaging-Guided Photothermal Cancer Therapy." *Acta Biomaterialia* 115: 358–70. https://doi.org/10.1016/j.actbio.2020.08.007.

Zhou, Changcheng, Liuhua Zhou, Jingyu Liu, Lutheyi Xu, Zheng Xu, Zaozao Chen, Yuzheng Ge, et al. 2020. "Kidney Extracellular Matrix Hydrogel Enhances Therapeutic Potential of Adipose-Derived Mesenchymal Stem Cells for Renal Ischemia Reperfusion Injury." *Acta Biomaterialia* 115: 250–63. https://doi.org/10.1016/j.actbio.2020.07.056.

Zhou, Liangyan, Hongshun Zhao, Kang Liang, Jinyuan Chen, Jianbin Li, Xiaobing Huang, Yanli Qi, and Yurong Ren. 2022. "Novel PETEA-Based Grafted Gel Polymer Electrolyte with Excellent High-Rate Cycling Performance for LiNi0.5Co0.2Ni0.2Mn0.3O2 Lithium-Ion Batteries." *Journal of Colloid and Interface Science* 613: 606–15. https://doi.org/10.1016/j.jcis.2021.12.147.

Zhou, Xiang, Xiuhong Huang, Bincheng Wang, Lihui Tan, Yi Zhang, and Yanpeng Jiao. 2021. "Light/Gas Cascade-Propelled Janus Micromotors That Actively Overcome Sequential and Multi-Staged Biological Barriers for Precise Drug Delivery." *Chemical Engineering Journal* 408 (September 2020): 127897. https://doi.org/10.1016/j.cej.2020.127897.

Zhu, Danzhu, Bin Liu, and Gang Wei. 2021. "Two-Dimensional Material-Based Colorimetric Biosensors: A Review." *Biosensors* 11: 259. https://doi.org/10.3390/bios11080259.

Zhu, Yangbin, Yang Liu, Hailong Hu, Zhongtheyi Xu, Chaoxing Wu, Dae Hun Kim, Tailiang Guo, Fushan Li, and Tae Whan Kim. 2021. "Intelligent, Biomimetic, Color-Tunable, Light-Emitting Artificial Skin with Memory Function." *Nano Energy* 90 (Part A): 106569. https://doi.org/10.1016/j.nanoen.2021.106569.

6 Mixed-Dimensional Heterostructures of Titanium-Based MXene with other 2D Layered Materials

6.1 INTRODUCTION

Unlike other 2D materials like transition metal dichalcogenides (TMDCs) and graphene, MXenes derivatives, especially $Ti_3C_2T_x$, show remarkably high metallic conductivities, usually in the range of about 6,000 to 8,000 S/cm. Additionally, they have a water-attracting surface (Anasori, Lukatskaya, and Gogotsi 2017). The intrinsic qualities of MXenes make them exceedingly attractive for a broad spectrum of applications, encompassing various networks for energy storage (Hantanasirisakul and Gogotsi 2018; Lukatskaya et al. 2013; Ghidiu et al. 2014b; Wang et al. 2015; Liang, Garsuch, and Nazar 2015), catalysis (Seh et al. 2016), electromagnetic interference shielding (Shahzad et al. 2016), transparent and conducting thin films (Halim et al. 2014; Hantanasirisakul et al. 2016; Dillon et al. 2016), water splitting (Ren et al. 2015), and a myriad of other domains.

However, even though many people want MXene-based materials around the world, there is still a lack of detailed research information on how stable MXene structures are, thinking about surface roughness and the progress made in making them. To confront the challenges associated with structural stability and surface smoothness, with the ultimate goal of enhancing the practical utility of MXenes, the concept of fabricating heterostructures involving MXene materials emerges as a promising avenue for exploration (Aïssa et al. 2016).

In recent times, there is a growing global interest in using MXene-based van der Waals heterostructures in various areas like photovoltaics, catalysis, and energy storage. This interest is rising similar to graphene back in 2006. Over the last few years, researchers made a significant advancement in producing MXene and the associated van der Waals heterostructures. This advancement has culminated in the proliferation of increasingly sophisticated materials and devices, thereby catering to an expanding community of research cohorts (Yuan and Li 2018).

There is a pressing need for the creation of adept heterostructures, integrating $Ti_3C_2T_x$ and Ti_2CT_x with other two-dimensional (2D) materials, with the intention of their application in energy storage devices. The frequently employed 2D materials

DOI: 10.1201/9781003319658-6

boast of a suite of distinctive attributes, including elevated electronic conductivity, substantial power density, remarkable flexibility, and robust mechanical integrity. However, in the context of graphene (GR)-based electrodes, despite their commendable electrical conductivity and mechanical resilience, their intrinsic potential is inherently circumscribed, primarily confined to surface-related functionalities. Furthermore, their chemical composition is susceptible to modulation by various surface functional groups, thereby precipitating a deleterious diminution in electrical conductivity (Stoller et al. 2008; Peng et al. 2016).

Conversely, transition metal dichalcogenides (TMDCs) exhibit noteworthy initial capacities; nonetheless, during the initial discharge cycle, a portion of their energy is dissipated as a result of a conversion reaction. When TMDC-based electrodes are confined within the scope of the reversible intercalation reaction, they demonstrate an enhanced cycling stability; however, it is noteworthy that their power output experiences a notable decline (Pumera, Sofer, and Ambrosi 2014).

Moreover, while certain TMDCs manifest commendable electronic conductivity, a significant proportion of their phases are semiconducting in nature (Chhowalla et al. 2013). Apart from their involvement in redox activity during intercalation reactions and their capability to function within a broad spectrum of working potentials, transition metal oxides (TMOs) are considered favorable contenders for cathodic uses, showcasing promise in various applications. Nevertheless, owing to their inherently low electronic conductivity, oxides necessitate amalgamation with conductive additives to improve their overall efficiency (Whittingham 2004).

To overcome these limitations, one viable strategy involves the construction of heterostructure electrodes, accomplished by layering alternating strata of diverse 2D materials. MXene phases, particularly $Ti_3C_2T_x$ and Ti_2CT_x, are distinguished by their significant electronic conductivity and mechanical resilience (Peng et al. 2016; Liu and Liu 2012). Moreover, they exhibit the ability to accommodate a substantial charge reservoir through a pseudocapacitive mechanism. The formation of heterostructures, coupled with the incorporation of supplementary 2D layered materials, serves to amplify the surface area, consequently enhancing the energy storage capability of the resultant device (Naguib et al. 2014). When skillfully arranged, these distinct layers form stacked configurations of 2D layered materials, leading to a harmonious enhancement of their properties.

This dynamic interaction emerges as a consequence of fusing the innate metallic electrical conductivity found in $Ti_3C_2T_x$ and Ti_2CT_x with that of different classes of 2D layered materials, encompassing graphene (GR), transition metal dichalcogenides (TMDCs), black phosphorus (BP), and transition metal oxides (TMO). This amalgamation works to concurrently enhance the electrical conductivity and surface area of the heterostructure, leading to a substantial boost in the overall performance (Naguib et al. 2014).

In this chapter, we have gathered the latest advancements in how we make 2D materials called $Ti_3C_2T_x$ and Ti_2CT_x MXene and combine them in different ways. Many review articles have already been written about how to make MXene from various starting materials, resulting in a wide range of 2D MXene materials that look different like tiny dots, small flakes, thin sheets, and long ribbons. These MXene materials have different shapes and can be used in many different ways (Naguib and Gogotsi 2015; Lei, Zhang, and Zhou 2015; Zhong et al. 2016; Khazaei et al. 2017; Alhabeb et al. 2017b; Anasori, Lukatskaya, and Gogotsi 2017; Huang et al. 2018; Pang et al. 2019; Cai et al. 2018; Verger et al. 2019; Zhan et al. 2020; Ronchi, Arantes, and Santos 2019; Kannan et al. 2020).

Our primary goal in this chapter is to showcase the latest advancements by highlighting the most pertinent and well-researched studies. This chapter presents a detailed exploration of the various methods for creating 2D Ti-based MXene materials, specifically $Ti_3C_2T_x$ and Ti_2CT_x. It reveals the process of combining these materials with a range of layered counterparts, such as graphene, transition metal dichalcogenides, boron nitride, and black phosphorus, resulting in the creation of innovative new structures.

6.2 CRAFTING MXENE HETEROSTRUCTURES VIA MORPHOLOGICAL ENGINEERING

Numerous methodologies have been advanced and elucidated in recent scholarly works concerning the production of heterostructures based on graphene (GR) (Reina et al. 2008). In the context of GR-0D heterojunctions, a standard procedure involves transferring synthetically produced graphene onto a chosen substrate, such as SiO_2 or a flexible polymer. Following this, there comes the epitaxial expansion of organic molecules or the step-by-step assembly of quantum dots utilizing dip or spin coating methodologies (Kim et al. 2014; Bie et al. 2011; Zhang et al. 2017; Deng et al. 2015). Conversely, for the development of GR-3D heterostructures, GR is generally transferred directly onto three-dimensional materials. Nevertheless, the fabrication of GR in conjunction with one-dimensional nanotubes or nanowires can pose an increased complexity due to the intricate transfer processes necessitated for both materials (Zhang et al. 2019). To streamline this process, precise transfer methodologies have been devised for individual nanowires or arrays of nanowires, facilitating their amalgamation with GR. This approach, reliant on transfer assistance, proves to be versatile and has found widespread application in the creation of diverse heterostructures centered around GR. Figure 6.1a–h provides an overview of distinct techniques employed in the preparation of heterostructures, utilizing various deposition methods. It is imperative to contemplate the underlying concept of graphene-based heterostructures (Pomerantseva and Gogotsi 2017).

The identical conceptual framework is utilized in the fabrication of heterostructures found of MXene materials, specifically $Ti_3C_2T_x$ and Ti_2CT_x, in conjunction with other two-dimensional (2D) layered materials. A diverse range of 2D materials, encompassing transition metal dichalcogenides (TMDCs), are employed in this construction (Li and Zhu 2015; Kou, Chen, and Smith 2015; Wang et al. 2012; Sahoo et al. 2013), various forms of graphene (GR) (Choi et al. 2010; Khan et al. 2015; Jiao et al. 2009; Han 2009; Akinwande, Petrone, and Hone 2014; Das et al. 2014), black phosphorus (BP) (Kou, Chen, and Smith 2015; Ospina et al. 2016; Li et al. 2014), metal oxides (MO) (Whittingham 2004; Peng et al. 2016), and various other forms of nanomaterials, exhibits advantageous (electrical and mechanical) properties. These properties make them highly suitable for developing efficient energy storage devices and highly responsive sensors. These 2D materials have found applications spanning interdisciplinary domains, encompassing materials science, chemistry, biology,

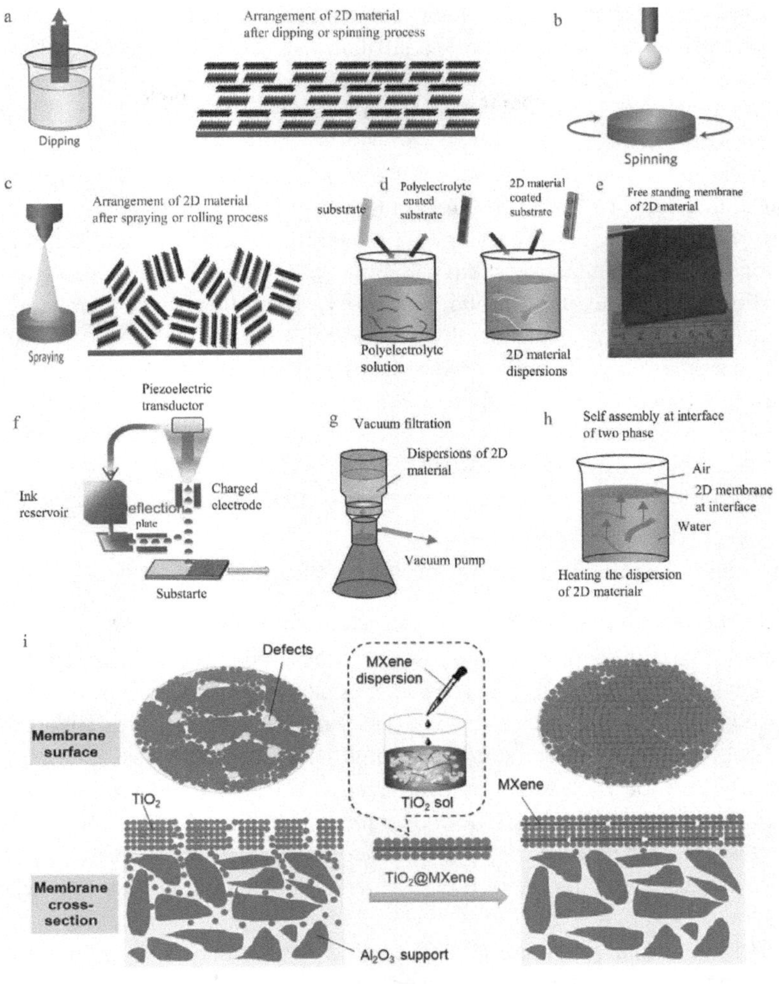

FIGURE 6.1 Layer-by-layer assembly represents a prevalent technique employed in the construction of two-dimensional (2D) heterostructure electrodes. (a–c) This methodology entails the sequential stacking of distinct layers, derived from disparate heterostructures, thereby engendering novel 2D heterostructures endowed with prospective applications. Permission awarded by Macmillan Publishers limited 2017 (see Pomerantseva and Gogotsi 2017). The fabrication of two-dimensional (2D) heterostructures can be realized through a multitude of techniques: (d) Layer-by-layer assembly entails the sequential stacking of distinct layers atop one another. (e) A self-assembled 2D material membrane emerges autonomously at the interface of water and air, standing free in its formation. (f) Inkjet printing offers precise material deposition onto a substrate. (g) Vacuum filtration involves the filtration of a solution containing the desired materials to produce a thin film. (h) Spontaneous material organization is facilitated through self-assembly going on at the interface of two distinct phases, where materials align themselves at the juncture of dissimilar realms. Permission awarded by Royal Society of Chemistry (see Cai et al. 2018). (i) The production of heterostructures involving TiO_2-MXene layers aims to mitigate potential defects.

Sources: (a–c) Adopted with permission from springer nature 2017. (d, e, f, g, h) or (d–h) Adapted with permission from the Royal Society of Chemistry. (i) Permission awarded by Elsevier Copy right 2018 (Xu et al. 2018).

and medicine. Synthetic methodologies have made substantial advancements in the precise control of the physicochemical attributes, surface functionalities, forms, and dimensions of 2D nanomaterials. The layered structure and super reactive surfaces of these nanomaterials make them crucial for uses in energy storage and catalysis in the optoelectronic field.

The 2D nanomaterial array recently expanded to include MXenes, which are carbides, nitrides, or carbonitrides of transition metals, and were first described by Professor Gogotsi and colleagues in 2011 (Naguib et al. 2011a; Anasori, Lukatskaya, and Gogotsi 2017; Hantanasirisakul and Gogotsi 2018). The layer-dependent optical properties, notable metallic conductivity, and hydrophilic surface characteristics of $Ti_3C_2T_x$ MXene bestow versatility for the design of plasmonic and electronic sensor detection systems. Significantly, MXene has demonstrated promising potential in the domain of biomedical sensing applications (Naguib et al. 2011a), positioning it as a valuable contender in the burgeoning field of biomedical sensors. Before embarking on the assembly of diverse heterostructures, it is vital to acquire a complete understanding of the intrinsic characteristics of various 2D materials, including MXene.

6.2.1 Heterostructure of MXene with Various 2D Materials

6.2.1.1 Heterostructures Relying on MXene and Metal Oxide

The layered structures in MXene do not work as conductive stabilizers for helping ion and electron transport. Instead, they act as a base material to prevent the clumping together of TMO nanocrystals. The TMO nanocrystals serve as linkers, alleviating the restacking of MXene layers during the assembly process and the lithium extraction/insertion cycle. This innovative approach fosters interfacial interactions and safeguards the active sites. The cooperative interplay between TMO and MXene substrates results in an amplified performance, rendering TMO/MXene heterostructures promising contenders for the next generation of anodic nanostructured energy storage materials (Liu et al. 2018).

In 2018, Cui and collaborators (Xu et al. 2018) introduced a novel method for fabricating mesoporous layers founded on TiO_2 by amalgamating TiO_2 nanoparticles with two-dimensional (2D) $Ti_3C_2T_X$ MXene nanosheets. This technique involves directly coating an aqueous solution onto the surface of a macroporous substrate.

Figure 6.1i visually represents the manner in which the nanolayers of MXene form a resilient two-dimensional framework during the deposition of TiO_2 nanoparticles.

Throughout the sol-gel coating process, TiO_2 nanoparticles seamlessly infiltrate the macroporous structure of the ceramic-like supporting nanosheet. At the same time, the micrometer-scale MXene nanolayers encapsulate potential imperfections in the macroporous ceramic support, eliminating the need for an additional mesoporous layer on the intended substrate. Both dip and spin coating techniques have been successfully applied to both hollow fiber structures and disc-shaped membranes.

To produce a meticulously crafted mesoporous TiO_2-MXene membrane, the stratified films underwent a drying and maturing procedure at 60°C when heated in a microwave for a duration of 12 hours. Following this, they underwent an open-air heating process, gradually reaching 400°C, with a temperature control system configured to increase or decrease by 0.5°C every minute.

The dip-coating technique was employed to construct TiO_2-MXene hollow fiber layers measuring approximately 7 cm in length and exhibiting an outer surface area of about 8.8 cm^2. Following specific coating procedures, including a withdrawal rate of 1 cm/s and drying time of 1 hour at 40°C, the blends naturally adhered to the exterior of the α-Al2O3 hollow fiber cores (Xu et al. 2018; Liu et al. 2018).

After four cycles of the protocols, the sol-coated hollow fiber was calcined in room-temperature air at 400°C for four hours, with a regulated heating and cooling rate of 0.5°C/min. The morphology of the MXene nanosheets underwent examination through scanning electron microscopy (SEM) and atomic force microscopy (AFM), as illustrated in Figure 6.2a–b. The nanosheets displayed an average dimension of approximately 500 nm and a thickness varying between 1 to 2 nm, signifying a configuration of MXene arranged in either a single or double layer (Lipatov et al. 2016).

Images captured using transmission electron microscopy (TEM) were used to evaluate the size, arrangement, and distribution of the bulk TiO_2 structures, as well as the TiO_2-MXene amalgamations in aqueous solutions. As depicted in Figure 6.2c–d, TEM examination unveiled that TiO_2 nanoparticles, averaging approximately 5 nm in size, were evenly distributed in the initial solution. Moreover, dynamic light scattering (DLS) analysis, representing an average size for the particles distribution of 15 nm, indicated that only a limited number of TiO_2 nanoparticles had aggregated due to weak attractive forces, leading to a consistent coating. When we added 1% MXene nanolayers to the TiO_2 mix, the TiO_2 particles consistently settled on the 2D MXene nanolayers' surface, without big changes in their shapes (see Figure 6.2e, f). This shows that TiO_2 and MXene were effectively combined in the heterostructures (Jiao et al. 2019).

Moreover, the fluorine on the MXene surface, mainly in the form of unblocked and oxygen-substituted (–O–) fluorine, goes through a temporary change when making TiO_2-MXene heterostructures. This transformation happens through the hydrothermal technique, briefly turning the MXene into TiO_2, as depicted in Figure 6.2e–f (Lipatov et al. 2016). The findings of this study highlight a scalable and highly effective approach for manufacturing mesoporous TiO_2-MXene heterostructure membranes. Leveraging the unique properties of MXene's two-dimensional nanolayers, precise control over the infiltration of TiO_2 nanoparticles was achieved, minimizing the occurrence of thin-layer imperfections and enhancing conductivity. The obtained hollow fiber membranes exhibited outstanding filtration performance, demonstrating a molecular weight cutoff of approximately 22,000 Da and an impressive pure water flow rate exceeding 90 L m^2 h^{-1} bar^{-1}. This underscores their considerable potential for practical applications in real-world scenarios.

Furthermore, the TiO_2-MXene heterostructure serves as a versatile catalyst, offering benefits such as enhanced polysulfide adsorption, improved conversion efficiency, and facilitated polysulfide migration.

The even spread of TiO_2 structures on MXene sheets serves as a main hub for grabbing polysulfides, and the interface between them makes sure polysulfides move quickly from TiO_2 to MXene, supporting efficient conversion. The oxygen-functionalized surface of MXene layers exhibits noteworthy catalytic activity in the transformation of polysulfides. Figure 6.2g illustrates the deliberate utilization

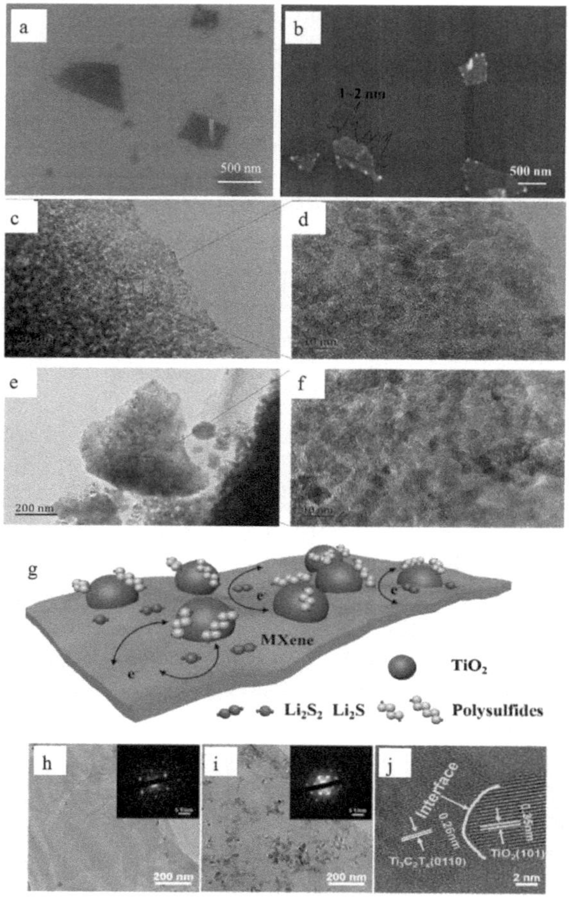

FIGURE 6.2 (a) The SEM image and (b) AFM photograph serve as visual representations and provide insight into the external structure characteristics of the MXene layers. TEM micrographs of the hydrosol provide valuable observations on TiO_2 nanoparticle size and dispersion. Panels (c, d) show the hydrosol containing TiO_2, while panels (e, f) show the hydrosol of TiO_2-MXene with a 1 wt.% concentration of MXene. These TEM images offer an in-depth insight into the dimensions, configuration, and spatial arrangement of the nanoparticles within their individual solutions. (g) A schematic diagram illustrating the process of transforming and trapping lithium polysulfides (LiPSs) on the TiO_2-$Ti_3C_2T_x$ heterostructure. This heterostructure, composed of both TiO_2 and $Ti_3C_2T_x$, serves as an effective platform for the conversion and trapping of LiPSs. The form and structural characteristics of the TiO_2-$Ti_3C_2T_x$ heterostructure have been examined using transmission electron microscopy (TEM). The TEM images show the original pristine structure of $Ti_3C_2T_x$ (panel h) and the configuration of the TiO_2-$Ti_3C_2T_x$ heterostructure (panel i). Insets in panels (h) and (i) correspond to selected area electron diffraction (SAED) patterns, providing additional insight into crystallographic orientation. (j) A high-resolution TEM (HRTEM) micrograph focusing on the TiO_2-$Ti_3C_2T_x$ heterostructure boundary is presented. This image provides a detailed view of the interface between the TiO_2 and $Ti_3C_2T_x$ components at higher magnification.

Sources: (a–f) Adopted with permission from Elsevier Ltd. (Xu et al. 2018). (g–j) Adopted with permission from Weinheim: Wiley-VCH Verlag GmbH & Co (Jiao et al. 2019).

of TiO_2 nanoparticles as adsorption sites for lithium polysulfides (LiPSs), with the hetero-interface facilitating their transfer from TiO_2 to MXene for subsequent conversion. Controlled partial oxidation of MXene layers occurs during the meticulous processing to generate TiO_2 nanoparticles, while preserving their two-dimensional geometry, conductivity, and surface area, thereby enabling efficient LiPS exchange. The TiO_2-$Ti_3C_2T_x$ heterostructure is prepared, using a hydrothermal method carried out at a temperature of 180°C.

The TiO_2-MXene heterostructures amalgamate several advantageous characteristics, encompassing an expansive surface area, robust LiPS adsorption capacity, excellent conductivity, and notable electrocatalytic performance.

Through the utilization of the hydrothermal oxidation technique, TiO_2 nanoparticles (NPs) are uniformly dispersed over the MXene surface.

The TEM images represented the original pristine structure of $Ti_3C_2T_x$ (panel h) and the existence of anatase TiO_2 is corroborated by SAED (selected area electron diffraction), as illustrated in the minor inset in Figure 6.2i. This can be seen by the peaks coinciding with the crystallographic planes of (101), (200), and (105) (Kasuga et al. 1998). Furthermore, the transmission electron microscopy with high-resolution (HRTEM) image in Figure 6.2j highlights the distinctive spindle-like morphology of the TiO_2 NPs, accentuating the interface formed by the (101) plane of TiO_2 and the (0110) plane of $Ti_3C_2T_x$ (Jiao et al. 2019).

The well-dispersed TiO_2 nanoparticles (NPs) on the MXene surface create numerous active sites conducive to efficient lithium polysulfide (LiPS) capture. Furthermore, the heterostructure interface facilitates the rapid transfer of LiPSs to the MXene structure, promoting swift conversion. The combination of TiO_2-MXene heterostructures and conductive graphene (GR) effectively hinders LiPS migration and enhances sulfur utilization, resulting in exceptional cycling stability. These findings offer valuable insights into catalysts for lithium-sulfur (Li-S) energy storage systems, aimed at enhancing performance, minimizing sulfur loss, and achieving long-term stability. These developments play a crucial role in the commercialization of lithium-sulfur (Li-S) batteries.

In addition, these combinations have shown a great promise in developing more effective catalysts for a range of chemical processes and their applications, including fuel cells and metal-air batteries.

Particularly within the domain of MXenes, $Ti_3C_2T_x$ distinguishes itself and has attracted considerable attention from the scientific community due to its exceptional properties.

These include an outstanding packing density of 4 g cm^{-3}, an impressive metallic conductivity of up to 8,000 S cm^{-1}, and the ability to carry out reversible oxidative and reductive reactions on its surface (Alhabeb et al. 2017a).

These characteristics provide high-rate performance (1,000–1,500 F/cm^3) (Lukatskaya et al. 2017), exceptional cycle stability even when electrolytes are acidic, and a significant specific volumetric capacitance (Lukatskaya et al. 2013; Ghidiu et al. 2014a).

The pseudocapacitive behavior exhibited by MXene allows it to function across a wide potential window, and, when combined with other pseudocapacitive nanomaterials, the operational voltage window can be further expanded (Lukatskaya et al. 2017; Anasori, Lukatskaya, and Gogotsi 2017).

Regarding pseudocapacitive materials, the author points out that ruthenium oxide (RuO_2) is a beneficial positive working electrode. This is because RuO_2 can create pseudocapacitance through proton-induced methods, it has good conductivity, and it operates within a favorable overpotential range to help with the oxygen evolution reaction in acidic environments (Lukatskaya et al. 2017). In the development of a fabric-based pseudocapacitor, the approach involves the interconnection of negative $Ti_3C_2T_x$ electrodes with positive RuO_2 electrodes within an electrolyte network characterized by its acidic nature (Hu et al. 2006; Liu, Pell, and Conway 1997).

This method has the potential to create high-performance pseudocapacitors by combining the unique features of MXene and RuO_2, allowing for effective energy storage and conversion. A three-dimensional carbon fabric (CF) structure shows outstanding electrical conductivity, making it a suitable base for different nanomaterials used in electrodes. Including conductive carbon fibers (CNFs) between the layers of 2D materials can improve the process of transporting charges, ultimately boosting efficiency and energy density in systems based on heterostructures (Liu, Liu et al. 2017; Liu et al. 2016). When CNFs are strategically positioned within the layers of 2D materials, several advantages can be realized.

To begin with, the presence of carbon nanofibers (CNFs) creates interconnected pathways that accelerate rapid electrolyte penetration and seamlessly transfer ions over longer distances, significantly improving battery efficiency (Liu et al. 2019). Furthermore, carbon nanofibers (CNFs) can serve as a supportive framework for the carbon layers, expediting the swift transfer of electrons. Hence, without surface preparation, conductive additives, or binders, RuO_2 and MXene may be directly deposited onto carbon fabric (CF) without any intermediate steps.

The exploration of new avenues for MXene-based applications can be pursued through the utilization of all-pseudocapacitive devices that incorporate 2D materials such as metal oxides (MO_2) and metal carbides (Jiang et al. 2018). The combination of electro-active nanomaterials is facilitated by the three-dimensional (3D) structure of carbon fiber (CF), which eliminates the need for binders. This allows for increased mass loading of active nanomaterials within a given area (as shown in Figure 6.3a, panel 1). In addition, CFs exhibit chemical stability and have been shown to be compatible with the lamination process of multidirectional nanomaterials. This demonstrates their superiority over metal-based current collectors. The outstanding properties of carbon fiber (CF) render it an optimal choice for the proposed network of irregular devices. For this research, CF was chosen as the electrically conductive base material. $Ti_3C_2T_x$ MXene was then deposited onto it using a drop-casting technique, as shown in Figure 6.3a (panel 2). The hydrothermal technique was used to uniformly deposit hydrous RuO_2 onto the surface of CF. The starting material employed was ruthenium chloride ($RuCl_3$) with different mass loadings, as shown in Figure 6.3a (panel 3).

As shown in Figure 6.3a (panel 4) (Jiang et al. 2018), to construct the all-pseudocapacitive asymmetric capacitor, the $Ti_3C_2T_x$/CF negative electrode system and the RuO_2/CF positive electrode were combined in an acidic electrolyte.

Throughout the integration process, both $Ti_3C_2T_x$ and RuO_2 were uniformly dispersed over the entire surface of the carbon fabric, as shown in Figure 6.3b and 6.3d.

In addition, the robust adhesion of the RuO2 units to the individual layers of the carbon-based fiber is further confirmed by the SEM image in Figure 6.3c.

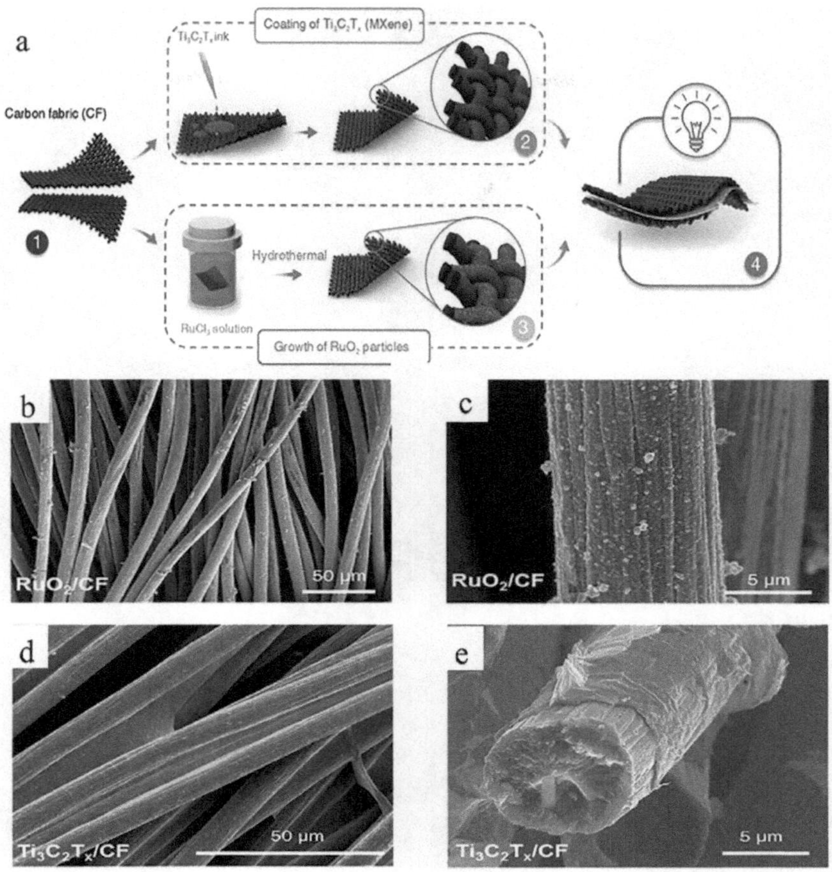

FIGURE 6.3 (a) Diagrammatic representation for explanation of the sequential assembly process of an asymmetric supercapacitor device. In the first panel, the three-dimensional surface of the CF material is shown. The deposition process of $Ti_3C_2T_x$ MXene coated upon carbon fiber (CF) through a drop-casting process and the preparation of powdered RuO_2 by a hydrothermal process are observed in panels 2 and 3, respectively. Panel 4 vividly illustrates the amalgamated structure of RuO_2//$Ti_3C_2T_x$ heterostructures, forming an irregular pseudocapacitor. (b) For a more comprehensive analysis of the RuO_2/CF composite, detailed field emission scanning electron microscopy (FESEM) images are presented, with additional magnification in part (c). (d) The low-magnification SEM image serves to validate the uniform $Ti_3C_2T_x$ MXene deposition process on the CFC surface. In addition, in part (e), he configuration of $Ti_3C_2T_x$ flakes enveloping different fiber surfaces within the material made of carbon fiber is shown. Permission awarded by Wiley 2018 (Jiang et al. 2018). SEM pictures of (f) Ti_3C_2 and (g) Ti_3C_2/ZnO composite are shown. The etched Ti_3C_2 has a layered microstructure. In contrast to the Ti_3AlC_2 powder, the thin layers are clearly separated. (h–i) The even dispersion of zinc oxide nanoparticles (ZnONPs) across the upper surface and amid the layers of the Ti_3C_2 nanosheets. This prevents the Ti_3C_2 layers from restacking, resulting in a sandwich-like arrangement. TEM and high-resolution TEM (HRTEM) images of the Ti_3C_2/ZnO composite structure are presented, illustrating the successful integration of MXene and nanostructures of 1D metal oxide in the composite structure.

Source: Permission awarded by Royal society of chemistry (Wang et al. 2016).

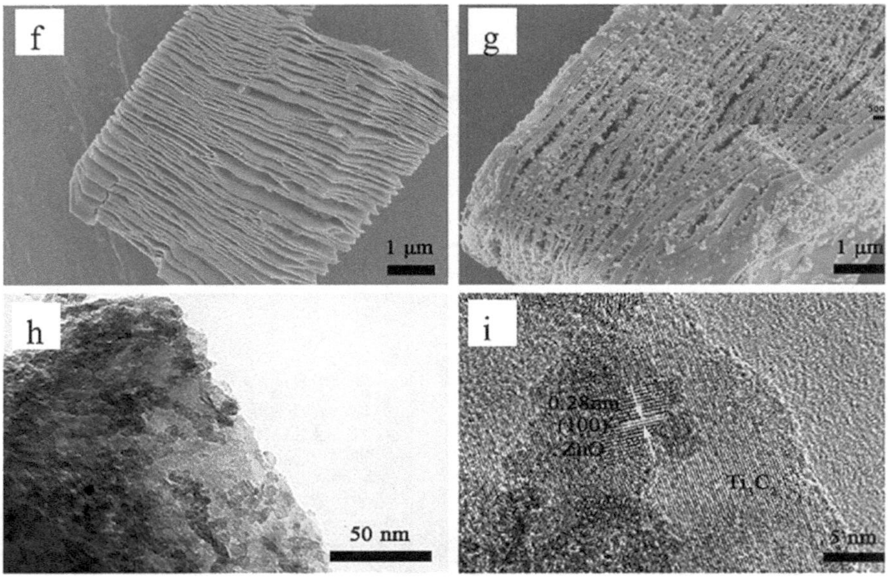

FIGURE 6.3 (Continued)

In summary, carbon fabric (CF) was found to be an effective substrate material for accommodating $Ti_3C_2T_x$ and RuO_2. This resulted in a well-distributed and securely bonded configuration on the carbon fabric surface.

In parallel, as shown by the SEM micrograph in Figure 6.3e (Jiang et al. 2018), the $Ti_3C_2T_x$ structures were consistently attached to the upper surface nanofibers based on carbon material.

Hydrous RuO_2 electrodes with $Ti_3C_2T_x$ MXene were integrated on the surface of carbon fabric to fabricate asymmetric pseudocapacitive supercapacitors. It is worth highlighting that the electrode incorporating MXene in an aqueous electrolyte demonstrated the widest voltage range for the electrochemical processes in capacitors. These results suggest that pseudocapacitive devices using MXene–RuO_2 combinations showed considerable potential for application with aqueous-based electrolytes. They allow for an extended voltage range and improved energy density.

Zinc oxide (ZnO) is a unique nanomaterial that has several uses due to its piezoelectric, semiconducting, and pyroelectric properties; these properties make it ideal for use in sensors, optics, and optoelectronics. Recently, the practical application of ZnO nanoparticles (NPs), such as in supercapacitor electrodes for nanostructures, has been supported due to their flexible electrochemical behavior, environmentally friendly properties, and easy availability (Aravinda et al. 2013; Kalpana et al. 2006).

Researchers have shown that conducting substrates like 2D graphene and its derivatives, when combined with zinc oxide (ZnO) nanoparticles, increase the surface area. This helps to enhance the interplay between the electrode and the electrolyte.

This, in turn, promotes effective ion diffusion and electron transport which ultimately lead to the improvement of the supercapacitive properties. Numerous efforts have been dedicated to the integration of zinc oxide (ZnO) and graphene (GR) for the fabrication of composite materials that are tailor-made for the development of highly efficient supercapacitors (Prakash and Bahadur 2014; Fang et al. 2014; Ramadoss and Kim 2013).

It is logical to devise a carefully planned method for constructing structures and explore the diverse applications of Ti_3C_2/ZnO composite structures in the world of supercapacitors, considering the trustworthy electrical conductivity of Ti_3C_2 nanosheets, their outstanding chemical durability, and their simple combination with ZnO (Wang et al. 2016).

The creation of the Ti_3C_2/ZnO composite network involved a direct precipitation technique, succeeded by a subsequent heat treatment process. Initially, a suspension is prepared comprising 100 mg of Ti_3C_2 in approximately 100 mL of deionized (DI) water under the influence of magnetic stirring for 30 minutes to achieve a uniform suspension. Subsequently, a blend containing 98 mg NaOH and 73 mg $Zn(NO_3)_2$-$6H_2O$ was gradually introduced into the Ti_3C_2 suspension, with continuous stirring. After 2 hours, the resulting mixture underwent washing with DI water until it reached a neutral pH of 7, followed by three rinses with ethanol. Lastly, the product underwent heating at 100 °C for 3 hours in an argon (Ar) atmosphere. The procedure for the synthesis of ZnO nanoparticles (NPs) was similar to that of the Ti_3C_2/ZnO composite, with the notable exception of the omission of Ti_3C_2 MXene.

The size and shape of the samples were examined using SEM and TEM images. Figure 6.3f represents the SEM image of Ti_3C_2 MXene. The etched multilayer microstructure is characterized by distinct layers.

Figure 6.3g shows the Ti_3C_2/ZnO composite structure. It illustrates the uniform distribution of ZnO NPs on the top surface and in the interstitial spaces between the Ti_3C_2 layers. This arrangement effectively mitigates the restacking of Ti_3C_2 nanosheets, forming a configuration similar to a sandwich. Meanwhile, the extensive and uniform dispersion of ZnO NPs on the Ti_3C_2 nanosheets is validated through TEM and HRTEM imagery depicting the Ti_3C_2/ZnO composite structure in Figure 6.3h–i. These results corroborate the SEM images and highlight the robust interaction between ZnO NPs and Ti_3C_2 nanosheets, even after an ultrasonication process (Wang et al. 2016).

The nanocomposite structures of Ti_3C_2/ZnO showcase remarkable attributes concerning specific capacitance, performance at varying rates, and sustained cycling stability when employed as electrodes in supercapacitors. Therefore, as electrode materials for supercapacitors, nanocomposites of metal oxides and 2D layered materials are expected to show a great promise. The Ti_3C_2/ZnO material, characterized by its sandwich-like structure, was effectively synthesized using a simple precipitation-based method followed by heat-treating. This methodology can be extrapolated to the fabrication of other composite structures in which transition metal oxides and 2D layered materials are amalgamated.

MXenes are characterized by their surface functionalities. They demonstrate advantageous dissolvability in an extensive array of solvents, encompassing polar organic solvents and water alike (Maleski, Mochalin, and Gogotsi 2017). This unique combination of properties makes MXenes an attractive class of materials. They are similar to a hybrid composite of graphene and graphene oxide. In addition, MXenes possess remarkable properties such as easy lithium-ion transport and mechanical flexibility. This renders them exceedingly appropriate for a diverse array of potential applications (Borysiuk, Mochalin, and Gogotsi 2015, 2018; Lukatskaya et al. 2013; Ghidiu et al. 2014a; Tang, Zhou, and Shen 2012; Eames and Islam 2014).

The amalgamation of nanostructures of transition metal oxide (TMO) with nanosheets of MXene in a coherent structure holds paramount significance, as it enables the synergistic incorporation of the exceptional properties and applications inherent in both constituents. Recent research endeavors have centered on the combination of TMO nanoparticles, such as Fe_3O_4 or MnO2, with robust MXene-based films. In this composite material, the primary placement of TMO nanoparticles occurs within the slender MXene layer, capitalizing on the distinctive dimensions and surface characteristics of MXene (Zhang et al. 2016). Xu et al. (Liu et al. 2018) have devised a method to integrate TMO nanocrystals, SnO2 nanowires, and TiO_2 nanorods onto MXene sheets by using van der Waals forces. This technique is distinguished by its simplicity, affordability, outstanding efficacy, and capacity to precisely control the packing density of TMO nanocrystals on the MXene surface. The MXene nanosheets were obtained using vacuum filtration from an aqueous solution and immediately transferred to tetrahydrofuran (THF) while still damp.

Subsequently, TiO_2 nanorods or SnO_2 nanowires were introduced into the solutions at varying mass ratios (40:60, 70:30, and 85:15). These mixtures underwent sonication under argon protection for a specified duration. Following sonication, the solid residues were filtered, subjected to THF washing, vacuum-dried, and preserved within an inert atmosphere within a glove box for subsequent utilization. The topmost layer of MXene nanolayers enables the deposition of transition metal oxide (TMO) nanostructures material under the influence of van der Waals forces in the presence of tetrahydrofuran (THF) solvent. This leads to a decrease in surface energy and stabilization of the heterostructures, analogous to micelles. The entire process of transition metal oxide (TMO) incorporation onto 2D surface materials is typically stochastic and guided by free energy reduction.

Significantly, no additional intervention is required. This process can be performed under ambient conditions. Figure 6.4a illustrates a diagrammatic depiction of the even distribution of MXene nanocrystals. A variety of heterostructures can be created by adjusting the mass ratios of SnO2 nanowires and TiO_2 nanorods in the presence of MXene nanosheets in THF solvent (Liu et al. 2018).

The nanoscale geometry of the transition metal oxide (TMO) is dispersed on the top surface of the MXene nanolayers at a comparatively modest TMO/MXene mass ratio such as 40:60. In addition, at an optimal mass ratio, e.g., 70:30, the TMO-based nanomaterials are uniformly and densely distributed. They cover the whole top surface portion of the MXene nanolayers. When a higher amount of TMOs is introduced, such as reflected in the ratio 85:15, they completely surround the MXene nanosheets.

A mass ratio of 70:30 is chosen for the transition metal oxide (TMO)/MXene heterostructures for subsequent electrochemical assessments. It is important to note that both SnO2 nanowires and TiO_2 nanorods are organized layer by layer on top of the MXene sheets, as shown in Figure 6.4b and 6.4e, in contrast to previous TMO/MXene composites that exhibited irregularities and challenges in controlling the distribution of TMOs.

The uniform dispersion of transition metal oxides (TMOs) on MXene nanolayers is achieved through the indiscriminate interactions of van der Waals forces. As shown in Figure 6.4c and 6.4f, this one-step assembly process demonstrates remarkable efficiency, culminating in the deposition of nearly all TMO nanostructures that are located on the upper surface of MXene layers.

Furthermore, the HRTEM representation confirms the existence of clear lattice fringe spacings that are in keeping with the crystalline characteristics of the transition metal oxide (TMO) nanostructures. Panels d and g of Figure 6.4 demonstrate that the (110) and (101) planes of rutile TiO_2 and anatase SnO2 have been assigned measurements of 0.34 nm and 0.35 nm, respectively (Liu et al. 2018).

In a distinct study carried out by Chen et al. (Zeng et al. 2019), they created a sophisticated method for converting solar energy into carbon dioxide using Ti_3C_2 quantum dots (QDs)/Cu_2O nanowires (NWs) and Cu heterostructures, based on electrostatic self-assembly.

It is noteworthy that MXene QDs, which are categorized as 0D nanostructures, have lately captivated significant research attention owing to their distinctive characteristics. Integrating Ti_3C_2 quantum dots (QDs) with Cu_2O nanowires (NWs) in a heterostructure shows enhanced photocatalytic prowess and durability of Cu_2O nanowires. The synergistic effects of 0D Ti_3C_2 QDs and 1D Cu_2O NWs were investigated using computational methods, in particular density functional theory (DFT), to gain insight into the photoinduced conversion mechanism (Zeng et al. 2019).

In Figure 6.4h, the fabrication process commences with the treatment of Ti_3AlC_2 powder using HF solvent to eliminate the aluminum layers, yielding Ti_3C_2 nanolayers (Naguib et al. 2011b). Subsequently, these nanolayers undergo functionalization with surface moieties (–O, –F, and –OH) to replace the aluminum (Al). Ultrasonication is then employed to disintegrate the Ti_3C_2 nanolayers into microsheets. Finally, functionalized Ti_3C_2 QDs are derived by hydrothermally segmenting the Ti_3C_2 sheets and enveloping them with polyethylenimine (PEI).

Semiconductor nanowires, such as Cu_2O nanowires, offer advantages in terms of efficient charge transport and light absorption, leading to improved light harvesting (Boettcher et al. 2010). The fabrication process involves the anodization of a copper network to generate $Cu(OH)_2$ nanowires, which are subsequently transformed into red Cu_2O nanowires through calcination in an inert atmosphere at 550°C (Figure 6.4h). These Cu_2O nanowires possess an approximate diameter of 500 nm and a porous surface (Figure 6.5a, b). In order to protect the reactive surface of Cu_2O nanowires, a layer of Poly (Sodium 4-styrene Sulfonate) (PSS) molecules is applied. PEI is used to organize positively charged Ti_3C_2 quantum dots (QDs) onto the Cu_2O nanowires, resulting in a hierarchical Ti_3C_2 QDs/Cu_2O nanowires/Cu heterostructure via electrostatic interactions.

Following calcination under an inert atmosphere, the PEI and PSS molecules are removed, leaving Ti_3C_2 QDs and Cu_2O nanowires in close proximity. The coating of Ti_3C_2 QDs minimally affects the size and shape of Cu_2O nanowires but serves to shield their porous surfaces (Figure 6.5c, d). Analogously, a Ti_3C_2 sheets/Cu_2O nanowires/Cu heterostructure is also fabricated by depositing Ti_3C_2 sheets (approximately 200 × 300 nm) onto the Cu_2O nanowires (Figure 6.5e, f) (Zeng et al. 2019). Subsequently, the Cu_2O nanowires/Cu network is treated with a PSS solution,

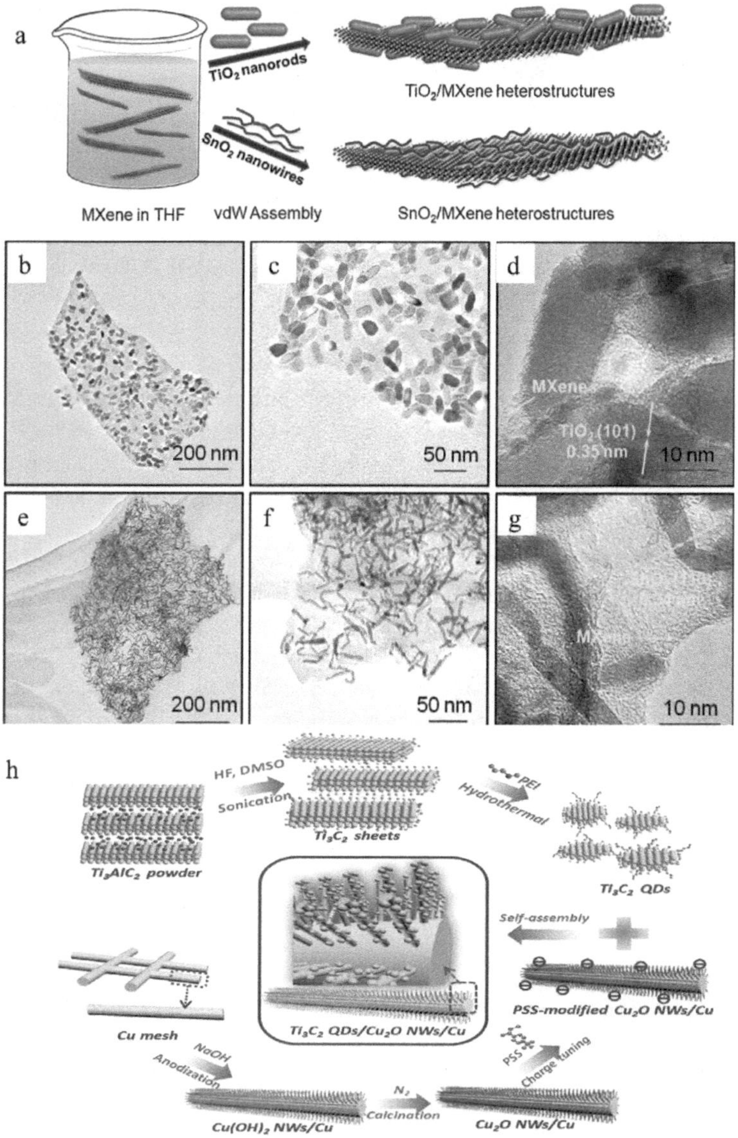

FIGURE 6.4 (a) Schematic demonstration of the streamlined assembly procedure of TMO nanostructures, SnO_2 nanowires, and TiO_2 nanorods on MXene sheets in THF solvent via van der Waals interactions. (b) low-magnification TEM images. (c) High-magnification TEM images. (d) HRTEM image showing detailed insights into the structure of TiO_2/MXene heterostructures. (e) A low-magnification TEM image. (f) A high-magnification TEM image. (g) HRTEM image showing the detailed structure of SnO_2/MXene heterostructures. Permission awarded by Wiley 2018 (Liu et al. 2018). (h) A diagram depicting the step-by-step process for the synthesis process of Ti_3C_2 quantum dots (QDs) and the subsequent formation of a Ti_3C_2 QDs/Cu_2O nanowires/Cu heterostructure.

Source: Permission awarded by Wiley 2018 (Zeng et al. 2019).

FIGURE 6.5 Field emission SEM images were captured for different heterostructures: (a, b) Cu2O NWs/Cu, (c, d) Ti_3C_2QDs/Cu_2O NWs/Cu, and (e, f) Ti_3C_2 sheets/Cu_2O NWs/Cu. Additionally, SEM images were taken of $Ti_3C_2T_x$ (g) and $Ti_3C_2T_x$/PtNP (h) at a magnification of 50,000×. The additional photo shows PtNPs formed by the reduction process of H_2PtCl_6 at 100,000× magnification.

Source: Permission awarded by Wiley 2018 (Zeng et al. 2019).

subjected to ethanol washing, and dried with nitrogen gas to eliminate any excess PSS.

To preserve the chemically reactive surface of Cu_2O nanowires, a protective layer of Poly (Sodium 4-styrene Sulfonate) (PSS) molecules is meticulously applied. Afterward, Ti_3C_2 quantum dots (QDs) covered with positively charged PEI are arranged onto the Cu_2O nanowires through electrostatic forces. This results in the creation of a layered structure described as Ti_3C_2 QDs/Cu_2O nanowires/Cu heterostructure. Following this stage, a calcination process is completed under inert atmosphere, resulting in the complete removal of the PEI and PSS molecules, thereby bringing Ti_3C_2 QDs and Cu_2O nanowires into close proximity. It is noteworthy that the introduction of Ti_3C_2 QDs does not significantly impact the dimensions or morphology of the Cu_2O nanowires; rather, it serves to safeguard their porous surfaces, as visually demonstrated in Figure 6.5c and 6.5d.

Alternatively, an alternative approach involves the creation of a Ti_3C_2 sheets/Cu_2O nanowires/Cu heterostructure by depositing Ti_3C_2 sheets (measuring approximately 200 × 300 nm) onto the Cu_2O nanowires, as visually depicted in the SEM images presented in Figure 6.5e and 6.5f. In the case of the Cu_2O nanowires/Cu network, a solution containing PSS (comprising 1% volume in ethanol solvent) is administered, followed by a 120-minute treatment at 90°C. Subsequently, the structure undergoes a thorough cleansing with ethanol solvent and is subjected to drying under a controlled flow of nitrogen gas to ensure the comprehensive removal of any excess PSS.

The Cu_2O nanowires/Cu structures, coated with PSS, underwent immersion in solutions containing positively charged PEI-coated Ti_3C_2 quantum dots (QDs) at varying concentrations for a duration of 24 hours under ambient conditions. Following this, a thorough rinse with deionized water was performed, and the structures were subsequently dried using a stream of nitrogen. Subsequent to these preparatory steps, the meticulously arranged heterostructure comprising Ti_3C_2 QDs/Cu_2O nanowires/Cu electrode material was realized through a calcination process conducted under an inert atmosphere at 200°C for a period of 60 minutes. Similarly, an electrode heterostructure of Ti_3C_2 sheets/Cu_2O nanowires/Cu was fabricated by employing a solution of Ti_3C_2 sheets in lieu of Ti_3C_2 QDs (Zeng et al. 2019).

In summary, the incorporation of Ti_3C_2 quantum dots (QDs) significantly improves the stability of Cu_2O nanowires. This improvement is expressed in terms of enhanced charge retention, increased carrier density, improved the charge transfer efficiency, optimized light absorption processes, and a reduction in the number of charge recombination events. In addition, band bending effects are mitigated.

6.2.1.2 Structures Derived from Semiconductor and Semimetallic Group IVA Elements

The importance of carbon nanostructures, such as graphene (Gr), carbon nanotubes (CNTs), and carbon nanotubes/silicon, in optoelectronics has been acknowledged by researchers, as shown in earlier studies (Hu et al. 1999; Li et al. 2010; Chen et al. 2011).

These substances have ushered in revolutionary advancements in the field. Specifically, the fabrication of vertical junction photoelectric devices involves incorporating MXene/Si van der Waals heterostructures, employing Ti_3C_2TX MXene on an n-type silicon (n-Si) substrate (Zhu et al. 2014; Li et al. 2016). This vertical configuration offers advantages for efficient separation and transport of photon-induced carriers.

In the fabrication process, a colloidal solution of $Ti_3C_2T_X$ is drop cast onto a substrate composed of a silicon wafer with n-type characteristics. The deposited $Ti_3C_2T_X$ is then dried under ambient conditions (see Figure 6.6a).

The resulting structure is a van der Waals heterostructure made of n-type silicon (n-Si) and $Ti_3C_2T_X$, as shown in Figure 6.6b. Upon contact, a Schottky connection forms between the n-Si and the $Ti_3C_2T_X$ metal carbide.

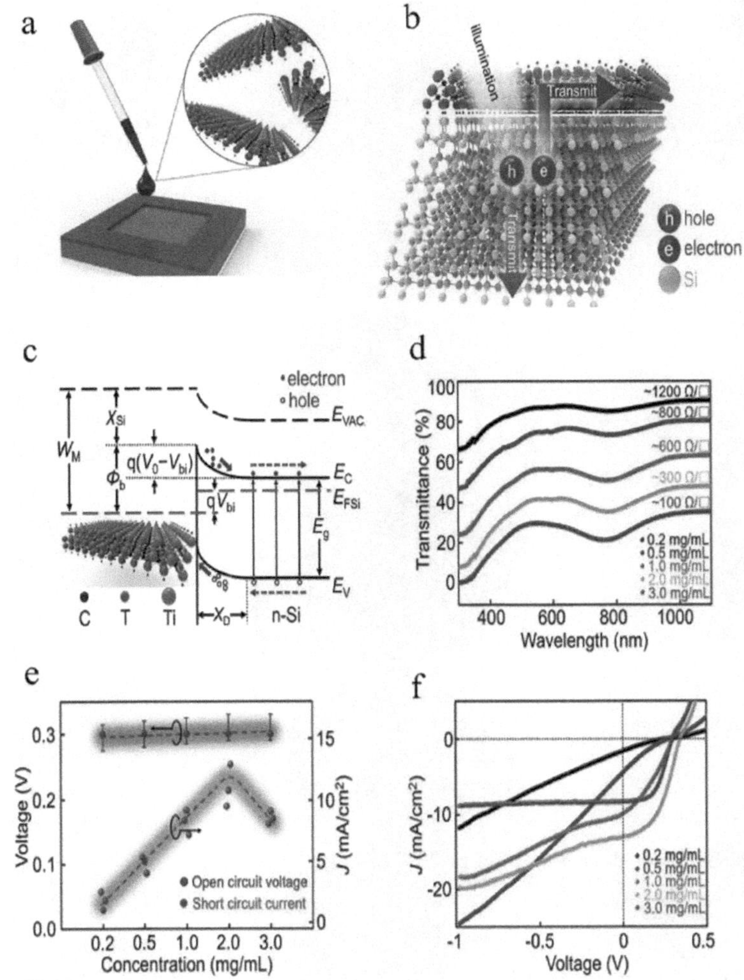

FIGURE 6.6 (a) The $Ti_3C_2T_X$/n-Si heterostructure network architecture was designed using a drop-casting methodology. (b) The subsequent heterostructure was created by merging $Ti_3C_2T_X$ and n-Si. (c) The illustration delineates the Schottky junction between $Ti_3C_2T_X$ and n-Si and elucidates their energy bandgap interplay. (d) Variations in the values for transmission range and sheet resistance were evaluated over $Ti_3C_2T_X$ MXene films of different thicknesses. (e) The investigation encompassed the short-circuit current (J_{SC}) and open-circuit voltage (V_{OC}) of the $Ti_3C_2T_X$/n-Si system across varying concentrations of the $Ti_3C_2T_X$ colloidal solution (f). For the $Ti_3C_2T_X$/n-Si composite structure units, the current–voltage (J–V) properties were investigated using a range of $Ti_3C_2T_X$ colloidal solution concentrations.

(Continued)

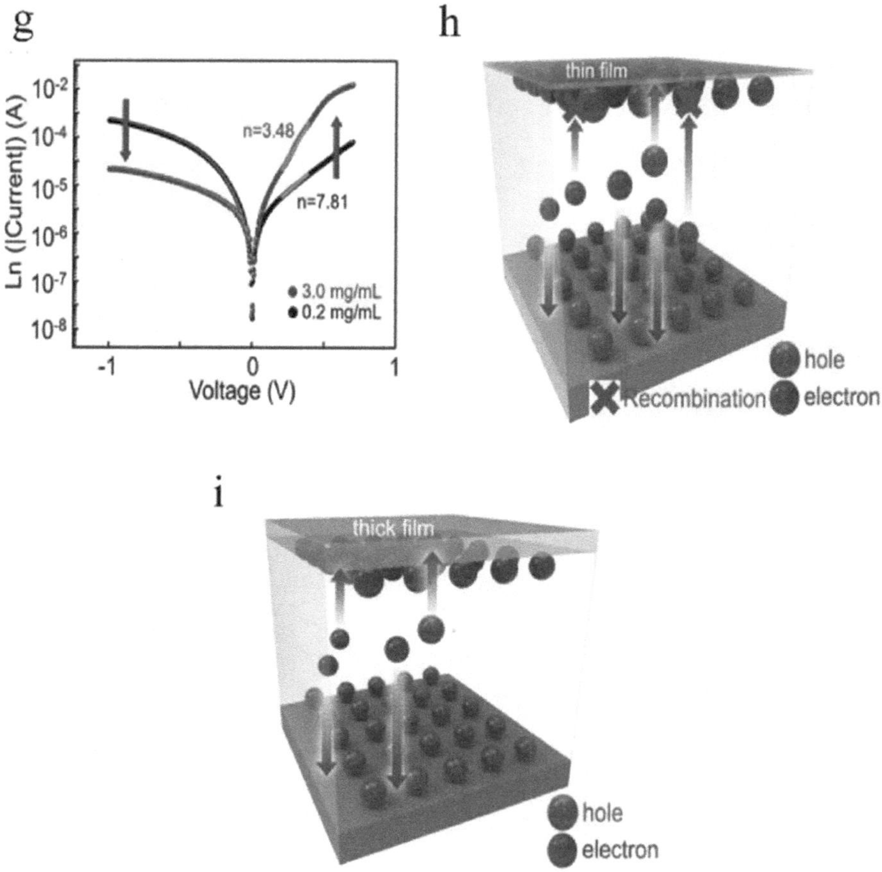

FIGURE 6.6 (Continued) (g) Additionally, the dark heterostructure's current–voltage characteristics of $Ti_3C_2T_X$/n-Si were assessed (h). A schematic representation and the functional apparatus of the $Ti_3C_2T_X$/n-Si heterostructure were presented, featuring both thin and thick $Ti_3C_2T_X$-MXene films.

Source: Permission awarded by Wiley 2017 (Kang et al. 2017).

This generates a robust electric field within the n-Si region. This electric field adeptly segregates photoinduced carriers upon illumination. The carriers are then recombined in an external circuit to generate electrical signals.

In Figure 6.6c, the illustration showcases the non-equilibrium Schottky junction's band energy arrangement within the $Ti_3C_2T_X$/n-Si heterostructure. It visually represents the intrinsic potential value (V_0) and barrier height (Φ_b). Remarkably, the variation in qVbi values along the $Ti_3C_2T_X$ and n-Si Fermi level (EF-Si) working functions enables efficient photon injection. Furthermore, the presence of an active electron–hole space-charge region facilitates the transport of both holes and electrons, directing them toward $Ti_3C_2T_X$ and the metal electrode backbone system, respectively, through V_0.

Figure 6.6d displays the sheet resistance and transmittance profiles of thin films made of $Ti_3C_2T_X$ with different concentrations of $Ti_3C_2T_X$-MXene colloidal solution. The $Ti_3C_2T_X$ thin film may achieve a balance between transparency and conductivity by adjusting the concentration of the colloidal solution.

Figure 6.6e illustrates the characteristics of short-circuit current (J_{SC}) and open-circuit voltage (V_{OC}) of the junction of the $Ti_3C_2T_X$/n-Si device using $Ti_3C_2T_X$ colloidal solutions at different concentrations. A marginal increase in VOC is observed with the elevation of the $Ti_3C_2T_X$ colloidal solution concentration. The $Ti_3C_2T_X$/n-Si configuration exhibits a notable improvement in photovoltaic performance, particularly when using a $Ti_3C_2T_X$ colloidal solution concentration of 10 μL (2 mg/mL), as shown in Figure 6.6f. The investigation revealed that the intensity of the intrinsic electric field increased proportionally with the thickness of the $Ti_3C_2T_X$ layer. Notably, the efficiency of the $Ti_3C_2T_X$/nSi-based device accomplished through the application of $Ti_3C_2T_X$-MXene colloidal solutions exceeded that of the comparable Schottky junction-based device circuit, as shown in Figure 6.6f (Kang et al. 2017).

Figure 6.6g illustrates the dark current–voltage characteristics of the manufactured $Ti_3C_2T_X$/n-Si devices, revealing negligible reverse current in the case of thicker films. The I–V curve in the range of 0.2 to 0.4 V (Figure 6.6g) was used to derive the ideality factor (n) associated with the $Ti_3C_2T_X$/n-Si-based van der Waals heterostructure approach.

The thinner $Ti_3C_2T_X$ film exhibited superior transmittance in comparison to denser films, requiring the trapping of additional photon species and facilitating charge transport, as illustrated in Figure 6.6h (Kang et al. 2017).

The resistance at the interface of the $Ti_3C_2T_X$/n-Si heterostructure network negatively affected both the junction at short circuit (J_{SC}) and the voltage at open-circuit (V_{OC}) within the context of the given scenario. This resistance impeded the efficient separation of the generated photon pairs and electrons by the built-in electric field, resulting in an overall performance reduction. However, at concentrations as high as 2.0 mg/mL, the $Ti_3C_2T_X$/n-Si heterostructure method showing $Ti_3C_2T_X$ colloidal solutions showed a significantly improved photovoltaic response potential.

A technique for the controlled distribution of Sn4+ ions on the conducting Ti_3C_2 MXene surface was devised by Zhang et al. (Luo et al. 2016) using a liquid-phase absorption procedure.

Utilizing electrostatic interactions and ion-exchange connections, this technique harnessed the Ti_3C_2 MXene as a buffering layer. It adeptly adapted to volume fluctuations during electrochemical reactions, furnishing ion-exchange sites and constructing a profoundly conductive network to amplify the rate performance of the nanocomposites. The indispensable role of the PVP stabilizer in the nanocomposite synthesis process ensured the even dispersion of nanoparticles (NPs) on the Ti_3C_2 matrix, preventing any tendency to clump together. Figure 6.7a illustrates the procedural steps involved in the preparation of PVP-Sn(IV)@Ti_3C_2 nanocomposites. It started with the immersion of Ti_3AlC_2 powder in a 40% HF solution for 24 hours.

This selectively removed the aluminum layers. They were replaced by other reactive functionalities such as fluorine and hydroxyl groups. Subsequently, there occurred an alkalinization phase, during which hydroxyl-functionalized Ti_3C_2 (alk-Ti_3C_2) underwent immersion in a 1 M LiOH solution for 24 hours, aiming to enlarge the interlayer spacing. Following extensive rinsing with deionized water to eliminate lingering alkali traces, the alk-Ti_3C_2 underwent dehydration at 80°C for a duration of 12 hours.

Following that, the alkylated Ti_3C_2-MXene underwent submersion in a 1 M $SnCl_4$ solution at 25°C to fashion PVP-Sn(IV)@Ti_3C_2 nanocomposites. Concurrently, 0.1 g of polyvinylpyrrolidone (PVP) was introduced into the solution and agitated for 24 hours. Ultimately, the PVP-Sn(IV)@Ti_3C_2 nanocomposites were acquired through centrifugation, and the residual solution was desiccated at 80°C for a span of 24 hours (refer to Figure 6.7a) (Luo et al. 2016).

SEM images were used to examine the size and morphology differences before and after Sn4+ loading. Figure 6.7b shows the SEM image of Sn(IV)@Ti_3C_2 composites in the absence of PVP.

The visual depiction showcases the irregular and clustered arrangement of nanoparticles (NPs) on the exterior of alkylated Ti_3C_2. This visual indicator indicates that there will be significant volume variations and a consequent decline in performance when used as anode material for LIBs. A common solution to this problem is the use of surfactants, which decrease particle size and prevent the amalgamation of composite materials (Wang et al. 2010; Ming et al. 2014).

To improve the synthesis process of PVP-Sn(IV)@Ti_3C_2 nanocomposites, polyvinylpyrrolidone (PVP) was used as a surfactant and stabilizer in the SnCl4 solution. The SEM images of the PVP-Sn(IV)@Ti_3C_2 system are shown in Figure 6.7c and d, demonstrating a uniform dispersion of numerous nanoparticles (NPs) over the surface of alk-Ti_3C_2. The introduction of PVP into the composite sample resulted in the dispersion of non-aggregated and flexible particles, in contrast to the clustered particles observed in Sn(IV)@Ti_3C_2 (Figure 6.7b). This morphological transformation can be attributed to the effective integration of PVP. PVP reduces the effects of interfacial tension forces and facilitates the even dispersion of nanoparticles (NPs) inside the alk-Ti_3C_2 matrix (Xu et al. 2014).

The TEM representation in Figure 6.7e affirms the even dispersion of the rod-shaped Sn(IV) nanocomplexes, aligning with the findings observed in SEM images (Figure 6.7c and 6.7d). The lower-left corner of Figure 6.7e has an inset that shows the size-distribution analysis of the attached Sn(IV) nanocomplex particles. These particles have diameters ranging from 5 to 10 nm. The HRTEM examination in Figure 6.7f provides more evidence of the amorphous structure of the Sn(IV) nanocomplex. It shows that the Sn(IV)-centered nanocomplex particles do not have any lattice fringes.

Evidence of fringes with interplanar spacings of 0.228 nm and 0.214 nm, corresponding to the (104) and (105) crystallographic planes of Ti_3C_2 MXene, is seen in the HRTEM image (Figure 6.7f). The identification of Ti_3C_2 MXene layers within the PVP-Sn(IV)@Ti_3C_2 composite is further supported by FFT patterns derived from Fourier analysis. Elemental analysis conducted using energy-dispersive spectroscopy (EDS) validates the composition of the composites. The STEM image

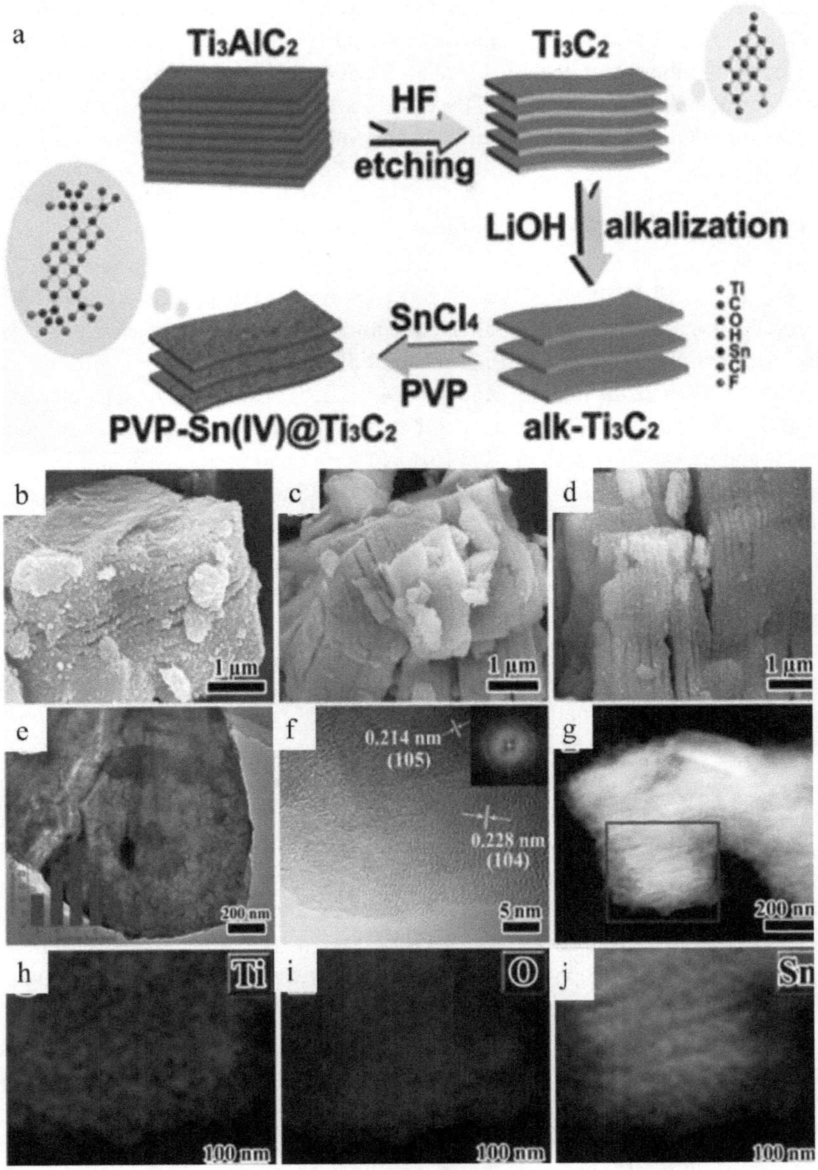

FIGURE 6.7 (a) The composites' synthesis of PVP-Sn(IV)@Ti_3C_2 is shown in the schematic diagram. (b) The SEM image delivers insight into the structural characteristics of Sn(IV)@ Ti_3C_2-based composites. (c and d) The SEM images show the changes in the morphology of PVP-Sn(IV)@Ti_3C_2. (e) TEM image reveals the internal structure of PVP-Sn(IV)@ Ti_3C_2, showing the distribution process of anchored nanocomplex nanostructure of Sn(IV). (f) HRTEM image lucidates the fringe spacing, accompanied by a Fast Fourier Transform (FFT) analysis, providing a detailed study. (g–h) STEM images and elemental mapping confirm the presence of Ti, O, and Sn elements.

Source: Permission awarded by American Chemical Society 2016 (Luo et al. 2016).

(Figure 6.7g) vividly displays the distinct stratified structure of the PVP-Sn(IV)@ Ti_3C_2 composite. Furthermore, elemental mapping of Ti, O, and Sn (Figure 6.7h, i, j) clearly demonstrates the uniform distribution of Sn(IV) nanocomplex on the matrix surface (Luo et al. 2016). These results underscore the effectiveness of the synthetic methods employed and offer promising prospects for advancing MXene-based materials.

6.2.1.3 MXene Structures Combined with Graphene

Two-dimensional (2D) nanomaterials, including TMO/hydroxides (Shan et al. 2016), graphene (GR) (El-Kady et al. 2012), MXenes (Ghidiu et al. 2014a), and TMDCs (Acerce, Voiry, and Chhowalla 2015), have undergone comprehensive investigation owing to their potential, prompting substantial research efforts in energy storage applications (Mendoza-Sánchez and Gogotsi 2016). MXenes, in particular, have garnered substantial consideration owing to their versatility in applications such as supercapacitors (Dall'Agnese et al. 2016), Na-ion (Zhao et al. 2017) and Li-ion batteries (Liu et al. 2015), electromagnetic shielding (Shahzad et al. 2016), ion sieving (Ren et al. 2015a), and cellular imaging (Xue et al. 2017). MXenes possess several advantages, such as metallic conductivity, thin-layer thickness, and hydrophilic properties, rendering them promising candidates for high-capacity supercapacitors with enhanced volumetric capacitance compared to existing devices. For instance, films of titanium carbide ($Ti_3C_2T_x$) have exhibited a notable volumetric capacitance value of 900 F cm^{-3}, along with exceptional cyclability range and rate performance (Ghidiu et al. 2014a). Nonetheless, one limitation of MXene electrodes lies in their relatively limited surface area available for electrolyte interactions.

Graphene (GR), a well-known 2D carbon material, is widely used as an electrode in supercapacitors and shows great potential for developing electronic devices (El-Kady et al. 2012). Its distinctive attributes, encompassing exceptional electrical conductivity, mechanical robustness, optical characteristics, singular energy band structure, and heightened electron mobility, render it well suited for diverse applications, including micro-supercapacitors, transistors, sensors, and biosensors.

In a recent study, graphene films of different thicknesses were produced by spraying varying volumes of graphene solutions onto glass substrates, with a specific focus on investigating the impact of temperature on these thin graphene films (Simon and Gogotsi 2010; Park and Ruoff 2009; Haditale, Dariani, and Ghasemian Lemraski 2019).

At temperatures up to 150°C, the current remains relatively stable, but beyond this range, it gradually increases while resistance sharply decreases. This phenomenon is particularly noticeable in samples with fewer graphene layers. One plausible explanation for this behavior lies in the evaporation of solvent trapped within the graphene flakes. The unique 2D structure of graphene, characterized by strong covalent σ bonds in a honeycomb pattern and weak van der Waals π bonds, allows for vertical bonding between the graphene layers and the solvent. As temperature rises, the solvent evaporates, causing the graphene layers to approach each other more closely, intensifying their interaction. Consequently, this results in enhanced current flow and reduced resistance as the flakes become densely packed and in closer proximity.

Another contributing factor to the increased current is the percolation effect, which describes fluid flow through porous materials. Below the percolation threshold, there is minimal interaction between the components. However, as the temperature exceeds 150°C, the graphene layers reach the percolation threshold, leading to stronger interaction between them. This is reflected in a notable jump in the current–temperature curve, signifying the occurrence of the percolation threshold. With rising temperature, the connection between the graphene layers strengthens, leading to heightened transient current (Newman and Ziff 2000; Chen et al. 2012). This effect is more noticeable in thinner samples because higher temperatures bring the graphene planes closer together, making it easier for the current to flow (Haditale, Dariani, and Ghasemian Lemraski 2019).

To combat the restacking challenge associated with 2D graphene nanosheets, an effective strategy involves creating layers with more open structures. This design provides additional room for the storage and movement of electrolyte ions, mitigating the problem of self-stacking (Yan et al. 2017).

A recent advancement in this realm is evidenced by the research conducted by Feng et al., wherein a vacuum-based filtration technique was employed to fabricate flexible thin-film layers comprising MXene and graphene (MXene/GR). Following this, these layers were integrated into flexible all-solid-state supercapacitors, demonstrating an impressive volumetric capacitance of 216 F cm^{-3}. This innovative approach not only prevents self-stacking but also enhances the overall capacitance of supercapacitors (Li et al. 2017).

Another noteworthy contribution by Gogotsi et al. (Yan et al. 2017) introduces a straightforward method for crafting flexible thin films using MXene/rGO through a self-assembly process.

This methodology leverages electrostatic interaction process between the positively charged graphene oxide (GO) and the negatively charged rich MXene, ensuring the well-ordered arrangement of MXene layers. Adding reduced graphene oxide (rGO) nanosheets effectively prevents MXene layers from restacking. This advancement ensures efficient dispersion and transportation of electrolytes within the material. The fabrication process, illustrated in Figure 6.8a, emphasizes the electrostatic-based uniform organization, starting with the reduction of GO nanolayers via hydrazine treatment at a temperature of 90°C for a time duration of 24 hours (Yan et al. 2010).

The MXene/rGO hybrid electrodes were created using an electrostatic self-assembly technique. The procedure included the incremental addition of a solution comprising poly(diallyl dimethyl ammonium chloride)-modified reduced graphene oxide (G-PDDA) to a 20-mL suspension of MXene at a concentration of 0.5 mg/mL while stirring continuously. The resulting mixture was then ultrasonicated (for 10 minutes), followed by filtration (using a polypropylene sheet and DI water) and dehydration (for 12 hours under vacuum) to acquire flexible and individual MXene/rGO heterostructure films. The composition of these hybrid films was varied on the basis of the volume of G-PDDA suspension added, resulting in different compositions such as M/G–1% (0.2 mL), M/G –5% (1 mL), and M/G–10% (2.2 mL). In addition, to establish a baseline, a pristine MXene thin film was arranged without the incorporation of reduced graphene oxide (rGO). This carefully controlled fabrication process facilitated the precise integration of rGO into the MXene films (Yan et al. 2017).

SEM and TEM imaging techniques were used to examine the morphology of the MXene/rGO hybrid films (Figure 6.8b, c). Both pristine MXene films and MXene/rGO hybrid films demonstrated a folded configuration. An examination of the cross-section through SEM of MXene films illustrated an orderly layered structure (Figure 6.8d), maintaining consistency even after the introduction of rGO nanosheets, as depicted in the SEM image (Figure 6.8e). The TEM picture (Figure 6.8f) confirmed the presence of well-arranged nanolayers, indicating the successful inhibition of the self-aggregation of the MXene nanolayers. XRD patterns were acquired for both the pristine MXene and the MXene/rGO hybrid architectures. They show a significant change in layer spacing (Figure 6.8g) (Yan et al. 2017).

A multilayer (ML) variant of $Ti_3C_2T_x$ particles was created by removing aluminum from Ti_3AlC_2 precursors using an aqueous HF solution in a separate investigation by Nedil and colleagues. Delaminated $Ti_3C_2T_x$ MXene particles were dispersed in an ethanol solvent through the process of sonication. The dispersion was then centrifuged. The supernatant containing different concentrations of graphene and $Ti_3C_2T_x$ MXene was collected. The application of electrohydrodynamic atomization (EHDA) deposition method facilitated the deposition of $Ti_3C_2T_x$/graphene thin films in a sandwich-like configuration onto glass substrates (refer to Figure 6.9a) (Naguib and Gogotsi 2015).

To prepare the colloidal solution, a specific quantity of delaminated $Ti_3C_2T_x$ MXene particles was measured and introduced into a solvent such as ethanol. The particles then underwent 60 minutes of sonication to ensure even dispersion. After sonication, the mixture was spun in a centrifuge at 3,500 rpm for 30 minutes at a temperature between 25 and 30 °C. This step led to the separation of the liquid above, which held the particles from the centrifuge.

Various quantities of the obtained liquid from above were blended with a set amount of pure defect-free (df) nanoplatelet graphene solution at a concentration of 0.05 mg/mL and $Ti_3C_2T_x$ MXene at a concentration of 1.16 mg/mL. The resulting mix was used as the deposition solution. The Electrohydrodynamic Atomization (EHDA) method was used to apply thin films of the $Ti_3C_2T_x$/graphene sandwich structure onto a glass surface.

The aforementioned liquid was stored in the syringe pump and employed in the course of the EHDA process. Several methods, including different flow rates and different potential ranges, were used to observe EHDA for both graphene flakes and $Ti_3C_2T_x$ dispersions. A stable meniscus was observed when the potential was set at zero. Increasing the voltage to 3.08 kV resulted in the achievement of stable cone-jet modes for depositing $Ti_3C_2T_x$, while a potential of 6.30 kV was employed for the graphene nanoflakes. A steady cone-jet mode was used to produce thin films of the $Ti_3C_2T_x$/GR composite (Figure 6.9b). Prior to deposition, the top surface of the substrate was cleaned with ethanol and dehydrated.

The stand-off setup was methodically adjusted within the range of 1 to 4 cm for graphene to ascertain the shielded deposition area for both $Ti_3C_2T_x$ MXene and graphene utilizing a stable cone-jet mode. The nozzle underwent lateral adjustment at a walking speed of 2.2 mm/s, while sustaining stand-off distances of 1.5 cm and 1 cm for graphene (GR) and $Ti_3C_2T_x$, respectively. This adjustment covered an area of (2 × 2) cm² (Ali et al. 2016; Choi, Ali, and Jo 2013).

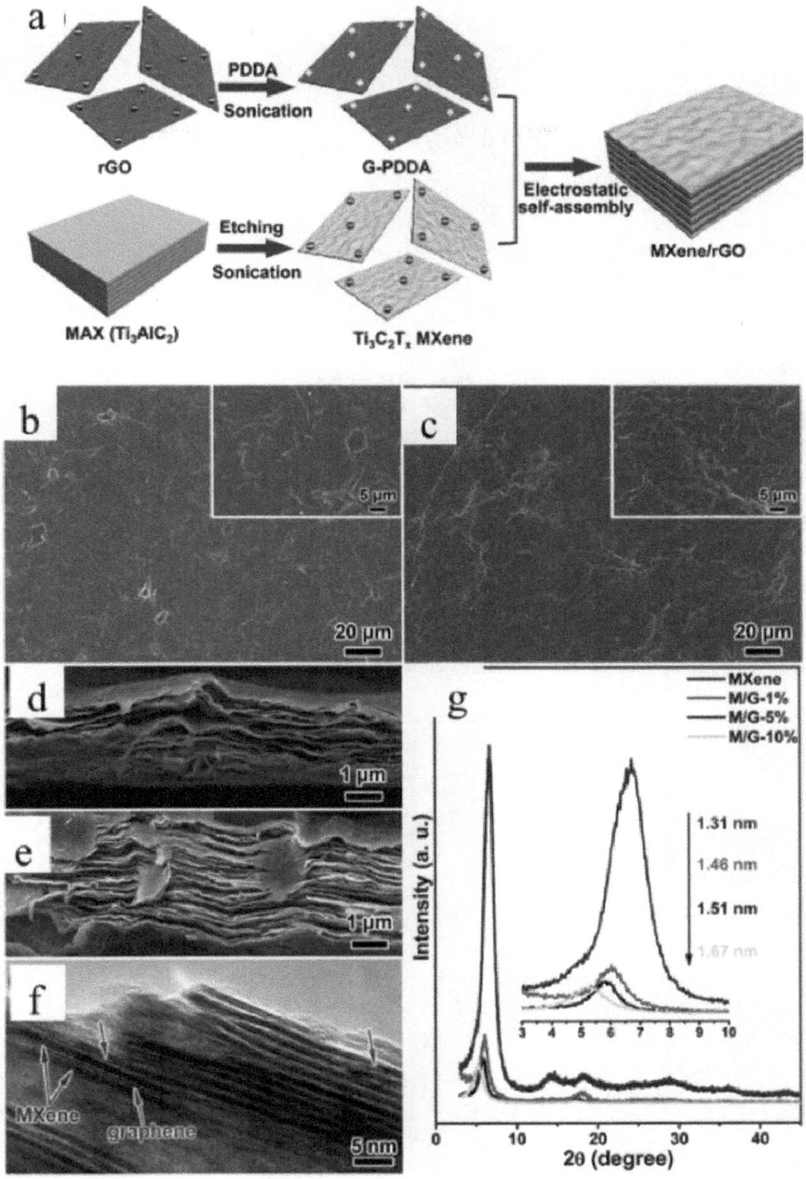

FIGURE 6.8 (a) Scheme showing the integration of MXene and reduced graphene oxide (rGO) within a hybrid framework. (b, c) SEM images from top view reveal the morphology of MXene/rGO hybrid structure. (d, e) SEM images from cross-section provide an insight into layered arrangement of MXene/rGO hybrid. (f) TEM picture offers a detailed and complex perspective of the microstructure present in the M/G–5% hybrid arrangement. (g) The XRD analysis shows diffraction patterns for both the processed MXene material and the hybrid composed of MXene and rGO.

Source: Permission awarded by Wiley 2017 (Yan et al. 2017).

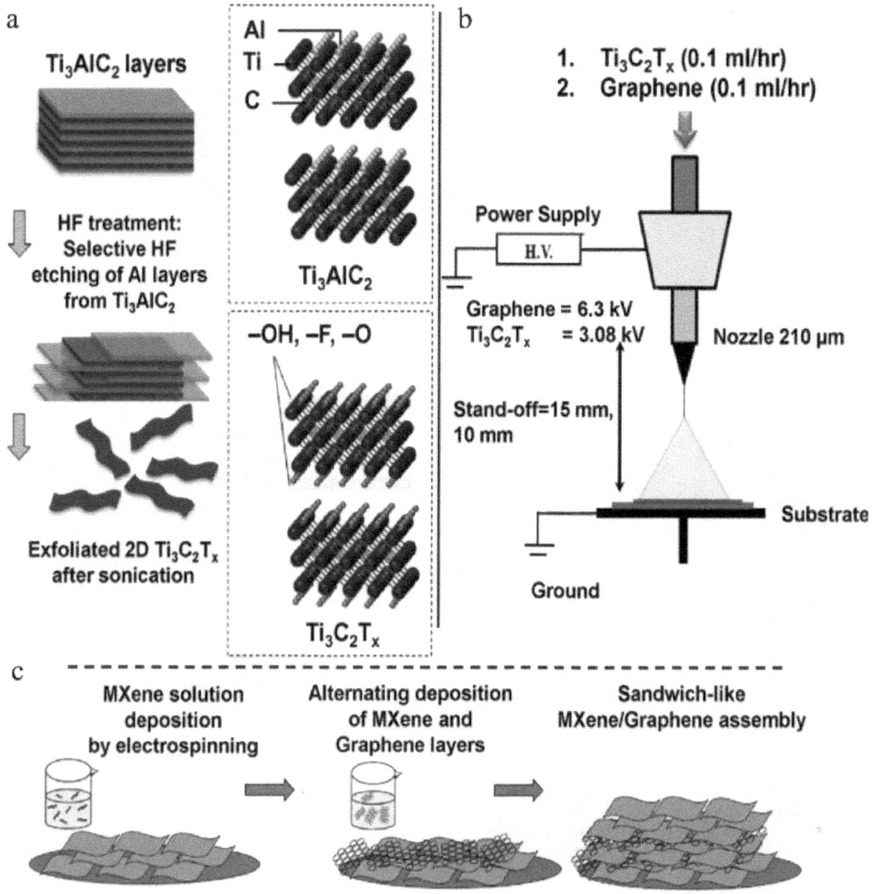

FIGURE 6.9 Basic procedures for fabricating MXene/Gr heterostructures: (a) The diagram provides an overview of the method for fabricating a 2D $Ti_3C_2T_x$ graphene membrane using MAX phase precursors, specifically Ti_3AlC_2. (b) The photograph illustrates the configuration of the electrospray technique used to deposit a composite thin film composed of $Ti_3C_2T_x$/Gr. (c) The schematic diagram delineates the production procedure of a sandwich-like nanocomposite of MXene and graphene.

Source: Permission awarded by AIP Publishing (Aïssa et al. 2016).

The deposition process involved multiple repetitions using various graphene concentrations to achieve films consistently comprising 5–10 layers of both GR and $Ti_3C_2T_x$ (Figure 6.9c). Following deposition, the films were strengthened through exposure to a vacuum at 103 mbar for 2 hours at 200°C. Figure 6.10a presents a TEM image revealing the df-GR nanoflakes, with a small inset confirming their crystalline nature via SAED patterns. A three-dimensional sectional view in Figure 6.10b from AFM illustrates the deposition of df-graphene nanoflakes on a SiO_2 substrate, with a film thickness of approximately 0.1 nm. Panels c and d of Figure 6.10 present

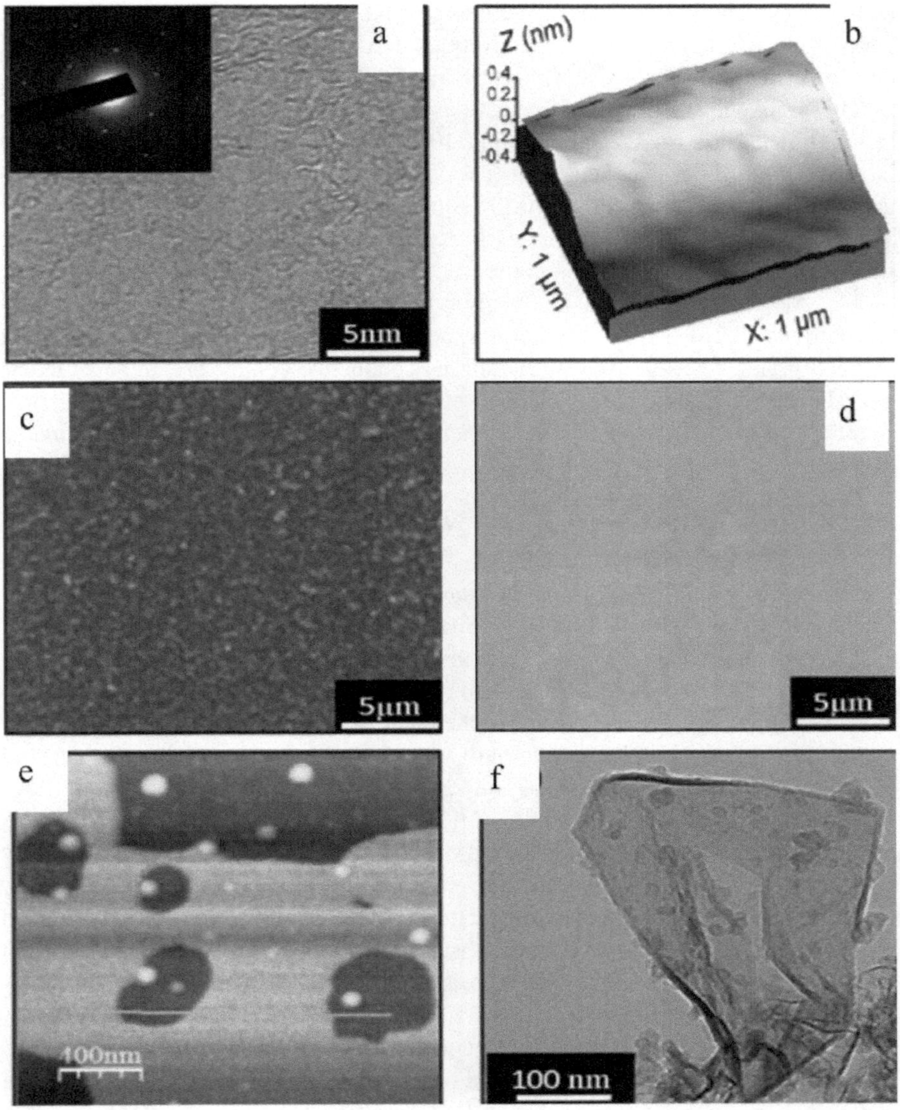

FIGURE 6.10 (a) A TEM image of df-GR nanoflakes, along with a corresponding selected area electron diffraction (SAED) pattern in the top-left inset. (b) A three-dimensional view of the atomic force microscopy (AFM) image illustrates the deposition of df-graphene nanoflakes on a SiO_2 substrate. (c) An SEM image provides insight into the morphological structure of $Ti_3C_2T_x$/graphene composite films, while (d) shows the morphology of the $Ti_3C_2T_x$ film. (e) An AFM image offers an illustrative view of the as-deposited $Ti_3C_2T_x$ MXene. (f) A low-magnification TEM image highlights the $Ti_3C_2T_x$ nanoflakes. The synthesis of heterostructures using MXene-silicene, MXene-germanene, and MXene-stanene is uncommon, making these discoveries very remarkable in the scientific field.

Source: Permission awarded by AIP Publishing (Aïssa et al. 2016).

distinct top-view SEM images of the fabricated thin films based on $Ti_3C_2T_x$ and the composite $Ti_3C_2T_x$/graphene, respectively. The composite film exhibits a consistent and uniformly even surface morphology, in stark contrast to the $Ti_3C_2T_x$ film alone.

The graphene integration fills surface basins or pinholes in the $Ti_3C_2T_x$ film, resulting in a smoother surface (Figure 6.10c). The presence of pinholes in the AFM image (Figure 6.10e) is notably reduced after the successful deposition of graphene nanoflakes. Furthermore, Figure 6.10f depicts a representative TEM image of the $Ti_3C_2T_x$ MXene nanoflakes extracted from the substrate and prepared for analysis, showing minor defects like nanopores.

The MXene/rGO composite meaningfully improves the value of electrical conductivity (2,261 S cm^{-1}) by merging both MXene and rGO. The nanolayers of reduced graphene oxide (rGO) function as conductive connectors among MXene nanosheets, widening the intervals between the MXene layers and creating distinct channels for electrolytes. This arrangement ensures a high-rate performance (Aïssa et al. 2016).

6.2.1.4 Heterostructures of MXene with Other Carbon-Based Layered Material

The growing popularity of electrochemical devices is driven by their exceptional energy density and extended lifespan (Simon, Gogotsi, and Dunn 2014; Simon and Gogotsi 2010), particularly in response to the rising demand for portable and wearable technology (Gogotsi and Simon 2011; Yang et al. 2013). However, the quest to enhance the energy density of electrochemical capacitors in relation to their volume is ongoing. Carbon-based alternatives, although having a volumetric potential of about 60 F cm^{-3} for commercial applications, come with limitations (Yang et al. 2013; Ghaffari et al. 2013; Jung et al. 2013). On the other hand, hydrated RuO_2 can yield high-capacitance capacitors (1,000–1,500 F cm^{-3}), but their broad applicability is hindered by substantial costs (Zheng and Rapp 1995).

In a recent study by Gogotsi et al., an innovative filtration method was introduced using a water-based mixture to form a distinctive sandwich-like structure of carbon nanotubes (CNTs) and MXenes. The resultant MXene/CNT paper exhibits attributes of being both self-supporting and exceptionally pliable.

When utilized as an electrode, this meticulously crafted paper demonstrated a significant augmentation in capacitance and overall efficiency in contrast to the sole use of pure MXene. The detailed procedure for fabricating these composite MXene/CNT paper electrodes is elucidated in Figure 6.11a. It encompasses the filtration of a solution containing MXene flakes with the aid of vacuum, leading to the creation of $Ti_3C_2T_x$ nanoflakes. Following this, a stratum of CNTs was applied atop the MXene flakes, and this sequence was iterated multiple times to assemble composite paper electrodes. All the resultant films made from this paper exhibited thicknesses ranging from 2 to 3 μm, as depicted in Figure 6.11b. Furthermore, these paper electrodes demonstrated remarkable flexibility and strength, as they could be wrapped around 5 mm diameter glass bars without visible damage, as shown in Figure 6.11c.

The paper's density, at 3.2 g cm^{-3}, indicates its high compactness. This dense nature somewhat limits the contact of electrolyte ions with $Ti_3C_2T_x$ nanoflakes, consequently restricting their electrochemical performance (Figure 6.11f). However,

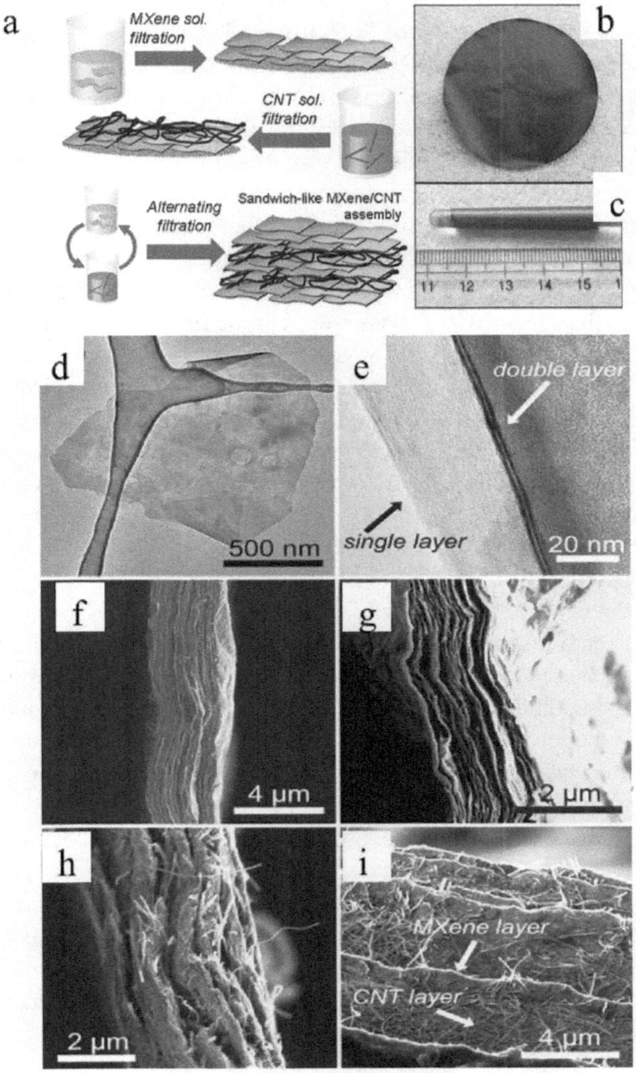

FIGURE 6.11 (a) The preparation process is depicted through a schematic representation, illustrating the creation of a sandwich-like MXene/CNT paper heterostructure. (b) Digital images vividly display the MXene/CNT paper, showcasing its remarkable flexibility and ability to stand freely in a sandwich-like heterostructure. (c) The flexibility and strength of the MXene/CNT paper are convincingly demonstrated as it is effortlessly wrapped around a glass rod with an approximate diameter of 5 mm. (d, e) Transmission electron microscopy (TEM) images provide detailed views of $Ti_3C_2T_x$ nanoflakes, offering a close-up examination of their structural characteristics. (f–i) SEM images capture the cross-sectional features of various samples, including pure $Ti_3C_2T_x$ MXene (f), the sandwich-like $Ti_3C_2T_x$/SWCNT heterostructure (g), and the $Ti_3C_2T_x$/MWCNT paper heterostructure. These images provide insights into the morphology of these materials.

Source: Permission awarded by 2014 Wiley (Zhao et al. 2015).

the integration of CNTs into the $Ti_3C_2T_x$ MXene structures leads to a noticeable improvement in the arrangement of $Ti_3C_2T_x$ nanoflakes (Figure 6.11g, h).

Upon comparing MXene/CNT composites with structures resembling sandwich-like paper, a discernible enhancement and well-organized configuration of MXene and CNT layers are evident (refer to Figure 6.11i). These papers, comprising a sandwich-like structure of MXene and single-walled carbon nanotube (SWCNT), showcase a significantly heightened capacitance, approximately measuring 390 F cm^{-3} at a scan rate of 2 mV s^{-1}. Remarkable is their enduring performance, remaining steadfast even after undergoing 10,000 cycles. Subsequent improvements in performance can be attained by precision refinement of the composite's geometry, adjustments in the MXene/CNT mass ratio, and optimization of the layer thickness (Zhao et al. 2015).

In recent scholarly investigations (Liang et al. 2017), lithium-sulfur-based (Li-S) energy storage arrangements have garnered substantial attention as prospective solutions to address the escalating market demands. These systems possess a theoretical energy density of approximately 2500 Wh kg^{-1}, representing a three to five times increase compared to contemporary Li-ion batteries. Elemental sulfur, functioning as a reactive nanomaterial, offers numerous advantages, encompassing cost-effectiveness, widespread availability, and environmental resilience.

Despite these benefits, the commercialization of Li-S batteries has encountered setbacks owing to enduring challenges, despite extensive years of research. A noteworthy discovery is the presence of thiosulfate/polythionate functional groups on the MXene surface, signifying the involvement of –OH groups in the redox activities of polysulfides through functionalization (Evers and Nazar 2013; Yang, Zheng, and Cui 2013).

This discovery aligns with analogous analyses conducted on graphene oxide and metal oxide materials.

To fabricate composite structures, multi-walled carbon nanotubes (CNTs), having an estimated diameter of 8 nm, were blended with exfoliated MXenes in chlorosulfonic acid. The combination underwent a filtration procedure utilizing an anodic aluminum oxide (AAO) membrane, succeeded by dehydration at 80°C. To establish the composite configuration, Ti2C-MXene and Ti_3C_2 were incorporated with approximately 10% CNTs, while Ti_3CNTs were blended with around 20% CNTs. The resulting composite structures, named sulfur S/CNT-MXene, were created using a simple melt-diffusion method at 155°C for 12 hours under normal air circumstances (Liang et al. 2017).

The MXene phases show less than ideal exterior organization compared to nanomaterials made of meso/microporous carbon, as seen in Figure 6.12a of the Brunauer–Emmett–Teller (BET) method.

However, an outstanding surface area can be achieved by utilizing delaminated multilayer MXene nanosheets (Wang et al. 2016). The restacking of nanolayers, induced by van der Waals interactions during the elimination of residual solvents from delamination, serves as a plausible explanation for this pivotal parameter. The surface area of the host material plays a crucial role in chemical interactions within the framework of Li-S batteries (Hart et al. 2015; Liang et al. 2015). To tackle these challenges, Nazar et al. (Liang et al. 2017) introduced a porous design by sandwiching MXene components with nanolayered carbon nanotubes (CNTs) (Figure 6.12b). The inclusion of CNTs enables efficient charge transport between the nanosheets,

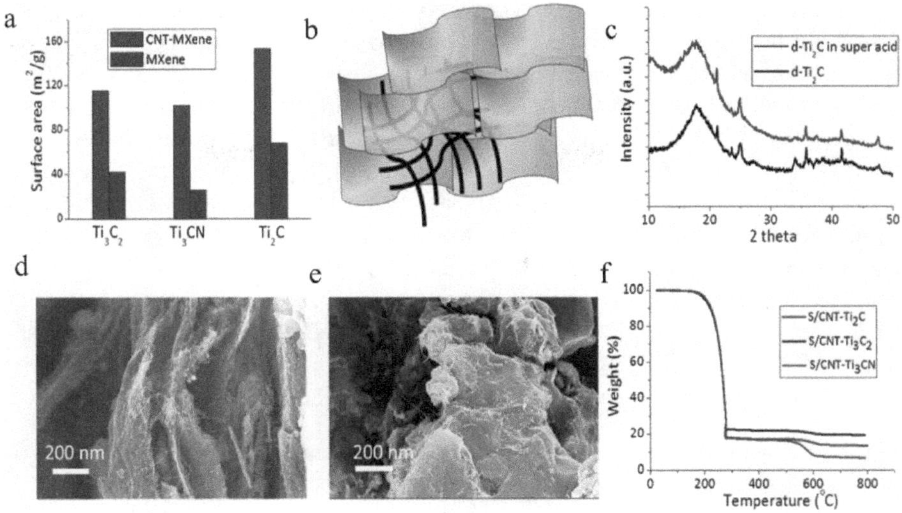

FIGURE 6.12 (a) Using the five-point BET analysis approach, the surface areas of the MXene nanolayer and the CNT-MXene composite structure were evaluated. This allowed for a comparison of their relative surface properties. (b) A schematic representation to illustrate the collective structure of the CNT-MXene nanomaterial, providing a visual depiction of its composition. (c) X-ray diffraction (XRD) patterns were employed to elucidate the structural distinctions in the Ti2C nanolayers both prior to and subsequent to exposure to chlorosulfonic acid. It is noteworthy that the appearance of prominent 18° peaks in the patterns can be attributed to the lubricating substance utilized during the sampling process. (d, e) SEM images are showcased, featuring the morphological attributes of both the CNT-Ti_3C_2 composite structure and the S/CNT-Ti_3C_2 composite structure, offering visual insights into their structural characteristics. (f) An investigation of the unique thermal behavior of S/CNT-MXene composites is elucidated through the examination of thermal gravimetric analysis (TGA) curves, with a focus on their thermal properties and decomposition profiles.

Source: Permission awarded by Wiley 2016 (Liang et al. 2017).

a crucial factor in the creation of highly effective Li-S batteries capable of accommodating sulfur loading.

The resistance value of the Ti2C MXene phase to acid was verified by immersing the nanolayers in acid for 12 hours, as depicted in Figure 6.12c. XRD analysis demonstrated that there was no noticeable change in crystallinity following this acid treatment. Furthermore, Figure 6.12d, a SEM image, clearly illustrates the even distribution of carbon nanotubes (CNTs) within the MXene layers. The inclusion of CNTs enhances electronic conductivity across the nanolayers, as evidenced by electrochemical impedance spectroscopy on the S/CNT-Ti_3C_2 electrode, which indicated a resistance level comparable to the CNT-free material. Through the employment of a melt dispersion technique, S/CNT-MXene achieved a significantly higher sulfur content, attributed to improved surface area and absorbency, as depicted in Figure 6.12f, all without notable sulfur aggregation, as seen in Figure 6.12e (Liang et al. 2017).

The introduction of trace amounts of carbon nanotubes (CNTs) between the layers of MXene generates a permeable and electronically conductive network characterized by an increased capacity for polysulfide adsorption. This facilitates sulfur's role as a host material, displaying exceptional long-term cycling efficiency and minimal capacity decay—as low as 0.043% per cycle over 1,200 cycles. In addition, this methodology has enabled reliable sulfur loading of up to 5.5 mg/cm^2, making it suitable for electrodes designed for high-energy density applications. These findings open doors for advancements in hybrid film structures using MXene materials, particularly in the field of electrochemical applications. These innovations are promising because compact, integrated energy storage devices require exceptional properties such as volumetric capacity and rate performance. In addition, the potential applications of these nanomaterials extend to a variety of thin-film-based electronics. This presents opportunities for several technological breakthroughs (Liang et al. 2017).

6.2.1.5 Heterostructures Based on MXene-Semiconductors

6.2.1.5.1 Heterostructures of MXene with Metal Chalcogenide

6.2.1.5.1.1 MXene with Elemental III–VI Group of Periodic Table Recently, there has been a remarkable upsurge of interest in thin-film semiconducting materials, especially those in the III–VI group. Materials such as GaTe, GaSe, InSe, and various others are included in this category. The increased attention toward them is mainly due to their exceptional electrical and optoelectronic characteristics. Unlike transition metal dichalcogenides (TMDs), which have indirect bandgaps, these III–VI-layered semiconductors have direct, precisely tunable bandgaps. This unique property positions them as highly promising candidates for various optoelectronic applications (Ho et al. 2017; Yang et al. 2018; Yao et al. 2020; Feng et al. 2018).

6.2.1.5.1.2 Heterostructures of MXene and InSe InSe, known for its exceptional attributes such as high chemical stability, wide spectral responsiveness, versatile charge carriers, and a narrow bandgap of 1.26 eV, has garnered significant attention in the realm of photoelectronic applications. Lee and colleagues (Yang et al. 2019) introduced an efficient photodetector (PD) based on a few-layer InSe, incorporating an electrode network using MXene (Ti_2CT_x). The fabrication of this electrode system involved several key steps. Initially, a solution containing exfoliated Ti_2CT_x nanoflakes was spin-coated onto a SiO_2/Si substrate, resulting in a thickness value of approximately 285 nm. After a drying process, few-layer Ti_2CT_x nanoflakes, approximately 5 nm wide, were attached to the surface of exfoliated InSe nanolayers, which were around 20 nm thick, on the SiO_2/Si substrate using a mechanical transfer technique.

The electrode device was designed through electron-beam lithography (EBL), followed by the deposition of a thin gold (Au) layer via electron-beam vaporization. These steps led to the creation of the InSe/Ti_2CT_x field-effect transistor (FET). The engineering manufacturing process of the InSe/Ti_2CT_x heterostructure electrode is visually depicted in Figure 6.13a, encompassing the utilization of EBL and mechanical transfer techniques (Yang et al. 2019).

In the initial phase, the upper surface of the InSe/Ti_2CT_x-based field-effect transistor (FET) was covered with a layer of poly methyl methacrylate (PMMA). Afterward,

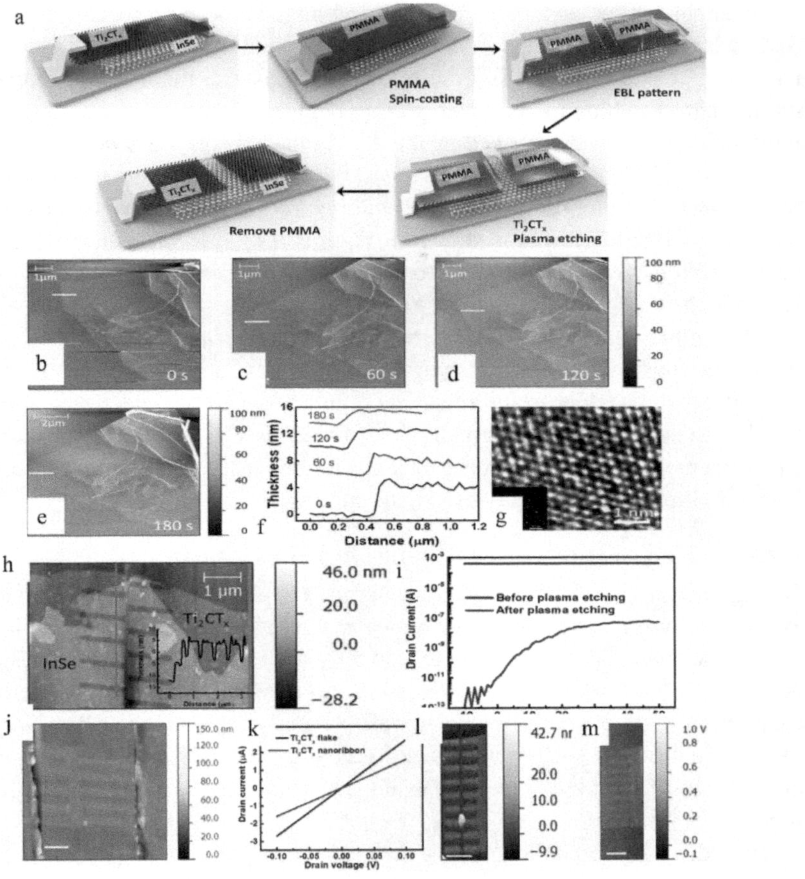

FIGURE 6.13 Fabrication and Characterization of Vertically Stacked InSe/Ti_2CT_x Heterostructure Photodetectors via Plasma-Etching and Lithographic Processing (a) The schematic sequence illustrates the fabrication process for the vertically stacked InSe/Ti_2CT_x heterostructure photodetector, with protective PMMA layers incorporated for device stability. The manufacturing protocol involves precise plasma etching and electron beam lithography (EBL) steps. (b, c) AFM characterization reveals the progressive Ti_2CT_x thinning process under different plasma etching durations, while (d, e) demonstrate the resulting surface topographies, with white reference lines indicating measurement positions. The etching process employed Ar/SF_6 gas mixture (40:1 ratio) at 25 W power and 30 mTorr pressure. (f) HRTEM analysis with SAED pattern (inset) confirms the structural properties of plasma-modified Ti_2CT_x flakes. (g) Subsequent AFM imaging shows the vertical alignment of InSe and Ti_2CT_x flakes post-etching, with the Ti_2CT_x flake protecting the underlying InSe and displaying characteristic periodic nanoribbon features. (h) Electrical characterization compares the drain current-gate voltage (I_4-V_9) relationship for the heterostructure before and after plasma treatment ($V_4 = 1$ V). (i, j) AFM micrographs (scale bar: 1 μm) detail the Ti_2CT_x nanoribbon devices fabricated via ICP plasma etching. (k) Current-voltage (I-V) profiles contrast the performance of pristine Ti_2CT_x flakes versus nanoribbon configurations. (l, m) Complementary AFM and KPFM images elucidate the electrode architecture and work function distribution across the Ti_2CT_x surface.

Source: Permission awarded by American Chemical Society. 2019 (Yang et al. 2019).

the channel region underwent EBL processing, while precise adjustments to the Ti_2CT_x layer thickness were made using a controlled minor gravure inductively coupled plasma (ICP) process (Figure 6.13b–g). To expose the Ti_2CT_x MXene layer, a plasma-based etching method was employed, ensuring the preservation of the underlying InSe layers, as established by the AFM image displayed in Figure 6.13h.

Figure 6.13i clearly shows the successful channel content transfer from Ti_2CT_x MXene to InSe. Importantly, the ICP etching process does not significantly alter the properties of Ti_2CT_x. This is shown in Figures 6.13j–m. Subsequently, the PMMA layers are removed. A photodetector made of InSe/Ti_2CT_x with a 250 nm channel length is the result of this process (Yang et al. 2019).

To summarize, 2D Ti_2CT_x MXene layers are advocated as highly efficient electrode materials with fast response times and excellent photoresponsivity. This breakthrough opens the door to future advances in integrated optoelectronic applications.

The exceptional metallic conductivity of 2D Ti_2CT_x positions it as a promising electrode material. It offers great potential for enhancing the performance of photodetectors through plasmonic-based configurations.

6.2.1.6 Heterostructures of MXene with Phosphorene

Owing to its impressive theoretical performance, phosphorus (P) has surfaced as an exceptionally effective option for anode materials in sodium-ion batteries (SIBs), giving rise to a Na3P alloy distinguished by a low electrode potential (Ni, Li, and Lu 2018; Ellis, Hatchard, and Obrovac 2012; Allan et al. 2016). This outperforms the capabilities of other anodic nanomaterials presently applicable in SIBs. Nonetheless, the electrochemical performance of P as an alloyed anode material encounters hurdles due to substantial volume expansion and insulating properties during the redox process.

To address these challenges, the strategic incorporation of phosphorus (P) into electrically conductive materials, such as carbon, has been identified as an effective approach to enhance structural stability and improve electrochemical kinetics (Qian et al. 2013; Li et al. 2013; Song et al. 2015; Hu et al. 2018). The authors have presented a novel electrostatic method for creating well-structured heterostructures at the molecular level (PDDA-BP/Ti_3C_2). In this method, the BP layers are altered using poly(diallyl dimethyl ammonium chloride) (PDDA), which acts as a positively charged polymeric surfactant (Xiong et al. 2018; Feng et al. 2018). This process of functionalization not only increases the tendency of BP nanolayers to attract electrons but also protects them from undergoing oxidation when exposed to water, guaranteeing their stability and even distribution.

The parallel arrangement of the 2D interlayer space in the PDDA-BP/Ti_3C_2 heterostructures facilitates efficient ion diffusion, leading to improved electrochemical performance.

The manufacturing process utilizes electrostatic attraction-driven molecular self-assembly. In this process, a mixture of exfoliated Ti_3C_2 and PDDA-BP with a specified feeding ratio is combined, sonicated, stirred, filtered, and freeze-dried to produce well-defined and stable heterostructures.

Yin et al. in 2019 conducted a study (Zhao et al. 2019), where they produced a film composed of PDDA-BP/ex-Ti_3C_2, laying the foundation for the development of heterostructures based on PDDA-BP/Ti_3C_2. These heterostructures were categorized into three types: PDDA-BP/Ti_3C_{2-1}, PDDA-BP/Ti_3C_{2-2}, and PDDA-BP/Ti_3C_{2-3}.

The differentiation among these configurations was attained through a precise control of the suspension of exfoliated Ti_3C_2 and PDDA-BP, maintaining feeding ratios of 1.2:1, 1.7:1, and 2.2:1, respectively. In contrast, hybrid BP-Ti_3C_2 structures were formed by directly blending suspensions of ex-Ti_3C_2 and ex-BP in comparable mass ratios for assessment. The creation of PDDA-BP/Ti_3C_2 heterostructures followed a flocculation-based method, bringing together exfoliated Ti_3C_2 (ex-Ti_3C_2) layers and PDDA-modified exfoliated black phosphorus (PDDA-BP) layers with distinct positions of charged distribution (Figure 6.14a). This strategy ensured the formation of well-organized heterostructures with controlled layer arrangements and surface properties (Zhao et al. 2019).

A monodispersed suspension of nanolayers undergoes a transformation, as depicted in Figure 6.14b (inset), resulting in a layered product that can be readily converted into a flexible film through a vacuum filtration process, as shown in Figure 6.14c (inset).

A monodisperse suspension of nanolayers is transformed, as shown in Figure 6.14c (inset), resulting in a layered product that is easily converted into a flexible film through a vacuum filtration process, as shown in Figure 6.14a (inset).

The cross-sectional SEM micrograph in Figure 6.14b shows a layered structure. This indicates a methodical process of assembling the combined nanolayers, consisting of PDDA-BP and exfoliated Ti_3C_2, in a layer-by-layer (LbL) manner. In the TEM image (Figure 6.14c), the exceedingly thin nanolayers of PDDA-BP are evenly dispersed and adorned with delicate wrinkles. They are juxtaposed with the distinctly crystalline ex-Ti_3C_2 nanosheets. The electron diffraction (ED) configuration (Figure 6.14d) reveals distinct in-plane diffraction patterns and discrete sharp deflection positions corresponding to the ex-Ti_3C_2 and PDDA-BP nanolayers. The atomic force microscopy (AFM) image (Figure 6.14e, f) presents a representative illustration of the assembly phase involving both types of thin-layered nanosheets (Sun et al. 2015; Meng et al. 2018).

Further confirmation of the lamellar structure inherent in the heterostructures is provided by the X-ray diffraction (XRD) spectrum shown in Figure 6.14g. The peak positions observed at 4.0°, 7.9°, 20.2°, 24.3°, and 28.1° correspond to the (001), (002), (005), (006), and (007) planes of the Ti_3C_2 layered geometry, respectively. Additional sharp peaks at 16.7°, 26.4°, 34.1°, 34.9°, and 52.1° correspond to the (020), (021), (040), (111), and (060) diffraction peaks of BP, respectively. This molecular heterostructure, with a well-organized design, ensures a close interaction between Ti_3C_2 and BP, resulting in enhanced performance of both components (Li et al. 2017).

6.2.1.7 MXene and Layered Structure of Transition Metal Dichalcogenide

6.2.1.7.1 MXene with MOS_2

The search for new and sustainable energy sources and their storage systems has acknowledged considerable attention due to the global increasing energy demand

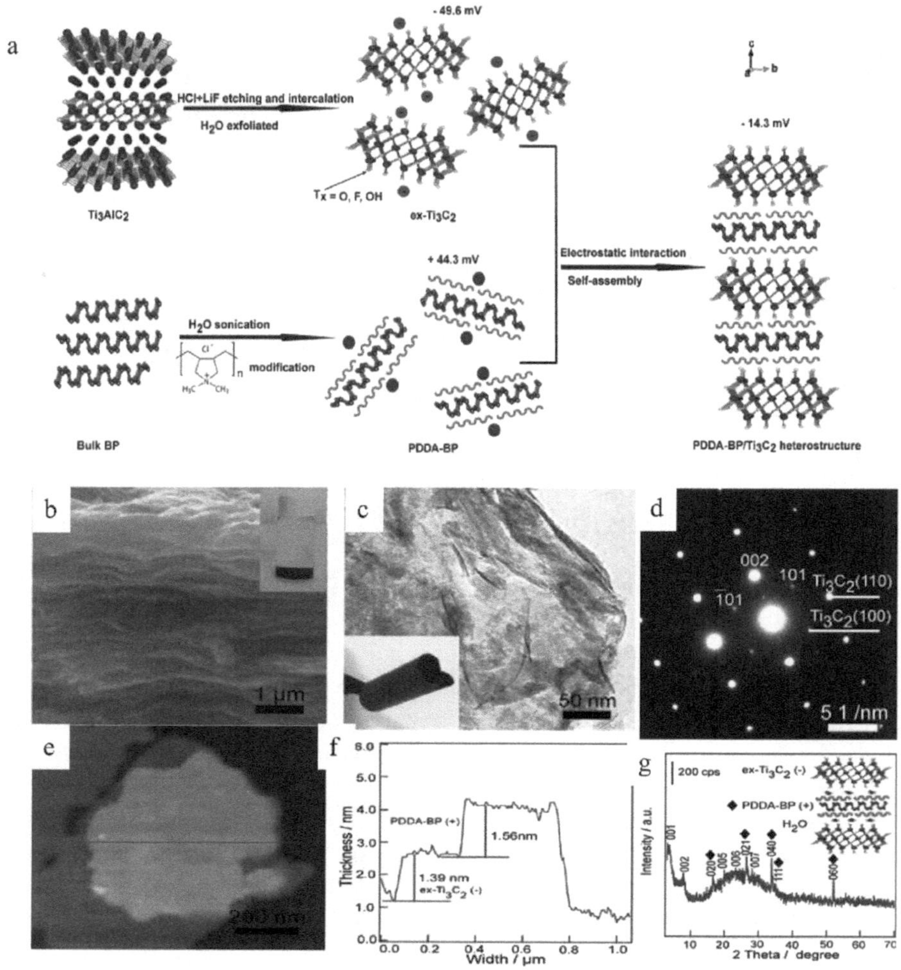

FIGURE 6.14 (a) The schematic diagram outlines the fabrication process used to create heterostructures comprising PDDA-BP/Ti_3C_2 layers. It is significant to note that the presence of water content, ensuring clarity, is not depicted in this illustration. Characterization is a crucial phase in constructing PDDA-BP/Ti_3C_2 layer heterostructures. (b) A SEM cross-sectional image provides a view of the PDDA-BP/Ti_3C_2 film's structure, with the flocculation of PDDA-BP/Ti_3C_2 shown in the inset. (c) A TEM photograph highlights the freestanding and flexible nature of the PDDA-BP/Ti_3C_2 film, with a visual representation in the inset. (d) The electron diffraction pattern of the PDDA-BP/Ti_3C_2 flocculation illustrates its specific arrangement at the microscopic level. (e–f) AFM images and width profiles offer a closer examination of the heterostructures composed of single-layer PDDA-BP/Ti_3C_2 nanolayers. (g) The XRD pattern of the PDDA-BP/Ti_3C_2 material provides insights into its crystalline structure and composition.

Source: Permission awarded by 2019 Elsevier Ltd (Zhao et al. 2019).

and environmental concerns (Wang, Zhang, and Zhang 2012; Zhang and Zhao 2009; Han et al. 2018).

2D-layered transition metal dichalcogenides have recently come to the forefront as potential electrocatalysts for the hydrogen evolution process (HER) because of their distinct structural and physical features (Wang et al. 2015; Yang and Shin 2014).

A prospective electrocatalyst for the hydrogen evolution reaction (HER) is molybdenum disulfide (MoS_2). This material has undergone thorough investigation due to its advantageous Gibbs free energy for hydrogen adsorption and its robust capabilities in accepting electrons (Hinnemann et al. 2005; Consider Liu et al. 2017). Nevertheless, MoS_2's charge transport and HER activity are hindered by its poor intrinsic conductivity.

To address this limitation, researchers have explored coupling MoS_2 with other electrically conductive nanomaterials (Yan et al. 2014; Qian et al. 2018). For instance, Qiu's group (Wu et al. 2017) developed hierarchical nanohybrid structures of $MoS_2/Ti_3C_2@C$, which exhibited enhanced HER performance. Nevertheless, the instability of Ti_3C_2 MXenes in the network required protection through carbon nanoplating.

To tackle these challenges, scientists have devised an in-situ synthesis method to create vertically aligned 2D MoS_2 nanolayers on the planar surface of Ti_3C_2 nanolayers, resulting in continuous 2D-2D graded heterostructures. These MoS2/Ti_3C_2 heterostructures leverage their layered geometry and strong internal bonding forces to significantly improve HER performance. The synthesis of these heterostructures involved a hydrothermal method, where specific amounts of reactants were mixed, heated, and precipitated. Among the prepared samples, MoS_2/Ti_3C_2–5% demonstrated the best electrocatalytic activity. Figure 6.15a schematically depicts the synthesis process of MoS2/Ti_3C_2 heterostructures, and SEM images provide insights into their morphology and size.

Figure 6.15b provides a comprehensive SEM image of untreated Ti_3C_2, illustrating the formation of closely packed nanolayers resulting from the HF-based etching process, effectively eliminating aluminum (Al) from the Ti_3AlC_2 layers. The resulting Ti_3C_2 displays a multilayered, accordion-like structure with a flat surface and noticeable interlayer spacing, resembling exfoliated lamellar graphite (Figure 6.15c).

Figure 6.15d shows an SEM image of the MoS_2/Ti_3C_2 heterostructures showing the previously uniform Ti_3C_2 surface changing to a rougher, patchy texture. A closer look reveals that the MoS_2 nanolayers are well protected and closely integrated with the Ti_3C_2 nanolayers (Figure 6.15e). This close 2D to 2D heterostructural configuration of MoS_2 and Ti_3C_2 facilitates direct charge transfer between MoS_2 and Ti_3C_2, leading to enhanced electrocatalytic efficiency.

The distribution of elements within the MoS_2/Ti_3C_2 heterostructure was investigated by element mapping. The elements Ti, Mo, C, S, and O show distinct segregation and even dispersion throughout the heterostructure as shown in Figure 6.15f. The inherent oxidation process of MoS_2 and the presence of surface-bound functional groups containing oxygen on the Ti_3C_2 surface explain the presence of oxygen (O) (Huang et al. 2019).

MoS_2/Ti_3C_2 heterostructures were fabricated using a simple hydrothermal technique. This methodology has enabled the direct cultivation of vertically oriented

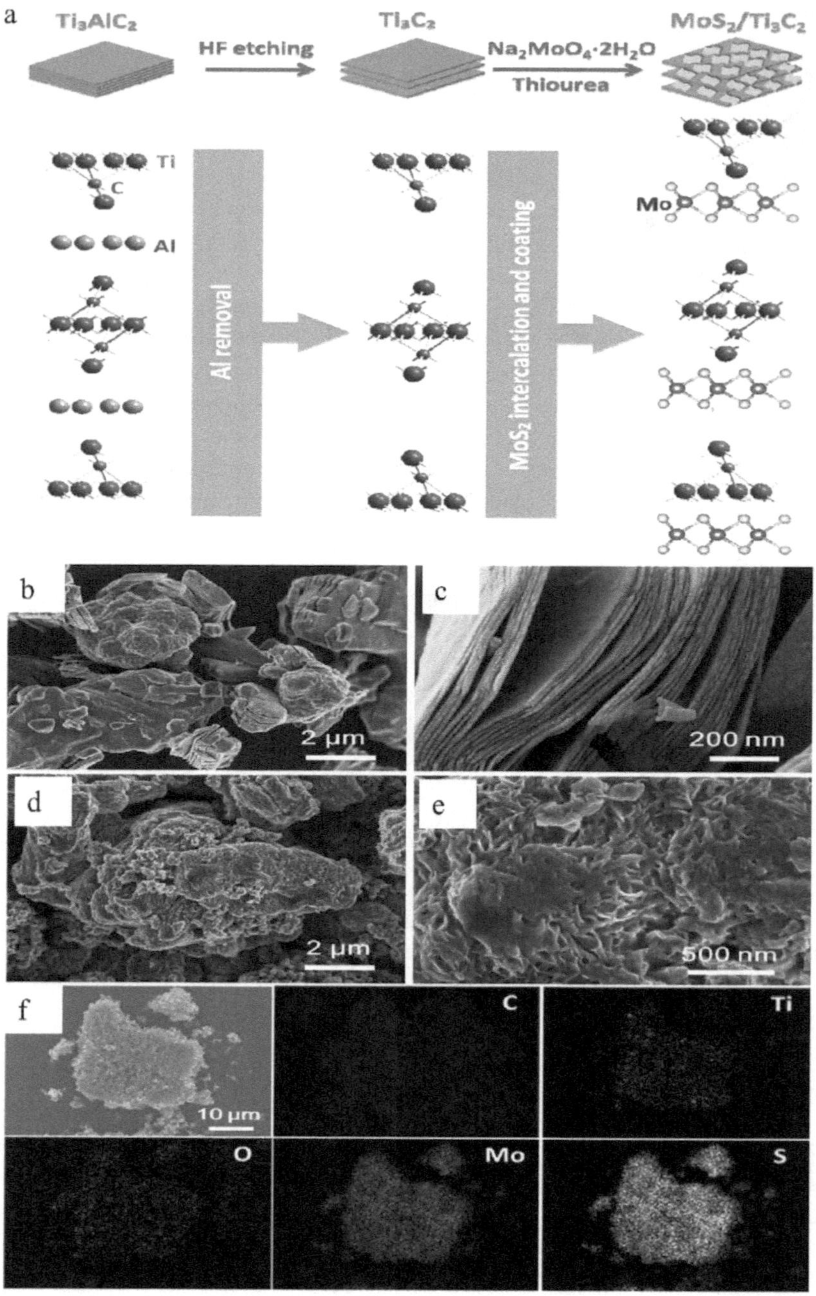

FIGURE 6.15 (a) A detailed schematic representation of the sequential synthesis process for MoS_2/Ti_3C_2 heterostructures. (b, c) SEM images reveal the morphology of the Ti_3C_2 nanosheets. (d, e) The MoS_2/Ti_3C_2 heterostructures. (f) Elemental mapping images, offering insights into the element distribution within the MoS_2/Ti_3C_2 heterostructures.

Source: Permission awarded by Elsevier Ltd. 2018 (Huang et al. 2019).

MoS_2 nanolayers on the planar surface of Ti_3C_2 nanolayers. These discoveries highlight the auspicious potential of MoS_2/Ti_3C_2 heterostructures as electrode materials in flexible and highly efficient metal-ion batteries.

6.2.1.8 Heterostructures of MXene and WS_2

The integration of MXene and transition metal dichalcogenides (TMDs) in heterostructures has garnered significant interest owing to their modifiable geometry and band structures, rendering them important for upcoming electronic devices (Li et al. 2015).

This particular arrangement serves as the electrode material, significantly increasing the capacity of the SIB (Ma et al. 2014; Wu et al. 2017). In another study, Xu et al. (2016) used Ti2C(OH)xFy precursors as an innovative electrode material coupled with WSe_2, resulting in a marked decrease in the Schottky barrier height between WSe_2 and the electrode.

However, $WS_2/Ti_3C_2T_x$ heterostructure fabrication is still in its infancy, and scientists are actively exploring the fundamental material properties of this newly developed configuration. Figure 6.16a shows a simple representation of the procedure for straight deposition and lamination of WS_2 on Si/SiO_2 substrates (300 nm), while Figure 6.16b shows an optical microscope image of a truncated triangular WS_2 nanoflake deposited autonomously without the presence of $Ti_3C_2T_x$.

The pale violet color signifies a nanoflake volume that is equal to 1 liter, which has been confirmed by the presence of a single excitonic photoluminescence (PL) intensity band, which will be further discussed in following sections.

The development of WS_2 flakes on the $Ti_3C_2T_x$ MXene surface is notably influenced by both the deposition pressure and the sample-base spacing (D). Panels c and d of Figure 6.16 show the variations in the surface morphology of the deposited WS_2 flakes with $Ti_3C_2T_x$ MXene at pressures of 200 Torr and 250 Torr, respectively.

In the diminished pressure conditions of approximately 200 Torr, the nanoflakes' growth rate around the MXene substrate is comparatively moderate. Conversely, under heightened pressures of 250 Torr (illustrated in Figure 6.16d) and 350 Torr (depicted in Figure 6.16e–h), the emergence of more substantial nanoflakes, distinguished by a few nanolayers, becomes increasingly evident. WS_2 nanoflake growth behavior is also significantly affected by sample spacing (D). Images of nanoflakes with D values of 9 cm, 9.5 cm, 10 cm, and 10.5 cm, respectively, are shown in Figure 6.16e–h.

The growth of WS_2 nanoflakes occurs rapidly when the sample-base distance (D) is shorter (9 cm) (Figure 6.16e). Conversely, for larger D values in a range from 10.5 to 11 cm, it is possible to achieve WS_2 nanoflakes of larger dimensions, with a volume measure of 1 L and a structure consisting of a few layers, as shown in Figure 6.16g and h. The growth behavior of WS_2 nanoflakes on the top surface of the MXene was also investigated. Three distinctive SEM images are shown in Figure 6.16i–k.

The formation of WS_2 flakes originating from a planar region of $Ti_3C_2T_x$ nanoflakes dispersed over the substrate is observed in Figure 6.16i. Over time, these WS_2 flakes uniformly envelop the top surface. This observation strongly indicates unhindered nucleation and growth of WS_2 flakes on the upper surface of the $Ti_3C_2T_x$ MXene flake.

Subsequently, in Figure 6.16j, the WS_2 material assumes a morphology characterized by terrace-like boundaries, prominently formed on the surface of the $Ti_3C_2T_x$

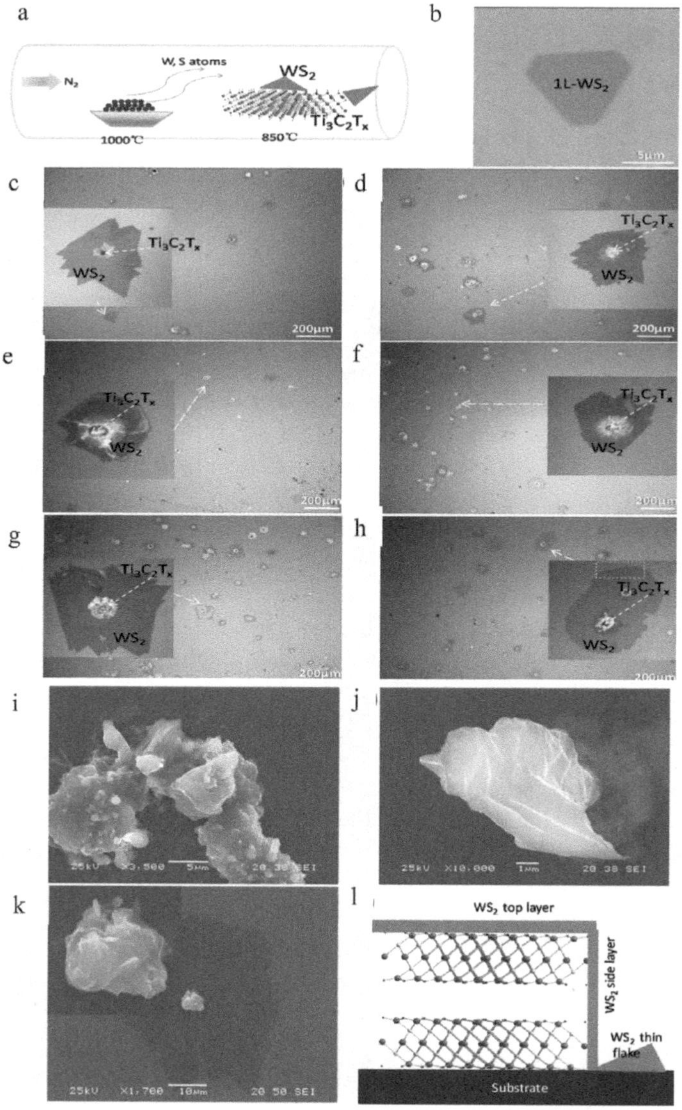

FIGURE 6.16 (a) Schematic representation of the vapor transport process for the formation of WS_2 flakes on the $Ti_3C_2T_x$ MXene surface. (b) A single WS_2 nanoflake without $Ti_3C_2T_x$ MXene is captured in an optical microscope photograph. Optical images show WS_2 nanoflakes grown at different pressures: (c) At 200 Torr and (d) at 250 Torr. (e) The source distance is varied at 10 cm, (f) 10.5 cm, (g) 11 cm, and (h) 11.5 cm. These images provide insight into the optical properties of WS_2 growth on different substrates. Magnified views of the visible flakes are shown in the supplementary images. (i–k) Type engineering of WS_2 nanoflakes grown on a variety of $Ti_3C_2T_x$ flakes. (l) A schematic outlines a potential fabrication approach for WS_2 on $Ti_3C_2T_x$ nanoflakes. A magnified SEM image shows WS_2 flakes on $Ti_3C_2T_x$. This is indicated by the diminutive inset in (k). It is noteworthy that the nanoflakes shown in (i–k) were selected from samples that underwent lamination at a pressure of 350 Torr.

Source: Permission awarded by Su et al. (2018).

MXene flake. Significantly, both the upper and neighboring facets of these terrace-like edges are embellished with slender layers of WS_2. Next, in Figure 6.16j, the WS2 material exhibits a morphology with terrace-like boundaries, which are prominently formed on the surface of the Ti_3C_2Tx MXene flake. It is noteworthy that both the upper and neighboring facets of these terrace-like edges are embellished with slender layers of WS_2. The absence of $Ti_3C_2T_x$ boundaries in Figure 6.16i and j signifies that the adjacent sides are entirely enveloped by WS_2. Figure 6.16k presents a contrasting feature compared to Figure 6.16j. In addition to the WS_2 layered flakes on the lateral side of the $Ti_3C_2T_x$ nanoflake, numerous smaller petals also grow on the top and adjacent sides.

A conceivable growth arrangement shown in Figure 6.16i–k is envisioned and schematically outlined in Figure 6.16l. In this suggested structure, there is a bilayer configuration of $Ti_3C_2T_x$ nanoflakes on the top surface of the target substrate. The $Ti_3C_2T_x$ nanoflakes function as nonuniform nucleation mediators, akin to other precursors like BN and graphene, which have demonstrated efficacy in generating nanoflakes of WS_2 and MoS_2 with a volume equivalent to 1 liter (Okada et al. 2014; Ling et al. 2014; Shi et al. 2012).

In conclusion, these intricately engineered 2D transition metal dichalcogenides (TMDs), specifically $Ti_3C_2T_x$ (MXenes), and their amalgamations demonstrate significant promise for application in solar energy conversion and energy storage devices.

6.2.1.9 MXene with WSe_2

The functionality of field-effect transistors (FETs) in transition metal dichalcogenide (TMD) networks depends heavily on the formation of Schottky junctions between channel and source/drain electrodes. The barrier heights at these junctions play a crucial role in influencing the presentation and the polarity of these devices, thus emphasizing the need for appropriate electrode networks in nanodevices based on TMDs (Chuang et al. 2014; Liu et al. 2013; Das et al. 2013).

In the field of 2D electronic devices, graphene, a highly valued and lightweight semimetallic material, plays a critical role (Eda et al. 2009; Lu et al. 2013).

A vertically stacked field-effect transistor (FET) integrating both graphene and molybdenum disulfide (MoS_2) was presented in a landmark study by Duan et al. (Yu et al. 2013). This FET, especially when graphene was used as a modifiable base electrode, exhibited exceptional properties, in particular a significant on-off ratio and current density.

The electrical performance of graphene-based electrode networks over other metal electrodes has been consistently confirmed in subsequent research.

In addition, elastic and transparent field-effect transistors (FETs) using molybdenum disulfide (MoS_2) have been developed by Mr. Lee and his research team (Yoon et al. 2013). In this design, a network of boron nitride (BN) acts as a thin dielectric layer, while graphene serves as the gate electrode. This innovative configuration allows FETs to integrate seamlessly with other two-dimensional (2D) electronic devices. It is worth noting that the work function of graphene is amenable to modulation by an electric field, thereby allowing for the fine tuning of the Schottky barrier heights (Yu et al. 2009).

The pronounced conical geometry that is inherent in the band structure of graphene guarantees a comparatively constant elevation of the existing barrier, even when it is subjected to high gate voltages. This phenomenon arises because graphene's Fermi level is located at a considerable distance from the Dirac point, in a region characterized by high-density states (Yoon et al. 2013). Consequently, graphene is primarily suited for use in n-channel network-based field-effect transistor (FET) electrode systems due to its relatively modest work function of about 4.6 eV, coupled with its limited tunability range (Yoon et al. 2013; Yu et al. 2009).

Shifting the emphasis, the introduction of MXene nanomaterials, a contemporary inclusion in the domain of two-dimensional (2D) materials, has attracted noteworthy considerations. MXenes are becoming increasingly acknowledged for their exceptional conductivity and remarkable performance in lithium-ion (Li-ion) battery applications (Naguib et al. 2014). At the same time, investigations are delving into their potential applicability in the field of two-dimensional field-effect transistor (2D FET) systems. Nevertheless, it is imperative to emphasize that the incorporation of MXenes in the field of 2D electrode networks is still an area that has yet to be subjected to a comprehensive investigation.

Halim et al. (2014) have highlighted the advantageous property of high light transmission, up to 90%, in thin films of Ti_3C_2. These films are produced by deposition and subsequent etching processes. This makes them suitable for use as translucent electrodes spanning the visible to infrared spectrum. However, there has been limited research on MXenes in the context of 2D electrode networks.

Figure 6.17a illustrates the comprehensive synthesis methodology for $Ti_2C(OH)$ xFy flakes derived from the precursor Ti_2AlC.

The process involves the treatment of the precursor of the MAX phase, Ti_2AlC, with HF, which leads to the effective removal of aluminum species from the Ti2C nanolayers by etching. This process introduces surface functionalities. As a result, mechanical exfoliation effectively produces Ti2C(OH)xFy nanoflakes. Afterwards, the particles are deposited onto SiO_2/Si substrates that have been coated with polyvinyl alcohol (PVA)/PMMA, while being seen using an optical microscope.

In this method of exfoliation, MoS_2 and WSe_2 nanoflowers were intricately handled on a 285 nm SiO_2/p+ doped substrate. To scrutinize the atomic structure and internal morphology of the synthesized Ti2C(OH)xFy nanoflakes, a Transmission Electron Microscopy (TEM) technique was utilized, visually depicted in Figure 6.17b. In this figure, a high-resolution TEM image is shown alongside an adjacent SAED image. Both images confirm the preserved hexagonal symmetry and the stable crystal structure. Throughout the synthesis process, the α lattice parameter is approximately 0.31 nm.

When energy-dispersive X-ray (EDX) spectroscopy is examined during the formation of Ti2C(OH)xFy nanoflakes, it is evident that there is no detectable Al signal. This observation is a clear indicator of the effective removal of aluminum atoms after RF treatment (Xu et al. 2016). However, identifying signals attributed to oxygen (O) and fluorine (F) implies the attendance of surface functional groups, particularly −F and −OH. These functional groups are generated as a result of the proficient etching process that is applied to aluminum (Naguib et al. 2011a).

In Figure 6.17c, an AFM image shows the synthesized Ti2C(OH)xFy nanoflake, accompanied by an optical microscopy photo inset showing the surface morphology of these flakes. The exfoliation process successfully yields thicker Ti2C(OH)xFy nanoflakes with sizes around 20 to 30 nm, making them promising for device fabrication.

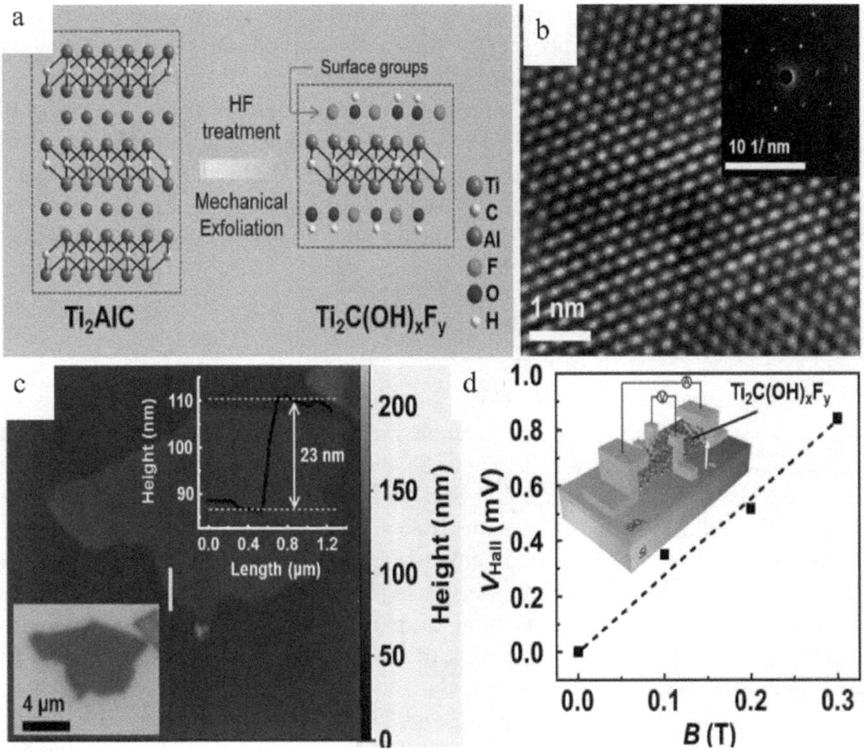

FIGURE 6.17 (a) The schematic depiction delineates the process of grounding Ti2C(OH) xFy nanoflakes. (b) The TEM image displays a Ti2C(OH)xFy nanoflake, accompanied by the corresponding Selected Area Electron Diffraction (SAED) pattern illustrated in the inset. (c) AFM image demonstrates the surface morphology of an exfoliated Ti2C(OH)xFy nanoflake on a SiO_2 surface, complemented by optical photographs in the inset. (d) The Ti2C(OH)xFy device is used for Hall voltage measurement, serving as a built-in component in the magnetic field source. A schematic diagram of the Ti2C(OH)xFy device is shown in the inset. Heterostructures that include semiconducting and insulating materials, such as MXene-HBN, have been documented in previous research.

Source: Permission awarded by 2016 Wiley-VCH (Xu et al. 2016).

As shown in Figure 6.17d, the electrical properties of Ti2C(OH)xFy nanoflakes are investigated for electrode-based applications by measuring the Hall voltage under different magnetic fields (B). These nanoflakes exhibit an impressive conductivity value of 0.026 S.

In addition, using a quadruple sampling method, both the Ti2C(OH)xFy nanoflakes and the chromium (Cr) entities exhibit a calculated dimension of approximately 0.243 μm for each species. This measurement is significantly reduced compared to the reported dimensions of graphene-chromium (graphene-Cr) bonds (Song and Cho 2013). This discrepancy underscores Ti2C(OH)xFy's significant potential as electrode material within transition metal dichalcogenide (TMD)-based interfacial networks. This potential is reinforced by its commendable power conductivity, its reduced metal contact resistance, and its reduced Schottky barrier values, which are evident in both molybdenum disulfide (MoS_2) and tungsten diselenide (WSe_2).

In contrast to graphene, which has a work function value of 4.6 eV, 2D electronic devices that incorporate Ti2C(OH)xFy electrodes in WSe_2 and MoS_2 FETs exhibit an elevated work function value of about 4.98 eV. In addition, Ti2C(OH)xFy is a compelling choice for 2D electrodes in MOS inverters because of its low Schottky barrier for both MoS_2 and WSe_2. This allows seamless integration of metal oxide semiconductors.

6.3 SUMMARY

In conclusion, the exploration of MXenes, particularly $Ti_3C_2T_x$ and Ti_2CT_x, opens a world of possibilities due to their remarkable metallic conductivities and hydrophilic surfaces. After years of study, $Ti_3C_2T_x$ and Ti_2CT_x MXene compounds have matured. The integration of $Ti_3C_2T_x$ and Ti_2CT_x MXene derivatives with diverse nanomaterials (0D, 1D, and 2D) has been proficiently achieved employing various deposition methodologies. This strategic amalgamation aims to augment prolonged stability and facilitate diverse functional applications. The resultant mixed-dimensional heterostructures proffer an alternative material paradigm for the semiconductor industry, adeptly tackling challenges inherent in the development of expansive ultrathin membranes for channels.

The distinctive attributes of $Ti_3C_2T_x$ and Ti_2CT_x, stemming from their surface functional groups, encompass a spectrum of desirable qualities. These encompass the facilitation of facile large-scale synthesis in aqueous environments, augmented electrical conductivity, manifestation process of both metallic and semiconducting characteristics, hydrophilic behavior, chemical resilience, environmental compatibility, and the propensity to engage in the formation of van der Waals heterostructures. Collectively, these attributes position $Ti_3C_2T_x$ and Ti_2CT_x MXene derivatives as promising materials with multifaceted applications and substantial potential within the semiconductor realm.

Graphene, carbon nanotubes, and black phosphorous are examples of additional dimensional semiconductors that may be combined with MXene derivatives to create novel materials that can take use of MXene's advantages as well as those of other substances. There is a lot of room for advancements in physics, nanotechnology, materials science, and commercialization in this subject. Studying the uses of MXene derivatives outside the $Ti_3C_2T_x/Ti_2CT_x$ framework may be aided by research on mixed-dimensional heterostructures based on $Ti_3C_2T_x/Ti_2CT_x$.

Within these heterostructures, the minimal interaction observed between MXene and companion materials, exemplified by diverse 2D sheets, nanoparticles, and nanowires, affords the amalgamation of respective advantages while concurrently mitigating inherent drawbacks. This synergistic amalgamation yields heightened mobility, modifiable narrow band gaps, augmented mechanical resilience, and the instigation of novel functionalities. The integration of 2D MXene nanosheets within these heterostructures establishes efficacious conduits for the transport of both electrons and ions, a pivotal facet for materials deployed in the areas of energy conversion and storage.

Moreover, the intrinsic attributes of MXene persist unfalteringly within these heterostructures, thereby furnishing robust safeguarding against environmental oxidation. The subsequent sections dedicated to applications meticulously scrutinize the utilitarian aspects of mixed-dimensional heterostructures, specifically rooted in $Ti_3C_2T_x$ and Ti_2CT_x MXene, within the realms of catalysis photovoltaics, catalysis, and energy storage devices. This comprehensive exploration serves to underscore the inherent significance of these heterostructures in the aforementioned domains.

REFERENCES

Acerce, Muharrem, Damien Voiry, and Manish Chhowalla. 2015. "Metallic 1T phase MoS2 nanosheets as supercapacitor electrode materials." *Nature Nanotechnology* 10 (4):313.

Aïssa, B, A Ali, KA Mahmoud, T Haddad, and M Nedil. 2016. "Transport properties of a highly conductive 2D Ti_3C_2Tx MXene/graphene composite." *Applied Physics Letters* 109 (4):043109.

Akinwande, Deji, Nicholas Petrone, and James Hone. 2014. "Two-dimensional flexible nanoelectronics." *Nature Communications* 5:5678.

Alhabeb, Mohamed, Kathleen Maleski, Babak Anasori, Pavel Lelyukh, Leah Clark, Saleesha Sin, and Yury Gogotsi. 2017a. "Guidelines for synthesis and processing of two-dimensional titanium carbide (Ti_3C_2Tx MXene)." *Chemistry of Materials* 29 (18):7633–7644.

Alhabeb, Mohamed, Kathleen Maleski, Babak Anasori, Pavel Lelyukh, Leah Clark, Saleesha Sin, and Yury Gogotsi. 2017b. "Guidelines for synthesis and processing of two-dimensional titanium carbide (Ti_3C_2Tx MXene)." *Chemistry of Materials* 29 (18):7633–7644.

Ali, Adnan, Abdelhak Belaidi, Shawkat Ali, Mohamed I Helal, and Khaled A Mahmoud. 2016. "Transparent and conductive Ti_3C_2Tx (MXene) thin film fabrication by electrohydrodynamic atomization technique." *Journal of Materials Science: Materials in Electronics* 27 (5):5440–5445.

Allan, Phoebe K, John M Griffin, Ali Darwiche, Olaf J Borkiewicz, Kamila M Wiaderek, Karena W Chapman, Andrew J Morris, Peter J Chupas, Laure Monconduit, and Clare P Grey. 2016. "Tracking sodium-antimonide phase transformations in sodium-ion anodes: Insights from operando pair distribution function analysis and solid-state NMR spectroscopy." *Journal of the American Chemical Society* 138 (7):2352–2365.

Anasori, Babak, Maria R Lukatskaya, and Yury Gogotsi. 2017. "2D metal carbides and nitrides (MXenes) for energy storage." *Nature Reviews Materials* 2:16098.

Aravinda, LS, KK Nagaraja, HS Nagaraja, K Udaya Bhat, and Badekai Ramachandra Bhat. 2013. "ZnO/carbon nanotube nanocomposite for high energy density supercapacitors." *Electrochimica Acta* 95:119–124.

Bie, Ya-Qing, Yang-Bo Zhou, Zhi-Min Liao, Kai Yan, Song Liu, Qing Zhao, Shishir Kumar, Han-Chun Wu, Georg S Duesberg, and Graham LW Cross. 2011. "Site-specific transfer-printing of individual graphene microscale patterns to arbitrary surfaces." *Advanced Materials* 23 (34):3938–3943.

Boettcher, Shannon W, Joshua M Spurgeon, Morgan C Putnam, Emily L Warren, Daniel B Turner-Evans, Michael D Kelzenberg, James R Maiolo, Harry A Atwater, and Nathan S Lewis. 2010. "Energy-conversion properties of vapor-liquid-solid–grown silicon wire-array photocathodes." *Science* 327 (5962):185–187.

Borysiuk, Vadym N, Vadym N Mochalin, and Yury Gogotsi. 2015. "Molecular dynamic study of the mechanical properties of two-dimensional titanium carbides Tin+ 1Cn (MXenes)." *Nanotechnology* 26 (26):265705.

Borysiuk, Vadym N, Vadym N Mochalin, and Yury Gogotsi. 2018. "Bending rigidity of two-dimensional titanium carbide (MXene) nanoribbons: A molecular dynamics study." *Computational Materials Science* 143:418–424.

Cai, Xingke, Yuting Luo, Bilu Liu, and Hui-Ming Cheng. 2018. "Preparation of 2D material dispersions and their applications." *Chemical Society Reviews* 47 (16):6224–6266. https://doi.org/10.1039/C8CS00254A.

Chen, Cheng-Meng, Jia-Qi Huang, Qiang Zhang, Wen-Zhao Gong, Quan-Hong Yang, Mao-Zhang Wang, and Yong-Gang Yang. 2012. "Annealing a graphene oxide film to produce a free standing high conductive graphene film." *Carbon* 50 (2):659–667. https://doi.org/10.1016/j.carbon.2011.09.022.

Chhowalla, Manish, Hyeon Suk Shin, Goki Eda, Lain-Jong Li, Kian Ping Loh, and Hua Zhang. 2013. "The chemistry of two-dimensional layered transition metal dichalcogenide nanosheets." *Nature Chemistry* 5 (4):263.

Choi, Kyung Hyun, Adnan Ali, and Jeongdai Jo. 2013. "Randomly oriented graphene flakes film fabrication from graphite dispersed in N-methyl-pyrrolidone by using electrohydrodynamic atomization technique." *Journal of Materials Science: Materials in Electronics* 24 (12):4893–4900.

Choi, Wonbong, Indranil Lahiri, Raghunandan Seelaboyina, and Yong Soo Kang. 2010. "Synthesis of graphene and its applications: A review." *Critical Reviews in Solid State and Materials Sciences* 35 (1):52–71.

Chuang, Steven, Corsin Battaglia, Angelica Azcatl, Stephen McDonnell, Jeong Seuk Kang, Xingtian Yin, Mahmut Tosun, Rehan Kapadia, Hui Fang, and Robert M Wallace. 2014. "MoS2 p-type transistors and diodes enabled by high work function MoO × contacts." *Nano Letters* 14 (3):1337–1342.

Dall'Agnese, Yohan, Patrick Rozier, Pierre-Louis Taberna, Yury Gogotsi, and Patrice Simon. 2016. "Capacitance of two-dimensional titanium carbide (MXene) and MXene/carbon nanotube composites in organic electrolytes." *Journal of Power Sources* 306:510–515. https://doi.org/10.1016/j.jpowsour.2015.12.036.

Das, Saptarshi, Hong-Yan Chen, Ashish Verma Penumatcha, and Joerg Appenzeller. 2013. "High performance multilayer MoS2 transistors with scandium contacts." *Nano Letters* 13 (1):100–105.

Das, Saptarshi, Wei Zhang, Marcel Demarteau, Axel Hoffmann, Madan Dubey, and Andreas Roelofs. 2014. "Tunable transport gap in phosphorene." *Nano Letters* 14 (10):5733–5739. https://doi.org/10.1021/nl5025535.

Deng, Bing, Po-Chun Hsu, Guanchu Chen, BN Chandrashekar, Lei Liao, Zhawulie Ayitimuda, Jinxiong Wu, Yunfan Guo, Li Lin, and Yu Zhou. 2015. "Roll-to-roll encapsulation of metal nanowires between graphene and plastic substrate for high-performance flexible transparent electrodes." *Nano Letters* 15 (6):4206–4213.

Dillon, Andrew D, Michael J Ghidiu, Alex L Krick, Justin Griggs, Steven J May, Yury Gogotsi, Michel W Barsoum, and Aaron T Fafarman. 2016. "Highly conductive optical quality solution-processed films of 2D titanium carbide." *Advanced Functional Materials* 26 (23):4162–4168. https://doi.org/10.1002/adfm.201600357.

Eames, Christopher, and M Saiful Islam. 2014. "Ion intercalation into two-dimensional transition-metal carbides: Global screening for new high-capacity battery materials." *Journal of the American Chemical Society* 136 (46):16270–16276.

Eda, Goki, Cecilia Mattevi, Hisato Yamaguchi, HoKwon Kim, and Manish Chhowalla. 2009. "Insulator to semimetal transition in graphene oxide." *The Journal of Physical Chemistry C* 113 (35):15768–15771.

El-Kady, Maher F, Veronica Strong, Sergey Dubin, and Richard B Kaner. 2012. "Laser scribing of high-performance and flexible graphene-based electrochemical capacitors." *Science* 335 (6074):1326–1330.

Ellis, Leah D, Timothy D Hatchard, and Mark N Obrovac. 2012. "Reversible insertion of sodium in tin." *Journal of the Electrochemical Society* 159 (11):A1801–A1805.

Evers, Scott, and Linda F Nazar. 2013. "New approaches for high energy density lithium–sulfur battery cathodes." *Accounts of Chemical Research* 46 (5):1135–1143.

Fang, Linxia, Baoliang Zhang, Wei Li, Jizhong Zhang, Kejing Huang, and Qiuyu Zhang. 2014. "Fabrication of highly dispersed ZnO nanoparticles embedded in graphene nanosheets for high performance supercapacitors." *Electrochimica Acta* 148:164–169.

Feng, Qingliang, Hongyan Liu, Meijie Zhu, Jing Shang, Dan Liu, Xiaoqi Cui, Diqin Shen, Liangzhi Kou, Dong Mao, and Jianbang Zheng. 2018. "Electrostatic functionalization and passivation of water-exfoliated few-layer black phosphorus by poly dimethyldiallyl ammonium chloride and its ultrafast laser application." *ACS Applied Materials & Interfaces* 10 (11):9679–9687.

Feng, Wei, Zehua Jin, Jiangtan Yuan, Jing Zhang, Shuai Jia, Liangliang Dong, Jongwon Yoon, Linan Zhou, Robert Vajtai, and James M Tour. 2018. "A fast and zero-biased photodetector based on GaTe–InSe vertical 2D p–n heterojunction." *2D Materials* 5 (2):025008.

Ghaffari, Mehdi, Yue Zhou, Haiping Xu, Minren Lin, Tae Young Kim, Rodney S Ruoff, and QM Zhang. 2013. "Retracted: High-volumetric performance aligned nano-porous microwave exfoliated graphite oxide-based electrochemical capacitors." *Advanced Materials* 25 (35):4879–4885.

Ghidiu, Michael, Maria R Lukatskaya, Meng-Qiang Zhao, Yury Gogotsi, and Michel W Barsoum. 2014a. "Conductive two-dimensional titanium carbide 'clay' with high volumetric capacitance." *Nature* 516 (7529):78–81.

Ghidiu, Michael, Maria R Lukatskaya, Meng-Qiang Zhao, Yury Gogotsi, and Michel W Barsoum. 2014b. "Conductive two-dimensional titanium carbide 'clay' with high volumetric capacitance." *Nature* 516:78. https://doi.org/10.1038/nature13970.

Gogotsi, Yury, and Patrice Simon. 2011. "True performance metrics in electrochemical energy storage." *Science* 334 (6058):917–918.

Haditale, M, RS Dariani, and E Ghasemian Lemraski. 2019. "Electrical behavior of graphene under temperature effect and survey of I–T curve." *Journal of Theoretical and Applied Physics* 13 (4):351–356. https://doi.org/10.1007/s40094-019-00349-1.

Halim, Joseph, Maria R Lukatskaya, Kevin M Cook, Jun Lu, Cole R Smith, Lars-Åke Näslund, Steven J May, Lars Hultman, Yury Gogotsi, Per Eklund, and Michel W Barsoum. 2014. "Transparent conductive two-dimensional titanium carbide epitaxial thin films." *Chemistry of Materials* 26 (7):2374–2381. https://doi.org/10.1021/cm500641a.

Han, Ning, Pengyun Liu, Jing Jiang, Lunhong Ai, Zongping Shao, and Shaomin Liu. 2018. "Recent advances in nanostructured metal nitrides for water splitting." *Journal of Materials Chemistry A* 6 (41):19912–19933.

Han, Wei-Qiang. 2009. "Anisotropic hexagonal boron nitride nanomaterials: Synthesis and applications." In *Mixed Metal Nanomaterials*. Wiley, Weinheim.

Hantanasirisakul, Kanit, and Yury Gogotsi. 2018. "Electronic and optical properties of 2D transition metal carbides and nitrides (MXenes)." *Advanced Materials* 30 (52):1804779. https://doi.org/10.1002/adma.201804779.

Hantanasirisakul, Kanit, Meng-Qiang Zhao, Patrick Urbankowski, Joseph Halim, Babak Anasori, Sankalp Kota, Chang E Ren, Michel W Barsoum, and Yury Gogotsi. 2016. "Fabrication of Ti_3C_2Tx MXene transparent thin films with tunable optoelectronic properties." *Advanced Electronic Materials* 2 (6):1600050. https://doi.org/10.1002/aelm.201600050.

Hart, Connor J, Marine Cuisinier, Xiao Liang, Dipan Kundu, Arnd Garsuch, and Linda F Nazar. 2015. "Rational design of sulphur host materials for Li–S batteries: Correlating lithium polysulphide adsorptivity and self-discharge capacity loss." *Chemical Communications* 51 (12):2308–2311.

Hinnemann, Berit, Poul Georg Moses, Jacob Bonde, Kristina P Jørgensen, Jane H Nielsen, Sebastian Horch, Ib Chorkendorff, and Jens K Nørskov. 2005. "Biomimetic hydrogen evolution: MoS2 nanoparticles as catalyst for hydrogen evolution." *Journal of the American Chemical Society* 127 (15):5308–5309.

Ho, Po-Hsun, Yih-Ren Chang, Yu-Cheng Chu, Min-Ken Li, Che-An Tsai, Wei-Hua Wang, Ching-Hwa Ho, Chun-Wei Chen, and Po-Wen Chiu. 2017. "High-mobility InSe transistors: The role of surface oxides." *Acs Nano* 11 (7):7362–7370.

Hu, Chi-Chang, Kuo-Hsin Chang, Ming-Champ Lin, and Yung-Tai Wu. 2006. "Design and tailoring of the nanotubular arrayed." *Nano Letters* 6 (12):2690–2695.

Hu, Jiangtao, Min Ouyang, Peidong Yang, and Charles M Lieber. 1999. "Controlled growth and electrical properties of heterojunctions of carbon nanotubes and silicon nanowires." *Nature* 399 (6731):48–51.

Hu, Ye, Bing Li, Xingxing Jiao, Chaofan Zhang, Xiaohan Dai, and Jiangxuan Song. 2018. "Stable cycling of phosphorus anode for sodium-ion batteries through chemical bonding with sulfurized polyacrylonitrile." *Advanced Functional Materials* 28 (23):1801010.

Huang, Kai, Zhongjun Li, Jing Lin, Gang Han, and Peng Huang. 2018. "Two-dimensional transition metal carbides and nitrides (MXenes) for biomedical applications." *Chemical Society Reviews* 47 (14):5109–5124. https://doi.org/10.1039/C7CS00838D.

Huang, Lan, Lunhong Ai, Mei Wang, Jing Jiang, and Shaobin Wang. 2019. "Hierarchical MoS2 nanosheets integrated Ti_3C_2 MXenes for electrocatalytic hydrogen evolution." *International Journal of Hydrogen Energy* 44 (2):965–976.

Jiang, Qiu, Narendra Kurra, Mohamed Alhabeb, Yury Gogotsi, and Husam N Alshareef. 2018. "All pseudocapacitive MXene-RUO2 asymmetric supercapacitors." *Advanced Energy Materials* 8 (13):1703043.

Jiao, Liying, Li Zhang, Xinran Wang, Georgi Diankov, and Hongjie Dai. 2009. "Narrow graphene nanoribbons from carbon nanotubes." *Nature* 458 (7240):877.

Jiao, Long, Chen Zhang, Chuannan Geng, Shichao Wu, Huan Li, Wei Lv, Ying Tao, Zijin Chen, Guangmin Zhou, Jia Li, Guowei Ling, Ying Wan, and Quan-Hong Yang. 2019. "Capture and catalytic conversion of polysulfides by in situ built TiO2-MXene heterostructures for lithium–sulfur batteries." *Advanced Energy Materials* 9 (19):1900219.

Jung, Naeyoung, Soongeun Kwon, Dongwook Lee, Dong-Myung Yoon, Young Min Park, Anass Benayad, Jae-Young Choi, and Jong Se Park. 2013. "Synthesis of chemically bonded graphene/carbon nanotube composites and their application in large volumetric capacitance supercapacitors." *Advanced Materials* 25 (47):6854–6858.

Kalpana, D, KS Omkumar, S Suresh Kumar, and NG Renganathan. 2006. "A novel high power symmetric ZnO/carbon aerogel composite electrode for electrochemical supercapacitor." *Electrochimica Acta* 52 (3):1309–1315.

Kang, Zhe, Yanan Ma, Xinyu Tan, Miao Zhu, Zhi Zheng, Nishuang Liu, Luying Li, Zhengguang Zou, Xueliang Jiang, Tianyou Zhai, and Yihua Gao. 2017. "MXene–silicon Van Der Waals heterostructures for high-speed self-driven photodetectors." *Advanced Electronic Materials* 3 (9):1700165. https://doi.org/10.1002/aelm.201700165.

Kannan, Karthik, Kishor Kumar Sadasivuni, Aboubakr M Abdullah, and Bijandra Kumar. 2020. "Current trends in MXene-based nanomaterials for energy storage and conversion system: A mini review." *Catalysts* 10 (5):495.

Kasuga, Tomoko, Masayoshi Hiramatsu, Akihiko Hoson, Toru Sekino, and Koichi Niihara. 1998. "Formation of titanium oxide nanotube." *Langmuir* 14 (12):3160–3163.

Khan, Majharul Haque, Zhenguo Huang, Feng Xiao, Gilberto Casillas, Zhixin Chen, Paul J Molino, and Hua Kun Liu. 2015. "Synthesis of large and few atomic layers of hexagonal boron nitride on melted copper." *Scientific Reports* 5:7743.

Khazaei, Mohammad, Ahmad Ranjbar, Masao Arai, Taizo Sasaki, and Seiji Yunoki. 2017. "Electronic properties and applications of MXenes: A theoretical review." *Journal of Materials Chemistry C* 5 (10):2488–2503. https://doi.org/10.1039/C7TC00140A.

Kim, Hyun Ho, Yoonyoung Chung, Eunho Lee, Seong Kyu Lee, and Kilwon Cho. 2014. "Water-free transfer method for CVD-grown graphene and its application to flexible air-stable graphene transistors." *Advanced Materials* 26 (20):3213–3217.

Kou, Liangzhi, Changfeng Chen, and Sean C Smith. 2015. "Phosphorene: Fabrication, properties, and applications." *The Journal of Physical Chemistry Letters* 6 (14):2794–2805. https://doi.org/10.1021/acs.jpclett.5b01094.

Lei, Jincheng, Xu Zhang, and Zhen Zhou. 2015. "Recent advances in MXene: Preparation, properties, and applications." *Frontiers of Physics in China* 10 (3):276–286.

Li, Hongyan, Yang Hou, Faxing Wang, Martin R Lohe, Xiaodong Zhuang, Li Niu, and Xinliang Feng. 2017. "Flexible all-solid-state supercapacitors with high volumetric capacitances boosted by solution processable MXene and electrochemically exfoliated graphene." *Advanced Energy Materials* 7 (4):1601847.

Li, Likai, Yijun Yu, Guo Jun Ye, Qingqin Ge, Xuedong Ou, Hua Wu, Donglai Feng, Xian Hui Chen, and Yuanbo Zhang. 2014. "Black phosphorus field-effect transistors." *Nature Nanotechnology* 9 (5):372–377. https://doi.org/10.1038/nnano.2014.35.

Li, Wei-Jie, Shu-Lei Chou, Jia-Zhao Wang, Hua-Kun Liu, and Shi-Xue Dou. 2013. "Simply mixed commercial red phosphorus and carbon nanotube composite with exceptionally reversible sodium-ion storage." *Nano Letters* 13 (11):5480–5484.

Li, Xiao, and Hongwei Zhu. 2015. "Two-dimensional MoS2: Properties, preparation, and applications." *Journal of Materiomics* 1 (1):33–44.

Li, Xinming, Hongwei Zhu, Kunlin Wang, Anyuan Cao, Jinquan Wei, Chunyan Li, Yi Jia, Zhen Li, Xiao Li, and Dehai Wu. 2010. "Graphene-on-silicon Schottky junction solar cells." *Advanced Materials* 22 (25):2743–2748.

Li, Xinming, Miao Zhu, Mingde Du, Zheng Lv, Li Zhang, Yuanchang Li, Yao Yang, Tingting Yang, Xiao Li, and Kunlin Wang. 2016. "High detectivity graphene-silicon heterojunction photodetector." *Small* 12 (5):595–601.

Li, Xinru, Ying Dai, Yandong Ma, Qunqun Liu, and Baibiao Huang. 2015. "Intriguing electronic properties of two-dimensional MoS2/TM2CO2 (TM=Ti, Zr, or Hf) hetero-bilayers: Type-II semiconductors with tunable band gaps." *Nanotechnology* 26 (13):135703.

Liang, Xiao, Arnd Garsuch, and Linda F Nazar. 2015. "Sulfur cathodes based on conductive MXene nanosheets for high-performance lithium–sulfur batteries." *Angewandte Chemie International Edition* 54 (13):3907–3911. https://doi.org/10.1002/anie.201410174.

Liang, Xiao, Connor Hart, Quan Pang, Arnd Garsuch, Thomas Weiss, and Linda F Nazar. 2015. "A highly efficient polysulfide mediator for lithium–sulfur batteries." *Nature Communications* 6 (1):1–8.

Liang, Xiao, Yverick Rangom, Chun Yuen Kwok, Quan Pang, and Linda F Nazar. 2017. "Interwoven MXene nanosheet/carbon-nanotube composites as Li–S cathode hosts." *Advanced Materials* 29 (3):1603040. https://doi.org/10.1002/adma.201603040.

Ling, Xi, Yi-Hsien Lee, Yuxuan Lin, Wenjing Fang, Lili Yu, Mildred S Dresselhaus, and Jing Kong. 2014. "Role of the seeding promoter in MoS2 growth by chemical vapor deposition." *Nano Letters* 14 (2):464–472.

Lipatov, Alexey, Mohamed Alhabeb, Maria R Lukatskaya, Alex Boson, Yury Gogotsi, and Alexander Sinitskii. 2016. "Effect of synthesis on quality, electronic properties and environmental stability of individual monolayer Ti_3C_2 MXene flakes." *Advanced Electronic Materials* 2 (12):1600255.

Liu, Jiehua, and Xue-Wei Liu. 2012. "Two-dimensional nanoarchitectures for lithium storage." *Advanced Materials* 24 (30):4097–4111.

Liu, Jinlong, Yao Zheng, Dongdong Zhu, Anthony Vasileff, Tao Ling, and Shi-Zhang Qiao. 2017. "Identification of pH-dependent synergy on Ru/MoS2 interface: A comparison of alkaline and acidic hydrogen evolution." *Nanoscale* 9 (43):16616–16621.

Liu, Mingkai, Yuqing Liu, Yan Yan, Fengsheng Wang, Jiahui Liu, and Tianxi Liu. 2017. "A highly conductive carbon–sulfur film with interconnected mesopores as an advanced cathode for lithium–sulfur batteries." *Chemical Communications* 53 (65):9097–9100.

Liu, Mingkai, Zhibin Yang, Hao Sun, Chao Lai, Xinsheng Zhao, Huisheng Peng, and Tianxi Liu. 2016. "A hybrid carbon aerogel with both aligned and interconnected pores as interlayer for high-performance lithium–sulfur batteries." *Nano Research* 9 (12): 3735–3746.

Liu, Mingkai, Peng Zhang, Zehua Qu, Yan Yan, Chao Lai, Tianxi Liu, and Shanqing Zhang. 2019. "Conductive carbon nanofiber interpenetrated graphene architecture for ultra-stable sodium ion battery." *Nature Communications* 10 (1):3917. https://doi.org/10.1038/s41467-019-11925-z.

Liu, Tongchang, Wendy G Pell, and Brian Evans Conway. 1997. "Self-discharge and potential recovery phenomena at thermally and electrochemically prepared RuO2 supercapacitor electrodes." *Electrochimica Acta* 42 (23–24):3541–3552.

Liu, Wei, Jiahao Kang, Deblina Sarkar, Yasin Khatami, Debdeep Jena, and Kaustav Banerjee. 2013. "Role of metal contacts in designing high-performance monolayer n-type WSe2 field effect transistors." *Nano Letters* 13 (5):1983–1990.

Liu, Yi-Tao, Peng Zhang, Ning Sun, Babak Anasori, Qi-Zhen Zhu, Huan Liu, Yury Gogotsi, and Bin Xu. 2018. "Self-assembly of transition metal oxide nanostructures on MXene nanosheets for fast and stable lithium storage." *Advanced Materials* 30 (23):1707334. https://doi.org/10.1002/adma.201707334.

Liu, Yu, Wei Wang, Yulong Ying, Yewu Wang, and Xinsheng Peng. 2015. "Binder-free layered Ti_3C_2/CNTs nanocomposite anodes with enhanced capacity and long-cycle life for lithium-ion batteries." *Dalton Transactions* 44 (16):7123–7126. https://doi.org/10.1039/C4DT02058H.

Luo, Jianmin, Xinyong Tao, Jun Zhang, Yang Xia, Hui Huang, Liyuan Zhang, Yongping Gan, Chu Liang, and Wenkui Zhang. 2016. "Sn4+ Ion Decorated Highly Conductive Ti_3C_2 MXene: Promising Lithium-Ion Anodes with Enhanced Volumetric Capacity and Cyclic Performance." *ACS Nano* 10 (2):2491–2499. https://doi.org/10.1021/acsnano.5b07333.

Lu, Ganhua, Kehan Yu, Zhenhai Wen, and Junhong Chen. 2013. "Semiconducting graphene: Converting graphene from semimetal to semiconductor." *Nanoscale* 5 (4):1353–1368.

Lukatskaya, Maria R, Sankalp Kota, Zifeng Lin, Meng-Qiang Zhao, Netanel Shpigel, Mikhael D Levi, Joseph Halim, Pierre-Louis Taberna, Michel W Barsoum, and Patrice Simon. 2017. "Ultra-high-rate pseudocapacitive energy storage in two-dimensional transition metal carbides." *Nature Energy* 2 (8):17105.

Lukatskaya, Maria R, Olha Mashtalir, Chang E Ren, Yohan Dall'Agnese, Patrick Rozier, Pierre Louis Taberna, Michael Naguib, Patrice Simon, Michel W Barsoum, and Yury Gogotsi. 2013. "Cation intercalation and high volumetric capacitance of two-dimensional titanium carbide." *Science* 341 (6153):1502. https://doi.org/10.1126/science.1241488.

Luo, Jianmin, Xinyong Tao, Jun Zhang, Yang Xia, Hui Huang, Liyuan Zhang, Yongping Gan, Chu Liang, and Wenkui Zhang. 2016. "Sn4+ ion decorated highly conductive Ti_3C_2 MXene: Promising lithium-ion anodes with enhanced volumetric capacity and cyclic performance." *ACS Nano* 10 (2):2491–2499. https://doi.org/10.1021/acsnano.5b07333.

Ma, Zhinan, Zhenpeng Hu, Xudong Zhao, Qing Tang, Dihua Wu, Zhen Zhou, and Lixin Zhang. 2014. "Tunable band structures of heterostructured bilayers with transition-metal dichalcogenide and MXene monolayer." *The Journal of Physical Chemistry C* 118 (10):5593–5599. https://doi.org/10.1021/jp500861n.

Maleski, Kathleen, Vadym N Mochalin, and Yury Gogotsi. 2017. "Dispersions of two-dimensional titanium carbide MXene in organic solvents." *Chemistry of Materials* 29 (4):1632–1640.

Mendoza-Sánchez, Beatriz, and Yury Gogotsi. 2016. "Synthesis of two-dimensional materials for capacitive energy storage." *Advanced Materials* 28 (29):6104–6135.

Meng, Ruijin, Jimei Huang, Yutong Feng, Lianhai Zu, Chengxin Peng, Lirong Zheng, Lei Zheng, Zhibin Chen, Guanglei Liu, and Bingjie Chen. 2018. "Black phosphorus quantum dot/Ti_3C_2 MXene nanosheet composites for efficient electrochemical lithium/sodium-ion storage." *Advanced Energy Materials* 8 (26):1801514.

Ming, Hai, Jun Ming, Seung-Min Oh, Shu Tian, Qun Zhou, Hui Huang, Yang-Kook Sun, and Junwei Zheng. 2014. "Surfactant-assisted synthesis of Fe_2O_3 nanoparticles and F-doped carbon modification toward an improved Fe_3O_4@ CF x/LiNi0. 5Mn1. 5O4 battery." *ACS Applied Materials & Interfaces* 6 (17):15499–15509.

Naguib, Michael, and Yury Gogotsi. 2015. "Synthesis of two-dimensional materials by selective extraction." *Accounts of Chemical Research* 48 (1):128–135. https://doi.org/10.1021/ar500346b.

Naguib, Michael, Murat Kurtoglu, Volker Presser, Jun Lu, Junjie Niu, Min Heon, Lars Hultman, Yury Gogotsi, and Michel W Barsoum. 2011a. "Two-dimensional nanocrystals produced by exfoliation of Ti_3AlC_2." *Advanced Materials* 23 (37):4248–4253.

Naguib, Michael, Murat Kurtoglu, Volker Presser, Jun Lu, Junjie Niu, Min Heon, Lars Hultman, Yury Gogotsi, and Michel W Barsoum. 2011b. "Two-dimensional nanocrystals: Two-dimensional nanocrystals produced by exfoliation of Ti_3AlC_2 (Adv. Mater. 37/2011)." *Advanced Materials* 23 (37):4207–4207.

Naguib, Michael, Vadym N Mochalin, Michel W Barsoum, and Yury Gogotsi. 2014. "25th anniversary article: MXenes: A new family of two-dimensional materials." *Advanced Materials* 26 (7):992–1005. https://doi.org/10.1002/adma.201304138.

Newman, MEJ, and Robert M Ziff. 2000. "Efficient Monte Carlo algorithm and high-precision results for percolation." *Physical Review Letters* 85 (19):4104.

Ni, Jiangfeng, Liang Li, and Jun Lu. 2018. "Phosphorus: An anode of choice for sodium-ion batteries." *ACS Energy Letters* 3 (5):1137–1144.

Okada, Mitsuhiro, Takumi Sawazaki, Kenji Watanabe, Takashi Taniguch, Hiroki Hibino, Hisanori Shinohara, and Ryo Kitaura. 2014. "Direct chemical vapor deposition growth of WS2 atomic layers on hexagonal boron nitride." *ACS Nano* 8 (8):8273–8277.

Ospina, DA, CA Duque, JD Correa, and Eric Suárez Morell. 2016. "Twisted bilayer blue phosphorene: A direct band gap semiconductor." *Superlattices and Microstructures* 97:562–568.

Pang, Jinbo, Rafael G Mendes, Alicja Bachmatiuk, Liang Zhao, Huy Q Ta, Thomas Gemming, Hong Liu, Zhongfan Liu, and Mark H Rummeli. 2019. "Applications of 2D MXenes in energy conversion and storage systems." *Chemical Society Reviews* 48 (1):72–133.

Park, Sungjin, and Rodney S Ruoff. 2009. "Chemical methods for the production of graphenes." *Nature Nanotechnology* 4 (4):217–224.

Peng, Lele, Yue Zhu, Dahong Chen, Rodney S Ruoff, and Guihua Yu. 2016. "Two-dimensional materials for beyond-lithium-ion batteries." *Advanced Energy Materials* 6 (11):1600025.

Pomerantseva, Ekaterina, and Yury Gogotsi. 2017. "Two-dimensional heterostructures for energy storage." *Nature Energy* 2 (7):17089.

Prakash, Anand, and D Bahadur. 2014. "The role of ionic electrolytes on capacitive performance of ZnO-reduced graphene oxide nanohybrids with thermally tunable morphologies." *ACS Applied Materials & Interfaces* 6 (3):1394–1405.

Pumera, Martin, Zdeněk Sofer, and Adriano Ambrosi. 2014. "Layered transition metal dichalcogenides for electrochemical energy generation and storage." *Journal of Materials Chemistry A* 2 (24):8981–8987.

Qian, Jiangfeng, Xianyong Wu, Yuliang Cao, Xinping Ai, and Hanxi Yang. 2013. "High capacity and rate capability of amorphous phosphorus for sodium ion batteries." *Angewandte Chemie International Edition* 52 (17):4633–4636.

Qian, Xingyue, Junfei Ding, Jianli Zhang, Yue Zhang, Yining Wang, Erjun Kan, Xin Wang, and Junwu Zhu. 2018. "Ultrathin molybdenum disulfide/carbon nitride nanosheets with abundant active sites for enhanced hydrogen evolution." *Nanoscale* 10 (4):1766–1773.

Ramadoss, Ananthakumar, and Sang Jae Kim. 2013. "Facile preparation and electrochemical characterization of graphene/ZnO nanocomposite for supercapacitor applications." *Materials Chemistry and Physics* 140 (1):405–411.

Reina, Alfonso, Hyungbin Son, Liying Jiao, Ben Fan, Mildred S Dresselhaus, ZhongFan Liu, and Jing Kong. 2008. "Transferring and identification of single-and few-layer graphene on arbitrary substrates." *The Journal of Physical Chemistry C* 112 (46):17741–17744.

Ren, Chang E, Kelsey B Hatzell, Mohamed Alhabeb, Zheng Ling, Khaled A Mahmoud, and Yury Gogotsi. 2015. "Charge- and size-selective ion sieving through Ti_3C_2Tx MXene membranes." *The Journal of Physical Chemistry Letters* 6 (20):4026–4031. https://doi.org/10.1021/acs.jpclett.5b01895.

Ronchi, Rodrigo Mantovani, Jeverson Teodoro Arantes, and Sydney Ferreira Santos. 2019. "Synthesis, structure, properties and applications of MXenes: Current status and perspectives." *Ceramics International* 45:18167–18188.

Sahoo, Satyaprakash, Anand PS Gaur, Majid Ahmadi, Maxime J-F Guinel, and Ram S Katiyar. 2013. "Temperature-dependent Raman studies and thermal conductivity of few-layer MoS2." *The Journal of Physical Chemistry C* 117 (17):9042–9047.

Seh, Zhi Wei, Kurt D Fredrickson, Babak Anasori, Jakob Kibsgaard, Alaina L Strickler, Maria R Lukatskaya, Yury Gogotsi, Thomas F Jaramillo, and Aleksandra Vojvodic. 2016. "Two-dimensional molybdenum carbide (MXene) as an efficient electrocatalyst for hydrogen evolution." *ACS Energy Letters* 1 (3):589–594. https://doi.org/10.1021/acsenergylett.6b00247.

Shahzad, Faisal, Mohamed Alhabeb, Christine Hatter, Babak Anasori, Soon Hong, Chong Min Koo, and Yury Gogotsi. 2016. "Electromagnetic interference shielding with 2D transition metal carbides (MXenes)." *Science* 353:1137–1140. https://doi.org/10.1126/science.aag2421.

Shan, Dandan, Jiao Yang, Wei Liu, Jun Yan, and Zhuangjun Fan. 2016. "Biomass-derived three-dimensional honeycomb-like hierarchical structured carbon for ultrahigh energy density asymmetric supercapacitors." *Journal of Materials Chemistry A* 4 (35): 13589–13602.

Shi, Yumeng, Wu Zhou, Ang-Yu Lu, Wenjing Fang, Yi-Hsien Lee, Allen Long Hsu, Soo Min Kim, Ki Kang Kim, Hui Ying Yang, and Lain-Jong Li. 2012. "Van Der Waals epitaxy of MoS2 layers using graphene as growth templates." *Nano Letters* 12 (6):2784–2791.

Simon, Patrice, and Yury Gogotsi. 2010. "Materials for electrochemical capacitors." In *Nanoscience and Technology: A Collection of Reviews From Nature Journals*, 320–329. World Scientific.

Simon, Patrice, Yury Gogotsi, and Bruce Dunn. 2014. "Where do batteries end and supercapacitors begin?" *Science* 343 (6176):1210–1211.

Song, Jiangxuan, Zhaoxin Yu, Mikhail L Gordin, Xiaolin Li, Huisheng Peng, and Donghai Wang. 2015. "Advanced sodium ion battery anode constructed via chemical bonding between phosphorus, carbon nanotube, and cross-linked polymer binder." *ACS Nano* 9 (12):11933–11941.

Song, Seung Min, and Byung Jin Cho. 2013. "Contact resistance in graphene channel transistors." *Carbon Letters* 14 (3):162–170.

Stoller, Meryl D, Sungjin Park, Yanwu Zhu, Jinho An, and Rodney S Ruoff. 2008. "Graphene-based ultracapacitors." *Nano Letters* 8 (10):3498–3502.

Su, Weitao, Shenguang Wang, Li Fu, Fei Chen, Kaixin Song, Xiwei Huang, and Li Yang. 2018. "Growth of WS2 flakes on Ti_3C_2Tx Mxene using vapor transportation routine." *Coatings* 8 (8):281.

Sun, Jie, Hyun-Wook Lee, Mauro Pasta, Hongtao Yuan, Guangyuan Zheng, Yongming Sun, Yuzhang Li, and Yi Cui. 2015. "A phosphorene–graphene hybrid material as a high-capacity anode for sodium-ion batteries." *Nature Nanotechnology* 10 (11):980.

Tang, Qing, Zhen Zhou, and Panwen Shen. 2012. "Are MXenes promising anode materials for Li ion batteries? Computational studies on electronic properties and Li storage capability of Ti_3C_2 and $Ti_3C_2X_2$ (X=F, OH) monolayer." *Journal of the American Chemical Society* 134 (40):16909–16916.

Verger, Louisiane, Chuan Xu, Varun Natu, Hui-Ming Cheng, Wencai Ren, and Michel W Barsoum. 2019. "Overview of the synthesis of MXenes and other ultrathin 2D transition metal carbides and nitrides." *Current Opinion in Solid State and Materials Science* 23 (3):149–163. https://doi.org/10.1016/j.cossms.2019.02.001.

Wang, Fen, Minjuan Cao, Yi Qin, Jianfeng Zhu, Lei Wang, and Yi Tang. 2016. "ZnO nanoparticle-decorated two-dimensional titanium carbide with enhanced supercapacitive performance." *RSC Advances* 6 (92):88934–88942. https://doi.org/10.1039/C6RA15384D.

Wang, Fengmei, Tofik Ahmed Shifa, Xueying Zhan, Yun Huang, Kaili Liu, Zhongzhou Cheng, Chao Jiang, and Jun He. 2015. "Recent advances in transition-metal dichalcogenide based nanomaterials for water splitting." *Nanoscale* 7 (47):19764–19788.

Wang, Guoping, Lei Zhang, and Jiujun Zhang. 2012. "A review of electrode materials for electrochemical supercapacitors." *Chemical Society Reviews* 41 (2):797–828.

Wang, Hsiu-Wen, Michael Naguib, Katharine Page, David J Wesolowski, and Yury Gogotsi. 2016. "Resolving the structure of Ti_3C_2Tx MXenes through multilevel structural modeling of the atomic pair distribution function." *Chemistry of Materials* 28 (1):349–359.

Wang, Xi, Xing-Long Wu, Yu-Guo Guo, Yeteng Zhong, Xinqiang Cao, Ying Ma, and Jiannian Yao. 2010. "Synthesis and lithium storage properties of Co_3O_4 nanosheet-assembled multishelled hollow spheres." *Advanced Functional Materials* 20 (10):1680–1686.

Wang, Xianfen, Satoshi Kajiyama, Hiroki Iinuma, Eiji Hosono, Shinji Oro, Isamu Moriguchi, Masashi Okubo, and Atsuo Yamada. 2015. "Pseudocapacitance of MXene nanosheets for high-power sodium-ion hybrid capacitors." *Nature Communications* 6 (1):6544. https://doi.org/10.1038/ncomms7544.

Whittingham, M Stanley. 2004. "Lithium batteries and cathode materials." *Chemical Reviews* 104 (10):4271–4302.

Wu, Xianhong, Zhiyu Wang, Mengzhou Yu, Luyang Xiu, and Jieshan Qiu. 2017. "Stabilizing the MXenes by carbon nanoplating for developing hierarchical nanohybrids with efficient lithium storage and hydrogen evolution capability." *Advanced Materials* 29 (24):1607017.

Wu, Yuting, Ping Nie, Jiangmin Jiang, Bing Ding, Hui Dou, and Xiaogang Zhang. 2017. "MoS2-nanosheet-decorated 2D titanium carbide (MXene) as high-performance anodes for sodium-ion batteries." *ChemElectroChem* 4 (6):1560–1565.

Xiong, Pan, Renzhi Ma, Nobuyuki Sakai, Leanddas Nurdiwijayanto, and Takayoshi Sasaki. 2018. "Unilamellar metallic MoS2/graphene superlattice for efficient sodium storage and hydrogen evolution." *ACS Energy Letters* 3 (4):997–1005.

Xu, Jiao, Jaewoo Shim, Jin-Hong Park, and Sungjoo Lee. 2016. "MXene electrode for the integration of WSe2 and MoS2 field effect transistors." *Advanced Functional Materials* 26 (29):5328–5334. https://doi.org/10.1002/adfm.201600771.

Xu, Yan, Xiaoshu Zhu, Xiaosi Zhou, Xia Liu, Yunxia Liu, Zhihui Dai, and Jianchun Bao. 2014. "Ge nanoparticles encapsulated in nitrogen-doped reduced graphene oxide as an advanced anode material for lithium-ion batteries." *The Journal of Physical Chemistry C* 118 (49):28502–28508.

Xu, Zhi, Yuqing Sun, Yongxiang Zhuang, Wenheng Jing, Hua Ye, and Zhanfeng Cui. 2018. "Assembly of 2D MXene nanosheets and TiO_2 nanoparticles for fabricating mesoporous TiO2-MXene membranes." *Journal of Membrane Science* 564:35–43.

Xue, Qi, Huijie Zhang, Minshen Zhu, Zengxia Pei, Hongfei Li, Zifeng Wang, Yang Huang, Yan Huang, Qihuang Deng, and Jie Zhou. 2017. "Photoluminescent Ti_3C_2 MXene quantum dots for multicolor cellular imaging." *Advanced Materials* 29 (15):1604847.

Yan, Jun, Zhuangjun Fan, Tong Wei, Weizhong Qian, Milin Zhang, and Fei Wei. 2010. "Fast and reversible surface redox reaction of graphene–MnO_2 composites as supercapacitor electrodes." *Carbon* 48 (13):3825–3833.

Yan, Jun, Chang E Ren, Kathleen Maleski, Christine B Hatter, Babak Anasori, Patrick Urbankowski, Asya Sarycheva, and Yury Gogotsi. 2017. "Flexible MXene/graphene films for ultrafast supercapacitors with outstanding volumetric capacitance." *Advanced Functional Materials* 27 (30):1701264. https://doi.org/10.1002/adfm.201701264.

Yan, Ya, BaoYu Xia, Zhichuan Xu, and Xin Wang. 2014. "Recent development of molybdenum sulfides as advanced electrocatalysts for hydrogen evolution reaction." *ACS Catalysis* 4 (6):1693–1705.

Yang, Hung-Wei, Ho-Feng Hsieh, Ruei-San Chen, Ching-Hwa Ho, Kuei-Yi Lee, and Liang-Chiun Chao. 2018. "Ultraefficient ultraviolet and visible light sensing and ohmic contacts in high-mobility InSe nanoflake photodetectors fabricated by the focused ion beam technique." *ACS Applied Materials & Interfaces* 10 (6):5740–5749.

Yang, Jieun, and Hyeon Suk Shin. 2014. "Recent advances in layered transition metal dichalcogenides for hydrogen evolution reaction." *Journal of Materials Chemistry A* 2 (17):5979–5985.

Yang, Xiaowei, Chi Cheng, Yufei Wang, Ling Qiu, and Dan Li. 2013. "Liquid-mediated dense integration of graphene materials for compact capacitive energy storage." *Science* 341 (6145):534–537.

Yang, Yajie, Jaeho Jeon, Jin-Hong Park, Mun Seok Jeong, Byoung Hun Lee, Euyheon Hwang, and Sungjoo Lee. 2019. "Plasmonic transition metal carbide electrodes for high-performance InSe photodetectors." *ACS Nano* 13 (8):8804–8810.

Yang, Yuan, Guangyuan Zheng, and Yi Cui. 2013. "Nanostructured sulfur cathodes." *Chemical Society Reviews* 42 (7):3018–3032.

Yao, Hui, Qiang Wang, Jianwei Li, Weishan Cai, Yadong Wei, Bin Wang, and Jian Wang. 2020. "Two-dimensional few-layered PC3 as a promising photocatalyst for overall water splitting." *Physical Chemistry Chemical Physics* 22 (17):9477–9486.

Yoon, Jongwon, Woojin Park, Ga-Yeong Bae, Yonghun Kim, Hun Soo Jang, Yujun Hyun, Sung Kwan Lim, Yung Ho Kahng, Woong-Ki Hong, and Byoung Hun Lee. 2013. "Highly flexible and transparent multilayer MoS_2 transistors with graphene electrodes." *Small* 9 (19):3295–3300.

Yu, Woo Jong, Zheng Li, Hailong Zhou, Yu Chen, Yang Wang, Yu Huang, and Xiangfeng Duan. 2013. "Vertically stacked multi-heterostructures of layered materials for logic transistors and complementary inverters." *Nature Materials* 12 (3):246–252.

Yu, Young-Jun, Yue Zhao, Sunmin Ryu, Louis E Brus, Kwang S Kim, and Philip Kim. 2009. "Tuning the graphene work function by electric field effect." *Nano Letters* 9 (10):3430–3434.

Yuan, Hao, and Zhenyu Li. 2018. "Interfacial properties of black phosphorus/transition metal carbide van der Waals heterostructures." *Frontiers of Physics* 13 (3):138103.

Zeng, Zhiping, Yibo Yan, Jie Chen, Ping Zan, Qinghua Tian, and Peng Chen. 2019. "Boosting the photocatalytic ability of Cu_2O nanowires for CO_2 conversion by MXene quantum dots." *Advanced Functional Materials* 29 (2):1806500. https://doi.org/10.1002/adfm.201806500.

Zhan, Xiaoxue, Chen Si, Jian Zhou, and Zhimei Sun. 2020. "MXene and MXene-based composites: Synthesis, properties and environment-related applications." *Nanoscale Horizons*. https://doi.org/10.1039/C9NH00571D.

Zhang, Jincan, Li Lin, Luzhao Sun, Yucheng Huang, Ai Leen Koh, Wenhui Dang, Jianbo Yin, Mingzhan Wang, Congwei Tan, and Tianran Li. 2017. "Clean transfer of large graphene single crystals for high-intactness suspended membranes and liquid cells." *Advanced Materials* 29 (26):1700639.

Zhang, Li, and XS Zhao. 2009. "Carbon-based materials as supercapacitor electrodes." *Chemical Society Reviews* 38 (9):2520–2531.

Zhang, Qingrui, Jie Teng, Guodong Zou, Qiuming Peng, Qing Du, Tifeng Jiao, and Jianyong Xiang. 2016. "Efficient phosphate sequestration for water purification by unique sandwich-like MXene/magnetic iron oxide nanocomposites." *Nanoscale* 8 (13):7085–7093.

Zhang, Zheng, Pei Lin, Qingliang Liao, Zhuo Kang, Haonan Si, and Yue Zhang. 2019. "Graphene-based mixed-dimensional van der Waals heterostructures for advanced optoelectronics." *Advanced Materials* 31 (37):1806411. https://doi.org/10.1002/adma.201806411.

Zhao, Meng-Qiang, Chang E Ren, Zheng Ling, Maria R Lukatskaya, Chuanfang Zhang, Katherine L Van Aken, Michel W Barsoum, and Yury Gogotsi. 2015. "Flexible MXene/carbon nanotube composite paper with high volumetric capacitance." *Advanced Materials* 27 (2):339–345. https://doi.org/10.1002/adma.201404140.

Zhao, Meng-Qiang, Xiuqiang Xie, Chang E Ren, Taron Makaryan, Babak Anasori, Guoxiu Wang, and Yury Gogotsi. 2017. "Hollow MXene spheres and 3D macroporous MXene frameworks for Na-ion storage." *Advanced Materials* 29 (37):1702410.

Zhao, Ruizheng, Zhao Qian, Zhongyuan Liu, Danyang Zhao, Xiaobin Hui, Guanzhong Jiang, Chengxiang Wang, and Longwei Yin. 2019. "Molecular-level heterostructures assembled from layered black phosphorene and Ti_3C_2 MXene as superior anodes for high-performance sodium ion batteries." *Nano Energy* 65:104037. https://doi.org/10.1016/j.nanoen.2019.104037.

Zheng, Xuejin, and Robert A Rapp. 1995. "Electrochemical impedance study of platinum electrode in fused $Na_2SO_{4\text{-}10}$ mole percent $NaVO_3$ melts." *Journal of the Electrochemical Society* 142 (1):142.

Zhong, Yu, Xinhui Xia, Fan Shi, Jiye Zhan, Jiangping Tu, and Hong Jin Fan. 2016. "Transition metal carbides and nitrides in energy storage and conversion." *Advanced Science* 3 (5):1500286. https://doi.org/10.1002/advs.201500286.

Zhu, Miao, Xinming Li, Yibo Guo, Xiao Li, Pengzhan Sun, Xiaobei Zang, Kunlin Wang, Minlin Zhong, Dehai Wu, and Hongwei Zhu. 2014. "Vertical junction photodetectors based on reduced graphene oxide/silicon Schottky diodes." *Nanoscale* 6 (9):4909–4914.

7 Mixed-Dimensional Titanium-Based MXene Heterostructures in Supercapacitors and Batteries-Based Energy Storage Network

7.1 INTRODUCTION

Nanostructures with small dimensions (2D, 1D, or 0D) have garnered significant attention due to their exceptional electrical properties and biocompatibility. Particularly, extensive development is currently underway for 2D nanomaterials, and their unique properties set them apart from other nanomaterials, like their zero-dimensional and one-dimensional counterparts, making them particularly intriguing for both industrial and academic research. Despite possessing remarkable mechanical, electrical, and thermal properties, graphene, as a two-dimensional (2D) material, is not particularly practical for the real world. This limitation arises mainly from its lack of inherent chemical reactivity and the absence of a discernible band gap, both of which restrict its usefulness in these applications. Overcoming this challenge requires diving deeper into the study of fundamental two-dimensional (2D) materials. In the last ten years, several materials have been used in various fields such as electronics, photonics, and biomedicine. These materials include layered metal dichalcogenides (LMDCs), metal oxides, boron nitride (BN), and multi-element hydroxides (Fuhrer and Hone 2013). Moreover, transition metal dichalcogenides (TMDCs) such as MoSe2, MoS_2, WSe_2, and WS_2 are distinguished by their remarkable and distinctive optical properties. They hold significant potential for advancements in sensing, optoelectronics, energy storage, and supercapacitors (Ding et al. 2017).

Additionally, molybdenum disulfide (MoS_2), a fascinating two-dimensional transition metal dichalcogenide (TMD), exhibits semiconductor properties with a direct bandgap. This distinctive quality makes MoS_2 a promising material for a range of cutting-edge applications. However, its stated carrier mobility value typically ranges from 60 to 200 $cm^2V^{-1}s^{-1}$ (Radisavljevic et al. 2011; Fuhrer and Hone 2013). When compared to graphene, single-layer molybdenum disulfide (MoS_2) exhibits a relatively lower value of nominal carrier mobility, typically varying from 60 to

DOI: 10.1201/9781003319658-7

200 $cm^2V^{-1}s^{-1}$ (Liao et al. 2010; Schwierz 2010). Due to its lower nominal carrier mobility value, ranging from 60 to 200 $cm^2V^{-1}s^{-1}$, single-layer molybdenum disulfide (MoS_2) is considered less advantageous for practical applications compared to graphene. In contrast, black phosphorus (BP) has a naturally viable band gap and increased carrier mobility value of 103 $cm^2V^{-1}s^{-1}$. Furthermore, it showcases a distinctive puckered structure, setting it apart from other materials (Zhang, Abbasi, and Wang 2021). However, the effectiveness of electronic devices based on black phosphorus (BP) deteriorates rapidly upon exposure to oxygen. This is primarily attributed to the material's chemical instability, posing a significant challenge to its practical implementation.

Given the limitations and shortcomings of existing two-dimensional (2D) materials, there is an increasing need, driven by market demands, to investigate 2D semiconductors with a diverse array of desirable properties. These characteristics include a moderate energy gap, excellent light-harvesting capabilities, stability in ambient conditions, high carrier mobility, and a straightforward, reproducible implementation process.

Recently, a novel class of two-dimensional (2D) nanostructures, referred to as MXenes, has emerged from transition metal carbides, nitrides, and carbonitrides. These innovative materials are paving the way for exciting advancements in various fields. These MXenes have been introduced to address the aforementioned challenges and provide innovative solutions in the field (Zhao et al. 2019). In recent times, the MXene family has garnered significant attention as a vibrant and fruitful area of research within the realm of two-dimensional (2D) materials. The MAX phases, serving as the precursor materials for MXenes, demonstrate robust bonding interactions that effectively hinder their cleavage process via mechanical means, such as shearing or other approaches. This distinctive attribute sets MXene apart from other 2D materials in a remarkable manner (Alhabeb et al. 2017; Anasori, Lukatskaya, and Gogotsi 2017; Sun et al. 2017; Ronchi, Arantes, and Santos 2019). As a result, the chemical exfoliation technique has pioneered the process of extracting two-dimensional (2D) structures, namely MXenes, from their precursor solids (MAX phases), marking a significant milestone in the field. For the first time, this breakthrough allows for the independent separation and isolation of MXene layers (Naguib et al. 2012; Libo et al. 2015).

In contrast to other two-dimensional stacked materials such as graphite, metal oxides, and transition metal dichalcogenides, MXenes are distinguished by their exceptional multidirectional properties. This unique versatility makes them highly promising for a wide range of applications. Unlike these materials, which are bound together by weak interactions, MXenes exhibit a unique structural arrangement that imparts enhanced stability and robustness (Anasori, Lukatskaya, and Gogotsi 2017; Kishor et al. 2020). Despite the growing global market for MXene-based products, there remains a notable scarcity of comprehensive and well-organized information pertaining to the study of structural stability, surface roughness, and strategies for enhancing the properties of MXenes. To meet the growing demand and enhance the quality of MXene materials, ongoing research and optimization are essential (Aïssa et al. 2016). The utilization of heterostructures of MXene presents a promising approach to overcome challenges related to stability and surface roughness, which have hindered the practical implementation of MXene in real-world applications. MXene-based van der Waals heterostructures have recently garnered significant attention from the photovoltaic, catalytic, and energy storage device communities.

These innovative applications are capturing the imagination of researchers and driving exciting advancements. This enthusiasm is growing rapidly, reminiscent of the enthusiasm seen in 2006. Recent progress in methodologies for synthesizing MXene and its van der Waals heterostructures has achieved substantial advancements.

As a result, an increasing number of research groups are now able to investigate complex devices and materials, contributing to the advancement of this exciting field (Yuan and Li 2018).

Numerous research laboratories are presently conducting intensive investigations into the integration of two-dimensional MXenes using a diverse array of van der Waals (vdW) heterostructures. Their objective is to overcome challenges such as long-term stability and surface irregularities.

Scientists are exploring mixed-dimensional heterostructures that incorporate 2D MXene derivatives with three different morphologies, hoping to discover solutions to these issues. These morphologies include the use of 2D MXene in conjunction with 0D, 1D, and 2D materials. Such novel combinations have the potential to alleviate stability concerns while also improving MXene's surface properties, paving the way for improved the functionality across several domains.

Developing energy storage device heterostructures using $Ti_3C_2T_x$ and Ti_2CT_x alongside other 2D materials is a challenging and meticulous process. The innovative strides being made in materials science to advance energy storage technologies are evident from the significant breakthroughs achieved in this demanding effort. Prominent 2D materials display properties such as excellent electrical conductivity, power generation capability, flexibility, and durability. Graphene (GR) electrodes, for example, are consistently robust both mechanically and electrically, but their influence is primarily superficial. Because graphene does not alter its composition significantly, the introduction of various surface functional groups tends to reduce conductivity.

This limitation prevents GR from being fully utilized in energy storage applications.

In contrast, transition metal dichalcogenides (TMDCs) exhibit significant initial capacities. However, a conversion reaction occurs during the first discharge cycle, resulting in energy waste. In the realm of reversible intercalation processes, TMDC-based electrodes demonstrate enhanced cycle stability. However, this stability comes at a cost, as there is a notable reduction in power output. This trade-off underscores the ongoing challenge of balancing performance and durability in advanced energy storage solutions (Pumera, Sofer, and Ambrosi 2014). Furthermore, although some transition metal dichalcogenides (TMDCs) demonstrate high electronic conductivity, most TMDC phases are inherently semiconducting. On the other hand, the high operational potential, redox activity during intercalation reactions, and other advantageous properties of transition metal oxides (TMOs) make them highly attractive candidates for making cathode materials (Chhowalla et al. 2013).

However, achieving success with TMOs necessitates the addition of a conductive material to offset their inherent poor electronic conductivity (Whittingham 2004).

Over the past decade, extensive research on $Ti_3C_2T_x$ and Ti_2CT_x as two-dimensional (2D) materials has unveiled their unique and remarkable properties. These materials stand out for their impressive conductivity, hydrophilicity, exceptional chemical stability, environmental robustness, and the convenience of large-scale synthesis in water. Moreover, they showcase both metallic and semiconducting properties, adding to their versatility and appeal in various applications.

Furthermore, their surfaces are rich in terminated functionalities, allowing for simple functionalization and the development of heterostructures (Sajid 2021). The electrochemical characteristics of $Ti_3C_2T_x$ and Ti_2CT_x MXene are significantly impacted by the synthesis conditions and surface chemistry. Consequently, it becomes imperative to devise methodologies for regulating the surface terminations, aiming to reduce irreversible energy losses. However, the fabrication of electrodes possessing a comprehensive array of desired attributes within a single structure poses a formidable challenge. The integration of monolayers from diverse 2D materials and the implementation of scalable assembly or self-assembly techniques in pursuit of this objective prove to be arduous endeavors (Zhao et al. 2016). In heterostructure electrodes, multiple layers of various two-dimensional materials are stacked alternately, blending distinct components into a unified structure (Akuzum et al. 2016; Liu and Liu 2012).

The phases of MXene, especially $Ti_3C_2T_x$ and Ti_2CT_x, are notable for their robust mechanical properties and excellent electronic conductivity. Pseudocapacitive materials like MXenes excel at storing substantial amounts of charge. Moreover, constructing heterostructures in combination with other two-dimensional layered materials enhances the surface area, thereby significantly boosting the energy storage capacity of the devices (Naguib et al. 2014). The metallic electrical conductivity of $Ti_3C_2T_x$ and Ti_2CT_x, when paired with layered 2D materials such as TMDCs, graphene (GR), black phosphorus (BP), and transition metal oxides (TMO), can synergistically enhance their properties. Achieving this involves stacking individual layers into heterostructures using these 2D materials, which not only boosts surface area but also enhances conductivity (Naguib et al. 2014).

In this chapter, we delve into the latest advancements pertaining to 2D $Ti_3C_2T_x$ and Ti_2CT_x MXene materials, as well as their heterostructures, which hold great promise for diverse applications. We are grateful to the diligent nanotechnology and materials research scientists around the world for their valuable contributions to the scientific literature in this field. There have been numerous review articles published that shed light on the synthesis and exfoliation techniques used to obtain MXene from various MAX phases. A diverse set of 2D nanomaterials based on MXenes, varying in morphologies such as nanoflakes, quantum dots, nanoribbons, and nanosheets have been developed and extensively researched, offering numerous applications supported by a wealth of published data (Naguib and Gogotsi 2015; Lei, Zhang, and Zhou 2015; Zhong et al. 2016; Khazaei et al. 2017; Alhabeb et al. 2017; Anasori, Lukatskaya, and Gogotsi 2017; Huang et al. 2018; Pang et al. 2019; Cai et al. 2018; Verger et al. 2019; Zhan et al. 2020; Ronchi, Arantes, and Santos 2019; Sadasivuni et al. 2020). This book chapter aims to consolidate recent advancements in the field and highlight exemplary articles that effectively encapsulate key concepts. The authors made deliberate efforts to ensure that this chapter's title stands apart from other reviews on the same topic. The work provides a comprehensive overview of diverse deposition techniques employed in creating heterostructures with 2D Ti-based MXene materials ($Ti_3C_2T_x$ and Ti_2CT_x), along with other layered materials like BN, graphene (GR), transition metal dichalcogenides (TMDs), and black phosphorus (BP). Furthermore, this chapter delves into the potential applications of macrostructures formed by functionalizing 2D Ti-based MXene materials in optics, catalysis, energy storage, and other fields of research.

7.2 DEVELOPMENT OF MXENE-BASED MIXED-DIMENSIONAL HETEROSTRUCTURES BY MORPHOLOGICAL ENGINEERING

A variety of transfer techniques are expected to emerge from the numerous published studies available on the subject (Reina et al. 2008). A notable approach in the case study of GR-0D heterojunctions involves using a flexible polymer or SiO2 as a substrate to transfer the graphene produced. Following this, quantum dots can be produced from a solution using dip/spin-coating techniques or through layer-by-layer epitaxial growth of small organic molecules (Kim et al. 2014; Bie et al. 2011; Zhang et al. 2017; Deng et al. 2015). Heterostructures based on GR-3D are commonly achieved by directly transferring graphene onto unpackaged 3D materials. However, when combining GR with 1D nanotubes/nanowires, the material transfer process becomes more intricate. The precise transfer of a single nanowire or arrays of nanowires has been established through various transfer methods, facilitating their integration with graphene. In general, the transfer-assisted construction method described earlier is adaptable and allows to create several graphene-based heterostructures during fabrication (Zhang et al. 2019). Figure 7.1 illustrates various techniques for the production of heterostructures by means of several deposition procedures. Keep in mind the concept of heterostructures based on graphene.

Assorted 2D layered materials like TMDCs offer intriguing properties for energy storage and sensing (Kou, Chen, and Smith 2015; Wang et al. 2012; Sahoo et al. 2013); metal oxides (MO) (Whittingham 2004; Zhu et al. 2016), GR and its oxides (Choi et al. 2010; Khan et al. 2015; Jiao et al. 2009; Han 2009; Akinwande, Petrone, and Hone 2014; Das et al. 2014), BP (Kou, Chen, and Smith 2015; Ospina et al. 2016; Li et al. 2014), and a few other 2D nanomaterials like Mxene (Li, Wang et al. 2015; Akuzum et al. 2016) exhibit exceptional material properties that make them attractive to craft effective power storage devices and sensitive sensors due to their inherent structures. In intriguing interdisciplinary fields such as materials science, chemistry, biology, and medicine, 2D materials are widely employed. These diverse 2D nanomaterials offer precise control over their physicochemical properties and surface characteristics, making them highly versatile for various applications. Diverse 2D nanomaterials with precise control over their physicochemical properties, surface burden, morphologies, and dimensions advance synthetic methodologies significantly. The importance of nanomaterial energy storage stems from its layered architecture and exceptionally reactive surface, as well as its improved physical, chemical, and biological characteristics.

In optoelectronics, the surface morphology and conductivity are critical factors for energy storage applications and network devices. Professor Gogotsi and his colleagues made a groundbreaking discovery in 2011 when they introduced the carbides of 2D transition metals, nitrides, and carbonitrides into the realm of 2D nanomaterials (Naguib et al. 2011; Anasori, Lukatskaya, and Gogotsi 2017; Hantanasirisakul and Gogotsi 2018).

Researchers have considerable flexibility in designing plasmonic and electrical sensor detection systems using $Ti_3C_2T_x$ MXene due to its hydrophilic surfaces, impressive metal conductivity (6,000–8,000 S/cm), and optical properties that vary with layer thickness. MXene has shown remarkable potential and adaptability as a biomedical sensing agent, establishing itself as a strong contender in the rapidly growing field of biomedical sensors. Understanding the fundamental characteristics of 2D materials is crucial before preparing various heterostructures. Table 7.1 provides a summary of the key characteristics of MXene and other 2D materials.

7.3 MIXED-DIMENSIONAL HETEROSTRUCTURES BASED ON MXENE AND THEIR ENERGY STORAGE APPLICATIONS

7.3.1 SUPERCAPACITORS

Many people are intrigued by supercapacitors (SCs) due to their potential for energy storage in future electronics. They surpass lithium-ion batteries in terms of effectiveness and durability, establishing them as the preferred choice (Xu et al. 2017; Lukatskaya, Dunn, and Gogotsi 2016). There is an increasing demand for energy storage solutions that can be integrated into electronic devices such as wearable sensors and the Internet of Things (IoT). This is driven by the shrinking size of these devices and the rising requirement for autonomous and intelligent systems (Wang 2010; Beidaghi and Gogotsi 2014; Kyeremateng, Brousse, and Pech 2017). In this age of microscopic marvels, when microsensors whisper secrets and nanorobots dance, it is critical to create flexible, bending device systems that can provide these electronic friends with vital electricity (Kyeremateng, Brousse, and Pech 2017; Beidaghi and Gogotsi 2014). Microbatteries or thin-film batteries have been used in energy storage systems on a smaller scale; nevertheless, their power density and cycle life are sometimes severely constrained. The need for advancements in this field is, therefore, increasing (Kyeremateng, Brousse, and Pech 2017; Xiong et al. 2014; Mendoza-Sánchez and Gogotsi 2016).

In the realm of energy storage, the allure of electrochemical double-layer capacitors and pseudocapacitors lies in their swift surface redox reactions, which have sparked their surge in popularity, ushering in the term "supercapacitor" into everyday parlance. These devices find their niche in applications demanding swift power delivery or absorption for electrical energy harvesting and storage, often standing in for or complementing traditional batteries. Noteworthy among their merits are their safety, enduring cycling life (over 10^4 cycles cycles), lightning-fast charge and discharge rates (within seconds), and impressive power density (10^3 W kg^{-1}, thousands of watts per kilogram).

Supercapacitors stand as invaluable assets in the realms of transportation, lighting, and elevators, showcasing superior power density compared to conventional

TABLE 7.1
Details on the Performance of Devices Based on Heterostructures Made of MXene and Layered Materials

a	$Ti_3C_2T_x$/rGO	Electrolyte capacitance of 1 M PVA/H_2SO_4 The formula for the volumetric capacitance is about 80 F cm^{-3}, where mF cm^{-2} is the value at a voltage of 2 mV s^{-1}.	Supercapacitors	(Couly et al. 2018)
b	3D micro-supercapacitors (MSCs) $Ti_3C_2T_x$/rGO	With a scan rate of 1 mV s^{-1}, the PVA–H_2SO_4 gel electrolyte has a specific capacitance of 34.6 mF cm^{-2}. Capacitance preservation during cycling efficiency of up to 91% Specific capacitance security at 81.7% for 15,000 cycles.	Supercapacitors	(Yue et al. 2018)
c	$Ti_3C_2T_x$/SWCNT	A volumetric capacitance of around 350 F cm^{-3}, a current density of 5 A g^{-1}, and a number of 10,000 cycles are all used with an electrolyte solution of 1 M $MgSO_4$.	Supercapacitors	(Zhao et al. 2015)
	$Ti_3C_2T_x$/CNTs	One millimeter of $TEABF_4$ in acetonitrile is the organic electrolyte. At 2 mV s^{-1}, the capacitance is 245 F cm^{-3}, or 85 F g^{-1}. The capacitance retention is 100 mV s^{-1}.	supercapacitor	(Dall'Agnese et al. 2016)
d	$Ti_3C_2T_x/RuO_2$	The electrolyte is 1 M H_2SO_4. Specific capacitance value: 388 F g^{-1} (10 mV s^{-1} scan rate value) for capacitance. The effective current density value over 20,000 cycles is 20 A g^{-1}.	Supercapacitor	(Jiang et al. 2018)
e	Ti_3C_2 Ti_3C_2/CNT	After 300 cycles, the reversible capacity is 96.2 mA h g^{-1} at 0.5 C. After 300 cycles, the reversible capacity is 428.1 mA h g^{-1} at 0.5C.	Li-ion batteries	(Liu et al. 2015)
f	Porous $Ti_3C_2T_x$ (p-$Ti_3C_2T_x$) p-$Ti_3C_2T_x$/CNTs	p-$Ti_3C_2T_x$ had a capacity value of 110 mAh g^{-1} when operated at a current rate of 0.5 C. Under conditions of 0.5 C current, the p-$Ti_3C_2T_x$/CNTs material exhibited a capacity value of 790 mAh g^{-1}. At a cycle rate of 0.1 C, the capacity value of p-$Ti_3C_2T_x$/CNTs was around 1,250 mAh g^{-1}.	Li-ion batteries	(Ren et al. 2016b)
g	Ti_3C_2/TiO_2	Based on various current densities, such as 500 or 1,000 mA g^{-1}, the capacity value after 200 cycles is 209 or 176 mA h g^{-1}.	Li-ion batteries	(Liu et al. 2018)
h	$Ti_3C_2T_x$-50 wt.% Fe_2O_3	With a rate performance of 100 mA h g^{-1} at 10 C and a specific capacity of around 270 mA h g^{-1} at 1 C (or roughly 203 mAh g^{-1} depending on the composite).	Li-ion batteries	(Ali et al. 2018)
i	MXene ($Ti_3C_2T_x$)/TiO_2	The lithium-sulfur battery can provide 800 mAh g^{-1} at 2 C after 1,000 cycles.	Lithium–sulfur (Li-S) batteries	(Jiao et al. 2019)
j	3D $Ti_3C_2T_x$	The ability to undo about 330 mAh g^{-1} after 1,000 cycles at a current pace of 2.5 C	Na-Ion batteries	(Zhao et al. 2017)
k	(Diallyl dimethyl ammonium chloride PDDA-BP/Ti_3C_2	A current density of around 0.1 A g^{-1} and a capacity value of 1,112 mA h g^{-1}, both within the 2,000 cycle range	Na-Ion batteries	(Zhao et al. 2019)

capacitors and batteries, as illustrated on the Ragone plot (Simon and Gogotsi 2010). Balancing this high power density with energy density remains a significant challenge for supercapacitors, as their energy storage capacities often fall short of those offered by batteries. However, supercapacitors have seen a tremendous change, providing improved performance and future promise, thanks to discoveries in charge storage processes and state-of-the-art nanostructured materials.

In comparison to single-layer capacitors made from inert carbon, electrochemical double-layer capacitors crafted from conductive carbon-based heterostructure electrodes with sub-nanometer pores exhibit significantly larger capacitance within the realm of battery storage technology.

This heightened capacitance arises because solvated ions shed their solvation shells in these tiny crevices. This breakthrough sets the stage for developing electrolytic devices with substantial energy densities.

Moreover, the landscape of electrochemical capacitors has evolved significantly. These capacitors now rival batteries in terms of energy density, thanks to the latest generation of nanostructured lithium electrodes that incorporate pseudocapacitive nanomaterials like oxides, nitrides, and polymers. This advancement marks a leap forward in the quest for compact, efficient energy storage solutions.

In this unfolding tale, the micro-scale electrochemical capacitors (MSCs) reign supreme with their unmatched power density and extended cycle life, surpassing the capabilities of traditional battery arrays. Their electrode components boast enduring longevity, coupled with rapid electron-ion kinetics. Across academic and industrial labs, intricate MSC devices have been meticulously crafted through diverse methodologies.

Researchers have unearthed a plethora of electrode materials, each bearing the potential to propel MSCs to new heights of performance and reliability. These discoveries pave the way for future advancements in MSC technology, promising novel solutions for energy storage in our ever-evolving world.

Activated carbon has been used to create carbon-based MSCs (Pech et al. 2010), carbon derived from carbide (Huang et al. 2016), nanotubes made of carbon (Xiao et al. 2014), graphene (Wu et al. 2013), and onion-like carbon (Pech et al. 2010). Additionally, a number of micro-pseudocapacitor substances, polypyrrole, polyaniline, and polyurethane, examples of conducting polymers (3,4-ethylene dioxythiophene) (Meng et al. 2014; Kurra et al. 2016) as well as transition metal oxides such as MnO_2 (Kurra, Jiang, and Alshareef 2015; Kurra, Alhebshi, and Alshareef 2015) have been extensively studied in the past decade.

Graphene, transition metal dichalcogenides (TMDCs), black phosphorus (BP), and titanium-based MXene materials ($Ti_3C_2T_x$ and Ti_2CT_x) could synergize perfectly when combined. Together, they weave intricate heterostructures designed to harness energy through ion adsorption and swift surface redox reactions. These materials boast a symphony of metallic and semiconducting properties, coupled with expansive surface areas that beckon electrons and ions alike to dance and mingle, unlocking new frontiers in energy storage technology. The incorporation of titanium-based MXenes in these heterostructures is set to revolutionize microelectrochemical capacitor technology, leading to the development of more flexible and adaptive devices. The subsequent sections provide an extensive array of typical examples of these advanced devices.

7.3.1.1 MXene and Graphene-Based Supercapacitors

In the world of materials, there exists a fascinating counterpart to graphene known as graphene oxide (GO). This variant boasts unique traits, crafted through a meticulous exfoliation process where functional oxygen groups are artfully introduced into graphite. This transformation imbues GO with a personality all its own, blending the foundational strength of graphene with enhanced chemical versatility, opening doors to new realms of innovation and application. GO offers significant versatility in electronics and optoelectronics due to its solubility in various solvents and its ability to form thin films over large areas. Graphene oxide and reduced graphene oxide both have great promise in these specific uses. They are used on platforms resembling plastic or paper due to GO's advantageous processing characteristics, including mechanical stability and electrical and optical tenability (Eda, Fanchini, and Chhowalla 2008).

Similar to carbon nanotube electronics, the conversion of GO/rGO sheets into layered network architectures enables the development of "polycrystalline" films, where the individual sheets serve as the individual "crystals." The adaptability and usefulness of GO/rGO in many applications are improved by this design. Two strategies may be used to increase competition and set exfoliated graphite or reduced graphene oxide (RGO) apart from their bulk equivalents. In order to produce high-quality RGO, it is first necessary to develop more effective methods of reducing graphene oxide (GO). Second, appropriate exfoliation techniques should be used to produce high-quality single-layer or even massive graphene layers (Kim and Rogers 2008).

Shape engineering has the potential to drastically alter GO's electrical and optical properties. At the molecular level, GO has high mechanical strength due to the covalent oxygen functional groups present (Dikin et al. 2007; Robinson 2008a, 2008b). These functional groups also change the electrical composition of graphene. It is possible to precisely modify the optoelectronic characteristics of GO by chemically, thermally, or electrochemically altering its chemical makeup (Eda, Fanchini, and Chhowalla 2008; Mattevi et al. 2009).

Even while GO and RGO do not have certain characteristics like relativistic charge-carrier transport that is seen in nearly perfect graphene, they nevertheless have many benefits. They are very intriguing for both basic study and possible applications in areas like photovoltaics, catalysis, and energy storage devices because of their accessibility, simplicity of processing, and adaptability.

Laser-based techniques for producing micro-supercapacitors have been developed (MSCs). El-Kady et al., for example, presented a large-scale synthesis method for laser-scribing-produced reduced graphene oxide (rGO) MSCs (El-Kady and Kaner 2013). Another approach involves using laser cutting to create asymmetric interlinking micro-supercapacitors from a thin film of rGO (Wu et al. 2016). In another study, researchers created symmetric interdigitated micro-supercapacitors by laser cutting a sheet of reduced graphene oxide (rGO).

In the fascinating world of materials science, symmetric devices crafted from reduced graphene oxide (rGO) frequently pose a challenge due to their rather unimpressive electrical conductivity, ranging from 15 to 30 S cm^{-1}, a thickness of approximately 1 μm, and a modest areal capacitance of 0.08 to 0.5 mF cm^{-2}. This combination challenges engineers and scientists alike to innovate and enhance the capabilities of rGO, seeking pathways to amplify its potential in future applications. Combining rGO with a more conductive substance may improve device performance overall (Wu et al. 2014; Niu et al. 2013).

To address the aforementioned issue, Gogotsi et al. (Couly et al. 2018) devised a current collector-free micro-supercapacitor (MSC) based on $Ti_3C_2T_x$ MXene. They employed a spray-coating technique combined with a direct laser patterning process (Peng, Akuzum et al. 2016). The hydrophilic surface and metallic conductivity of $Ti_3C_2T_x$ MXene (about 6700 S cm1) enable effective electron transport without the need for plasmonic noble metal current collectors. Additionally, $Ti_3C_2T_x$ MXene's high density of packing up to around 4.0 g cm^3 enhances its capacity for volumetric energy storage (Lukatskaya et al. 2013). Due to $Ti_3C_2T_x$ MXene's propensity to undergo oxidation at higher anode potentials, the $Ti_3C_2T_x$ MXene-based MSCs are only operable within a narrow voltage window (0.6 V) in an aqueous electrolyte system (e.g., in an asymmetric 2-electrode configuration, 0.8 V). To avoid limiting the voltage window associated with $Ti_3C_2T_x$ MXene as a positive electrode material, an unusual electrode material is used to overcome this restriction. Furthermore, by fabricating an asymmetric sandwich-type MSC structure from spray-treated rGO and $Ti_3C_2T_x$ MXene, the researchers examined the advantages of an interdigitated electrode design (Anasori, Lukatskaya, and Gogotsi 2017).

When compared to symmetric device designs, the irregular micro-supercapacitor (MSC) design displays extraordinary flexibility even after being twisted and wrapped at a 90° angle and enduring up to 10,000 charge–discharge cycles. rGO and $Ti_3C_2T_x$ nanosheets were both coated using the spray-coating approach with a 1 M PVA/H_2SO_4 solution over a transparent flexible polyethylene terephthalate (PET) substrate. A hot air pistol was used to properly dry each sprayed layer, preventing the production of coffee ring patterns, in order to provide a smooth process and prevent the formation of nonuniform coatings. To establish a homogeneous spray for each material throughout the coating process was the primary goal as it allowed for the construction of an asymmetric interdigitated structure (Couly et al. 2018).

Figure 7.1a shows the two device designs that were considered. One design had 2D nanosheets of rGO and $Ti_3C_2T_x$ sandwiched together, while the other had interdigitated 2D nanosheets arranged in a coplanar pattern. Nanosheets of MXene and rGO occur side by side on a same plane when they are interdigitated, but when they are arranged in a sandwich-like pattern, their thin layers are facing each other on opposing sides (Figure 7.1b). Figure 7.1c depicts the created 14-finger MSC-based device system numerically, with the different electrode network colors denoting the increased number of fingers (Figure 7.1d). When the same mass is deposited per unit area, $Ti_3C_2T_x$ MXene looks brighter than rGO-based thin films because of its increased density relative to rGO. The electrode network also shows a structured morphology/architecture with consistent spacing, demonstrating the accuracy of the manufacturing process.

The 14-finger MSC's interlocking configuration and width separation are shown in a top-surface view SEM image (Figure 7.1f). As captured by the scanning electron micrograph, the fingers stretch out to approximately 900-μm wide, with spaces between them spanning about 400 μm. The spray-coating technology's resolution

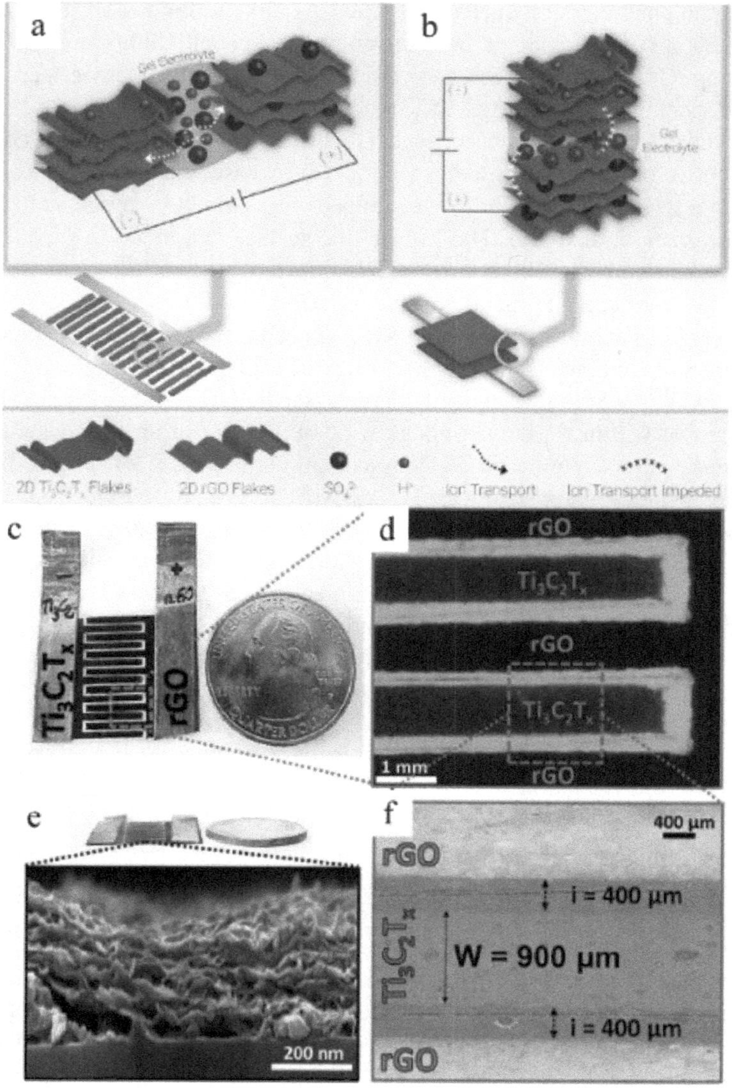

FIGURE 7.1 Combined layered structure of rGO and $Ti_3C_2T_x$. (a) The structure's interdigitation ions may spread out easily across the electrodes. (b) the sandwich-type construction makes it difficult for ions to move between the electrode nanomaterials' thin layers. (c) Size measurements may be determined from a digital picture of a quarter-dollar coin's top view using an asymmetric MSC device. (d) An enlarged view of the electrode interspace and the rGO and $Ti_3C_2T_x$ combinations may be shown in a different image. (e) A unit featuring a $Ti_3C_2T_x$ electrode, roughly the size of a quarter-dollar coin, is depicted through cross-sectional imaging and a side view in a scanning electron microscope (SEM) photograph. (f) The surface morphology of a $Ti_3C_2T_x$ electrode that is sandwiched between two rGO electrodes is seen in an SEM image.

Source: With permission adopted from Couly et al. (2018). Copyright© 2017 Wiley-VCH Verlag GmbH & Co. KGaA, Weinheim.

may be constrained by the mask employed during manufacturing. The conventional method for crafting asymmetric micro-supercapacitors relies on a rigid mask, which imposes constraints on the electrode system's dimensions. This intricate mix of size and precision challenges researchers to innovate beyond these limits, seeking novel approaches to enhance performance and functionality in micro-scale energy storage devices.

The desired feature dimensions can be achieved by integrating sputtering-based lamination with electron beam lithography or photolithography processes. The electrodes have a thickness of around 300 nm based on cross-sectional SEM scans (Figure 7.1e, bottom).

Wide-ranging applications for these thin electrode layers, which do not need current collectors at the top or bottom, show their suitability, intelligence, and distinctive conductive qualities. This distinguishes them from other MSCs, which often depend on current collectors (Pech et al. 2010a, 2010b; Chmiola et al. 2010).

Additionally, the micro-supercapacitor's (MSC) electrochemical performance was assessed using a two-electrode setup, with a negative electrode made of $Ti_3C_2T_x$-MXene and a positive electrode of reduced graphene oxide (rGO).

The MSC's rectangular cyclic voltammetry (CV) curve within a voltage range of 1 volt(V) is depicted in Figure 7.2a, demonstrating the device's successful capacitive behavior. Figure 7.2b illustrates that at a scan rate of 2 mV s^{-1}, the network exhibited an areal capacitance of approximately 2.4 mF cm^{-2} and an estimated volumetric capacitance of around 80 F cm^{-3}. The negative electrode operated within a cyclic voltage window of 0 to 0.5 V, while the positive electrode (rGO) functioned between 0.5 and 1 V ($Ti_3C_2T_x$). Figure 7.2c presents the results from cyclic voltammetry (CV) measurements, illustrating that $Ti_3C_2T_x$ underwent reduction at 0 V and oxidation at 0.5 V under a controlled scan rate of 2 mV s^{-1}.

The balanced configuration phase's potential range was discovered to be bound by the symmetric $Ti_3C_2T_x$ interdigitated MSC system, which had a narrow range of voltages around 0.6 V. This restriction can be removed by using rGO as the positive electrode material, increasing the viable voltage window to 1 V (Peng, Akuzum et al. 2016). The irregularly connected MSC, in contrast, has a higher energy density value than its symmetric counterpart and operates over a wider voltage range (Figure 7.2d).

Comparing the symmetric device structure to the asymmetric version, greater scan rates show a small increase in efficiency (Couly et al. 2018). The asymmetric micro-supercapacitor utilizing MXene operates effectively within a voltage range of 1 V and impressively retains 97% of its initial capacitance after 10,000 cycles.

It boasts of a power density of 0.2 W cm^{-3} and an energy density of 8.6 mW h cm^{-3}. Moreover, these micro-supercapacitors exhibit impressive flexibility, maintaining performance even when subjected to mechanical bending.

An asymmetrical design utilizing $Ti_3C_2T_x$ as the negative electrode and reduced graphene oxide (rGO) as the positive electrode has a great promise for increasing the energy and power densities of micro-supercapacitors, specifically because $Ti_3C_2T_x$-MXene electrodes can function efficiently at negative potentials in water-based electrolytes.

In a separate investigation conducted by Gao et al. (Yue et al. 2018), to improve energy storage capacity, they concentrated on creating extremely effective

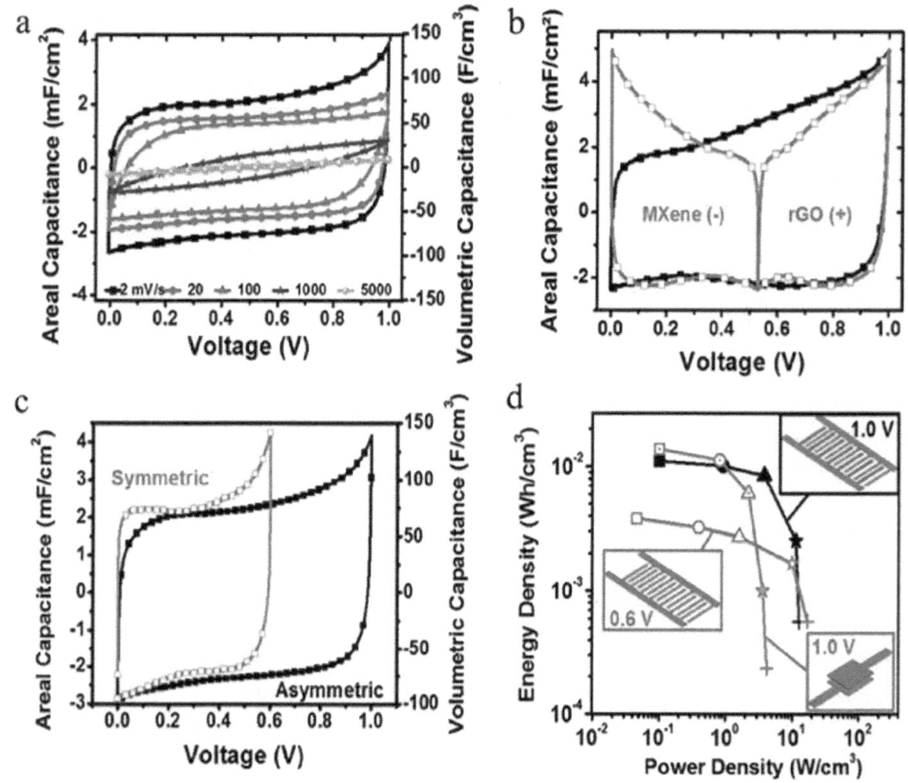

FIGURE 7.2 (a) Cyclic voltammograms (CVs), with a range from 2 to 5,000 mV s^{-1}, were recorded for the asymmetrically constructed micro-supercapacitor (MSC) at various scan speeds. (b) CV curves were obtained using a three-electrode arrangement with a scan rate of 2 mV s^{-1} for both the MSC and a distant electrode device. (c) An asymmetric interdigitated MSC and an all-MXene symmetric interdigitated MSC were compared at a scan rate of 2 mV s^{-1}. (d) The grey, middle-left, sandwich-like MSC structure, along with the grey,bottom right, and black, top-right Ragone curves, illustrates the performance of the symmetric interdigitated all-MXene MSC structure. (e) Diagrams that show how to create aerogels utilizing MXene-rGO heterostructures are supplied. (f) A schematic illustration shows how the self-healing MSC system, which makes use of MXene-rGO heterostructures, is made. (g) A picture depicts the stable heterostructure that results from the dispersion of rGO, MXene, and MXene-rGO in water. (h) An analysis and presentation of the size distribution of MXene flakes. (i) Pressure–strain curves are utilized to compare the behavior of composite aerogels based on rGO and MXene-rGO heterostructure-based aerogels.

Source: (d) Adopted with permission from Wiley-VCH Verlag GmbH & Co. KGaA, Weinheim (Couly et al. 2018). (i) Adopted with permission adopted from Yue et al. (2018) Copyright. 2018, American Chemical Society.

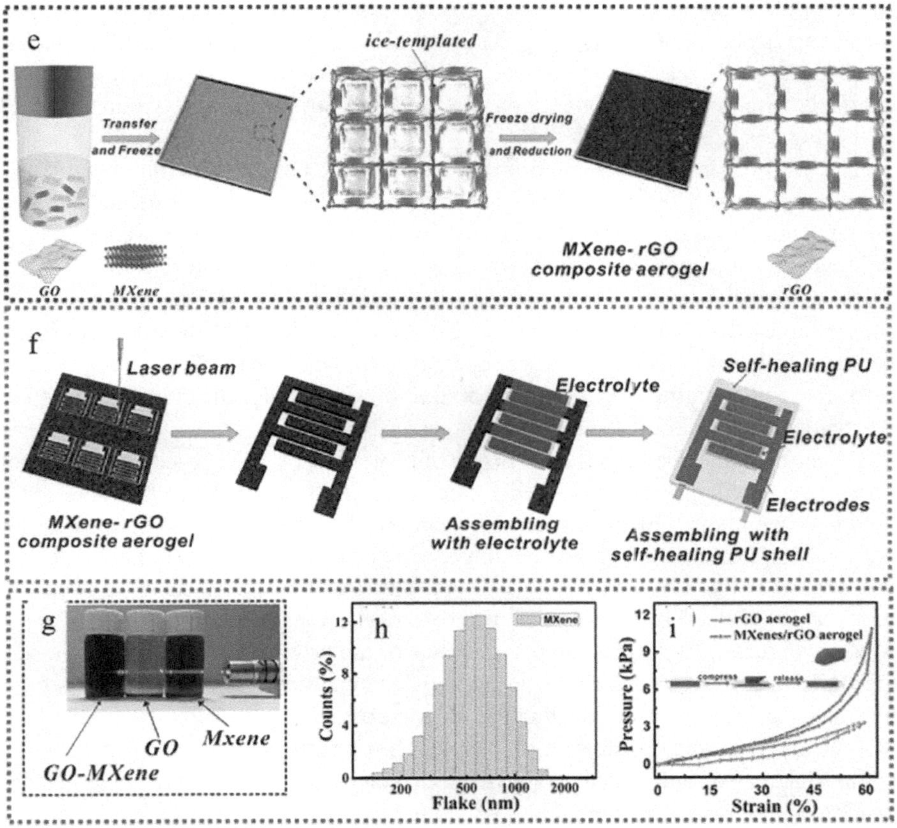

FIGURE 7.2 (Continued)

three-dimensional (3D) micro-supercapacitors (MSCs). To address the issue of mechanical deformation, which often leads to the breakdown of electrode combinations with 3D shapes in real applications, they developed a self-healing 3D micro-supercapacitor (MSC) network. This innovative solution involves incorporating aerogel-based network electrode heterostructures composed of $Ti_3C_2T_x$ and reduced graphene oxide (rGO). An exterior shell made of self-healing polyurethane was employed to shield this structure (Figure 7.2e). Figure 7.2f shows a schematic of the 3D MSC device with self-healing capabilities.

Laser cutting was used throughout the 3D system's creation to build interlink patterns inside the MXene-rGO heterostructure aerogel. The constructed MXene-rGO heterostructure network was then coupled with a PVA-H_2SO_4 gel solid-phase electrolyte. A 3D MSC network was designed with self-healing properties by encasing it in a carboxylated polyurethane (PU) surfactant, which is well-known for its compatibility with industrial procedures. The interfacial hydrogen bonding networks in the PU supermolecular network provide the substance its capacity for self-healing (Cho, White, and Braun 2009; Ahn et al. 2014). Additionally, the PVA-based gel itself has some degree of self-healing ability and self-adhesive qualities.

Notably, the accurate reconnection of disconnected electrodes is made possible by the 3D electrode geometry (Zhang, Xia, and Zhao 2012; Wang et al. 2017). The PU shell successfully recovers the device's performance at the necessary areas by successfully directing the reconnection of the 3D electrode network's disconnected parts (Figure 7.2g).

The well-organized dispersion process of MXene-GO heterostructures enables the construction of heterostructure-based aerogels using an ice-template method. As depicted in Figure 7.2h, the size distribution of MXene flakes predominantly falls within the nanometer range, which is slightly smaller than that of graphene oxide (GO) nanosheets. This enhanced dispersion allows MXene nanosheets to adhere more effectively to the GO surface. Figure 7.2i illustrates the pressure–strain behavior of the original rGO aerogel compared to the MXene-rGO heterostructure aerogel, demonstrating that the incorporation of MXene significantly enhances the mechanical properties of the aerogel (Yue et al. 2018).

Because of the well-organized dispersion process of MXene-GO heterostructures, heterostructure-based aerogels can be created using an ice-template technique. When compared to the size distribution of GO nanosheets, MXene flakes are concentrated in the nanometer range, as shown in Figure 7.2h. This distribution promotes the adhesion of MXene nanosheets to the GO surface. To illustrate how the mechanical characteristics of the aerogel are enhanced by the addition of MXene, Figure 7.2i presents the performance of the original rGO aerogel alongside the MXene-rGO heterostructure aerogel under conditions of pressure and strain, highlighting the enhanced mechanical characteristics imparted by the addition of MXene (Conway and Supercapacitors 1999). The ongoing advancements in wearable and portable electronics, such as handheld displays and flexible sensor systems, are driving significant interest in all-solid-state flexible power sources as crucial energy solutions. The rGO/($Ti_3C_2T_x$) heterostructures, as well as the presence of incomplete surface functional groups on the $Ti_3C_2T_x$ surface, have an effect on the electrochemical performance-based electrode system construction.

7.3.1.2 MXene and CNT-Based Supercapacitors

The rising popularity of electrochemical capacitors can be attributed to their prolonged cycle lifespans and impressive power densities (Simon, Gogotsi, and Dunn 2014). In response to the growing demand for wearable and portable electronics, advancements are being made to enhance the energy density per unit volume of electrochemical capacitors (Gogotsi and Simon 2011; Yang et al. 2013). Nevertheless, carbon-based electrode materials have restricted volumetric capacitance values. For low-density permeable carbon (around 0.5–1 g/cm^3), the range for commercial devices is from roughly 60 F/cm^3 to 300 F/cm^3 (Ghaffari et al. 2013; Jung et al. 2013). Hydrous ruthenium oxide (RuO_2) has higher capacitance values in the 1,000–1,500 F cm^3 range, but its high cost limits its widespread use (Zheng, Cygan, and Jow 1995).

Due to their straightforward synthesis, large-scale production capabilities, and efficient graded geometry, MXenes have emerged as a promising material for tackling these challenges. However, there are currently only a limited number of MXene-based composites documented in the literature. Gogotsi and colleagues (Zhao et al. 2015) developed an innovative technique to regulate the filtration and dispersion processes of carbon nanotubes (CNTs) and MXenes, resulting in the creation of flexible and elastic sandwich-like heterostructures for MXene/CNT electrodes. They also investigated the concentrations of CNTs in thin films with reduced graphene oxide (rGO) and onion-like carbon (OLC) formations. The sandwich-like heterostructure demonstrated superior electrochemical performance compared to both the original MXene structure and the randomly mixed MXene/CNT heterostructures.

In summary, the high power densities and longevity of electrochemical capacitors make them highly desirable. However, carbon-based electrodes are limited by their relatively low range of volumetric capacitance values. While hydrous ruthenium oxide offers higher capacitance, it is prohibitively expensive. MXenes show considerable promise; yet, the existing literature on MXene-based composites remains sparse. The flexible MXene/CNT sandwich-like heterostructures proposed by Gogotsi and colleagues have demonstrated superior electrochemical performance. Further research is necessary to explore and fully realize the potential of MXenes in practical applications.

As operational electrodes, $Ti_3C_2T_x$ and $Ti_3C_2T_x$/CNT heterostructure-based sheets were tested in an aqueous electrolyte solution containing 1 M $MgSO_4$.

For this, a conventional Ag/AgCl electrode was employed in a three-electrode irregular network arrangement. The CV curves of the original $Ti_3C_2T_x$ MXene-based electrode system and the heterostructure network are shown in Figures 7.3a and b, respectively. Separators were made of polypropylene membranes, and the electrolyte was a 1 M $MgSO_4$ aqueous solution.

The findings show that at a potential of 20 mV/s, the CV curves of the $Ti_3C_2T_x$/CNT paper heterostructures had a more rectangular form compared to the original $Ti_3C_2T_x$ paper electrode network. In the articles based on heterostructures, this predicts enhanced capacitive behavior and decreased resistance values. In contrast to randomly mixed articles, the composite network framework incorporating both multi-walled carbon nanotubes (MWCNT) and single-walled carbon nanotubes (SWCNT) greatly increased the electrochemical efficiency of the sandwich-like $Ti_3C_2T_x$/CNT paper heterostructure.

In conclusion, a three-electrode setup was used to assess the electrochemical base efficiency of papers based on $Ti_3C_2T_x$ and $Ti_3C_2T_x$/CNT heterostructures. The CV curves showed that the $Ti_3C_2T_x$/CNT heterostructures outperformed the original $Ti_3C_2T_x$ paper electrodes, offering better capacitive performance and lower resistance. Furthermore, when MWCNT and SWCNT were included, the sandwich-like heterostructure made of $Ti_3C_2T_x$/CNT papers showed better electrochemical efficiency in comparison to randomly mixed articles.

As shown in Figure 7.3c, the electrode network topologies' rate efficiency had a similar trend. In comparison to pure $Ti_3C_2T_x$ papers, the $Ti_3C_2T_x$/CNT-based heterostructure papers had noticeably better gravimetric capacitance values and outstanding output rates, with the sandwich-like structure demonstrating the highest performance. At a scan rate of 2 mV/s, the $Ti_3C_2T_x$/MWCNT heterostructure exhibited a capacitance of 150 F/g. Even at a higher scan rate of 200 mV/s, it maintained a capacitance value of 117 F/g, which is approximately 130% higher than that of $Ti_3C_2T_x$ paper at the same scan rate. The $Ti_3C_2T_x$/CNT heterostructure papers demonstrated high gravimetric capacitance values and improved performance compared to pure $Ti_3C_2T_x$ sheets.

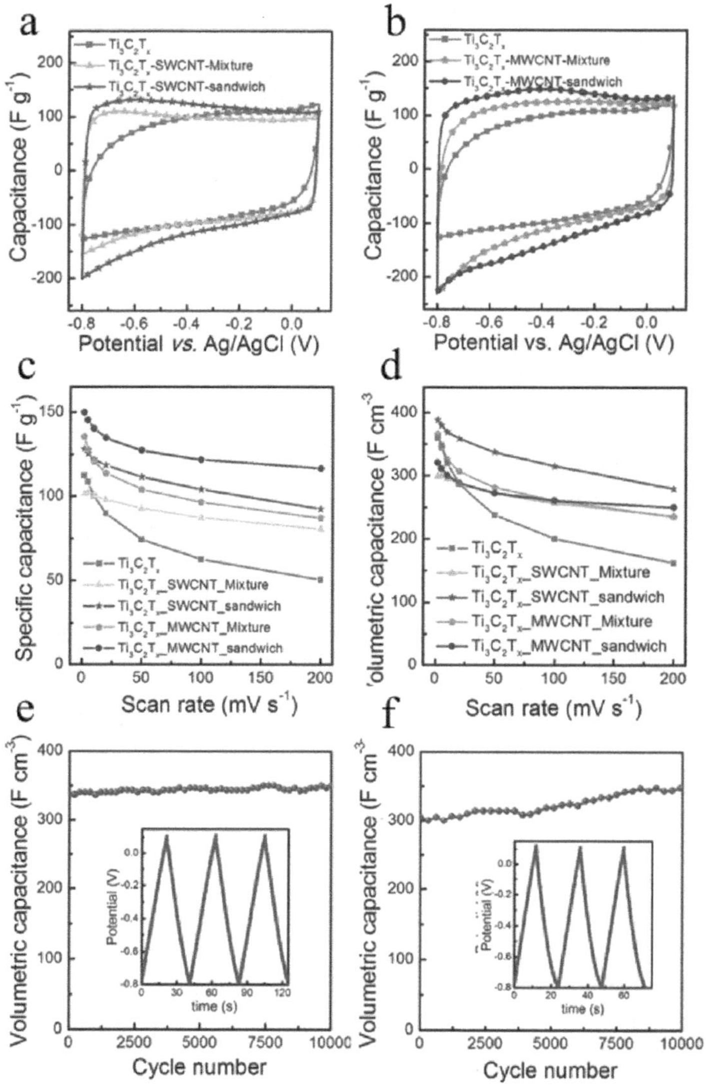

FIGURE 7.3 At a scan rate of 20 mV/s, the CV curves of the $Ti_3C_2T_x$/CNT electrode framework (panels a and b) reveal the electrochemical behavior of the new $Ti_3C_2T_x$ structure and heterostructure. (b) The gravimetric and volumetric capacitance values of pure $Ti_3C_2T_x$ and the $Ti_3C_2T_x$/CNT electrode heterostructure were computed at different scan speeds. (c) The specific capacitance value measured by weight. (d) The volumetric capacitance value of the pristine $Ti_3C_2T_x$ material, along with the capacitance of the $Ti_3C_2T_x$/CNT electrode heterostructure, evaluated at different scan rates. (e, f) At current densities of 5 A g^{-1} (e) and 10 A g^{-1} (f), respectively, the cycle stability of the $Ti_3C_2T_x$/SWCNT-based and $Ti_3C_2T_x$/MWCNT-based sandwich-like heterostructure electrode networks was assessed. We used an aqueous solution of 1 M $MgSO_4$ for all of our tests.

Source: Adopted with consent of Zhao et al. (2015). Copyright© 2014 Wiley-VCH Verlag GmbH & Co. KGaA, Weinheim.

The sandwich-like architecture showed the best performance at a scan rate of 2 mV/s, achieving a capacitance of 150 F/g. Even at a higher scan rate of 200 mV/s, these heterostructure sheets maintained a capacitance value of 117 F/g, highlighting their superior performance over $Ti_3C_2T_x$ paper. The volumetric capacitance of the pristine $Ti_3C_2T_x$ material, as well as the capacitance of the $Ti_3C_2T_x$/CNT composite electrode structure, were measured and compared across a range of scan rates (Figure 7.3d). This analysis provides insights into the electrochemical performance and energy storage capabilities of both the standalone $Ti_3C_2T_x$ and its hybrid heterostructure with CNTs under varying conditions. The results highlight how the integration of CNTs enhances the overall capacitance and stability of the electrode material (Zhao et al. 2015).

The effectiveness of galvanostatic cycling on sandwich-like $Ti_3C_2T_x$/CNT heterostructure sheets is illustrated in Figures 7.3e and 7.3f. The $Ti_3C_2T_x$/SWCNT paper preserved its structural integrity over 10,000 cycles at a specific density of 5 A/g, achieving a volumetric capacitance of approximately 350 F/cm³. Similarly, the sandwich-like $Ti_3C_2T_x$/MWCNT heterostructures maintained a steady volumetric capacitance of around 300 F/cm³, even at a high current density of 10 A/g. Remarkably, this value increased to about 350 F/cm³ after 10,000 cycles, indicating an enhancement in capacitance over time. This improvement is attributed to the cycling process, which optimizes the distribution of slit pore widths within the MXene nanoflakes. Overall, these figures demonstrate the galvanostatic cycling efficiency of the sandwich-like $Ti_3C_2T_x$/CNT heterostructure papers, highlighting their robust performance and capacity retention (Lukatskaya et al. 2013).

Alternately, there is still opportunity for this framework to be improved. The obtained findings show that $Ti_3C_2T_x$/CNT heterostructures in sandwich-like electrodes significantly increase volumetric capacitance and perform very well. The volumetric capacitance values obtained with $Ti_3C_2T_x$/CNT paper in the sandwich-like structure are much greater than those of carbon-based electrode systems. For instance, the usual activated carbon range is 60 to 100 F/cm³ (Ghaffari et al. 2013), carbide-derived carbon networks reach around 180 F/cm³ (Heon et al. 2011), graphene-based conductive electrode networks achieve approximately 260 F/cm³c (Yang et al. 2013), and the Gr/CNT composite system reaches 165 F/cm³ (Jung et al. 2013), Especially at such high rates, for building sandwich-like structured nanostructures, the layer-by-layer (LbL) organizational process is widely used (Byon et al. 2013).

The goal of producing and examining uniform structures composed of $Ti_3C_2T_x$ MXene nanoflakes, combined with 0D onion-like carbon (OLC) and 2D reduced graphene oxide (rGO), was to demonstrate the effectiveness of a streamlined filtration technique. This method successfully fabricates sandwich-like structures with exceptional electrochemical properties. Figure 7.3d illustrates the rate performance of these nanostructures, emphasizing their outstanding effectiveness.

Overall, MXene and CNT sandwich-like arrays were synthesized from water-based suspensions by means of a straightforward alternative filtering procedure. A freestanding, very flexible, and exceptionally fluid MXene/CNT sheet is the ultimate product. After comparing rate performance and volumetric capacitance, supercapacitor devices that used sandwich-like heterostructure papers based on MXene/CNT as electrode materials performed better than those that utilized pristine MXene or randomly dispersed MXene/CNT heterostructure papers.

MXenes, which are 2D carbides that were discovered relatively recently, have demonstrated astounding capacitance values in aqueous electrolyte systems. However, due to their small potential window, both power and energy density are constrained. To make the composite material, delaminated Ti_3C_2 MXene with a calculated surface area of 175 m2/g was combined with 20 weight percent multi-walled carbon nanotubes (MWCNT C100). A glass tube and polypropylene-based membranes were used to filter the manufactured $Ti_3C_2T_x$ MXene, delaminated $Ti_3C_2T_x$ MXene, and CNT/$Ti_3C_2T_x$ composite. Thin-film electrode materials were then created on Teflon membranes. The resulting films, including freestanding $Ti_3C_2T_x$, d-$Ti_3C_2T_x$, and CNT-$Ti_3C_2T_x$, could be easily peeled off from the membrane surface after drying without any folding.

All things considered, the streamlined filtering procedure was effective in creating sandwich-like $Ti_3C_2T_x$/CNT heterostructures with remarkable electrochemical characteristics.

Carefully, thin sheets of electrode material of around 25 mm^2 in size were cut into rectangular shapes using a tiny razor blade. The thicknesses of the films were estimated on the basis of the density values of the pure $Ti_3C_2T_x$, delaminated $Ti_3C_2T_x$ (d-$Ti_3C_2T_x$), and mixed CNT-$Ti_3C_2T_x$ structures. These densities were determined to be approximately 2.3 g/cm^3, 3.0 g/cm^3, and 2.9 g/cm^3, respectively. Figure 7.4a shows the produced MXene clay, also known as $Ti_3C_2T_x$, and Figure 7.4 displays cross-sectional scanning electron micrographs of the electrode materials (Dall'Agnese et al. 2016).

With some shaping and trimming, the MXene nanosheets may be made into a flexible and freestanding electrode network. Delaminated MXene electrodes, known as d-MXene, were developed to enhance the electrochemical performance (Figure 7.4b) (Ghidiu et al. 2014; Dall'Agnese et al. 2014). The goal of this delamination method was to increase the SSA and, by extension, the electrochemical efficiency.

Earlier research using nitrogen gas sorption analysis reported the specific surface area (SSA) of multilayered Ti_3C_2 to be 23 m^2/g, while delaminated Ti_3C_2 had a much higher SSA of 98 m^2/g (Mashtalir et al. 2013). The CNT-$Ti_3C_2T_x$ electrode network is shown in a cross-sectional scanning electron microscope (SEM) in Figure 7.4c and 7.4d, revealing CNTs dispersed among thin layers of @Ti_3C_2. The addition of small amounts of CNTs not only prevents the layers from restacking but also improves the overall electrical conductivity value of the electrode material (Liu et al. 2015; Zhao et al. 2015). At different voltage levels, in-situ XRD patterns were obtained to investigate the energy storage process (Figure 7.4e). According to the study of various patterns, there are two unique sets of Bragg peaks with changing intensities that rely on the applied voltage, and the cyclic voltammetry (CV) plot clearly shows redox peaks that correspond to phase changes (Figure 7.4f).

Nanomaterials based on MXene for metal-ion capacitors and water supercapacitors usually exhibit a continuous intercalation process, and the deintercalation/intercalation processes follow a variable two-phase mechanism (Come et al. 2012; Dall'Agnese et al. 2015). Ghidiu et al. (2014) used the Ti_3C_2 MXene (001) diffraction planes, which correspond to the gaps between the layers, to index each pair of detected Bragg peaks. According to the deflection angle of the (002) Bragg peaks, the potential range of 0.8 V to 0.5 V corresponds to an insertion and removal of intercalated materials phase, when ions from the electrolyte species enter and leave the $Ti_3C_2T_x$ MXene species (Figure 7.4g).

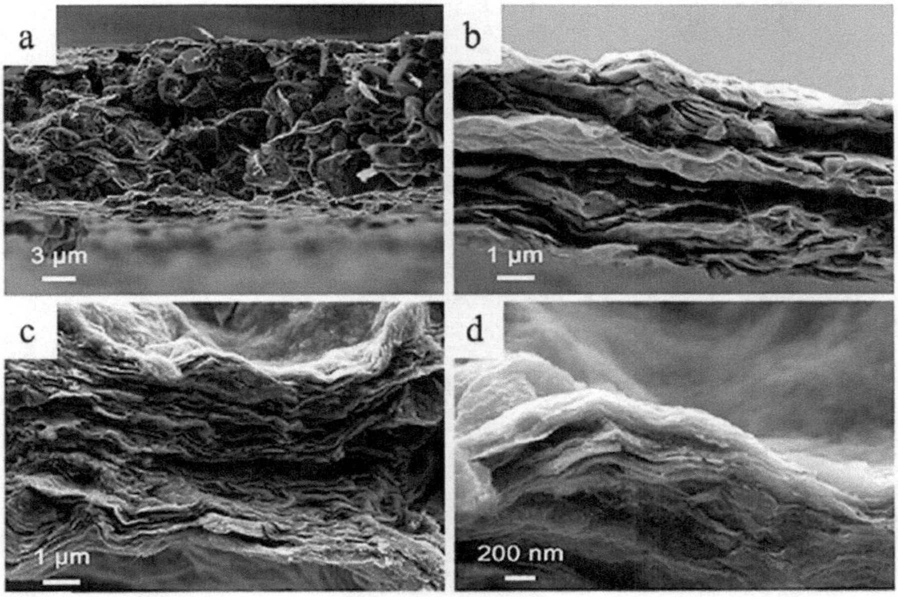

FIGURE 7.4 To document the morphology of several electrode materials, SEM photos were acquired. Panel (a) shows a SEM image of rolled $Ti_3C_2T_x$ "clay," and panel (b) shows a structure made of delaminated $Ti_3C_2T_x$. Panels (c) and (d) also show thin films of electrodes made from CNT-$Ti_3C_2T_x$ heterostructures. (e) The CNT-$Ti_3C_2T_x$ heterostructure was subjected to in-situ XRD patterns at a range of potential values between 5 and 422q. Purple diamond markings indicate the peaks produced by the cell during the in-situ measuring procedure. Furthermore, an inset (f) shows the improvement seen in the 5–72q range. Last but not least, panel (g) shows a measurement made using CV at a scan rate of 20 mV s^{-1}. The electrochemical behavior of the RuO_2/CF heterostructure in a 1 M H2SO4 electrolyte was investigated using a three-electrode setup. These results were achieved: (h) Cyclic voltammetry (CV) curves demonstrating the electrochemical performance of the RuO_2/CF electrode at different scan rates were generated. (i) Galvanostatic charge–discharge curves for the RuO_2/CF electrode at various current densities were observed, highlighting the heterostructure's charge storage capabilities. (j) To assess the impact of scan rate on the capacitive properties of the RuO_2/CF electrode, both single and areal capacitances were compared. This comparison provided insight into how scan rate influences the electrode's performance. (k) In a 1 M H2SO4 electrolyte, 20,000 cycles of long-term cycling stability experiments were performed at an extreme current density value of around 20 A g^{-1}. The inset of the data depicts the resistance fluctuation shown on the Nyquist plot, demonstrating the exceptional stability of the RuO_2/CF electrode.

Source: (a–g) Adopted with consent from Elsevier (Dall'Agnese et al. 2016). (h–k) Adopted with permission, Wiley-VCH Verlag GmbH & Co. KGaA, Weinheim (Jiang et al. 2018).

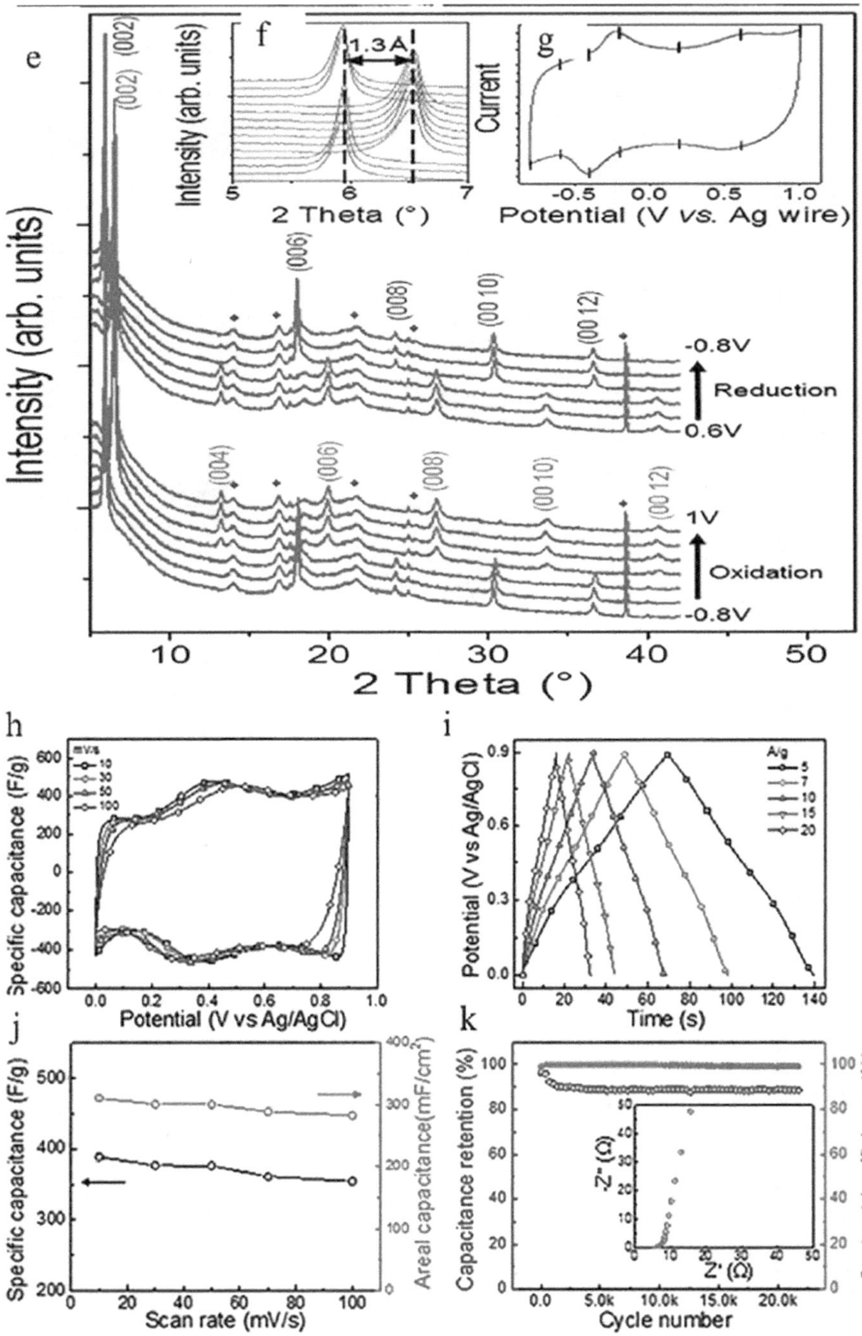

FIGURE 7.4 (Continued)

7.3.1.3 Supercapacitors Based on MXene and Metal Oxide

In recent years, electrochemical energy storage systems have captured global attention due to the increasing demands in electronics, smart textiles, renewable energy sources, and electric vehicles (Simon and Gogotsi 2010; Winter and Brodd 2005). The efficiency and power density of energy storage devices depend significantly on the kinetics of charge storage and the interactions occurring at the electrolyte-electrode interface. For instance, batteries utilize reversible redox reactions involving mobile metal ions within the electrode material, which allows for high energy density. However, this process often leads to lower power density due to slower diffusion processes (Simon, Gogotsi, and Dunn 2014; Gogotsi and Simon 2011). Electrochemical double-layer capacitors (EDLCs), often called supercapacitors, have a number of advantages over traditional batteries when it comes to storing large amounts of power. These include high power densities; effective charge–discharge rates; and consistent, durable cycling performance over time (Gogotsi 2014; Zhang and Zhao 2009).

Electrolyte ions are adsorbed onto the large surface area of carbonaceous materials in EDLCs, which function according to the electrostatic adsorption principle. Unlike other energy storage technologies, EDLCs store charges (electrons and holes) directly on the electrode surface rather of relying on charge transfer channels. The accessible surface area for the ionic species in the electrolyte determines the overall charge storage capacity. Pseudocapacitive substances, such as metal oxides or conducting polymers, provide a greater capacity for charge–discharge via reversible surface redox processes (Augustyn, Simon, and Dunn 2014). These materials have a power density comparable to EDLCs and an energy density comparable to batteries. However, conventional pseudocapacitive nanomaterials' poor cycle stability and low electrical conductivity limit their performance.

One of the promising applications for the newly discovered 2D metal nanomaterial MXene is in the field of energy storage. MXene proves to be an excellent material for highly efficient pseudocapacitive energy storage systems because of its hydrophilic nature, redox surface interactions, and metallic conductivity. Due to concerns about oxidation at high anodic potentials, generally about 0.6 V, MXene symmetric supercapacitors have constrained voltage windows. Recently, MXene, notably titanium carbide ($Ti_3C_2T_x$), has been effectively combined with positive electrodes like ruthenium oxide (RuO_2) in acidic electrolytes to create all-pseudocapacitive asymmetric devices. The operating voltage window for this asymmetric design is around 1.5 V, essentially tripling the power output.

The entire system efficiency is greatly enhanced by carefully adjusting the operating potential windows of both RuO_2 and MXene. Figure 7.4h's experimental findings illustrate the typical pseudocapacitive behavior of RuO_2/CF electrodes, which has discrete peaks within the chosen voltage range. Additionally, the capacitive-type kinetics of the $Ti_3C_2T_x$/CF electrodes show effective surface-based redox processes. At a voltage of 0.3 V, the symmetrical triangular arcs on the galvanostatic charge/discharge curves shown in Figure 7.4i with noticeable redox characteristics demonstrate the well-balanced performance of the RuO_2/CF heterostructure.

The electrodes of the RuO_2/CF heterostructure exhibit minimal Ohmic drop values even at high current densities, such 20 A g^{-1}, demonstrating their superior conductivity and effective energy transfer. This proves that energy storage systems based on asymmetric MXene may provide high-performance results.

The RuO_2 electrode exhibited a specific capacitance of 388 F/g at a scan rate of 10 mV/s. However, when the scan rate was increased to 100 mV/s, the capacitance dropped to 355 F/g, still retaining an impressive 91% of its initial value. Figure 7.4j provides detailed information on these results. Even after 20,000 cycles under real-world conditions with a current density of 20 A/g in a 1 M H_2SO_4 electrolyte, the electrode continues to demonstrate exceptional performance, as depicted in Figure 43d, highlighting its excellent long-term cycling capabilities. The results show that 88% of the original capacitance is still present, demonstrating the electrode's exceptional cycle stability and 100% coulombic efficiency.

Taking into consideration the impact of the electrolyte, current collectors, and electrode materials, the total resistance in the high-frequency regions measures approximately 6 ohms, as indicated in the inset of Figure 7.4k, based on Nyquist plot analysis. These findings strongly suggest that in asymmetric electrochemical capacitors, carbon-based negative electrodes can be replaced with pseudocapacitive MXene-based negative electrodes to achieve higher energy densities.

The RuO_2 electrode exhibits excellent retention rates, long-term cycle stability, and high specific capacitance values, positioning it as a robust contender for energy storage applications. The findings imply that adding electrodes made of MXene to asymmetric capacitors may greatly increase the total energy density.

All-pseudocapacitive supercapacitors are made by fusing hydrous RuO_2 electrodes with $Ti_3C_2T_x$ MXene over a surface made of carbon-based fabric, in summary. This represents the widest voltage range achieved to date in aqueous solutions using electrode networks incorporating MXene. The $RuO_2/Ti_3C_2T_x$ electrode system achieves energy densities of 37 Wh cm^{-2} at a power density of 40 mW cm^{-2} and 45 Wh cm^{-2} at a power density of 6 mW cm^{-2}, operating within a 1.5 V voltage window. ZnO is a further intriguing contender for use with MXene in supercapacitor applications. The semiconducting properties of ZnO, which also include piezoelectric and pyroelectric capabilities, make it useful in optics, optoelectronics, and sensors.

Thanks to the excellent chemical stability and high electrical conductivity of Ti_3C_2 sheets, along with the straightforward fabrication of ZnO, the Ti_3C_2/ZnO composite emerges as a promising candidate for supercapacitors. This composite material holds the potential to significantly enhance supercapacitor performance. The device performance of various 2D MXene-based layered materials is detailed in Table 7.1(a–d).

7.3.2 Heterostructures Made of MXene for Use in Energy Conversion and Storage

7.3.2.1 Li-Ion Batteries

During last decade, there has been a notable rise in competition within the sector of lithium-ion batteries (LIBs), leading to widespread recognition and praise globally. They have established themselves as the go-to choice for employment in compact electric vehicles and consumer electronic devices due to their exceptional power density, energy density, and long cycle life. However, to further elevate the capabilities of Li-ion batteries including other energy storage devices, a plethora of electrode materials have been proffered as potential solutions. Among these materials is a fascinating class known as two-dimensional (2D)-layered nanomaterials, which includes graphene, WS_2, and SnS_2. These materials have an enormous potential for supplementing or replacing the role of Li^+ ions in LIBs (Chang and Chen 2011; Wang et al. 2009). These materials present a host of benefits, notably encompassing expeditious transport of ions and electrons, which prove paramount in attaining elevated power density. Nonetheless, traditional cathode materials for batteries, characterized by their oxide composition, persistently grapple with inherent predicaments, including diminished energy density, protracted charging duration, abbreviated lifespan, and apprehensions regarding safety (Whittingham 2004). 2D materials are appealing in LIBs for a variety of reasons, including their wide spacing between layers, high conductivity for efficient transport, and better capacity for storing lithium due to structural engineering possibilities and large surface area (Pomerantseva and Gogotsi 2017).

7.3.2.1.1 $Ti_3C_2T_x$/CNT-Based Heterostructures

Ti_3C_2 MXene has emerged as an attractive alternative to graphite used in LIB anodes due to its easy availability, metallic properties, and distinct thin-layered structure. Computational investigations have unveiled that Ti_3C_2 MXene showcases a considerably lower diffusion barrier (0.07 eV) for Li+ ions in contrast to graphite interlayers (0.3 eV), endowing it with a favorable advantage in facilitating efficient diffusion of Li+ ions (Tang, Zhou, and Shen 2012; Mashtalir et al. 2013) Nevertheless, it is worth noting that Ti_3C_2 MXene layers are frequently subjected to functionalization with hydroxyl (OH) and/or fluorine (F) groups, a practice that inadvertently contributes to escalated diffusion barriers and compromised conductivity (Naguib et al. 2011; Tang, Zhou, and Shen 2012). In order to tackle this challenge, CNTs have been harnessed as conductive materials as a result of their exceptional electronic transport properties and remarkable aspect ratio. Layered composite structures, such as WS_2/CNTs, MoS_2/CNTs, and graphene/CNTs, have been cleverly engineered to enhance stability and discharge capacity as materials used as anodes in lithium-ion batteries (LIBs) (Wang et al. 2013; Liu et al. 2014; Zhang et al. 2012; Cohn et al. 2014). The electrochemical effectiveness of pure Ti_3C_2 and Ti_3C_2/CNT composites was evaluated using coin cells, and cyclic voltammetry (CV) results showed distinct voltage plateaus at 0.65 V and 1.1 V during the initial cycling (Figure 7.5a).

The initial peak observed in the cyclic voltammetry results is likely due to the formation of a somewhat reversible solid electrolyte interphase (SEI) layer. The subsequent peak is most likely the result of the formation of titanium oxides or the interaction of Li+ ions with Ti_3C_2 (Zeng et al. 2008). The presence of two broad peaks at 2.0 V and 1.8 V throughout the subsequent cycles reveals the reversible oxidation of active Li+ ions as well as their subsequent interaction with the titanium v(IV)-based composite. We can articulate the electrochemical process mechanistically as follows (Naguib et al. 2012)

$$Ti_3C_2 + xLi^+ + xe^- \leftrightarrow Ti_3C_2Li_x$$

Figure 7.5b shows the potential–capacity patterns of composite structures made from Ti_3C_2 and CNTs in equal proportions. These profiles were recorded during multiple cycles (first, fifth, tenth, and twentieth) during a 0.5 C value of the current rate. Within the initial discharge curve, the profiles show the presence of two distinct plateaus at 0.7 V and 1.2 V, respectively. It is worth noting that the initial discharge and charge capacities are 403.5 and 642.5 $mAhg^{-1}$, respectively. The gradual decrease in reversible capacity is caused by irreversible processes, which do not include factors such as electrolyte degradation or the formation of the solid

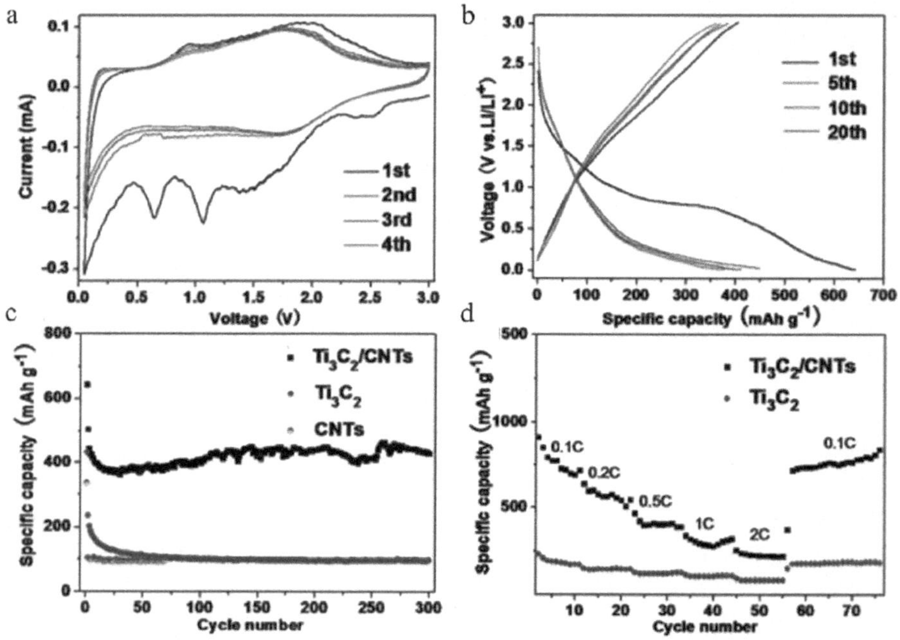

FIGURE 7.5 (a) Using a scan rate of 0.1 mV s^{-1}, the cyclic voltammetry (CV) curves vividly depict the redox processes of the Ti_3C_2/CNT heterostructure. (b) A load/decloak cycle diagram can provide additional information. (c) A comprehensive comparison of cycle performance is also presented, with the original Ti_3C_2 MXene, Ti_3C_2/CNT heterostructures (with a 1:1 mass ratio), and CNTs all tested at 0.5 C. (d) Cycling efficiency of delaminated Ti_3C_2 and heterostructures of Ti_3C_2/CNT at various current rates, including 0.1 C and 2 C, is investigated. It is important to note that the equimolar mass ratio of $Ti_3C_2T_x$ and CNTs remains constant throughout the heterostructure formation process. Permission has been obtained from Royal Society of Chemistry (Liu et al. 2015). The 5 mm thick p-$Ti_3C_2T_x$/CNT films that were created are used for Li-ion storage. (e) The redox process of the p-$Ti_3C_2T_x$/CNT paper electrode is demonstrated using cyclic voltammetry (CV) curves at a scanning speed of 0.2 mV s^{-1}. (f) The cycling stability of various electrode configurations, including the original $Ti_3C_2T_x$, p-$Ti_3C_2T_x$, $Ti_3C_2T_x$/CNT heterostructure, and p-$Ti_3C_2T_x$/CNT film electrodes, is tested before and after pre-cycling at 0.5 C. The coulombic efficiency is depicted on the right-hand y-axis for the p-$Ti_3C_2T_x$/CNT films. (g) The cycling rate influences the discharge capacity both before and after cycling. (h) The voltage curves of pre-cycled films at various cycling rates provide further insights into potential irreversible interactions on the surface of p-$Ti_3C_2T_x$ nanoflakes, particularly involving Li ions and surface functional groups such as –F and –OH.

Source: Adopted with Permission Wiley-VCH Verlag GmbH & Co. KGaA Weinheim (Ren et al. 2016).

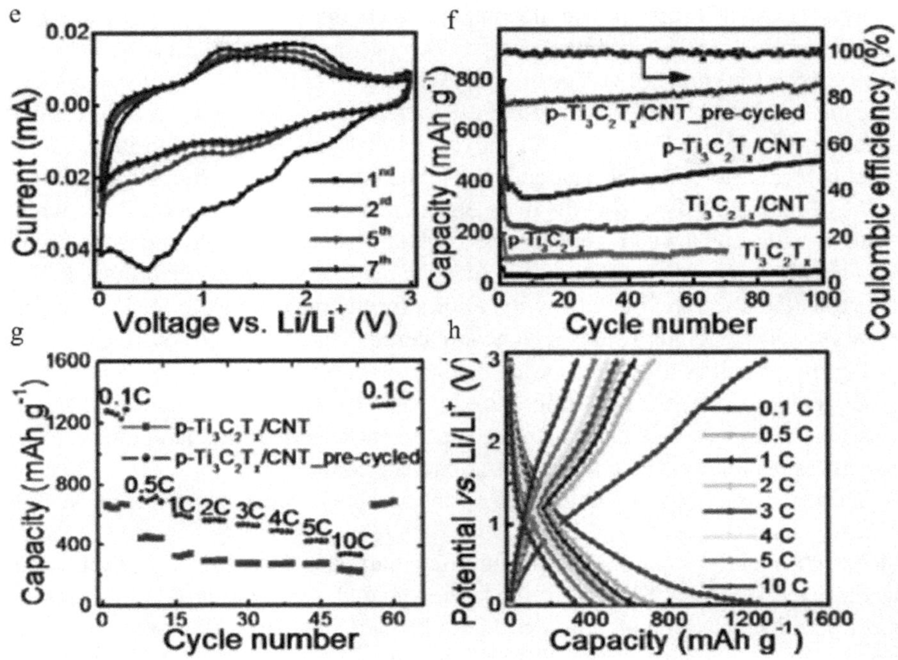

FIGURE 7.5 (Continued)

electrolyte interphase (SEI) layer (Allcorn and Manthiram 2014; Liu et al. 2013). The convergence of the discharge and charge curves indicates the electrode's commendable stability throughout the aforementioned cycles. The Ti_3C_2/CNT heterostructure with a weight ratio of 1:1 exhibits the best power performance, as evidenced by its sustained efficacy at a rate of 1 C for 200 cycles.

It is important to note that higher proportions of CNTs or Ti_3C_2 in the composites result in decreased capacity or compromised electrical conductivity, resulting in a trade-off effect on the electrochemical properties.

Corroborating evidence from additional published sources substantiates that the optimal mass ratio between Ti_3C_2 MXene nanolayers and CNTs is established at 1:1 (Liu et al. 2014). In Figure 7.5c, the video demonstrates how a Ti_3C_2/CNTs (1:1) composite paper and a pure Ti_3C_2 film perform during cycling at a rate of 0.5 C. Surprisingly, even after 300 cycles, the discharge capacities of the original Ti_3C_2 and the Ti_3C_2/CNTs composite remain above 96.2 and 428.1 mA h g^{-1}, respectively. Thanks to the combined effect of Ti_3C_2 and CNTs, the discharge rate of the Ti_3C_2/CNTs composite saw a marked improvement. In particular, the electrical conductivity of the composite was significantly improved by the CNTs.

In comparison, the discharge capacity of the CNT material alone at 0.5 C is only 94.2 mAhg^{-1}, highlighting its significant inferiority when compared to the combined Ti_3C_2/CNT composite. Even after 70 cycles, the composite paper maintains a discharge capacity of 428.1 mA h gl, highlighting its long-lasting performance.

The Ti_3C_2/CNT heterostructure outperformed the original Ti_3C_2 MXene film in maintaining cyclic capacity across various current rates (0.1–2 C), as shown in Figure 7.5d (Liu et al. 2015). The discharge capacities of the Ti_3C_2/CNT heterostructure are 713.2, 543.1, 407.7, 320.6, and 218.2 mA h g^{-1}, respectively, at different current rates (0.1 C, 0.2 C, 0.5 C, 1 C, and 2 C). Interestingly, the composite electrode quickly regains its reversible characteristics when the current density is reduced back to 0.1 C, indicating a high degree of reversibility. The incorporation of CNTs, which effectively occupy the gaps between the Ti_3C_2 layers, is primarily responsible for the enhanced cycling stability, increased capacity, and notable rate performance. This improves the electrode's conductivity, allows for the rapid transport of Li ions, and allows for efficient interactions with nanomaterials, resulting in the desired capacity and cycling stability.

Gogotsi et al. (Ren et al. 2016b) introduced a novel type of mesoporous MXene, termed p-$Ti_3C_2T_x$, which exhibits more precise surface areas and larger, more open structures in comparison to their pristine MXene counterparts.

These p-$Ti_3C_2T_x$ electrodes showed a considerable improvement in Li-ion storage capacity compared to $Ti_3C_2T_x$/CNT films with reduced porosity. Figure 7.5e shows the results of electrochemical testing on Li-ion half-cells using p-$Ti_3C_2T_x$ films as the working electrode and Li metal as the functional and counter electrodes. The results displayed distinct lithiation and delithiation peaks, indicating improved capacitive performance. At a current rate of 0.5 C, the reversible baseline lithium capacity of p-$Ti_3C_2T_x$ was measured at 110 mAh/g, a significant increase from the 35 mAh/g observed in the original $Ti_3C_2T_x$ films. Furthermore, adding 10% wt. CNTs resulted in a first-cycle capacity of 790 $mAhg^{-1}$, which gradually increased to 500 $mAhg^{-1}$ after 100 cycles (as illustrated in Figure 7.5f).

The capacity of approximately 500 $mAhg^{-1}$ is significantly higher than the 110 $mAhg^{-1}$ capacity of p-$Ti_3C_2T_x$ without CNTs, as depicted in Figure 7.5f, following 100 cycles at a current rate of 0.5 C. Furthermore, it outperforms the $Ti_3C_2T_x$/CNT film's capacity of about 220 $mAhg^{-1}$, which lacks the porosity found in MXene structures (as shown in Figure 7.5f). This means that combining the porous structures of p-$Ti_3C_2T_x$ with CNTs significantly increases the Li capacities of the MXene material. However, because of their inadequate substance and inaccurate precision ability, CNTs have a negligible impact on the overall capacity of the composite paper (De las Casas and Li 2012). The electrodes made from the combination of p-$Ti_3C_2T_x$ and CNTs have a remarkable and consistent specific capacity even at high value of current rates, as illustrated in Figure 7.5g and h. These electrodes achieve an impressive capacity of approximately 650 milliampere-hours per gram when cycling at 0.1 C ($mAhg^{-1}$). Over time, the capacity declines due to cycling rate increases, but in a gradual and progressive manner. For instance, at 60 C, the capacity is about 110 $mAhg^{-1}$, while at 10 C, it increases to around 230 $mAhg^{-1}$.

At a cycling rate of 0.1 C, the combined electrodes of p-$Ti_3C_2T_x$ and CNTs exhibit an extraordinary capacity of approximately 1250 $mAhg^{-1}$ during the pre-cycling process. When the cycling rate is increased to 10 C, the capacity drops to approximately 330 $mAhg^{-1}$. These capacity values highlight the superior Li-ion storage capabilities of p-$Ti_3C_2T_x$ compared to lithium titanate materials under similar cycling conditions (Landi et al. 2009; Naguib et al. 2014; Mashtalir et al. 2013; Naguib et al. 2013; Kim et al. 2015; Anasori et al. 2015). The capacity values reported for the p-$Ti_3C_2T_x$/CNT heterostructure films exceed any previously documented in the literature across all current rates. This remarkable capacity enhancement is primarily due to the improved Li-ion adsorption and storage capabilities facilitated by the porous p-$Ti_3C_2T_x$ flakes.

The charge/discharge profiles presented clearly show that the cycling behavior is similar to that of supercapacitor-based devices rather than conventional Li-ion energy storage devices. This property must be considered when designing circuits that use the specified electrodes, such as Li-ion capacitors or hybrid devices (Kim et al. 2015).

Even under high current densities and when cycled at a rate of 0.1 C, the p-$Ti_3C_2T_x$/CNT composite electrode retains an impressive capacity of 1250 mAh g^{-1}. Furthermore, newly synthesized MXenes such as Nb_2CTx and V_2CTx can undergo similar chemical etching processes as explored in this study, paving the way for a multitude of novel MXene-based structures. Overall, the findings offer a simple and scalable method for improving the physical and chemical properties of two-dimensional materials such as MXenes.

7.3.2.1.2 MXene and Metal Oxides Based Heterostructures

The various layers of MXenes, contrary to their conductive nature, serve a crucial role as stabilizers, facilitating the smooth flow of ions and electrons. Moreover, they act as an essential starting material, preventing the aggregation of nanocrystals based on transition metal oxides (TMOs). These TMO-based nanocrystals, in turn, serve as linkers, effectively isolating the MXene layers to avoid stacking during assembly and throughout the extraction/insertion process of lithium ions. This novel method harnesses the power of interfacial interactions while preserving the sites of activity, resulting in an exceptional synergistic effect between the two substrates. As a result, a heterostructure involving TMO and MXene performs significantly better, establishing themselves as promising and dependable candidates for anodic nanostructured materials for use into upcoming thermal energy storage devices (Liu et al. 2018).

The intentional combination of MXene nanolayers and transition metal oxide (TMO) nanostructures in a specific configuration is critical because it allows for the synergistic integration of the unique properties offered by both contributors. By effectively combining thick MXene films with TMO materials such as Fe_3O_4 or MnO_2, several successful outcomes have been obtained (Zhang et al. 2016; Tang et al. 2016).

The deposition of transition metal oxide (TMO) nanoparticles is limited to the outer surfaces of thick MXene composites within a hybrid structural model, resulting in a significant portion of MXene's active areas remaining unutilized. Zhao et al. conducted another study (Zhao et al. 2016). Researchers employed a hydrothermal method to effectively combine $NiCo_2O_4$ nanoplatelet-like structures with MXene nanosheets. In this case, however, the precursor cations of the MXene nanolayers must be stored within the defects of MXene via electrostatic interactions. This causes the oxide nanoplatelets to nucleate, crystallize, and grow in an asymmetric morphology. Hence, controlling the morphology (size, shape) and arrangement of oxide nanoparticles proves to be a challenging task. Moreover, MXenes undergo oxidation during hydrothermal reactions in solutions containing dissolved oxygen (Wu et al. 2017).

7.3.2.1.3 MXene/TiO$_2$ Nanoparticles

In their 2018 study, Xu et al. (Liu et al. 2018) employed a van der Waals-based bonding technique to create self-assembled transition metal oxide (TMO) nanostructures atop MXene (Ti_3C_2) nanolayers. However, there is currently a shortage of data in the literature regarding the controlled fabrication of two-dimensional heterostructures consisting of single MXene nanolayers decorated with various types of TMO nanomaterials, such as TiO_2 nanorods. Figure 7.6a displays the results from cyclic voltammetry (CV) tests performed on TiO_2/MXene heterostructures to investigate their lithium-ion storage capabilities. The experiments were carried out within the voltage range of 0.01–3 V versus Li/Li+.

In the initial scanning cycle, there was a sharp decrease in peak intensity at 1.71 V, accompanied by the appearance of an oxidation peak at 2.05 V. The inclusion or removal of lithium ions from the TiO_2 anatase crystal structure is responsible for this reduction. It is important to note that subsequent scan curves had lower peak intensities, indicating the occurrence of irreversible reactions. The variations in peak intensities, on the other hand, suggest conversion reactions, pointing to a two-phase mechanical mechanism during the lithium-ion insertion and extraction process, as described by the following equation.

$$TiO_2 + xLi + +xe- \leftrightarrow LixTiO_2$$

Additionally, a minor anodic peak was detected around 1.9 V, accompanied by two broad cathodic peaks at approximately 1.5 V and 0.75 V, along with two narrow peaks near 0.75 V in the initial scan region. The tiny anodic peak indicates the removal of MXene nanolayers due to lithium-based species interacting with the surface. Figure 7.6b illustrates the cyclic testing of the TiO_2/MXene heterostructure, evaluating its long-term stability, particularly under rigorous cycling conditions. After 200 cycles, the TiO_2/MXene heterostructure exhibited impressive performance, maintaining capacities of 209 mA h g^{-1} at 500 mA g^{-1} and 176 mA h g^{-1} at 1000 mA g^{-1}. Still, even when the current density is large, i.e., 2,000 mA g^{-1} during fast cycling, it retained a notable capacity of 138 mA h g^{-1}. Figure 7.6c offers a comparative analysis of the cycling performance among multilayer MXene, nanorods, and TiO_2/(Ti_3C_2) MXene heterostructures.

While the two control anodes exhibited capacities of 48 mA h g^{-1} after 200 cycles, the TiO_2/MXene heterostructure demonstrated a much higher capacity of 167 mA h g^{-1} at a current density of 500 mA g^{-1}.

Figure 7.6d provides a comparison and graphical representation of their rate capabilities. The TiO_2/MXene heterostructures displayed enhanced capacities of 272, 250, 233, 204, 175, and 140 mA h g^{-1} after cycling at different current densities of 50, 100, 200, 500, 1,000, and 2,000 mA g^{-1}. The results obtained from these combined components (MXene nanolayers plus TiO_2 nanorods) significantly surpass those

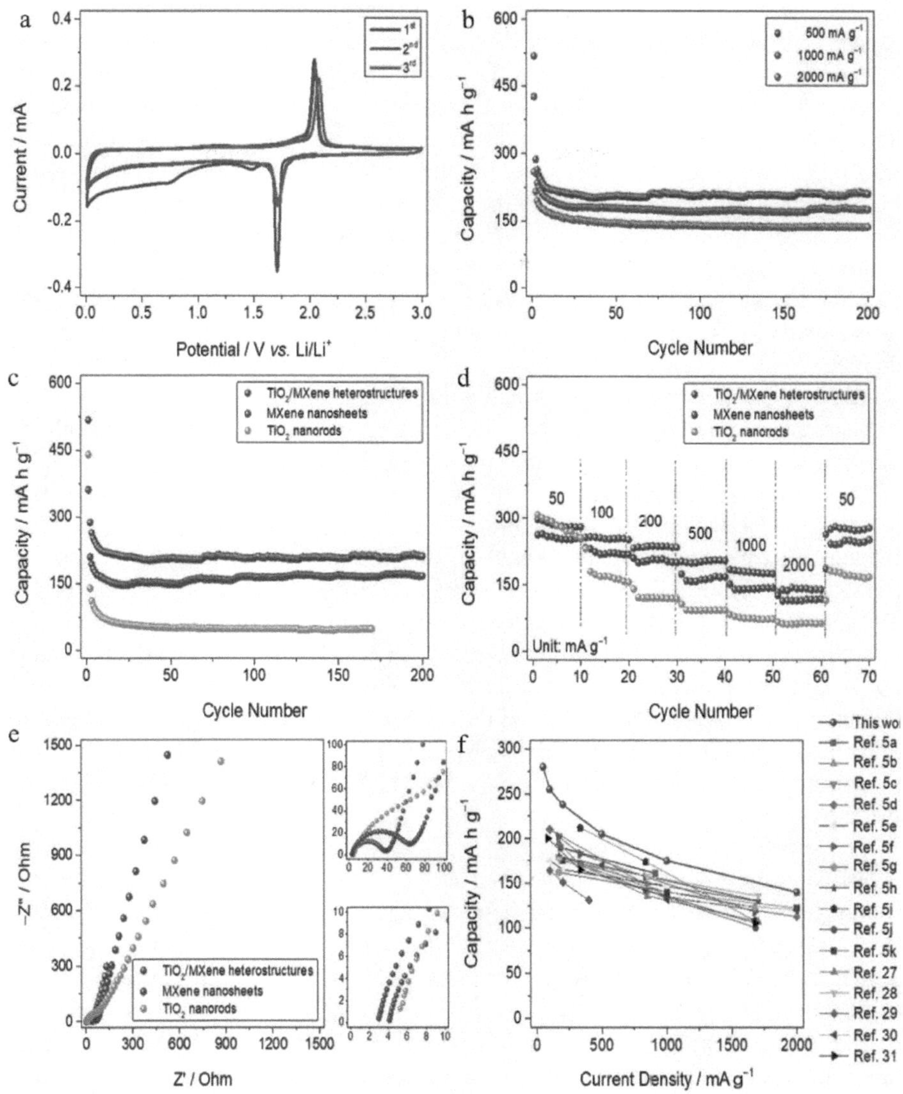

FIGURE 7.6 (a) The scan rate of 0.1 millivolts per second (mV s^{-1}) was employed to record the cyclic voltammetry (CV) curves. (b) The heterostructure composed of TiO_2/MXene demonstrates a collective cycling performance. (c) The cycling performance was assessed at a current density of approximately 500 milliamperes per gram (mA g^{-1}). (d) The TiO_2/MXene heterostructures' rate capabilities were investigated. (e) TiO_2 nanorods, MXene nanocaps, and TiO_2/MXene heterostructures all had Nyquist curves. (f) The TiO_2/MXene heterostructures were compared to other TiO_2-based anode structures, such as TiO_2/graphene, walnut-like TiO_2, and mesoporous TiO_2, 3D network of TiO_2–δ nanomembranes, and anatase TiO_2 mesocrystals.

Source: Permission has been obtained for the inclusion of this information from Liu et al. (2018b). Copyright© 2018 Wiley-VCH Verlag GmbH & Co. KGaA, Weinheim.

achieved with their individual components. A slight increase of 50 mA g^{-1} yields a capacity of 277 mA h g^{-1}. The remarkable electrochemical stability observed in TiO_2/MXene heterostructures arises from the synergistic interplay between their components and the robustness of their structural integrity (Liu et al. 2018a). There are noticeable characteristics in the Nyquist curves that are produced by Electrochemical Impedance Spectroscopy (EIS) as shown in Figure 7.6e. A single arc is observed at high frequencies, a representation of the charge-transfer resistance (Rct) mechanism. An inclined line is observed at low frequencies, indicating lithium solid-state diffusion. The TiO_2/MXene heterostructures have an Rct value of 83, which is greater than the Rct value of the MXene nanolayers (51). As a result of charge transfer, the resistance is higher, as the semicircle's diameter is the same as the TiO_2 nanorods' diameter.

When TiO_2 nanorods are combined with MXene nanosheets, they form a TiO_2/MXene heterostructure with an intermediate electronic resistance (Rs), highlighting the enhancement of electrical conductivity by the MXene nanolayers.

Depositing TiO_2 nanomaterials on the surface of the MXene nanolayer establishes a conductive network, enhancing the efficient transfer of ions and electrons. Because of this, the Rct value—now at 30—is significantly affected (Liu et al. 2018a). Furthermore, the inclusion of various TiO_2-based anode materials, such as TiO_2/graphene, influences the electrochemical efficiency of TiO_2/MXene heterostructures (Mo et al. 2014; Xin et al. 2012; Fei et al. 2015; Qiu, Xing, and Zhang 2014) and other recently reported TiO_2 nanomaterials with different structures (Huang et al. 2017; Cai et al. 2017; Hu et al. 2016). Figure 7.6f highlights the durability and high efficiency of the TiO_2/MXene heterostructures, highlighting their potential as excellent anode structures and materials for highly efficient lithium-ion storage.

The electrochemical processes of SnO_2/MXene-based heterostructures were investigated in order to determine their efficiency. The remarkable high-rate efficiency observed in SnO_2/MXene heterostructures validates the efficacy of the structural approach that incorporates MXene nanolayers as a promising two-dimensional (2D) material with diverse and potentially useful applications (Liu et al. 2018b).

Due to their exceptional high-rate performance, the TiO_2/MXene heterostructures surprised researchers by not requiring surface encapsulation or optimization of the binder/electrolyte.

The potential applications of these heterostructures in LIBs of the future are rather exciting, with high efficiency and exceptional energy storage capabilities.

7.3.2.1.4 Fe_2O_3/MXene Nanoparticles

LIBs have gained significant attention in recent years (Tarascon and Armand 2011; Kang and Ceder 2009; Dunn, Kamath, and Tarascon 2011; Zhao et al. 2012), particularly for applications such as rechargeable batteries and electric vehicles, due to their long lifespan and high energy density (Armand and Tarascon 2008; Wagner, Lakshmanan, and Mathias 2010). As a result, much work has been put into finding and creating new electrode materials for LIBs (Poizot et al. 2000). In the domain of anode materials, the development of MXene/transition metal oxide (TMO) heterostructures is of paramount importance. These heterostructures aim to enhance electrical conductivity and promote enhanced diffusion properties. $Ti_3C_2T_x$ Mxene's distinctive 2D sheet-like structure makes it an excellent substrate for supporting metal oxide (MO) materials or can be integrated into a sandwich-like configuration. It is now possible to overcome the difficulties linked with volume changes observed in the MO species during the lithium insertion and extraction method by combining the MO material with $Ti_3C_2T_x$ MXene. This integration eventually leads to improved rate capabilities and cyclability, advancing anode material performance (Anasori, Lukatskaya, and Gogotsi 2017).

Wang et al. (2014) introduced ball milling method to prepare a composite structure based on Fe_2O_3 and graphite as an anode material for LIBs (Ali et al. 2018). The combination of MXene's high conductivity and stable cyclability with Fe_2O_3's significant lithium storage capacity bodes well for the development of anode materials in LIBs.

Regarding this matter, a heterostructure material composed of $Ti_3C_2T_x$-25 wt. percent 2D Fe_2O_3/MXene nanoparticles shows promise. Notably, using a Fe_2O_3 content of Ti_3C_2Tx-50 wt. percent in the ball milling (BM) process results in a well-dispersed distribution of Fe_2O_3 nanoparticles on the surface of $Ti_3C_2T_x$ nanolayers while minimizing MXene oxidation.

Furthermore, using a minimum restacking method to combine $Ti_3C_2T_x$ and Fe_2O_3 yields a higher degree of surface area accuracy. This method improves the $Ti_3C_2T_x/Fe_2O_3$ heterostructure's overall performance and stability, making it a viable candidate for advanced LIB anode materials (Ali et al. 2018). Figure 7.7a displays the CV profiles for cyclic voltammetry of the $Ti_3C_2T_x/Fe_2O_3$ heterostructures, obtained at a scan rate of 0.2 millivolts per second (mV/s). The curves exhibit distinctive redox peaks observed at voltage levels of 0.85 and 1.9 volts versus Li+/Li. These peaks correspond to the reduction of Fe(III) to Fe(0) and the subsequent insertion of lithium into the Fe_2O_3 structure. The charge and discharge curves captured within the voltage plateau (Figure 7.7b) provide additional evidence supporting the presence of the observed redox peaks in the CV results.

The procedures for charging and discharging of the prepared heterostructures are administered by the distinctive voltage profile associated with the Fe_2O_3 conversion reaction. When the charging and discharging processes are complete, this voltage profile shows sloping voltage changes and distinct voltage plateaus (Cabana et al. 2010). Furthermore, even after 50 cycles, the properties shown in Figure 7.7c and d for BM25 and BM50 remain unchanged. Although the voltage profiles differ from those of $Ti_3C_2T_x$ MXene alone, this indicates that the both materials make a contribution to the overall capacity value.

These findings strongly propose that the $Ti_3C_2T_x/Fe_2O_3$-based heterostructure prepared using the BM method provides mutual benefits. This heterostructure demonstrates outstanding cycling stability, rate capability, and Li-ion storage capacity. As a result, these findings provide a straightforward and efficient method for developing composite structures integrating transition metal oxides and MXene as anodes in lithium-ion batteries (Ali et al. 2018).

The $Ti_3C_2T_x/Fe_2O_3$ combined structure found a way to reverse the effects of around 270 milliampere-hours per gram (mA h g^{-1}) at a current rate of 1 C and an output capacity of around 100 mA g^{-1} at a current rate of 10 C. These promising outcomes showcase the significant potential for advancing high-performance and stable LIBS through the utilization of heterostructures comprising MXene and transition metal oxide materials. Table 7.1(c) in the catalysis section provides detailed performance metrics for MXene-based 2D heterostructures in laser-induced breakdown spectroscopy (LIBS) applications.

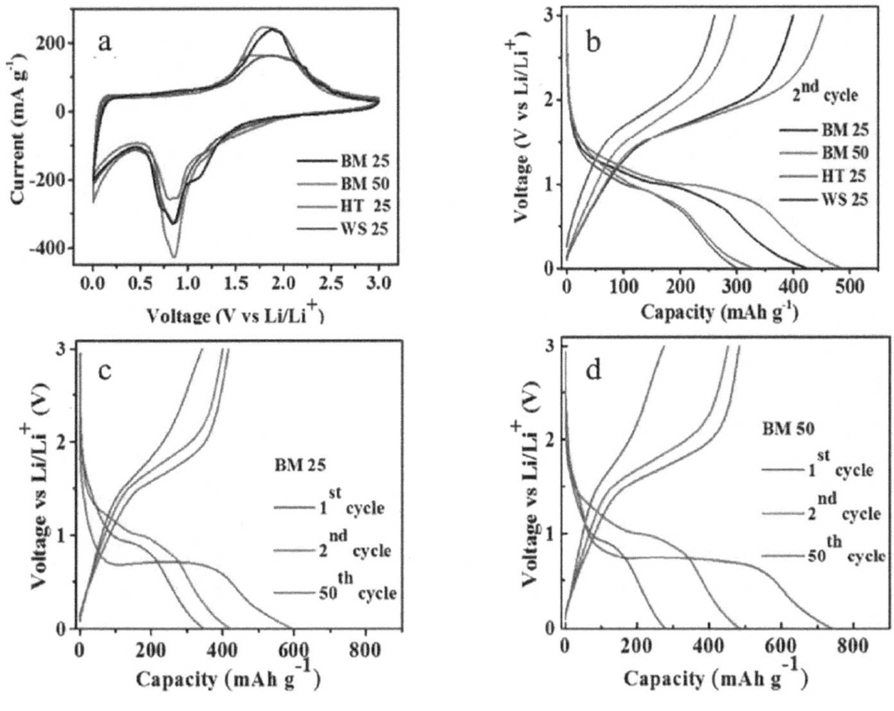

FIGURE 7.7 (a) CV curve was acquired by scanning the $Ti_3C_2T_x/Fe_2O_3$ combined structure at a rate of 0.2 millivolts per second (mVs^{-1}). (b) The voltage behavior of the combined structure during the second cycle of charging and discharging is depicted. (c) The voltage patterns of the BM25 sample are illustrated for the initial, second, and fiftieth cycles. (d) Voltage distributions along the BM50 sample are demonstrated for the first, second, and fiftieth cycles. Panels e–i depict SEM images illustrating the progressive change in color, indicating the transformation of $Ti_3C_2T_x$ nanosheets. In (e), the $Ti_3C_2T_x$ nanosheets exhibit a distinct color, and as the oxidation time increases, the formation of small nanoparticles begins, resulting in a color change (f). With further oxidation, the nanoparticles continue to grow, leading to a more pronounced color change (g, h). After approximately 8 hours of oxidation, the $Ti_3C_2T_x$ nanosheets completely transform into irregularly shaped nanoparticles, which are reflected in a different color (i). In panel j, the XRD patterns provide information about the crystal structure. The XRD pattern in panel j displays distinct diffraction peaks, indicating the transformation of $Ti_3C_2T_x$. These findings have been made public with permission (see Jiao et al. 2019).

Sources: (a–d) Adopted with permission from American Chemical Society (Ali et al. 2018). (e–j) Adopted with permission from Weinheim: Wiley-VCH Verlag GmbH & Co.

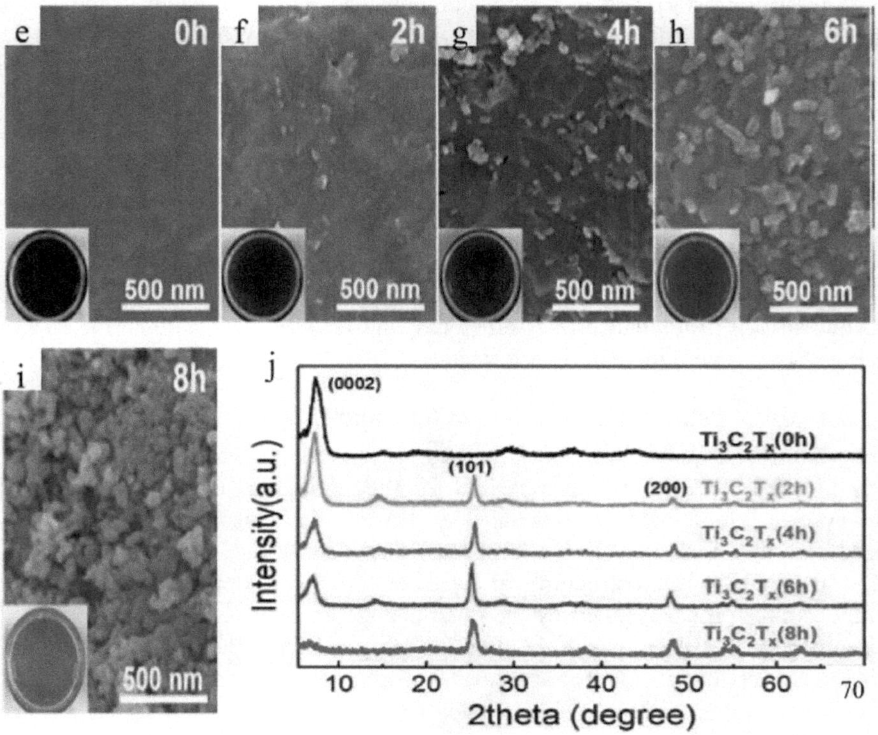

FIGURE 7.7 (Continued)

7.3.3 Li-S Batteries

Recently, a new generation of energy storage devices based on lithium-sulfur (Li-S) has emerged as a highly attractive alternative. This is primarily due to sulfur's low-cost, high theoretical energy density, and minimal environmental impact (Bruce et al. 2012; Manthiram et al. 2014; Liang et al. 2016). Nonetheless, significant obstacles have hampered the widespread adoption of Li-S batteries. These impediments include sulfur's insulative properties, the formation of lithium sulfides (Li2S) species, and the well-known shuttle effect caused by lithium polysulfides (LiPSs) during battery cycling. The shuttle effect contributes to reactive nanomaterial degradation and anode corrosion, resulting in rapid capacity deterioration and decreased faradaic efficiency or current efficiency (Huang et al. 2013; Peng et al. 2017).

Carbon materials with superior electrical conductivity, appropriate porous structure, and precisely controlled particle size have been extensively researched in the search for sulfur host materials or interlayers. These carbon materials are designed to effectively capture and confine lithium polysulfides (LiPSs). (Ji, Lee, and Nazar 2009; Wang et al. 2011) within the cathode region (Su and Manthiram 2012; Zhou et al. 2015). Nevertheless, the inadequate binding strength between polar LiPSs and nonpolar carbon materials presents a hindrance to achieving long-term cycling performance. To overcome this limitation, a viable strategy involves the chemical entrapment of LiPSs through the utilization of TiO_2 and other transition metal oxides, MnO_2, and other analogous materials.

These transition metal oxides engage in polar-polar interactions to effectively restrict the mobility of LiPSs (Seh et al. 2013; Chen et al. 2018; Liang et al. 2015).

MXene has garnered significant attention in the field of energy storage due to its exceptional conductivity, two-dimensional structure, and unique acidic surface properties (Ghidiu et al. 2014; Naguib et al. 2013; Wang et al. 2015). However, the stacking behavior of MXene nanolayers and the abundance of F-terminal functionalities on the surface present obstacles in efficiently capturing and catalytically converting lithium polysulfides (LiPSs) (Li et al. 2017). These factors limit the overall performance of MXenes, particularly their reactive functional surface. Hence, it is crucial to explore approaches that can improve the operational characteristics of MXene surfaces to enhance their efficiency and activity in lithium-sulfur energy storage systems with increased performance.

7.3.3.1 Heterostructures Based on MXene and TiO_2

The solubility of polysulfide intermediates, their mobility, and their slow conversion kinetics are primarily responsible for the limited transport characteristics observed in lithium-sulfur (Li-S) batteries. Efforts are being made by Yang et al. to address this challenge (Jiao et al. 2019). The in-situ functionalization of TiO_2-MXene ($Ti_3C_2T_x$) heterostructures allowed for the creation of a versatile catalyst with desirable properties. During the functionalization process, $Ti_3C_2T_x$ underwent hydrothermal oxidation, where the duration of oxidation could be adjusted to control the conversion of $Ti_3C_2T_x$ to TiO_2. The resulting $Ti_3C_2T_x$ samples (denoted as x h) exhibited a range of colors from cloudy-white to black, as shown in the inset of Figure 7.7e–i. In Figure 7.7e, the $Ti_3C_2T_x$ nanosheets exhibited a well-defined surface, and after a two-hour oxidation period, small nanoparticles began to form on the surface (Figure 7.7f). With increasing oxidation time, the growth of nanoparticles became more pronounced, leading to a significant increase in their size (Figure 7.7g, h). After approximately 8 hours of oxidation, the $Ti_3C_2T_x$ nanosheets completely disintegrated, resulting in the formation of irregularly shaped nanoparticles (Figure 7.7i). These compositional changes in $Ti_3C_2T_x$ were also evident in the XRD patterns (Figure 7.7j), where the $Ti_3C_2T_x$ phase showed a prominent (0002) diffraction peak, indicating the fractional stacking of Ti4C2Tx nanosheets.

The XRD patterns reveal the emergence of diffraction peaks at 25.3° and 47°, which correspond to the crystal planes 101 and (200) of TiO_2 nanoparticles (NPs). These peaks indicate the transformation of $Ti_3C_2T_x$ MXene into TiO_2 NPs, leading to a shift toward lower angles for the (0002) peak of $Ti_3C_2T_x$. This shift indicates a restacking phenomenon of $Ti_3C_2T_x$ nanosheets, which is regulated by the presence of TiO_2 NPs on its surface (Ren et al. 2016a).

To assess the adsorption capacity of the TiO_2-$Ti_3C_2T_x$ heterostructure toward lithium polysulfides (LiPSs), an adsorption experiment was conducted. The vials containing the Li2S6 solution were supplemented with 5 mg of equal weight after diluting it with 5 mL of 1,2-dimethoxyethane (DME) at a concentration of 0.1 M.

Figure 7.8a shows that as the oxidation time interval of $Ti_3C_2T_x$ extended from 0 to 8 hours, the rate of growth in LiPSs' adsorption capacity was constant. The sample oxidized for 4 hours showed a significant color change in the LiPS solution, while longer oxidation times showed minimal fluctuations. To further investigate the

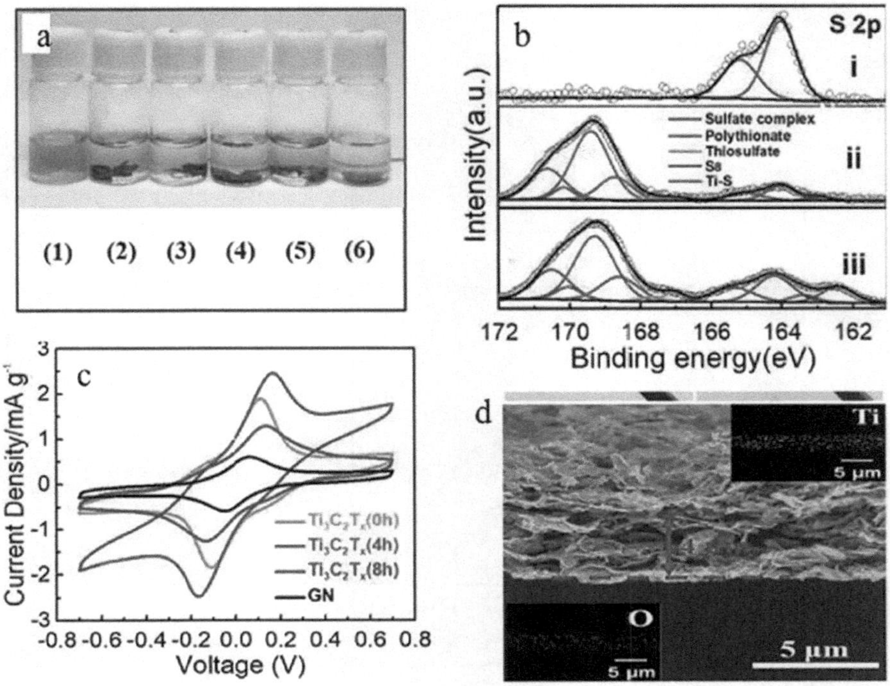

FIGURE 7.8 (a) The photograph illustrates the adsorption of Li2S6 in various DME solutions samples after 24 hours. The samples include Null Li2S6, Blank $Ti_3C_2T_x$ (0 hour), $Ti_3C_2T_x$ (2 hours), Blank $Ti_3C_2T_x$ (4 hours), Blank $Ti_3C_2T_x$ (6 hours), and Blank $Ti_3C_2T_x$ (8 hours). (b) The S2p XPS spectra of elemental sulfur for $Ti_3C_2T_x$ (0 h)-GN and $Ti_3C_2T_x$ (4 h)-GN are displayed after undergoing 200 cycles at 2.35 V. (c) Cyclic voltammetry (CV) curves for the Li2S6-Li2S6 symmetrical cell configuration, utilizing various heterostructures, are illustrated at a scan rate of 10 mV/s. (d) The cross-sectional SEM photographs of $Ti_3C_2T_x$ (4 h)-GN, along with the interlaced SEM images of $Ti_3C_2T_x$ (4 h), are displayed. The insets in the upper right and bottom left corners provide elemental mapping for titanium (Ti) and oxygen (O), respectively (see Jiao et al. 2019). (e) As electrode materials, 3D macroporous structures based on MXene films were used. The cyclic voltammetry (CV) curves were acquired at a scan rate of 0.1 mV/s and revealed the electrode materials' Na^+ storing capacity. (f) The scan rate dependency of the CV curves can provide information about the effectiveness of the 3D macroporous MXene films as Na+ storage electrodes. (g) The charging and discharging patterns of the $Ti_3C_2T_x$-based film electrode with a 3D structure were observed at various current rates. (h) When comparing the Na+ storage efficiency of the 3D microporous MXene thin-film electrodes, other electrode forms were taken into account. (i) The long-term cycling performance of 3D macroporous MXene-based film electrodes was evaluated at a high current rate of 2.5 C, confirming their stability and endurance over numerous charge–discharge cycles (see Zhao et al. 2017).

Source: (d and i) Adopted with permission from Wiley-VCH Verlag GmbH & Co. KGaA, Weinheim.

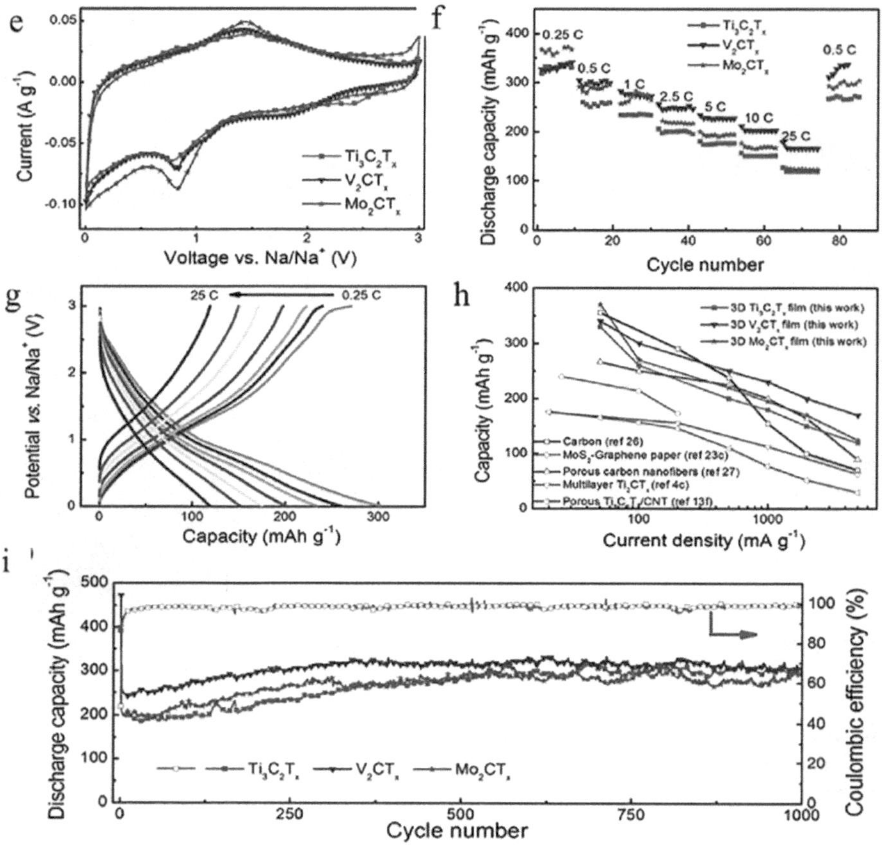

FIGURE 7.8 (Continued)

chemical interactions between the materials and lithium polysulfides (LiPSs), cells were fabricated using sulfur as the cathode material, with varying immersion times: Specifically, $Ti_3C_2T_x$ for 0 hours and $Ti_3C_2T_x$ for 4 hours. The performance of these cells was subsequently analyzed (Figure 7.8b).

After cycling, the S2p XPS spectrum of the cathode using $Ti_3C_2T_x$ (0 h) as the material showed a high intensity of elemental S8 and the presence of sulfate/polythio nate/thiosulfate compounds formed during cycling. In contrast, the cathode configuration with $Ti_3C_2T_x$ (4 h) showed different characteristics (Tang et al. 2018). In addition, two additional peaks at different potentials (163.4 and 162.3 eV) were observed in the S2p XPS spectrum, indicating the interaction between sulfur and TiO_2 as well as the Lewis acidic surface region of $Ti_3C_2T_x$.

To get a better understanding of how lithium polysulfides (LiPSs) are converted on the surface of the TiO_2-$Ti_3C_2T_x$ heterostructure, symmetric cells were built with the TiO_2-$Ti_3C_2T_x$ heterostructure and graphite nanosheets (GN) as the electrode material. Li2S6 was present in the electrolyte. Figure 7.8c showcases the cyclic voltammetry (CV) curves obtained at a scan rate of 10 mV/s within a voltage range of −0.7 to 0.7 V.

$Ti_3C_2T_x$ (0 h) exhibits more pronounced redox peaks at different time intervals than $Ti_3C_2T_x$ (8 h) on the symmetrical cell. Thanks to its well-structured composition and surface features, graphene (GR) offers an effective platform for sulfur material utilization. It allows for a balanced adsorption process of LiPS and its surface conversion (Jiao et al. 2019).

The TiO_2-$Ti_3C_2T_x$ heterostructure improves trapping and kinetics for the conversion of lithium polysulfides (LiPSs). The cross-sectional scanning electron micrograph (Figure 7.8d) reveals a 4 μm wide space between the layers. The TiO_2-$Ti_3C_2T_x$ heterostructure is evenly dispersed inside the interlayer, as seen by the tiny insets at the top left and bottom right, which are energy-dispersive X-ray (EDX) maps of titanium (Ti) and oxygen (O). The well-dispersed TiO_2 nanostructures within the MXene layers play a crucial role in effectively confining the polysulfide content. Due to the oxygen-terminated surface of the MXene and the hetero-interface, the transfer of polysulfide species from TiO_2 to MXene is accelerated, resulting in an environment with strong catalytic activity (Jiao et al. 2019).

Heterostructures composed of TiO_2-MXene were developed through the partial oxidation of $Ti_3C_2T_x$ nanosheets (NSs). These heterostructures offer numerous advantages, including high conductivity, excellent electrocatalytic activity, a large surface area, and a significant capacity to adsorb lithium polysulfides (LiPSs). The uniformly distributed TiO_2 nanoparticles on the MXene surface serve as effective sites for lithium polysulfide (LiPS) capture. The boundaries of the heterostructure facilitate the effective catalytic activity, enabling successful LiPS conversion. The cyclic stability is enhanced by the interlayer formed by the TiO_2-MXene and graphene (GN) heterostructures, which actively inhibit LiPS migration while allowing the utilization of sulfur components.

At a 2C cycling rate, the improved lithium-sulfur energy storage batteries demonstrate a capacity of 800 $mAhg^{-1}$, with a capacity degradation of only 0.028% per cycle after 1,000 cycles at the same rate. Even with a sulfur loading of 5.1 mg/cm^2, the battery retains 93% of its initial capacity after 200 cycles. These results demonstrate that the polysulfide content in lithium-sulfur batteries can be effectively controlled using catalyst materials with meticulously engineered surface properties and chemical compositions. This paves the way for novel approaches in developing advanced energy storage systems. For further details on the performance of 2D heterostructures based on MXene in lithium-ion batteries, please refer to Table 7.1(i) in the catalysis section.

7.3.4 Na-Ion Batteries

The sodium-ion (Na-ion) battery has garnered significant attention as a potential replacement for the more conventional lithium-based energy storage batteries because of its enormous storage capacity.

Na-ion batteries have a good and dependable energy density, are inexpensive, and have a lot of sodium salts (Slater et al. 2013; Yabuuchi et al. 2014). The high conductivity and three-dimensional (3D) macroporous structure of MXenes make them attractive materials for use in electrochemistry and sodium-ion (Na-ion) storage. The total Na-ion capacity of MXenes is reduced due to their pseudocapacitive Na-ion storage mechanism, in contrast to materials used for double-layer electrodes.

However, when compared to bulk Faradic electrode materials, they provide a significantly higher rate of efficiency (Wang et al. 2015; Wang et al. 2015).

7.3.4.1 $Ti_3C_2T_x$-Based Na-Ion Batteries

In their study, Gogotsi et al. (Zhao et al. 2017) produced films that can be used as anodes without the need for additional current collectors and are binder-free and stand-alone. An electrochemical test cell employing a sodium (Na) foil counter electrode was utilized to assess the electrochemical performance of the films. In the cyclic voltammetry (CV) profiles of the MXene-based films, the cathodic peak occurred at approximately 0.8 V, while the anodic peak was around 1.4 V, relative to Na/Na^+.

These peaks indicate the insertion and extraction of sodium ions (Na^+) from the electrode surfaces. These peaks point to a non-diffusion-limited pseudocapacitive method of storing Na+ on the surface of MXene nanoflakes. This observation is consistent with previous research on MXene compounds (Wang et al. 2015; Zhao et al. 2016).

During the early CV cycles, the MXene-based film electrodes showed irreversible cathodic peaks, indicating the presence of a solid-state electrolyte interface. Sodium ions (Na^+) interact irreversibly with reactive sites in two specific locations: The interlayer gaps of the surface MXene material and with water molecules, specifically fluorine (F), hydroxyl (OH), and oxygen (O) groups (Zhao et al. 2016; Xie et al. 2014). As depicted in Figure 7.8e, the charge–discharge plateaus at 0.8/1.4 V closely align with the strengths of the redox peaks observed on the CV curve under a current flow of 0.25 C. Three distinct 3D-shaped MXene-based films are compared in terms of rate performance in Figure 7.8f. Electrodes made of $Ti_3C_2T_x$ and V_2CTx films showed capacities of around 330 and 340 $mAhg^{-1}$, respectively, at a current rate of 0.25 C, whereas the Mo_2CTx film showed a reversible capacity of about 370 $mAhg^{-1}$.

Both films effectively utilized available resources, demonstrating an excellent capacity retention over time.

The $Ti_3C_2T_x$, Mo_2CTx, and V_2CTx films demonstrated specific capacity values of 120, 125, and 170 $mAhg^{-1}$, respectively, at a high current rate of 25 C. Because of its wider interlayer spacing, the V_2CTx film demonstrated higher rate capabilities. Figure 7.8g depicts the charge and discharge patterns of the 3D $Ti_3C_2T_x$ film electrode at various current rates. According to these profiles, MXenes have pseudocapacitive Na+ storage (Wang et al. 2015), which is characterized by its distinct charge/discharge curves. The 3D V_2CTx and Mo_2CTx film electrodes behaved similarly, making them promising candidates for Na-ion storage applications. Figure 7.8h compares the ability of 3D microporous MXene film structures to store Na+ to that of other nanostructured materials. In comparison to other carbon-based materials, such as banana peels, the gravimetric efficiency of the 3D MXene thin films was higher (Lotfabad et al. 2014) for porous carbon nanofibers (Li et al. 2014) and graphene/MoS_2 papers (Makaryan et al. 2016). The cycling stability of an MXene macroporous film electrode during 1,000 cycles at the present rate of 2.5 C is shown in Figure 7.8i. $Ti_3C_2T_x$, V_2CTx, and Mo_2CTx all have initial discharge capacities of around 390, 470, and 400 $mAhg^{-1}$, respectively. For these thin-film electrodes, the capacity retention values were around 210, 260, and 210 $mAhg^{-1}$. In comparison to the $Ti_3C_2T_x$/CNT film electrode system, which had an efficiency of about 40%, $Ti_3C_2T_x$, V_2CTx, and Mo_2CTx had coulombic efficiencies of 53.8%, 55.3%, and 51.2%, respectively. The forward interaction between MXene nanoflakes and electrolyte species, in the end, was principally responsible for the first irreversible capabilities (Xie et al. 2014).

The thermal annealing method on MXene partially removes surface functions, leading to an improved Coulombic efficiency. MXenes' surface chemical properties can be improved further by optimizing them. Because of their 3D macroporous structure, $Ti_3C_2T_x$ films exhibit constant gravimetric storage performance, as seen in multilayered $Ti_3C_2T_x$ particulate structures and hybrid thin films like $Ti_3C_2T_x$/CNT composites. The macroporous structure improves electrolyte accessibility and cycle efficiency at active sites by widening the interlayer gap (Ren et al. 2016a; Zhao et al. 2016).

The larger spacing between its layers is responsible for the V_2CTx film's outstanding stability. All of the aforementioned 3D MXene-based film topologies showed exceptional stability after 1000 full cycles. While maintaining their pseudocapacitive Na-ion storage behavior, which is characterized by high capacitance and high-rate capability, they attained a high Coulombic efficiency that was over 100%. For $Ti_3C_2T_x$, V_2CTx, and Mo_2CTx, the residual capacity values after 1,000 cycles were around 295, 310, and 290 $mAhg^{-1}$, respectively. In conclusion, a uniform distribution of 2D $Ti_3C_2T_x$ MXene nanoflakes on the polymer surface allowed for the production of 3D macroporous structures utilizing a sacrificial template-based technique.

The hollow spherical $Ti_3C_2T_x$ structures' stability and ease of detachment in ethanol and water suggested their potential use in the fields of medicine and ecology. The robust interaction between the spherical structure and the metallic conductivity of $Ti_3C_2T_x$ MXene resulted in the formation of independent, elastic, and highly conductive 3D macroporous films. Other MXene types, such as V_2CTx and Mo_2CTx, have also been used to successfully create comparable 3D structures. The 3D MXene sheets served as anode materials for storing Na ions in the absence of binders and current collectors. Electrodes fabricated from 3D Ti_3C_2Tx, V_2CTx, and Mo_2CTx demonstrated impressive reversible capacities of approximately 330, 340, and 370 $mAhg^{-1}$, respectively, showcasing excellent performance and cycle stability at a lower charging rate of 0.25 C. Even after 1,000 cycles at a higher current rate of 2.5 C, these electrodes maintained capacities of around 295 $mAhg^{-1}$ for $Ti_3C_2T_x$, 310 mAh g^{-1} for V_2CTx, and 290 mAh g^{-1} for Mo_2CTx. These results represent a substantial improvement over hybrid MXene/CNT structures and multilayer MXene structures (Zhao et al. 2017).

7.3.4.2 BP and Ti_3C_2-Based Sodium-Ion Batteries

Phosphorus (P) is highly appealing as an anode material for sodium-ion batteries due to its relatively low electrode potential and impressive theoretical capacity value of 2,596 mA h g^{-1} when alloyed as Na_3P (Ni, Li, and Lu 2018; Ellis, Hatchard, and Obrovac 2012; Allan et al. 2016). P, on the other hand, has been less effective electrochemically as an anode material for SIBs due to its poor structural stability and insulating properties. In recent studies, researchers have made significant progress in solving these problems by adding P species to electrically conductive matrices.

Yang and colleagues' pioneering contributions (Qian et al. 2013; Li et al. 2013) have resulted in extensive research into the use of P as an anode material in SIBs (Song et al. 2015; Hu et al. 2018; Li et al. 2017).

The important advantages of interaction between phosphorus (P) and carbon (C) materials, which results in increased electrochemical stability and decreased potential, have been emphasized in several research papers. One example is the work of Song et al., who improved the interaction between the P and CNT species by forming P-O-C bonds, resulting in a long-lasting hybrid structure of P and CNTs. After 100 cycles, the conductivity was much enhanced and the environment was favorable to accept the significant volume expansion that occurs during cycling. This led to a constant capacity retention of about 91% (Song et al. 2015). Despite these developments, reaching even larger energy storage capacities is still difficult since as inclination to disintegrate poses problems that require the attention of experts all over the world in order to strengthen the framework and increase the stability of the cycle.

The most stable form of phosphorus, known as black phosphorus (BP), may be readily exfoliated to form nanosheets composed of phosphorene, which are structures with just a few atomic layers (Liao et al. 2019). The single-layer form of Ti_3C_2 nanosheets has been found to exhibit a high conductivity of approximately 6.76 × 105 S m^{-1} (Sang et al. 2016). On the other hand, Ti_3C_2 MXene's terminal functional groups (Fgs) (–F, –O, and –OH) have been shown to have Na+ ion diffusion barriers that are around 0.19 eV, 0.2 eV, and 0.013 eV, respectively. Due of the reactive Fgs on the Ti_3C_2 MXene surface, these findings suggest favorable Na+ ion transport kinetics (Yu 2016).

With their excellent conductivity, theoretical capacity, and thin-layered architectures, Ti_3C_2 MXene and black phosphorus (BP) offer a lot of potential. The multilayer structure of BP, which has a high range value of theoretical capacity, and Ti_3C_2 MXene, which has a high electrical conductivity and flexible features, are combined to construct molecular-level heterostructures (PDDA-BP/Ti_3C_2) via a unique method based on electrostatic interactions. The electronegativity of BP nanolayers is increased, and the BP layers are shielded from oxidation in water by functionalizing the surface of BP with a cationic organic molecule termed poly(diallyl dimethyl ammonium chloride) (PDDA), which improves stability and dispersibility (Xiong et al. 2018; Feng et al. 2018). The two components' direct contact causes the 2D interlayer distance to align in a parallel orientation, creating a good channel for ion diffusion and improving the effectiveness of the electrochemical process. As an electrode, the PDDA-BP/Ti_3C_2 heterostructure material performs remarkably well with regard to load capacity and cycle reliability.

The phosphorus (P) anode exhibits remarkable cycling stability, with only a 0.05% capacity loss per cycle over 2,000 cycles at a current density of 1.0 A g^{-1}. It also delivers a high capacity of 1,112 mA h g^{-1} at a lower current density of 0.1 A g^{-1}. Even after repeated cycling, the heterostructure electrode retains its ordered structure and maintains the Ti_3C_2-based BP quantum dot monodispersed structure.

Research suggests that the strong connections between BP and Ti_3C_2 might speed up sodiation and lower the binding energy. Extensive research has been conducted on the redox phenomena of sodiation/desodiation in Ti_3C_2 MXene, including atomic-scale mechanistic investigations and the role of reactive functionalities. The synergistic impact of this PDDA-BP/Ti_3C_2 heterostructure should benefit high-performance sodium-ion batteries significantly. For a detailed examination of how MXene-based 2D heterostructures perform in various batteries, refer to Table 7.1(j–k), which showcases the performance of MXene-based materials in this specific application.

7.4 SUMMARY

In conclusion, the journey through the development and applications of MXene-based mixed-dimensional heterostructures has been both enlightening and promising. Our exploration began with an understanding of the unique properties of two-dimensional (2D) materials, particularly focusing on the recent advancements in MXenes. These novel materials, derived from transition metal carbides, nitrides, and carbonitrides, have shown immense potential in overcoming the limitations of traditional 2D materials like graphene and molybdenum disulfide (MoS_2).

The integration of MXenes with other materials such as graphene, CNTs, and metal oxides has led to the creation of advanced supercapacitors and energy storage devices, demonstrating significant improvements in performance metrics.

Our discussion on supercapacitors showcased the remarkable synergy achieved by combining MXenes with graphene and CNTs, resulting in devices with enhanced electrical conductivity and energy storage capacity. Similarly, the exploration of MXene-metal oxide composites revealed their potential in developing high-performance supercapacitors with robust stability and efficiency.

Moving forward, the application of MXene-based heterostructures in energy conversion and storage was thoroughly examined. The incorporation of MXenes in lithium-ion and sodium-ion batteries has opened new avenues for creating batteries with higher capacities, faster charge–discharge cycles, and improved longevity. The development of MXene-TiO_2 and MXene-Fe_2O_3 nanoparticles has further expanded the scope of these materials in energy storage, showcasing their versatility and adaptability.

Throughout this journey, it has become evident that the field of MXene research is rapidly evolving, driven by the relentless efforts of scientists worldwide. The synthesis and exfoliation techniques for MXenes continue to advance, enabling the production of high-quality materials with tailored properties. The integration of MXenes into mixed-dimensional heterostructures holds immense promise for a wide range of applications, from optoelectronics to biomedicine.

In essence, the development of MXene-based mixed-dimensional heterostructures represents a significant leap forward in the realm of nanomaterials. The unique properties of MXenes, combined with innovative engineering approaches, are paving the way for next-generation devices that are more efficient, durable, and versatile. As we look to the future, continued research and collaboration will be crucial in unlocking the full potential of these remarkable materials, ultimately leading to groundbreaking advancements in technology and industry.

REFERENCES

Ahn, B Kollbe, Dong Woog Lee, Jacob N Israelachvili, and J Herbert Waite. 2014. "Surface-initiated self-healing of polymers in aqueous media." *Nature Materials* 13 (9):867–872.

Aïssa, B, A Ali, KA Mahmoud, T Haddad, and M Nedil. 2016. "Transport properties of a highly conductive 2D Ti_3C_2Tx MXene/graphene composite." *Applied Physics Letters* 109 (4):043109.

Akinwande, Deji, Nicholas Petrone, and James Hone. 2014. "Two-dimensional flexible nano-electronics." *Nature Communications* 5:5678.

Alhabeb, Mohamed, Kathleen Maleski, Babak Anasori, Pavel Lelyukh, Leah Clark, Saleesha Sin, and Yury Gogotsi. 2017. "Guidelines for synthesis and processing of two-dimensional titanium carbide (Ti_3C_2Tx MXene)." *Chemistry of Materials* 29 (18):7633–7644.

Ali, Adnan, Kanit Hantanasirisakul, Ahmed Abdala, Patrick Urbankowski, Meng-Qiang Zhao, Babak Anasori, Yury Gogotsi, Brahim Aïssa, and Khaled A Mahmoud. 2018. "Effect of synthesis on performance of MXene/iron oxide anode material for lithium-ion batteries." *Langmuir* 34 (38):11325–11334.

Allan, Phoebe K, John M Griffin, Ali Darwiche, Olaf J Borkiewicz, Kamila M Wiaderek, Karena W Chapman, Andrew J Morris, Peter J Chupas, Laure Monconduit, and Clare P Grey. 2016. "Tracking sodium-antimonide phase transformations in sodium-ion anodes: Insights from operando pair distribution function analysis and solid-state NMR spectroscopy." *Journal of the American Chemical Society* 138 (7):2352–2365.

Allcorn, Eric, and Arumugam Manthiram. 2014. "NiSb–Al2O3–C nanocomposite anodes with long cycle life for Li-ion batteries." *The Journal of Physical Chemistry C* 118 (2):811–822.

Anasori, Babak, Maria R Lukatskaya, and Yury Gogotsi. 2017. "2D metal carbides and nitrides (MXenes) for energy storage." *Nature Reviews Materials* 2:16098.

Anasori, Babak, Yu Xie, Majid Beidaghi, Jun Lu, Brian C. Hosler, Lars Hultman, Paul RC Kent, Yury Gogotsi, and Michel W Barsoum. 2015. "Two-dimensional, ordered, double transition metals carbides (MXenes)." *ACS Nano* 9 (10):9507–9516. https://doi.org/10.1021/acsnano.5b03591.

Armand, Michel, and J-M Tarascon. 2008. "Building better batteries." *Nature* 451 (7179):652–657.

Augustyn, Veronica, Patrice Simon, and Bruce Dunn. 2014. "Pseudocapacitive oxide materials for high-rate electrochemical energy storage." *Energy & Environmental Science* 7 (5):1597–1614.

Beidaghi, Majid, and Yury Gogotsi. 2014. "Capacitive energy storage in micro-scale devices: Recent advances in design and fabrication of micro-supercapacitors." *Energy & Environmental Science* 7 (3):867–884.

Bie, Ya-Qing, Yang-Bo Zhou, Zhi-Min Liao, Kai Yan, Song Liu, Qing Zhao, Shishir Kumar, Han-Chun Wu, Georg S Duesberg, and Graham LW Cross. 2011. "Site-specific transfer-printing of individual graphene microscale patterns to arbitrary surfaces." *Advanced Materials* 23 (34):3938–3943.

Bruce, Peter G, Stefan A Freunberger, Laurence J Hardwick, and Jean-Marie Tarascon. 2012. "Li–O2 and Li–S batteries with high energy storage." *Nature Materials* 11 (1):19–29.

Byon, Hye Ryung, Betar M Gallant, Seung Woo Lee, and Yang Shao-Horn. 2013. "Role of oxygen functional groups in carbon nanotube/graphene freestanding electrodes for high performance lithium batteries." *Advanced Functional Materials* 23 (8):1037–1045.

Cabana, Jordi, Laure Monconduit, Dominique Larcher, and M Rosa Palacin. 2010. "Beyond intercalation-based Li-ion batteries: The state of the art and challenges of electrode materials reacting through conversion reactions." *Advanced Materials* 22 (35):E170–E192.

Cai, Xingke, Yuting Luo, Bilu Liu, and Hui-Ming Cheng. 2018. "Preparation of 2D material dispersions and their applications." *Chemical Society Reviews* 47 (16):6224–6266. https://doi.org/10.1039/C8CS00254A.

Cai, Yi, Hong-En Wang, Xu Zhao, Fei Huang, Chao Wang, Zhao Deng, Yu Li, Guozhong Cao, and Bao-Lian Su. 2017. "Walnut-like porous core/shell TiO2 with hybridized phases enabling fast and stable lithium storage." *ACS Applied Materials & Interfaces* 9 (12):10652–10663.

Chang, Kun, and Weixiang Chen. 2011. "L-cysteine-assisted synthesis of layered MoS2/graphene composites with excellent electrochemical performances for lithium ion batteries." *ACS Nano* 5 (6):4720–4728.

Chen, Yi, Sinho Choi, Dawei Su, Xiaochun Gao, and Guoxiu Wang. 2018. "Self-standing sulfur cathodes enabled by 3D hierarchically porous titanium monoxide-graphene composite film for high-performance lithium-sulfur batteries." *Nano Energy* 47:331–339.

Chhowalla, Manish, Hyeon Suk Shin, Goki Eda, Lain-Jong Li, Kian Ping Loh, and Hua Zhang. 2013. "The chemistry of two-dimensional layered transition metal dichalcogenide nanosheets." *Nature Chemistry* 5 (4):263.

Chmiola, John, Celine Largeot, Pierre-Louis Taberna, Patrice Simon, and Yury Gogotsi. 2010. "Monolithic carbide-derived carbon films for micro-supercapacitors." *Science* 328 (5977):480–483.

Cho, Soo Hyoun, Scott R White, and Paul V Braun. 2009. "Self-healing polymer coatings." *Advanced Materials* 21 (6):645–649.

Choi, Wonbong, Indranil Lahiri, Raghunandan Seelaboyina, and Yong Soo Kang. 2010. "Synthesis of graphene and its applications: A review." *Critical Reviews in Solid State and Materials Sciences* 35 (1):52–71.

Cohn, Adam P, Landon Oakes, Rachel Carter, Shahana Chatterjee, Andrew S Westover, Keith Share, and Cary L Pint. 2014. "Assessing the improved performance of freestanding, flexible graphene and carbon nanotube hybrid foams for lithium ion battery anodes." *Nanoscale* 6 (9):4669–4675.

Come, Jérémy, Michael Naguib, Patrick Rozier, Michel W Barsoum, Yury Gogotsi, P-L Taberna, Mathieu Morcrette, and Patrice Simon. 2012. "A non-aqueous asymmetric cell with a Ti2C-based two-dimensional negative electrode." *Journal of the Electrochemical Society* 159 (8):A1368–A1373.

Conway, BE, and Electrochemical Supercapacitors. 1999. *Scientific Fundamentals and Technological Applications.* KluwerAcademic/Plenum Publishers, New York, 528.

Couly, Cedric, Mohamed Alhabeb, Katherine L Van Aken, Narendra Kurra, Luisa Gomes, Adriana M Navarro-Suárez, Babak Anasori, Husam N Alshareef, and Yury Gogotsi. 2018. "Asymmetric flexible MXene-reduced graphene oxide micro-supercapacitor." *Advanced Electronic Materials* 4 (1):1700339. https://doi.org/10.1002/aelm.201700339.

Dall'Agnese, Yohan, Maria R Lukatskaya, Kevin M Cook, Pierre-Louis Taberna, Yury Gogotsi, and Patrice Simon. 2014. "High capacitance of surface-modified 2D titanium carbide in acidic electrolyte." *Electrochemistry Communications* 48:118–122.

Dall'Agnese, Yohan, Patrick Rozier, Pierre-Louis Taberna, Yury Gogotsi, and Patrice Simon. 2016. "Capacitance of two-dimensional titanium carbide (MXene) and MXene/carbon nanotube composites in organic electrolytes." *Journal of Power Sources* 306:510–515. https://doi.org/10.1016/j.jpowsour.2015.12.036.

Dall'Agnese, Yohan, Pierre-Louis Taberna, Yury Gogotsi, and Patrice Simon. 2015. "Two-dimensional vanadium carbide (MXene) as positive electrode for sodium-ion capacitors." *The Journal of Physical Chemistry Letters* 6 (12):2305–2309.

Das, Saptarshi, Wei Zhang, Marcel Demarteau, Axel Hoffmann, Madan Dubey, and Andreas Roelofs. 2014. "Tunable transport gap in phosphorene." *Nano Letters* 14 (10):5733–5739. https://doi.org/10.1021/nl5025535.

De las Casas, Charles, and Wenzhi Li. 2012. "A review of application of carbon nanotubes for lithium ion battery anode material." *Journal of Power Sources* 208:74–85.

Deng, Bing, Po-Chun Hsu, Guanchu Chen, BN Chandrashekar, Lei Liao, Zhawulie Ayitimuda, Jinxiong Wu, Yunfan Guo, Li Lin, and Yu Zhou. 2015. "Roll-to-roll encapsulation of metal nanowires between graphene and plastic substrate for high-performance flexible transparent electrodes." *Nano Letters* 15 (6):4206–4213.

Dikin, Dmitriy A, Sasha Stankovich, Eric J Zimney, Richard D Piner, Geoffrey HB Dommett, Guennadi Evmenenko, SonBinh T Nguyen, and Rodney S Ruoff. 2007. "Preparation and characterization of graphene oxide paper." *Nature* 448 (7152):457–460.

Ding, Li, Yanying Wei, Yanjie Wang, Hongbin Chen, Jürgen Caro, and Haihui Wang. 2017. "A two-dimensional lamellar membrane: MXene nanosheet stacks." *Angewandte Chemie International Edition* 56 (7):1825–1829. https://doi.org/10.1002/anie.201609306.

Dunn, Bruce, Haresh Kamath, and Jean-Marie Tarascon. 2011. "Electrical energy storage for the grid: A battery of choices." *Science* 334 (6058):928–935.

Eda, Goki, Giovanni Fanchini, and Manish Chhowalla. 2008. "Large-area ultrathin films of reduced graphene oxide as a transparent and flexible electronic material." *Nature Nanotechnology* 3 (5):270–274.

El-Kady, Maher F, and Richard B Kaner. 2013. "Scalable fabrication of high-power graphene micro-supercapacitors for flexible and on-chip energy storage." *Nature Communications* 4:1475.

Ellis, Leah D, Timothy D Hatchard, and Mark N Obrovac. 2012. "Reversible insertion of sodium in tin." *Journal of the Electrochemical Society* 159 (11):A1801–A1805.

Feng, Qingliang, Hongyan Liu, Meijie Zhu, Jing Shang, Dan Liu, Xiaoqi Cui, Diqin Shen, Liangzhi Kou, Dong Mao, and Jianbang Zheng. 2018. "Electrostatic functionalization and passivation of water-exfoliated few-layer black phosphorus by poly dimethyldiallyl ammonium chloride and its ultrafast laser application." *ACS Applied Materials & Interfaces* 10 (11):9679–9687.

Fuhrer, Michael S., and James Hone. 2013. "Measurement of mobility in dual-gated MoS2 transistors." *Nature Nanotechnology* 8 (3):146–147. https://doi.org/10.1038/nnano.2013.30.

Ghaffari, Mehdi, Yue Zhou, Haiping Xu, Minren Lin, Tae Young Kim, Rodney S Ruoff, and QM Zhang. 2013. "Retracted: High-volumetric performance aligned nano-porous microwave exfoliated graphite oxide-based electrochemical capacitors." *Advanced Materials* 25 (35):4879–4885.

Ghidiu, Michael, Maria R Lukatskaya, Meng-Qiang Zhao, Yury Gogotsi, and Michel W Barsoum. 2014. "Conductive two-dimensional titanium carbide 'clay' with high volumetric capacitance." *Nature* 516 (7529):78–81.

Gogotsi, Yury. 2014. "Materials science: Energy storage wrapped up." *Nature* 509 (7502): 568–570.

Gogotsi, Yury, and Patrice Simon. 2011. "True performance metrics in electrochemical energy storage." *Science* 334 (6058):917–918.

Han, Wei-Qiang. 2009. "Anisotropic hexagonal boron nitride nanomaterials: Synthesis and applications." In *Mixed Metal Nanomaterials*. Wiley, Weinheim.

Hantanasirisakul, Kanit, and Yury Gogotsi. 2018. "Electronic and optical properties of 2D transition metal carbides and nitrides (MXenes)." *Advanced Materials* 30 (52):1804779. https://doi.org/10.1002/adma.201804779.

Heon, Min, Samuel Lofland, James Applegate, Robert Nolte, Emma Cortes, Jeffrey D Hettinger, Pierre-Louis Taberna, Patrice Simon, Peihua Huang, and Magali Brunet. 2011. "Continuous carbide-derived carbon films with high volumetric capacitance." *Energy & Environmental Science* 4 (1):135–138.

Hu, Ye, Bing Li, Xingxing Jiao, Chaofan Zhang, Xiaohan Dai, and Jiangxuan Song. 2018. "Stable cycling of phosphorus anode for sodium-ion batteries through chemical bonding with sulfurized polyacrylonitrile." *Advanced Functional Materials* 28 (23):1801010.

Huang, Jia-Qi, Xiao-Fei Liu, Qiang Zhang, Cheng-Meng Chen, Meng-Qiang Zhao, Shu-Mao Zhang, Wancheng Zhu, Wei-Zhong Qian, and Fei Wei. 2013. "Entrapment of sulfur in hierarchical porous graphene for lithium–sulfur batteries with high rate performance from –40 to 60°C." *Nano Energy* 2 (2):314–321.

Huang, Kai, Zhongjun Li, Jing Lin, Gang Han, and Peng Huang. 2018. "Two-dimensional transition metal carbides and nitrides (MXenes) for biomedical applications." *Chemical Society Reviews* 47 (14):5109–5124. https://doi.org/10.1039/C7CS00838D.

Huang, Peihua, Christophe Lethien, Sébastien Pinaud, Kevin Brousse, Raphaël Laloo, Viviane Turq, Marc Respaud, Arnaud Demortiere, Barbara Daffos, and Pierre-Louis Taberna. 2016. "On-chip and freestanding elastic carbon films for micro-supercapacitors." *Science* 351 (6274):691–695.

Huang, Shaozhuan, Lin Zhang, Xueyi Lu, Lifeng Liu, Lixiang Liu, Xiaolei Sun, Yin Yin, Steffen Oswald, Zhaoyong Zou, and Fei Ding. 2017. "Tunable pseudocapacitance in 3D TiO2– δ nanomembranes enabling superior lithium storage performance." *ACS Nano* 11 (1):821–830.

Ji, Xiulei, Kyu Tae Lee, and Linda F Nazar. 2009. "A highly ordered nanostructured carbon–sulphur cathode for lithium–sulphur batteries." *Nature Materials* 8 (6):500–506.

Jiang, Qiu, Narendra Kurra, Mohamed Alhabeb, Yury Gogotsi, and Husam N Alshareef. 2018. "All pseudocapacitive MXene-Ruo2 asymmetric supercapacitors." *Advanced Energy Materials* 8 (13):1703043.

Jiao, Liying, Li Zhang, Xinran Wang, Georgi Diankov, and Hongjie Dai. 2009. "Narrow graphene nanoribbons from carbon nanotubes." *Nature* 458 (7240):877.

Jiao, Long, Chen Zhang, Chuannan Geng, Shichao Wu, Huan Li, Wei Lv, Ying Tao, Zijin Chen, Guangmin Zhou, Jia Li, Guowei Ling, Ying Wan, and Quan-Hong Yang. 2019. "Capture and catalytic conversion of polysulfides by in situ built TiO2-MXene heterostructures for lithium–sulfur batteries." *Advanced Energy Materials* 9 (19):1900219.

Jung, Naeyoung, Soongeun Kwon, Dongwook Lee, Dong-Myung Yoon, Young Min Park, Anass Benayad, Jae-Young Choi, and Jong Se Park. 2013. "Synthesis of chemically bonded graphene/carbon nanotube composites and their application in large volumetric capacitance supercapacitors." *Advanced Materials* 25 (47):6854–6858.

Kang, Byoungwoo, and Gerbrand Ceder. 2009. "Battery materials for ultrafast charging and discharging." *Nature* 458 (7235):190–193.

Kannan, Karthik, Kishor Kumar Sadasivuni, Aboubakr M Abdullah, and Bijandra Kumar. 2020. "Current trends in MXene-based nanomaterials for energy storage and conversion system: A mini review." *Catalysts* 10 (5):495. https://doi.org/10.3390/catal10050495

Kannan, Karthik, Mostafa H Sliem, Aboubakr M Abdullah, Kishor Kumar Sadasivuni, and Bijandra Kumar. 2020. "Fabrication of ZnO-Fe-MXene based nanocomposites for efficient CO2 reduction." *Catalysts* 10 (5):549. https://doi.org/10.3390/catal10050549

Khan, Majharul Haque, Zhenguo Huang, Feng Xiao, Gilberto Casillas, Zhixin Chen, Paul J Molino, and Hua Kun Liu. 2015. "Synthesis of large and few atomic layers of hexagonal boron nitride on melted copper." *Scientific Reports* 5:7743.

Khazaei, Mohammad, Ahmad Ranjbar, Masao Arai, Taizo Sasaki, and Seiji Yunoki. 2017. "Electronic properties and applications of MXenes: A theoretical review." *Journal of Materials Chemistry C* 5 (10):2488–2503. https://doi.org/10.1039/C7TC00140A.

Kim, Dae-Hyeong, and John A Rogers. 2008. "Stretchable electronics: Materials strategies and devices." *Advanced Materials* 20 (24):4887–4892.

Kim, Hyun Ho, Yoonyoung Chung, Eunho Lee, Seong Kyu Lee, and Kilwon Cho. 2014. "Water-free transfer method for CVD-grown graphene and its application to flexible air-stable graphene transistors." *Advanced Materials* 26 (20):3213–3217.

Kim, Seon Joon, Michael Naguib, Mengqiang Zhao, Chuanfang Zhang, Hee-Tae Jung, Michel W Barsoum, and Yury Gogotsi. 2015. "High mass loading, binder-free MXene anodes for high areal capacity Li-ion batteries." *Electrochimica Acta* 163:246–251.

Kou, Liangzhi, Changfeng Chen, and Sean C Smith. 2015. "Phosphorene: Fabrication, properties, and applications." *The Journal of Physical Chemistry Letters* 6 (14):2794–2805. https://doi.org/10.1021/acs.jpclett.5b01094.

Kurra, Narendra, Nuha A Alhebshi, and Husam N Alshareef. 2015. "Microfabricated pseudocapacitors using Ni(OH)2 electrodes exhibit remarkable volumetric capacitance and energy density." *Advanced Energy Materials* 5 (2):1401303.

Kurra, Narendra, Qiu Jiang, and Husam N Alshareef. 2015. "A general strategy for the fabrication of high performance microsupercapacitors." *Nano Energy* 16:1–9.

Kurra, Narendra, Qiu Jiang, Ahad Syed, Chuan Xia, and Husam N Alshareef. 2016. "Micro-pseudocapacitors with electroactive polymer electrodes: Toward AC-line filtering applications." *ACS Applied Materials & Interfaces* 8 (20):12748–12755.

Kyeremateng, Nana Amponsah, Thierry Brousse, and David Pech. 2017. "Microsupercapacitors as miniaturized energy-storage components for on-chip electronics." *Nature Nanotechnology* 12 (1):7.

Landi, Brian J, Matthew J Ganter, Cory D Cress, Roberta A DiLeo, and Ryne P Raffaelle. 2009. "Carbon nanotubes for lithium ion batteries." *Energy & Environmental Science* 2 (6):638–654.

Lei, Jincheng, Xu Zhang, and Zhen Zhou. 2015. "Recent advances in MXene: Preparation, properties, and applications." *Frontiers of Physics in China* 10 (3):276–286.

Li, Jian, Xiaotao Yuan, Cong Lin, Yanquan Yang, Le Xu, Xin Du, Jinglin Xie, Jianhua Lin, and Junliang Sun. 2017. "Achieving high pseudocapacitance of 2D titanium carbide (MXene) by cation intercalation and surface modification." *Advanced Energy Materials* 7 (15):1602725.

Li, Likai, Yijun Yu, Guo Jun Ye, Qingqin Ge, Xuedong Ou, Hua Wu, Donglai Feng, Xian Hui Chen, and Yuanbo Zhang. 2014. "Black phosphorus field-effect transistors." *Nature Nanotechnology* 9 (5):372–377. https://doi.org/10.1038/nnano.2014.35.

Li, Wei, Fei Wang, Yupu Liu, Jinxiu Wang, Jianping Yang, Lijuan Zhang, Ahmed A Elzatahry, Daifallah Al-Dahyan, Yongyao Xia, and Dongyuan Zhao. 2015. "General strategy to synthesize uniform mesoporous TiO2/graphene/mesoporous TiO2 sandwich-like nanosheets for highly reversible lithium storage." *Nano Letters* 15 (3):2186–2193.

Li, Wei-Jie, Shu-Lei Chou, Jia-Zhao Wang, Hua-Kun Liu, and Shi-Xue Dou. 2013. "Simply mixed commercial red phosphorus and carbon nanotube composite with exceptionally reversible sodium-ion storage." *Nano Letters* 13 (11):5480–5484.

Li, Xiao, and Hongwei Zhu. 2015. "Two-dimensional MoS2: Properties, preparation, and applications." *Journal of Materiomics* 1 (1):33–44.

Li, Zhengyang, Libo Wang, Dandan Sun, Yude Zhang, Baozhong Liu, Qianku Hu, and Aiguo Zhou. 2015. "Synthesis and thermal stability of two-dimensional carbide MXene Ti_3C_2." *Materials Science and Engineering: B* 191:33–40. https://doi.org/10.1016/j.mseb.2014.10.009.

Liang, Ji, Zhen-Hua Sun, Feng Li, and Hui-Ming Cheng. 2016. "Carbon materials for Li–S batteries: Functional evolution and performance improvement." *Energy Storage Materials* 2:76–106.

Liang, Xiao, Connor Hart, Quan Pang, Arnd Garsuch, Thomas Weiss, and Linda F Nazar. 2015. "A highly efficient polysulfide mediator for lithium–sulfur batteries." *Nature Communications* 6 (1):1–8.

Liao, Guangfu, Yan Gong, Li Zhang, Haiyang Gao, Guan-Jun Yang, and Baizeng Fang. 2019. "Semiconductor polymeric graphitic carbon nitride photocatalysts: The "holy grail" for the photocatalytic hydrogen evolution reaction under visible light." *Energy & Environmental Science* 12 (7):2080–2147.

Liao, Lei, Yung-Chen Lin, Mingqiang Bao, Rui Cheng, Jingwei Bai, Yuan Liu, Yongquan Qu, Kang L. Wang, Yu Huang, and Xiangfeng Duan. 2010. "High-speed graphene transistors with a self-aligned nanowire gate." *Nature* 467 (7313):305–308. https://doi.org/10.1038/nature09405.

Liu, Jiehua, and Xue-Wei Liu. 2012. "Two-dimensional nanoarchitectures for lithium storage." *Advanced Materials* 24 (30):4097–4111.

Liu, Yi-Tao, Peng Zhang, Ning Sun, Babak Anasori, Qi-Zhen Zhu, Huan Liu, Yury Gogotsi, and Bin Xu. 2018. "Self-assembly of transition metal oxide nanostructures on MXene nanosheets for fast and stable lithium storage." *Advanced Materials* 30 (23):1707334. https://doi.org/10.1002/adma.201707334.

Liu, Yu, Wei Wang, Lin Gu, Yewu Wang, Yulong Ying, Yiyin Mao, Luwei Sun, and Xinsheng Peng. 2013. "Flexible CuO nanosheets/reduced-graphene oxide composite paper: Binder-free anode for high-performance lithium-ion batteries." *ACS Applied Materials & Interfaces* 5 (19):9850–9855.

Liu, Yu, Wei Wang, Hubiao Huang, Lin Gu, Yewu Wang, and Xinsheng Peng. 2014. "The highly enhanced performance of lamellar WS2 nanosheet electrodes upon intercalation of single-walled carbon nanotubes for supercapacitors and lithium ions batteries." *Chemical Communications* 50 (34):4485–4488.

Liu, Yu, Wei Wang, Yewu Wang, and Xinsheng Peng. 2014. "Homogeneously assembling like-charged WS2 and GO nanosheets lamellar composite films by filtration for highly efficient lithium ion batteries." *Nano Energy* 7:25–32.

Liu, Yu, Wei Wang, Yulong Ying, Yewu Wang, and Xinsheng Peng. 2015. "Binder-free layered Ti_3C_2/CNTs nanocomposite anodes with enhanced capacity and long-cycle life for lithium-ion batteries." *Dalton Transactions* 44 (16):7123–7126. https://doi.org/10.1039/C4DT02058H.

Lotfabad, Elmira Memarzadeh, Jia Ding, Kai Cui, Alireza Kohandehghan, W Peter Kalisvaart, Michael Hazelton, and David Mitlin. 2014. "High-density sodium and lithium ion battery anodes from banana peels." *ACS Nano* 8 (7):7115–7129.

Lukatskaya, Maria R, Bruce Dunn, and Yury Gogotsi. 2016. "Multidimensional materials and device architectures for future hybrid energy storage." *Nature Communications* 7 (1):1–13.

Lukatskaya, Maria R, Olha Mashtalir, Chang E Ren, Yohan Dall'Agnese, Patrick Rozier, Pierre Louis Taberna, Michael Naguib, Patrice Simon, Michel W Barsoum, and Yury Gogotsi. 2013. "Cation intercalation and high volumetric capacitance of two-dimensional titanium carbide." *Science* 341 (6153):1502. https://doi.org/10.1126/science.1241488.

Manthiram, Arumugam, Yongzhu Fu, Sheng-Heng Chung, Chenxi Zu, and Yu-Sheng Su. 2014. "Rechargeable lithium–sulfur batteries." *Chemical Reviews* 114 (23):11751–11787.

Mashtalir, Olha, Michael Naguib, Vadym N Mochalin, Yohan Dall'Agnese, Min Heon, Michel W Barsoum, and Yury Gogotsi. 2013. "Intercalation and delamination of layered carbides and carbonitrides." *Nature Communications* 4 (1):1–7.

Mattevi, Cecilia, Goki Eda, Stefano Agnoli, Steve Miller, K Andre Mkhoyan, Ozgur Celik, Daniel Mastrogiovanni, Gaetano Granozzi, Eric Garfunkel, and Manish Chhowalla. 2009. "Evolution of electrical, chemical, and structural properties of transparent and conducting chemically derived graphene thin films." *Advanced Functional Materials* 19 (16):2577–2583.

Mendoza-Sánchez, Beatriz, and Yury Gogotsi. 2016. "Synthesis of two-dimensional materials for capacitive energy storage." *Advanced Materials* 28 (29):6104–6135.

Meng, Chuizhou, Jimin Maeng, Simon WM John, and Pedro P Irazoqui. 2014. "Ultrasmall integrated 3D micro-supercapacitors solve energy storage for miniature devices." *Advanced Energy Materials* 4 (7):1301269.

Mo, Runwei, Zhengyu Lei, Kening Sun, and David Rooney. 2014. "Facile synthesis of anatase TiO2 quantum-dot/graphene-nanosheet composites with enhanced electrochemical performance for lithium-ion batteries." *Advanced Materials* 26 (13):2084–2088.

Naguib, Michael, and Yury Gogotsi. 2015. "Synthesis of two-dimensional materials by selective extraction." *Accounts of Chemical Research* 48 (1):128–135. https://doi.org/10.1021/ar500346b.

Naguib, Michael, Joseph Halim, Jun Lu, Kevin M. Cook, Lars Hultman, Yury Gogotsi, and Michel W. Barsoum. 2013. "New two-dimensional niobium and vanadium carbides as promising materials for Li-ion batteries." *Journal of the American Chemical Society* 135 (43):15966–15969. https://doi.org/10.1021/ja405735d.

Naguib, Michael, Murat Kurtoglu, Volker Presser, Jun Lu, Junjie Niu, Min Heon, Lars Hultman, Yury Gogotsi, and Michel W Barsoum. 2011. "Two-dimensional nanocrystals produced by exfoliation of Ti3AlC2." *Advanced Materials* 23 (37):4248–4253.

Naguib, Michael, Olha Mashtalir, Joshua Carle, Volker Presser, Jun Lu, Lars Hultman, Yury Gogotsi, and Michel W Barsoum. 2012. "Two-dimensional transition metal carbides." *ACS Nano* 6 (2):1322–1331. https://doi.org/10.1021/nn204153h.

Naguib, Michael, Vadym N Mochalin, Michel W Barsoum, and Yury Gogotsi. 2014. "25th anniversary article: MXenes: A new family of two-dimensional materials." *Advanced Materials* 26 (7):992–1005.

Ni, Jiangfeng, Liang Li, and Jun Lu. 2018. "Phosphorus: An anode of choice for sodium-ion batteries." *ACS Energy Letters* 3 (5):1137–1144.

Niu, Zhiqiang, Li Zhang, Lili Liu, Bowen Zhu, Haibo Dong, and Xiaodong Chen. 2013. "All-solid-state flexible ultrathin micro-supercapacitors based on graphene." *Advanced Materials* 25 (29):4035–4042.

Ospina, DA, CA Duque, JD Correa, and Eric Suárez Morell. 2016. "Twisted bilayer blue phosphorene: A direct band gap semiconductor." *Superlattices and Microstructures* 97:562–568.

Pang, Jinbo, Rafael G Mendes, Alicja Bachmatiuk, Liang Zhao, Huy Q Ta, Thomas Gemming, Hong Liu, Zhongfan Liu, and Mark H Rummeli. 2019. "Applications of 2D MXenes in energy conversion and storage systems." *Chemical Society Reviews* 48 (1):72–133.

Pech, David, Magali Brunet, Hugo Durou, Peihua Huang, Vadym Mochalin, Yury Gogotsi, Pierre-Louis Taberna, and Patrice Simon. 2010a. "Ultrahigh-power micrometre-sized supercapacitors based on onion-like carbon." *Nature Nanotechnology* 5 (9):651–654.

Pech, David, Magali Brunet, Pierre-Louis Taberna, Patrice Simon, Norbert Fabre, Fabien Mesnilgrente, Véronique Conédéra, and Hugo Durou. 2010b. "Elaboration of a microstructured inkjet-printed carbon electrochemical capacitor." *Journal of Power Sources* 195 (4):1266–1269.

Peng, Hong-Jie, Jia-Qi Huang, Xin-Bing Cheng, and Qiang Zhang. 2017. "Review on high-loading and high-energy lithium–sulfur batteries." *Advanced Energy Materials* 7 (24):1700260.

Peng, Lele, Yue Zhu, Dahong Chen, Rodney S Ruoff, and Guihua Yu. 2016. "Two-dimensional materials for beyond-lithium-ion batteries." *Advanced Energy Materials* 6 (11):1600025.

Peng, You-Yu, Bilen Akuzum, Narendra Kurra, Meng-Qiang Zhao, Mohamed Alhabeb, Babak Anasori, Emin Caglan Kumbur, Husam N Alshareef, Ming-Der Ger, and Yury Gogotsi. 2016. "All-MXene (2D titanium carbide) solid-state microsupercapacitors for on-chip energy storage." *Energy & Environmental Science* 9 (9):2847–2854. https://doi.org/10.1039/C6EE01717G.

Poizot, PLSG, S Laruelle, S Grugeon, L Dupont, and JM Tarascon. 2000. "Nano-sized transition-metal oxides as negative-electrode materials for lithium-ion batteries." *Nature* 407 (6803):496.

Pomerantseva, Ekaterina, and Yury Gogotsi. 2017. "Two-dimensional heterostructures for energy storage." *Nature Energy* 2 (7):17089.

Pumera, Martin, Zdeněk Sofer, and Adriano Ambrosi. 2014. "Layered transition metal dichalcogenides for electrochemical energy generation and storage." *Journal of Materials Chemistry A* 2 (24):8981–8987.

Qian, Jiangfeng, Xianyong Wu, Yuliang Cao, Xinping Ai, and Hanxi Yang. 2013. "High capacity and rate capability of amorphous phosphorus for sodium ion batteries." *Angewandte Chemie International Edition* 52 (17):4633–4636.

Qiu, Bocheng, Mingyang Xing, and Jinlong Zhang. 2014. "Mesoporous TiO2 nanocrystals grown in situ on graphene aerogels for high photocatalysis and lithium-ion batteries." *Journal of the American Chemical Society* 136 (16):5852–5855.

Radisavljevic, B, A Radenovic, J Brivio, V Giacometti, and A Kis. 2011. "Single-layer MoS2 transistors." *Nature Nanotechnology* 6 (3):147–150. https://doi.org/10.1038/nnano.2010.279.

Reina, Alfonso, Hyungbin Son, Liying Jiao, Ben Fan, Mildred S Dresselhaus, ZhongFan Liu, and Jing Kong. 2008. "Transferring and identification of single-and few-layer graphene on arbitrary substrates." *The Journal of Physical Chemistry C* 112 (46):17741–17744.

Ren, Chang E, Meng-Qiang Zhao, Taron Makaryan, Joseph Halim, Muhammad Boota, Sankalp Kota, Babak Anasori, Michel W Barsoum, and Yury Gogotsi. 2016. "Porous two-dimensional transition metal carbide (MXene) flakes for high-performance Li-ion storage." *ChemElectroChem* 3 (5):689–693. https://doi.org/10.1002/celc.201600059.

Robinson, Jeremy T, F Keith Perkins, Eric S Snow, Zhongqing Wei, and Paul E Sheehan. 2008a. "Reduced graphene oxide molecular sensors." *Nano Letters* 8 (10):3137–3140.

Robinson, Jeremy T, Maxim Zalalutdinov, Jeffrey W Baldwin, Eric S Snow, Zhongqing Wei, Paul Sheehan, and Brian H Houston. 2008b. "Wafer-scale reduced graphene oxide films for nanomechanical devices." *Nano Letters* 8 (10):3441–3445.

Ronchi, Rodrigo Mantovani, Jeverson Teodoro Arantes, and Sydney Ferreira Santos. 2019. "Synthesis, structure, properties and applications of MXenes: Current status and perspectives." *Ceramics International* 45:18167–18188.

Sahoo, Satyaprakash, Anand PS Gaur, Majid Ahmadi, Maxime J-F Guinel, and Ram S Katiyar. 2013. "Temperature-dependent Raman studies and thermal conductivity of few-layer MoS2." *The Journal of Physical Chemistry C* 117 (17):9042–9047.

Sajid, Muhammad. 2021. "MXenes: Are they emerging materials for analytical chemistry applications?—A review." *Analytica Chimica Acta* 1143:267–280. https://doi.org/10.1016/j.aca.2020.08.063.

Sang, Xiahan, Yu Xie, Ming-Wei Lin, Mohamed Alhabeb, Katherine L Van Aken, Yury Gogotsi, Paul RC Kent, Kai Xiao, and Raymond R Unocic. 2016. "Atomic defects in monolayer titanium carbide (Ti_3C_2Tx) MXene." *ACS Nano* 10 (10):9193–9200.

Schwierz, Frank. 2010. "Graphene transistors." *Nature Nanotechnology* 5 (7):487–496. https://doi.org/10.1038/nnano.2010.89.

Seh, Zhi Wei, Weiyang Li, Judy J Cha, Guangyuan Zheng, Yuan Yang, Matthew T McDowell, Po-Chun Hsu, and Yi Cui. 2013. "Sulphur–TiO2 yolk–shell nanoarchitecture with internal void space for long-cycle lithium–sulphur batteries." *Nature Communications* 4 (1):1–6.

Simon, Patrice, and Yury Gogotsi. 2010. "Materials for electrochemical capacitors." In *Nanoscience and Technology: A Collection of Reviews From Nature Journals*, 320–329. World Scientific.

Simon, Patrice, Yury Gogotsi, and Bruce Dunn. 2014. "Where do batteries end and supercapacitors begin?" *Science* 343 (6176):1210–1211.

Slater, Michael D, Donghan Kim, Eungje Lee, and Christopher S Johnson. 2013. "Sodium-ion batteries." *Advanced Functional Materials* 23 (8):947–958.

Song, Jiangxuan, Zhaoxin Yu, Mikhail L Gordin, Xiaolin Li, Huisheng Peng, and Donghai Wang. 2015. "Advanced sodium ion battery anode constructed via chemical bonding between phosphorus, carbon nanotube, and cross-linked polymer binder." *ACS Nano* 9 (12):11933–11941.

Su, Yu-Sheng, and Arumugam Manthiram. 2012. "Lithium–sulphur batteries with a microporous carbon paper as a bifunctional interlayer." *Nature Communications* 3 (1):1–6.

Sun, Renhui, Hao-Bin Zhang, Ji Liu, Xi Xie, Rui Yang, Yue Li, Song Hong, and Zhong-Zhen Yu. 2017. "Highly conductive transition metal carbide/carbonitride(MXene)@polystyrene nanocomposites fabricated by electrostatic assembly for highly efficient electromagnetic interference shielding." *Advanced Functional Materials* 27 (45):1702807. https://doi.org/10.1002/adfm.201702807.

Tang, Huan, Wenlong Li, Limei Pan, Conor P Cullen, Yu Liu, Amir Pakdel, Donghui Long, Jian Yang, Niall McEvoy, and Georg S Duesberg. 2018. "In situ formed protective barrier enabled by Sulfur@ titanium carbide (MXene) ink for achieving high-capacity, long lifetime Li-S batteries." *Advanced Science* 5 (9):1800502.

Tang, Qing, Zhen Zhou, and Panwen Shen. 2012. "Are MXenes promising anode materials for Li ion batteries? Computational studies on electronic properties and Li storage capability of Ti_3C_2 and $Ti_3C_2X_2$ (X=F, OH) monolayer." *Journal of the American Chemical Society* 134 (40):16909–16916.

Tang, Yi, JianFeng Zhu, ChenHui Yang, and Fen Wang. 2016. "Enhanced supercapacitive performance of manganese oxides doped two-dimensional titanium carbide nanocomposite in alkaline electrolyte." *Journal of Alloys and Compounds* 685:194–201.

Tarascon, J-M, and Michel Armand. 2011. "Issues and challenges facing rechargeable lithium batteries." In *Materials for Sustainable Energy: A Collection of Peer-Reviewed Research and Review Articles From Nature Publishing Group*, 171–179. World Scientific.

Verger, Louisiane, Chuan Xu, Varun Natu, Hui-Ming Cheng, Wencai Ren, and Michel W Barsoum. 2019. "Overview of the synthesis of MXenes and other ultrathin 2D transition metal carbides and nitrides." *Current Opinion in Solid State and Materials Science* 23 (3):149–163. https://doi.org/10.1016/j.cossms.2019.02.001.

Wagner, Frederick T, Balasubramanian Lakshmanan, and Mark F Mathias. 2010. "Electrochemistry and the future of the automobile." *The Journal of Physical Chemistry Letters* 1 (14):2204–2219.

Wang, Caiyun, Dan Li, Chee O Too, and Gordon G Wallace. 2009. "Electrochemical properties of graphene paper electrodes used in lithium batteries." *Chemistry of Materials* 21 (13):2604–2606.

Wang, Hailiang, Yuan Yang, Yongye Liang, Joshua Tucker Robinson, Yanguang Li, Ariel Jackson, Yi Cui, and Hongjie Dai. 2011. "Graphene-wrapped sulfur particles as a rechargeable lithium–sulfur battery cathode material with high capacity and cycling stability." *Nano Letters* 11 (7):2644–2647.

Wang, Jia-Zhao, Lin Lu, Mustafa Lotya, Jonathan N Coleman, Shu-Lei Chou, Hua-Kun Liu, Andrew I Minett, and Jun Chen. 2013. "Development of MoS2–CNT composite thin film from layered MoS2 for lithium batteries." *Advanced Energy Materials* 3 (6):798–805.

Wang, Qing Hua, Kourosh Kalantar-Zadeh, Andras Kis, Jonathan N Coleman, and Michael S Strano. 2012. "Electronics and optoelectronics of two-dimensional transition metal dichalcogenides." *Nature Nanotechnology* 7:699. https://doi.org/10.1038/nnano.2012.193.

Wang, Siliang, Nishuang Liu, Jun Su, Luying Li, Fei Long, Zhengguang Zou, Xueliang Jiang, and Yihua Gao. 2017. "Highly stretchable and self-healable supercapacitor with reduced graphene oxide based fiber springs." *ACS Nano* 11 (2):2066–2074.

Wang, Xianfen, Satoshi Kajiyama, Hiroki Iinuma, Eiji Hosono, Shinji Oro, Isamu Moriguchi, Masashi Okubo, and Atsuo Yamada. 2015. "Pseudocapacitance of MXene nanosheets for high-power sodium-ion hybrid capacitors." *Nature Communications* 6 (1):6544. https://doi.org/10.1038/ncomms7544.

Wang, Xuefeng, Xi Shen, Yurui Gao, Zhaoxiang Wang, Richeng Yu, and Liquan Chen. 2015. "Atomic-scale recognition of surface structure and intercalation mechanism of Ti3C 2X." *Journal of the American Chemical Society* 137 (7):2715–2721.

Wang, Yukun, Lichun Yang, Renzong Hu, Liuzhang Ouyang, and Min Zhu. 2014. "Facile synthesis of Fe2O3-graphite composite with stable electrochemical performance as anode material for lithium ion batteries." *Electrochimica Acta* 125:421–426.

Wang, Zhong Lin. 2010. "Toward self-powered sensor networks." *Nano Today* 5 (6):512–514.

Whittingham, M Stanley. 2004. "Lithium batteries and cathode materials." *Chemical Reviews* 104 (10):4271–4302.

Winter, Martin, and Ralph J Brodd. 2005. "What are batteries, fuel cells, and supercapacitors?(Chem. Rev. 2003, 104, 4245–4269. Published on the Web 09/28/2004.)." *Chemical Reviews* 105 (3):1021–1021.

Wu, Mingmao, Yingru Li, Bowen Yao, Ji Chen, Chun Li, and Gaoquan Shi. 2016. "A high-performance current collector-free flexible in-plane micro-supercapacitor based on a highly conductive reduced graphene oxide film." *Journal of Materials Chemistry A* 4 (41):16213–16218.

Wu, Xianhong, Zhiyu Wang, Mengzhou Yu, Luyang Xiu, and Jieshan Qiu. 2017. "Stabilizing the MXenes by carbon nanoplating for developing hierarchical nanohybrids with efficient lithium storage and hydrogen evolution capability." *Advanced Materials* 29 (24):1607017.

Wu, Zhong-Shuai, Khaled Parvez, Xinliang Feng, and Klaus Müllen. 2013. "Graphene-based in-plane micro-supercapacitors with high power and energy densities." *Nature Communications* 4 (1):1–8.

Wu, Zhong-Shuai, Khaled Parvez, Xinliang Feng, and Klaus Müllen. 2014. "Photolithographic fabrication of high-performance all-solid-state graphene-based planar micro-supercapacitors with different interdigital fingers." *Journal of Materials Chemistry A* 2 (22):8288–8293.

Xiao, Xu, Tianqi Li, Zehua Peng, Huanyu Jin, Qize Zhong, Qiyi Hu, Bin Yao, Qiuping Luo, Chuanfang Zhang, and Li Gong. 2014. "Freestanding functionalized carbon nanotube-based electrode for solid-state asymmetric supercapacitors." *Nano Energy* 6:1–9.

Xie, Huiqi, Linfeng Hu, Feilong Wu, Min Chen, and Limin Wu. 2016. "Self-templated synthesis of ultrathin nanosheets constructed TiO2 hollow spheres with high electrochemical properties." *Advanced Science* 3 (11):1600162.

Xie, Xiuqiang, Taron Makaryan, Mengqiang Zhao, Katherine L Van Aken, Yury Gogotsi, and Guoxiu Wang. 2016. "MoS2 nanosheets vertically aligned on carbon paper: A freestanding electrode for highly reversible sodium-ion batteries." *Advanced Energy Materials* 6 (5):1502161.

Xie, Xiuqiang, Meng-Qiang Zhao, Babak Anasori, Kathleen Maleski, Chang E Ren, Jingwen Li, Bryan W Byles, Ekaterina Pomerantseva, Guoxiu Wang, and Yury Gogotsi. 2016. "Porous heterostructured MXene/carbon nanotube composite paper with high volumetric capacity for sodium-based energy storage devices." *Nano Energy* 26:513–523.

Xie, Yu, Michael Naguib, Vadym N Mochalin, Michel W Barsoum, Yury Gogotsi, Xiqian Yu, Kyung-Wan Nam, Xiao-Qing Yang, Alexander I Kolesnikov, and Paul RC Kent. 2014. "Role of surface structure on Li-ion energy storage capacity of two-dimensional transition-metal carbides." *Journal of the American Chemical Society* 136 (17):6385–6394.

Xin, Xing, Xufeng Zhou, Jinghua Wu, Xiayin Yao, and Zhaoping Liu. 2012. "Scalable synthesis of TiO2/graphene nanostructured composite with high-rate performance for lithium ion batteries." *Acs Nano* 6 (12):11035–11043.

Xiong, Guoping, Chuizhou Meng, Ronald G Reifenberger, Pedro P Irazoqui, and Timothy S Fisher. 2014. "A review of graphene-based electrochemical microsupercapacitors." *Electroanalysis* 26 (1):30–51.

Xiong, Pan, Renzhi Ma, Nobuyuki Sakai, Leanddas Nurdiwijayanto, and Takayoshi Sasaki. 2018. "Unilamellar metallic MoS2/graphene superlattice for efficient sodium storage and hydrogen evolution." *ACS Energy Letters* 3 (4):997–1005.

Xu, Shuaikai, Guodong Wei, Junzhi Li, Wei Han, and Yury Gogotsi. 2017. "Flexible MXene–graphene electrodes with high volumetric capacitance for integrated co-cathode energy conversion/storage devices." *Journal of Materials Chemistry A* 5 (33):17442–17451.

Yabuuchi, Naoaki, Kei Kubota, Mouad Dahbi, and Shinichi Komaba. 2014. "Research development on sodium-ion batteries." *Chemical Reviews* 114 (23):11636–11682.

Yang, Xiaowei, Chi Cheng, Yufei Wang, Ling Qiu, and Dan Li. 2013. "Liquid-mediated dense integration of graphene materials for compact capacitive energy storage." *Science* 341 (6145):534–537.

Yu, Yang-Xin. 2016. "Prediction of mobility, enhanced storage capacity, and volume change during sodiation on interlayer-expanded functionalized Ti_3C_2 MXene anode materials for sodium-ion batteries." *The Journal of Physical Chemistry C* 120 (10):5288–5296. https://doi.org/10.1021/acs.jpcc.5b10366.

Yuan, Hao, and Zhenyu Li. 2018. "Interfacial properties of black phosphorus/transition metal carbide van der Waals heterostructures." *Frontiers of Physics* 13 (3):138103.

Yue, Yang, Nishuang Liu, Yanan Ma, Siliang Wang, Weijie Liu, Cheng Luo, Hang Zhang, Feng Cheng, Jiangyu Rao, Xiaokang Hu, Jun Su, and Yihua Gao. 2018. "Highly self-healable 3D microsupercapacitor with MXene–graphene composite aerogel." *ACS Nano* 12 (5):4224–4232.

Zeng, ZY, JP Tu, YZ Yang, JY Xiang, XH Huang, F Mao, and M Ma. 2008. "Nanostructured Si/TiC composite anode for Li-ion batteries." *Electrochimica Acta* 53 (6):2724–2728.

Zhan, Xiaoxue, Chen Si, Jian Zhou, and Zhimei Sun. 2020. "MXene and MXene-based composites: Synthesis, properties and environment-related applications." *Nanoscale Horizons* 5 (2):235–258. https://doi.org/10.1039/C9NH00571D.

Zhang, Dengsong, Tingting Yan, Liyi Shi, Zheng Peng, Xiaoru Wen, and Jianping Zhang. 2012. "Enhanced capacitive deionization performance of graphene/carbon nanotube composites." *Journal of Materials Chemistry* 22 (29):14696–14704.

Zhang, Han, Nasir Mahmood Abbasi, and Bing Wang. 2021. *Semiconducting Black Phosphorus: From 2D Nanomaterial to Emerging 3D Architecture*. CRC Press.

Zhang, Hongji, Hesheng Xia, and Yue Zhao. 2012. "Poly (vinyl alcohol) hydrogel can autonomously self-heal." *ACS Macro Letters* 1 (11):1233–1236.

Zhang, Jincan, Li Lin, Luzhao Sun, Yucheng Huang, Ai Leen Koh, Wenhui Dang, Jianbo Yin, Mingzhan Wang, Congwei Tan, and Tianran Li. 2017. "Clean transfer of large graphene single crystals for high-intactness suspended membranes and liquid cells." *Advanced Materials* 29 (26):1700639.

Zhang, Li, and XS Zhao. 2009. "Carbon-based materials as supercapacitor electrodes." *Chemical Society Reviews* 38 (9):2520–2531.

Zhang, Qingrui, Jie Teng, Guodong Zou, Qiuming Peng, Qing Du, Tifeng Jiao, and Jianyong Xiang. 2016. "Efficient phosphate sequestration for water purification by unique sandwich-like MXene/magnetic iron oxide nanocomposites." *Nanoscale* 8 (13):7085–7093.

Zhang, Zheng, Pei Lin, Qingliang Liao, Zhuo Kang, Haonan Si, and Yue Zhang. 2019. "Graphene-based mixed-dimensional van der Waals heterostructures for advanced optoelectronics." *Advanced Materials* 31 (37):1806411. https://doi.org/10.1002/adma.201806411.

Zhao, Meng-Qiang, Chang E Ren, Zheng Ling, Maria R Lukatskaya, Chuanfang Zhang, Katherine L Van Aken, Michel W Barsoum, and Yury Gogotsi. 2015. "Flexible MXene/carbon nanotube composite paper with high volumetric capacitance." *Advanced Materials* 27 (2):339–345. https://doi.org/10.1002/adma.201404140.

Zhao, Meng-Qiang, Michelle Torelli, Chang E Ren, Michael Ghidiu, Zheng Ling, Babak Anasori, Michel W Barsoum, and Yury Gogotsi. 2016. "2D titanium carbide and transition metal oxides hybrid electrodes for Li-ion storage." *Nano Energy* 30:603–613.

Zhao, Meng-Qiang, Xiuqiang Xie, Chang E Ren, Taron Makaryan, Babak Anasori, Guoxiu Wang, and Yury Gogotsi. 2017. "Hollow MXene spheres and 3D macroporous MXene frameworks for Na-ion storage." *Advanced Materials* 29 (37):1702410.

Zhao, Ruizheng, Zhao Qian, Zhongyuan Liu, Danyang Zhao, Xiaobin Hui, Guanzhong Jiang, Chengxiang Wang, and Longwei Yin. 2019. "Molecular-level heterostructures assembled from layered black phosphorene and Ti_3C_2 MXene as superior anodes for high-performance sodium ion batteries." *Nano Energy* 65:104037. https://doi.org/10.1016/j.nanoen.2019.104037.

Zhao, Yunlong, Lin Xu, Liqiang Mai, Chunhua Han, Qinyou An, Xu Xu, Xue Liu, and Qingjie Zhang. 2012. "Hierarchical mesoporous perovskite La0. 5Sr0. 5CoO2. 91 nanowires with ultrahigh capacity for Li-air batteries." *Proceedings of the National Academy of Sciences* 109 (48):19569–19574.

Zheng, JP, PJ Cygan, and TR Jow. 1995. "Hydrous ruthenium oxide as an electrode material for electrochemical capacitors." *Journal of the Electrochemical Society* 142 (8):2699–2703.

Zhong, Yu, Xinhui Xia, Fan Shi, Jiye Zhan, Jiangping Tu, and Hong Jin Fan. 2016. "Transition metal carbides and nitrides in energy storage and conversion." *Advanced Science* 3 (5):1500286. https://doi.org/10.1002/advs.201500286.

Zhou, Guangmin, Lu Li, Da-Wei Wang, Xu-yi Shan, Songfeng Pei, Feng Li, and Hui-Ming Cheng. 2015. "A flexible sulfur-graphene-polypropylene separator integrated electrode for advanced Li–S batteries." *Advanced Materials* 27 (4):641–647.

8 Mixed-Dimensional Titanium-Based Structures in Photovoltaics and Catalysis

8.1 INTRODUCTION

Researchers have recently investigated MXenes, a fascinating area of new materials. The vast capabilities of these 2D materials, which originate from the broader group of transition metal carbides and nitrides, have been shown in several technological uses. Their unique attributes, such as their ability to attract water, high ability to conduct electricity, and adjustable surface chemistry, make them excellent contenders for making progress in solar cells (Kang et al. 2017), photodetectors (Montazeri et al. 2019), and catalysis (Guo et al. 2016; Seh et al. 2016). Photodetectors play a crucial role in several applications such as environmental monitoring and medical diagnostics.

The introduction of MXene-based photodetectors has revolutionized the field due to their exceptional electronic properties, resulting in superior performance. $Ti_3C_2T_x$ is particularly remarkable when used in combination with other materials to form heterostructures, making it stand out among the other MXenes (Yang et al. 2019). Heterostructures, specifically those made with n-Si, have remarkable light-detecting characteristics, facilitating the advancement of highly sensitive and efficient photodetectors. Similarly, the integration of MXenes with materials like InSe has resulted in devices that have improved sensitivity and velocity. MXene technology has provided significant benefits to the solar cell industry. By integrating MXenes, the drawbacks of conventional silicon-based solar cells can be mitigated, enhancing their effectiveness (Yu et al. 2019a; Guo et al. 2018; Fu et al. 2019; Yang et al. 2019; Xu et al. 2017). For instance, the combination of $Ti_3C_2T_x$ and SiO2 in heterostructures has shown promise in enhancing the durability and effectiveness of solar cells.

This combination capitalizes on the robust and stable nature of silicon and the exceptional conductive properties of MXenes, potentially resulting in significant advancements in solar energy harvesting (Fu et al. 2019).

MXenes are also exceptional in the field of catalysis, which is a fundamental component of environmental science and chemical engineering.

MXene-based heterostructures have demonstrated superior effectiveness in various catalytic processes (Guo et al. 2016; Seh et al. 2016). MXene-based catalysts significantly enhance durability and activity, particularly in water splitting for electrocatalysis (Brown et al. 2012; Wang et al. 2011). Moreover, MXenes are crucial in

DOI: 10.1201/9781003319658-8

the field of photocatalysis for converting CO_2 to methanol, which is vital for developing sustainable energy solutions (Chu, Cui, and Liu 2017; Mikkelsen, Jørgensen, and Krebs 2010; Habisreutinger, Schmidt-Mende, and Stolarczyk 2013; Sun et al. 2017). The study of MXenes has revealed new possibilities in multiple scientific fields. In addition to transforming catalytic processes and improving photodetectors and solar cells, these materials are indispensable for a future defined by sustainable solutions and technological progress.

The full extent of the potential of MXenes has not yet been fully recognized, and there are promising advancements on the horizon as research progresses. This book chapter explores the many roles of MXenes in photodetectors, solar cells, and catalytic processes. The material researcher analyzed the notable progress achieved in photodetectors using MXene materials, with a special emphasis on the exceptional capabilities of $Ti_3C_2T_x$ when integrated with n-Si and InSe in heterostructures. Scientists are now investigating the potential for enhancing the effectiveness of solar cells by integrating MXenes with silicon in $Ti_3C_2T_x$-SiO2 heterostructures in the realm of solar energy research. Moreover, the exceptional catalytic ability of MXenes is emphasized, demonstrating their effectiveness in electrocatalysis for the process of water splitting and in the photocatalytic transformation of CO_2 into methanol (Yu et al. 2019a; Guo et al. 2018; Fu et al. 2019; Yang et al. 2019; Xu et al. 2017). This book chapter highlights the significance of findings influence of MXenes in improving the photovoltaics and catalysis technology and advancing sustainable solutions.

8.2 PHOTODETECTORS

The exponential growth in the demands for information, data transmission, and distribution has led to a dramatic rise in the need for photodetectors, which have found extensive use in many different areas, but notably simplified fiber and data communications. The need for efficient data center operations and the realization of high-value bandwidth are significantly contributed by key areas for instance, the Internet of Things (IoT), virtual reality, artificial intelligence networks, and other internet-driven demands. Photodetectors play a vital part in enabling these advancements and ensuring the seamless flow of data in modern communication systems.

P-type, intrinsic n-type (PIN) photodiodes are commonly used (Burrus, Bowers, and Tucker 1985) along with metal-semiconductor-metal (MSM) photodiodes (Sugeta et al. 1980; Assefa et al. 2010) and avalanche photodiodes (APDs) (Lenox et al. 1999). Alternatively, metal-graphene-metal photodetectors (Koppens et al. 2014; Liu et al. 2014; Mueller, Xia, and Avouris 2010) are employed for fast and sensitive detection. Photodetectors are used for fast and sensitive identification. Even though there are many methods to reduce the dark current in MSM-based photodetectors, such as using semiconductors with larger bandgap, such as that of AlGaAs, to reduce Schottky crossover at metal-semiconductor junctions, these photodetectors still exhibit a broader noise spectrum compared to PIN photodiodes and APDs (Liou and Nabet 1996). A combination of optimized semiconductor conditions and a compact final size, along with industry-regulated transport gases, effectively blocks the entry of species with a negative charge from metal surfaces (Nabet et al. 2014).

All of the aforementioned strategies require larger surface areas for effective absorption, leading to higher responsiveness values (Shiue et al. 2015). On the other hand, the increased surface area of the system leads to longer electrical assembly times, which restrict the bandwidth due to rising RC time constants and power source transit times. Consequently, various applications require a compromise (Nabet 2015).

8.2.1 Photodetectors Using MXene Bases

The MXene family of 2D materials is regarded as a relatively recent innovation. The unique combination of hydrophilic qualities and high conductivity largely accounts for the many recently discovered technological applications of MXenes. Additionally, the ability to quickly and easily produce large volumes of stable aqueous colloidal suspensions is another significant advantage. This study isolated five unique MXene nanolayers—V_2CTx, T_2CTx, Nb_2CTx, $Ti_3C_2T_x$, and Mo_2CTx—from the parent MAX phase precursors. Among these MXene derivatives, $Ti_3C_2T_x$ exhibits the highest free carrier density and electrical conductivity, establishing it as the gold standard. This research presents a detailed examination of the morphological, mechanical, and steady-state optical characteristics of these materials (Anasori, Lukatskaya, and Gogotsi 2017; Li et al. 2018; El-Demellawi et al. 2018).

In general, nanostructure materials that have the same conductivity as metal are not good for photodetector uses because they do not have high dark currents. In the last ten years, MXenes have been used in many different ways and have reached their best values in many ways (Bahram et al. (Montazeri et al.). The successful fabrication of highly organized patterning is demonstrated in Figure 8.1, even though photolithography has its imperfections. The process involves using spin-coating to apply the MXene suspension onto GaAs plates and later removing it with an organic solvent like acetone. The resulting devices, depicted in the drawings, show images of the energy bands and were achieved at room temperature with features with sizes in the 2 m range. The functional gadget exhibited improved contact gaps compared to previous designs.

This low-cost, simple, robust, and scalable method produces photodetectors that are comparable to, and in many aspects superior to, materials like electrode networks made of Ti/Au. The choice of Ti_3C_2Tz MXene for the process is due to its extensive study and investigation.

To create the system shown in Figure 8.1b, the Ti_3AlC_2 MAX stage predecessors (shown in Figure 8.1a) were printed and separated. A patent is written on the surface of the GaAs substrate, as depicted in Figure 8.1c. The photoresist and contact lithography method, shown in Figure 8.1, are employed to translate the patent into a practical application in the real world. The colloidal solution is then put on the surface of the target suspender using a spin-coating process (Figure 8.1e).

Finally, the completed device is immersed in acetone solvent to dissolve and remove the photoresist. This process results in two MXene pads separated by a 2 m gap, as shown in Figure 8.1g's SEM image. It is observed that the corners and edges exhibit rapid movement, indicating that the lithography technique imposes limitations on the resolution power of the approach. The results obtained in this study demonstrate higher efficiency and better organization compared to previous work (Hu and Hua 2017; Xu et al. 2016a).

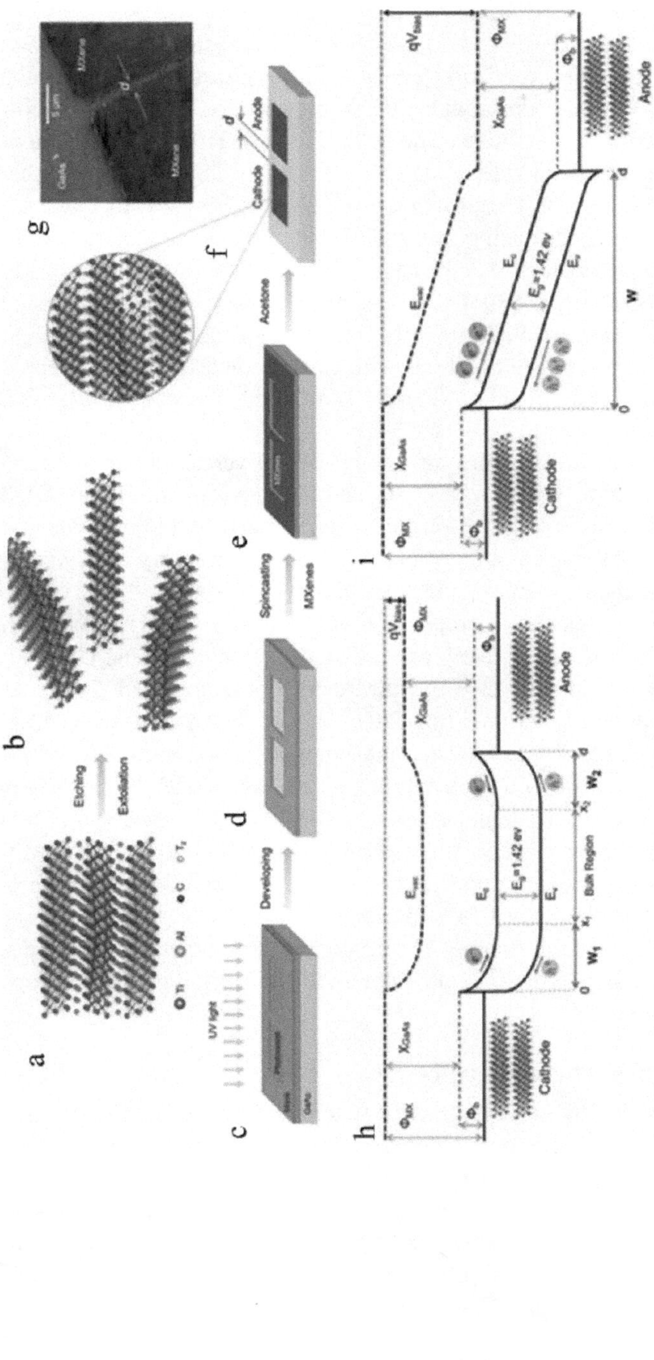

FIGURE 8.1 The MXene-GaAs-MXene heterostructure-based photodetector system manufacturing method and energy band diagrams (MX-S-MX). (a, b) The growth process of a colloidal suspension based on MXene using the classic photolithography technique. This results in exposed contact areas. (c, d) Photolithography method is performed, exposing zones of touch. (e) Aqueous colloidal MXene spin-coating suspension is applied. (f) Acetone is eliminated. (g) SEM pictures of the completed construction show the separation distance between the cathode and anode. (h) The energy band diagram sketch illustrates the device's energy band system under a mild bias, highlighting regions such as the depletion zone, intrinsic area, built-in voltage, barrier height, and Schottky junction areas. However, as shown in panel (h), applying a strong biased voltage results in flat band conditions.

Source: Permission awarded by Wiley 2019 (Montazeri et al. 2019).

This chapter demonstrate that this cost-effective, straightforward, and scalable technique yielded efficient and reliable photodetectors that outperformed other materials that were made up using Ti/Au-based electrodes system in terms of maximal statistics of distinction (FOM). The significance of this achievement cannot be overstated as it eliminates the need for complex equipment, apart from a spin-coater and a mask aligner, making the process simpler and more accessible.

Figure 8.1h illustrates the developed MX-S-MX device system's energy band representation under the influence of a little bias, while Figure 8.1i illustrates the same thing under the effect of a substantial external predilection. In order to facilitate side-by-side comparisons, a manufactured MSM network Ti/Au-GaAs-Ti/Au of the MX-based device system was developed using a well-established lithographic technique and mask arrangement. Panels a and g of Figure 8.2 present the current–voltage (I–V) behavior of devices using electrodes made of MXene and AU under various lighting conditions. In order to emphasize contact clarity, wavelengths of 532 nm and 780 nm are chosen, while 830 nm is important for short-range optical tele/data infrastructure. At various wavelengths, the $Ti_3C_2T_x$ MXene displays different dark powers. The dark current (I(dark)) varies with gap distance and is not affected by the applied potential. At an optical power of 0.34 W and a bias of 8 V, there is a significant variation in photocurrent.

Figure 8.2 illustrates an investigation of optical power versus photocurrent at various biases for wavelength values of 830 nm and 532 nm. Substituting MXene for Au significantly enhances the optical response, particularly at 830 nm, where a notable improvement is observed at an 8 V bias. The unique property of MXenes, such as Ti_3C_2Tz, lies in their tunable work function ranging from 2.14 to 5.65 eV, enabling surface terminal elemental composition with –O–, –F–, (Khazaei et al. 2015; Schultz et al. 2019; Hart et al. 2019) or –Cl, as well as hole doping (Li et al. 2019). Such a wide range is unparalleled in MSM detectors (Kang et al. 2017). At 830 nm, the detected value of quantum efficiency for MXene-GaAs-MXene is 42%, compared to 11% for Ti/Au MSM, as reported in a study at a maximum bias of 8 V. In contrast, the Ti/Au-responsivity device exhibits a value of 72 mA W^{-1}, whereas the MXene-GaAs-MXene devices demonstrate a responsivity value of 280 mA W^{-1} under an 8 V bias. It has been observed that photodetectors based on MX-S-MX structures often exhibit lower quantum efficiencies (QEs) compared to similar heterostructures using Ti/Au, which are 1.5–3 times higher.

Additionally, the potential for shaping ohmic connections utilizing the proper MXene surface characteristics process is attractive to the micro and nanoelectronics sector.

8.2.1.1 MXene and InSe Photodetectors

A thin-layer photodetector (PD) leveraging the carrier-friendly replication effect of Ti_2CT_x (MXene) is optimized for high-value systems. The avalanche-based photodetector utilizing InSe/Ti_2CT_x heterostructures can benefit from the structure of nanoribbon networks and an array of plasmonic gratings geometry enabled by Ti_2CT_x, thereby enhancing the photodetector's light-absorbing capacity. Lee et al. (Yang et al. 2019) fabricated InSe/Ti_2CT_xb heterostructures and evaluated their photodetector application. The InSe/Ti_2CT_x electrode functions as a photodetector (PD) whose performance is influenced by light intensity, as demonstrated in

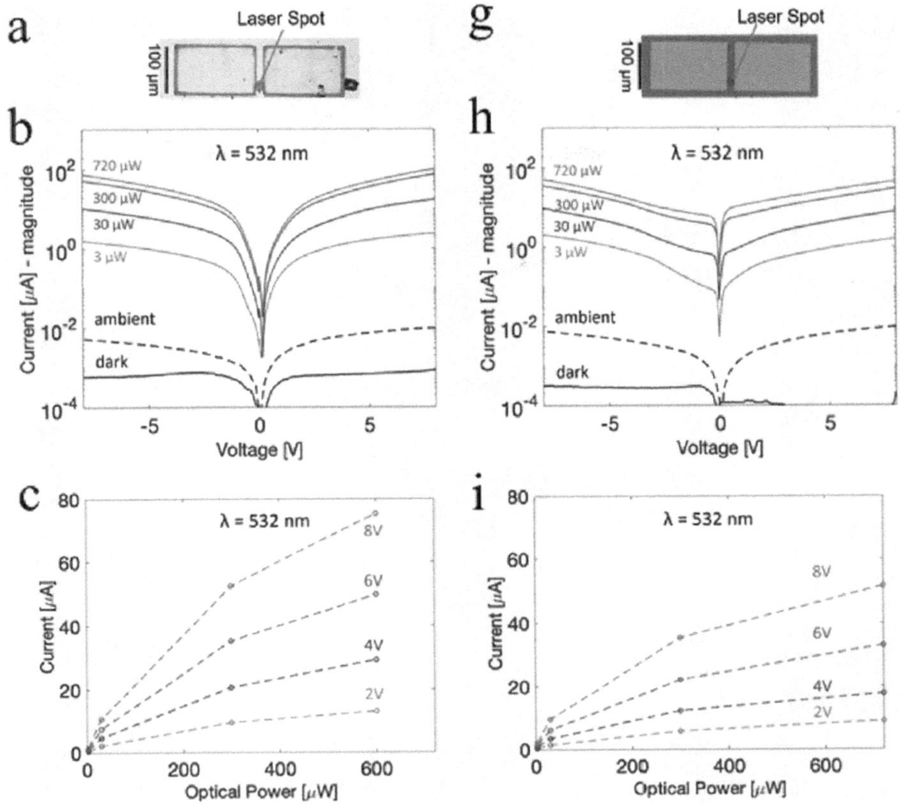

FIGURE 8.2 I–V characteristics of two photodetectors were compared: A sandwich-like A Ti/Au-GaAs-Ti/Au (MSM) detector and an MXene-GaAs-MXene (MX-S-MX) detector. (a) Micrographs of MX-S-MX photodetectors show overlapping laser spots. (g) The same micrographs but with the MSM photodetector. (b–e) Current–voltage curves (I–V characteristics) of the MX-S-MX photodetector at various wavelengths (532 nm, 780 nm, and 830 nm). (h–j) An MSM-based photodetector with I–V curves similar to (b)–(d) but at different wavelengths. (b, h) The reaction to 532 nm light. Photocurrent detectors with a gap of 10 m and a bias range under 532 nm light. Lower row: (d, j) Light reaction at 830 nm. When compared to the Ti/Au-based detector, the MX-S-MX detector has up to 300% higher values at all measured wavelengths, particularly at 830 nm photocurrent. (e, k) Response to 830 nm light with various biases and a 10 m gap.

Source: Adopted with permission from Wiley 2019 (Montazeri et al.).

Figure 8.3a and 8.3b. The voltage component of the dark current is used to analyze the disparity of heterostructures as a primary wastewater. The dark current value gradually increases with increasing discharge voltage and reaches saturation at 1.3 V of potential. As the drain voltage was increased further, the value of the dark current increased as a result of the applied high voltage's effect on the carrier increase.

The saturation current level, which is linearly accumulated, is a key measure of the ionization of heated carriers in InSe channels. Additionally, the researchers explore

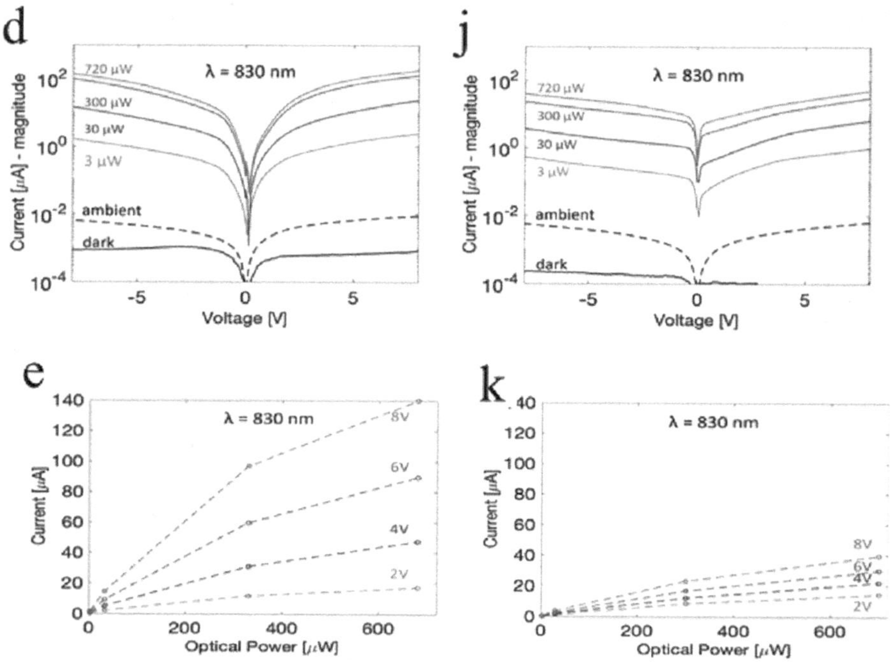

FIGURE 8.2 (Continued)

the correlation between ln(V) and ln(1–1/M), where M represents the multiplication factor of the charge carriers, defined as $M = I/I_{sat}$, with I_{sat} being the observed current value at a potential of 1.3 V. As the avalanche transport progresses, the inset in Figure 8.3b (top right) indicates the presence of electron–hole pairs due to the increasing linearity in the process of interacting with ln(1–1/M) and ln(V). With an electric field value approaching 52 kV/cm and a drain voltage exceeding 1.3 V, the carrier multiplication effect diminishes. Figure 8.3c clearly illustrates the influence of blue light, with a wavelength of 405 nm, on the interpretation of the drain current.

The avalanche value rises significantly, reaching a maximum of 500. Figure 8.3d depicts the time-resolved measurements' short rise time (2 ms) and decay time (25 ms).

The reaction time is longer compared to other photodetectors based on 2D materials, as previously discussed, possibly due to transmission and patterning technologies used during device fabrication, which introduce defects at material boundaries (Lei et al. 2015). Extending the reaction time window may be necessary during the optimization of device manufacturing processes. The fabrication of InSe/Ti_2CT_x APDs and the Fermi level modulation of Ti_2CT_x and InSe are monitored and controlled using Kelvin Probe Force Microscopy (KPFM) (Figure 8.3e). The calculated Fermi levels for Ti_2CT_x and InSe are 4.9 eV and 4.4 eV, respectively, consistent with literature references (Xu et al. 2016b).

On the other hand, MXenes demonstrate remarkable efficiency in converting light into heat (Chaudhuri et al. 2018). Further research is needed to explore the impact of photothermoelectric effects on photocurrent generation and heat dissipation efficiency in InSe and Ti_2CT_x MXene, despite the lack of direct relevance to

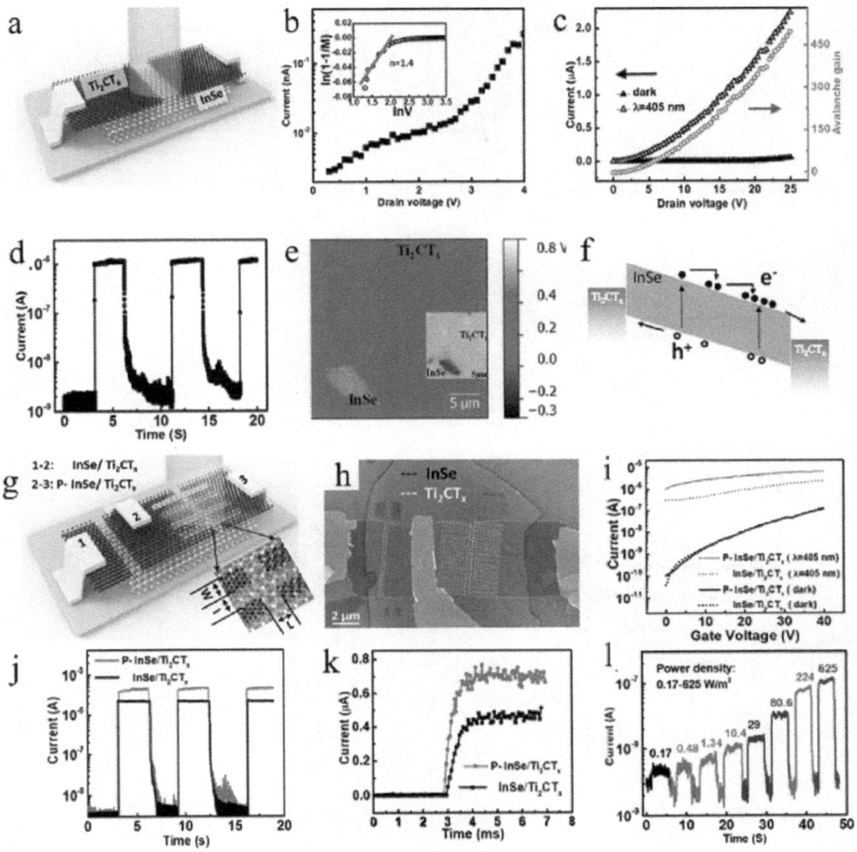

FIGURE 8.3 (a) Under light illumination, the heterostructure arrangement of an $InSe/Ti_2CT_x$ photodetector is depicted. The channel material is InSe, and the electrode material is Ti_2CT_x. (b) A logarithmic plot of the photodetector's dark current output at a gate voltage of 0 V reveals a direct proportionality between the natural logarithm of the inverse of the multiplication factor (ln(1/M)) and the natural logarithm of the voltage (lnV). (c) The photodetector's output curve under various light illuminations, including without light and with light at a wavelength of 405 nm and brightness of 625 W/m^2. (d) Time-resolved photoresponse measurements show a fast response time of less than 2 ms and a decay time of about 25 milliseconds. (e) Kelvin Probe Force Microscopy (KPFM) InSe and Ti_2CT_x nanoflakes can be seen on a silicon substrate in a potential mapping image. (f) The avalanche phenomenon is seen schematically in the $InSe/Ti_2CT_x$ heterostructure-based photodetector. (g) Unadorned and embellished $InSe/Ti_2CT_x$ heterostructure-based photodetectors are shown schematically and in a SEM image. (h) The light current and dark current characteristics of $InSe/Ti_2CT_x$ heterostructure-based photodetectors that are decorated and unadorned as a function of gate voltage. (i) High-resolution time-resolved photoresponse measurements and time-resolved photoresponse measurements. (j) The ornamented $InSe/Ti_2CT_x$ heterostructure avalanche photodetector's photoresponse curves under various lighting conditions. (k) Comparison of photocurrent values, responsivity values, and detectivity values for the decorated and undecorated $InSe/Ti_2CT_x$ heterostructures. (l) The plasmonic $InSe/Ti_2CT_x$ avalanche photodetector (7.3×10^{12} Jones) has a high detectivity, a high responsivity (1×10^5 AW1), and a low dark current (3 nA).

(Continued)

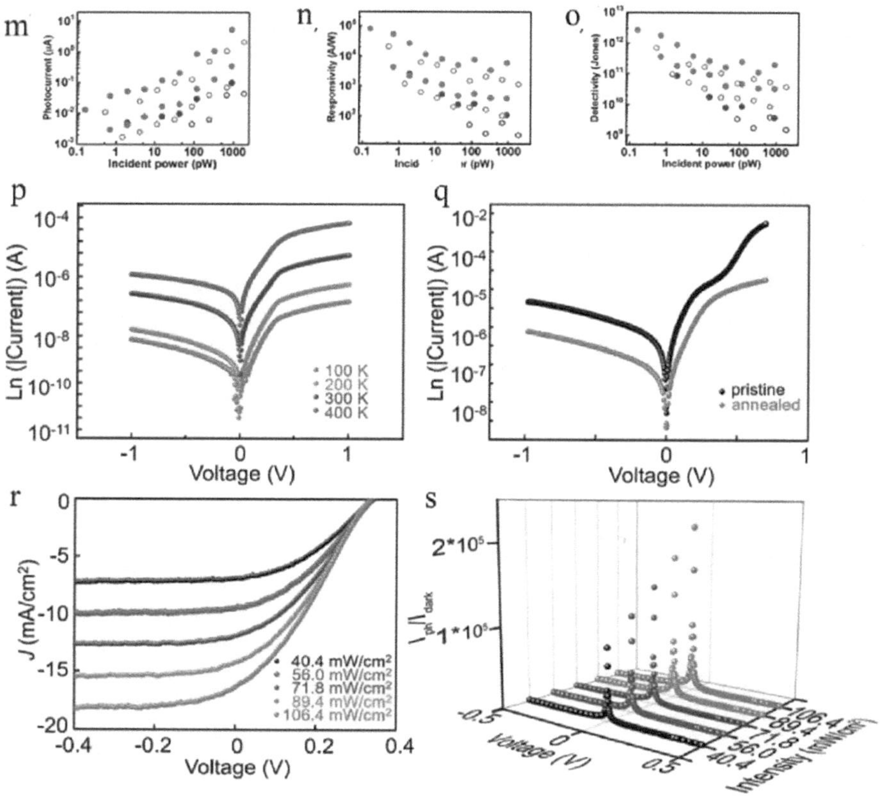

FIGURE 8.3 (Continued) (m) The variation in photocurrent values under a variety of conditions. (n) The calculated responsivities (R) from panel (m). Under 405 nm light illumination, the decorated $InSe/Ti_2CT_x$ APD has the highest detectivity (7.3×10^{12} Jones) and the shortest photoresponse time (0.8 ms), as shown in panel o. (p) At 100 K, 200 K, 300 K, and 350 K, the dark J–V properties of a $Ti_3C_2T_x$/n-Si heterostructure were measured and displayed on a logarithmic scale. (q) The J–V characteristics of the $Ti_3C_2T_x$/n-Si heterostructure were measured and plotted on a logarithmic scale before and after a vacuum process. (r) J–V curves of the $Ti_3C_2T_x$/n-Si electrode heterostructure were recorded under various light illuminations based on energy density. (s) The importance of photoresponsiveness and photocurrent (I_{ph}) to dark current (I_{dark}) ratio were calculated using the power dependence analysis in panel.

Sources: (n–o) Adopted with permission from American Chemical Society (Yang et al. 2019). (s) Permission awarded by Wiley 2017 (Kang et al. 2017).

avalanche multiplication devices (Yang et al. 2019). The Ti_2CT_x electrode network comprises MXene and InSe nanoflakes separated by a 0.7 eV thick Schottky barrier. This Schottky junction results in lower dark current values, and the avalanche effect can be controlled by applying a high bias voltage. The band alignment of the $InSe/Ti_2CT_x$ heterostructure-based photodetector has been successfully characterized, as depicted in Figure 8.3f.

The photoexcited carriers are accelerated to high energy levels by the high electric field range, resulting in the generation of additional electron–hole pairs and an increase in photocurrent value (Luo et al. 2015; Tamalampudi et al. 2014).

In Figure 8.3g, the visually striking unpatterned APD electrodes (InSe/Ti_2CT_x on the left) and the patterned APD (P-InSe/Ti_2CT_x on the right) both utilize equal amounts of Ti_2CT_x and InSe nanoflakes. The SEM images of the fabricated devices are presented in Figure 8.3h. The transfer characteristics, illustrating the variation of the drain current as a function of the gate voltage, are depicted in Figure 8.3i. Both the decorated and unadorned InSe/Ti_2CT_x heterostructure electrodes exhibit similar dark current levels in avalanche photodetectors (APDs). This similarity underscores their equivalent performance characteristics under low-light conditions. However, when illuminated, the decorated APD exhibits approximately three times the current of the undecorated APD, indicating enhanced photoresponse due to the Ti_2CT_x MXene nanoribbon morphology arrays. At 0 V gate voltage, the decorated APD exhibits more than three orders of magnitude of current amplification from light to dark. The time-resolved measurements in Figure 8.3j show that the decorated P-InSe/Ti_2CT_x heterostructure APD has a faster photoresponse behavior, with response times of 0.5 and 0.8 ms, as shown in Figure 8.3k. Figure 8.3l depicts the decorated electrode's progressive potential within the visible to near-infrared spectrum.

Due to its lower dark current, the coated APD exhibits an unconventional current response when illuminated, indicating increased light sensitivity. The incorporation of MXene Ti_2CT_x with a plasmonic-based geometry can substantially enhance the photoresponse behavior in the near-infrared (NIR) range. In the visible range, the InSe has a high photoresponse. In comparison to the InSe/Ti_2CT_x heterostructure electrode that does not exhibit plasmonic reaction due to geometry, the P-InSe/Ti_2CT_x heterostructure APD device (Figure 8.3l) exhibits strong photoresponse and detects a laser with extremely low power density. Figure 8.3m depicts the variation in photocurrent values under a variety of conditions.

Figure 8.3n presents the calculated responsivities (R) from Figure 8.3m. As incident intensity increases, responsivity values decrease due to the saturation of InSe nanolayers with light. Increased photocurrent leads to higher responsiveness of the P-InSe/Ti_2CT_x APD. The patterned InSe/Ti_2CT_x APD achieves a maximum photoresponsivity value of 1×10^5 AW when exposed to blue light (–1). Under illumination with 405 nm light, the decorated InSe/Ti_2CT_x APD demonstrates the highest detectivity of 7.3×10^{12} Jones and the shortest photoresponse time of 0.8 ms, as illustrated in Figure 8.3o. These measurements highlight the exceptional sensitivity and rapid response of the device (Yang et al. 2019). Table 8.1(1a) (photovoltaics column) summarizes the MXene-based 2D heterostructure-based device performance. The graph demonstrates that when incident light power rises, photocurrent values rise along with it, with the heterostructure layers becoming decorated and with InSe/Ti_2CT_x generally exhibiting higher photocurrents compared to the unmodified counterparts.

8.2.1.2 Photo Detectors Based on Heterostructures of $Ti_3C_2T_X$ and n-Si

MXenes have been mentioned as a potential new 2D material in the field of optoelectronics research on rare occasions. Previously, scientists investigated the properties of carbon nanotubes and graphene (Gr and CNTs) arranged on silicon in an effort to enhance comprehension of the potential applications that may be found in the field of graphene (Gr and CNTs) research (Hu et al. 1999; Li et al. 2010; Chen et al. 2011; Chen et al. 2017).

Silicon is highly valued for its commendable and dependable traits, including stable physical and chemical properties, an ideal band gap, affordability, and ease of use, all without compromising environmental friendliness. In a study led by Gao et al. (Kang et al. 2017), the researchers investigated van der Waals heterostructures using $Ti_3C_2T_x$ MXene on n-type silicon (n-Si) as the preferred substrate. Through a thorough examination and analysis of the properties of $Ti_3C_2T_x$/n-Si heterostructures, this research offered substantial potential for advancing the development of vertical junction photoelectrical devices (Zhu et al. 2014; Li et al. 2016b). Based on the $Ti_3C_2T_X$/n-Si crossing effect, $Ti_3C_2T_X$ MXene has a disproportionate effect on the van der Waals heterostructure. Despite a decrease in $Ti_3C_2T_X$ transmittance with increasing thin-layer concentration, $Ti_3C_2T_X$/n-Si heterostructure devices showed enhanced junction effects and conductivity.

The outcome of the junction is influenced by temperature, bias voltage, and curing conditions. Leveraging insights into the Schottky crossover effects in $Ti_3C_2T_X$/n-Si heterostructures, researchers have engineered a self-sustaining vertical junction system. They noted a gradual enhancement in the Schottky effect with rising temperatures. Nonetheless, as shown in Figure 8.3p, the inductive current increases progressively as the applied voltage is reversed and the temperature rises.

Instead of the Schottky junction effect, it is strongly advised that a thermally based stimulation mechanism be used to regulate the current concentration. Figure 8.3q shows the I–V curve relationship created by an air-cooled Schottky junction diode that was annealed at 200°C for 1,200 minutes at a temperature of 25 to 30°C. Before and after Schottky, the optimum impacts and barriers for $Ti_3C_2T_X$/n-Si heterostructures are 3.10 and 0.48 eV and 2.53 and 0.53 eV. After the annealing procedure, the $Ti_3C_2T_X$/n-Si film's boundary flaw was removed, as indicated by the optimal element and diminished current value as well as annealing method. And the $Ti_3C_2T_X$ MXene annealing process under the influence of Ar was able to eradicate the responsive capabilities of (F) in the model (Rakhi et al. 2015). According to the published scientific study, the work function of the unmodified Ti_3C_2 surface is greater than that of the Ti_3C_2 surface functionalized with fluorine (F) (Liu, Xiao, and Goddard III 2016). Therefore, annealing enhances the Schottky barriers of the $Ti_3C_2T_X$/n-Si heterostructure. The ability to enhance the Schottky barriers by rearrangement of reactive terminal groups on Si and MXene surfaces is demonstrated by this finding. At luminance intensities ranging from 40.4 to 106.4 mW cm^{-2}, the I–V topographies of the $Ti_3C_2T_X$/n-Si heterostructure system are examined, as illustrated in Figure 8.3r.

The photocurrent value has been steadily increasing as the system's brightness has been increased by a robust illuminating technique. Figure 8.3 depicts the extreme I_{ph}/I_{dark} ratio of the $Ti_3C_2T_X$/n-Si-based heterostructure system at zero-bias potential using a variety of illumination intensity levels. The I_{ph}/I_{dark} ratio significantly influences the zero-bias potential value, which is measured at 10^5. By leveraging heterostructures formed through van der Waals forces between $Ti_3C_2T_x$ and n-Si integrated circuits, highly reliable and self-sufficient photoelectric detectors can be developed (Kang et al. 2017).

The arrangement in the vertical plane is adequate for photon-mediated transporters to exit and transmit. Utilizing van der Waals forces, a Schottky contact can be formed between $Ti_3C_2T_X$ MXenes at temperatures ranging from 25 to 30°C, as demonstrated in dimensional analysis. The material consistently exhibits a work function value of 4.37 eV. An open-circuit voltage of 0.34 V and a short-circuit current density of 12.9 mA cm² were produced by this $Ti_3C_2T_X$/n-Si heterostructure device under 100 mW cm² of illumination. A self-powered photodetector leveraging $Ti_3C_2T_x$/n-Si heterostructures has been successfully developed, exhibiting an impressive I_{ph}/I_{dark} ratio of 10^5. This advanced device boasts a high responsiveness of 26.95 mA/W, a swift recovery time of 1.67 ms, and an exceptionally fast response time of 0.84 ms.

8.3 HETEROSTRUCTURES FOR SOLAR CELLS BASED ON MXENES

Solar energy has emerged as one of the cleanest and most sustainable energy sources. However, its effective utilization relies on access to sunlight. To overcome the limitations of solar energy availability, the integration of solar cells with energy storage devices has proven to be a viable solution. This approach allows for the efficient harnessing and storage of solar energy, enabling its utilization for various everyday tasks even during periods of limited sunlight (Yu et al. 2019a; Guo et al. 2018; Fu et al. 2019; Yang et al. 2019; Xu et al. 2017).

8.3.1 $Ti_3C_2T_x$-SiO_2

A heterojunction consisting of a transparent layer of conductivity made of an electrode and an n-type silicone (n-Si) wafer composed of MXene is used to create a new type of solar cell. The incorporation of an extremely thin layer of native silicon dioxide (SiO_2) is critical for facilitating efficient charge-carrier recombination. A solution-processable solar cell employing a $Ti_3C_2T_x$-on-Si heterojunction has achieved initial efficiencies of 5% under repeated AM 1.5 full sun illumination, making it highly desirable and cost-effective. Incorporating a straightforward chemical treatment involving HCl and $AuCl_3$ could further enhance photovoltaic (PV) efficiency. The use of HCl treatment to intensify surface doping effects on MXene layers significantly enhances the charge transport capabilities of fabricated solar cells, resulting in a 9% efficiency improvement.

By appropriately utilizing a thin anti-reflection layer, the solar cells demonstrate a typical efficiency exceeding 10% based on measurements from over 15 devices. Furthermore, these solar cells have a high power conversion efficiency (PCE) of around 11%, indicating a repeatable and valuable scheme for this type of solar cell.

This section describes a new class of solar cells that make use of a heterojunction of 2D $Ti_3C_2T_x$ (MXene) and Si materials (Yu et al. 2019a). The floating technique (Yu et al. 2019a) can be used to create the heterojunction with a negligible viscous SiO2 nanolayer (as depicted in Figure 8.4a). After the thermal oxide layer is removed, the $Ti_3C_2T_x$ films are easily detached from the Si substrate.

After exposing an Si substrate with a wet $Ti_3C_2T_x$/MCE (metal composite electrode) to 75°C in a vacuum oven, a thin layer of SiO_2 is formed due to oxidation

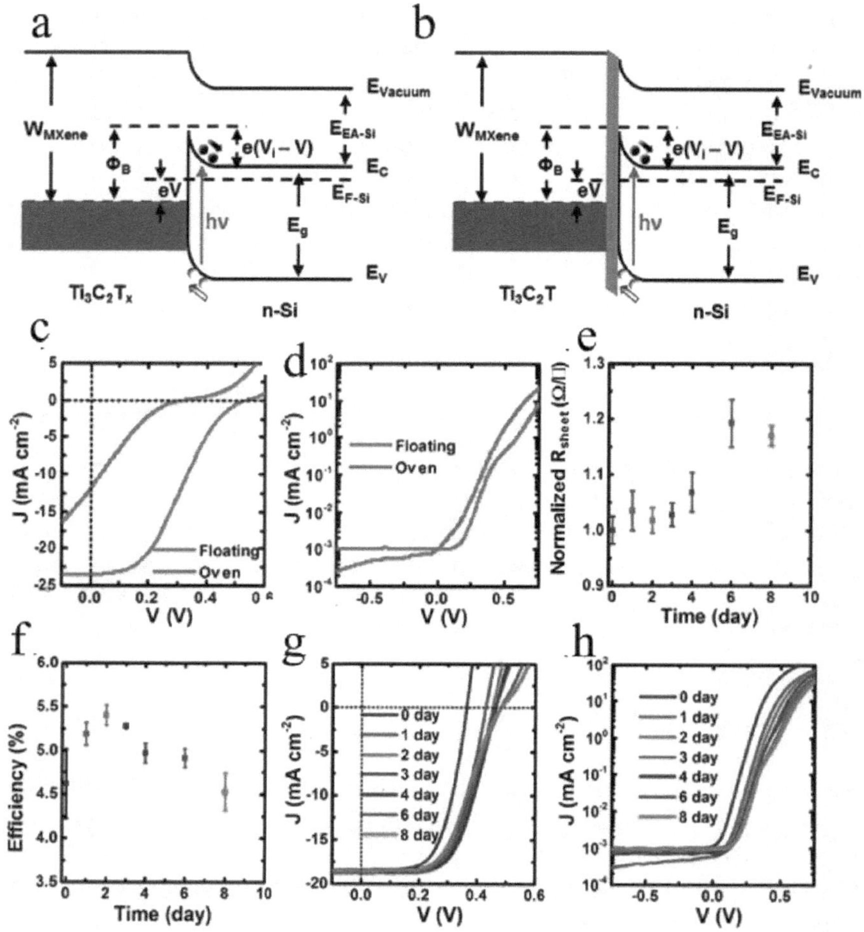

FIGURE 8.4 (a) In the $Ti_3C_2T_x$/Si heterojunction band diagram, WMXene represents the work function value of $Ti_3C_2T_x$ MXene, B represents the barrier height, V represents the output voltage of the solar cell, eVi represents the built-in potential, Eg represents the bandgap of Si, EEA-Si represents the electron affinity of silicon, and EF-Si represents the Fermi level of Si. EVacuum, EC, and EV represent the bottom, vacuum, and top levels of Si's conduction band and valence band, respectively. The presence of both negative and positive charges is represented by black and white zigzag dots, respectively. (b) The band diagram illustrates the $Ti_3C_2T_x$/SiO2/Si heterojunction, where the thin layer is depicted by the grey rectangle of SiO2 between the $Ti_3C_2T_x$ layer and the Si substrate. (c) J–V curves show the current–voltage characteristics of the solar cell under illumination conditions. (d) J–V curves display the solar cell's current–voltage characteristics under dark conditions using both techniques that use oven transmission and floating. (e) Measurement of uniform sheet resistance is performed on thin films of $Ti_3C_2T_x$ after subjecting them to 2% HF treatment. (f) The productivity curve (J–V) shows the current–voltage characteristics of the solar cell under light conditions. (g) J–V curve represents the current–voltage characteristics of the solar cell under light conditions. (h) J–V curve illustrates the current–voltage characteristics of the solar cell under dark conditions using the oven transfer process.

Source: Permission awarded by Wiley 2019 (Yu et al. 2019a).

caused by moisture presence. Figure 8.4b illustrates the potential formation of a $Ti_3C_2T_x/SiO_2/Si$ junction boundary at 25°C.

Panels c and d of Figure 8.4 depict typical J–V curves for both the floating and oven transfer techniques, along with the overall solar output of the fabricated devices. Introducing a SiO2 layer (0.17–0.35) between the Si and $Ti_3C_2T_x$ electrodes resulted in a decrease in the short-circuit current density (J_{sc}) from 10.86 to 22.77 mA/cm². However, both the fill factor (FF) and open-circuit voltage (V_{oc}) showed significant increases, rising from 0.319 to 0.525 V. Solar cells lacking an oxide layer exhibited an efficiency of 0.28%, while those with a thin SiO2 layer achieved an efficiency of 4.20%. The transparency value of HF-treated electrodes cannot be told apart after 8 days, but the sheet resistance value keeps going up (as shown in Figure 8.4e). As shown in Figure 8.4f, the effectiveness of the device went from 4.6% to 5.2% in 2 days during the production process. After 8 days in the open air at 25oC, it went down to 4.5%.

TiO_2 nanolayers were formed after 2 days, increasing competence. Nanolayers based on MO could be used in carbon nanotube/Si based solar cells as carrier dopants and antireflective coatings (Wang et al. 2015; Shi et al. 2012). However, higher SiO_2, TiO_2, and carbon formation as a result of the oxidation process reduces device efficiency over 8 days (VahidMohammadi et al. 2019). Previously, as-fabricated $Ti_3C_2T_x$/Si solar cells demonstrated efficiencies comparable to Gr/Si (1.5%) (Li et al. 2010) as well as perovskite-based solar cells (3.8%) (Kojima et al. 2009).

The J–V arcs of a typical oven-based device in both light and dark situations are shown in panels g and h of Figure 8.4. Electrodes placed between transparent electrically conducting silicon (Si) and a thin layer of silicon oxide (SiO2) can act as a barrier, preventing charge recombination. This oxide layer improves junction quality by preventing the recombination of electron–hole pairs generated by light absorption. Solar cells with reduced charge recombination exhibit superior overall performance, featuring higher open-circuit voltage (V_{oc}) and greater fill factor (FF). The utilization of chemical treatments like HCl and $AuCl_3$ to boost the solar cell's power conversion efficiency (PCE) from 5% to over 9%, coupled with the incorporation of a thin anti-reflection layer achieving over 10% efficiency, highlights the potential for future advancements in this emerging solar cell technology. The initial output of 11% efficiency is already quite high for newly developed solar cell, and the fact that these results can be replicated is a promising sign of the reliability and repeatability of the technology. With continued advancements and enhancements to different components of these solar cells, there is hope that they can be effectively utilized in commercial settings, further promoting the adoption of this cutting-edge technology for solar energy generation.

8.3.2 Solar Cells Made of Silicon and MXenes

MXene-based derivatives in solar system engineering have received little attention. A van der Waals heterostructure consisting of $Ti_3C_2T_x$ MXene and silicon was developed by Kang et al. to propose ultrafast perpendicular junction-based photodetectors (Kang et al. 2017). Figure 8.5a illustrates the Schottky contact at the interface between $Ti_3C_2T_x$ MXene and n-Si. $Ti_3C_2T_x$ MXene/n-Si heterostructures not only function as transparent electrodes but also regulate the generation and transport of photogenerated charge carriers. Furthermore, 2D $Ti_3C_2T_x$ MXene has significantly

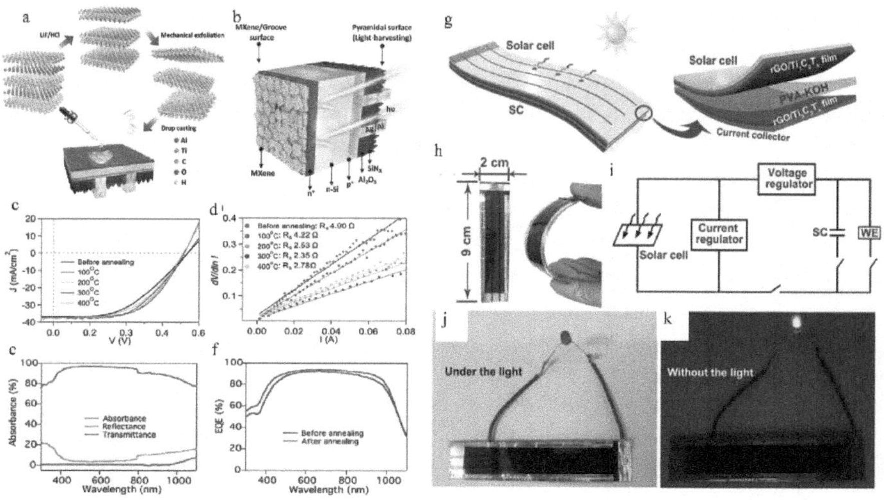

FIGURE 8.5 (a) The preparation process, which included the deposition of delaminated $Ti_3C_2T_x$ MXene nanosheets onto n+-Si surfaces followed by drop-casting, is depicted schematically. A method for fabricating a $Ti_3C_2T_x$ solar cell based on an MXene/n+np+-Si heterostructure is also detailed. (b) A $Ti_3C_2T_x$ MXene/n+np+-Si solar cell structure, emphasizing its layered configuration, is depicted. (c) The current density–voltage (J–V) characteristics of the $Ti_3C_2T_x$ MXene/n+np+-Si solar cell were analyzed both before and after rapid thermal annealing (RTA) at temperatures ranging from 100 to 400 degrees Celsius under AM 1.5 illumination. (d) The serial resistance (Rs) values were determined from dV/dln I versus I curves obtained before and after RTA treatment. The solar cell structure comprises layers including n+np+-Si for absorption, reference, and transmission. (e) The n+np+-Si material demonstrates favorable light absorption behavior across the 400 to 1000 nm wavelength range under sunlight. (f) The external quantum efficiency (EQE) spectra of the $Ti_3C_2T_x$ MXene/n+np+-Si solar cell were measured before and after RTA. (Fu et al. 2019). (g) A graphic illustration is given to show how the flexible coupled ECSD looks. (h) The produced integrated ECSD's surface form is seen in the digital picture. (i) A circuit illustrating the coordinated operation of the ECSD based on the co-cathode approach is shown. (j and k) Photosensitive pictures show how the ECSD is used in conjunction with supercapacitors to power an LED both with and without light.

Source: Adopted with permission from Wiley 2019.

higher conductivity than graphene produced from solutions (Lipatov et al. 2018). Its capability makes it a useful contact material for solar cells. Aside from that, the 2D $Ti_3C_2T_x$ MXene is significantly superior to that of graphene. The n+np+-Si-based solar cell undergoes meticulous modification with deep-doped n+ surfaces utilizing delaminated $Ti_3C_2T_x$ MXene as the foundation of its heterostructure. The resulting solar cell structure exhibits outstanding properties, including the fill factor (FF), short-circuit current density (J_{sc}), and high open-circuit voltage (V_{oc}). Expertly reducing contact resistance and fostering synergies that effectively eliminate alternate charge carriers in the high-contact area are pivotal to achieving these remarkable advancements. See Figure 8.5b for details. The current density–voltage (J–V) characteristics of $Ti_3C_2T_x$ MXene/n+np+-Si heterostructure solar cells under

AM 1.5G illumination at temperatures ranging from 100 to 400°C are illustrated in Figure 8.5c. In terms of power conversion efficiency (PCE), short-circuit current density (J_{sc}), open-circuit voltage (V_{oc}), and fill factor (FF), the $Ti_3C_2T_x$ MXene/n+ np+-Si-based solar heterostructure cell exhibited excellent performance metrics, achieving a noteworthy 9.53% prior to rapid thermal annealing (RTA).

However, as RTA temperatures increased, the device's performance gradually declined, reaching its lowest point at 400°C. Notably, the best results were obtained when the apparatus was heated to 300°C, yielding a PCE of about 11.5%, J_{sc} of 36.70 mA cm^2, V_{oc} of 0.54 V, and FF of 57.99%. The electrical conductivity of the $Ti_3C_2T_x$ MXene film increased to 300°C during the RTA procedure. Nevertheless, the conductivity experienced a precipitous decline above 300°C, suggesting that the MXene $Ti_3C_2T_x$ film is unstable at elevated temperatures.

Furthermore, the $Ti_3C_2T_x$ MXene-Si modeling of the solar cell system's contact resistance (Rs) was determined by analyzing the slope values obtained from the before and after RTA dV/dln I versus I plots. The Rs values for the system at 100, 200, 300, and 400°C were 4.90, 4.22, 2.53, 2.38, and 2.78, respectively (Figure 8.5d). Among all the temperatures tested, the device treated at 300°C had the lowest sheet resistance (Rs).

It should be noted that the interaction of the metal with Si, as well as the strain between the Si and metal layers influence the resistance (R) of a solar cell. High Rs values can have a negative impact on short-circuit current while having no effect on open-circuit voltage. Optical characterizations were used to achieve the overarching goal of developing highly efficient $Ti_3C_2T_x$ MXene-interacted Si solar cells.

According to Figure 8.5e, the n+np+-Si material exhibits a favorable light absorption behavior within the range of 400 to 1000 nm in sunlight. This could be attributed to the pyramidal-shaped surfaces of the material, combined with the presence of SiNx anti-reflection coatings (ARCs), which effectively capture and redirect light, preventing it from being reflected away. This effect allows the n+np+-Si material to absorb light efficiently. Applying a coating of $Ti_3C_2T_x$ MXene to a delaminated and coated Si solar cell improves absorption even more by reducing internal reflections. This speeds up the absorption process by making the light's conversion to electrical energy inside the Si material more efficient. Power conversion efficiency (PCE) of around 11.5% is achieved by the solar cell made of a few-layered Si coated with $Ti_3C_2T_x$ MXene.

The decrease in contact resistance is a result of the enhancement in the Ohmic contact between the highly doped silicon-based surface and the $Ti_3C_2T_x$ MXene metal. The increase in solar cell efficiency is a result of the decreased influence of junction charge carriers, leading to improvements in both the cell's open-circuit voltage (V_{oc}) and short-circuit current density (J_{sc}).

Rapid thermal annealing (RTA) is used to enhance the electrical contact between the Si and $Ti_3C_2T_x$ MXene layers. As a result, the resistance of both the series of cells and the $Ti_3C_2T_x$ MXene thin films (Rs) is reduced, resulting in a significant improvement in the overall performance of the solar cell. The fill factor (FF) has the potential to rise by a maximum of 20%, while the power conversion efficiency (PCE) may be enhanced by up to 11.5%.

The enhancements in PCE (Power Conversion Efficiency) and FF (Fill Factor) serve as evidence of the solar cell's exceptional quality and efficacy (refer to Figure 8.5f).

Moreover, the discoveries provide insight into the fundamental physics that drives the electrical connection between MXene and Si materials at their interfaces. Having a clear comprehension of this concept is essential for making progress in the development of MXene-based touch devices, as it enables the identification of possible obstacles and areas for improvement.

By addressing the intricacies of the MXene–Si interface, these findings pave the way for enhancing the performance of similar devices that rely on MXene-based contacts (Fu et al. 2019).

8.3.3 MXene and Graphene-Based Solar Cell

To meet these requirements, solid-state SCs based on rGO/$Ti_3C_2T_x$ heterostructures and thin-film solar cells were combined to create all-solid-state flexible energy conversion and storage devices (ECSDs) (Xu et al. 2017). Figure 8.5g shows how to build a flexible, bendable static SC combined using a thin foil-based solar cell, which, when exposed to sunlight, can provide an open-circuit voltage of 2.5 V. The PVA/KOH gel-based solid-state electrolyte is essential in enabling the energy conversion and storage process. Figure 8.5h shows digital photos that demonstrate how the integrated ECSD is set up.

To ensure effective operation and energy storage adaptation, a single working circuit was created that included the co-cathode-dependent ECSD, voltage regulator, and current regulator, as shown in Figure 8.5i. This integrated system enables group energy conversion and storage, which improves the efficiency of the devices.

In integrated co-cathode ECSDs, the performance of the thin-film solar cells is unaffected by a noticeable drop in operating voltage or overall functionality. This demonstrates that adding ECSDs does not degrade the solar cell's performance. The researchers have successfully developed flexible, integrated co-cathode ECSDs by seamlessly merging thin-film SCs with a thin-layer solar cell substrate. This innovative system of thin-film solar cells effectively powers both the red LED and the supercapacitor, as depicted in Figure 8.5j.

The SC supplies power to the LEDs, so they continue to work even when the power is turned off (Figure 8.5k) (Xu et al. 2017). The rGO/($Ti_3C_2T_x$ few-layer) combination structure improves the electrochemical efficacy of the electrodes without removing the functional groups present on the surface of $Ti_3C_2T_x$. Fabricating flexible supercapacitors (SCs) using the co-cathode technique included combining thin-film solar cell technology with rGO/$Ti_3C_2T_x$ thin films. The supercapacitors achieved an impressive volumetric energy density of 63 mWh/cm^3 and a high power density of 67 mW/cm^3. Additionally, they demonstrated a low self-discharge rate, maintaining 30 mW h cm^{-3} even at a higher power density of 2786 mW cm^{-3}. The fabricated ECSDs exhibited excellent durability, enduring 10,000 cycles at a current density of 5 A g-1 without any degradation. A 6 M aqueous solution of KOH was used as the electrolyte in these experiments. The integrated ECSD arrangement provides numerous potential applications to facilitate the creation of mobile and wearable electronic devices.

Table 8.1 from the photovoltaics section provides a summary of the device performance for solar cells built on 2D heterostructures made of MXene.

TABLE 8.1
Performance Comparison of MXene-Based Heterostructures in Photovoltaics and Catalysis

Serial Number	Major Field	Layer Structure	Efficiency of the Device	Application	Reference
1a	Photovoltaics	MXene-GaAs-MXene	The responsiveness value is 280 mA W^{-1}, and the quantum efficiency is 42%.	Photodetectors	(Montazeri et al. 2019)
		Ti/Au MSM	The efficiency of quantum systems is 11%; value of responsiveness approximating 72 mA W^{-1}	Photodetectors	(Montazeri et al. 2019)
		Ti_3C_2TX/n-Si	Current density in a short circuit 12.9 mA, an open-circuit voltage of 0.34 V, an I_{ph}/I_{dark} ratio of around 10^5, and a current-to-wattage ratio of 26.95 mA-W. Response time: 0.84 milliseconds, responsivity: 1.67 milliseconds	Photodetectors	(Kang et al. 2017)
		InSe/Ti_2CTx	Decreasing dark current (3 nA); the value of responsiveness is 1×10^5 AW^{-1}. The highest range of detectivity is 7.3×10^{12} Jones. Minimal photoresponse time is about 0.8 ms	Photodetectors	(Yang et al. 2019)
		$Ti_3C_2T_x$-Si heterojunction	Efficiency in converting power 11%	Solar cell	(Yu et al. 2019a)
		$Ti_3C_2T_x$ MXene/n+np+-Si solar cell	Effectiveness of power conversion 11.5%	Solar cell	(Fu et al. 2019)
1b	Catalysis	TiO_2/Ti_3C_2	Daily averages for O_2 production range from 0.07 mL to 0.15 mL, with H_2 emission ranging from 0.07 mL to 0.15 mL for each second. The ratio of the mixture of hydrogen and oxygen is 2:1.	Water splitting	(Yang et al. 2019)
		Ti_3C_2NPs/CdS	The activity of producing H_2 at 420 nm is 14,342 mmol h^{-1} g^{-1}, and the quantum efficiency is 40.1%. Amazing factor ratio is 136.6	Water splitting	(Ran et al. 2017)
		Ti_3C_2 QDs/Cu_2O NWs/Cu	The production value of methanol is 153.38 ppm cm^{-2}. Cu_2O nanowires/Cu have a carrier density ND of 1.09×10^{-3} cm^{-3}. The nuclear density (ND) of Ti_3C_2 sheets/Cu_2O NWs/Cu is 1.89×10^4 cm^{-3}, whereas the ND of Ti_3C_2 quantum dots/Cu_2O NWs/Cu is 2.75×10^6 cm^{-3}	Reduction of CO_2 to Methanol	(Zeng et al. 2019)

8.4 CATALYSIS BASED ON HETEROSTRUCTURES OF MXENE

MXene derivatives are ideally suited for a variety of uses because of their amazing chemical methodology and distinctive morphological engineering such as energy storage (Naguib et al. 2012; Lukatskaya et al. 2013), sensors (Yu et al. 2015), water purification (Peng et al. 2014), and catalysis (Guo et al. 2016; Seh et al. 2016). Particularly, rapid ion transport and storage are made possible by the thin-layered structure of MXenes, making them highly utilized in gadgets that store energy using metal ions (Zhou et al. 2017; Tang, Zhou, and Shen 2012) and supercapacitor electrodes (Lukatskaya et al. 2013). The oxygen reduction reaction (ORR), oxygen evolution reaction (OER), and hydrogen evolution reaction (HER) are essential electrochemical processes for renewable energy technologies, including metal-air batteries, fuel cells, and water electrolysis (Lee et al. 2016; Anantharaj et al. 2016; Nie, Li, and Wei 2015). Scalable synthesis techniques are being investigated in order to create inexpensive and effective catalysts in order to replace pricey noble metal catalysts.

To leverage the synergistic effects, transition metal and N-doped graphitic carbon composite materials have become a significant class of electrocatalysts for OER, HER, and ORR (Lee et al. 2016; Nie, Li, and Wei 2015; Deng et al. 2014). Given the excellent charge transfer kinetics and metallic properties of MXenes, the potential for developing a new class of catalysts made from composites of N-doped carbon materials and MXenes is quite promising. In the following sections, we will explore some of the most common methods for generating and storing energy.

8.4.1 Electrocatalysis

8.4.1.1 Water Splitting

The creation of a cutting-edge, environmentally responsible clean energy network is urgently required, alongside efficient energy storage systems (Hisatomi, Kubota, and Domen 2014; Ran et al. 2015; Fujishima and Honda 1972). Many researchers believe that hydrogen is an excellent energy source for the future because it is renewable, environmentally friendly, and highly efficient (Chen et al. 2010; Kudo and Miseki 2009). A promising and eco-friendly solution to these challenges is water electrolysis, which converts and stores renewable energy sources like wind and solar into hydrogen. Platinum (Pt)-based nanomaterials have proven to be highly effective in catalyzing the electrochemical hydrogen evolution reaction (HER), achieving significant success with minimal overpotentials. Nonetheless, the widespread utilization of Pt-based catalysts is impeded by their high market price and limited availability.

Despite advancements in the development of earth-abundant co-catalysts, there are still certain limitations associated with their inherent characteristics. One such limitation is the insufficient surface functionalities to enable strong interactions with photocatalysts. Problems like poor long-term stability and sluggish interfacial charge transfer may emerge from this. Additionally, these co-catalysts, such as graphene oxide (GO) (Hernandez et al. 2008), suffer from ineffective semiconducting/insulating conductivity and degradation of the p-conjugated system, which lead to inefficient electron flow. Additionally, there are challenges related to the need for a moisture-free environment, poor interaction with water molecules due to insufficient hydrophilic properties, and the unfavorable Gibbs energy for H_2 production (Brown et al. 2012; Wang et al. 2011). Therefore, investigating different co-catalysts that can get beyond these restrictions is essential.

MXene, a new class of more than 60 2D materials, has emerged in the last ten years including metal carbides, nitrides (Khazaei et al. 2013; Gao et al. 2015) or carbonitrides and has emerged as a promising material for various applications, including for making supercapacitors (Lukatskaya et al. 2013) and electrodes in lithium-ion batteries (Naguib et al. 2013).

MXene possesses remarkable properties that make it highly suitable for addressing the aforementioned issues. One key advantage is the existence of functions for hydrophilic terminal groups, like –OH and –O, on its surface. These functionalities allow for strong interactions with a wide range of semiconductor materials, facilitating improved performance and addressing issues related to insufficient surface functionalities. Second, MXene's excellent metallic conductivity facilitates efficient charge transport. Third, other surfaces, the exposed metal sites in MXene (such as Nb, Ti, or V), are more reactive to redox than carbon materials (Ghidiu et al. 2014). Fourth, MXene's hydrophilic surface features promote robust interaction with aqueous solutions, making it stable in such environments (Ran et al. 2017). Additionally, MXene demonstrates promising potential as a photocatalytic material due to its unique properties.

In 2019 (Yu et al. 2019), Ti_3C_2 MXene nanoparticles were synthesized and used in the fabrication of TiO_2/Ti_3C_2 heterostructured water splitting photoanodes for photoelectrochemical (PEC) processes by Zhang et al. Due to increased interfacial contact, the heterostructure displayed significantly better photoactivity and charge separation. Before and after loading Ti_3C_2, Raman spectroscopy study was done to look at the TiO_2 band structure (Yu et al. 2019a). The presence of Ti_3C_2 MXene was verified by the presence of a new peak at 143 cm^{-1}, in addition to the characteristic peaks of rutile TiO_2 (Figure 8.6a) (Liu et al. 2014). Electron spin resonance (ESR) measurements showed that unpaired electrons are present in Ti_3C_2, and the disappearance of these peaks after coupling with TiO_2 indicated charge transfer at the heterostructure boundaries (Figure 8.6) (Liu et al. 2015).

Utilizing X-ray photoelectron spectroscopy (XPS), the surface elemental composition, and oxidation state, states of TiO_2/Ti_3C_2 heterostructures were investigated (Xue et al. 2017). Peak intensities of $Ti_2p_{3/2}$ and $Ti_2p_{1/2}$ migrated to lower binding energies when compared to pristine TiO_2, indicating a reduction process (Figure 8.6c). The presence of Ti^{3+} species suggested that Ti_3C_2 and TiO_2 had a strong interaction (Liu et al. 2015). The presence of Ti-OH species at 532.2 eV in the O 1s spectrum confirmed TiO_2/enhanced Ti_3C_2's hydrophilicity, which is crucial for PEC water splitting (Figure 8.6d) (Zuo et al. 2010).

In summary, the unique properties of MXene, as demonstrated by the TiO_2/Ti_3C_2 heterostructure study, offer great potential for various applications, especially those in the field of photocatalysis.

Figure 8.6e illustrates TiO_2 photocurrent–voltage (I–V) curves and various Ti_3C_2 NP concentrations. TiO_2 alone exhibits a modest photocurrent under AM 1.5 illumination. However, all samples containing Ti_3C_2 NPs have increased photoactivity. An impressively high photocurrent density of 1.94 mA/cm^2 at 1.23 V versus RHE was

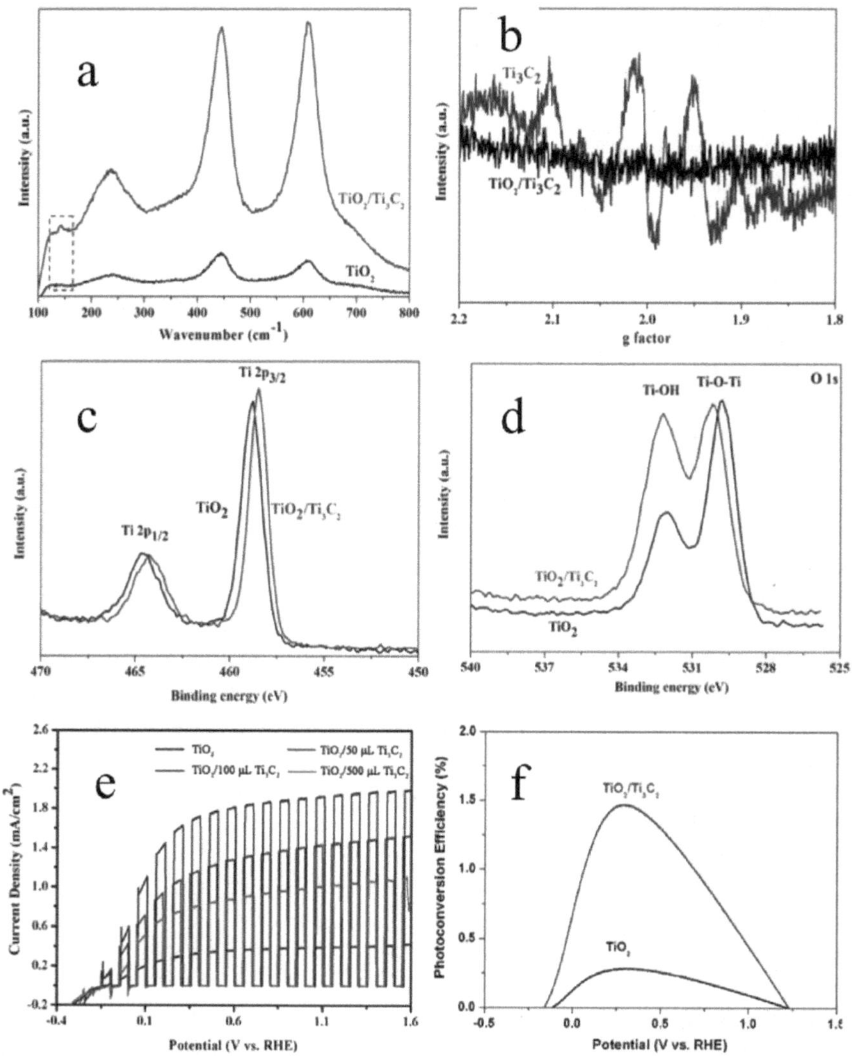

FIGURE 8.6 (a) Comparing the original and Raman spectra of TiO_2 nanoparticles (TiO_2NPs) with TiO_2/Ti_3C_2 heterostructures. (b) Electron Spin Resonance (ESR) spectra of Ti_3C_2 and TiO_2/Ti_3C_2 heterostructures. (c) Spectra with high resolution of Ti2p for TiO_2NPs and TiO_2/Ti_3C_2 heterostructures. (d) O 1s high-resolution spectrum for TiO_2NPs and TiO_2/Ti_3C_2 heterostructures. (e) The current density–voltage characteristics of TiO_2 nanoparticles (TiO_2NPs) and TiO_2/Ti_3C_2 composite electrodes were examined under AM 1.5 simulated solar irradiation. (f) TiO_2NPs and TiO_2/Ti_3C_2 photoanodes' photon-to-current conversion efficiencies at various applied voltages. Photoanodes based on TiO_2/Ti_3C_2 heterostructures were created using a simple and readily available method that included ultrasonic treatment and drop-casting. The resulting TiO_2/Ti_3C_2 photoanode demonstrated greater stability and improved photoactivity when exposed to AM 1.5 radiation.

Source: With permission, adopted. Permission awarded by Elsevier 2018 (Yang et al. 2019).

achieved with a 100 mL Ti_3C_2 lamination, which is six times greater than that of the original TiO_2 arrays at a concentration of 0.3 mg/mL.

The photocatalytic water splitting process produces H_2 and O_2, with the TiO_2/Ti_3C_2 photoanode producing 0.15 mL H_2 and 0.07 mL O_2 on a daily basis. The 2:1 H_2:O_2 ratio confirms successful water splitting. The applied bias photon-to-current efficiency (ABPE) of the TiO_2/Ti_3C_2 heterostructures is 1.47%, a significant improvement over the original TiO_2's 0.27% (Figure 8.6f) (Yang et al. 2019).

A study by Qiao et al. (Ran et al. 2017) explored, using density functional theory (DFT), the possibility of using Ti_3C_2 MXene as a co-catalyst for the generation of hydrogen (H2). Theoretical investigations, coupled with experimental techniques, investigated the efficacy of Ti_3C_2 quantum dots in conjunction with the photocatalyst CdS to enhance H_2 production. However, it was observed that pristine Ti_3C_2 and F-terminated Ti_3C_2 exhibited inadequate hydrogen evolution reaction (HER) activity and unfavorable electrochemical properties, rendering them unsuitable for H_2 production.

Another study found that the O-terminal groups on Ti_3C_2 were suitable co-catalyst candidates because of their good metallic conductivity, positive HER activity, and expected Fermi level (EF). During the experiment, Ti_3C_2 nanoparticles (NPs) were submerged in an aqueous solution of Cd(Ac)2, facilitating the adsorption of Cd^{2+} cations onto the O-terminal groups. The resultant suspension was then exposed to thiourea, an organic sulfur source, which then interacted with the Cd^{2+} ions. Hydrothermal treatment was then applied to the prepared combination or suspension. The breakdown of thiourea released S^{2-} anions into the solution, which then replaced most of the F-terminal species on the Ti_3C_2 nanoparticles with reactive functional groups like O/OH. As a result of interactions between these S^{2-} anions and the accumulating Cd^{2+} cations on their surface, Ti_3C_2 nanoparticles facilitated the formation and growth of CdS nanoparticles (CdSNPs).

As a result of the brief contact between the S^{2-} anions, excess Cd2+ cations were formed, which made it easier to use a similar nucleation technique to form pure CdS nanoparticles (CdSNPs).

The combination of CdS and Ti_3C_2 leads to the formation of well-organized structures, specifically sub-microsphere (SMS) structures resembling a cauliflower-like morphology (Ran et al. 2017). The connection between CdS and Ti_3C_2 nanoparticles significantly boosts the overall photocatalytic process. According to Figure 8.7a, bulk CdS (CT0) demonstrates a modest photocatalytic activity of 10^5 mmol h^{-1} g^{-1}.

In contrast, adding just a small amount of Ti_3C_2 nanoparticles (0.05 wt. percent) significantly enhances the photocatalytic activity to 993 mmol h^{-1} g^{-1}, as shown by CT0.05.

The photocatalytic performance of the Ti_3C_2-based CdS system improves with increasing Ti_3C_2 NPs. CT2.5 exhibits a remarkable 14,342 mmol h^{-1} g^{-1} photocatalytic H_2-generating activity, which is 136.6 times greater than CT0.

Under identical conditions (2.5 wt.% co-catalyst loading), the photocatalytic hydrogen evolution rates for Ni-CdS, NiS-CdS, and MoS_2-CdS heterostructures are 12,953, 8,649, and 6,183 mmol h^{-1} g^{-1}, respectively (Figure 8.7a). These results support the use of Ti_3C_2 nanoparticles as the most efficient and widely available co-catalysts. Multiple experiments were conducted to explore the impact of varying Ti_3C_2 content on the photocatalytic activity of the CdS/Ti_3C_2 composite.

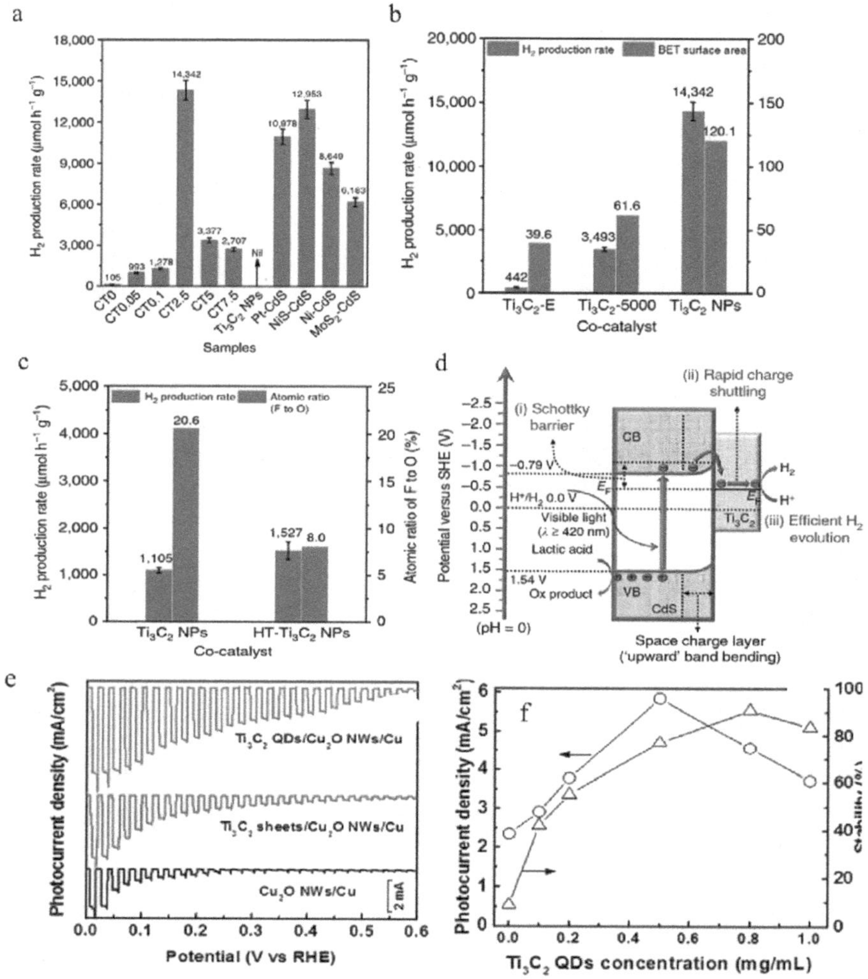

FIGURE 8.7 The performance of the photocatalyst and its enhanced form in the CdS/Ti_3C_2 configuration. (a) It was determined how photocatalytic H_2-generating behavior of NPs Ti_3C_2, PtCdS, NiS-CdS, Ni-CdS, and MoS_2-CdS performed. Standard deviation is indicated by error bars. (b) The area of the co-surface catalyst has a significant impact on photocatalytic activity. (c) The photocatalytic activity is affected by the co-F catalyst's to O atomic ratio. (d) Exposure to visible light (420 nm) radiation excites electrons from CdS' valence band (VB) to its conduction band (CB). With permission, adopted. (see Ran et al. 2017) (e) The performance of Cu_2O NWs/Cu, Ti_3C_2 sheets/Cu_2O NWs/Cu, and photoelectrochemical Ti_3C_2 QDs/Cu_2O NWs Cu was assessed. (f) Performance values were associated with photostability and photo current density of Ti_3C_2 QDs/Cu_2O NWs/Cu at various Ti_3C_2 QD concentrations (measured at 0 V versus RHE on the left axis). (g) The load transfer resistance of Ti_3C_2 QDs/Cu_2O NWs/Cu was calculated by examining the diameter of the previous arc in the electrochemical impedance spectroscopy (EIS) spectrum. (h) Photoelectrochemical measurements under simulated solar light illumination yielded the Mott–Schottky (M-S) curve.

Source: (h) Permission awarded by Wiley 2018 (Zeng et al. 2019).

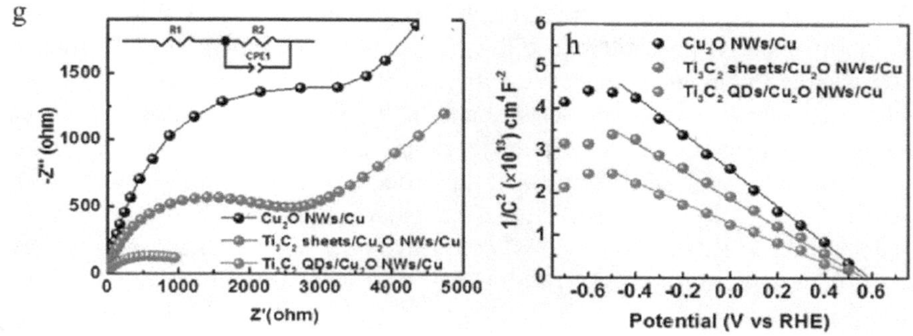

FIGURE 8.7 (Continued)

are suitable for use as high-performance photocatalysts and photoelectrodes that are both common on Earth and efficient at producing hydrogen through photocatalytic water splitting. For further information on how well 2D heterostructures based on MXene exhibit performance during water splitting, refer to Table 8.1(1b) in the catalysis section.

8.4.1.2 Photocatalysis/Conversion of CO_2 to Methanol

The excessive emission and accumulation of CO_2 pose a significant threat to sustainable development. There have been significant efforts to address this issue by converting carbon dioxide into sustainable fuels.

A promising class of materials for effective photoelectrochemical (PEC) and photocatalytic reduction includes semiconductor-based catalysts of CO_2. (Chu, Cui, and Liu 2017; Mikkelsen, Jørgensen, and Krebs 2010; Habisreutinger, Schmidt-Mende, and Stolarczyk 2013; Sun et al. 2017). Among these, p-type Cu_2O nanostructures have received a lot of attention because of their favorable conduction band position, which is more negative than CO_2's redox potential and a band gap (2.2 eV) that allows for efficient light absorption. (Li, Zhang, and MacFarlane 2015; Li and Kanan 2012; Li et al. 2016a; Schreier et al. 2016; Ghadimkhani et al. 2013). However, a major drawback of Cu_2O-based nanomaterials is their rapid recombination of photo-induced charge carriers during the redox reaction. Various tactics have been created over the past ten years to safeguard the surface of Cu_2O by incorporating protective and conductive layers, such as carbon-based materials like CNTs (Zhang et al. 2013; Kecsenovity et al. 2017), noble metals, or other semiconducting materials (Paracchino et al. 2011; Morales-Guio et al. 2014; Luo et al. 2016). These approaches, however, often require complex or exclusive fabrication processes. Recently, the potential of MXenes in photocatalysis has been explored (Naguib et al. 2014; Wang et al. 2018). For instance, Ti_3C_2 nanosheets have shown promising results as effective co-catalysts when coupled with TiO_2, CdS, C3N4, ZnS, or Bi2WO6, enhancing their photocatalytic activities (Ran et al. 2017; Low et al. 2018).

Peng et al. (Zeng et al. 2019) demonstrated the fabrication of Ti_3C_2 quantum dot (QD)/Cu_2O nanostructures (nanowires) through an electrostatic assembly method for solar-driven CO_2 conversion. The Ti_3C_2 QDs were found to significantly enhance the photocatalytic performance and stability of Cu_2O nanowires. Photocurrent

The size and surface area of the co-catalyst were first investigated. When Ti_3C_2-E, Ti_3C_2-5000, and Ti_3C_2 NP are loaded, it was discovered that increasing the surface area gradually improves the photocatalytic activity, as shown in Figure 8.7b. This can be attributed to Ti_3C_2's smaller size and higher visible energy position, which allow for faster reactant access and stronger coupling with CdS. We also investigated how the co-active catalyst's functional groups affected the efficiency of photocatalysis using the CdS/Ti_3C_2 heterostructure. In comparison to Ti_3C_2 NPs (Figure 8.7c), which had a surface area of 120.1 m^2 g^{-1} before hydrothermal treatment and a surface area of 56.7 m^2 g^{-1} after, HT-Ti_3C_2 NPs had an increased capacity for photocatalysis of 1,527 mmol h^{-1} g^{-1}.

The increased density of active sites, specifically the –O/–OH end groups that replaced the –F on Ti_3C_2 NPs, could explain the observed increase in photocatalytic activity. In contrast, after hydrothermal therapy, the floor space is reduced (Ran et al. 2017). Figure 8.7d depicts a photocatalytic mechanistic approach to explaining CT2.5's remarkably improved H2-generating activity based on these experimental findings. In CT2.5, there is a close interaction between CdS and Ti_3C_2, allowing electrons to transfer from CdS to Ti_3C_2 (somewhat below its conduction band position of 0.91 V versus standard hydrogen electrode (SHE)). The O-ended Ti_3C_2 has an original Fermi level (EF) of 1.88 V versus the SHE, which is less negative than that of n-type CdS. As a result, the CdS/Ti_3C_2 heterostructure network maintains a stable EF, since the Ti_3C_2 EF value is higher than the hydrogen evolution potential (0.00 V against SHE).

Jakob, Levanon, and Kamat (2003) described a similar phenomenon, where an electric charge layer is formed near the CdS/Ti_3C_2 interface due to the presence of CdS, which has a limited positive ion capacity. A consequence of this is that the CdS valence band (VB) and conduction band (CB) are aligned upwards. Consequently, between the CdS and Ti_3C_2 structures, a Schottky junction boundary is formed. Under visible light at 420 nanometers, CdS is excited from its lower valence band (VB) level to its higher conduction band (CB) position. However, the compact thickness of the space-charge layer, consisting of primary CdS nanoparticles, limits the vertical bending between the upper CB and lower VB locations of the CdS layer (Figure 8.7d) (Osterloh 2013; Leung et al. 2010; Yan et al. 2009; Yang et al. 2013).

In 2018, Wang and colleagues (Huang et al. 2019) developed a simple hydrothermal method to fabricate vertically aligned MoS_2 nanolayers on planar Ti_3C_2 nanolayers, forming hierarchical heterostructures. The hydrogen evolution reaction (HER) activity was significantly enhanced in these MoS_2/Ti_3C_2 heterostructures compared to bulk MoS_2 nanosheets, thanks to their strong interfacial coupling and unique layer structure. The charge transfer resistance of MoS_2/Ti_3C_2 heterostructures was measured at 14.74 U, much lower than that of pure MoS_2 (187.3 U), indicating faster interfacial charge transfer (182, 183). This efficient charge transfer was facilitated by the conductive Ti_3C_2 nanosheets, which acted as promoters. At an overpotential of 400 mV, the MoS_2/Ti_3C_2 heterostructure exhibited a catalytic current density around 6.2 times more than the bulk MoS_2 nanosheets (Xie et al. 2013; Xue et al. 2017a). These results highlight the potential of Ti_3C_2 MXene nanosheets as promising electrocatalysts for water splitting, offering a cost-effective and efficient alternative to scarce platinum-based catalysts. Overall, these results show that MXene materials measurements revealed that the Ti_3C_2 QDs/Cu_2O NWs/Cu heterostructure outperformed the other heterostructures in terms of light absorption, charge separation efficiency, and charge transport (Figure 8.7e). Moreover, the photostability of Ti_3C_2 QDs/Cu_2O NWs/Cu was exceptionally high, addressing the issue of Cu_2O NWs' poor stability in an open atmosphere due to oxidation. The experimental results also indicated that Cu^2O NWs had better interaction with Ti_3C_2 QDs compared to Ti_3C_2 sheets, highlighting the protective nature of Ti_3C_2. Additionally, the concentration of Ti_3C_2 QDs influenced the photocurrent and photostability, with an optimal concentration observed to promote favorable interactions without hindering light absorption (Figure 8.7f). The Ti_3C_2 QDs/Cu^2O NWs/Cu heterostructure produced the most methanol at a concentration of 153.38 ppm cm^2, outperforming the Cu_2O NWs/Cu heterostructure by 8.25 times and the Ti_3C_2 nanosheets/Cu_2O NWs/Cu heterostructure by 2.15 times. Ti_3C_2 QDs/Cu_2O NWs/Cu (2.75×10^{19} cm^3) had a higher carrier density (ND) than Ti_3C_2 sheets/Cu_2O NWs/Cu (1.89×10^{19} cm^3) and Cu_2O NWs/Cu (1.09×10^{19} cm3), indicating that conductive Ti_3C_2 nanosheets contributed to the improved performance.

The effectiveness of using Ti_3C_2 quantum dots (QDs) alongside Cu_2O nanowires (NWs) was examined using electrochemical impedance spectroscopy (EIS). As shown in Figure 8.7g, the Ti_3C_2 QDs/Cu_2O NWs/Cu heterostructure exhibited the lowest charge transfer resistance among the tested configurations, which included Cu_2O NWs/Cu and Ti_3C_2 sheets/Cu_2O NWs/Cu.

This is evident from the smaller semicircle radius in the EIS plot for Ti_3C_2 QDs/Cu_2O NWs/Cu. Additionally, Mott–Schottky (M-S) curves obtained from photoelectrochemical measurements under simulated solar light illumination were presented in Figure 8.7h.

Ti_3C_2 quantum dots (QDs) have been unearthed to bestow an exquisite augmentation upon the stability and photocatalytic efficiency of Cu_2O nanowires (NWs). Notably, the conspicuous descent observed in the graph signifies an intimate inverse association amidst the reciprocal of the square of the capacitance (expressed as 1/C2) and the momentous potential of Cu^2O as a p-type semiconductor. Intriguingly, when the band potentials of Cu_2O NWs merged with Cu, Ti_3C_2 sheets positioned atop Cu_2O NWs and Cu, and Ti_3C_2 QDs intercalated within Cu_2O NWs and Cu are compared, they are 0.59, 0.56, and 0.52 V, respectively (versus RHE, or reversible hydrogen electrode). As a result, the presence of Ti_3C_2 QDs at the interface reduces the occurrence of band bending when combined with Cu_2O NWs (Zeng et al. 2019).

The meticulous particulars concerning the efficacy of MXene-derived two-dimensional (2D) heterostructures in catalyzing the reduction of carbon dioxide (CO_2) to methanol can be readily perused within the confines of Table 8.1 as expounded in the corresponding section.

The notable improvement of efficiency of CO_2 reduction can be ascribed to the remarkable prowess exhibited by Ti_3C_2 quantum dots (QDs) in fostering the swift transmission of charges, mitigating the occurrence of band bending, and amplifying the density of carriers. These Ti_3C_2 QDs manifest themselves as an innovative zero-dimensional (0D) substance, replete with distinctive characteristics that proffer unprecedented prospects for forthcoming applications (Zeng et al. 2019).

8.5 SUMMARY

In conclusion, this chapter has emphasized the remarkable potential and a wide range of applications of MXene-based heterostructures in fields including optoelectronics, electronics, solar cells, and catalysis. For example, the integration of MXene materials such as $Ti_3C_2T_x$ with n-Si and InSe in photodetectors has resulted in substantial advancements in light detection technology, including increased sensitivity and efficiency that have the potential to revolutionize optical sensing.

The efficiency of solar cells has been demonstrated to be enhanced by the combination of $Ti_3C_2T_x$, SiO2, and silicon in solar cell technology. This approach provides a sustainable alternative to conventional materials and paves the way for the development of next-generation energy solutions.

Electrocatalysis for water splitting and photocatalysis for the conversion of CO_2 to methanol are two applications in which MXene-based heterostructures have shown exceptional potential in the field of catalysis. These applications underscore the versatility and efficacy of MXene materials in enabling critical chemical reactions, thereby furthering green energy technologies and promoting environmental sustainability.

As a whole, this chapter offers a thorough examination of the most recent advancements and potential applications of MXene-based heterostructures. MXene materials exhibit substantial potential for a diverse array of sophisticated technological applications, which emphasizes the necessity of ongoing research and development in this field. It is anticipated that the ongoing advancements in the synthesis, characterization, and application of MXene will stimulate additional innovations, thereby establishing their significance in the future of materials science and engineering.

REFERENCES

Anantharaj, Sengeni, Sivasankara Rao Ede, Kuppan Sakthikumar, Kannimuthu Karthick, Soumyaranjan Mishra, and Subrata Kundu. 2016. "Recent trends and perspectives in electrochemical water splitting with an emphasis on sulfide, selenide, and phosphide catalysts of Fe, Co, and Ni: A review." *ACS Catalysis* 6 (12):8069–8097.

Anasori, Babak, Maria R. Lukatskaya, and Yury Gogotsi. 2017. "2D metal carbides and nitrides (MXenes) for energy storage." *Nature Reviews Materials* 2:16098.

Assefa, Solomon, Fengnian Xia, Stephen W Bedell, Ying Zhang, Teya Topuria, Philip M Rice, and Yurii A Vlasov. 2010. "CMOS-integrated high-speed MSM germanium waveguide photodetector." *Optics Express* 18 (5):4986–4999.

Brown, Katherine A, Molly B Wilker, Marko Boehm, Gordana Dukovic, and Paul W King. 2012. "Characterization of photochemical processes for H_2 production by CdS nanorod–[FeFe] hydrogenase complexes." *Journal of the American Chemical Society* 134 (12):5627–5636.

Burrus, CA, JE Bowers, and RS Tucker. 1985. "Improved very-high-speed packaged InGaAs PIN punch-through photodiode." *Electronics Letters* 21 (7):262–263.

Chaudhuri, Krishnakali, Mohamed Alhabeb, Zhuoxian Wang, Vladimir M Shalaev, Yury Gogotsi, and Alexandra Boltasseva. 2018. "Highly broadband absorber using plasmonic titanium carbide (MXene)." *ACS Photonics* 5 (3):1115–1122.

Chen, Chun-Chung, Mehmet Aykol, Chia-Chi Chang, AFJ Levi, and Stephen B Cronin. 2011. "Graphene-silicon Schottky diodes." *Nano Letters* 11 (5):1863–1867.

Chen, Xiaobo, Shaohua Shen, Liejin Guo, and Samuel S Mao. 2010. "Semiconductor-based photocatalytic hydrogen generation." *Chemical Reviews* 110 (11):6503–6570.

Chen, Zefeng, Xinming Li, Jiaqi Wang, Li Tao, Mingzhu Long, Shi-Jun Liang, Lay Kee Ang, Chester Shu, Hon Ki Tsang, and Jian-Bin Xu. 2017. "Synergistic effects of plasmonics and electron trapping in graphene short-wave infrared photodetectors with ultrahigh responsivity." *ACS Nano* 11 (1):430–437.

Chu, Steven, Yi Cui, and Nian Liu. 2017. "The path towards sustainable energy." *Nature Materials* 16 (1):16–22.

Deng, Jiao, Pengju Ren, Dehui Deng, Liang Yu, Fan Yang, and Xinhe Bao. 2014. "Highly active and durable non-precious-metal catalysts encapsulated in carbon nanotubes for hydrogen evolution reaction." *Energy & Environmental Science* 7 (6):1919–1923.

El-Demellawi, Jehad K, Sergei Lopatin, Jun Yin, Omar F Mohammed, and Husam N Alshareef. 2018. "Tunable multipolar surface plasmons in 2D Ti_3C_2 T_x MXene flakes." *ACS Nano* 12 (8):8485–8493.

Fu, Hui-Chun, Vinoth Ramalingam, Hyunho Kim, Chun-Ho Lin, Xiaosheng Fang, Husam N Alshareef, and Jr-Hau He. 2019. "MXene-contacted silicon solar cells with 11.5% efficiency." *Advanced Energy Materials* 9 (22):1900180.

Fujishima, Akira, and Kenichi Honda. 1972. "Electrochemical photolysis of water at a semiconductor electrode." *Nature* 238 (5358):37–38.

Gao, Yupeng, Libo Wang, Aiguo Zhou, Zhengyang Li, Jingkuo Chen, Hari Bala, Qianku Hu, and Xinxin Cao. 2015. "Hydrothermal synthesis of TiO_2/Ti_3C_2 nanocomposites with enhanced photocatalytic activity." *Materials Letters* 150:62–64.

Ghadimkhani, Ghazaleh, Norma R de Tacconi, Wilaiwan Chanmanee, Csaba Janaky, and Krishnan Rajeshwar. 2013. "Efficient solar photoelectrosynthesis of methanol from carbon dioxide using hybrid CuO–Cu2O semiconductor nanorod arrays." *Chemical Communications* 49 (13):1297–1299.

Ghidiu, Michael, Maria R Lukatskaya, Meng-Qiang Zhao, Yury Gogotsi, and Michel W Barsoum. 2014. "Conductive two-dimensional titanium carbide 'clay' with high volumetric capacitance." *Nature* 516 (7529):78–81.

Guo, Zhanglin, Liguo Gao, Zhenhua Xu, Siowhwa Teo, Chu Zhang, Yusuke Kamata, Shuzi Hayase, and Tingli Ma. 2018. "High electrical conductivity 2D MXene serves as additive of perovskite for efficient solar cells." *Small* 14 (47):1802738. https://doi.org/10.1002/smll.201802738.

Guo, Zhonglu, Jian Zhou, Linggang Zhu, and Zhimei Sun. 2016. "MXene: A promising photocatalyst for water splitting." *Journal of Materials Chemistry A* 4 (29):11446–11452.

Habisreutinger, Severin N, Lukas Schmidt-Mende, and Jacek K Stolarczyk. 2013. "Photocatalytic reduction of CO_2 on TiO_2 and other semiconductors." *Angewandte Chemie International Edition* 52 (29):7372–7408.

Hart, James L, Kanit Hantanasirisakul, Andrew C Lang, Babak Anasori, David Pinto, Yevheniy Pivak, J Tijn van Omme, Steven J May, Yury Gogotsi, and Mitra L Taheri. 2019. "Control of MXenes' electronic properties through termination and intercalation." *Nature Communications* 10 (1):522. https://doi.org/10.1038/s41467-018-08169-8.

Hernandez, Yenny, Valeria Nicolosi, Mustafa Lotya, Fiona M Blighe, Zhenyu Sun, Sukanta De, IT McGovern, Brendan Holland, Michele Byrne, and Yurii K Gun'Ko. 2008. "High-yield production of graphene by liquid-phase exfoliation of graphite." *Nature Nanotechnology* 3 (9):563.

Hisatomi, Takashi, Jun Kubota, and Kazunari Domen. 2014. "Recent advances in semiconductors for photocatalytic and photoelectrochemical water splitting." *Chemical Society Reviews* 43 (22):7520–7535.

Hu, Haibo, and Tao Hua. 2017. "An easily manipulated protocol for patterning of MXenes on paper for planar micro-supercapacitors." *Journal of Materials Chemistry A* 5 (37):19639–19648.

Hu, Jiangtao, Min Ouyang, Peidong Yang, and Charles M Lieber. 1999. "Controlled growth and electrical properties of heterojunctions of carbon nanotubes and silicon nanowires." *Nature* 399 (6731):48–51.

Huang, Lan, Lunhong Ai, Mei Wang, Jing Jiang, and Shaobin Wang. 2019. "Hierarchical MoS_2 nanosheets integrated Ti_3C_2 MXenes for electrocatalytic hydrogen evolution." *International Journal of Hydrogen Energy* 44 (2):965–976.

Jakob, Manuela, Haim Levanon, and Prashant V Kamat. 2003. "Charge distribution between UV-irradiated TiO2 and gold nanoparticles: Determination of shift in the Fermi level." *Nano Letters* 3 (3):353–358.

Kang, Zhe, Yanan Ma, Xinyu Tan, Miao Zhu, Zhi Zheng, Nishuang Liu, Luying Li, Zhengguang Zou, Xueliang Jiang, Tianyou Zhai, and Yihua Gao. 2017. "MXene–silicon Van Der Waals heterostructures for high-speed self-driven photodetectors." *Advanced Electronic Materials* 3 (9):1700165. https://doi.org/10.1002/aelm.201700165.

Kecsenovity, Egon, Balázs Endrődi, Péter S Tóth, Yuqin Zou, Robert AW Dryfe, Krishnan Rajeshwar, and Csaba Janáky. 2017. "Enhanced photoelectrochemical performance of cuprous oxide/graphene nanohybrids." *Journal of the American Chemical Society* 139 (19):6682–6692.

Khazaei, Mohammad, Masao Arai, Taizo Sasaki, Chan-Yeup Chung, Natarajan S Venkataramanan, Mehdi Estili, Yoshio Sakka, and Yoshiyuki Kawazoe. 2013. "Novel electronic and magnetic properties of two-dimensional transition metal carbides and nitrides." *Advanced Functional Materials* 23 (17):2185–2192.

Khazaei, Mohammad, Masao Arai, Taizo Sasaki, Ahmad Ranjbar, Yunye Liang, and Seiji Yunoki. 2015. "OH-terminated two-dimensional transition metal carbides and nitrides as ultralow work function materials." *Physical Review B* 92 (7):075411.

Kojima, Akihiro, Kenjiro Teshima, Yasuo Shirai, and Tsutomu Miyasaka. 2009. "Organometal halide perovskites as visible-light sensitizers for photovoltaic cells." *Journal of the American Chemical Society* 131 (17):6050–6051.

Koppens, FHL, T Mueller, Ph Avouris, AC Ferrari, MS Vitiello, and M Polini. 2014. "Photodetectors based on graphene, other two-dimensional materials and hybrid systems." *Nature Nanotechnology* 9 (10):780.

Kudo, Akihiko, and Yugo Miseki. 2009. "Heterogeneous photocatalyst materials for water splitting." *Chemical Society Reviews* 38 (1):253–278.

Lee, Dong Un, Pan Xu, Zachary P Cano, Ali Ghorbani Kashkooli, Moon Gyu Park, and Zhongwei Chen. 2016. "Recent progress and perspectives on bi-functional oxygen electrocatalysts for advanced rechargeable metal–air batteries." *Journal of Materials Chemistry A* 4 (19):7107–7134.

Lei, Sidong, Fangfang Wen, Liehui Ge, Sina Najmaei, Antony George, Yongji Gong, Weilu Gao, Zehua Jin, Bo Li, and Jun Lou. 2015. "An atomically layered InSe avalanche photodetector." *Nano Letters* 15 (5):3048–3055.

Lenox, C, H Nie, P Yuan, G Kinsey, AL Homles, BG Streetman, and JC Campbell. 1999. "Resonant-cavity InGaAs-InAlAs avalanche photodiodes with gain-bandwidth product of 290 GHz." *IEEE Photonics Technology Letters* 11 (9):1162–1164.

Leung, Dennis YC, Xianliang Fu, Cuifang Wang, Meng Ni, Michael KH Leung, Xuxu Wang, and Xianzhi Fu. 2010. "Hydrogen production over titania-based photocatalysts." *ChemSusChem* 3 (6):681–694.

Li, Christina W, and Matthew W Kanan. 2012. "CO_2 reduction at low overpotential on Cu electrodes resulting from the reduction of thick Cu_2O films." *Journal of the American Chemical Society* 134 (17):7231–7234.

Li, Fei, Li Zhang, Jincheng Tong, Yingliang Liu, Shengang Xu, Yan Cao, and Shaokui Cao. 2016a. "Photocatalytic CO_2 conversion to methanol by Cu_2O/graphene/TNA heterostructure catalyst in a visible-light-driven dual-chamber reactor." *Nano Energy* 27:320–329.

Li, Guangjiang, Kateryna Kushnir, Yongchang Dong, Sergii Chertopalov, Apparao M Rao, Vadym N Mochalin, Ramakrishna Podila, and Lyubov V Titova. 2018. "Equilibrium and non-equilibrium free carrier dynamics in 2D $Ti_3C_2T_x$ MXenes: THz spectroscopy study." *2D Materials* 5 (3):035043.

Li, Haitao, Xinyi Zhang, and Douglas R MacFarlane. 2015. "Carbon quantum dots/Cu_2O heterostructures for solar-light-driven conversion of CO_2 to methanol." *Advanced Energy Materials* 5 (5):1401077.

Li, Mian, Jun Lu, Kan Luo, Youbing Li, Keke Chang, Ke Chen, Jie Zhou, Johanna Rosen, Lars Hultman, and Per Eklund. 2019. "Element replacement approach by reaction with Lewis acidic molten salts to synthesize nanolaminated MAX phases and MXenes." *Journal of the American Chemical Society* 141 (11):4730–4737.

Li, Xinming, Hongwei Zhu, Kunlin Wang, Anyuan Cao, Jinquan Wei, Chunyan Li, Yi Jia, Zhen Li, Xiao Li, and Dehai Wu. 2010. "Graphene-on-silicon Schottky junction solar cells." *Advanced Materials* 22 (25):2743–2748.

Liou, Liann-Chern, and Bahram Nabet. 1996. "Simple analytical model of bias dependence of the photocurrent of metal–semiconductor–metal photodetectors." *Applied Optics* 35 (1):15–23.

Lipatov, Alexey, Haidong Lu, Mohamed Alhabeb, Babak Anasori, Alexei Gruverman, Yury Gogotsi, and Alexander Sinitskii. 2018. "Elastic properties of 2D $Ti_3C_2T_x$ MXene monolayers and bilayers." *Science Advances* 4 (6):eaat0491.

Liu, Chang-Hua, You-Chia Chang, Theodore B Norris, and Zhaohui Zhong. 2014a. "Graphene photodetectors with ultra-broadband and high responsivity at room temperature." *Nature Nanotechnology* 9 (4):273–278.

Liu, Yuanxu, Bingsen Zhang, Liangfeng Luo, Xuanye Chen, Zhonglei Wang, Erlong Wu, Dangsheng Su, and Weixin Huang. 2015. "TiO_2/Cu_2O core/ultrathin shell nanorods as efficient and stable photocatalysts for water reduction." *Angewandte Chemie International Edition* 54 (50):15260–15265.

Liu, Yuanyue, Hai Xiao, and William A Goddard III. 2016. "Schottky-barrier-free contacts with two-dimensional semiconductors by surface-engineered MXenes." *Journal of the American Chemical Society* 138 (49):15853–15856.

Low, Jingxiang, Liuyang Zhang, Tong Tong, Baojia Shen, and Jiaguo Yu. 2018. "TiO_2/MXene Ti_3C_2 composite with excellent photocatalytic CO_2 reduction activity." *Journal of Catalysis* 361:255–266.

Lukatskaya, Maria R, Olha Mashtalir, Chang E Ren, Yohan Dall'Agnese, Patrick Rozier, Pierre Louis Taberna, Michael Naguib, Patrice Simon, Michel W Barsoum, and Yury Gogotsi. 2013. "Cation intercalation and high volumetric capacitance of two-dimensional titanium carbide." *Science* 341 (6153):1502. https://doi.org/10.1126/science.1241488.

Luo, Jingshan, Ludmilla Steier, Min-Kyu Son, Marcel Schreier, Matthew T Mayer, and Michael Grätzel. 2016. "Cu_2O nanowire photocathodes for efficient and durable solar water splitting." *Nano Letters* 16 (3):1848–1857.

Luo, Wengang, Yufei Cao, Pingan Hu, Kaiming Cai, Qi Feng, Faguang Yan, Tengfei Yan, Xinhui Zhang, and Kaiyou Wang. 2015. "Gate tuning of high-performance InSe-based photodetectors using graphene electrodes." *Advanced Optical Materials* 3 (10):1418–1423.

Mikkelsen, Mette, Mikkel Jørgensen, and Frederik C Krebs. 2010. "The teraton challenge. A review of fixation and transformation of carbon dioxide." *Energy & Environmental Science* 3 (1):43–81.

Montazeri, Kiana, Marc Currie, Louisiane Verger, Pouya Dianat, Michel W Barsoum, and Bahram Nabet. 2019. "Beyond gold: Spin-coated Ti_3C_2-based MXene photodetectors." *Advanced Materials* 31 (43):1903271. https://doi.org/10.1002/adma.201903271.

Morales-Guio, Carlos G, S David Tilley, Heron Vrubel, Michael Grätzel, and Xile Hu. 2014. "Hydrogen evolution from a copper (I) oxide photocathode coated with an amorphous molybdenum sulphide catalyst." *Nature Communications* 5 (1):1–7.

Mueller, Thomas, Fengnian Xia, and Phaedon Avouris. 2010. "Graphene photodetectors for high-speed optical communications." *Nature Photonics* 4 (5):297.

Nabet, Bahram. 2015. *Photodetectors: Materials, Devices and Applications in Communications and Imaging Technologies*. Elsevier Science & Technology.

Nabet, Bahram, Marc Currie, Pouya Dianat, Fabio Quaranta, and Adriano Cola. 2014. "High-speed, high-sensitivity optoelectronic device with bilayer electron and hole charge plasma." *ACS Photonics* 1 (7):560–569.

Naguib, Michael, Jérémy Come, Boris Dyatkin, Volker Presser, Pierre-Louis Taberna, Patrice Simon, Michel W Barsoum, and Yury Gogotsi. 2012. "MXene: A promising transition metal carbide anode for lithium-ion batteries." *Electrochemistry Communications* 16 (1):61–64.

Naguib, Michael, Joseph Halim, Jun Lu, Kevin M Cook, Lars Hultman, Yury Gogotsi, and Michel W Barsoum. 2013. "New two-dimensional niobium and vanadium carbides as promising materials for Li-ion batteries." *Journal of the American Chemical Society* 135 (43):15966–15969. https://doi.org/10.1021/ja405735d.

Naguib, Michael, Vadym N Mochalin, Michel W Barsoum, and Yury Gogotsi. 2014. "25th anniversary article: MXenes: A new family of two-dimensional materials." *Advanced Materials* 26 (7):992–1005.

Nie, Yao, Li Li, and Zidong Wei. 2015. "Recent advancements in Pt and Pt-free catalysts for oxygen reduction reaction." *Chemical Society Reviews* 44 (8):2168–2201.

Osterloh, Frank E. 2013. "Inorganic nanostructures for photoelectrochemical and photocatalytic water splitting." *Chemical Society Reviews* 42 (6):2294–2320.

Paracchino, Adriana, Vincent Laporte, Kevin Sivula, Michael Grätzel, and Elijah Thimsen. 2011. "Highly active oxide photocathode for photoelectrochemical water reduction." *Nature Materials* 10 (6):456–461.

Peng, Qiuming, Jianxin Guo, Qingrui Zhang, Jianyong Xiang, Baozhong Liu, Aiguo Zhou, Riping Liu, and Yongjun Tian. 2014. "Unique lead adsorption behavior of activated hydroxyl group in two-dimensional titanium carbide." *Journal of the American Chemical Society* 136 (11):4113–4116.

Rakhi, Raghavan B, Bilal Ahmed, Mohamed N Hedhili, Dalaver H Anjum, and Husam N Alshareef. 2015. "Effect of postetch annealing gas composition on the structural and electrochemical properties of Ti_2CT_x MXene electrodes for supercapacitor applications." *Chemistry of Materials* 27 (15):5314–5323.

Ran, Jingrun, Guoping Gao, Fa-Tang Li, Tian-Yi Ma, Aijun Du, and Shi-Zhang Qiao. 2017. "Ti_3C_2 MXene co-catalyst on metal sulfide photo-absorbers for enhanced visible-light photocatalytic hydrogen production." *Nature Communications* 8 (1):13907. https://doi.org/10.1038/ncomms13907.

Ran, Jingrun, Tian Yi Ma, Guoping Gao, Xi-Wen Du, and Shi Zhang Qiao. 2015. "Porous P-doped graphitic carbon nitride nanosheets for synergistically enhanced visible-light photocatalytic H_2 production." *Energy & Environmental Science* 8 (12):3708–3717.

Schreier, Marcel, Jingshan Luo, Peng Gao, Thomas Moehl, Matthew T Mayer, and Michael Grätzel. 2016. "Covalent immobilization of a molecular catalyst on Cu_2O photocathodes for CO_2 reduction." *Journal of the American Chemical Society* 138 (6):1938–1946.

Schultz, Thorsten, Nathan C Frey, Kanit Hantanasirisakul, Soohyung Park, Steven J May, Vivek B Shenoy, Yury Gogotsi, and Norbert Koch. 2019. "Surface termination dependent work function and electronic properties of $Ti_3C_2T_x$ MXene." *Chemistry of Materials* 31 (17):6590–6597.

Seh, Zhi Wei, Kurt D Fredrickson, Babak Anasori, Jakob Kibsgaard, Alaina L Strickler, Maria R Lukatskaya, Yury Gogotsi, Thomas F Jaramillo, and Aleksandra Vojvodic. 2016. "Two-dimensional molybdenum carbide (MXene) as an efficient electrocatalyst for hydrogen evolution." *ACS Energy Letters* 1 (3):589–594. https://doi.org/10.1021/acsenergylett.6b00247.

Shi, Enzheng, Luhui Zhang, Zhen Li, Peixu Li, Yuanyuan Shang, Yi Jia, Jinquan Wei, Kunlin Wang, Hongwei Zhu, and Dehai Wu. 2012. "TiO_2-coated carbon nanotube-silicon solar cells with efficiency of 15%." *Scientific Reports* 2:884.

Shiue, Ren-Jye, Yuanda Gao, Yifei Wang, Cheng Peng, Alexander D Robertson, Dmitri K Efetov, Solomon Assefa, Frank HL Koppens, James Hone, and Dirk Englund. 2015. "High-responsivity graphene–boron nitride photodetector and autocorrelator in a silicon photonic integrated circuit." *Nano Letters* 15 (11):7288–7293.

Sugeta, Takayuki, Tsuneo Urisu, Seizo Sakata, and Yoshihiko Mizushima. 1980. "Metal-semiconductor-metal photodetector for high-speed optoelectronic circuits." *Japanese Journal of Applied Physics* 19 (S1):459.

Sun, Zhuxing, Julia Melisande Theresa Agatha Fischer, Qian Li, Jing Hu, Qijun Tang, Haiqiang Wang, Zhongbiao Wu, Marlies Hankel, Debra J Searles, and Lianzhou Wang. 2017. "Enhanced CO_2 photocatalytic reduction on alkali-decorated graphitic carbon nitride." *Applied Catalysis B: Environmental* 216:146–155.

Tamalampudi, Srinivasa Reddy, Yi-Ying Lu, Rajesh Kumar U, Raman Sankar, Chun-Da Liao, Karukanara Moorthy B, Che-Hsuan Cheng, Fang Cheng Chou, and Yit-Tsong Chen. 2014. "High performance and bendable few-layered InSe photodetectors with broad spectral response." *Nano Letters* 14 (5):2800–2806.

Tang, Qing, Zhen Zhou, and Panwen Shen. 2012. "Are MXenes promising anode materials for Li ion batteries? Computational studies on electronic properties and Li storage capability of Ti_3C_2 and $Ti_3C_2X_2$ (X=F, OH) monolayer." *Journal of the American Chemical Society* 134 (40):16909–16916.

VahidMohammadi, Armin, Mehrnaz Mojtabavi, Nuala Mai Caffrey, Meni Wanunu, and Majid Beidaghi. 2019. "Assembling 2D MXenes into highly stable pseudocapacitive electrodes with high power and energy densities." *Advanced Materials* 31 (8):1806931.

Wang, Feijiu, Daichi Kozawa, Yuhei Miyauchi, Kazushi Hiraoka, Shinichiro Mouri, Yutaka Ohno, and Kazunari Matsuda. 2015. "Considerably improved photovoltaic performance of carbon nanotube-based solar cells using metal oxide layers." *Nature Communications* 6 (1):1–7.

Wang, Feng, Wen-Guang Wang, Xiao-Jun Wang, Hong-Yan Wang, Chen-Ho Tung, and Li-Zhu Wu. 2011. "A highly efficient photocatalytic system for hydrogen production by a robust hydrogenase mimic in an aqueous solution." *Angewandte Chemie International Edition* 50 (14):3193–3197.

Wang, Hou, Yan Wu, Xingzhong Yuan, Guangming Zeng, Jin Zhou, Xin Wang, and Jia Wei Chew. 2018. "Clay-inspired MXene-based electrochemical devices and photo-electrocatalyst: State-of-the-art progresses and challenges." *Advanced Materials* 30 (12):1704561.

Xie, Xiaohong, Siguo Chen, Wei Ding, Yao Nie, and Zidong Wei. 2013. "An extraordinarily stable catalyst: Pt NPs supported on two-dimensional $Ti_3C_2X_2$ (X=OH, F) nanosheets for oxygen reduction reaction." *Chemical Communications* 49 (86):10112–10114.

Xu, Bingzhe, Minshen Zhu, Wencong Zhang, Xu Zhen, Zengxia Pei, Qi Xue, Chunyi Zhi, and Peng Shi. 2016a. "Ultrathin MXene-micropattern-based field-effect transistor for probing neural activity." *Advanced Materials* 28 (17):3333–3339.

Xu, Jiao, Jaewoo Shim, Jin-Hong Park, and Sungjoo Lee. 2016b. "MXene electrode for the integration of WSe_2 and MoS_2 field effect transistors." *Advanced Functional Materials* 26 (29):5328–5334. https://doi.org/10.1002/adfm.201600771.

Xu, Shuaikai, Guodong Wei, Junzhi Li, Wei Han, and Yury Gogotsi. 2017. "Flexible MXene–graphene electrodes with high volumetric capacitance for integrated co-cathode energy conversion/storage devices." *Journal of Materials Chemistry A* 5 (33):17442–17451.

Xue, Qi, Zengxia Pei, Yan Huang, Minshen Zhu, Zijie Tang, Hongfei Li, Yang Huang, Na Li, Haiyan Zhang, and Chunyi Zhi. 2017a. "Mn_3O_4 nanoparticles on layer-structured Ti_3C_2 MXene towards the oxygen reduction reaction and zinc–air batteries." *Journal of Materials Chemistry A* 5 (39):20818–20823.

Yan, Hongjian, Jinhui Yang, Guijun Ma, Guopeng Wu, Xu Zong, Zhibin Lei, Jingying Shi, and Can Li. 2009. "Visible-light-driven hydrogen production with extremely high quantum efficiency on Pt–PdS/CdS photocatalyst." *Journal of Catalysis* 266 (2):165–168.

Yang, Jinhui, Donge Wang, Hongxian Han, and Can Li. 2013. "Roles of cocatalysts in photocatalysis and photoelectrocatalysis." *Accounts of Chemical Research* 46 (8):1900–1909.

Yang, Yajie, Jaeho Jeon, Jin-Hong Park, Mun Seok Jeong, Byoung Hun Lee, Euyheon Hwang, and Sungjoo Lee. 2019. "Plasmonic transition metal carbide electrodes for high-performance InSe photodetectors." *ACS Nano* 13 (8):8804–8810.

Yu, LePing, Abdulaziz SR Bati, Tom SL Grace, Munkhbayar Batmunkh, and Joseph G Shapter. 2019a. "$Ti_3C_2T_x$ (MXene)-silicon heterojunction for efficient photovoltaic cells." *Advanced Energy Materials* 9 (31):1901063.

Yu, Xue-Fang, Yan-Chun Li, Jian-bo Cheng, Zhen-bo Liu, Qing-zhong Li, Wen-zuo Li, Xin Yang, and Bo Xiao. 2015. "Monolayer Ti_2CO_2: A promising candidate for NH_3 sensor or capturer with high sensitivity and selectivity." *ACS Applied Materials & Interfaces* 7 (24):13707–13713.

Yu, Xuelian, Tao Wang, Wenchao Yin, and Yihe Zhang. 2019. "Ti_3C_2 MXene nanoparticles modified metal oxide composites for enhanced photoelectrochemical water splitting." *International Journal of Hydrogen Energy* 44 (5):2704–2710.

Zeng, Zhiping, Yibo Yan, Jie Chen, Ping Zan, Qinghua Tian, and Peng Chen. 2019. "Boosting the photocatalytic ability of Cu_2O nanowires for CO_2 conversion by MXene quantum dots." *Advanced Functional Materials* 29 (2):1806500. https://doi.org/10.1002/adfm.201806500.

Zhang, Zhonghai, Rubal Dua, Lianbin Zhang, Haibo Zhu, Hongnan Zhang, and Peng Wang. 2013. "Carbon-layer-protected cuprous oxide nanowire arrays for efficient water reduction." *ACS Nano* 7 (2):1709–1717.

Zhou, Jie, Xianhu Zha, Xiaobing Zhou, Fanyan Chen, Guoliang Gao, Shuwei Wang, Cai Shen, Tao Chen, Chunyi Zhi, Per Eklund, Shiyu Du, Jianming Xue, Weiqun Shi, Zhifang Chai, and Qing Huang. 2017. "Synthesis and electrochemical properties of two-dimensional hafnium carbide." *ACS Nano* 11 (4):3841–3850. https://doi.org/10.1021/acsnano.7b00030.

Zhu, Miao, Xinming Li, Yibo Guo, Xiao Li, Pengzhan Sun, Xiaobei Zang, Kunlin Wang, Minlin Zhong, Dehai Wu, and Hongwei Zhu. 2014. "Vertical junction photodetectors based on reduced graphene oxide/silicon Schottky diodes." *Nanoscale* 6 (9):4909–4914.

Zuo, Fan, Le Wang, Tao Wu, Zhenyu Zhang, Dan Borchardt, and Pingyun Feng. 2010. "Self-doped Ti3+ enhanced photocatalyst for hydrogen production under visible light." *Journal of the American Chemical Society* 132 (34):11856–11857.

9 Preparation of 3D Assemblies/Structures of Titanium-Based MXenes

9.1 INTRODUCTION

Nanomaterials have experienced substantial progress in recent decades, showcasing captivating optical, electrical, and magnetic characteristics at the nanoscale (Z. L. Wang, Jiang, and Xu 2017; Lin, Pan, and Wang 2018; Veloso, Videira, and Oliveira 2022; Wan et al. 2021). These remarkable qualities have spurred the development of useful gadgets using these nanomaterials for real-world applications. Film materials made from nanoparticles have garnered considerable interest due to their capacity to combine nanoscience principles with practical applications on a larger scale. These applications include portable power supply, water purification, and electromagnetic interference shielding (Guohao Li et al. 2021; Xiran Li et al. 2020; Y. Luo et al. 2022). More functionality and better device performance are anticipated with the unique nanosized building blocks of film materials, which is what fuels the continued development of numerous contemporary technologies (Uchino and Debus 2012; Y. Zou, Raveendran, and Chen 2020; Y. Yu and Wang 2016; J. Xu et al. 2020).

Ever since the groundbreaking discovery of graphene in 2004, two-dimensional (2D) materials have emerged as a prominent area of scientific investigation in materials science due to its exceptional physicochemical properties and unique 2D structure. The rapid advancement of 2D materials in various applications over the past decade has expanded the innovation of the scientific community. (Rasheed 2022; Choudhury et al. 2020; Louwen Zhang et al. 2021; Bian et al. 2021). Since Drexel University researchers initially reported it in 2011, MXenes are a collection of carbides made from transition metals, and nitrides is a recent addition to the collection of 2D materials and has attracted significant academic interest. (Anasori, Lukatskaya, and Gogotsi 2017; Wychowaniec et al. 2020; Lim et al. 2021; Guohao Li et al. 2021; Prabhakar et al. 2022). Al is frequently used as an A-group element layer. These layers are selectively etched to create 2D MXenes from their corresponding 3D bulk MAX phase predecessors. M is a metal in the early transition (e.g., Ta, Nb, Ti, V, Zr, or Mo). The surface functionalities that result from the etching process, such as −(OH), = −Cl, −F, and, O, are represented by T_x, where X is carbon and nitrogen (Anasori, Lukatskaya, and Gogotsi 2017; Etesami et al. 2022; Li Li et al. 2020; G. Ma et al. 2022). The expression of MXenes can be denoted as Mn+1XnTx (n = 1–4). MXenes are the focus of research because of their distinct structure and surface chemistry, which also provide them with metallic conductivity, high hydrophilicity,

DOI: 10.1201/9781003319658-9

and outstanding flexibility (Qin, Su, and Ji 2022; Z. Xiao et al. 2020; W. Liu, Ning, Li, et al. 2021; Jaya Prakash and Kandasubramanian 2021).

More than 30 distinct MXenes, including V_2CT_x, $Nb_4C_3T_x$, Ti_2CT_x, Mo_2CT_x, Nb_2CT_x, $Mo_2TiC_2T_x$, $Ti_3C_2T_x$, and $Mo_2Ti_2C_3Tx$ have been synthesized by experimental methods. $Ti_3C_2T_x$ has attracted considerable interest among MXene compounds because of its comparatively low diffusion of ions barrier, strong electrical conductivity, negatively surface charges in liquids (measured 40 mV in water), flexibility in mechanics, low voltage of operation, and ecologically favorable properties (Jaya Prakash and Kandasubramanian 2021; J. S. Park et al. 2022; Ungai-Salánki et al. 2021; Z. Jing et al. 2022). These characteristics, along with $Ti_3C_2T_x$'s abundance of surface functions, high electrical conductivity, considering redox-active transition metals' potential for incorporation, have shown the material's intriguing potential for use in various disciplines, including wireless communications and EMI shielding. $Ti_3C_2T_x$ stands out among other 2D materials as having beneficial features for electrochemical energy storage (E. Kim et al. 2021; Sun et al. 2021; Z. Niu et al. 2022; Aïssa et al. 2021). These exceptional electrochemical, mechanical, and optoelectronic capabilities can also be applied to further broaden the range of flexible devices for which they can be used. $Ti_3C_2T_x$ films, which are flexible and self-supporting, can function as electrodes in storage of energy systems without requiring a metal current receiver. These films are appropriate for utilization in smart and lightweight gadgets where simplicity is crucial (Swain et al. 2022; Z. Huang et al. 2022; Wulong Li et al. 2022; Pei et al. 2021; Peng Liu, Liu, Zhang, et al. 2021). Xu et al. utilized an improved electrophoretic plating technique to create supercapacitor membranes composed of flexible $Ti_3C_2T_x$. These electrodes exhibited reliable cycling capability. By combining the two components of the $Ti_3C_2T_x$ film on conducting fabric material, an elastic symmetric all-solid-state capacitor is created. This capacitor demonstrates outstanding adaptability and may be utilized to produce batteries in the shape of watchbands, which are capable of providing electricity to an LED (S. Xu et al. 2017; Y. Lei et al. 2020; Xinyu Wu et al. 2020; Xiaochen Yang et al. 2021).

Nevertheless, there are still various challenges in the production and creation of films based on $Ti_3C_2T_x$. These obstacles include the limited stability of the element in environments with high humidity or oxygen levels above the temperature of the room. Additionally, the $Ti_3C_2T_x$ nanosheets experience a direct and close restacking procedure when they are assemble into sheets (P. Zhang et al. 2020; Lv et al. 2021; Xue Yang et al. 2022; Y. Deng et al. 2019; G. Zhou et al. 2022). These obstacles can seriously impair the material's physicochemical properties and prevent its further use. The performance of $Ti_3C_2T_x$-based layered structures has been improved by the development of surface modification and structural design techniques to counteract these effects. During the past five years, researchers have worked very hard on the intricate design of porous MXenes, drawn from the remarkable results of the permeable arrangement in tackling the difficulties faced by other 2D materials (W. Chen et al. 2022; Z. Shen et al. 2020; Jung et al. 2021; Ying et al. 2021; Hyeonho Lee et al. 2022). For instance, using a straightforward sulfur-template technique, Zhao and colleagues transformed 2D $Ti_3C_2T_x$ flakes into 3D structures. The $Ti_3C_2T_x$ films' abundance of active sites, open 3D macroporous structure, and flexibility makes them highly conductive, freestanding, flexible, and highly free of self-restacking

(K. Li et al. 2022; Lulu Li et al. 2021; Xing Wu et al. 2022; Gusnawan et al. 2018). The $Ti_3C_2T_x$ sheets possess a notable and stable capacity of 314.9 mA h g^{-1} at 50 mA h g^{-1} after 300 cycles, thanks to their 3D macroporous composition. The capacity of this material is almost quadruple of that of films made just from layered $Ti_3C_2T_x$. In this chapter, numerous $Ti_3C_2T_x$ and Ti_2CT_x MXene structures, such as honeycomb, lantern, 3D-printed scaffolds, sponge morphology, nanoflakes, and more, are described in detail with the addition of cellulose and palladium (F. Li et al. 2022; Y. Wu et al. 2018; Tao et al. 2018; Xingxing Li et al. 2022; Xiang Zhang et al. 2022; Y. Jin et al. 2021).

9.2 3D STRUCTURES BASED ON 2D TI_3C_2TX MXENE NANOMATERIAL

9.2.1 HONEYCOMB-LANTERN-INSPIRED 3D STRUCTURES OF 2D TI_3C_2TX MXENE NANOMATERIAL

Since ancient times, people have admired and pondered about the honeybee's hexagonal comb. According to Greek mythology, Daedalus was tasked with establishing the first honeycomb by lost wax casting gold more than 3,000 years ago. Marcus Varro claims that the Greek mathematicians Euclid and Zeno Dorus realized that the hexagon design maximizes building space and resources (C. Xu et al. 2018; J. Chen et al. 2013a; Chu et al. 2016; Y. Li et al. 2021; Z. L. Wang and Wu 2012). On the inside, the dome of the Pantheon in Rome features ribs and secret compartments. This honeycomb structure extends back through time. The geometric structure of a honeycomb enables the reduction of material usage to achieve both low weight and cost (Qi et al. 2020; Jian Tang et al. 2020; Das and Dhar 2020; Meng et al. 2022; Guzel et al. 2022). Honeycomb structures can be either natural or man-made. Honeycomb structures exhibit diverse geometries, although they uniformly possess an arrangement of empty cells formed by thin vertical walls. The cells have a regular pattern of hexagonal shape and are elongated in a columnar arrangement. A material with a honeycomb-shaped structure has low density and relatively strong resistance to compression and shear forces in directions perpendicular to its plane (Rafiq et al. 2022; Britz, Strutwolf, and Østerby 2022; Y. O. Kim et al. 2020; Zhenhua Xu et al. 2022; J. L. Yang et al. 2022).

MXene materials with 3D pores can ensure good EMI shielding. The generated 3D porous MXene structures at lower MXene loading typically lead to poor control of cell size and distribution. Al_2O_3 honeycomb plates were able to offer regular architectures, opening the door to controllable cell size, shape, and distribution designs (P. Song et al. 2020; Hongyan Li et al. 2021; S. K. Singh et al. 2022; P. T. M. Nguyen et al. 2021; A. P. Singh, Pandey, and Sharma 2021). The electromagnetic wave channels in the materials might also be extended because of the distinct honeycomb arrangements, enhancing the polymer composites' EMI shielding capabilities. rGO and MXene's additive effects can be fully utilized by 3D rGMH skeleton structures, according to P. Song et al., which greatly increases the electromagnetic interference shielding effectiveness (EMI SE) values of the rGMH/epoxy nanocomposites are shown in Figure 9.1a–d. Furthermore, the electromagnetic interference (EMI)

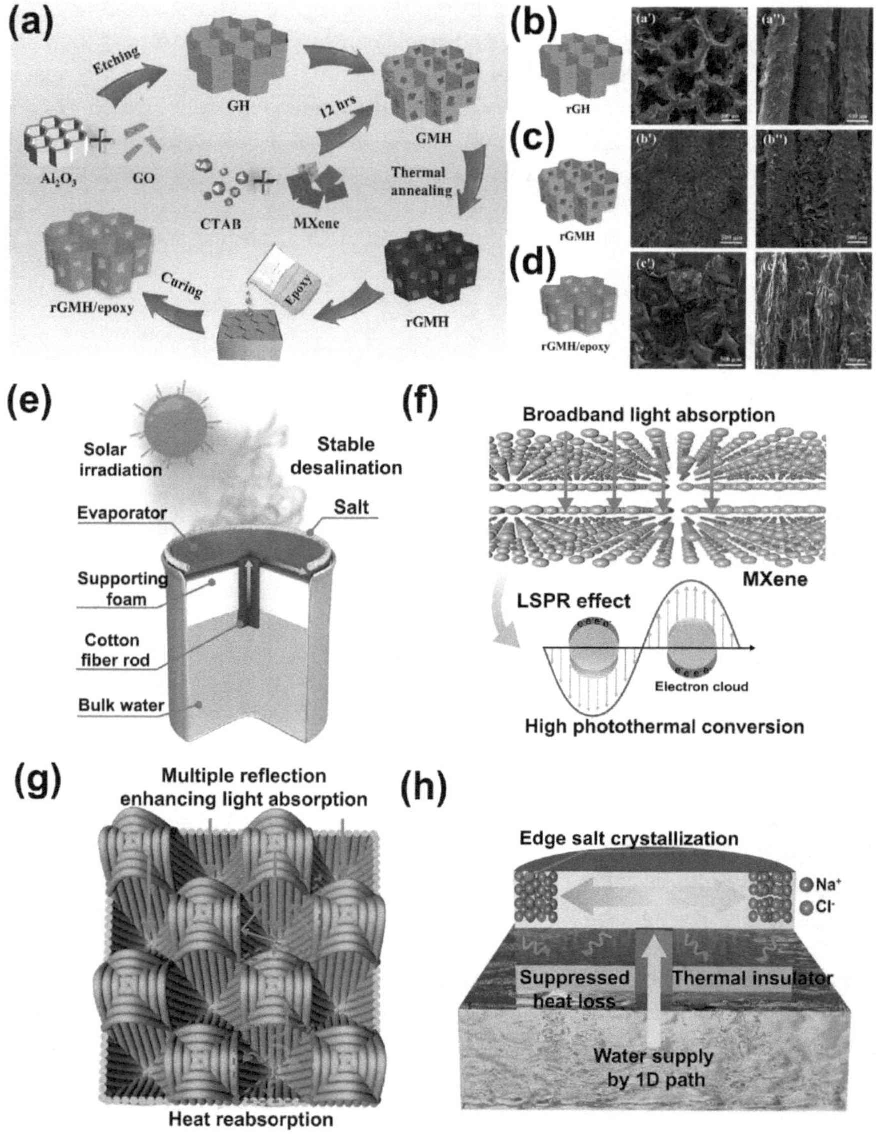

FIGURE 9.1 (a) A schematic showing the synthesis of rGMH/epoxy nanocomposites. SEM images captured in the axial and radial directions of rGMH, rGH, and rGMH/epoxy nanocomposites, respectively, are presented in b–d. (e) A schematic representation of an MXene-adorned solar evaporator for desalination that utilizes a fabric made of 3D honeycomb structure. (f) The photothermal conversion mechanism is based on the LSPR effect in the MXene. (g) Schematic representation of a fabric with a 3D honeycomb structure that absorbs light and dissipates heat. (h) A picture of cotton fiber rods with polystyrene foam acting as insulation and carrying water to stop heat loss, salt ring formation, and a 1D water channel.

Sources: (a–h) Adopted with permission from Elsevier (P. Song et al. 2020). (a–h) Permission awarded by Z. Lei et al. (2022).

shielding effectiveness (SE) of the rGMH/epoxy nanocomposites decreases as the honeycomb cell dimension grows, while the amount of weight of MXene remains unchanged (H. Ji et al. 2018; Y. Zou, Raveendran, and Chen 2020; Y. Yu and Wang 2016; J. Xu et al. 2020; Tyler et al. 2020). When the honeycomb cell size is 0.5 mm and the loading of reduced graphene oxide (rGO) and MXene is 1.2 weight percent (wt.%) and 3.3 wt.%, respectively, the rGMH/epoxy nanocomposites are 2,978 and 5 times more efficient than the rGM/epoxy nanocomposites (with a specific conductivity of 0.13 S/m and electromagnetic interference shielding effectiveness (EMI SE) of 11 dB). In the shielding mechanism of rGMH/epoxy nanocomposites, absorption is the dominant factor. The tensile modulus (Es), glass transition temperature (T_g), and thermal stability (THRI) values of the rGMH/epoxy nanocomposites are 7565.6 MPa, 150.9°C, and 184.1°C, respectively. The values obtained represent an increase of 25% for the epoxy resin's readings, namely 13.9°C and 10.6°C, compared to the initial values of 6041.2 MPa, 137.0°C, and 173.5°C (Mondal et al. 2020; Waseem and Ryu 2020; T. Kim et al. 2020; Amici et al. 2022; W. Deng et al. 2022; Chai et al. 2022; Gaâbel et al. 2020).

Z. Lei et al. designed an efficient and easily expandable technique for constructing a solar steam evaporator using a 3D honeycomb-structured fabric adorned with MXene. This method attained a solar power efficiency that was up to 93.5% and a rapid evaporating rate of 1.62 kilograms per square meter per hour under one sun irradiance condition (Figure 9.1e–h). The salt crystals may also be temporarily separated from the evaporator's surface thanks to a confined salt concentration transferring channel, which guarantees the evaporator's long-term usage durability and salt harvesting (J. Lee et al. 2020; Y. Zou, Raveendran, and Chen 2020; Y. Yu and Wang 2016; J. Xu et al. 2020; Tyler et al. 2020; Medina-Franco 2016; Byun et al. 2016). Additionally, the MXene/3D honeycomb fabric evaporator's evaporation rate stayed constant over 25 cycles and 35 nonstop hours of evaporating salt water. The remarkable solar efficiency is ascribed to multifunctional layers' combined optimization and synergistic effects, utilizing a 3D honeycomb fabric surface with concave intervals covered with MXene as efficient light absorbers, a thermal barrier made of a honeycomb fabric with low heat conductivity reinforced by insulating foam, and a1D highly hydrophilic honeycomb fabric with significant capillary force in the center and hydrophilic fibrous rods acting as water transport (T. T. Li et al. 2020; Panzera et al. 2020; Lal et al. 2020; zhao et al. 2019; M. Kim et al. 2019; Taikar 2020; Khalili and Chenari 2020). The MXene/3D honeycomb fabric solar evaporator has novel possibilities for desalination, wastewater treatment, and the retrieval of valuable salts from water. It is cost-effective, adaptable, capable of being expanded, and easily transportable (Hyeseong Lee et al. 2019; T. G. Kim et al. 2020; Yi Han et al. 2022; Y. Zheng et al. 2022; B. Yang et al. 2019).

Extremely permeable MMMs were produced after the in-situ loading of Zn/Co-ZIF nanoparticles into a semi-permeable PEO-based membrane was carried out (Figure 9.2a–g) (Cheng et al. 2019; L. Jin et al. 2021; Yao et al. 2022; Y. Nie et al. 2020; E. Kim et al. 2021; Sun et al. 2021; Z. Niu et al. 2022). Better CO_2 solubility and diffusion in MMMs and the enhanced permeability to CO_2 arise from the dipole–quadrupole coupling between the PEO group and CO_2 within the membrane,

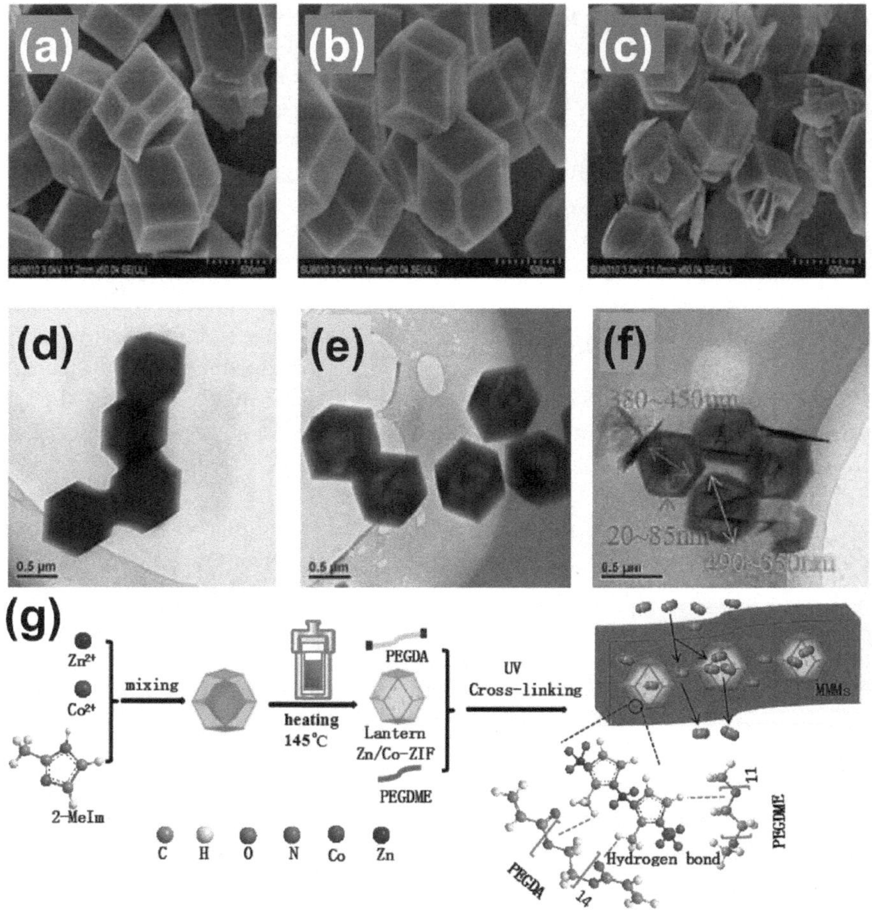

FIGURE 9.2 (a–f) TEM images of temperature-dependent Zn/Co-ZIF nanoparticles were taken at 40°C and 120°C, whereas SEM images were taken at 40°C and 145°C. (g) Diagram showing the manufacture of MMMs and the route taken by the gas. *Notes:* The abbreviation XLPEO stands for cross-linked poly (ethylene glycol) diacrylate and poly (ethylene glycol) dimethyl ether. MMMs, also known as $XLPEO^{-5}$, $XLPEO^{-10}$, $XLPEO^{-2}$, and $XLPEO^{-15}$, are membranes made of the mixed matrix, indicating the lantern Zn/Co-ZIF nanoparticle mass fractions in MMMs.

Source: Permission awarded by Cheng et al. (2019).

along with the remarkable performance of permeable lantern Zn/Co-ZIF nanostructures possessing thin outer shells. The nanoparticles function as inorganic doping agents, engaging in the complexation with duplex metal ions and CO_2, so generating many routes with little resistance for the movement of gases (Ding et al. 2021; Kazmina, Borovoy, and Semenova 2021; Hsia et al. 2021; Uchino and Debus 2012; Slabov et al. 2020; Zeng and Xuan 2018; Hedong Chen et al. 2021; J. Chen et al. 2013b; Roman et al. 2020; Y. Yin et al. 2020).

S. Yue et al. designed a highly responsive and adaptable piezoresistive sensor with an arrangement like a honeycomb by employing femtosecond filamentation pulse processing and drip-coating MXene nanosheets. Because of the exceptional qualities of nanosheets made of $Ti_3C_2T_x$ and the distinct shape of the filament-processed microscopic structure, the MXene hybrid sensor has a sensitivity of 0.61 kPa^{-1}, exceeding a two-fold improvement (Figure 9.3a–g) (Yue Su et al. 2022; J. Gao et al. 2022; Xinru Zhang et al. 2018; Yonghong Wang et al. 2022; Hassan et al. 2021). The device also displays excellent repeatability throughout loading and unloading cycles and an extraordinarily fast response time of only 160 ms. The established device's sensing method is based on modifying the electrically conducting surface of the interface states that exist between interdigitated electrodes and the MXene/PDMS conductivity layer (G. Yin et al. 2020; T. Hou et al. 2021; J. J. Liu et al. 2022; Cui et al. 2021; Y. Z. Cai et al. 2021; L. Yang et al. 2018). They were able to identify a wide range of sensing inputs, from minor muscle contractions to substantial human motions including artery pulse, joint motion, and facial emotions, thanks to these remarkable sensing performance features. Then, they created a smart glove that was effective in recognizing complicated and deft gestures. To keep an eye on healthcare and immediately identify human activity, it is anticipated that these manufactured sensing devices would be utilized in upcoming wearable electronic products (Yao et al. 2022; Y. Nie et al. 2020; E. Kim et al. 2021; Sun et al. 2021; Z. Niu et al. 2022; Aïssa et al. 2021).

Numerous benefits include the honeycomb structure's remarkable ratio of energy to mass, high toughness, corrosion resistance, high temperature performance, fire and fungal resistance, lack of moisture absorption, ease of machining and formability, and cost-effective advantages including energy savings. A structural core material with one of the best strengths to weight ratios on the market, aluminum honeycomb offers the maximum amount of rigidity (Goyat et al. 2021; Zhongliao Wang et al. 2022; S. Liu, Liu, Liu, et al. 2021; Jun et al. 2020; G. W. Lee et al. 2021; Z. Fang et al. 2019; Demin et al. 2021). The honeycomb structure is utilized in various applications due to its numerous advantages. These applications include skis, LED technology, rocket substructures, snowboards, jet aircraft, automobile structures, clean room panels, loudspeaker technology, air and water directionalities, heating and air-conditioning equipment, fluid and light directionalities, electric shielding enclosures, exterior architectural curtain wall panels, wind turbine blades, ventilation, and energy absorption and recovery (Q. T. Nguyen and Ahn 2021; Y. Zou, Raveendran, and Chen 2020; Y. Yu and Wang 2016; J. Xu et al. 2020; Tyler et al. 2020; Medina-Franco 2016; Byun et al. 2016). Due to its benefits and practical uses, the honeycomb structure holds great promise for materials science and engineering applications that aim to improve and innovate upon material qualities (Y. Xiao et al. 2020; D. Deng et al. 2018; Trabocchi 2019; L. Huang et al. 2022; R. Zhou et al. 2019; Makurvet 2021; A. Yu et al. 2019).

9.2.2 3D-Printed Scaffolds from 2D Ti_3C_2Tx MXene Nanomaterial

The 2D nanomaterial family now includes new members known as MXenes, which are carbonitrided, carbides, and transition metal nitrides. MXenes have garnered significant attention due to their amazing electrical, optical, and mechanical characteristics.

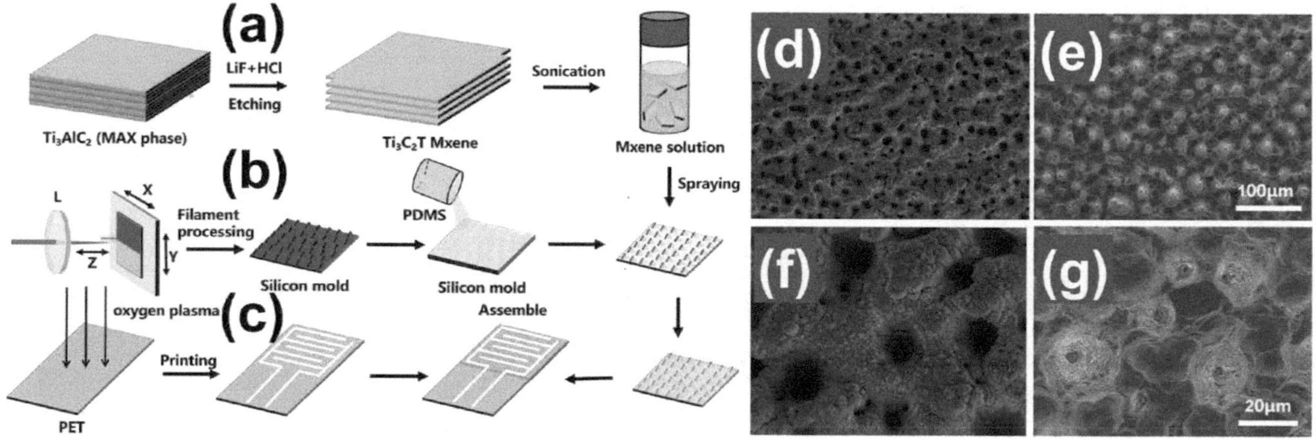

FIGURE 9.3 (a) $Ti_3C_2T_x$ MXene with micro-patterned MXene/PDMS conducting layer fabrication. (b–c) Assembling the piezoresistive sensor based on MXene. (d–e) Top-view SEM pictures at low and high magnifications of a silicon mold surface that has undergone filament processing. (f) SEM images of the manufactured MXene-based PDMS film surface at a lower magnification, revealing the overall surface morphology and distribution of MXene within the PDMS matrix. (g) SEM images of the manufactured MXene-based PDMS film surface at a higher magnification, highlighting the detailed structure and dispersion of MXene nanosheets within the PDMS composite, and the quality of the MXene and PDMS adhesion.

Source: Permission awarded by Yue Su et al. (2022).

Materials based on MXene are widely used in a variety of industries, including energy storage, light-to-heat conversion, electronics, biomedicine, and so forth (B. Huang et al. 2021; W. Kong et al. 2021; Yazdani-Kachoei and Jalali-Asadabadi 2020; C. Gao et al. 2022; Trzciński et al. 2022; Jensen et al. 2021; Bhatnagar et al. 2022). Moreover, MXene sheets have favorable hydrophilicity as a result of the plentiful presence of terminal groups such as –O and –OH on their surface. Consequently, several hydrophilic polymers exhibit enhanced compatibility with MXene sheets. The cause of this phenomenon is the localized surface plasmon resonance (LSPR) effect. Additionally, MXene nanosheets demonstrate efficient light absorption (Hirai et al. 2021; Padhan et al. 2022; Hui Wang, Huang, and Su 2020; Chang et al. 2021; Gupta et al. 2019; Rozman et al. 2022; Yeo et al. 2022). It is anticipated that MXenes will produce a perfect practical photo blocker for stereolithography due to their electrical conductivity, exceptional biological safety, dispersibility, and light absorption capacity (Rodríguez-Peña et al. 2021; Wen Li et al. 2020; H. Y. Yang et al. 2021; Franco et al. 2021; Geng et al. 2022; J. Hu et al. 2021; Ahmad et al. 2022; Zhuowen Wang et al. 2022).

The $Ti_3C_2T_x$ MXene was used by B. Huang et al. to produce functional gels with various geometries using CLIP 3D printing as a photo blocker (Figure 9.4a–d) (B. Huang et al. 2021). By incorporating MXene nanosheets into a CLIP aqueous ink, it is possible to precisely regulate the photopolymerization of the ink within a specific region using light projection. This controlled process ensures the desired thickness and enhances the overall quality of the printed material (Xiran Li et al. 2020; Y. Luo et al. 2022; X. Song et al. 2022; S. Xu et al. 2017; Y. Lei et al. 2020; Xinyu Wu et al. 2020; Xiaochen Yang et al. 2021). The hydrogel produced using CLIP exhibited a smooth surface devoid of the usual macroscopic defects ("staircase effect") associated with 3D printing. Additionally, MXene improves the hydrogel's mechanical performance by acting in the role of a non-covalent cross-linking agent and reinforcement. The MXene/SA/PAM organogel that results from a solvent exchange with glycerin has improved structural stability and acceptable adherence to a variety of surfaces (Jiqi Wang et al. 2022; J. Pan et al. 2022; Xuan Yang et al. 2022; R. Cheng et al. 2022; Eskandari and Ahmadi 2015; Zhi et al. 2022; H. Ma, Zhang, and Feng 2022; D. Yin et al. 2021; T. Yang et al. 2022). The organogel can therefore function as a strain or pressure sensor that uses adhesive properties to attach to surfaces. The organogel exhibits a fast rate of converting light into heat and a high efficiency in absorbing moisture. In comparison to a regular cuboid, an organogel that has a hollow structure created using CLIP with a meticulously engineered 3D structure exhibits enhanced environmental harvesting of water capabilities, and because of its larger particular area of surface, it is able to pull in a greater quantity of water molecules. Overall, the research has demonstrated that the utilization of MXene functions as an optical barrier in conjunction with CLIP printing technology and holds promise as a viable approach for manufacturing a diverse array of advanced hydrogel devices (F. Kong et al. 2018; Jinbu Su et al. 2015; Elibol 2020; H. Niu et al. 2022; H. Liu, Wang, and Zhang 2022; Hongbing Wang et al. 2015; Xiaomeng Li and Zhou 2022; L. Cai et al. 2022; Z. Li et al. 2021; Shu, Li, et al. 2022; Y. Zhan et al. 2018; Dai et al. 2022; Feng et al. 2022; Peijiang Liu et al. 2018; Shu, Xu, et al. 2022; Y. Deng et al. 2019; X. Jin et al. 2020).

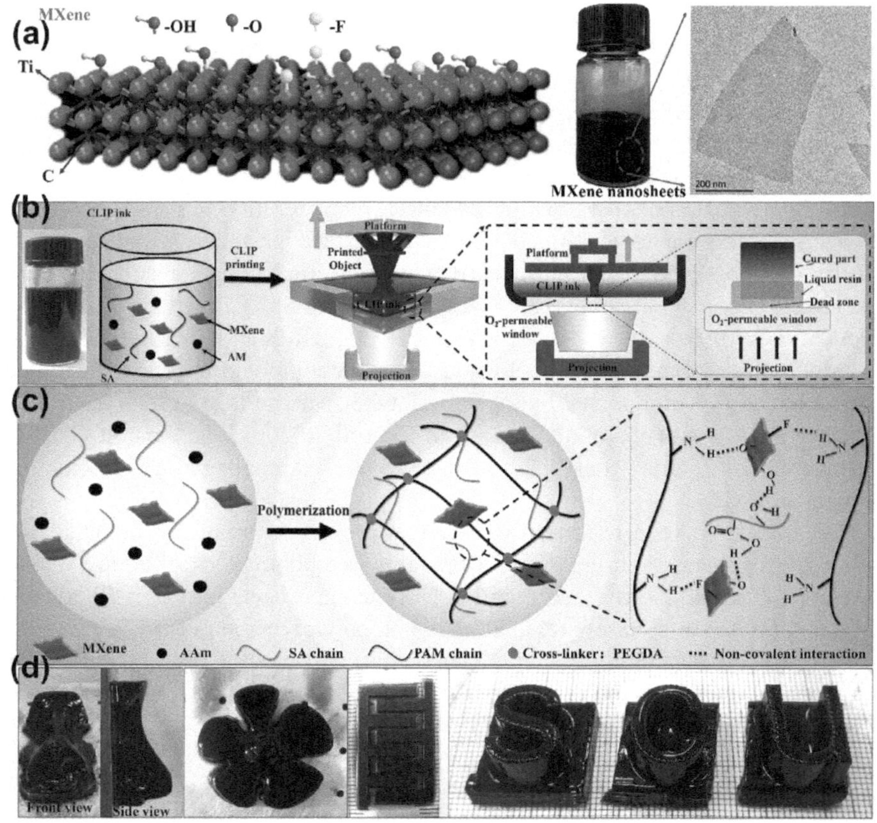

FIGURE 9.4 (a) A photograph of the $Ti_3C_2T_x$ MXene aqueous dispersion, along with an illustration of its chemical composition. (b) A schematic outlining the CLIP procedure. (c) Schematic representation demonstrating the chemical composition of the hydrogel created using CLIP. (d) letters with a nose, flower, electrode, and SCU printing on CLIPs.

Source: Permission awarded (B. Huang et al. 2021).

Li Xiran et al. developed a procedure that uses unidirectional freezing and 3D printing to produce MSCs that are naturally flexible (Figure 9.5) (Xiran Li et al. 2020). MXene, MnONWs, AgNWs, and C60 were combined to create a thick, porous structure resembling a honeycomb, with a gel-like nanocomposite ink. The MSC device, which was created using 3D printing technology, had exceptional electrochemical characteristics, achieving an output power density of 58.3 mW cm^2, an energy density of 19.2 Wh cm^2, an aerial capacitance of 216.2 mF cm^2, long-term cycling stability, and excellent rate capability (Chenyang Yu et al. 2020; Y. Zou, Raveendran, and Chen 2020; Y. Yu and Wang 2016; J. Xu et al. 2020; Tyler et al. 2020; Medina-Franco 2016; Byun et al. 2016; Y. Xiao et al. 2020; D. Deng et al. 2018; Trabocchi 2019; L. Huang et al. 2022; R. Zhou et al. 2019). This was made possible by the interplay between the electrode architecture and the nanocomponents. The 3D-printed MSC, with its structured honeycomb-like porosity framework and

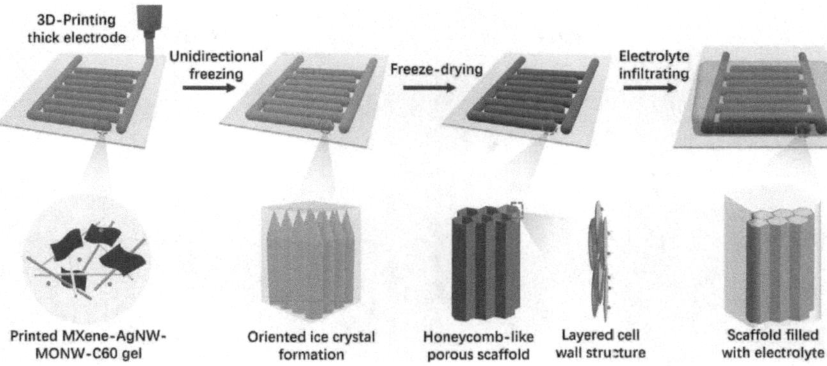

FIGURE 9.5 The schematic diagram illustrates the process of creating inherently stretchable MSCs using longitudinal cooling and 3D printing. The 3D-printed interdigitated sensors possess a layered cell wall architecture and an impermeable framework that closely resembles honeycomb structure.

Source: Permission awarded by Xiran Li et al. (2020).

the capability of C60 to migrate between MXene layers, demonstrated the capacity to preserve its electrochemical characteristics under a load of as much as 50% and after undergoing 1,000 stretch/release sessions. This approach shows great potential as an initial step in integrating adaptable battery backups into wearable and transportable technological devices (Mokhtari, Latifi, and Shamshirsaz 2016; Blachowicz and Ehrmann 2020; Shi, He, and Lee 2019; X. S. Zhang et al. 2018a; R. Zhou et al. 2019; Trabocchi 2019; L. Huang et al. 2022; Y. Xiao et al. 2020; D. Deng et al. 2018; Byun et al. 2016; Bafandeh and Lee 2021; Duan et al. 2021).

H. Wei et al. effectively employed vacuum-assisted filtration to create extremely thin, pliable films of composites composed of d-$Ti_3C_2T_x$/ANFs. These films demonstrated outstanding physical characteristics and proved extraordinary proficiency in insulating against electromagnetic wave interference (EMI). The composite film contained layers of $Ti_3C_2T_x$ that were impregnated with one-dimensional (1D) aligned nanofibers (ANFs), creating a "brick and cement" structure (H. Wei et al. 2020) (Figure 9.6). Within this configuration, the MXene sheets functioned as the fundamental building blocks, while the ANFs served as the binding agent, offering simultaneous linkage and reinforcement to the building blocks. Amorphous nanofillers (ANFs) have a capacity to greatly enhance the mechanical durability of composite films. The composite film, consisting of d-$Ti_3C_2T_x$/ANFs, demonstrated a maximum elastic strength of 116.71 MPa and a fracture pressure of 2.64%. The combination of films demonstrated exceptional electromagnetic interference (EMI) shielding capabilities, as reported by multiple studies (H. Wei et al. 2020; Zhen Wang et al. 2021; Xiaohan Wang et al. 2022; J. Cheng et al. 2020; Guo et al. 2021; R. Liu et al. 2018; F. Li et al. 2022; Tao et al. 2018; Y. Wu et al. 2018; Xingxing Li et al. 2022; Xiang Zhang et al. 2022). A 12 m thick composite film, composed of d-$Ti_3C_2T_x$ and ANFs, exhibited an electromagnetic interference (EMI SE) shielding efficacy of 34.71 dB at a frequency of 8.2 GHz. Furthermore, the composite material exhibited a specific

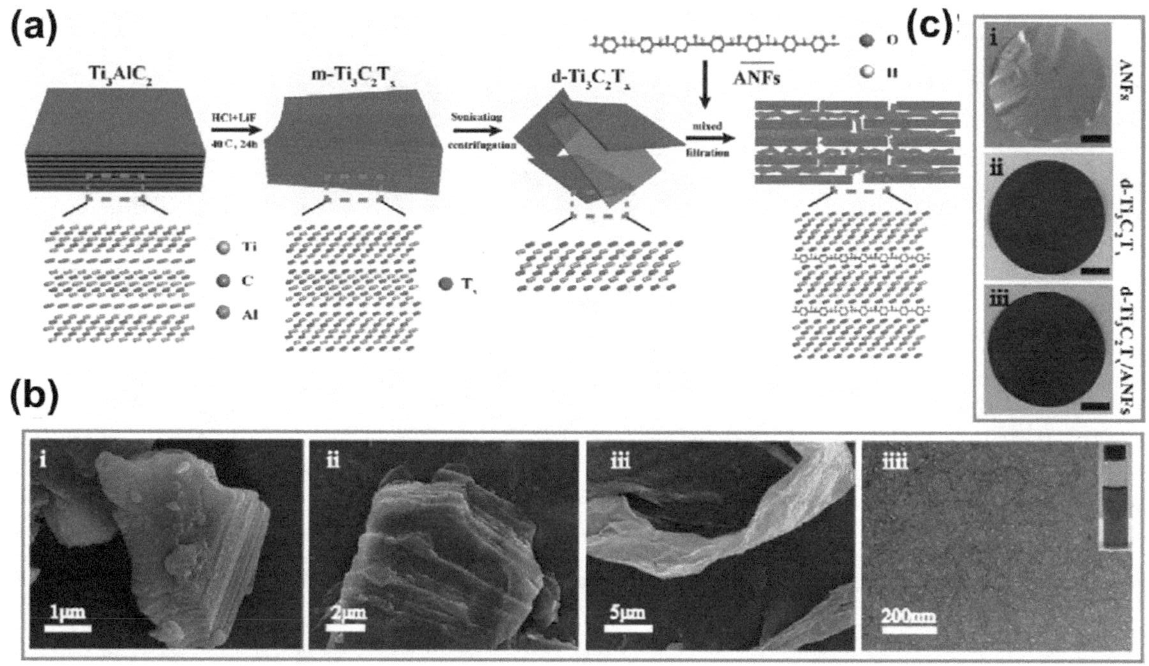

FIGURE 9.6 (a) The diagram illustrates the process of producing composite films consisting of d-$Ti_3C_2T_x$ and ANFs. (b) The morphology of (i) Ti_3AlC_2, (ii) m-$Ti_3C_2T_x$, (iii) d-$Ti_3C_2T_x$, and (iv) TEM of ANFs; the suspension of ANFs was shown as the inset of the greyscale solution. (c) A digital image of the films for the ANFs, d-$Ti_3C_2T_x$, and d-$Ti_3C_2T_x$/ANFs, where the bar represents 10 mm.

Source: Permission awarded by H. Wei et al. (2020).

surface energy per unit thickness of 21971.37 dB cm^2 g^{-1}. The composite films made from d-$Ti_3C_2T_x$/ANFs exhibit a range of desired attributes, including being remarkably thin, flexible, and possessing remarkable mechanical properties. Furthermore, they demonstrate exceptional electromagnetic interference (EMI) shielding capacity. Such films have significant potential applications in various fields, such as electronic watches, robotic parts, machine joints, and protecting delicate equipment (Hong Chen et al. 2022; Jiang et al. 2018; Z. Zhan et al. 2022; J. Xiao et al. 2020; R. Zheng et al. 2019; Wandi Wang et al. 2020; Tan et al. 2022; Guohao Li et al. 2021; Xiran Li et al. 2020).

The fabrication technique of 3D printing has a lot of potential for making scaffolds for material engineering. The range of materials that can be utilized in different materials restricts the uses of 3D printing in the field of material engineering. To make new biomaterials and compositions suitable for 3D printing, many researchers have created them. The capacity to produce various porosities and geometries, co-culture multiple cells, and include growth factors are just a few of the benefits of constructing scaffolds utilizing 3D printing. In general, the MXene via 3D printing technique described earlier has great promise for the creation of scaffolds and the design and functionalization of materials (Heck et al. 2022; Manousiouthakis et al. 2022; Pucci et al. 2022; Wei Wang et al. 2020; Jingwen Zhang et al. 2021; Xie et al. 2020; Long et al. 2021; Ghaemi, Slade, and Amini Horri 2021; Besisa, Ewais, and Ahmed 2021).

9.2.3 2D Ti_3C_2Tx MXene Nanomaterial Sponges by the Modified Electrochemical Approach

Chemical engineering's electrochemical approach is the area that involves the technological uses of electrochemical phenomena, such as metal interference patterns and refining, corrosion, flow batteries, fuel cells, electrodeposition surface modification, and electrochemical separations. Chemical engineering and electrochemistry both have applications in this field. Membrane technology is put to a great task by the rapidly developing sector that produces a multi-component pollutant–oil–water emulsion that poses a substantial hazard to both the well-being of humans and the surroundings. Recently, 2D MXene materials have shown great promise as viable choices for functional separating surfaces (Saqib et al. 2021; H. Wu et al. 2021; Kayahan et al. 2021; Zhenkai Wang et al. 2021; Cong Chen, Hang, et al. 2021; Bae et al. 2021; Y. Park et al. 2021; Xue et al. 2021; Zhong et al. 2020; Yan et al. 2020).

S. He et al. (2020) employed a facing technique to fabricate a multifunctional composite bilayer, which consists of an ordered framework of MXene@UIO-66-$(COOH)^2$. This membrane is the material is backed by a nylon 66 microporous matrix and exhibits exceptionally the stability of chemicals, along with successful segregation of complex pollutant–oil–water fluid emulsion (Figure 9.7a–g) (S. He et al. 2020; Miao et al. 2020; M. Hu et al. 2022; Qureshi and Panesar 2020; Q. Chen 2020; Vladimirova et al. 2020; Peng et al. 2020; Z. Wu et al. 2020; Xiaoyun Liu et al. 2022; Knozowski et al. 2021). The membrane's possibility for enhanced separation efficiency of diverse oil-in-water emulsions and dyes is attributed to its excellent

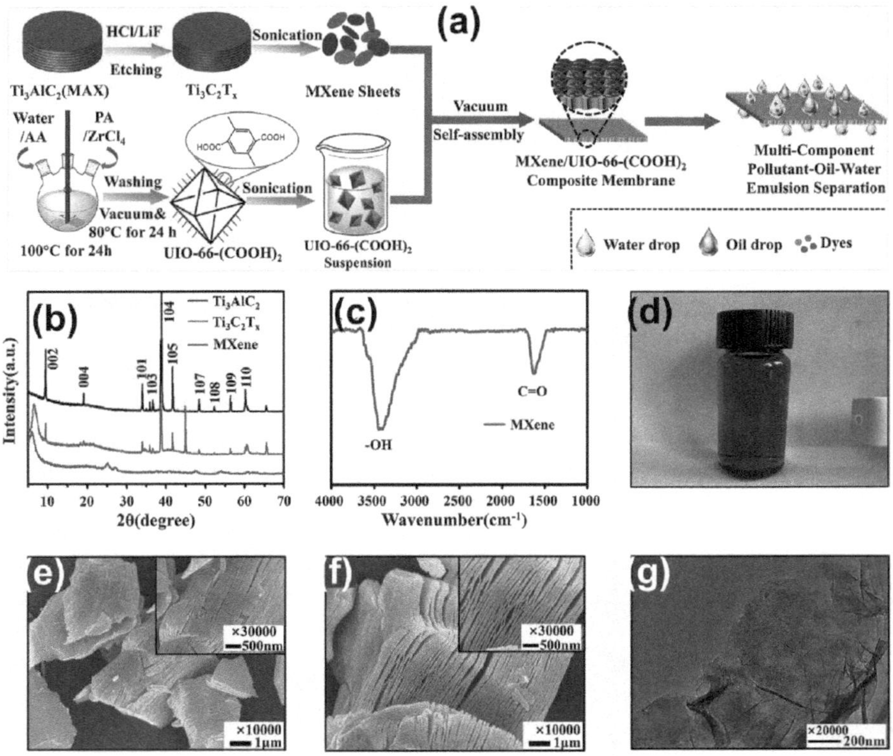

FIGURE 9.7 (a) MXene@UIO-66-(COOH)$_2$ composite membrane fabrication and separation technique for multi-component pollutant–oil–water emulsion. (b) Ti_3AlC_2, $Ti_3C_2T_x$, and MXene XRD patterns. (c) MXene's FTIR spectra. (d) A digital image of the MXene suspension Tyndall phenomena. (e) Ti_3AlC_2 and (f) $Ti_3C_2T_x$ SEM images. (g) A TEM picture of nanosheets made of MXene.

Source: Permission awarded by S. He et al. (2020).

antifouling resistance, low oil adhesion characteristic, inherent hydrophilicity, hierarchical intercalation structure, and good recyclability. Furthermore, the MXene@ UIO-66-(COOH)^2 composite membrane exhibited an exceptional resistance to corrosion in highly acidic (3 M HCl), alkaline (1 M NaOH), and saline (Saturated NaCl) environments, thanks to its chemical stability. This study provides a solid framework for the development of robust and practical 2D membrane materials for the separation of several components from pollutant–oil–water emulsions under difficult chemical circumstances (He et al. 2020; Nandal et al. 2022; S. Zhang et al. 2021; J. K. Chen et al. 2020; Waseem and Ryu 2020; T. Kim et al. 2020; Amici et al. 2022; Chai et al. 2022; Amorim et al. 2020; X. Zheng et al. 2021; Y. Hou et al. 2020; Paschalidou et al. 2022).

The qualities of sponge structure are highly encouraging, including compressibility, tensile strength, cleaning abilities, high liquid absorption and retention capacity, biodegradability, chemical resistance, biocompatibility, electrical conductivity, filtering ability, sterilization, and washability. The sponge form is particularly promising for usage in MXene and other sectors of material engineering due to its many characteristics (khalilifard and Javadian 2021; Zehao Wang et al. 2021; Xiqing Wang et al. 2021; Y. Zhang et al. 2022; Y. Xu et al. 2021; Gill et al. 2021; Sharma et al. 2021; Ateka et al. 2021).

9.2.4 Graphene Oxide/2D Ti_3C_2Tx MXene 3D Nanoflakes Aerogel

Graphite oxide (GO), also referred to as graphitic oxide or graphitic acid, is a carbon-based substance that is formed when powerful oxidizers and acids are used to eliminate surplus metals from graphite, including carbon, oxygen, and hydrogen. The extensively oxidized bulk material is a yellow solid with a carbon-to-oxygen ratio ranging from 2:1 to 2:9. The material retains the layered composition of graphite, but, the interlayer gaps are significantly larger and exhibit a greater degree of randomness (K. Liu et al. 2021; Altaf et al. 2020; F. Chen et al. 2022; Shao et al. 2022; P. Zhou et al. 2020; W. W. Kong et al. 2021; Eslami et al. 2022; Si et al. 2020; Hung et al. 2021; Jun Tang et al. 2019; Saha et al. 2022).

To generate monomolecular sheets, similar to graphene, the single-layer form of graphite, the bulk material can be subjected to sonication in a solvent with a specific polarity or dispersed naturally in alkaline solutions. Graphene oxide sheets have been utilized to fabricate robust paper-like materials, thin films, membranes, and composite materials. Initially, there was significant enthusiasm about graphene oxide as a potential intermediary in the production of graphene (W. Yang and Wang 2016; Sultana et al. 2019; G. Hu et al. 2018; Z. Liu et al. 2019; Sinha Ray 2013; Xiaoting Liu, Shen, Sa, et al. 2021; S. Wang et al. 2020; Lei Zhang et al. 2015; S. Pan et al. 2020; K. Dong et al. 2018). Numerous chemical and structural imperfections remain in the graphene created by decreasing graphene oxide, which may be helpful in some applications but troublesome in others. By analogy to graphene, the single-layer version of graphite, to produce monomolecular sheets, the bulk material can be sonicated in solvent polarity or disseminated spontaneously in basic solutions (Xi et al. 2017; Trinh and Chung 2018; X. Lu, Li, and Wang 2011; Zi and Wang 2017; Wahajuddin and Arora 2012; Liang, Tong, and Pei 2016; Dagdeviren, Li, and Wang 2017; Yuanjie Su et al. 2021; Firdous et al. 2021; Y. Dong et al. 2018). Graphene oxide strips have been utilized to fabricate robust paper-like substances, thin sheets, barriers, and composite materials. At first, there was a great deal of interest in graphene oxide as a possible intermediary in the synthesis of graphene. When graphene oxide is broken down, many chemical and structural flaws remain in the graphene that is made. These flaws may be useful in some situations, but they can be a problem in others (H. Zou et al. 2017; Helin Wang et al. 2017; Haoyi Li and Yang 2016; J. Lu et al. 2017; L. Xu et al. 2018; Getzlaff et al. 2019; C. X. Li et al. 2018; P. C. Huang et al. 2021).

A nanoflake is a particle with at least one nanometric dimension (between 1 and 100 nm, in other words). A flake can have a structure or form that resembles a plate; however, it need not be fully flat (Haotian Chen et al. 2019; Xun Han et al. 2017; Ferreira et al. 2019; Sheldon and Tokranov 2016; K. Cheng et al. 2022; Barhoum 2019; Xia et al. 2018; C. Liu, Neale, and Cao 2016; Yaojie Han et al. 2020). There are nanoflakes made of several substances. Given that the final product is still in the prototype stage, nanoflakes are a type of semiconductor that shows potential for solar energy generation. Because of their crystalline form, the crystals can absorb light and retain 30% of the solar energy that is shining on them (Y. Gao et al. 2020; H. Yang et al. 2021; Matsusaka et al. 2010; Jie Wang et al. 2015; Xingfu Wang et al. 2017; X. Yin et al. 2019; Karan et al. 2018; Ataeefard, Ghasemi, and Ebadi 2014; H. Jing et al. 2021; L. Zheng et al. 2017; Cao et al. 2018; K. Dong et al. 2017).

The therapeutic use of tissue-engineered cardiac patches' potential for myocardial infarction (MI) treatment is quite encouraging. For the patch to integrate well with the surrounding tissue and for the cells in the patch to function well, it is important that the extracellular matrix closely resembles the organized native matrix of cells, has a similar structure to that of the human heart, and also shares the same conductivity of electricity. (Basara et al. 2022; Esmaeili et al. 2022; S. Wu et al. 2022; Moreira et al. 2022; Barroca et al. 2022; Chaojie Yu, Yao, and Li 2022; Kiyotake, Martin, and Detamore 2022; X. Zhou et al. 2022). G. Basara et al. have developed a new composite material that provides conductivity and topographical stimuli for the differentiation of human-induced pluripotent stem cells into heart muscle cells (iCMs) for use in cardiac tissue engineering applications. The structures are fabricated through the utilization of aerosol jet printing to deposit pre-designed patterns of conductive titanium carbide ($Ti_3C_2T_x$) MXene onto polyethylene glycol (PEG) hydrogels, employing 3D printing technology. Next, these structures are implanted with induced cardiomyocytes (iCMs) and cultured for a week without any harmful effects on cell viability (Heck et al. 2022; Manousiouthakis et al. 2022; Pucci et al. 2022; Wei Wang et al. 2020; Jingwen Zhang et al. 2021; Long et al. 2021; Ghaemi, Slade, and Amini Horri 2021; Xie et al. 2020). The results of this work highlight the crucial importance of using 3D-printed $Ti_3C_2T_x$ MXene to align induced cardiomyocytes (iCMs). This is evidenced by the significantly increased levels of MYH7, SERCA2, and TNNT2, as well as the enhanced synchronized contraction and speed of electrical conduction. This study demonstrates that 3D-printed $Ti_3C_2T_x$ MXene can be used to create cardiac patches for myocardial infarction (MI) therapy in a manner that is suitable for the physiological conditions. The findings are supported by the works of Panda, Biswas, and Paul (2021), Zhixiang Xu et al. (2020), Taşdemir et al. (2021), Bulut, Oktem, and Yilmaz (2020), Pirmoradi and Kastner (2021), G. Xu et al. (2021), Dong Wang et al. (2021), Chenglong Yu et al. (2020), Jun feng Su et al. (2020), and S. Dong et al. (2020).

Artificially created cardiac chambers show significant promise as treatments for damaged areas of the heart following a myocardial infarction (MI) due to their ability to stimulate the regrowth of cardiac tissue. There have been numerous studies that have examined the uses of various conductive materials for creating cardiac patches (Figure 9.8a–f) (Basara et al. 2022). However, MXene/hydrogel composites and 2D $Ti_3C_2T_x$ MXene nanoflakes have yet to reach their full potential. In a recent study, Ye and coworkers combined solid titanium carbide MXene (Ti_2CT_x) particles distributed in water with polyethylene glycol diacrylate (PEGDA), a crosslinker solution, to form conductive nanocomposite hydrogels. To produce cardiac patches, rat cardiomyocytes were subsequently injected into these gels (Heck et al. 2022;

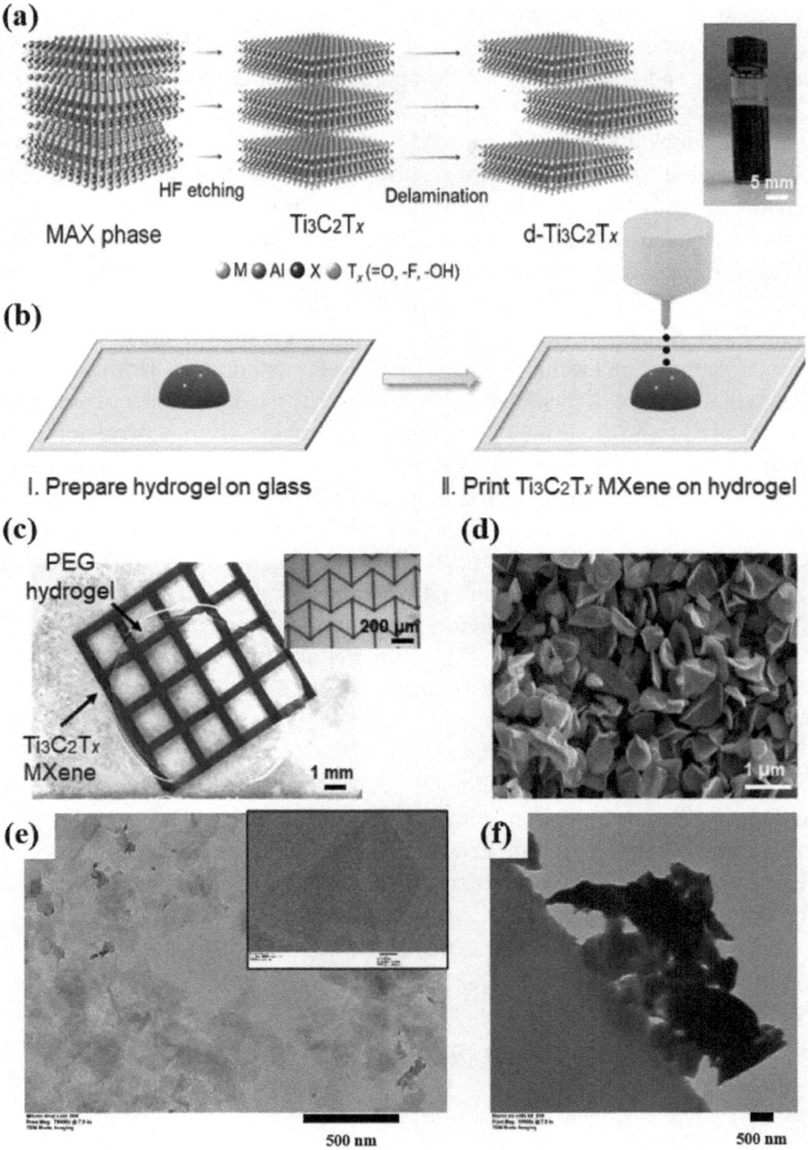

FIGURE 9.8 (a) $Ti_3C_2T_x$ MXene is printed in three dimensions using an aerosol jet over PEG gels. Layers of Ti_3AlC_2 MAX phases are selectively etched with hydrofluoric acid (HF) to produce $Ti_3C_2T_x$ MXene. (b) Using an aerosol jet printer, a diagrammatic representation of the process of printing $Ti_3C_2T_x$ MXene–hydrogel composite is shown, along with the procedures for making PEG hydrogel on glass and printing Ti_3C_2Tx MXene on the hydrogel. (c) Square (81 mm^2) $Ti_3C_2T_x$ MXene pattern printed on PEG hydrogel. $Ti_3C_2T_x$ on PEG is printed with a bowtie design in the inset. (d) $Ti_3C_2T_x$ MXene nanoflakes in the top view of an SEM picture. (e) $Ti_3C_2T_x$ MXene TEM pictures. (f) $Ti_3C_2T_x$ MXene printed on PEG hydrogel is shown in a TEM image.

Source: Permission awarded by Basara et al. (2022).

Manousiouthakis et al. 2022; Pucci et al. 2022; Wei Wang et al. 2020; Jingwen Zhang et al. 2021; Xie et al. 2020; Long et al. 2021; Ghaemi, Slade, and Amini Horri 2021; Besisa, Ewais, and Ahmed 2021) as compared to what G. Basara et al. performed, which was to merely implant the $Ti_3C_2T_x$ MXene nanoparticles in the hydrogel. They created continuous filaments by using AJP, and the objective was to apply a layer of 2D MXene nanoflakes onto a hydrogel substrate in order to enhance cell alignment and conductivity. By introducing iCMs onto the composite structures, they successfully showcased the alignment of cells with exceptional viability. (Zhixiang Xu et al. 2020; Taşdemir et al. 2021; Bulut, Oktem, and Yilmaz 2020; Pirmoradi and Kastner 2021; G. Xu et al. 2021; Dong Wang et al. 2021; Chenglong Yu et al. 2020; Jun feng Su et al. 2020; S. Dong et al. 2020). Studies that combine the fibers' structure and electroconductivity of the natural heart have been conducted. However, with the exception of one, none of them included printing in three dimensions. A current work utilized 3D printing using the extrusion method, which offers a precision of a few hundred micrometres, to generate a gel with a hydrogel incorporating gold nanorods and cells. An ambient field of electricity was utilized to orient the cells and nanorods. A recent study investigated the possible application of Ti_3C_2 MXene in combination with hyaluronic acid-alginate hydrogels as a bio-ink for three-dimensional printing using the extrusion method utilizing Human Embryonic Kidney 293 (HEK-293) cells (Dorishetty, Dutta, and Choudhury 2020; Y. Zou, Raveendran, and Chen 2020; Y. Yu and Wang 2016; J. Xu et al. 2020; Tyler et al. 2020; Medina-Franco 2016; Y. Xiao et al. 2020; Trabocchi 2019; L. Huang et al. 2022). This study improves the field by printing a cell-friendly conductive material with cell-level resolution directly on a 3D conformal scaffold. (It calls for a high sheath flow and a procedure of printing parameter optimization.) Additionally, AJP enables the creation of materials in desired forms faster than other conventional fabrication techniques like infusion molding and without the need for product-specific molds. In conclusion, the findings presented previously demonstrate that the nanoflakes technology is very promising and could boost the quality of the material (A. Yu et al. 2019; Zaarour et al. 2019; Mokhtari, Latifi, and Shamshirsaz 2016; Blachowicz and Ehrmann 2020; Y. Yu and Wang 2016; Yongzhi Wang et al. 2021).

9.2.5 Cellulose/2D Ti_3C_2Tx MXene-Based Hydrogel Possesses 3D Networks

Cellulose is an organic polysaccharide composed of a linear chain of many D-glucose units, with the chemical formula $(C_6H_{10}O_5)n$. Various species of algae and oomycetes possess fundamental cell walls that consist of cellulose, which serves as a crucial component for structural integrity. It is excreted by certain bacterial species within biofilms. Cellulose is the predominant organic polymer utilized on a global scale. Wood typically contains between 40 to 50% cellulose, while cotton fiber consists of cellulose. Dry hemp, on the other hand, contains approximately cellulose. Cellulose is used to produce two primary products: Paper and paperboard. Various derivative goods, including rayon and cellophane, are produced from lower quantities. Cellulosic ethanol and other biofuels derived from cellulose obtained from energy crops are presently being developed as a sustainable fuel option. Cotton and wood pulp are the primary sources of cellulose used in industrial applications (MOORE 1964; B. D. Chen et al. 2018; Zhu et al. 2018; Vasconcelos et al. 2016; To and Tip 1997; Rodrigues et al. 2007; P. Xu et al. 2018; X. S. Zhang et al. 2018b; Zhao et al. 2018).

Hydrogels made from bioinspired materials have unique qualities like those of biological tissues. They possess a diverse array of uses in wearable smart devices due to their extraordinary sensitivity to diverse environmental inputs. The integration of these multifaceted hydrogels with customizable patterns, responsive to stimuli, capable of self-repair, and compatible with living organisms, often poses significant challenges. The authors Wei et al. synthesized an MXene-based composite hydrogel that is self-healing, printable, and biocompatible. This was achieved by incorporating $Ti_3C_2T_x$ MXene nanosheets through a chelation process, where calcium ions (Ca^{2+}) bind to polyacrylic acid and cellulose nanofibers in an alkaline environment (L. Wang et al. 2021; J. Wei et al. 2022; K. Yu et al. 2022; Cong Chen, Liang, et al. 2021; Fabien et al. 2022; Yew et al. 2021; Cheng Chen, Su, et al. 2021). Mussels' secretions served as their inspiration. The biocompatible conductive hydrogel is an excellent choice for wearable multifunctional sensors. It may be used for strain monitoring, speech sensing, signature identification, and transmission of Morse code. The hydrogel demonstrates high sensitivity (2.16-gauge factor), strong adhesion, quick self-healing (within 1 second), and reliable recognition. The multifunctional hydrogel, with a thickness of 2.0 mm, provided significant protection against electromagnetic radiation and pollutants, reducing their levels by over 30 dB. Hence, this study presents a versatile method for creating conducting hydrogels that are with environmentally friendly properties, showcasing the varied applications of intelligent hydrogels (X. Xu et al. 2022; Russo et al. 2021; García-Soriano et al. 2022; Bromberg et al. 2021; Gibson et al. 2021; Mazzaglia et al. 2022; Peng Liu, Liu, Hu, et al. 2021; C. Zhou et al. 2020).

Wei et al. provided a simple method for creating a hydrogel composed of calcium ions, PAA, CNF, and MXene that is inspired by biological systems. This hydrogel displayed a wide range of useful qualities, such as electrical conductivity, biocompatibility, stretchability, plasticity, adhesion, and 3D printing capability (Figure 9.9a–d) (J. Wei et al. 2022). The hydrogel shows promise for use in multifunctional intelligent devices based on these characteristics (Ungai-Salánki et al. 2021; Sagar and Rao 2021; Jaya Prakash and Kandasubramanian 2021; J. S. Park et al. 2022; Z. Jing et al. 2022; Palomo-Briones et al. 2022; Neves et al. 2022). Additionally, the screen-printed layer made of Ca-PAA-CNF-MXene hydrogel on electrodes made of copper can work as a strain sensor, speech sensor, signature sensor, and Morse code receiver. It also has other capabilities. Moreover, the intelligent hydrogel's porous configuration is notable for its efficient electromagnetic interference (EMI) shielding capability, which has the potential to provide protection against electromagnetic radiation for individuals. This bioinspired multifunctional hydrogel shows great potential for advancing intelligent wearable technology due to its impressive properties (H. Su et al. 2022; Carbonati et al. 2021; G. Lei et al. 2021; Wright, Beach, and McCarron 2022; Gang Li et al. 2022; Lai et al. 2020; Jiao et al. 2021; Z. Yang et al. 2021; Mohseni et al. 2022; J. H. Shen et al. 2021).

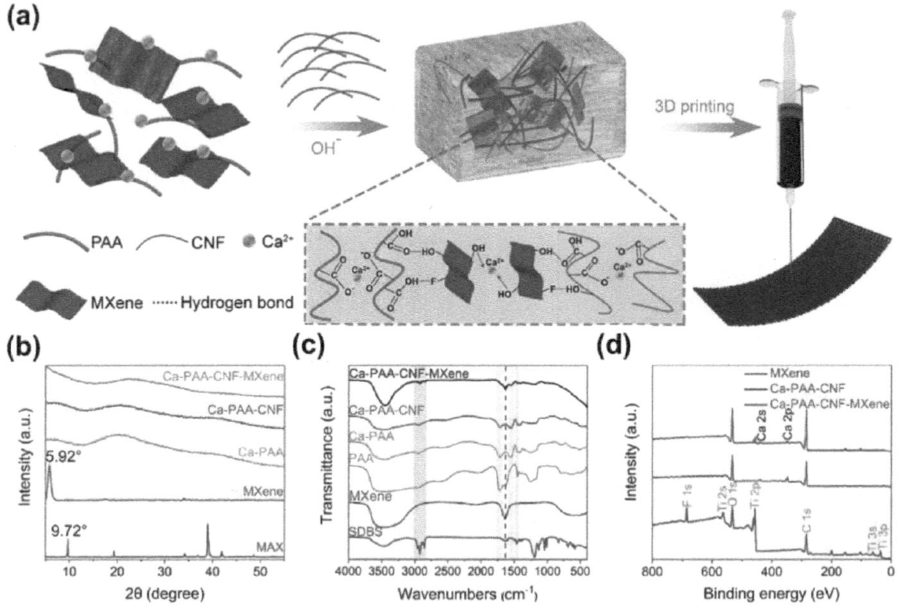

FIGURE 9.9 (a) The making and characterization of Ca, PAA, CNF, and MXene composite hydrogels. Ca-PAA-CNF-MXene hydrogel preparation and 3D printing are shown schematically. (b) Ca, PAA, CNF, and MXene hydrogels and the associated reference materials' XRD patterns, FTIR spectra. (c–d) XPS surveys. *Abbreviations:* XPS, X-ray photoelectron spectroscopy; FTIR, Fourier transform infrared; PAA, polyacrylic acid; XRD, X-ray diffraction.

Source: Permission awarded by J. Wei et al. (2022).

Cellulose differs from the other polysaccharides according to its properties. Cellulose's exceptional properties are due to its unique structure. The amount of glucose subunits in cellulose also has an impact on them. It has the following characteristics: Cellulose is the form of carbohydrate that occurs most frequently in nature. It is insoluble in water (Srasri et al. 2018; Y. Zou, Raveendran, and Chen 2020; Y. Yu and Wang 2016; J. Xu et al. 2020; Tyler et al. 2020; Medina-Franco 2016; Byun et al. 2016; Y. Xiao et al. 2020). It is a crystalline substance that looks like white powder. The remarkably high tensile strength of cellulose microfibrils is due to the strong bonds of hydrogen that exist between its individual strands. Cellulose fibrils possess a tensile strength that is equivalent to that of steel. The diverse arrangements of the sugar molecules also enhance the cellulose's exceptional tensile strength. It exhibits solubility in many solvents. The inherent properties of cellulose make it highly advantageous for application in several domains of materials science and engineering, with the aim of enhancing the properties of the materials (K. Wei et al. 2021; Ling et al. 2021; Nigam et al. 2021; Maletsky et al. 2021; Xiulan Wu et al. 2021; F. Fang et al. 2020).

9.2.6 Palladium Nanoparticles and 2D Ti_3C_2Tx MXene Hybrid 3D Structure

Palladium is a natural element with an atomic number of 46 and is represented by the symbol Pd. The British scientist William Hyde Wollaston discovered this valuable and lustrous white in color metal in 1803. He gave it the name Pallas in honor of the asteroid Pallas, which was given that name by the Greek goddess Athena after she killed Pallas. A group of substances known as the platinum group metals is composed of the elements platinum, palladium, ruthenium, iridium, rhodium, and osmium (PGMs) (Wellmer et al. 2020; Y. Zou, Raveendran, and Chen 2020; Y. Yu and Wang 2016; J. Xu et al. 2020; Tyler et al. 2020; Medina-Franco 2016; Byun et al. 2016; Y. Xiao et al. 2020). Palladium exhibits the lowest density and possesses the lowest melting point within the group, while yet displaying comparable chemical properties. Over 50% of the palladium and platinum supply is used by catalytic converters, which transform up to 90% of the noxious chemicals found in car exhaust (such as hydrocarbons, carbon monoxide, and nitrogen dioxide) into harmless substances (nitrogen, carbon dioxide, and water vapor). In addition, palladium is used in several industries such as jewelry, dentistry, electronics, medicine, chemical processes, hydrogen purification, and groundwater remediation. Fuel cells facilitate the chemical reaction between hydrogen and oxygen, resulting in the generation of electricity, heat, and water. Palladium is essential in these technologies (Wellmer et al. 2020; Y. Zou, Raveendran, and Chen 2020; Y. Yu and Wang 2016; J. Xu et al. 2020; Tyler et al. 2020; Medina-Franco 2016; Byun et al. 2016; Y. Xiao et al. 2020; S. Luo et al. 2020; T. Yu et al. 2021; Shin and Son 2021; Jinze Zhang et al. 2021; Y. H. Chen, Hu, et al. 2021; Mao et al. 2021; D. Yu, Tan, and Zhang 2021; Q. Ji et al. 2021; Go et al. 2021; Shawky et al. 2021).

According to research by E. Satheeshkumar et al. in a single step, gold, silver, and palladium nanoparticles were combined with liquids and then bonded to exfoliated two-dimensional (2D) Ti_3C_2 titanium carbide (MXene) nanosheets. This process is seen in Figure 9.10a–c. Surface-enhanced Raman spectroscopy (SERS) can be conducted on hybrid materials that have been synthesized. An estimated analytical technique can be used to compute the surface plasmon resonance (SPR) frequency of nanoparticles in a medium around the interface of two dielectric medium that have distinct constants of dielectric (Jinze Zhang et al. 2021; Y. H. Chen, Hu, et al. 2021; G. Xiao et al. 2021; J. Yin et al. 2021; Mao et al. 2021; Huan Wang et al. 2020; Q. Ji et al. 2021; Tong et al. 2021; Shawky et al. 2021; D. Li et al. 2021; S. Jin et al. 2020; C. Wang et al. 2021; Y. K. Park et al. 2020). For the silver and gold nanoparticles, respectively, SPR wavelengths of 440 nm and 558 nm were able to successfully match the experimental findings. When coupled with 2D MXene, palladium displays a broad, noticeably red-shifted SPR band with a peak at 230 nm, based on the computed SPR wavelength for palladium's planar shape of 160 nm (Satheeshkumar et al. 2016). The authors provide a possible mechanism for the hybridization of plasmonic nanoparticles with MXene. When coupled with 2D MXene, palladium displays a broad, noticeably red-shifted SPR band with a peak at 230 nm, compared to the computed SPR wavelength of 160 nm for palladium's planar shape. They provide a possible mechanism for the

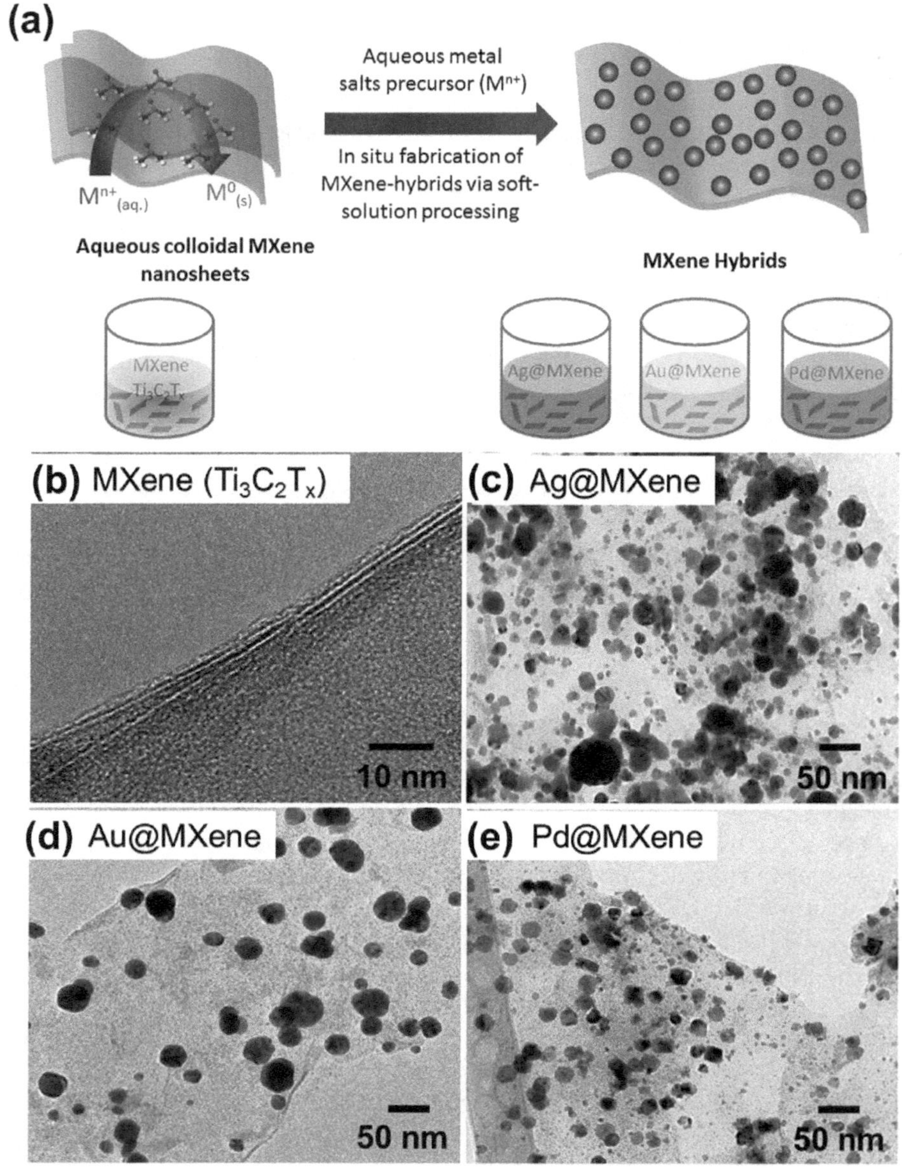

FIGURE 9.10 a) Visual representation of the in-situ one-step solution-processed synthesis of Ag, Au, and Pd@MXene ($Ti_3C_2T_x$) hybrids, utilizing a soft-solution processing method combined with sonochemistry. Morphology studies of (b) $Ti_3C_2T_x$, (c) Ag@MXene, (d) Au@MXene, and (e) Pd@MXene hybrid nanosheets are shown in low-resolution TEM images, along with high-resolution TEM images of MXene nanosheets.

Source: Permission awarded by E. Satheeshkumar et al. (2016).

hybridization of plasmonic nanoparticles with MXene. The noble metal nanoparticles on MXene, as they are initially synthesized, exhibit highly sensitive surface-enhanced Raman scattering (SERS) detection of methylene blue. The anticipated improvement factors are approximately 10^5. These discoveries create opportunities for the advancement of new 2D hybrid materials that can be used in visible-range SERS applications in the fields of sensors, catalysis, and biology (Dongming Wang, Wu, and Gao 2021; Shin and Son 2021; T. Yu et al. 2021; Xiaoyu Han et al. 2020; H. Nie et al. 2021; El-Fiqi and Kim 2021; Peera et al. 2021; M. Chen and Liu 2021; Son, Kim, and Ahn 2021; Ishak et al. 2020; Gavrila et al. 2020; Peña-Garcia et al. 2021).

The benefits of the hybrid material for producing 3D printed structures include the ability to cast complicated shapes that are challenging to make conventionally, as well as low prices and time savings for prototype or small series production.

9.3 SUMMARY

Out all the several types of MXene, $Ti_3C_2T_x$ has garnered the highest level of attention, specifically in the fields of storing energy and electromagnetic interference (EMI) shielding. Significantly, there is a presence of both possibilities and limitations for the relevant future research. Due to the demonstrated correlation between the quality of $Ti_3C_2T_x$ nanoflakes and their electrical conductivity, there is a strong demand for reliable and secure synthetic methods. Even while the generated $Ti_3C_2T_x$ flakes show that the LiF-HCl etching process produces considerably fewer nanometer-size flaws than HF etching, the interlayers' natural properties could be harmed by excess water or cations. Further research is required to better understand the characteristics of the influence of few- or monolayer sheets on films made from $Ti_3C_2T_x$. This is particularly important following the etching process, as it is vital to develop methods for separating and collecting the multilayered $Ti_3C_2T_x$. Additional work should be focused into this area to safely simplify the entire synthesis process and produce $Ti_3C_2T_x$ with large flake sizes and a low defect concentration.

Nevertheless, the process of oxidizing MXene films is a significant barrier that hinders their practical use. While the process of producing films by arranging $Ti_3C_2T_x$ flakes together does enhance the environmental stability of the flakes, it is important to note that these films still exhibit a moderate susceptibility to humid conditions. In order to maximize the potential of $Ti_3C_2T_x$-based films for practical applications in the future, it would be beneficial to chemically alter the edges and defective areas of the film while preserving its in-plane physiochemistry. The performance in transport applications may suffer because of the restack of $Ti_3C_2T_x$ flakes during the assembly process. To counteract the inclination to restacking and prepare for the discovery of more aggressive strategies, we have summarized some doable and efficient approaches. Previous researchers have mostly concentrated on reducing the amount of $Ti_3C_2T_x$ flakes that are stacked directly on top of each other during face-to-face interactions. The deliberate creation of vertically aligned holes and the process of chemically bonding $Ti_3C_2T_x$ flakes together in films by side-to-side connections have not been extensively studied, despite their significant relevance to electrochemical applications.

REFERENCES

Ahmad, Rabia, Naseem Iqbal, Tayyaba Noor, Ghulam Ali, Majid Ali, Nadia Shahzad, and Muhammad Arslan Raza. 2022. "Zeolitic Imidazolate Frameworks Derived Co-Zn-Nanoporous Carbon-Sulfide Material for Supercapacitors." *Electrochimica Acta* 404. Elsevier Ltd: 139739. doi:10.1016/j.electacta.2021.139739.

Aïssa, B., A. Sinopoli, A. Ali, Y. Zakaria, A. Zekri, M. Helal, M. Nedil, F. Rosei, S. Mansour, and K. A. Mahmoud. 2021. "Nanoelectromagnetic of a Highly Conductive 2D Transition Metal Carbide (MXene)/Graphene Nanoplatelets Composite in the EHF M-Band Frequency." *Carbon* 173. Elsevier Ltd: 528–39. doi:10.1016/j.carbon.2020.11.024.

Altaf, Cigdem Tuc, Mehrdad Faraji, Alihan Kumtepe, Nazrin Abdullayeva, Nazmi Yilmaz, Emine Karagoz, Ali Bozbey, Hamza Kurt, Mehmet Sankir, and Nurdan Demirci Sankir. 2020. "Highly Efficient 3D-ZnO Nanosheet Photoelectrodes for Solar-Driven Water Splitting: Chalcogenide Nanoparticle Sensitization and Mathematical Modeling." *Journal of Alloys and Compounds* 828. Elsevier B.V: 154472. doi:10.1016/j.jallcom.2020.154472.

Amici, Julia, Cecilia Andrea Calderón, Daniele Versaci, Guillermina Luque, Daniel Barraco, Ezequiel Leiva, Carlotta Francia, and Silvia Bodoardo. 2022. "Composite Polymer Electrolyte with High Inorganic Additive Contents to Enable Metallic Lithium Anode." *Electrochimica Acta* 404. Elsevier Ltd: 139772. doi:10.1016/j.electacta.2021.139772.

Amorim, C. O., K. Sloyan, M. R. Correia, A. A. C. S. Lourenço, M. S. Dahlem, and V. S. Amaral. 2020. "Constrained Titanohematite Formation at BTO/Fe Interfaces Deposited by RF-Sputtering." *Journal of Alloys and Compounds* 828. Elsevier B.V: 154244. doi:10.1016/j.jallcom.2020.154244.

Anasori, Babak, Maria R. Lukatskaya, and Yury Gogotsi. 2017. *2D Metal Carbides and Nitrides (MXenes) for Energy Storage. Nature Reviews Materials*. Vol. 2. doi:10.1038/natrevmats.2016.98.

Ataeefard, Maryam, Ebrahim Ghasemi, and Mona Ebadi. 2014. "Effect of Micro- and Nanomagnetite on Printing Toner Properties." *The Scientific World Journal* 2014. doi:10.1155/2014/706367.

Ateka, Ainara, Pablo Rodriguez-Vega, Tomás Cordero-Lanzac, Javier Bilbao, and Andrés T. Aguayo. 2021. "Model Validation of a Packed Bed LTA Membrane Reactor for the Direct Synthesis of DME from CO/CO_2." *Chemical Engineering Journal* 408 (September 2020). doi:10.1016/j.cej.2020.127356.

Bae, Ho Sub, Mahadeo A. Mahadik, Young Seok Seo, Won Sik Chae, Hee Suk Chung, Hyeon Ih Ryu, Min Cho, Patrick J. Shea, Sun Hee Choi, and Jum Suk Jang. 2021. "Palladium Metal Oxide/Hydroxide Clustered Cobalt Oxide Co-Loading on Acid Treated TiO_2 Nanorods for Degradation of Organic Pollutants and Salmonella Typhimurium Inactivation Under Simulated Solar Light." *Chemical Engineering Journal* 408 (September 2020). Elsevier B.V.: 127260. doi:10.1016/j.cej.2020.127260.

Bafandeh, Mohammad Reza, and Jae Shin Lee. 2021. "Two-Step Sintering of $0.93Bi0.5Na0.5TiO_{3-0}.07BaTiO_3$ Lead-Free Piezoelectric Material." *Ceramics International* 47 (20). Elsevier Ltd: 28723–28. doi:10.1016/j.ceramint.2021.07.031.

Barhoum, Ahmed. 2019. *Handbook of Nanofibers. Handbook of Nanofibers*. doi:10.1007/978-3-319-53655-2.

Barroca, Nathalie, Liam Collins, Brian J. Rodriguez, M. Helena V. Fernandes, and Paula M. Vilarinho. 2022. "Mechanical Writing of Electrical Polarization in Poly (L-Lactic) Acid." *Acta Biomaterialia* 139: 249–58. doi:10.1016/j.actbio.2021.05.057.

Basara, Gozde, Mortaza Saeidi-Javash, Xiang Ren, Gokhan Bahcecioglu, Brian C. Wyatt, Babak Anasori, Yanliang Zhang, and Pinar Zorlutuna. 2022. "Electrically Conductive 3D Printed Ti_3C_2Tx MXene-PEG Composite Constructs for Cardiac Tissue Engineering." *Acta Biomaterialia* 139: 179–89. doi:10.1016/j.actbio.2020.12.033.

Besisa, Dina H. A., Emad M. M. Ewais, and Yasser M. Z. Ahmed. 2021. "A Comparative Study of Thermal Conductivity and Thermal Emissivity of High Temperature Solar Absorber of ZrO_2/Fe_2O_3 and Al_2O_3/CuO Ceramics." *Ceramics International* 47 (20). Elsevier Ltd: 28252–59. doi:10.1016/j.ceramint.2021.06.240.

Bhatnagar, Priyanka, Joonpyo Hong, Malkeshkumar Patel, and Joondong Kim. 2022. "Transparent Photovoltaic Skin for Artificial Thermoreceptor and Nociceptor Memory." *Nano Energy* 91 (November 2021). Elsevier Ltd: 106676. doi:10.1016/j.nanoen.2021.106676.

Bian, Jiyong, Xiaoqiang An, Wei Jiang, Ruiping Liu, Chengzhi Hu, and Huijuan Liu. 2021. "Defect-Enhanced Activation of Carbon Nitride/Horseradish Peroxidase Nanohybrids for Visible-Light-Driven Photobiocatalytic Water Purification." *Chemical Engineering Journal* 408 (September 2020). Elsevier B.V.: 127231. doi:10.1016/j.cej.2020.127231.

Blachowicz, Tomasz, and Andrea Ehrmann. 2020. "Most Recent Developments in Electrospun Magnetic Nanofibers: A Review." *Journal of Engineered Fibers and Fabrics* 15. doi:10.1177/1558925019900843.

Britz, Dieter, Jörg Strutwolf, and Ole Østerby. 2022. "Rectangular Electrodes: Simulation of Accurate Steady State Currents and the Behaviour of Square Electrode Arrays." *Electrochimica Acta* 404. Elsevier Ltd: 139750. doi:10.1016/j.electacta.2021.139750.

Bromberg, Lev, Nil Ozbek, Kai Jher Tan, Xiao Su, Lokesh P. Padhye, and T. Alan Hatton. 2021. "Iron Phosphomolybdate Complexes in Electrocatalytic Reduction of Aqueous Disinfection Byproducts." *Chemical Engineering Journal* 408 (October 2020). Elsevier B.V.: 127354. doi:10.1016/j.cej.2020.127354.

Bulut, Onur, Huseyin Avni Oktem, and M. Deniz Yilmaz. 2020. "A Highly Substituted and Fluorescent Aromatic-Fused Imidazole Derivative That Shows Enhanced Antibacterial Activity against Methicillin-Resistant Staphylococcus Aureus (MRSA)." *Journal of Hazardous Materials* 399 (March). Elsevier: 122902. doi:10.1016/j.jhazmat.2020.122902.

Byun, Kyung Eun, Yeonchoo Cho, Minsu Seol, Seongsu Kim, Sang Woo Kim, Hyeon Jin Shin, Seongjun Park, and Sungwoo Hwang. 2016. "Control of Triboelectrification by Engineering Surface Dipole and Surface Electronic State." *ACS Applied Materials and Interfaces* 8 (28): 18519–25. doi:10.1021/acsami.6b02802.

Cai, Lei, Fei Pan, Xiaojie Zhu, Yanyan Dong, Yuyang Shi, Zhen Xiang, Jie Cheng, Haojie Jiang, Zhong Shi, and Wei Lu. 2022. "Etching Engineering and Electrostatic Self-Assembly of N-Doped MXene/Hollow Co-ZIF Hybrids for High-Performance Microwave Absorbers." *Chemical Engineering Journal* 434 (November 2021). Elsevier B.V.: 133865. doi:10.1016/j.cej.2021.133865.

Cai, Yong Zhu, Yong Sheng Fang, Wen Qiang Cao, Peng He, and Mao Sheng Cao. 2021. "MXene-CNT/PANI Ternary Material with Excellent Supercapacitive Performance Driven by Synergy." *Journal of Alloys and Compounds* 868. Elsevier: 159159. doi:10.1016/j.jallcom.2021.159159.

Cao, Ran, Xianjie Pu, Xinyu Du, Wei Yang, Jiaona Wang, Hengyu Guo, Shuyu Zhao, Zuqing Yuan, Chi Zhang, Congju Li, and Zhong Lin Wang. 2018. "Screen-Printed Washable Electronic Textiles as Self-Powered Touch/Gesture Tribo-Sensors for Intelligent Human-Machine Interaction." *ACS Nano* 12 (6). American Chemical Society: 5190–96. doi:10.1021/acsnano.8b02477.

Carbonati, Tecla, Carolina Cionti, Ewoud Cosaert, Ben Nimmegeers, Daniela Meroni, and Dirk Poelman. 2021. "NIR Emitting $GdVO_4$:Nd Nanoparticles for Bioimaging: The Role of the Synthetic Pathway." *Journal of Alloys and Compounds* 862. doi:10.1016/j.jallcom.2020.158413.

Chai, Bin, Kunming Shi, Haiyang Zou, Pingkai Jiang, Zixiang Wu, and Xingyi Huang. 2022. "Conductive Interlayer Modulated Ferroelectric Nanocomposites for High Performance Triboelectric Nanogenerator." *Nano Energy* 91 (September 2021). Elsevier Ltd: 106668. doi:10.1016/j.nanoen.2021.106668.

Chang, Ming, Zirui Jia, Shuangqiao He, Jixi Zhou, Shuo Zhang, Mengli Tian, Bingbing Wang, and Guanglei Wu. 2021. "Two-Dimensional Interface Engineering of $NiS/MoS_2/Ti_3C_2Tx$ Heterostructures for Promoting Electromagnetic Wave Absorption Capability." *Composites Part B: Engineering* 225 (July). Elsevier Ltd: 109306. doi:10.1016/j.compositesb.2021.109306.

Chen, Bao Dong, Wei Tang, Chuan He, Chao Ran Deng, Lei Jing Yang, Lai Pan Zhu, Jian Chen, Jia Jia Shao, Long Liu, and Zhong Lin Wang. 2018. "Water Wave Energy Harvesting and Self-Powered Liquid-Surface Fluctuation Sensing Based on Bionic-Jellyfish Triboelectric Nanogenerator." *Materials Today* 21 (1): 88–97. doi:10.1016/j.mattod.2017.10.006.

Chen, Cheng, Hang Su, Li Na Lu, Yang Shan Hong, Yunzhen Chen, Kang Xiao, Ting Ouyang, Yanlin Qin, and Zhao Qing Liu. 2021. "Interfacing Spinel $NiCo_2O_4$ and NiCo Alloy Derived N-Doped Carbon Nanotubes for Enhanced Oxygen Electrocatalysis." *Chemical Engineering Journal* 408 (July 2020). doi:10.1016/j.cej.2020.127814.

Chen, Cong, Weiyuan Hang, Yiwei Li, Mingjian Zhang, Kaiqi Nie, Jiaou Wang, Wenguang Zhao, Rui Qi, Changjian Zuo, Zhibo Li, Haocong Yi, and Feng Pan. 2021. "P2/O3 Biphasic Fe/Mn-Based Layered Oxide Cathode with Ultrahigh Capacity and Great Cyclability for Sodium Ion Batteries." *Nano Energy* 90 (PA). Elsevier Ltd: 106504. doi:10.1016/j.nanoen.2021.106504.

Chen, Cong, Jiwei Liang, Junjun Zhang, Xinxing Liu, Xinxing Yin, Hongsen Cui, Haibing Wang, Chen Wang, Zaifang Li, Junbo Gong, Qianqian Lin, Weijun Ke, Chen Tao, Bo Da, Zejun Ding, Xudong Xiao, and Guojia Fang. 2021. "Interfacial Engineering of a Thiophene-Based 2D/3D Perovskite Heterojunction for Efficient and Stable Inverted Wide-Bandgap Perovskite Solar Cells." *Nano Energy* 90 (July): 1–10. doi:10.1016/j.nanoen.2021.106608.

Chen, Fang, Chenglong Ma, Chongjie Zou, Limei Cao, and Ji Yang. 2022. "Highly Efficient and Robust Iridium Based Anodes Prepared via Thermal Decomposition for Significant Degradation of Carbamazepine." *Electrochimica Acta* 404. Elsevier Ltd: 139722. doi:10.1016/j.electacta.2021.139722.

Chen, Haotian, Yu Song, Xiaoliang Cheng, and Haixia Zhang. 2019. "Self-Powered Electronic Skin Based on the Triboelectric Generator." *Nano Energy* 56 (November 2018). Elsevier Ltd: 252–68. doi:10.1016/j.nanoen.2018.11.061.

Chen, Hedong, Peng Wang, Xingyu Wang, Xingfu Wang, Lujia Rao, Yinping Qian, Hongjie Yin, Xianhua Hou, Huapeng Ye, Guofu Zhou, and Richard Nötzel. 2021. "3D InGaN Nanowire Arrays on Oblique Pyramid-Textured Si (311) for Light Trapping and Solar Water Splitting Enhancement." *Nano Energy* 83 (October 2020): 1–10. doi:10.1016/j.nanoen.2021.105768.

Chen, Hong, Jianyu Zhou, Shuai Liu, Sheng Wang, and Xinglong Gong. 2022. "A Novel Anti-Impact and Flame Retardant Gel towards Human Protection and High-Temperature Alarm." *Composites Part A: Applied Science and Manufacturing* 158 (January). Elsevier Ltd: 106994. doi:10.1016/j.compositesa.2022.106994.

Chen, Jun, Guang Zhu, Weiqing Yang, Qingshen Jing, Peng Bai, Ya Yang, Te-Chien Hou, and Zhong Lin Wang. 2013a. "Harmonic-Resonator-Based Triboelectric Nanogenerator as a Sustainable Power Source and a Self-Powered Active Vibration Sensor." *Advanced Materials* 25: 6094–99. doi:10.1002/adma.201302397.

Chen, Jun, Guang Zhu, Weiqing Yang, Qingshen Jing, Peng Bai, Ya Yang, Te Chien Hou, and Zhong Lin Wang. 2013b. "Harmonic-Resonator-Based Triboelectric Nanogenerator as a Sustainable Power Source and a Self-Powered Active Vibration Sensor." *Advanced Materials* 25 (42): 6094–99. doi:10.1002/adma.201302397.

Chen, Mingming, and Qinghui Liu. 2021. "Multi-Scale Modelling of Progressive Damage and Failure Behaviour of 2D Woven SiC/SiC Composites." *Ceramics International* 47 (20). Elsevier Ltd: 28821–36. doi:10.1016/j.ceramint.2021.07.043.

Chen, Qiuling. 2020. "Nanobismuth Enhanced Plasmonic, Emission and Faraday Rotation Properties of Diamagnetic Tellurite Glasses." *Journal of Alloys and Compounds* 828. Elsevier B.V: 154448. doi:10.1016/j.jallcom.2020.154448.

Chen, Wenhua, Pengju Liu, Yuan Liu, and Zhuoxin Liu. 2022. "Recent Advances in Two-Dimensional Ti_3C_2Tx MXene for Flame Retardant Polymer Materials." *Chemical Engineering Journal* 446 (P3). Elsevier B.V.: 137239. doi:10.1016/j.cej.2022.137239.

Cheng, Jun, Yali Wang, Leiqing Hu, Niu Liu, Junchen Xu, and Junhu Zhou. 2020. "Using Lantern Zn/Co-ZIF Nanoparticles to Provide Channels for CO_2 Permeation through PEO-Based MMMs." *Journal of Membrane Science* 597 (November 2019). Elsevier B.V.: 117644. doi:10.1016/j.memsci.2019.117644.

Cheng, Kuan, Samuel Wallaert, Haleh Ardebili, and Alamgir Karim. 2022. "Advanced Triboelectric Nanogenerators Based on Low-Dimension Carbon Materials: A Review." *Carbon* 194 (March). Elsevier Ltd: 81–103. doi:10.1016/j.carbon.2022.03.037.

Cheng, Runrun, Yan Wang, Xiaochuang Di, Zhao Lu, Ping Wang, Mingliang Ma, and Jinrui Ye. 2022. "Construction of MOF-Derived Plum-like NiCo@C Composite with Enhanced Multi-Polarization for High-Efficiency Microwave Absorption." *Journal of Colloid and Interface Science* 609. Elsevier Inc.: 224–34. doi:10.1016/j.jcis.2021.11.197.

Choudhury, Deepak, Marcus Yee, Zach Lee Jia Sheng, Ahmad Amirul, and May Win Naing. 2020. "Decellularization Systems and Devices: State-of-the-Art." *Acta Biomaterialia* 115. Elsevier Ltd: 51–59. doi:10.1016/j.actbio.2020.07.060.

Chu, Hyenwoo, Houk Jang, Yongjun Lee, Youngcheol Chae, and Jong-hyun Ahn. 2016. "Nano Energy for Self-Powered Wearable Electronics." *Nano Energy* 27. Elsevier: 298–305. doi:10.1016/j.nanoen.2016.07.009.

Cui, Ce, Ronghui Guo, Erhui Ren, Hongyan Xiao, Mi Zhou, Xiaoxu Lai, Qin Qin, Shouxiang Jiang, and Wenfeng Qin. 2021. "MXene-Based RGO/Nb_2CTx/Fe_3O_4 Composite for High Absorption of Electromagnetic Wave." *Chemical Engineering Journal* 405 (March 2020). Elsevier: 126626. doi:10.1016/j.cej.2020.126626.

Dagdeviren, Canan, Zhou Li, and Zhong Lin Wang. 2017. "Energy Harvesting from the Animal/Human Body for Self-Powered Electronics." *Annual Review of Biomedical Engineering* 19 (1): 85–108. doi:10.1146/annurev-bioeng-071516-044517.

Dai, Bo, Yong Ma, Shixuan Feng, Haowen Wang, Mingliang Ma, Jianxu Ding, Xunqian Yin, and Tingxi Li. 2022. "Journal of Colloid and Interface Science Fabrication of One-Dimensional M (Co, Ni)@ Polyaniline Nanochains with Adjustable Thickness for Excellent Microwave Absorption Properties." *Journal of Colloid And Interface Science* 627. Elsevier Inc.: 113–25. doi:10.1016/j.jcis.2022.06.137.

Das, Krishna Ch, and Siddhartha S. Dhar. 2020. "Rapid Catalytic Degradation of Malachite Green by $MgFe_2O_4$ Nanoparticles in Presence of H_2O_2." *Journal of Alloys and Compounds* 828. Elsevier B.V: 154462. doi:10.1016/j.jallcom.2020.154462.

Demin, Alexander M., Aleksei I. Maksimovskikh, Alexander V. Mekhaev, Dmitry K. Kuznetsov, Artyom S. Minin, Alexandra G. Pershina, Mikhail A. Uimin, Vladimir Ya

Shur, and Victor P. Krasnov. 2021. "Silica Coating of Fe_3O_4 Magnetic Nanoparticles with PMIDA Assistance to Increase the Surface Area and Enhance Peptide Immobilization Efficiency." *Ceramics International* 47 (16). Elsevier Ltd: 23078–87. doi:10.1016/j.ceramint.2021.04.310.

Deng, Dan, Yang Yang, Wenjun Zou, Yajie Zhang, Zhen Wang, Zaiyu Wang, Jianqi Zhang, Kun Lu, Wei Ma, and Zhixiang Wei. 2018. "Aromatic End-Capped Acceptor Effects on Molecular Stacking and the Photovoltaic Performance of Solution-Processable Small Molecules." *Journal of Materials Chemistry A* 6 (44). Royal Society of Chemistry: 22077–85. doi:10.1039/C8TA04102D.

Deng, Weili, Alberto Libanori, Xiao Xiao, Jun Fang, Xun Zhao, Yihao Zhou, Guorui Chen, Song Li, and Jun Chen. 2022. "Computational Investigation of Ultrasound Induced Electricity Generation via a Triboelectric Nanogenerator." *Nano Energy* 91 (October 2021). Elsevier Ltd: 106656. doi:10.1016/j.nanoen.2021.106656.

Deng, Yaqian, Tongxin Shang, Zhitan Wu, Ying Tao, Chong Luo, Jiachen Liang, Daliang Han, Ruiyang Lyu, Changsheng Qi, Wei Lv, Feiyu Kang, and Quan-Hong Yang. 2019. "Fast Gelation of Ti_3C_2Tx MXene Initiated by Metal Ions." *Advanced Materials* 31 (43): 1–7. doi:10.1002/adma.201902432.

Ding, Rui, Yan Dai, Wenji Zheng, Xiangcun Li, Xiaoming Yan, Yi Liu, Xuehua Ruan, Shaojie Li, Xiaochen Yang, Kai Yang, and Gaohong He. 2021. "Vesicles-Shaped MOF-Based Mixed Matrix Membranes with Intensified Interfacial Affinity and CO_2 Transport Freeway." *Chemical Engineering Journal* 414 (October 2020). Elsevier B.V.: 128807. doi:10.1016/j.cej.2021.128807.

Dong, Kai, Jianan Deng, Wenbo Ding, Aurelia C. Wang, Peihong Wang, Chaoyu Cheng, Yi Cheng Wang, Limin Jin, Bohong Gu, Baozhong Sun, and Zhong Lin Wang. 2018. "Versatile Core–Sheath Yarn for Sustainable Biomechanical Energy Harvesting and Real-Time Human-Interactive Sensing." *Advanced Energy Materials* 8 (23): 1–12. doi:10.1002/aenm.201801114.

Dong, Kai, Yi Cheng Wang, Jianan Deng, Yejing Dai, Steven L. Zhang, Haiyang Zou, Bohong Gu, Baozhong Sun, and Zhong Lin Wang. 2017. "A Highly Stretchable and Washable All-Yarn-Based Self-Charging Knitting Power Textile Composed of Fiber Triboelectric Nanogenerators and Supercapacitors." *ACS Nano* 11 (9): 9490–99. doi:10.1021/acsnano.7b05317.

Dong, Shuying, Lingfang Cui, Yujia Tian, Longji Xia, Yuhan Wu, Jianjia Yu, David M. Bagley, Jianhui Sun, and Maohong Fan. 2020. "A Novel and High-Performance Double Z-Scheme Photocatalyst $ZnO\text{-}SnO_2\text{-}Zn_2SnO_4$ for Effective Removal of the Biological Toxicity of Antibiotics." *Journal of Hazardous Materials* 399 (March). Elsevier: 123017. doi:10.1016/j.jhazmat.2020.123017.

Dong, Yongchang, Sai Sunil, Kumar Mallineni, Kathleen Maleski, and Herbert Behlow. 2018. "Nano Energy Metallic MXenes : A New Family of Materials for Fl Exible Triboelectric Nanogenerators." *Nano Energy* 44 (October 2017). Elsevier Ltd: 103–10. doi:10.1016/j.nanoen.2017.11.044.

Dorishetty, Pramod, Naba K. Dutta, and Namita Roy Choudhury. 2020. "Bioprintable Tough Hydrogels for Tissue Engineering Applications." *Advances in Colloid and Interface Science* 281. Elsevier B.V.: 102163. doi:10.1016/j.cis.2020.102163.

Duan, Junjie, Menghang Zhang, Pengju Chen, Zhuan Li, Liang Pang, Peng Xiao, and Yang Li. 2021. "Tribological Behavior and Applications of Carbon Fiber Reinforced Ceramic Composites as High-Performance Frictional Materials." *Ceramics International* 47 (14). Elsevier Ltd: 19271–81. doi:10.1016/j.ceramint.2021.02.187.

El-Fiqi, Ahmed, and Hae Won Kim. 2021. "Nano/Micro-Structured Poly(ϵ-Caprolactone)/Gelatin Nanofibers with Biomimetically-Grown Hydroxyapatite Spherules: High Protein Adsorption, Controlled Protein Delivery and Sustained Bioactive Ions Release

Designed as a Multifunctional Bone Regenerative Mem." *Ceramics International* 47 (14). Elsevier Ltd: 19873–85. doi:10.1016/j.ceramint.2021.04.003.

Elibol, Erdem. 2020. "Effects of Different Counter Electrodes on Performance of CdSeTe Alloy QDSSC." *Solar Energy* 197 (January): 519–26. doi:10.1016/j.solener.2020.01.035.

Eskandari, M., and V. Ahmadi. 2015. "Copper Selenide as a New Counter Electrode for Zinc Oxide Nanorod Based Quantum Dot Solar Cells." *Materials Letters* 142. Elsevier: 308–11. doi:10.1016/j.matlet.2014.12.071.

Eslami, Reza, Nahid Azizi, Seyed Reza Ghaffarian, Mehrab Mehrvar, and Hadis Zarrin. 2022. "Highly Sensitive and Selective Non-Enzymatic Measurement of Glucose Using Arraying of Two Separate Sweat Sensors at Physiological PH." *Electrochimica Acta* 404. Elsevier Ltd: 139749. doi:10.1016/j.electacta.2021.139749.

Esmaeili, Hamid, Alejandra Patino-Guerrero, Masoud Hasany, Mohammad Omaish Ansari, Adnan Memic, Alireza Dolatshahi-Pirouz, and Mehdi Nikkhah. 2022. "Electroconductive Biomaterials for Cardiac Tissue Engineering." *Acta Biomaterialia* 139: 118–40. doi:10.1016/j.actbio.2021.08.031.

Etesami, Mohammad, Mai Thanh Nguyen, Tetsu Yonezawa, Adisorn Tuantranont, Anongnat Somwangthanaroj, and Soorathep Kheawhom. 2022. "3D Carbon Nanotubes-Graphene Hybrids for Energy Conversion and Storage Applications." *Chemical Engineering Journal* 446 (P3). Elsevier B.V.: 137190. doi:10.1016/j.cej.2022.137190.

Fabien, Aymeric, Guillaume Lefebvre, Brice Calvignac, Pierre Legout, Elisabeth Badens, and Christelle Crampon. 2022. "Interfacial Tension of Ethanol, Water, and Their Mixtures in High Pressure Carbon Dioxide: Measurements and Modeling." *Journal of Colloid and Interface Science* 613. Elsevier Inc.: 847–56. doi:10.1016/j.jcis.2022.01.058.

Fang, Fan, Peng Zhao, Nengjie Feng, Hui Wan, and Guofeng Guan. 2020. "Surface Engineering on Porous Perovskite-Type $La_{0.6}Sr_{0.4}CoO_{3-\delta}$ Nanotubes for an Enhanced Performance in Diesel Soot Elimination." *Journal of Hazardous Materials* 399 (March). Elsevier: 123014. doi:10.1016/j.jhazmat.2020.123014.

Fang, Zhitang, Bangchuan Zhao, Jiafeng Zhou, Jin Bai, Kunzhen Li, Hongyang Ma, and Shuai Lin. 2019. "Composited with Ti_3C_2Tx MXene Nanosheets." *Journal of Energy Chemistry*: 1419–28. https://doi.org/10.1016/j.jechem.2020.01.006.

Feng, Shiyi, Zeying Zhan, Ya Yi, Zehang Zhou, and Canhui Lu. 2022. "Facile Fabrication of MXene/Cellulose Fiber Composite Film with Homogeneous and Aligned Structure via Wet Co-Milling for Enhancing Electromagnetic Interference Shielding Performance." *Composites Part A: Applied Science and Manufacturing* 157 (January). doi:10.1016/j.compositesa.2022.106907.

Ferreira, Filipe V., Luciana S. Cividanes, Rubia F. Gouveia, and Liliane M. F. Lona. 2019. "An Overview on Properties and Applications of Poly(Butylene Adipate-Co-Terephthalate)–PBAT Based Composites." *Polymer Engineering and Science* 59 (s2): E7–15. doi:10.1002/pen.24770.

Firdous, Irum, Muhammad Fahim, Lingyun Wang, Wen Jung Li, Yunlong Zi, and Walid A. Daoud. 2021. "Boosting Current Output of Triboelectric Nanogenerator by Two Orders of Magnitude via Hindering Interfacial Charge Recombination." *Nano Energy* 89 (PA). Elsevier Ltd: 106315. doi:10.1016/j.nanoen.2021.106315.

Franco, Jefferson Honorio, Matteo Grattieri, Adalgisa R. de Andrade, and Shelley D. Minteer. 2021. "Unveiling Complete Lactate Oxidation Through a Hybrid Catalytic Cascade." *Electrochimica Acta* 376. Elsevier Ltd: 138044. doi:10.1016/j.electacta.2021.138044.

Gaâbel, F., M. Khlifi, N. Hamdaoui, K. Taibi, and J. Dhahri. 2020. "Conduction Mechanisms Study in $CaCu_2.8Ni_{0.2}Ti_4O_{12}$ Ceramics Sintered at Different Temperatures." *Journal of Alloys and Compounds* 828. doi:10.1016/j.jallcom.2020.154373.

Gao, Caifang, Qianfan Nie, Che Yi Lin, Fanming Huang, Liangjun Wang, Wei Xia, Xiang Wang, Zhigao Hu, Mengjiao Li, Hong-Wei Lu, Ying-Chih Lai, Yen-Fu Lin, Junhao Chu, and Wenwu Li. 2022. "Touch-Modulated van Der Waals Heterostructure with Self-Writing Power Switch for Synaptic Simulation." *Nano Energy* 91 (September 2021). Elsevier Ltd: 106659. doi:10.1016/j.nanoen.2021.106659.

Gao, Jiaojiao, Jinpeng Song, Yao Wang, Zhuo Wang, Jianping Song, and Xuedong Zhao. 2022. "Microstructures and Mechanical Properties of Functionally Graded TiCN–TaC Ceramics Prepared by a Novel Layer Processing Strategy." *Ceramics International* 48 (12). Elsevier Ltd: 16990–96. doi:10.1016/j.ceramint.2022.02.254.

Gao, Yuyu, Cheng Yan, Haichao Huang, Tao Yang, Guo Tian, Da Xiong, Ningjun Chen, Xiang Chu, Shen Zhong, Weili Deng, Yong Fang, and Weiqing Yang. 2020. "Microchannel-Confined MXene Based Flexible Piezoresistive Multifunctional Micro-Force Sensor." *Advanced Functional Materials* 30 (11): 1–8. doi:10.1002/adfm.201909603.

García-Soriano, David, Paula Milán-Rois, Nuria Lafuente-Gómez, Cristina Navío, Lucía Gutiérrez, Lorena Cussó, Manuel Desco, Daniel Calle, Álvaro Somoza, and Gorka Salas. 2022. "Iron Oxide-Manganese Oxide Nanoparticles with Tunable Morphology and Switchable MRI Contrast Mode Triggered by Intracellular Conditions." *Journal of Colloid and Interface Science* 613: 447–60. doi:10.1016/j.jcis.2022.01.070.

Gavrila, Ana Mihaela, Anamaria Zaharia, Lisa Paruch, Francois Xavier Perrin, Andrei Sarbu, Andreea Gabriela Olaru, Adam Mariusz Paruch, and Tanta Verona Iordache. 2020. "Molecularly Imprinted Films and Quaternary Ammonium-Functionalized Microparticles Working in Tandem against Pathogenic Bacteria in Wastewaters." *Journal of Hazardous Materials* 399 (February). Elsevier: 123026. doi:10.1016/j.jhazmat.2020.123026.

Geng, Zeyang, Yu Chuan Chien, Matthew J. Lacey, Torbjörn Thiringer, and Daniel Brandell. 2022. "Validity of Solid-State Li+ Diffusion Coefficient Estimation by Electrochemical Approaches for Lithium-Ion Batteries." *Electrochimica Acta* 404. Elsevier Ltd: 139727. doi:10.1016/j.electacta.2021.139727.

Getzlaff, M., M. Leifels, P. Weber, Ü. Kökcam-Demir, and Ch. Janiak. 2019. "Nanoparticles in Toner Material." *SN Applied Sciences* 1 (5). Springer International Publishing. doi:10.1007/s42452-019-0501-9.

Ghaemi, Nasrin, Robert C. T. Slade, and Bahman Amini Horri. 2021. "A Benzoate Coprecipitation Route for Synthesizing Nanocrystalline GDC Powder with Lowered Sintering Temperature." *Ceramics International* 47 (14). Elsevier Ltd: 20009–18. doi:10.1016/j.ceramint.2021.04.011.

Gibson, Rebecca L., Mark J. H. Simmons, E. Hugh Stitt, John West, Sam K. Wilkinson, and Robert W. Gallen. 2021. "Kinetic Modelling of Thermal Processes Using a Modified Sestak-Berggren Equation." *Chemical Engineering Journal* 408 (September 2020). Elsevier B.V.: 127318. doi:10.1016/j.cej.2020.127318.

Gill, Kirandeep K., Rachel Gibson, Kam Ho Chester Yiu, Patrick Hester, and Nuno M. Reis. 2021. "Microcapillary Film Reactor Outperforms Single-Bore Mesocapillary Reactors in Continuous Flow Chemical Reactions." *Chemical Engineering Journal* 408 (September 2020). Elsevier B.V.: 127860. doi:10.1016/j.cej.2020.127860.

Go, Dohyun, Jaehyeong Lee, Jeong Woo Shin, Sungje Lee, Wangu Kang, Jeong Hwan Han, and Jihwan An. 2021. "Phase-Gradient Atomic Layer Deposition of TiO_2 Thin Films by Plasma-Induced Local Crystallization." *Ceramics International* 47 (20). Elsevier Ltd: 28770–77. doi:10.1016/j.ceramint.2021.07.037.

Goyat, M. S., Amrita Hooda, Tejendra K. Gupta, Kaushal Kumar, Sudipta Halder, P. K. Ghosh, and Brijnandan S. Dehiya. 2021. "Role of Non-Functionalized Oxide Nanoparticles on Mechanical Properties and Toughening Mechanisms of Epoxy Nanocomposites." *Ceramics International* 47 (16). Elsevier Ltd: 22316–44. doi:10.1016/j.ceramint.2021.05.083.

Guo, Siyao, Hailong Guan, Ying Li, Yunfeng Bao, Dongyi Lei, Tiejun Zhao, Baomin Zhong, and Zhihong Li. 2021. "Dual-Loss $Ti_3C_2T_x$ MXene/$Ni_{0.6}Zn_{0.4}Fe_2O_4$ Heterogeneous Nanocomposites for Highly Efficient Electromagnetic Wave Absorption." *Journal of Alloys and Compounds* 887. Elsevier: 161298. doi:10.1016/j.jallcom.2021.161298.

Gupta, Shivam, Sanjeev Kumar Sharma, Debabrata Pradhan, and Nyan Hwa Tai. 2019. "Ultra-Light 3D Reduced Graphene Oxide Aerogels Decorated with Cobalt Ferrite and Zinc Oxide Perform Excellent Electromagnetic Interference Shielding Effectiveness." *Composites Part A: Applied Science and Manufacturing* 123 (January). Elsevier: 232–41. doi:10.1016/j.compositesa.2019.05.025.

Gusnawan, Pri Januar, Shangwen Zha, Lusi Zou, Guoyin Zhang, and Jianjia Yu. 2018. "Soybean and Moringa Based Green Biosolvents for Low-Concentration CO2 Capture via a Hollow Fiber Membrane Contactor." *Chemical Engineering Journal* 335 (November 2017). Elsevier: 631–37. doi:10.1016/j.cej.2017.11.023.

Guzel, Merve, Su Ryong Ha, Hyosung Choi, and Metin Ak. 2022. "Rational Design of an 'All-in-One' Monomer to Obtain Black-to-Highly Transmissive Electrochromic Polymer." *Electrochimica Acta* 404. Elsevier Ltd: 139761. doi:10.1016/j.electacta.2021.139761.

Han, Xiaoyu, Youpeng Qu, Jing Wu, Da Li, Nanqi Ren, and Yujie Feng. 2020. "Nitric Oxide Reduction by Microbial Fuel Cell with Carbon Based Gas Diffusion Cathode for Power Generation and Gas Purification." *Journal of Hazardous Materials* 399 (February). doi:10.1016/j.jhazmat.2020.122878.

Han, Xun, Weiming Du, Mengxiao Chen, Xiandi Wang, Xiaojia Zhang, Xiaoyi Li, Jing Li, Zhengchun Peng, Caofeng Pan, and Zhong Lin Wang. 2017. "Visualization Recording and Storage of Pressure Distribution Through a Smart Matrix Based on the Piezotronic Effect." *Advanced Materials* 29 (26): 1–6. doi:10.1002/adma.201701253.

Han, Yaojie, Yufeng Han, Xiaopan Zhang, Lin Li, Chengwu Zhang, Jinhua Liu, Gang Lu, Hai Dong Yu, and Wei Huang. 2020. "Fish Gelatin Based Triboelectric Nanogenerator for Harvesting Biomechanical Energy and Self-Powered Sensing of Human Physiological Signals." *ACS Applied Materials and Interfaces* 12 (14): 16442–50. doi:10.1021/acsami.0c01061.

Han, Yi, Yun Shan Jiang, Yang Xia, Liang Deng, Lan Fang Que, Fu Da Yu, and Zhen Bo Wang. 2022. "Suppressed Phase Separation in Spinel $LiNi_{0.5}Mn_{1.5}O_4$ Cathode via Interstitial Sites Modulation." *Nano Energy* 91 (92): 1–9. doi:10.1016/j.nanoen.2021.106636.

Hassan, Ali, Muhammad Adnan Aslam, Muhammad Bilal, Muhammad Salman Khan, Sajid ur Rehman, Kun Ma, Junfeng Wang, and Zhigao Sheng. 2021. "Modulating Dielectric Loss of MoS2@Ti_3C_2Tx Nanoarchitectures for Electromagnetic Wave Absorption with Radar Cross Section Reduction Performance Verified through Simulations." *Ceramics International* 47 (14). Elsevier Ltd: 20706–16. doi:10.1016/j.ceramint.2021.04.014.

He, Shuangjiang, Yingqing Zhan, Jiaxin Hu, Guiyuan Zhang, Shumei Zhao, Qingying Feng, and Wei Yang. 2020. "Chemically Stable Two-Dimensional MXene@UIO-66-(COOH)2 Composite Lamellar Membrane for Multi-Component Pollutant-Oil-Water Emulsion Separation." *Composites Part B: Engineering* 197 (February). Elsevier Ltd: 108188. doi:10.1016/j.compositesb.2020.108188.

Heck, Julian, Josef Goding, Roberto Portillo Lara, and Rylie Green. 2022. "The Influence of Physicochemical Properties on the Processibility of Conducting Polymers: A Bioelectronics Perspective." *Acta Biomaterialia* 139: 259–79. doi:10.1016/j.actbio.2021.05.052.

Hirai, Takayuki, Kenichi Yagi, Kyoko Nakai, Kazuo Okamoto, Takuro Matsunaga, and Hirotaka Okamoto. 2021. "Transparent Thermoplastic Composite from a Refractive Index-Adjustable Polymer Blend." *Composites Part B: Engineering* 225 (August). Elsevier Ltd: 109258. doi:10.1016/j.compositesb.2021.109258.

Hou, Tianqi, Zirui Jia, Bingbing Wang, Hanbin Li, Xuehua Liu, Lei Bi, and Guanglei Wu. 2021. "MXene-Based Accordion 2D Hybrid Structure with Co_9S_8/C/Ti_3C_2Tx as Efficient Electromagnetic Wave Absorber." *Chemical Engineering Journal* 414 (December 2020). Elsevier B.V.: 128875. doi:10.1016/j.cej.2021.128875.

Hou, Yuliang, Ying Tie, Cheng Li, Thaneshan Sapanathan, and Mohamed Rachik. 2020. "Corrigendum to 'Low-Velocity Impact Behaviors of Repaired CFRP Laminates: Effect of Impact Location and External Patch Configurations' [Composites Part B: Engineering 163(2019):669–680] (Composites Part B (2019) 163 (669–680), (S1359836818323114), (10.1016)." *Composites Part B: Engineering* 197 (June). Elsevier Ltd: 108202. doi:10.1016/j.compositesb.2020.108202.

Hsia, Hao Hsuan, Irish Valerie B. Maggay, Rasu Muruganantham, and Wei Ren Liu. 2021. "$ZnIn_2S_4$: A Promising Anode Material with High Electrochemical Performance for Sodium-Ion Batteries." *Ceramics International* 47 (20). Elsevier Ltd: 28634–41. doi:10.1016/j.ceramint.2021.07.022.

Hu, Gongwei, Yan Zhang, Lijie Li, and Zhong Lin Wang. 2018. "Piezotronic Transistor Based on Topological." doi:10.1021/acsnano.7b07996.

Hu, Jiaoli, Fei Yan, Hao Liu, Lianshuang Zhao, Zheng Li, Yu Liu, Yuhui Ao, and Liu Liu. 2021. "Water-Based PEKC-COOH Sizing Agent for Enhancing the Interfacial Adhesion of Carbon Fiber/Polyether-Ether-Ketone Composites." *Composites Part B: Engineering* 225 (September). Elsevier Ltd: 109279. doi:10.1016/j.compositesb.2021.109279.

Hu, Meihua, Yingyu Lin, Xin Li, Wenguang Zhang, Zhuoyue Chen, Yuxian Yang, Guanjie Li, Yikeng Lu, and Weishan Li. 2022. "Nano-Fe_3C@2D-NC@CC as Anode for Improving Extracellular Electron Transfer and Electricity Generation of Microbial Fuel Cells." *Electrochimica Acta* 404. Elsevier Ltd: 139618. doi:10.1016/j.electacta.2021.139618.

Huang, Bingxue, Zehang Zhou, Lingfei Wei, Quancheng Song, Weiye Yu, Yi Zhou, Rui Hu, Wei Zhang, and Canhui Lu. 2021. "Ti_3C_2Tx MXene as a Novel Functional Photo Blocker for Stereolithographic 3D Printing of Multifunctional Gels via Continuous Liquid Interface Production." *Composites Part B: Engineering* 225 (July). Elsevier Ltd: 109261. doi:10.1016/j.compositesb.2021.109261.

Huang, Lingling, Jianmiao Yang, Tiantian Wang, Jianqing Gao, and Donghang Xu. 2022. "Engineering of Small-Molecule Lipidic Prodrugs as Novel Nanomedicines for Enhanced Drug Delivery." *Journal of Nanobiotechnology* 20 (1). BioMed Central: 1–15. doi:10.1186/s12951-022-01257-4.

Huang, Po Cheng, Shao Ku Huang, Ting Chun Lai, Min Chuan Shih, Hung Chang Hsu, Chun Hsiang Chen, Cheng Chieh Lin, Chun-Hao Chiang, Chi-Ying Lin, Kazuhito Tsukagoshi, Chun-Wei Chen, Ya-Ping Chiu, Shiow-Fon Tsay, and Ying-Chiao Wang. 2021. "Visualizing Band Alignment Across 2D/3D Perovskite Heterointerfaces of Solar Cells with Light-Modulated Scanning Tunneling Microscopy." *Nano Energy* 89 (PA). Elsevier Ltd: 106362. doi:10.1016/j.nanoen.2021.106362.

Huang, Zhixiong, Zongyi Deng, Chuang Dong, Jiamin Fan, and Yilin Ren. 2022. "A Closed-Loop Recycling Process for Carbon Fiber Reinforced Vinyl Ester Resin Composite." *Chemical Engineering Journal* 446 (P3). Elsevier B.V.: 137254. doi:10.1016/j.cej.2022.137254.

Hung, H. T., Z. D. Ma, P. S. Shih, J. H. Huang, L. Y. Kao, C. Y. Yang, Vengudusamy Renganathan, C. L. Kao, Y. C. Hung, and C. R. Kao. 2021. "Highly Uniform Microfluidic Electroless Interconnections for Chip Stacking Applications." *Electrochimica Acta* 376. doi:10.1016/j.electacta.2021.138032.

Ishak, Mohd Adil Iman, Khairulazhar Jumbri, Shaari Daud, Mohd Basyaruddin Abdul Rahman, Roswanira Abdul Wahab, Hiroshi Yamagishi, and Yohei Yamamoto. 2020.

"Molecular Simulation on the Stability and Adsorption Properties of Choline-Based Ionic Liquids/IRMOF-1 Hybrid Composite for Selective H_2S/CO_2 Capture." *Journal of Hazardous Materials* 399 (March). Elsevier: 123008. doi:10.1016/j.jhazmat.2020.123008.

Jaya Prakash, Niranjana, and Balasubramanian Kandasubramanian. 2021. "Nanocomposites of MXene for Industrial Applications." *Journal of Alloys and Compounds* 862. Elsevier: 158547. doi:10.1016/j.jallcom.2020.158547.

Jensen, S. M., B. L.V. Bak, J. J. Bender, L. Carreras, and E. Lindgaard. 2021. "Transient Delamination Growth in GFRP Laminates with Fibre Bridging under Variable Amplitude Loading in G-Control." *Composites Part B: Engineering* 225 (September). Elsevier Ltd: 109296. doi:10.1016/j.compositesb.2021.109296.

Ji, He, Rui Zhao, Nan Zhang, Changxian Jin, Xiaofeng Lu, and Ce Wang. 2018. "Lightweight and Flexible Electrospun Polymer Nanofiber/Metal Nanoparticle Hybrid Membrane for High-Performance Electromagnetic Interference Shielding." *NPG Asia Materials* 10 (8): 749–60. doi:10.1038/s41427-018-0070-1.

Ji, Qiuyi, Xinying Cheng, Dunyu Sun, Yijie Wu, Xiangcheng Kong, Huan He, Zhe Xu, Chenmin Xu, Chengdu Qi, Yazi Liu, Shiyin Li, Limin Zhang, Shaogui Yang, and Cheng Sun. 2021. "Persulfate Enhanced Visible Light Photocatalytic Degradation of Iohexol by Surface-Loaded Perylene Diimide/Acidified Biochar." *Chemical Engineering Journal* 414 (February). Elsevier B.V.: 128793. doi:10.1016/j.cej.2021.128793.

Jiang, Hanmei, Zegao Wang, Qian Yang, Muhammad Hanif, Zhiming Wang, Lichun Dong, and Mingdong Dong. 2018. "A Novel MnO2/Ti_3C_2Tx MXene Nanocomposite as High Performance Electrode Materials for Flexible Supercapacitors." *Electrochimica Acta* 290. Elsevier Ltd: 695–703. doi:10.1016/j.electacta.2018.08.096.

Jiao, X. Y., C. F. Liu, Z. P. Guo, Harood Nishat, G. D. Tong, S. L. Ma, Y. Bi, Y. F. Zhang, Stuart Wiesner, and S. M. Xiong. 2021. "On the Characterization of Primary Iron-Rich Phase in a High-Pressure Die-Cast Hypoeutectic Al-Si Alloy." *Journal of Alloys and Compounds* 862. doi:10.1016/j.jallcom.2020.158580.

Jin, Le, Jiqi Wang, Fei Wu, Yanan Yin, and Baoliang Zhang. 2021. "MXene@Fe_3O_4 Microspheres/Fibers Composite Microwave Absorbing Materials: Optimum Composition and Performance Evaluation." *Carbon* 182. Elsevier Ltd: 770–80. doi:10.1016/j.carbon.2021.06.073.

Jin, Shicun, Kuang Li, Xiaowei Zhang, Qiang Gao, Ling Zeng, Sheldon Q. Shi, and Jianzhang Li. 2020. "Phytic Acid-Assisted Fabrication for Soybean Meal/Nanofiber Composite Adhesive via Bioinspired Chelation Reinforcement Strategy." *Journal of Hazardous Materials* 399 (March). Elsevier: 123064. doi:10.1016/j.jhazmat.2020.123064.

Jin, Xiuxiu, Jianfeng Wang, Lunzhi Dai, Xiaoya Liu, Lei Li, Y. Yang, Yanxia Cao, Wanjie Wang, Hong Wu, and Shaoyun Guo. 2020. "Flame-Retardant Poly(Vinyl Alcohol)/MXene Multilayered Films with Outstanding Electromagnetic Interference Shielding and Thermal Conductive Performances." *Chemical Engineering Journal* 380 (April 2019). Elsevier: 122475. doi:10.1016/j.cej.2019.122475.

Jing, Hongyu, Wei Liu, Zhengyan Zhao, Jiangwei Zhang, Chao Zhu, Yantao Shi, Dingsheng Wang, and Yadong Li. 2021. "Electronics and Coordination Engineering of Atomic Cobalt Trapped by Oxygen-Driven Defects for Efficient Cathode in Solar Cells." *Nano Energy* 89 (PA). Elsevier Ltd: 106365. doi:10.1016/j.nanoen.2021.106365.

Jing, Ziwei, Qiuzheng Du, Xiaojian Zhang, and Yu Zhang. 2022. "Nanomedicines and Nanomaterials for Cancer Therapy: Progress, Challenge and Perspectives." *Chemical Engineering Journal* 446 (P3). Elsevier B.V.: 137147. doi:10.1016/j.cej.2022.137147.

Jun, Byung Moon, Min Jang, Chang Min Park, Jonghun Han, and Yeomin Yoon. 2020. "Selective Adsorption of Cs+ by MXene (Ti_3C_2Tx) from Model Low-Level Radioactive

Wastewater." *Nuclear Engineering and Technology* 52 (6). Elsevier Ltd: 1201–7. doi:10.1016/j.net.2019.11.020.

Jung, Jeki, Joo Hee Kang, Su Hyeon Kim, Jiwon Park, Won Kyoung Kim, Yun Soo Lee, Cha Yong Lim, and Yong Ho Park. 2021. "Redox-Reaction Phenomenon in Cenosphere Reinforced Aluminum Alloy Matrix Syntactic Foam." *Journal of Alloys and Compounds* 862. Elsevier: 158686. doi:10.1016/j.jallcom.2021.158686.

Karan, Sumanta Kumar, Sandip Maiti, Sarbaranjan Paria, Anirban Maitra, Suman Kumar Si, Jin Kon Kim, and Bhanu Bhusan Khatua. 2018. "A New Insight towards Eggshell Membrane as High Energy Conversion Efficient Bio-Piezoelectric Energy Harvester." *Materials Today Energy* 9. Elsevier Ltd: 114–25. doi:10.1016/j.mtener.2018.05.006.

Kayahan, Emine, Daniele Urbani, Paolo Dambruoso, Alessandro Massi, Leen Braeken, Tom Van Gerven, and M. Enis Leblebici. 2021. "Overcoming Mass and Photon Transfer Limitations in a Scalable Reactor: Oxidation in an Aerosol Photoreactor." *Chemical Engineering Journal* 408 (September 2020). Elsevier B.V.: 127357. doi:10.1016/j.cej.2020.127357.

Kazmina, O., V. Borovoy, and V. Semenova. 2021. "White Vitreous Enamel for Ferrous Metals with Preliminary Thermal Activation of Frit." *Ceramics International* 47 (20). Elsevier Ltd: 28471–78. doi:10.1016/j.ceramint.2021.07.001.

Khalili, Saba, and Hossein Mahmoudi Chenari. 2020. "Successful Electrospinning Fabrication of ZrO_2 Nanofibers: A Detailed Physical—Chemical Characterization Study." *Journal of Alloys and Compounds* 828. Elsevier B.V: 154414. doi:10.1016/j.jallcom.2020.154414.

Khalilifard, Mahsa, and Soheila Javadian. 2021. "Magnetic Superhydrophobic Polyurethane Sponge Loaded with Fe_3O_4@oleic Acid@graphene Oxide as High Performance Adsorbent Oil from Water." *Chemical Engineering Journal* 408 (September 2020). Elsevier B.V.: 127369. doi:10.1016/j.cej.2020.127369.

Kim, Eunyoung, Hongming Zhang, Jeng Hun Lee, Haomin Chen, Heng Zhang, Muhammad Humza Javed, Xi Shen, and Jang Kyo Kim. 2021. "MXene/Polyurethane Auxetic Composite Foam for Electromagnetic Interference Shielding and Impact Attenuation." *Composites Part A: Applied Science and Manufacturing* 147 (April). Elsevier Ltd: 106430. doi:10.1016/j.compositesa.2021.106430.

Kim, Myounguk, Tae Won Lee, Sun Min Park, and Young Gyu Jeong. 2019. "Structures, Electrical and Mechanical Properties of Epoxy Composites Reinforced with MWCNT-Coated Basalt Fibers." *Composites Part A: Applied Science and Manufacturing* 123 (April). Elsevier: 123–31. doi:10.1016/j.compositesa.2019.05.011.

Kim, Tae Gun, Edmund Samuel, Chan Woo Park, Bhavana Joshi, Min Woo Kim, Mark T. Swihart, and Sam S. Yoon. 2020. "Supersonically Sprayed Zn_2SnO_4/SnO_2/Carbon Nanotube Films for High-Efficiency Water Splitting Photoanodes." *Journal of Alloys and Compounds* 828. doi:10.1016/j.jallcom.2020.154374.

Kim, Young O., Jaehyun Cho, Young Nam Kim, Kun Won Kim, Byoung Wan Lee, Jae Woo Kim, Minkook Kim, and Yong Chae Jung. 2020. "Recyclable, Flame-Retardant and Smoke-Suppressing Tannic Acid-Based Carbon-Fiber-Reinforced Plastic." *Composites Part B: Engineering* 197 (May). Elsevier Ltd: 108173. doi:10.1016/j.compositesb.2020.108173.

Kiyotake, Emi A., Michael D. Martin, and Michael S. Detamore. 2022. "Regenerative Rehabilitation with Conductive Biomaterials for Spinal Cord Injury." *Acta Biomaterialia* 139: 43–64. doi:10.1016/j.actbio.2020.12.021.

Knozowski, Dominik, Magdalena Graczyk-Zając, Dragoljub Vrankovic, Grzegorz Trykowski, Mirosław Sawczak, Dario M. De Carolis, and Monika Wilamowska-Zawłocka. 2021. "New Insights on Lithium Storage in Silicon Oxycarbide/Carbon Composites: Impact of Microstructure on Electrochemical Properties." *Composites Part B: Engineering* 225 (September): 1–11. doi:10.1016/j.compositesb.2021.109302.

Kong, Fanyu, Xiaodong He, Qianqian Liu, Xinxin Qi, Yongting Zheng, Rongguo Wang, and Yuelei Bai. 2018. "Effect of Ti_3AlC_2 Precursor on the Electrochemical Properties of the Resulting MXene Ti_3C_2 for Li-Ion Batteries." *Ceramics International* 44 (10). Elsevier Ltd and Techna Group S.r.l.: 11591–96. doi:10.1016/j.ceramint.2018.03.223.

Kong, Weiqian, Xiaofan Zhang, Yanzhen Guo, Guangli He, Huili Liu, Shouren Zhang, and Baocheng Yang. 2021. "TiO_2@PCN Core-Shell Nanoarrays Decorated with Au Nanoparticles for Enhanced Photoelectrochemical Performance." *Electrochimica Acta* 376. Elsevier Ltd: 138036. doi:10.1016/j.electacta.2021.138036.

Kong, Wei Wei, Chang Ge Zhou, Kun Dai, Li Chuan Jia, Ding Xiang Yan, and Zhong Ming Li. 2021. "Highly Stretchable and Durable Fibrous Strain Sensor with Growth Ring-like Spiral Structure for Wearable Electronics." *Composites Part B: Engineering* 225 (September). Elsevier Ltd: 109275. doi:10.1016/j.compositesb.2021.109275.

Lai, Fang I., Jui Fu Yang, Yu Chao Hsu, and Shou Yi Kuo. 2020. "Omnidirectional Light Harvesting Enhancement of Dye-Sensitized Solar Cells Decorated with Two-Dimensional ZnO Nanoflowers." *Journal of Alloys and Compounds* 815. Elsevier B.V: 152287. doi:10.1016/j.jallcom.2019.152287.

Lal, Ganesh, Khushboo Punia, Satya Narain Dolia, Parvez A. Alvi, Banwari L. Choudhary, and Sudhish Kumar. 2020. "Structural, Cation Distribution, Optical and Magnetic Properties of Quaternary $Co_{0.4}$+$xZn_{0.6}$-XFe_2O_4 (x = 0.0, 0.1 and 0.2) and Li Doped Quinary $Co_{0.4}$+$xZn_{0.5}$–$XLi_{0.1}Fe_2O_4$ (x = 0.0, 0.05 and 0.1) Nanoferrites." *Journal of Alloys and Compounds* 828. doi:10.1016/j.jallcom.2020.154388.

Lee, Gye Won, Tae Woo Kim, Willem G. Sloof, and Kee Sung Lee. 2021. "Self-Healing Capacity of Mullite-Yb_2SiO_5 Environmental Barrier Coating Material with Embedded Ti_2AlC MAX Phase Particles." *Ceramics International* 47 (16). Elsevier Ltd: 22478–86. doi:10.1016/j.ceramint.2021.04.257.

Lee, Hyeonho, Sung Ju Im, Chang min Kim, and Am Jang. 2022. "Relative Selectivity of Short- and Medium-Chain Fatty Acids in Hydrophobic Membrane for Resource Recovery in Wastewater: Effects of Supported Liquid, Feed PH, and Membrane Pores." *Chemical Engineering Journal* 446 (P3). Elsevier B.V.: 137258. doi:10.1016/j.cej.2022.137258.

Lee, Hyeseong, Doojin Lee, Jaehyun Cho, Young O. Kim, Soonho Lim, Sang Jun Youn, Yong Chae Jung, Seong Yun Kim, and Dong Gi Seong. 2019. "Super-Insulating, Flame-Retardant, and Flexible Poly(Dimethylsiloxane) Composites Based on Silica Aerogel." *Composites Part A: Applied Science and Manufacturing* 123 (February). Elsevier: 108–13. doi:10.1016/j.compositesa.2019.05.004.

Lee, Jaehyeon, Kiwoong Kim, Sung Ho Park, Gun Young Yoon, Jeongju Kim, and Sang Joon Lee. 2020. "Macroporous Photothermal Bilayer Evaporator for Highly Efficient and Self-Cleaning Solar Desalination." *Nano Energy* 77. Elsevier Ltd: 105130. doi:10.1016/j.nanoen.2020.105130.

Lei, Guangyu, Jingwen Ma, Mengyao Zhao, Shun Wu, Hongwei He, Hui Qi, Wenchao Peng, Xiaobin Fan, Guoliang Zhang, Fengbao Zhang, and Yang Li. 2021. "Nitrogen–Carbon Materials Base on Pyrolytic Graphene Hydrogel for Oxygen Reduction." *Journal of Colloid and Interface Science* 602. Elsevier Inc.: 274–81. doi:10.1016/j.jcis.2021.06.036.

Lei, Yiming, Zhengjun Yao, Shuzhou Li, Jintang Zhou, Azhar Ali Haidry, and Peijiang Liu. 2020. "Broadband High-Performance Electromagnetic Wave Absorption of Co-Doped NiZn Ferrite/Polyaniline on MXenes." *Ceramics International* 46 (8). Elsevier Ltd and Techna Group S.r.l.: 10006–15. doi:10.1016/j.ceramint.2019.12.189.

Lei, Zhiwei, Xuantong Sun, Shifeng Zhu, Kai Dong, Xuqing Liu, Lili Wang, Xiansheng Zhang, Lijun Qu, and Xueji Zhang. 2022. "Nature Inspired MXene-Decorated 3D Honeycomb-Fabric Architectures Toward Efficient Water Desalination and Salt

Harvesting." *Nano-Micro Letters* 14 (1). Springer Singapore: 1–16. doi:10.1007/s40820-021-00748-7.

Li, Chun Xiao, Shuang Yang Kuang, Yang Hui Chen, Zhong Lin Wang, Congju Li, and Guang Zhu. 2018. "In Situ Active Poling of Nanofiber Networks for Gigantically Enhanced Particulate Filtration." *ACS Applied Materials and Interfaces* 10 (29): 24332–38. doi:10.1021/acsami.8b07203.

Li, Dong, Shanshan Chen, Ruirui Huang, Chaorui Xue, Pengfei Li, Yangsen Li, Qing Chang, Huiqi Wang, Ning Li, Suping Jia, Shengliang Hu, and Jinlong Yang. 2021. "Photothermal, Photocatalytic, and Anti-Bacterial Ti-Ag-O Nanoporous Powders for Interfacial Solar Driven Water Evaporation." *Ceramics International* 47 (14). Elsevier Ltd: 19800–808. doi:10.1016/j.ceramint.2021.03.319.

Li, Fei, Yi lin Liu, Gui Gen Wang, Si Ying Zhang, Da Qiang Zhao, Kan Fang, Hua Yu Zhang, and Hui Ying Yang. 2022. "3D Porous H-Ti_3C_2Tx Films as Free-Standing Electrodes for Zinc Ion Hybrid Capacitors." *Chemical Engineering Journal* 435 (P3). Elsevier B.V.: 135052. doi:10.1016/j.cej.2022.135052.

Li, Gang, Kang Wang, Dan Wang, and Ling Lin. 2022. "Noninvasive Blood Glucose Detection System Based on Dynamic Spectrum and "M+N″ Theory." *Analytica Chimica Acta* 1201. Elsevier B.V.: 339635. doi:10.1016/j.aca.2022.339635.

Li, Guohao, Brian C. Wyatt, Fei Song, Changqiang Yu, Zhenjun Wu, Xiuqiang Xie, Babak Anasori, and Nan Zhang. 2021. "2D Titanium Carbide (MXene) Based Films: Expanding the Frontier of Functional Film Materials." *Advanced Functional Materials* 31 (46): 1–30. doi:10.1002/adfm.202105043.

Li, Haoyi, and Weimin Yang. 2016. "Electrospinning Technology in Non-Woven Fabric Manufacturing." *Non-Woven Fabrics*. doi:10.5772/62200.

Li, Hongyan, Qi Li, Hongli Liu, Xurui Wang, Kai Cao, Jing Li, Bo Zhang, and Wenjin Yuan. 2021. "Ice-Resistant Surface with Three Dimensional Spherical Halloysite Aerogel: Construction and Anti-Icing Mechanism." *Ceramics International* 47 (16): 22976–84. doi:10.1016/j.ceramint.2021.05.011.

Li, Kangle, Jiapeng Li, Qizhen Zhu, and Bin Xu. 2022. "Three-Dimensional MXenes for Supercapacitors: A Review." *Small Methods* 6 (4): 1–19. doi:10.1002/smtd.202101537.

Li, Li, Hanlu Wang, Zhengjun Xie, Cuihua An, Gaoxue Jiang, and Yijing Wang. 2020. "3D Graphene-Encapsulated Nearly Monodisperse Fe_3O_4 Nanoparticles as High-Performance Lithium-Ion Battery Anodes." *Journal of Alloys and Compounds* 815. doi:10.1016/j.jallcom.2019.152337.

Li, Lulu, Sai Zhao, Xin Jie Luo, Hao Bin Zhang, and Zhong Zhen Yu. 2021. "Smart MXene-Based Janus Films with Multi-Responsive Actuation Capability and High Electromagnetic Interference Shielding Performances." *Carbon* 175. Elsevier Ltd: 594–602. doi:10.1016/j.carbon.2020.10.090.

Li, Ting Ting, Wenna Dai, Shih Yu Huang, Hongyang Wang, Qi Lin, Jia Horng Lin, and Ching Wen Lou. 2020. "Spring-like Sandwich Foam Composites Reinforced by 3D Concave–Convex Structured Fabric: Manufacturing and Low-Velocity Cushion Response." *Composites Part B: Engineering* 197 (January). Elsevier Ltd: 108171. doi:10.1016/j.compositesb.2020.108171.

Li, Wen, Wen Yu, Qian Xu, Jianxin Zhou, Hai Nan, Yajun Yin, and Xu Shen. 2020. "Understanding the Atomistic Deformation Mechanisms of Polycrystalline γ-TiAl under Nanoindentation: Effect of Lamellar Structure." *Journal of Alloys and Compounds* 828. Elsevier B.V: 154443. doi:10.1016/j.jallcom.2020.154443.

Li, Wulong, Kexin Liu, Yaoxin Zhang, Shuai Guo, Zhanxiong Li, and Swee Ching Tan. 2022. "A Facile Strategy to Prepare Robust Self-Healable Superhydrophobic Fabrics with

Self-Cleaning, Anti-Icing, UV Resistance, and Antibacterial Properties." *Chemical Engineering Journal* 446 (P3). Elsevier B.V.: 137195. doi:10.1016/j.cej.2022.137195.

Li, Xiaomeng, and Shuxue Zhou. 2022. "Epoxy-Functionalized Ti_3C_2 Nanosheet for Epoxy Coatings with Prominent Anticorrosion Performance." *Progress in Organic Coatings* 162 (October 2021). Elsevier B.V.: 106559. doi:10.1016/j.porgcoat.2021.106559.

Li, Xingxing, Yanan Ma, Yang Yue, Guosheng Li, Chuankun Zhang, Minglei Cao, Yongchen Xiong, Jintao Zou, Yongheng Zhou, and Yihua Gao. 2022. "A Flexible Zn-Ion Hybrid Micro-Supercapacitor Based on MXene Anode and V_2O_5 Cathode with High Capacitance." *Chemical Engineering Journal* 428 (April 2021). Elsevier B.V.: 130965. doi:10.1016/j.cej.2021.130965.

Li, Xiran, Hongpeng Li, Xiangqian Fan, Xinlei Shi, and Jiajie Liang. 2020. "3D-Printed Stretchable Micro-Supercapacitor with Remarkable Areal Performance." *Advanced Energy Materials* 10 (14): 1–12. doi:10.1002/aenm.201903794.

Li, Yahui, Qi Zhang, Yang Liu, Penglei Zhang, Chao Ren, Haodong Zhang, Han Cai, Guifu Ding, Zhuoqing Yang, and Chi Zhang. 2021. "Regulation of Nanocrystals Structure for High-Performance Magnetic Triboelectric Nanogenerator." *Nano Energy* 89 (PA). Elsevier Ltd: 106390. doi:10.1016/j.nanoen.2021.106390.

Li, Zijiong, Dongfang Guo, Dianzhang Wang, Min Sun, and Haibin Sun. 2021. "Exploration of Metal/Ti_3C_2 MXene-Derived Composites as Anode for High-Performance Zinc-Ion Supercapacitor." *Journal of Power Sources* 506 (May). Elsevier B.V.: 230197. doi:10.1016/j.jpowsour.2021.230197.

Liang, Jiajie, Kwing Tong, and Qibing Pei. 2016. "A Water-Based Silver-Nanowire Screen-Print Ink for the Fabrication of Stretchable Conductors and Wearable Thin-Film Transistors." *Advanced Materials*: 5986–96. doi:10.1002/adma.201600772.

Lim, Gim Pao, Chin Fhong Soon, Agnieszka Maria Jastrzębska, Nyuk Ling Ma, Anita Rozmysłowska Wojciechowska, Aleksandra Szuplewska, Wan Ibtisam Wan Omar, Marlia Morsin, Nafarizal Nayan, and Kian Sek Tee. 2021. "Synthesis, Characterization and Biophysical Evaluation of the 2D Ti_2CT_x MXene Using 3D Spheroid-Type Cultures." *Ceramics International* 47 (16): 22567–77. doi:10.1016/j.ceramint.2021.04.268.

Lin, P., C. Pan, and Z. L. Wang. 2018. "Two-Dimensional Nanomaterials for Novel Piezotronics and Piezophototronics." *Materials Today Nano* 4. Elsevier Ltd: 17–31. doi:10.1016/j.mtnano.2018.11.006.

Ling, Yeqing, Xiandong Hao, Yuxi Gui, Hongju Qiu, Qiannan Li, Hewen Zheng, Mamdouh Omran, Lei Gao, Jin Chen, and Guo Chen. 2021. "Stability Properties and Microstructure Properties of Microwave-Sintered CeO_2 Doped Zirconia Ceramics." *Ceramics International* 47 (20). Elsevier Ltd: 28210–17. doi:10.1016/j.ceramint.2021.06.234.

Liu, Chaofeng, Zachary G. Neale, and Guozhong Cao. 2016. "Understanding Electrochemical Potentials of Cathode Materials in Rechargeable Batteries." *Biochemical Pharmacology* 19 (2). Elsevier Ltd.: 109–23. doi:10.1016/j.mattod.2015.10.009.

Liu, Hongfei, Yongjie Wang, and Zhiping Zhang. 2022. "Enhanced Negative Thermal Expansion of (KMg)3+-Substituted $Fe_2Mo_3O_{12}$ Ceramics." *Ceramics International* 48 (12). Elsevier Ltd: 16554–61. doi:10.1016/j.ceramint.2022.02.199.

Liu, Jun Jie, Wen Jie Yang, Yu Xu, Anthony Chun Yin Yuen, Timothy Bo Yuan Chen, Chun Xiang Wei, San E. Zhu, Guan Heng Yeoh, Wei Yang, and Hong Dian Lu. 2022. "MXene-Based Films via Scalable Fabrication with Improved Mechanical and Antioxidant Properties for Electromagnetic Interference Shielding." *Composites Communications* 31 (January): 3–7. doi:10.1016/j.coco.2022.101112.

Liu, Ke, Jie Hu, Zhengqing Kong, Jiawei Hu, Zhishuai Tian, Jiarui Hou, Jinli Qin, Chengshun Liu, Shuai Liang, Haipeng Wu, Jianming Zhang, Lu Zong, and Yongxin Duan. 2021. "High-Yield, High-Conductive Graphene/Nanocellulose Hybrids Prepared by

Co-Exfoliation of Low-Oxidized Expanded Graphite and Microfibrillated Cellulose." *Composites Part B: Engineering* 225 (May). Elsevier Ltd: 109250. doi:10.1016/j.compositesb.2021.109250.

Liu, Peijiang, Zhengjun Yao, Vincent Ming Hong Ng, Jintang Zhou, Ling Bing Kong, and Kan Yue. 2018. "Facile Synthesis of Ultrasmall Fe_3O_4 Nanoparticles on MXenes for High Microwave Absorption Performance." *Composites Part A: Applied Science and Manufacturing* 115 (October). Elsevier: 371–82. doi:10.1016/j.compositesa.2018.10.014.

Liu, Peng, Chao Liu, Ting Hu, Junjie Shi, Libo Zhang, Bingguo Liu, and Jinhui Peng. 2021. "Kinetic Study of Microwave Enhanced Mercury Desorption for the Regeneration of Spent Activated Carbon Supported Mercuric Chloride Catalysts." *Chemical Engineering Journal* 408 (September 2020). Elsevier B.V.: 127355. doi:10.1016/j.cej.2020.127355.

Liu, Peng, Huiling Liu, Shun Zhang, Jie Wang, and Cheng Wang. 2021. "A General Strategy for Obtaining BiOX Nanoplates Derived Bi Nanosheets as Efficient CO_2 Reduction Catalysts by Enhancing CO_2—Adsorption and Electron Transfer." *Journal of Colloid and Interface Science* 602. Elsevier Inc.: 740–47. doi:10.1016/j.jcis.2021.06.010.

Liu, Ruiting, Miao Miao, Yahui Li, Jianfeng Zhang, Shaomei Cao, and Xin Feng. 2018. "Ultrathin Biomimetic Polymeric Ti_3C_2Tx MXene Composite Films for Electromagnetic Interference Shielding." *ACS Applied Materials and Interfaces* 10 (51). Elsevier Ltd: 44787–95. doi:10.1021/acsami.8b18347.

Liu, Shiyuan, Jieyuan Liu, Xiaofang Liu, Jiaxiang Shang, Li Xu, Ronghai Yu, and Jianglan Shui. 2021. "Hydrogen Storage in Incompletely Etched Multilayer Ti_2CTx at Room Temperature." *Nature Nanotechnology* 16 (3): 331–36. doi:10.1038/s41565-020-00818-8.

Liu, Wei, Liang Min Ning, Shao Qing Li, Wei Xing Liu, Qiang Zhang, Jia Shao, and Jin Lei Tian. 2021. "N-Rich MOFs Derived N-Doped Carbon Nanotubes Encapsulating Cobalt Nanoparticles as Efficient and Magnetic Recoverable Catalysts for Nitro Aromatics Reduction." *Journal of Alloys and Compounds* 862. Elsevier: 158333. doi:10.1016/j.jallcom.2020.158333.

Liu, Xiaoting, Xiaofang Shen, Baisheng Sa, Yinggan Zhang, Xin Li, and Hao Xue. 2021. "Piezotronic-Enhanced Photocatalytic Performance of Heterostructured $BaTiO_3/SrTiO_3$ Nanofibers." *Nano Energy* 89 (May): 1–9. doi:10.1016/j.nanoen.2021.106391.

Liu, Xiaoyun, Fang Yuan, Chunrong Zhu, Jingrui Li, Xingqiang Lv, Guichuan Xing, Qi Wei, Gang Wang, Jinfei Dai, Hua Dong, Jie Xu, Bo Jiao, and Zhaoxin Wu. 2022. "Near-Unity Blue Luminance from Lead-Free Copper Halides for Light-Emitting Diodes." *Nano Energy* 91 (October 2021). Elsevier Ltd: 106664. doi:10.1016/j.nanoen.2021.106664.

Liu, Zhuo, Ye Ma, Han Ouyang, Bojing Shi, Ning Li, Dongjie Jiang, Feng Xie, Dan Qu, Yang Zou, Yue Huang, Hu Li, Chaochao Zhao, Puchuan Tan, Min Yu, Yubo Fan, Hao Zhang, Zhong Lin Wang, and Zhou Li. 2019. "Transcatheter Self-Powered Ultrasensitive Endocardial Pressure Sensor." *Advanced Functional Materials* 1807560. doi:10.1002/adfm.201807560.

Long, Yun, Yuzhang Wang, Xiaohu Chen, and Runxuan Sang. 2021. "3D Numerical Reconstruction and Microstructure Regulation of Thermal Barrier Coatings for High Temperature Components of Gas Turbine." *Ceramics International* 47 (14). Elsevier Ltd: 20126–40. doi:10.1016/j.ceramint.2021.04.019.

Lu, Jing, Yong Wang, Jianfeng Huang, Jie Fei, Liyun Cao, and Cuiyan Li. 2017. "In Situ Synthesis of Mesoporous C-Doped TiO_2 Single Crystal with Oxygen Vacancy and Its Enhanced Sunlight Photocatalytic Properties." *Dyes and Pigments* 144. Elsevier Ltd: 203–11. doi:10.1016/j.dyepig.2017.05.033.

Lu, Xiaoyan, Hui Li, and Biao Wang. 2011. "Theoretical Analysis of Electric, Magnetic and Magnetoelectric Properties of Nano-Structured Multiferroic Composites."

Journal of the Mechanics and Physics of Solids 59 (10). Elsevier: 1966–77. doi:10.1016/j.jmps.2011.07.007.

Luo, Shuang, Xingyan Liu, Xiangjun Wei, Min Fu, Peng Lu, Xuan Li, Yiming Jia, Qiao Ren, and Youzhou He. 2020. "Noble-Metal-Free Cobaloxime Coupled with Metal-Organic Frameworks NH2-MIL-125: A Novel Bifunctional Photocatalyst for Photocatalytic NO Removal and H_2 Evolution Under Visible Light Irradiation." *Journal of Hazardous Materials* 399 (February). Elsevier: 122824. doi:10.1016/j.jhazmat.2020.122824.

Luo, Yonghao, Weiye Zhang, Chenhong Ning, Xiaowu Zhu, Yueliang Wang, Bingwen Lu, Yanhui Li, and Xinyue Zhang. 2022. "Alpha-to-Gamma Phase Transformation Self-Reinforcing of Alumina Ceramic Slit Array in Nd:YAG Laser Cutting Based on Vapor-to-Melt Ratio Controlling." *Ceramics International* 48 (12). Elsevier Ltd: 16977–89. doi:10.1016/j.ceramint.2022.02.252.

Lv, Zepeng, Wansen Ma, Meng Wang, Jie Dang, Kailiang Jian, Dong Liu, and Dejun Huang. 2021. "Co-Constructing Interfaces of Multiheterostructure on MXene (Ti_3C_2Tx)-Modified 3D Self-Supporting Electrode for Ultraefficient Electrocatalytic HER in Alkaline Media." *Advanced Functional Materials* 31 (29): 1–8. doi:10.1002/adfm.202102576.

Ma, Ge, Xiaoyong Zhao, Mengzhe Guo, Yaqin Liu, Keqiang Shi, Cheng Guo, and Yuanjiang Pan. 2022. "6-Glycosylaminoquinoline-Assisted LDI MS for Detection and Imaging of Small Molecules with Enhanced Detection Selectivity and Sensitivity." *Analytica Chimica Acta* 1201. doi:10.1016/j.aca.2022.339620.

Ma, Hongqiang, Shaochen Zhang, and Jingjing Feng. 2022. "Early Hydration Properties and Microstructure Evolutions of MgO-Activated Slag Materials at Different Curing Temperatures." *Ceramics International* 48 (12). Elsevier Ltd: 17104–15. doi:10.1016/j.ceramint.2022.02.266.

Makurvet, Favour Danladi. 2021. "Biologics vs. Small Molecules: Drug Costs and Patient Access." *Medicine in Drug Discovery* 9: 100075. doi:10.1016/j.medidd.2020.100075.

Maletsky, A. V., D. R. Belichko, T. E. Konstantinova, G. K. Volkova, A. S. Doroshkevich, A. I. Lyubchyk, V. V. Burkhovetskiy, V. A. Aleksandrov, D. Mardare, C. Mita, D. Chicea, and L. H. Khiem. 2021. "Structure Formation and Properties of Corundum Ceramics Based on Metastable Aluminium Oxide Doped with Stabilized Zirconium Dioxide." *Ceramics International* 47 (14). Elsevier Ltd: 19489–95. doi:10.1016/j.ceramint.2021.03.286.

Manousiouthakis, Eleana, Junggeon Park, John G. Hardy, Jae Young Lee, and Christine E. Schmidt. 2022. "Towards the Translation of Electroconductive Organic Materials for Regeneration of Neural Tissues." *Acta Biomaterialia* 139: 22–42. doi:10.1016/j.actbio.2021.07.065.

Mao, Fengqi, Yaran Niu, Xin Zhong, Yawen Wang, Xiao Zhu, Le Zhang, Qilian Li, and Xuebin Zheng. 2021. "Oxidation Behaviors and Mechanisms of Yb_2O_3-Doped Silicon Coatings Fabricated by Vacuum Plasma Spray." *Ceramics International* 47 (14). Elsevier Ltd: 19906–13. doi:10.1016/j.ceramint.2021.04.005.

Matsusaka, S., H. Maruyama, T. Matsuyama, and M. Ghadiri. 2010. "Triboelectric Charging of Powders: A Review." *Chemical Engineering Science* 65 (22). Elsevier: 5781–5807. doi:10.1016/j.ces.2010.07.005.

Mazzaglia, Antonino, Giuseppe Di Natale, Rita Tosto, Angela Scala, Giuseppe Sortino, Anna Piperno, Maria Pia Casaletto, Alberto Riminucci, Maria Laura Giuffrida, Placido G. Mineo, Valentina Villari, Norberto Micali, and Giuseppe Pappalardo. 2022. "KLVFF Oligopeptide-Decorated Amphiphilic Cyclodextrin Nanomagnets for Selective Amyloid Beta Recognition and Fishing." *Journal of Colloid and Interface Science* 613. Elsevier Inc.: 814–26. doi:10.1016/j.jcis.2022.01.051.

Medina-Franco, José L. 2016. *Epi-Informatics: Discovery and Development of Small Molecule Epigenetic Drugs and Probes*. doi:10.1016/C2014-0-03789-6.

Meng, Lingxin, Haicheng Xuan, Guohong Zhang, Rui Wang, Jie Wang, Yayu Guan, Xin Yu, Xiaohong Liang, Yuping Li, and Peide Han. 2022. "Rational Construction of Uniform $CoS/NiFe_2O_4$ Heterostructure as Efficient Bifunctional Electrocatalysts for Hydrogen Evolution and Oxygen Evolution Reactions." *Electrochimica Acta* 404. Elsevier Ltd: 139596. doi:10.1016/j.electacta.2021.139596.

Miao, Jinkang, Daoping Cai, Junhui Si, Qianting Wang, and Hongbing Zhan. 2020. "Multi-Component Hierarchical Hollow Co–Mo–O Nanocages Anchored on Reduced Graphene Oxide with Strong Interfacial Interaction for Lithium-Ion Batteries." *Journal of Alloys and Compounds* 828. Elsevier B.V: 154379. doi:10.1016/j.jallcom.2020.154379.

Mohseni, Mojtaba, Wibke Zängler, Kristof Demeestere, Gijs Du Laing, Sabita Bhandari, Anna K. Mechler, Süleyman Yüce, Robert G. Keller, and Matthias Wessling. 2022. "One-Pot Synthesized, Fe-Incorporated Self-Standing Carbons with a Hierarchical Porosity Remove Carbamazepine and Sulfamethoxazole Through Heterogeneous Electro-Fenton." *Chemical Engineering Journal* 446 (P3). Elsevier B.V.: 137006. doi:10.1016/j.cej.2022.137006.

Mokhtari, F., M. Latifi, and M. Shamshirsaz. 2016. "Electrospinning/Electrospray of Polyvinylidene Fluoride (PVDF): Piezoelectric Nanofibers." *Journal of the Textile Institute* 107 (8). Taylor & Francis: 1037–55. doi:10.1080/00405000.2015.1083300.

Mondal, Subhadip, Revathy Ravindren, Poushali Bhawal, Beomsu Shin, Sayan Ganguly, Changwoon Nah, and Narayan Ch Das. 2020. "Combination Effect of Carbon Nanofiber and Ketjen Carbon Black Hybrid Nanofillers on Mechanical, Electrical, and Electromagnetic Interference Shielding Properties of Chlorinated Polyethylene Nanocomposites." *Composites Part B: Engineering* 197 (February). Elsevier Ltd: 108071. doi:10.1016/j.compositesb.2020.108071.

Moore, W. R. 1964. *An Introduction to Polymer Chemistry. Soil Science*. Vol. 97. doi:10.1097/00010694-196404000-00016.

Moreira, Joana, Margarida M. Fernandes, Estela O. Carvalho, Ana Nicolau, Vesna Lazic, Jovan M. Nedeljković, and Senentxu Lanceros-Mendez. 2022. "Exploring Electroactive Microenvironments in Polymer-Based Nanocomposites to Sensitize Bacterial Cells to Low-Dose Embedded Silver Nanoparticles." *Acta Biomaterialia* 139: 237–48. doi:10.1016/j.actbio.2021.07.067.

Nandal, Neha, Pankaj K. Prajapati, B. Moses Abraham, and Suman L. Jain. 2022. "CO_2 to Ethanol: A Selective Photoelectrochemical Conversion Using a Ternary Composite Consisting of Graphene Oxide/Copper Oxide and a Copper-Based Metal-Organic Framework." *Electrochimica Acta* 404. Elsevier Ltd: 139612. doi:10.1016/j.electacta.2021.139612.

Neves, Tauany de Figueiredo, Natália Gabriele Camparotto, Everton Augusto Rodrigues, Valmor Roberto Mastelaro, Renato Falcão Dantas, and Patrícia Prediger. 2022. "New Graphene Oxide-Safranin Modified@polyacrylonitrile Membranes for Removal of Emerging Contaminants: The Role of Chemical and Morphological Features." *Chemical Engineering Journal* 446 (March). doi:10.1016/j.cej.2022.137176.

Nguyen, Phuong T. M., Tai Nguyen, Minh D. Nguyen, and Thu Hien Vu. 2021. "Impact of Electrode Materials on Microstructure, Leakage Current and Dielectric Tunable Properties of Lead-Free BSZT Thin Films." *Ceramics International* 47 (16). Elsevier Ltd: 23214–21. doi:10.1016/j.ceramint.2021.05.033.

Nguyen, Quang Tan, and Kyoung Kwan K. Ahn. 2021. "Fluid-Based Triboelectric Nanogenerators: A Review of Current Status and Applications." *International Journal of Precision Engineering and Manufacturing—Green Technology* 8 (3). Korean Society for Precision Engineering: 1043–60. doi:10.1007/s40684-020-00255-x.

Nie, Hongqi, Li Ping Tan, Sreekumar Pisharath, and Huey Hoon Hng. 2021. "Nanothermite Composites with a Novel Cast Curable Fluoropolymer." *Chemical Engineering Journal* 414 (December 2020). Elsevier B.V.: 128786. doi:10.1016/j.cej.2021.128786.

Nie, Yan, Jiankun Huang, Shengyan Ma, Zhichao Li, Yongkang Shi, Xiaofang Yang, Xin Fang, Jingbin Zeng, Pengyu Bi, Jiantao Qi, Shengjie Wang, Yongqing Xia, Tifeng Jiao, Dongxiang Li, and Meiwen Cao. 2020. "MXene-Hybridized Silane Films for Metal Anticorrosion and Antibacterial Applications." *Applied Surface Science* 527 (January). Elsevier: 146915. doi:10.1016/j.apsusc.2020.146915.

Nigam, Abhishek, Sheetal Saini, Ambak Kumar Rai, and S. J. Pawar. 2021. "Structural, Optical, Cytotoxicity, and Antimicrobial Properties of MgO, ZnO and MgO/ZnO Nanocomposite for Biomedical Applications." *Ceramics International* 47 (14). Elsevier Ltd: 19515–25. doi:10.1016/j.ceramint.2021.03.289.

Niu, Huanhuan, Xueyang Tu, Shuang Zhang, Yaya Li, Hailong Wang, Gang Shao, Rui Zhang, Hongxia Li, Biao Zhao, and Bingbing Fan. 2022. "Engineered Core-Shell SiO_2@Ti_3C_2Tx Composites: Towards Ultra-Thin Electromagnetic Wave Absorption Materials." *Chemical Engineering Journal* 446 (P3). Elsevier B.V.: 137260. doi:10.1016/j.cej.2022.137260.

Niu, Zhenhua, Wenjia Luo, Peng Mu, and Jian Li. 2022. "Nanoconfined CO_2-Philic Ionic Liquid in Laminated g-C_3N_4 Membrane for the Highly Efficient Separation of CO_2." *Separation and Purification Technology* 297 (April). Elsevier B.V.: 121513. doi:10.1016/j.seppur.2022.121513.

Padhan, Aneeta Manjari, Sugato Hajra, Sanjib Nayak, Jagadish Kumar, Manisha Sahu, Hoe Joon Kim, and Perumal Alagarsamy. 2022. "Triboelectrification Based on NiO-Mg Magnetic Nanocomposite: Synthesis, Device Fabrication, and Energy Harvesting Performance." *Nano Energy* 91 (September 2021). Elsevier Ltd: 106662. doi:10.1016/j.nanoen.2021.106662.

Palomo-Briones, Rodolfo, Jiajie Xu, Catherine M. Spirito, Joseph G. Usack, Lauren H. Trondsen, Juan J. L. Guzman, and Largus T. Angenent. 2022. "Near-Neutral PH Increased n-Caprylate Production in a Microbiome with Product Inhibition of Methanogenesis." *Chemical Engineering Journal* 446 (P3). Elsevier B.V.: 137170. doi:10.1016/j.cej.2022.137170.

Pan, Junjie, Xin Sun, Zhengzhi Jin, Tao Wang, Qiliang Zhao, Hongjiao Qu, and Jianping He. 2022. "Constructing Two-Dimensional Lamellar Monometallic Carbon Nanocomposites by Sodium Chloride Hard Template for Lightweight Microwave Scattering and Absorption." *Composites Part B: Engineering* 228 (October 2021). Elsevier Ltd: 109422. doi:10.1016/j.compositesb.2021.109422.

Pan, Shuang, Yihuang Chen, Zewei Wang, Yeu Wei Harn, Jiwoo Yu, Aurelia Wang, Marcus J. Smith, Zili Li, Vladimir V. Tsukruk, Juan Peng, and Zhiqun Lin. 2020. "Strongly-Ligated Perovskite Quantum Dots with Precisely Controlled Dimensions and Architectures for White Light-Emitting Diodes." *Nano Energy* 77 (July). Elsevier Ltd: 105043. doi:10.1016/j.nanoen.2020.105043.

Panda, Sikta, Chandan Kumar Biswas, and Subhankar Paul. 2021. "A Comprehensive Review on the Preparation and Application of Calcium Hydroxyapatite: A Special Focus on Atomic Doping Methods for Bone Tissue Engineering." *Ceramics International* 47 (20). Elsevier Ltd: 28122–44. doi:10.1016/j.ceramint.2021.07.100.

Panzera, Tulio Halak, Thomas Jeannin, Xavier Gabrion, Vincent Placet, Chrystel Remillat, Ian Farrow, and Fabrizio Scarpa. 2020. "Static, Fatigue and Impact Behaviour of an Autoclaved Flax Fibre Reinforced Composite for Aerospace Engineering."

Composites Part B: Engineering 197 (January). Elsevier Ltd: 108049. doi:10.1016/j.compositesb.2020.108049.

Park, Jin Sung, Sung Eun Wang, Dae Soo Jung, Jung Kul Lee, and Yun Chan Kang. 2022. "Nanoconfined Vanadium Nitride in 3D Porous Reduced Graphene Oxide Microspheres as High-Capacity Cathode for Aqueous Zinc-Ion Batteries." *Chemical Engineering Journal* 446 (P3). Elsevier B.V.: 137266. doi:10.1016/j.cej.2022.137266.

Park, Y. K., Kyong Hwan Chung, In Soo Park, Sang Chai Kim, Sun Jae Kim, and Sang Chul Jung. 2020. "Photocatalytic Degradation of 1,4-Dioxane Using Liquid Phase Plasma on Visible Light Photocatalysts." *Journal of Hazardous Materials* 399 (June). Elsevier: 123087. doi:10.1016/j.jhazmat.2020.123087.

Park, Yongha, Jun Ho Kang, Dong Kyu Moon, Young Suk Jo, and Chang Ha Lee. 2021. "Parallel and Series Multi-Bed Pressure Swing Adsorption Processes for H_2 Recovery from a Lean Hydrogen Mixture." *Chemical Engineering Journal* 408 (October 2020). Elsevier B.V.: 127299. doi:10.1016/j.cej.2020.127299.

Paschalidou, Eirini Maria, Rebecka Lindblad, Leon Zendejas Medina, Dennis Karlsson, Ulf Jansson, and Leif Nyholm. 2022. "Corrosion Studies on Multicomponent CoCrFeMnNi(C) Thin Films in Acidic Environments." *Electrochimica Acta* 404. Elsevier Ltd: 139756. doi:10.1016/j.electacta.2021.139756.

Peera, Shaik Gouse, Chao Liu, Jaesool Shim, Akhila Kumar Sahu, Tae Gwan Lee, Manickam Selvaraj, and Ravindranadh Koutavarapu. 2021. "MXene (Ti_3C_2Tx) Supported Electrocatalysts for Methanol and Ethanol Electrooxidation: A Review." *Ceramics International* 47 (20). Elsevier Ltd: 28106–21. doi:10.1016/j.ceramint.2021.07.075.

Pei, Zhen, Qiang Zhang, Qiang Li, Chao Ji, Yan Liu, Kun Yang, Kai Zhuo, Wendong Zhang, and Shengbo Sang. 2021. "A Fully 3D Printed Electronic Skin with Bionic High Resolution and Air Permeable Porous Structure." *Journal of Colloid and Interface Science* 602. Elsevier Inc.: 452–58. doi:10.1016/j.jcis.2021.06.041.

Peña-Garcia, R., Y. Guerra, S. Castro-Lopes, Y. M. Camejo, João M. Soares, A. Franco, E. Padrón-Hernández, and M. Cabrera-Baez. 2021. "Morphological, Magnetic and EPR Studies of ZnO Nanostructures Doped and Co-Doped with Ni and Sr." *Ceramics International* 47 (20): 28714–22. doi:10.1016/j.ceramint.2021.07.030.

Peng, Yong, Haibin Wang, Chong Zhao, Huaxin Hu, Xuemei Liu, and Xiaoyan Song. 2020. "Nanocrystalline WC-Co Composite with Ultrahigh Hardness and Toughness." *Composites Part B: Engineering* 197 (March). Elsevier Ltd: 108161. doi:10.1016/j.compositesb.2020.108161.

Pirmoradi, Maryam, and James R. Kastner. 2021. "A Kinetic Model of Multi-Step Furfural Hydrogenation over a Pd-TiO_2 Supported Activated Carbon Catalyst." *Chemical Engineering Journal* 414 (February). Elsevier B.V.: 128693. doi:10.1016/j.cej.2021.128693.

Prabhakar, Nikhil, Rajapriya A, N Ponpandian, and C. Viswanathan. 2022. "Influence on Effective and Ineffective Delamination of Mxene (Ti_3C_2Tx) by Tightly Anchoring Tin Oxide Nanocomposite for Boosting the Specific Capacitance of Supercapacitor." *SSRN Electronic Journal* 921. Elsevier: 166092. doi:10.2139/ssrn.4112771.

Pucci, Carlotta, Attilio Marino, Özlem Şen, Daniele De Pasquale, Martina Bartolucci, Nerea Iturrioz-Rodríguez, Nicoletta di Leo, Giuseppe de Vito, Doriana Debellis, Andrea Petretto, and Gianni Ciofani. 2022. "Ultrasound-Responsive Nutlin-Loaded Nanoparticles for Combined Chemotherapy and Piezoelectric Treatment of Glioblastoma Cells." *Acta Biomaterialia* 139: 218–36. doi:10.1016/j.actbio.2021.04.005.

Qi, Chang, Feng Jiang, Alex Remennikov, Lian Zheng Pei, Jun Liu, Jun Shan Wang, Xiang Wei Liao, and Shu Yang. 2020. "Quasi-Static Crushing Behavior of Novel Re-Entrant Circular Auxetic Honeycombs." *Composites Part B: Engineering* 197 (January). Elsevier Ltd: 108117. doi:10.1016/j.compositesb.2020.108117.

Qin, Zuzeng, Tongming Su, and Hongbing Ji. 2022. *MXene-Based Photocatalysts: Fabrication and Applications. MXene-Based Photocatalysts: Fabrication and Applications.* doi:10.1201/9781003156963.

Qureshi, Tanvir S., and Daman K. Panesar. 2020. "Nano Reinforced Cement Paste Composite with Functionalized Graphene and Pristine Graphene Nanoplatelets." *Composites Part B: Engineering* 197 (March). Elsevier Ltd: 108063. doi:10.1016/j.compositesb.2020.108063.

Rafiq, Madiha, Xiaozhen Hu, Zhiliang Ye, Abdul Qayum, Hong Xia, Liangsheng Hu, Fushen Lu, and Paul K. Chu. 2022. "Recent Advances in Structural Engineering of 2D Hexagonal Boron Nitride Electrocatalysts." *Nano Energy* 91 (October 2021). Elsevier Ltd: 106661. doi:10.1016/j.nanoen.2021.106661.

Rasheed, Tahir. 2022. "3D MXenes as Promising Alternatives for Potential Electrocatalysis Applications: Opportunities and Challenges." *Journal of Materials Chemistry C*. Royal Society of Chemistry: 9669–90. doi:10.1039/d2tc01542k.

Rodrigues, C., Claudete Alves, Alvaro J. Santos-Neto, Christian Fernandes, and Fernando M. Lan. 2007. "Analysis of Tricyclic Antidepressant Drugs in Plasma by Means of Solid-Phase Microextraction-Liquid Chromatography-Mass Spectrometry." *Journal of Mass Spectrometry*, no. October: 1342–47. doi:10.1002/jms.

Rodríguez-Peña, M., J. A. Barrios Pérez, J. Llanos, C. Saez, C. E. Barrera-Díaz, and M. A. Rodrigo. 2021. "Understanding Ozone Generation in Electrochemical Cells at Mild PHs." *Electrochimica Acta* 376. doi:10.1016/j.electacta.2021.138033.

Roman, Julien, Wilfrid Neri, Vanessa Fierro, Alain Celzard, Ahmed Bentaleb, Isabelle Ly, Jing Zhong, Alain Derré, and Philippe Poulin. 2020. "Lignin-Graphene Oxide Inks for 3D Printing of Graphitic Materials with Tunable Density." *Nano Today* 33. doi:10.1016/j.nantod.2020.100881.

Rozman, Martin, Andraž Mavrič, Gregor Kravanja, Matjaž Valant, and Amirhossein Pakseresht. 2022. "Ultra-Low-Cost, Flexible and Durable Electrochromic Tape Device Based on Aluminum Foil." *Electrochimica Acta* 404. doi:10.1016/j.electacta.2021.139760.

Russo, Vincenzo, Francesco Taddeo, Tommaso Cogliano, Rosa Vitiello, Roberto Esposito, Riccardo Tesser, Tapio Salmi, and Martino Di Serio. 2021. "Investigation of the Intrinsic Reaction Kinetics and the Mass Transfer Phenomena of Nonanoic Acid Esterification with 2-Ethylhexanol Promoted by Sulfuric Acid or Amberlite IR120." *Chemical Engineering Journal* 408 (September 2020). Elsevier B.V.: 127236. doi:10.1016/j.cej.2020.127236.

Sagar, Raghavendra, and Asha Rao. 2021. "Nanoscale TiO_2 and Ta_2O_5 as Efficient Antireflection Coatings on Commercial Monocrystalline Silicon Solar Cell." *Journal of Alloys and Compounds* 862. Elsevier: 158464. doi:10.1016/j.jallcom.2020.158464.

Saha, Sudip, Jie Yang, S. Shayan Mousavi Masouleh, G. A. Botton, and Leyla Soleymani. 2022. "Hot Hole Direct Photoelectrochemistry of Au NPs: Interband versus Intraband Hot Carriers." *Electrochimica Acta* 404. Elsevier Ltd: 139746. doi:10.1016/j.electacta.2021.139746.

Saqib, Muhammad, Yunshan Fan, Rui Hao, and Bo Zhang. 2021. "Optical Imaging of Nanoscale Electrochemical Interfaces in Energy Applications." *Nano Energy* 90 (PA). Elsevier Ltd: 106539. doi:10.1016/j.nanoen.2021.106539.

Satheeshkumar, Taron Makaryan, Armen Melikyan, Hayk Minassian, Yury Gogotsi, and Masahiro Yoshimura. 2016. "One-step Solution Processing of Ag, Au and Pd@MXene Hybrids for SERS." *Scientific Reports* 6: 32049. doi:10.1038/srep32049.

Shao, Guangwei, Chuanli Su, Chenyang Shi, Zijie Xu, Meidan Ye, Nanliang Chen, Rui Yu, and Xiang Yang Liu. 2022. "Highly Flexible and High Energy Density Fiber Supercapacitors

Based upon Spiral Silk Composite Membranes Encapsulation." *Electrochimica Acta* 404. Elsevier Ltd: 139611. doi:10.1016/j.electacta.2021.139611.

Sharma, Surbhi, Soumen Basu, Nagaraj P. Shetti, Mallikarjuna N. Nadagouda, and Tejraj M. Aminabhavi. 2021. "Microplastics in the Environment: Occurrence, Perils, and Eradication." *Chemical Engineering Journal* 408 (October 2020). Elsevier B.V.: 127317. doi:10.1016/j.cej.2020.127317.

Shawky, Ahmed, Soha M. Albukhari, Nada Y. Tashkandi, and Z. I. Zaki. 2021. "Photoactivity Enhancement of La-Doped $NaTaO_3$ Nanocrystals by CuO Decoration Toward Fast Oxidation of Ciprofloxacin Under Visible Light." *Ceramics International* 47 (20). Elsevier Ltd: 28884–91. doi:10.1016/j.ceramint.2021.07.049.

Sheldon, Brian W., and Anton Tokranov. 2016. "Internal Stress Due to Solvent Co-Intercalation in Graphite Electrodes for Li Ion Batteries." *Extreme Mechanics Letters* 9. Elsevier Ltd: 379–85. doi:10.1016/j.eml.2016.05.001.

Shen, Jyun Hong, Yong Hsuan Tang, Zhu Wu Jiang, Dong Qi Liao, and Jao Jia Horng. 2021. "Optimized Preparation and Characterization of Co-N Codoped TiO_2 with Enhanced Visible Light Activity: An Insight into Effect of Dopants on Surface Redox Reactions of Photogenerated Charge Carriers for Hydroxyl Radical Formation." *Journal of Alloys and Compounds* 862. Elsevier: 158697. doi:10.1016/j.jallcom.2021.158697.

Shen, Zhouzhou, Junhong Chen, Bin Li, Guangqi Li, Zhijiao Zhang, and Xinmei Hou. 2020. "Recent Progress in SiC Nanowires as Electromagnetic Microwaves Absorbing Materials." *Journal of Alloys and Compounds* 815. Elsevier B.V: 152388. doi:10.1016/j.jallcom.2019.152388.

Shi, Qiongfeng, Tianyiyi He, and Chengkuo Lee. 2019. "More than Energy Harvesting—Combining Triboelectric Nanogenerator and Flexible Electronics Technology for Enabling Novel Micro-/Nano-Systems." *Nano Energy* 57 (January). Elsevier Ltd: 851–71. doi:10.1016/j.nanoen.2019.01.002.

Shin, Hyun Wook, and Jong Yeog Son. 2021. "Nonvolatile-Memory Current Characteristics of $BiFeO_3$ Nanodots Switched by Applying External Bias and Force." *Ceramics International* 47 (20). Elsevier Ltd: 28449–54. doi:10.1016/j.ceramint.2021.06.263.

Shu, Ruiwen, Xiaohui Li, Konghu Tian, and Jianjun Shi. 2022. "Fabrication of Bimetallic Metal-Organic Frameworks Derived Fe_3O_4/C Decorated Graphene Composites as High-Efficiency and Broadband Microwave Absorbers." *Composites Part B: Engineering* 228 (October 2021). Elsevier Ltd: 109423. doi:10.1016/j.compositesb.2021.109423.

Shu, Ruiwen, Jing Xu, Lijuan Nie, and Jianjun Shi. 2022. "Facile Construction of Three-Dimensional Porous Netlike Reduced Graphene Oxide/Zinc Oxide Composite Aerogels as the Lightweight, Flame Retardant, Compression Resilience and High-Performance Electromagnetic Wave Absorbers." *Composites Part A* 160 (July). Elsevier Ltd: 107068. doi:10.1016/j.compositesa.2022.107068.

Si, Wan Jie, Hao Zhang, Yi Dong Li, Caili Huang, Yun Xuan Weng, and Jian Bing Zeng. 2020. "Highly Toughened and Heat Resistant Poly(L-Lactide)/Poly(ε-Caprolactone) Blends via Engineering Balance between Kinetics and Thermodynamics of Phasic Morphology with Stereocomplex Crystallite." *Composites Part B: Engineering* 197 (March). Elsevier Ltd. doi:10.1016/j.compositesb.2020.108155.

Singh, Anoop Pratap, O. P. Pandey, and Puneet Sharma. 2021. "Impedance Spectroscopy and Magneto-Dielectric Analysis of La3+ Substituted Co_2Z Hexaferrite." *Ceramics International* 47 (16). Elsevier Ltd: 22550–57. doi:10.1016/j.ceramint.2021.04.266.

Singh, Suraj Kumar, Anshuman Singh, Vinod Kumar, Jalaj Gupta, Sima Umrao, Manoj Kumar, Devojit Kumar Sarma, Marcis Leja, Manohar Prasad Bhandari, and Vinod Verma. 2022. "Nanosheets Based Approach to Elevate the Proliferative and Differentiation Efficacy of

Human Wharton's Jelly Mesenchymal Stem Cells." *International Journal of Molecular Sciences* 23 (10): 5816. doi:10.3390/ijms23105816.

Sinha Ray, Suprakas. 2013. "Clay-Containing Polymer Nanocomposites: From Fundamentals to Real Applications." *Clay-Containing Polymer Nanocomposites: From Fundamentals to Real Applications*: 1–387. doi:10.1016/C2011-0-07134-X.

Slabov, Vladislav, Svitlana Kopyl, Marco P. Soares dos Santos, and Andrei L. Kholkin. 2020. "Natural and Eco-Friendly Materials for Triboelectric Energy Harvesting." *Nano-Micro Letters* 12 (1). Springer Singapore. doi:10.1007/s40820-020-0373-y.

Son, Hong Jin, Min Ji Kim, and Sung Hoon Ahn. 2021. "Monolithic Co-N-C Membrane Integrating Co Atoms and Clusters as a Self-Supporting Multi-Functional Electrode for Solid-State Zinc–Air Batteries and Self-Powered Water Splitting." *Chemical Engineering Journal* 414 (November 2020). Elsevier B.V.: 128739. doi:10.1016/j.cej.2021.128739.

Song, Ping, Hua Qiu, Lei Wang, Xinyu Liu, Yali Zhang, Junliang Zhang, Jie Kong, and Junwei Gu. 2020. "Honeycomb Structural RGO-MXene/Epoxy Nanocomposites for Superior Electromagnetic Interference Shielding Performance." *Sustainable Materials and Technologies* 24. Elsevier B.V.: e00153. doi:10.1016/j.susmat.2020.e00153.

Srasri, K., M. Thongroj, P. Chaijiraaree, S. Thiangtham, H. Manuspiya, P. Pisitsak, and S. Ummartyotin. 2018. "Recovery Potential of Cellulose Fiber from Newspaper Waste: An Approach on Magnetic Cellulose Aerogel for Dye Adsorption Material." *International Journal of Biological Macromolecules* 119 (July): 662–68. doi:10.1016/j.ijbiomac.2018.07.123.

Su, Hui, Jing Jiang, Ning Li, Yangqin Gao, and Lei Ge. 2022. "NiCu Alloys Anchored Defect-Rich NiFe Layered Double-Hydroxides as Efficient Electrocatalysts for Overall Water Splitting." *Chemical Engineering Journal* 446 (P3). Elsevier B.V.: 137226. doi:10.1016/j.cej.2022.137226.

Su, Jinbu, Wancheng Zhou, Yi Liu, Yuchang Qing, Fa Luo, and Dongmei Zhu. 2015. "Effect of Ti_3SiC_2 Addition on Microwave Absorption Property of Ti_3SiC_2/Cordierite Coatings." *Surface and Coatings Technology* 270. Elsevier B.V.: 39–46. doi:10.1016/j.surfcoat.2015.03.021.

Su, Jun feng, Zi zhen Wu, Ting lin Huang, Han Zhang, and Jia Wei Li. 2020. "A New Technology for Simultaneous Calcium–Nitrate and Fluoride Removal in the Biofilm Reactor." *Journal of Hazardous Materials* 399 (January). Elsevier: 122846. doi:10.1016/j.jhazmat.2020.122846.

Su, Yuanjie, Guorui Chen, Chunxu Chen, Qichen Gong, Guangzhong Xie, Mingliang Yao, Huiling Tai, Yadong Jiang, and Jun Chen. 2021. "Self-Powered Respiration Monitoring Enabled by a Triboelectric Nanogenerator." *Advanced Materials* 33 (35). doi:10.1002/adma.202101262.

Su, Yue, Kainan Ma, Fang Yuan, Jun Tang, Ming Liu, and Xu Zhang. 2022. "High-Performance Flexible Piezoresistive Sensor Based on Ti_3C_2Tx MXene with a Honeycomb-Like Structure for Human Activity Monitoring." *Micromachines* 13 (6): 821. doi:10.3390/mi13060821.

Sultana, Ayesha, Sujoy Kumar Ghosh, Mehebub Alam, Priyabrata Sadhukhan, Krittish Roy, Mengying Xie, Chris R. Bowen, Subrata Sarkar, Sachindranath Das, and Tapas Ranjan Middya. 2019. "Methylammonium Lead Iodide Incorporated Poly(Vinylidene Fl Uoride) Nano Fi Bers for Flexible Piezoelectric – Pyroelectric Nanogenerator." doi: 10.1021/acsami.9b04812.

Sun, Yan, Ruonan Ding, Sung Yong Hong, Jinwoo Lee, You Kyong Seo, Jae Do Nam, and Jonghwan Suhr. 2021. "MXene-Xanthan Nanocomposite Films with Layered Microstructure for Electromagnetic Interference Shielding and Joule Heating." *Chemical Engineering Journal* 410 (September 2020). Elsevier B.V.: 128348. doi:10.1016/j.cej.2020.128348.

Swain, Nilimapriyadarsini, Alekhika Tripathy, Arun Thirumurugan, B. Saravanakumar, Lukas Schmidt-Mende, and Ananthakumar Ramadoss. 2022. "A Brief Review on Stretchable, Compressible, and Deformable Supercapacitor for Smart Devices." *Chemical Engineering Journal* 446 (P3). Elsevier B.V.: 136876. doi:10.1016/j.cej.2022.136876.

Taikar, D. R. 2020. "Study of Energy Transfer from Bi3+ to Tb3+ in Y_2O_3 Phosphor and Its Application for W-LED." *Journal of Alloys and Compounds* 828. Elsevier B.V: 154405. doi:10.1016/j.jallcom.2020.154405.

Tan, Dongchen, Nan Sun, Long Chen, Jingyuan Bu, and Chengming Jiang. 2022. "Piezoelectricity in Monolayer and Multilayer Ti_3C_2Tx MXenes: Implications for Piezoelectric Devices." *ACS Applied Nano Materials* 5 (1): 1034–46. doi:10.1021/acsanm.1c03696.

Tang, Jian, Wenhao Xing, Kaijin Kang, Tixian Zeng, Wenlong Yin, and Bin Kang. 2020. "Quaternary Rare-Earth Sulfide LaSnGa3S7: Synthesis, Structure, Thermal and Optical Properties." *Journal of Alloys and Compounds* 828: 154380. doi:10.1016/j.jallcom.2020.154380.

Tang, Jun, Yentl Swolfs, Marco L. Longana, Ha Na Yu, Michael R. Wisnom, Stepan V. Lomov, and Larissa Gorbatikh. 2019. "Hybrid Composites of Aligned Discontinuous Carbon Fibers and Self-Reinforced Polypropylene Under Tensile Loading." *Composites Part A: Applied Science and Manufacturing* 123 (April). Elsevier: 97–107. doi:10.1016/j.compositesa.2019.05.003.

Tao, Mengli, Youquan Zhang, Renming Zhan, Bingshu Guo, Qiuju Xu, and Maowen Xu. 2018. "A Chemically Bonded $CoNiO_2$ Nanoparticles/MXene Composite as Anode for Sodium-Ion Batteries." *Materials Letters* 230. Elsevier B.V.: 173–76. doi:10.1016/j.matlet.2018.07.110.

Taşdemir, A., R. Aydin, A. Akkaya, N. Akman, Y. Altınay, H. Çetin, B. Şahin, A. Uzun, and E. Ayyıldız. 2021. "A Green Approach for the Preparation of Nanostructured Zinc Oxide: Characterization and Promising Antibacterial Behaviour." *Ceramics International* 47 (14): 19362–73. doi:10.1016/j.ceramint.2021.03.273.

To, Applied, and A Single Tip. 1997. "Electrostatic Forces Acting on the Tip in Atomic Force Microscopy: Modelization and Comparison with Analytic Expressions." *Journal of Applied Phyicss* 81: 1023–1030. https://doi.org/10.1063/1.363884

Tong, Yimin, Gansheng Shi, Gaowei Hu, Xiaoyou Hu, Lin Han, Xiaofeng Xie, Yongfen Xu, Rong Zhang, Jing Sun, and Jin Zhong. 2021. "Photo-Catalyzed TiO_2 Inactivates Pathogenic Viruses by Attacking Viral Genome." *Chemical Engineering Journal* 414 (October 2020). Elsevier B.V.: 128788. doi:10.1016/j.cej.2021.128788.

Trabocchi, Andrea. 2019. *Small Molecule Drug Discovery: Methods, Molecules and Applications*. Elsevier. https://shop.elsevier.com/books/small-molecule-drug-discovery/trabocchi/978-0-12-818349-6.

Trinh, V. L., and C. K. Chung. 2018. "Harvesting Mechanical Energy, Storage, and Lighting Using a Novel PDMS Based Triboelectric Generator with Inclined Wall Arrays and Micro-Topping Structure." *Applied Energy* 213 (June 2017). Elsevier: 353–65. doi:10.1016/j.apenergy.2018.01.039.

Trzciński, Konrad, Mariusz Szkoda, Zuzanna Zarach, Mirosław Sawczak, and Andrzej P. Nowak. 2022. "Towards Spectroscopic Monitoring of Photoelectrodes: In-Situ Raman Photoelectrochemistry of a TiO_2/Prussian Blue Photoanode." *Electrochimica Acta* 404. Elsevier Ltd: 139774. doi:10.1016/j.electacta.2021.139774.

Tyler, Andrew R., Ronnie Ragbirsingh, Charles J. McMonagle, Paul G. Waddell, Sarah E. Heaps, Jonathan W. Steed, Paul Thaw, Michael J. Hall, and Michael R. Probert. 2020. "Encapsulated Nanodroplet Crystallization of Organic-Soluble Small Molecules." *Chem* 6 (7). Elsevier Inc.: 1755–65. doi:10.1016/j.chempr.2020.04.009.

Uchino, K., and J. C. Debus. 2012. *Applications of ATILA FEM Software to Smart Materials: Case Studies in Designing Devices. Applications of ATILA FEM Software to Smart Materials: Case Studies in Designing Devices*. doi:10.1533/9780857096319.

Ungai-Salánki, Rita, Benjamin Csippa, Tamás Gerecsei, Beatrix Péter, Robert Horvath, and Bálint Szabó. 2021. "Nanonewton Scale Adhesion Force Measurements on Biotinylated Microbeads with a Robotic Micropipette." *Journal of Colloid and Interface Science* 602: 291–99. doi:10.1016/j.jcis.2021.05.180.

Vasconcelos, Luize Scalco de, Rong Xu, Jianlin Li, and Kejie Zhao. 2016. "Grid Indentation Analysis of Mechanical Properties of Composite Electrodes in Li-Ion Batteries." *Extreme Mechanics Letters* 9. Elsevier Ltd: 495–502. doi:10.1016/j.eml.2016.03.002.

Veloso, A. D., R. A. Videira, and M. C. Oliveira. 2022. "An Electrochemical Strategy for Synthesising Carbon-Based Nanomaterials with Tuned Redox Properties." *Electrochimica Acta* 404. Elsevier Ltd: 139734. doi:10.1016/j.electacta.2021.139734.

Vladimirova, S. A., K. Ya Prikhodko, M. N. Rumyantseva, E. A. Konstantinova, A. S. Chizhov, N. O. Khmelevsky, and A. M. Gaskov. 2020. "Nanocrystalline Complex Oxides $NixCo_3$-XO_4: Cations Distribution Impact on Electrical and Gas Sensor Behaviour." *Journal of Alloys and Compounds* 828. Elsevier B.V: 154420. doi:10.1016/j.jallcom.2020.154420.

Wahajuddin, and Sumit Arora. 2012. "Superparamagnetic Iron Oxide Nanoparticles: Magnetic Nanoplatforms as Drug Carriers." *International Journal of Nanomedicine* 7: 3445–71. doi:10.2147/IJN.S30320.

Wan, Keming, Yalin Li, Yan Wang, and Gang Wei. 2021. "Recent Advance in the Fabrication of 2d and 3d Metal Carbides-Based Nanomaterials for Energy and Environmental Applications." *Nanomaterials* 11 (1): 1–34. doi:10.3390/nano11010246.

Wang, Chunjie, Chao Liang, Yu Hao, Ziliang Dong, Yujie Zhu, Quguang Li, Zhuang Liu, Liangzhu Feng, and Meiwan Chen. 2021. "Photodynamic Creation of Artificial Tumor Microenvironments to Collectively Facilitate Hypoxia-Activated Chemotherapy Delivered by Coagulation-Targeting Liposomes." *Chemical Engineering Journal* 414 (October 2020). Elsevier B.V.: 128731. doi:10.1016/j.cej.2021.128731.

Wang, Dong, Lingxia Li, Mingkun Du, and Yu Zhan. 2021. "A Low-Sintering Temperature Microwave Dielectric Ceramic for 5G LTCC Applications with Ultralow Loss." *Ceramics International* 47 (20). Elsevier Ltd: 28675–84. doi:10.1016/j.ceramint.2021.07.027.

Wang, Dongming, Xiaomei Wu, and Qiang Gao. 2021. "Novel Energy-Saving Window Coating Based on F Doped TiO2 Nanocrystals with Enhanced NIR Shielding Performance." *Ceramics International* 47 (20). Elsevier Ltd: 28557–65. doi:10.1016/j.ceram int.2021.07.013.

Wang, Helin, Lingcheng Chen, Zhenbo Zhao, and Yi Xiao. 2017. "Aryl-Bisalkynyl Bridged Perylene Diimides Dimers: Efficient Synthesis, Properties and Improved Electron Mobilities." *Dyes and Pigments* 144. Elsevier Ltd: 184–89. doi:10.1016/j.dyepig.2017.05.030.

Wang, Hongbing, Yuping Wu, Jianfeng Zhang, Gaiye Li, Huajie Huang, Xin Zhang, and Quanguo Jiang. 2015. "Enhancement of the Electrical Properties of MXene Ti_3C_2 Nanosheets by Post-Treatments of Alkalization and Calcination." *Materials Letters* 160. Elsevier: 537–40. doi:10.1016/j.matlet.2015.08.046.

Wang, Huan, Huiyang Fan, Huijun Liu, Mingkang Jin, Shaoting Du, Dexiao Li, Ping Zhang, Songlin Ruan, and Jieren Qiu. 2020. "Oxidative Stress Response Mechanism of Scenedesmus Obliquus to Ionic Liquids with Different Number of Methyl-Substituents." *Journal of Hazardous Materials* 399 (January). Elsevier: 122847. doi:10.1016/j.jhazmat.2020.122847.

Wang, Hui, Hai You Huang, and Yan Jing Su. 2020. "Tuning the Operation Temperature Window of the Elastocaloric Effect in Cu–Al–Mn Shape Memory Alloys by

Composition Design." *Journal of Alloys and Compounds* 828. Elsevier B.V: 154265. doi:10.1016/j.jallcom.2020.154265.

Wang, Jie, Xiuhan Li, Yunlong Zi, Sihong Wang, Zhaoling Li, Li Zheng, Fang Yi, Shengming Li, and Zhong Lin Wang. 2015. "A Flexible Fiber-Based Supercapacitor-Triboelectric-Nanogenerator Power System for Wearable Electronics." *Advanced Materials* 27 (33): 4830–36. doi:10.1002/adma.201501934.

Wang, Jiqi, Fangrong Zhang, Yulong Li, Mudasir Ahmad, Pei Liu, Qiuyu Zhang, and Baoliang Zhang. 2022. "Constructing TCNFs/MXene/TiO_2 Microspheres with Wrinkled Surface for Excellent Electromagnetic Wave Absorption." *Journal of Alloys and Compounds* 918. Elsevier: 165623. doi:10.1016/j.jallcom.2022.165623.

Wang, Lipeng, Zheng Yan, Jianhang Qiu, Jinbo Wu, Chao Zhen, Kaiping Tai, Xin Jiang, and Shihe Yang. 2021. "Interface Regulation Enables Hysteresis Free Wide-Bandgap Perovskite Solar Cells with Low VOC Deficit and High Stability." *Nano Energy* 90 (PA). Elsevier Ltd: 106537. doi:10.1016/j.nanoen.2021.106537.

Wang, Sheng, Fang Yuan, Shuai Liu, Jianyu Zhou, Shouhu Xuan, Yu Wang, and Xinglong Gong. 2020. "A Smart Triboelectric Nanogenerator with Tunable Rheological and Electrical Performance for Self-Powered Multi-Sensors." *Journal of Materials Chemistry C* 8 (11). Royal Society of Chemistry: 3715–23. doi:10.1039/c9tc05969e.

Wang, Wandi, Demin Jiang, Xing Chen, Kun Xie, Yinhua Jiang, and Yuqiao Wang. 2020. "A Sandwich-like Nano-Micro LDH-MXene-LDH for High-Performance Supercapacitors." *Applied Surface Science* 515 (January). Elsevier: 145982. doi:10.1016/j.apsusc.2020.145982.

Wang, Wei, Wenwen Gong, Shuangxi Zhou, Jiajun Han, Delin Qi, and Han Qu. 2020. "β-Cyclodextrin Improve the Tolerant of Freshwater Algal Spiny Scenedesmus to Chiral Drugs Venlafaxine and Its Metabolite." *Journal of Hazardous Materials* 399 (15). Elsevier: 123076. doi:10.1016/j.jhazmat.2020.123076.

Wang, Xiaohan, Shen Bao, Feiyue Hu, Siyang Shang, Yongqiang Chen, Na Zhao, Rui Zhang, Biao Zhao, and Bingbing Fan. 2022. "The Effect of Honeycomb Pore Size on the Electromagnetic Interference Shielding Performance of Multifunctional 3D Honeycomb-like Ag/Ti_3C_2Tx Hybrid Structures." *Ceramics International* 48 (12). Elsevier Ltd: 16892–900. doi:10.1016/j.ceramint.2022.02.244.

Wang, Xingfu, Yejing Dai, Ruiyuan Liu, Xu He, Shuti Li, and Zhong Lin Wang. 2017. "Light-Triggered Pyroelectric Nanogenerator Based on a Pn-Junction for Self-Powered Near-Infrared Photosensing." *ACS Nano* 11 (8): 8339–45. doi:10.1021/acsnano.7b03560.

Wang, Xiqing, Atif Muhmood, Tao Lyu, Renjie Dong, Hongtao Liu, and Shubiao Wu. 2021. "Mechanisms of Genuine Humic Acid Evolution and Its Dynamic Interaction with Methane Production in Anaerobic Digestion Processes." *Chemical Engineering Journal* 408 (September 2020). Elsevier B.V.: 127322. doi:10.1016/j.cej.2020.127322.

Wang, Yonghong, Zhiwei Ma, Xinru Zhang, Jinping Li, Yi Zhou, Zhuo Jin, and Nanwen Li. 2022. "Mixed-Matrix Membranes Consisting of Pebax and Novel Nitrogen-Doped Porous Carbons for CO2 Separation." *Journal of Membrane Science* 644 (November 2021). Elsevier B.V.: 120182. doi:10.1016/j.memsci.2021.120182.

Wang, Yongzhi, Guorong Hu, Zhongdong Peng, Ke Du, Baichao Zhang, and Yanbing Cao. 2021. "Self-Assembly Synthesis of Dumbbell-like $MnPO_4 \cdot H_2O$ Hierarchical Particle and Its Application in $LiMnPO_4$/C Cathode Materials." *Ceramics International* 47 (14): 19687–93. doi:10.1016/j.ceramint.2021.03.306.

Wang, Zehao, Li Yuan, Guozheng Liang, and Aijuan Gu. 2021. "Mechanically Durable and Self-Healing Super-Hydrophobic Coating with Hierarchically Structured KH570 Modified SiO2-Decorated Aligned Carbon Nanotube Bundles." *Chemical Engineering Journal* 408 (September 2020). Elsevier B.V.: 127263. doi:10.1016/j.cej.2020.127263.

Wang, Zhen, Zhi Cheng, Li Xie, Xianliang Hou, and Changqing Fang. 2021. "Flexible and Lightweight Ti_3C_2Tx MXene/Fe_3O_4@PANI Composite Films for High-Performance Electromagnetic Interference Shielding." *Ceramics International* 47 (4). Elsevier Ltd: 5747–57. doi:10.1016/j.ceramint.2020.10.161.

Wang, Zhenkai, Kaicheng Zhu, Jiali Chen, Guicheng Zhang, Winston Duo Wu, and Sheng Peng Sun. 2021. "Oxone Activation by UVA-Irradiated FeIII-NTA Complex: Efficacy, Radicals Formation and Mechanism on Crotamiton Degradation." *Chemical Engineering Journal* 408 (September 2020). Elsevier B.V.: 127324. doi:10.1016/j.cej.2020.127324.

Wang, Zhong Lin, and Wenzhuo Wu. 2012. "Nanotechnology-Enabled Energy Harvesting for Self-Powered Micro-/Nanosystems." *Angewandte Chemie—International Edition* 51 (47): 11700–721. doi:10.1002/anie.201201656.

Wang, Zhong Lin, Tao Jiang, and Liang Xu. 2017. "Toward the Blue Energy Dream by Triboelectric Nanogenerator Networks." *Nano Energy* 39 (June). Elsevier Ltd: 9–23. doi:10.1016/j.nanoen.2017.06.035.

Wang, Zhongliao, Bei Cheng, Liuyang Zhang, Jiaguo Yu, Youji Li, S. Wageh, and Ahmed A. Al-Ghamdi. 2022. "S-Scheme 2D/2D Bi_2MoO_6/BiOI van Der Waals Heterojunction for CO_2 Photoreduction." *Chinese Journal of Catalysis* 43 (7). Dalian Institute of Chemical Physics, the Chinese Academy of Sciences: 1657–66. doi:10.1016/S1872-2067(21)64010-X.

Wang, Zhuowen, Luluan Feng, Chao Deng, and Shengping Wang. 2022. "2,5-Dimercapto-1,3,4-Thiadiazole/Acetylene Black@polypropylene Separator for Inhibiting the Shuttle Effect and Electrocatalyzing Electrode Reactions in Li-S Batteries." *Chemical Engineering Journal* 446 (P3). Elsevier B.V.: 137153. doi:10.1016/j.cej.2022.137153.

Waseem, Owais Ahmed, and Ho Jin Ryu. 2020. "Combinatorial Synthesis and Analysis of AlxTayVz-Cr20Mo20Nb20Ti20Zr10 and Al10CrMoxNbTiZr10 Refractory High-Entropy Alloys: Oxidation Behavior." *Journal of Alloys and Compounds* 828. Elsevier B.V: 154427. doi:10.1016/j.jallcom.2020.154427.

Wei, Huawei, Mingqiang Wang, Wenhui Zheng, Zaixing Jiang, and Yudong Huang. 2020. "2D Ti_3C_2Tx MXene/Aramid Nanofibers Composite Films Prepared via a Simple Filtration Method with Excellent Mechanical and Electromagnetic Interference Shielding Properties." *Ceramics International* 46 (5). Elsevier Ltd and Techna Group S.r.l.: 6199–6204. doi:10.1016/j.ceramint.2019.11.087.

Wei, Jingjiang, Chenglong Zhu, Zhihui Zeng, Fei Pan, Fuqiang Wan, Liwen Lei, Gustav Nyström, and Zhengyi Fu. 2022. "Bioinspired Cellulose-Integrated MXene-based Hydrogels for Multifunctional Sensing and Electromagnetic Interference Shielding." *Interdisciplinary Materials*, no. February: 1–12. doi:10.1002/idm2.12026.

Wei, Kexin, Yousef Faraj, Gang Yao, Ruzhen Xie, and Bo Lai. 2021. "Strategies for Improving Perovskite Photocatalysts Reactivity for Organic Pollutants Degradation: A Review on Recent Progress." *Chemical Engineering Journal* 414 (February). Elsevier B.V.: 128783. doi:10.1016/j.cej.2021.128783.

Wellmer, Friedrich W., Peter Buchholz, Jens Gutzmer, Peter Herzig, Christian Hagelüken, Ralf Littke, Rudolf K. Thauer, and Jenny Mash. 2020. *"Raw Materials for Future Energy Supply." Johnson Matthey Technology Review*. Vol. 64. doi:10.1595/205651320x15735482848588.

Wright, Elliott J., Daniel G. Beach, and Pearse McCarron. 2022. "Non-Target Analysis and Stability Assessment of Reference Materials Using Liquid Chromatography–High-Resolution Mass Spectrometry." *Analytica Chimica Acta* 1201. Elsevier Ltd: 339622. doi:10.1016/j.aca.2022.339622.

Wu, Hao, Mingyu Wang, Linwei Huai, Wenxiang Wang, Jiachi Zhang, and Yuhua Wang. 2021. "Optical Storage and Operation Based on Photostimulated Luminescence." *Nano Energy* 90 (PA). Elsevier Ltd: 106546. doi:10.1016/j.nanoen.2021.106546.

Wu, Shaohua, Ye Qi, Wen Shi, Mitchell Kuss, Shaojuan Chen, and Bin Duan. 2022. "Electrospun Conductive Nanofiber Yarns for Accelerating Mesenchymal Stem Cells Differentiation and Maturation into Schwann Cell-Like Cells Under a Combination of Electrical Stimulation and Chemical Induction." *Acta Biomaterialia* 139: 91–104. doi:10.1016/j.actbio.2020.11.042.

Wu, Xing, Wenxuan Hu, Jiahuan Qiu, Biao Geng, Miao Du, and Qiang Zheng. 2022. "Solvent-Assisted Self-Assembly to Fabricate a Ternary Flexible Free-Standing Polyaniline@ MXene-CNTs Electrode for High-Performance Supercapacitors." *Journal of Alloys and Compounds* 921. Elsevier: 166062. doi:10.1016/j.jallcom.2022.166062.

Wu, Xinyu, Bingyong Han, Hao Bin Zhang, Xi Xie, Tingxiang Tu, Yu Zhang, Yang Dai, Rui Yang, and Zhong Zhen Yu. 2020. "Compressible, Durable and Conductive Polydimethylsiloxane-Coated MXene Foams for High-Performance Electromagnetic Interference Shielding." *Chemical Engineering Journal* 381 (August 2019). Elsevier: 122622. doi:10.1016/j.cej.2019.122622.

Wu, Xiulan, Liang Du, Qiang Ren, and Ou Hai. 2021. "Study on the Color Tunability and Energy Transfer Mechanism in Tm3+/Dy3+ Co-Doped LiLaSiO4 Phosphors." *Ceramics International* 47 (20). Elsevier Ltd: 28384–99. doi:10.1016/j.ceramint.2021.06.256.

Wu, Yuting, Ping Nie, Langyuan Wu, Hui Dou, and Xiaogang Zhang. 2018. "2D MXene/SnS_2 Composites as High-Performance Anodes for Sodium Ion Batteries." *Chemical Engineering Journal* 334 (October 2017). Elsevier: 932–38. doi:10.1016/j.cej.2017.10.007.

Wu, Zhenhua, Erzhen Mu, Zhanxun Che, Yang Liu, Fangyuan Sun, Xinwei Wang, and Zhiyu Hu. 2020. "Nanoporous (00l)-Oriented Bi_2Te_3 Nanoplate Film for Improved Thermoelectric Performance." *Journal of Alloys and Compounds* 828. Elsevier B.V: 154239. doi:10.1016/j.jallcom.2020.154239.

Wychowaniec, Jacek K., Jagoda Litowczenko, Krzysztof Tadyszak, Varun Natu, Claudia Aparicio, Barbara Peplińska, Michel W. Barsoum, Michal Otyepka, and Błażej Scheibe. 2020. "Unique Cellular Network Formation Guided by Heterostructures Based on Reduced Graphene Oxide—Ti_3C_2Tx MXene Hydrogels." *Acta Biomaterialia* 115: 104–15. doi:10.1016/j.actbio.2020.08.010.

Xi, Yi, Jie Wang, Yunlong Zi, Xiaogan Li, Changbao Han, Xia Cao, Chenguo Hu, and Zhonglin Wang. 2017. "High Efficient Harvesting of Underwater Ultrasonic Wave Energy by Triboelectric Nanogenerator." *Nano Energy* 38 (April): 101–8. doi:10.1016/j.nanoen.2017.04.053.

Xia, Mengyang, Jinhui Nie, Zailei Zhang, Xianmao Lu, and Zhong Lin. 2018. "Nano Energy Suppressing Self-Discharge of Supercapacitors via Electrorheological Effect of Liquid Crystals." *Nano Energy* 47 (February). Elsevier Ltd: 43–50. doi:10.1016/j.nanoen.2018.02.022.

Xiao, Guodong, Biao Zhao, Wenfeng Ding, and Haixiang Huan. 2021. "On the Grinding Performance of Metal-Bonded Aggregated CBN Grinding Wheels Based on Open-Pore Structures." *Ceramics International* 47 (14). Elsevier Ltd: 19709–15. doi:10.1016/j.cer amint.2021.04.001.

Xiao, Junpeng, Jing Wen, Jiabao Zhao, Xinzhi Ma, Hong Gao, and Xitian Zhang. 2020. "A Safe Etching Route to Synthesize Highly Crystalline Nb_2CTx MXene for High Performance Asymmetric Supercapacitor Applications." *Electrochimica Acta* 337. Elsevier: 135803. doi:10.1016/j.electacta.2020.135803.

Xiao, Yao, Nasir Mahmood Abbasi, Yan Fang Zhu, Shi Li, Shuang Jie Tan, Wei Ling, Ling Peng, Tingqiang Yang, Lude Wang, Xiao-Dong Guo, Ya-Xia Yin, Han Zhang, and Yu-Guo Guo. 2020. "Layered Oxide Cathodes Promoted by Structure Modulation Technology for Sodium-Ion Batteries." *Advanced Functional Materials* 30 (30): 1–28. doi:10.1002/adfm.202001334.

Xiao, Zhuohao, Shuangchen Ruan, Ling Bing Kong, Wenxiu Que, Kun Zhou, Yin Liu, and Tianshu Zhang. 2020. *Synthesis and Properties of MXenes. Engineering Materials.* doi:10.1007/978-3-030-59373-5_2.

Xie, Yanluo, Yao Luo, Mingping Sheng, He Peng, Yuanming Gu, Heng Xu, and Yun Zhao. 2020. "24-Epibrassinolide Combined with Heavy Metal Resistant Bacteria Enhancing Phytoextraction of Amaranthus Hypochondriacus L. in Cd-Contaminated Soil." *Journal of Hazardous Materials* 399 (May). Elsevier: 123031. doi:10.1016/j.jhazmat.2020.123031.

Xu, Cheng, Yunlong Zi, Aurelia Chi Wang, Haiyang Zou, Yejing Dai, Xu He, Peihong Wang, Yi-Cheng Wang, Peizhong Feng, Dawei Li, and Zhong Lin Wang. 2018. "On the Electron-Transfer Mechanism in the Contact-Electrification Effect." *Advanced Materials* 1706790: 1–9. doi:10.1002/adma.201706790.

Xu, Guangda, Na Li, Ye Sun, Ce Gao, Liping Ma, Peng Song, and Lixin Xia. 2021. "A Label-Free, Rapid, Sensitive and Selective Technique for Detection of Fe2+ Using SERRS with 2,2′-Bipyridine as a Probe." *Chemical Engineering Journal* 414 (October 2020). doi:10.1016/j.cej.2021.128741.

Xu, Jing, Yongjiu Zou, Ardo Nashalian, and Jun Chen. 2020. "Leverage Surface Chemistry for High-Performance Triboelectric Nanogenerators." *Frontiers in Chemistry* 8 (November): 1–23. doi:10.3389/fchem.2020.577327.

Xu, Liang, Hao Wu, Guo Yao, Libo Chen, Xiaodan Yang, Baodong Chen, Xin Huang, Wei Zhong, Xiangyu Chen, Zhouping Yin, and Zhong Lin Wang. 2018. "Giant Voltage Enhancement via Triboelectric Charge Supplement Channel for Self-Powered Electroadhesion." *ACS Nano.* doi:10.1021/acsnano.8b05359.

Xu, Pei, Weijia Fu, Yadong Hu, and Yunsheng Ding. 2018. "Effect of Annealing Treatment on Crystalline and Dielectric Properties of PVDF/PEG-Containing Ionic Liquid Composites." *Composites Science and Technology* 158. Elsevier Ltd: 1–8. doi:10.1016/j.compscitech.2018.01.039.

Xu, Shuaikai, Guodong Wei, Junzhi Li, Yuan Ji, Nickolai Klyui, Vladimir Izotov, and Wei Han. 2017. "Binder-Free Ti_3C_2Tx MXene Electrode Film for Supercapacitor Produced by Electrophoretic Deposition Method." *Chemical Engineering Journal* 317: 1026–36. doi:10.1016/j.cej.2017.02.144.

Xu, Xiaodong, Ting Wang, Yanliang Wen, Xin Wen, Xuecheng Chen, Chuncheng Hao, Qingquan Lei, and Ewa Mijowska. 2022. "Intumescent Flame Retardants Inspired Template-Assistant Synthesis of N/P Dual-Doped Three-Dimensional Porous Carbons for High-Performance Supercapacitors." *Journal of Colloid and Interface Science* 613. Elsevier Inc.: 35–46. doi:10.1016/j.jcis.2022.01.018.

Xu, You, Tianlun Ren, Kaili Ren, Shanshan Yu, Mengying Liu, Ziqiang Wang, Xiaonian Li, Liang Wang, and Hongjing Wang. 2021. "Metal-Organic Frameworks-Derived Ru-Doped Co_2P/N-Doped Carbon Composite Nanosheet Arrays as Bifunctional Electrocatalysts for Hydrogen Evolution and Urea Oxidation." *Chemical Engineering Journal* 408 (September 2020). Elsevier B.V.: 127308. doi:10.1016/j.cej.2020.127308.

Xu, Zhenhua, Linxiang Zeng, Jinlong Hu, Zhen Wang, Putao Zhang, Christoph J. Brabec, Karen Forberich, Yaohua Mai, and Fei Guo. 2022. "Reducing Energy Barrier of δ-to-α Phase Transition for Printed Formamidinium Lead Iodide Photovoltaic Devices." *Nano Energy* 91 (October 2021). Elsevier Ltd: 106658. doi:10.1016/j.nanoen.2021.106658.

Xu, Zhixiang, Jichang Lu, Xianyao Zheng, Bo Chen, Yongming Luo, Muhammad Nauman Tahir, Bin Huang, Xueshan Xia, and Xuejun Pan. 2020. "A Critical Review on the Applications and Potential Risks of Emerging MoS_2 Nanomaterials." *Journal of Hazardous Materials* 399 (727). Elsevier: 123057. doi:10.1016/j.jhazmat.2020.123057.

Xue, Yudong, Nguyet N.T. Pham, Gwiwoong Nam, Jaemin Choi, Yong Yoon Ahn, Hongshin Lee, Jinho Jung, Seung Geol Lee, and Jaesang Lee. 2021. "Persulfate Activation by ZIF-67-Derived Cobalt/Nitrogen-Doped Carbon Composites: Kinetics and Mechanisms Dependent on Persulfate Precursor." *Chemical Engineering Journal* 408 (June 2020). Elsevier B.V.: 127305. doi:10.1016/j.cej.2020.127305.

Yan, Jianglong, Dandan Xia, Wenhao Zhou, Yangyang Li, Pan Xiong, Qiyao Li, Pei Wang, Ming Li, Yufeng Zheng, and Yan Cheng. 2020. "PH-Responsive Silk Fibroin-Based CuO/Ag Micro/Nano Coating Endows Polyetheretherketone with Synergistic Antibacterial Ability, Osteogenesis, and Angiogenesis." *Acta Biomaterialia* 115. Elsevier Ltd: 220–34. doi:10.1016/j.actbio.2020.07.062.

Yang, Bing, Yinghong Chen, Mingdan Zhang, and Gaowei Yuan. 2019. "Synergistic and Compatibilizing Effect of Octavinyl Polyhedral Oligomeric Silsesquioxane Nanoparticles in Polypropylene/Intumescent Flame Retardant Composite System." *Composites Part A: Applied Science and Manufacturing* 123 (April). Elsevier: 46–58. doi:10.1016/j.compositesa.2019.04.032.

Yang, Hong Yu, Yi Fan Yan, Tian Shu Liu, Bai Xin Dong, Liang Yu Chen, Shi Li Shu, Feng Qiu, Qi Chuan Jiang, and Lai Chang Zhang. 2021. "Unprecedented Enhancement in Strength-Plasticity Synergy of (TiC+Al6MoTi+Mo)/Al Cermet by Multiple Length-Scale Microstructure Stimulated Synergistic Deformation." *Composites Part B: Engineering* 225 (5988). Elsevier Ltd: 109265. doi:10.1016/j.compositesb.2021.109265.

Yang, Jin Lin, Da Qian Cai, Qiaowei Lin, Xin Yu Wang, Zou Qiang Fang, Ling Huang, Zhi Jie Wang, Xiao-Ge Hao, Shi-Xi Zhao, Jia Li, Guo-Zhong Cao, and Wei Lv. 2022. "Regulating the Li2S Deposition by Grain Boundaries in Metal Nitrides for Stable Lithium-Sulfur Batteries." *Nano Energy* 91 (September 2021). Elsevier Ltd: 106669. doi:10.1016/j.nanoen.2021.106669.

Yang, L., W. Zheng, P. Zhang, J. Chen, W. B. Tian, Y. M. Zhang, and Z. M. Sun. 2018. "MXene/CNTs Films Prepared by Electrophoretic Deposition for Supercapacitor Electrodes." *Journal of Electroanalytical Chemistry* 830–831 (October). Elsevier: 1–6. doi:10.1016/j.jelechem.2018.10.024.

Yang, Ting-Ting, Yue Shui, Cheng-Sha Wei, Li-Zhao Huang, and Chao-Wen Yang. 2022. "Effect of Cyclic Straining with Various Rates on Stress Softening/Hysteresis and Structural Evolution of Filled Rubber : A Time-Resolved SANS Study." *Composites Part B* 242 (July). Elsevier Ltd: 110100. doi:10.1016/j.compositesb.2022.110100.

Yang, Wendong, and Changhai Wang. 2016. "Graphene and the Related Conductive Inks for Flexible Electronics." *Journal of Materials Chemistry C* 4 (30). Royal Society of Chemistry: 7193–7207. doi:10.1039/c6tc01625a.

Yang, Xiaochen, Wenji Zheng, Yuan Xi, Weixin Guan, Xiaoming Yan, Xuehua Ruan, Canghai Ma, Yan Dai, and Gaohong He. 2021. "Constructing Low-Resistance and High-Selectivity Transport Multi-Channels in Mixed Matrix Membranes for Efficient CO_2 Separation." *Journal of Membrane Science* 624 (December 2020). Elsevier B.V.: 119046. doi:10.1016/j.memsci.2020.119046.

Yang, Xuan, Yuping Duan, Shuqing Li, Lingxi Huang, Huifang Pang, Ben Ma, and Tongmin Wang. 2022. "Constructing Three-Dimensional Reticulated Carbonyl Iron/Carbon Foam Composites to Achieve Temperature-Stable Broadband Microwave Absorption Performance." *Carbon* 188. Elsevier Ltd: 376–84. doi:10.1016/j.carbon.2021.12.044.

Yang, Xue, Yiwei Yao, Qian Wang, Kai Zhu, Ke Ye, Guiling Wang, Dianxue Cao, and Jun Yan. 2022. "3D Macroporous Oxidation-Resistant Ti_3C_2Tx MXene Hybrid Hydrogels for Enhanced Supercapacitive Performances with Ultralong Cycle Life." *Advanced Functional Materials* 32 (10): 1–10. doi:10.1002/adfm.202109479.

Yang, Zhouping, Yang Liu, Changfang Lu, Guizhou Yue, Yanying Wang, Hanbing Rao, Wei Zhang, Zhiwei Lu, and Xianxiang Wang. 2021. "One-Pot Synthesis of CeO2-Carbon Dots with Enhanced Peroxidase-like Activity and Carbon Dots for Ratiometric Fluorescence Detection of H_2O_2 and Cholesterol." *Journal of Alloys and Compounds* 862. Elsevier: 158323. doi:10.1016/j.jallcom.2020.158323.

Yao, Yuanyuan, Shaohua Jin, Manman Wang, Feng Gao, Bolin Xu, Xijuan Lv, and Qinghai Shu. 2022. "Mxene Hybrid Polyvinyl Alcohol Flexible Composite Films for Electromagnetic Interference Shielding." *Applied Surface Science* 578 (November 2021). Elsevier B.V.: 152007. doi:10.1016/j.apsusc.2021.152007.

Yazdani-Kachoei, M., and S. Jalali-Asadabadi. 2020. "Topological Analysis of Electron Density in Half-Heusler ZrXBi (X = Co, Rh) Compounds: A Density Functional Theory Study Accompanied by Bader's Quantum Theory of Atoms in Molecules." *Journal of Alloys and Compounds* 828. Elsevier B.V: 154287. doi:10.1016/j.jallcom.2020.154287.

Yeo, Taehan, Byungseok Seo, Jaeho Lee, Seounghyun Park, Kyungmin Kim, and Wonjoon Choi. 2022. "Ultrafast Extreme Thermal–Electrical Fabrication of Volcano-Shape-like Core-Shell Ag-MnxOy Branches Anchored on Carbon as High-Performance Electrochemical Electrodes." *Nano Energy* 91 (September 2021). Elsevier Ltd: 106663. doi:10.1016/j.nanoen.2021.106663.

Yew, Guo Yong, Xuefei Tan, Kit Wayne Chew, Jo Shu Chang, Yang Tao, Ning Jiang, and Pau Loke Show. 2021. "Thermal-Fenton Mechanism with Sonoprocessing for Rapid Non-Catalytic Transesterification of Microalgal to Biofuel Production." *Chemical Engineering Journal* 408 (September 2020). Elsevier B.V.: 127264. doi:10.1016/j.cej.2020.127264.

Yin, Dezheng, Yangfang Deng, Yuzhao Ma, and Xiaomeng Fan. 2021. "Effect of SiC Content on the Mechanical Behaviour of a Three-Dimensional Needled C/SiC Composite." *Ceramics International* 47 (17). Elsevier Ltd: 25067–73. doi:10.1016/j.ceramint.2021.05.237.

Yin, Guang, Yu Wang, Wei Wang, and Dan Yu. 2020. "Multilayer Structured PANI/MXene/CF Fabric for Electromagnetic Interference Shielding Constructed by Layer-by-Layer Strategy." *Colloids and Surfaces A: Physicochemical and Engineering Aspects* 601 (March). Elsevier: 125047. doi:10.1016/j.colsurfa.2020.125047.

Yin, Jinhui, Di Gao, Xiangmiao Zhu, Xiang Liu, and Hexing Li. 2021. "One-Pot Synthesis of 3D Porous Bi7O9I3/N-Doped Graphene Aerogel with Enhanced Photocatalytic Activity for Organic Dye Degradation in Wastewater." *Ceramics International* 47 (14). Elsevier Ltd: 19556–66. doi:10.1016/j.ceramint.2021.03.293.

Yin, Xing, Di Liu, Linglin Zhou, Xinyuan Li, Chunlei Zhang, Ping Cheng, Hengyu Guo, Weixing Song, Jie Wang, and Zhong Lin Wang. 2019. "Structure and Dimension Effects on the Performance of Layered Triboelectric Nanogenerators in Contact-Separation Mode." *ACS Nano* 13 (1): 698–705. doi:10.1021/acsnano.8b07935.

Yin, Yihua, Mengdan Yang, Jing Xi, Weiquan Cai, Ying Yi, Guanghua He, Yue Dai, Ting Zhou, and Mengqing Jiang. 2020. "A Sodium Alginate-Based Nano-Pesticide Delivery System for Enhanced in Vitro Photostability and Insecticidal Efficacy of Phloxine B." *Carbohydrate Polymers* 247 (January). Elsevier: 116677. doi:10.1016/j.carbpol.2020.116677.

Ying, Hao, Tingting Chen, Chenyun Zhang, Jiahui Bi, Zhonghao Li, and Jingcheng Hao. 2021. "Regeneration of Porous Fe_3O_4 Nanosheets from Deep Eutectic Solvent for

High-Performance Electrocatalytic Nitrogen Reduction." *Journal of Colloid and Interface Science* 602. Elsevier Inc.: 64–72. doi:10.1016/j.jcis.2021.05.185.

Yu, Aifang, Yaxing Zhu, Wei Wang, and Junyi Zhai. 2019. "Progress in Triboelectric Materials: Toward High Performance and Widespread Applications." *Advanced Functional Materials* 29 (41): 1–27. doi:10.1002/adfm.201900098.

Yu, Chaojie, Fanglian Yao, and Junjie Li. 2022. "Rational Design of Injectable Conducting Polymer-Based Hydrogels for Tissue Engineering." *Acta Biomaterialia* 139: 4–21. doi:10.1016/j.actbio.2021.04.027.

Yu, Chenglong, Dan Hou, Bichun Huang, Meijuan Lu, Ruosi Peng, and Zhiyong Zhong. 2020. "A MnOx@Eu-CeOx Nanorod Catalyst with Multiple Protective Effects: Strong SO2-Tolerance for Low Temperature DeNOx Processes." *Journal of Hazardous Materials* 399 (February). doi:10.1016/j.jhazmat.2020.123011.

Yu, Chenyang, Jianing An, Qiang Chen, Jinyuan Zhou, Wei Huang, Young Jin Kim, and Gengzhi Sun. 2020. "Recent Advances in Design of Flexible Electrodes for Miniaturized Supercapacitors." *Small Methods* 4 (6): 1–31. doi:10.1002/smtd.201900824.

Yu, Dan, Yongqiang Tan, and Haibin Zhang. 2021. "Oxidation Behaviors of (Ti1-XNbx)C Ceramic Solid Solutions." *Ceramics International* 47 (20). Elsevier Ltd: 28968–75. doi:10.1016/j.ceramint.2021.07.057.

Yu, Kai, Huagui Zhang, Suparit Tangparitkul, Jiatong Jiang, Chris Hodges, and David Harbottle. 2022. "Interfacial Behavior of Core–Shell Composite Nanoparticles under Compression and Shear: Influence of Polymer Shell Thickness." *Journal of Colloid and Interface Science* 613. Elsevier Inc.: 827–35. doi:10.1016/j.jcis.2022.01.069.

Yu, Tao, Xiaotong Chen, Weihua Zhuang, Yuan Tian, Zhen Liang, Qunshou Kong, Cheng Hu, Gaocan Li, and Yunbing Wang. 2021. "Nonglutaraldehyde Treated Porcine Pericardium with Good Biocompatibility, Reduced Calcification and Improved Anti-Coagulation for Bioprosthetic Heart Valve Applications." *Chemical Engineering Journal* 414 (October 2020). Elsevier B.V.: 128900. doi:10.1016/j.cej.2021.128900.

Yu, Yanhao, and Xudong Wang. 2016. "Chemical Modification of Polymer Surfaces for Advanced Triboelectric Nanogenerator Development." *Extreme Mechanics Letters* 9. Elsevier Ltd: 514–30. doi:10.1016/j.eml.2016.02.019.

Zaarour, Bilal, Lei Zhu, Chen Huang, Xiang Yu Jin, Hadeel Alghafari, Jian Fang, and Tong Lin. 2019. "A Review on Piezoelectric Fibers and Nanowires for Energy Harvesting." *Journal of Industrial Textiles* 0 (0): 1–44. doi:10.1177/1528083719870197.

Zeng, Jia, and Yimin Xuan. 2018. "Tunable Full-Spectrum Photo-Thermal Conversion Features of Magnetic-Plasmonic Fe_3O_4/TiN Nanofluid." *Nano Energy* 51 (July). Elsevier Ltd: 754–63. doi:10.1016/j.nanoen.2018.07.034.

Zhan, Yanhu, Jian Wang, Kaiye Zhang, Yuchao Li, Yanyan Meng, Ning Yan, Wenkang Wei, Fubin Peng, and Hesheng Xia. 2018. "Fabrication of a Flexible Electromagnetic Interference Shielding Fe_3O_4@reduced Graphene Oxide/Natural Rubber Composite with Segregated Network." *Chemical Engineering Journal* 344 (October 2017). Elsevier: 184–93. doi:10.1016/j.cej.2018.03.085.

Zhan, Zhaoshun, Li Wang, Ran Duan, Qixing Shi, Jinfeng Shen, Shihan Weng, Xing Wang, Lina Yan, Baoyong Zhang, and Lixin Li. 2022. "A Novel Strategy to Improve the Flame Retardancy and Electrical Conductivity of Polymethyl Methacrylate by Controlling the Configuration of Phosphorus-Containing Polyaniline@needle Coke with Magnetic Field." *Chemical Engineering Journal* 448 (April). Elsevier B.V.: 137642. doi:10.1016/j.cej.2022.137642.

Zhang, Jingwen, Chuchu Cheng, Fang Shu Xing, Cheng Chen, and Caijin Huang. 2021. "0D β-Ni(OH)2 Nanoparticles/1D Mn0.3Cd0.7S Nanorods with Rich S Vacancies for Improved Photocatalytic H2 Production." *Chemical Engineering Journal* 414 (December 2020). Elsevier B.V.: 129157. doi:10.1016/j.cej.2021.129157.

Zhang, Jinze, Zhongyan Sun, Yue Liu, Jiaxi Sui, Bo Zhong, Yingying Wang, and Xiaoxiao Huang. 2021. "Novel Synthesis Method and Microstructure Evolution of Ti_3C_2(OH)2/K2Ti8O17 Nanocomposites as an Effective Surface Enhanced Raman Scattering Substrate." *Ceramics International* 47 (14). Elsevier Ltd: 19864–72. doi:10.1016/j.ceramint.2021.03.325.

Zhang, Lei, Long Jin, Binbin Zhang, Weili Deng, Hong Pan, Junfeng Tang, Minhao Zhu, and Weiqing Yang. 2015. "Multifunctional Triboelectric Nanogenerator Based on Porous Micro-Nickel Foam to Harvest Mechanical Energy." *Nano Energy* 16. Elsevier: 516–23. doi:10.1016/j.nanoen.2015.06.012.

Zhang, Louwen, Meng Zhu, Yuan Sun, Jing Zhang, Mingming Zhang, Hanhong Zhang, Feifan Zhou, Junle Qu, and Jun Song. 2021. "Deep-Red Emissive Colloidal Lead-Based Triiodide Perovskite/Telluride Nanoscale Heterostructures with Reduced Surface Defects and Enhanced Stability for Indoor Lighting Applications." *Nano Energy* 90 (PA). Elsevier Ltd: 106506. doi:10.1016/j.nanoen.2021.106506.

Zhang, Peng, Qizhen Zhu, Razium A. Soomro, Shiyu He, Ning Sun, Ning Qiao, and Bin Xu. 2020. "In Situ Ice Template Approach to Fabricate 3D Flexible MXene Film-Based Electrode for High Performance Supercapacitors." *Advanced Functional Materials* 30 (47): 1–10. doi:10.1002/adfm.202000922.

Zhang, Shuaishuai, Liang Lu, Sheng Wang, Fang Yuan, Shouhu Xuan, and Xinglong Gong. 2021. "Coaxial Direct Ink Writing of Shear Stiffening Gel/Ecoflex Composite for Customized Insoles." *Composites Part B: Engineering* 225 (May). Elsevier Ltd: 109268. doi:10.1016/j.compositesb.2021.109268.

Zhang, Xiang, Jie Cheng, Zhen Xiang, Lei Cai, and Wei Lu. 2022. "A Hierarchical Co @ Mesoporous C/Macroporous C Sheet Composite Derived from Bimetallic MOF and Oroxylum Indicum for Enhanced Microwave Absorption." *Carbon* 187. Elsevier Ltd: 477–87. doi:10.1016/j.carbon.2021.11.044.

Zhang, Xiao Sheng, Mengdi Han, Beomjoon Kim, Jing Fu Bao, Juergen Brugger, and Haixia Zhang. 2018a. "All-in-One Self-Powered Flexible Microsystems Based on Triboelectric Nanogenerators." *Nano Energy* 47 (January). Elsevier Ltd: 410–26. doi:10.1016/j.nanoen.2018.02.046.

Zhang, Xiao Sheng, Mengdi Han, Beomjoon Kim, Jing Fu Bao, Juergen Brugger, and Haixia Zhang. 2018b. "All-in-One Self-Powered Flexible Microsystems Based on Triboelectric Nanogenerators." *Nano Energy* 47 (March). Elsevier Ltd: 410–26. doi:10.1016/j.nanoen.2018.02.046.

Zhang, Xinru, Tao Zhang, Yonghong Wang, Jinping Li, Chengcen Liu, Nanwen Li, and Jiayou Liao. 2018. "Mixed-Matrix Membranes Based on Zn/Ni-ZIF-8-PEBA for High Performance CO2 Separation." *Journal of Membrane Science* 560 (May). Elsevier B.V.: 38–46. doi:10.1016/j.memsci.2018.05.004.

Zhang, Yating, Peng Wang, Juan Yang, Keke Li, Nana Zhang, Guoyang Liu, Yingfeng Duan, and Jieshan Qiu. 2022. "Mechanochemical Coordination Self-Assembly for Cobalt-Based Metal-Organic Framework-Derived Bifunctional Oxygen Electrocatalysts." *Journal of Colloid and Interface Science* 613. Elsevier Inc.: 733–46. doi:10.1016/j.jcis.2021.12.191.

Zhao, Che, Luquan Ren, Zhengyi Song, Linhong Deng, and Qingping Liu. 2019. "Structure-Driven Biomimetic Self-Morphing Composites Fabricated by Multi-Process 3-D Printing." *Composites Part A: Applied Science and Manufacturing* 123 (April). Elsevier: 1–9. doi:10.1016/j.compositesa.2019.04.030.

Zhao, Xue Jiao, Shuang Yang Kuang, Zhong Lin Wang, and Guang Zhu. 2018. "Highly Adaptive Solid-Liquid Interfacing Triboelectric Nanogenerator for Harvesting Diverse Water Wave Energy." *ACS Nano* 12 (5): 4280–85. doi:10.1021/acsnano.7b08716.

Zheng, Li, Yali Wu, Xiangyu Chen, Aifang Yu, Liang Xu, Yongsheng Liu, Hexing Li, and Zhong Lin Wang. 2017. "Self-Powered Electrostatic Actuation Systems for Manipulating the Movement of Both Microfluid and Solid Objects by Using Triboelectric Nanogenerator." *Advanced Functional Materials* 27 (16). doi:10.1002/adfm.201606408.

Zheng, Ruixin, Chaozhu Shu, Zhiqian Hou, Anjun Hu, Peng Hei, Tingshuai Yang, Jiabao Li, Ranxi Liang, and Jianping Long. 2019. "In Situ Fabricating Oxygen Vacancy-Rich TiO2 Nanoparticles via Utilizing Thermodynamically Metastable Ti Atoms on Ti_3C_2Tx MXene Nanosheet Surface to Boost Electrocatalytic Activity for High-Performance Li-O2 Batteries." *ACS Applied Materials and Interfaces* 11 (50): 46696–704. doi:10.1 021/acsami.9b14783.

Zheng, Xiaoguang, Gen Wang, Yanhong Zhao, Liyuan Wu, Yuanjiang Wang, Yanyong Song, Peijun Tian, and Xiaojing Wang. 2021. "Controllable Morphology of Pd-Loaded Potassium Tantalates with High Catalytic Performance for Ethylene Glycol Electrooxidation." *ElectrochimicaActa* 376. ElsevierLtd: 137978. doi:10.1016/j.electacta. 2021.137978.

Zheng, Youbin, Shaochen Ma, Enrico Benassi, Yange Feng, Shiwei Xu, Ning Luo, Ying Liu, Li Cheng, Yong Qin, Miaomiao Yuan, Zuankai Wang, Daoai Wang, and Feng Zhou. 2022. "Surface Engineering and On-Site Charge Neutralization for the Regulation of Contact Electrification." *Nano Energy* 91 (November 2021). Elsevier Ltd: 106687. doi:10.1016/j.nanoen.2021.106687.

Zhi, Dandan, Tian Li, Zhaohui Qi, Jinzhe Li, Yingrui Tian, Wenting Deng, and Fanbin Meng. 2022. "Core-Shell Heterogeneous Graphene-Based Aerogel Microspheres for High-Performance Broadband Microwave Absorption via Resonance Loss and Sequential Attenuation." *Chemical Engineering Journal* 433 (P1). Elsevier B.V.: 134496. doi:10.1016/j.cej.2022.134496.

Zhong, Yaping, Yibiao Zou, Lingyan Liu, Ruohan Li, Fengfeng Xue, and Tao Yi. 2020. "PH-Responsive Ag2S Nanodots Loaded with Heat Shock Protein 70 Inhibitor for Photoacoustic Imaging-Guided Photothermal Cancer Therapy." *Acta Biomaterialia* 115. Elsevier Ltd: 358–70. doi:10.1016/j.actbio.2020.08.007.

Zhou, Changcheng, Liuhua Zhou, Jingyu Liu, Luwei Xu, Zheng Xu, Zaozao Chen, Yuzheng Ge, Feng Zhao, Ran Wu, Xinning Wang, Nan Jiang, Liang Mao, and Ruipeng Jia. 2020. "Kidney Extracellular Matrix Hydrogel Enhances Therapeutic Potential of Adipose-Derived Mesenchymal Stem Cells for Renal Ischemia Reperfusion Injury." *Acta Biomaterialia* 115. Elsevier Ltd: 250–63. doi:10.1016/j.actbio.2020.07.056.

Zhou, Guoqiang, Mei Chun Li, Chaozheng Liu, Qinglin Wu, and Changtong Mei. 2022. "3D Printed Ti_3C_2Tx MXene/Cellulose Nanofiber Architectures for Solid-State Supercapacitors: Ink Rheology, 3D Printability, and Electrochemical Performance." *Advanced Functional Materials* 32 (14): 1–11. doi:10.1002/adfm.202109593.

Zhou, Pengfei, Yanbai Shen, Wei Lu, Sikai Zhao, Tingting Li, Xiangxi Zhong, Baoyu Cui, Dezhou Wei, and Yunhai Zhang. 2020. "Highly Selective NO_2 Chemiresistive Gas Sensor Based on Hierarchical In_2O_3 Microflowers Grown on Clinoptilolite Substrates." *Journal of Alloys and Compounds* 828 (2). Elsevier B.V: 154395. doi:10.1016/j.jallcom.2020.154395.

Zhou, Ruimin, Zhaoyan Jiang, Chen Yang, Jianwei Yu, Jirui Feng, Muhammad Abdullah Adil, Dan Deng, Wenjun Zou, Jianqi Zhang, Kun Lu, Wei Ma, Feng Gao, and Zhixiang Wei. 2019. "All-Small-Molecule Organic Solar Cells with over 14% Efficiency by

Optimizing Hierarchical Morphologies." *Nature Communications* 10 (1). Springer US: 1–9. doi:10.1038/s41467-019-13292-1.

Zhou, Xin, Ashna Rajeev, Arunprabaharan Subramanian, Yang Li, Nicolò Rossetti, Giovanniantonio Natale, Gregory A. Lodygensky, and Fabio Cicoira. 2022. "Self-Healing, Stretchable, and Highly Adhesive Hydrogels for Epidermal Patch Electrodes." *Acta Biomaterialia* 139: 296–306. doi:10.1016/j.actbio.2021.07.069.

Zhu, Laipan, Yan Zhang, Pei Lin, Ying Wang, Leijing Yang, Libo Chen, Longfei Wang, Baodong Chen, and Zhong Lin Wang. 2018. "Piezotronic Effect on Rashba Spin-Orbit Coupling in a ZnO/P3HT Nanowire Array Structure." *ACS Nano* 12 (2): 1811–20. doi: 10.1021/acsnano.7b08618.

Zi, Yunlong, and Zhong Lin Wang. 2017. "Nanogenerators: An Emerging Technology towards Nanoenergy." *APL Materials* 5 (7). doi:10.1063/1.4977208.

Zou, Haiyang, Xiaogan Li, Wenbo Peng, Wenzhuo Wu, Ruomeng Yu, Changsheng Wu, Wenbo Ding, et al. 2017. "Piezo-Phototronic Effect on Selective Electron or Hole Transport through Depletion Region of Vis–NIR Broadband Photodiode." *Advanced Materials* 29 (29): 1–10. doi:10.1002/adma.201701412.

Zou, Yongjiu, Vidhur Raveendran, and Jun Chen. 2020. "Wearable Triboelectric Nanogenerators for Biomechanical Energy Harvesting." *Nano Energy* 77 (November). Elsevier Ltd. doi:10.1016/j.nanoen.2020.105303.

10 Application of 3D Assemblies/Structures of Titanium-Based MXenes in Sensing, Photovoltaics, and Catalysis

10.1 INTRODUCTION

MXenes have garnered significant attention for their diverse range of uses in areas such as catalysts, gas storage medium, ion batteries, and sensors. Their significant surface area can be attributed to their exceptional adsorption capability, the hydrophilicity hydrophile nature, and robust surface reactivity. Solar cells have garnered significant interest from researchers due to their ability to enable various applications such as energy storage, catalysis, and electrochemical sensors (Qin et al. 2021). It is important to look into how the surface functionalization is done and synthetic procedures affect the performance diversity; modularity; and chemical, thermal, and physical properties of MXenes. MXenes are an interesting basis for good biosensors because of their unique electrochemical redox properties. As the lone determining element for excellent selectivity, dynamic ranges, sensitivity, repeatability, and ultrasensitive detection limits of sensing devices, electrode modification is a vital component of the sensor (Qin et al. 2021).

Three-dimensional (3D) materials have great prospects for numerous electronic and energy uses because of their unusual electrical architectures and great surface areas. The creation, characteristics, or potential uses of innovative 3D-materials are now among most fascinating fields of study in science and technology. Successful synthesis of boron nitride (BN), graphene, MoS_2, and phosphorene can be obtained by isolating individual layers from the massive quantities of van der Waals layered frameworks (Wei Zhang, Huang, and Bernstein 2022; J. Long et al. 2021). Combining chemical exfoliation with sonication has been demonstrated recently to allow synthesis and mass production of 2D materials from three-dimensional (3D) layered compounds with chemical bonding between the layers.

In this regard, research has shown that HF concentrations and ultrasound treatment can effectively separate several MAX phase materials into individual layers. These layers, known as 2D MXenes, belong to a family of transition metal carbides and nitrides. A combination of O, F, or the surface of MXenes is terminated by OH groups during exfoliation based on the specific chemical conditions. Not only is HF

DOI: 10.1201/9781003319658-10

not the only etchant, exfoliating the skin has also been accomplished with LiF + HCl and NH_4HF_2 (Cao et al. 2019).

The superior mechanical properties, metallic stability at high temperatures, and the restricted applications of solid MAX phases are their features. However, because of their several compositional decisions, the phases of MAX consist of several crystalline forms and composites components. The phases of MAX, which are rarely seen in their original MAX phases, exhibit considerably different optical, electrical, magnetic, and electrochemical properties when the material undergoes exfoliation, resulting in the formation of 2D MXenes. The specific transition metals present on the surface functioning and the thickness of the material influence this process. Many theoretical investigations reported in the literature predict MXenes to have remarkable electrical characteristics and a broad range of device uses. Most of these forecasts (R. Li et al., 2021) rely on density functional theory (DFT), which has limitations in accurately predicting the chemical and physical properties of MXenes. DFT predictions depend on energy differences between the highest occupied and lowest unoccupied electronic states, as well as intermolecular forces like van der Waals interactions or strongly correlated system. First, we thoroughly address in this chapter the structural aspects of MXene-based photovoltaics, batteries, photothermal process, tissue engineering, drug delivery, and photodynamic treatment.

This chapter can be seen as a supplement to and an update of the previous chapter on the preparation of 3D assemblies/structures of $Ti_3C_2T_x$ and Ti_2CT_x as well as some typical 2D assemblies' structures and architectures of graphene and transition metallic dichalcogenides (TMDCs) and the characteristics and uses of MXenes (Fu et al. 2022; Wei et al. 2021; Jiang et al. 2020). In this chapter, we provide the most recent methods for making MXenes, their characteristics, microstructural layout, production methods, and uses in wearable electronics, energy storage, and miniature electrochemical sensors. Our main goal is to focus on microelectrochemical sensors based on MXene for various critical analytes, redox mediators, electrode modification processes, sensing mechanisms, and analytical parameters. We conclude with a summary of the current state and anticipated future uses of disposable sensors.

10.2 MXENE-BASED HETEROSTRUCTURE SENSORS

Simple and dependable sensing methods can quickly categorize the demand for chemical or biological analytes that can identify and quantify gas molecules, hazardous ions, and macromolecules with high responsiveness, specificity, and stability under different environmental circumstances. These analytes are crucial for many different kinds of practical applications. The recognition approach has several applications in daily life, including the successful detection of diseases, correction of gas leaks, and health care facilities (P. Xu et al. 2018; P. Liu et al. 2018; Bai et al. 2022).

10.2.1 Gas Sensing

It has long been known that tracking human breathing can help with disease diagnosis. Thanks to recent developments in nanotechnology, gas sensors can today identify, forecast, and track various human breath problems. Patients must improve their performance and cut expenditures for care in the early stages of all diseases, from diabetes to cancer. Fast and nonintrusive sensors are therefore a possible alternative to traditional techniques for meeting that need. Gas sensors built specifically to analyze human breath provide tempting alternatives to current conventional diagnostic techniques such as slow and invasive blood tests, enabling a real-time, accurate, and rapid disease diagnosis (Q. Wang et al. 2021).

C. Winston et al. (W. Y. Chen et al. 2020) stated that, recently, there has been a growing interest in the use of two-dimensional transition metallic nitrides and carbides for gas monitoring purposes (Figure 10.1). At the moment, there is limited study on the hybridization of MXenes and two-dimensional transition metallic dichalcogenides for use as gas-sensing substances. Here, we choose WSe_2 and $Ti_3C_2T_x$ which are utilized as exemplary substances for hybridization and application in the identification of various volatile chemical compounds. The $WSe_2/Ti_3C_2T_x$ multifunctional detector offers low noise levels, great adaptability for a broad spectrum of volatile organic chemicals, and very quick reaction and recovery times. Compared to pristine $Ti_3C_2T_x$, the hybrid, mix sensor's sensitivity to ethanol has increased almost 12-fold. Additionally, the hybridization procedure successfully provides a barrier by impeding the relationship between molecules of water at the margins of $Ti_3C_2T_x$; $Ti_3C_2T_x/WSe_2$ heterostructure materials are proposed to gain from a better method for the very sensitive and focused detection of volatile organic compounds including oxygen.

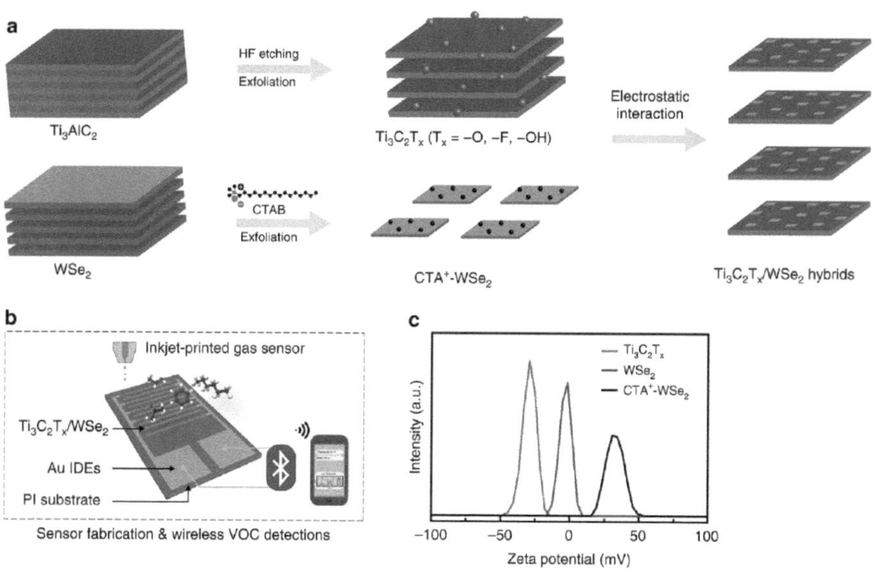

FIGURE 10.1 Fabrication of sensors using hybrid $Ti_3C_2T_x/WSe_2$. (a) Process preparation for Ti_3C_2Tx/WSe_2 nanohybrids is shown schematically. (b) A wireless monitoring system can identify volatile organic chemicals using gas sensors that are inkjet printed. (c) Distributions of the zeta potential for the $Ti_3C_2T_x$, WSe_2, and CTA+-WSe_2 dispersions.

Source: Adopted with permission (W. Y. Chen et al. 2020).

Future research into field-deployable sensors of the next generation may be guided by the scientific findings of this work (B. Fan et al. 2020; Selamneni et al. 2022). MXene gas sensors provide an essential means to ensure the safety of manufacturing, and it is necessary to monitor the concentration of gas and data related to the environment. They are one of the most important tools for detecting hazardous gases. Therefore, investigation into the characteristics of elevated sensitivity, exceptional selectivity, and outstanding consistency has gained popularity.

10.2.2 Humidity Sensors

Many industries, including electronics, food, chemicals, metallurgy, pharmaceuticals, climatology, and cosmetics, depend on humidity sensing. Still, it is interesting for getting an instantaneous response to dynamic humidity variation, much as the human respiratory system does (Q. Lu et al. 2022).

According to A. Hyosung et al. (H. An et al. 2019), the demand for quick humidity sensors is being driven by the fast growing fields of non-contact medical diagnosis, environmental monitoring, and noninvasive epidermal sensing (Figure 10.2). Quick

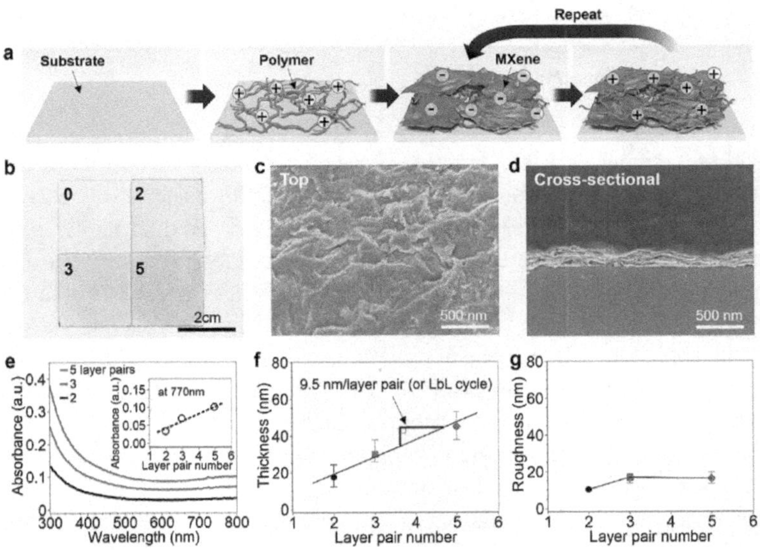

FIGURE 10.2 Characterization of polyelectrolyte and MXene multilayers produced utilizing LbL: (a) A PDAC/MXene combo schematic. (b) A technological design of the MXene/ polyelectrolyte multiple layered films produced by alternately adsorbing PDAC and MXene on glass for 15 minutes with the number of layer pairs varying (0, 3, 4, and 5). (c) From the top down, a cross-sectional SEM photograph. (d) Multiple-layer films of MXene and polyelectrolyte are shown (ten pairings). (e) Ultraviolet (UV) spectrum of absorption for multilayered films made of MXene and polyelectrolytes . The absorbance values at 770 nm are plotted against the number of layer pairings in the inset figure. (f) Relationship between the thickness and the number of MXene/ polyelectrolyte layer pairings. (g) The number of layer pairings, as indicated by the root-mean-square, influences roughness. Utilizing profilometry, thickness and roughness are measured.

Source: Adopted with permission from H. An et al. (2019).

responses to dynamic variations in humidity, like those brought on by respiration, are challenging to come up with. This is because the sensing method, water sorption in the material, and water diffusion all have the potential to limit the response. For ultrafast humidity sensing, they propose multilayer assemblies constructed of MXene nanosheets and polyelectrolytes, along with details on their water sorption and response mechanisms. Ti_3C_2 belongs to the MXene classification of two-dimensional transition metallic carbides, with metal-like conductivity and hydrophilicity. They demonstrate that the MXene/polyelectrolyte multiple layers films, assembled using a layer-by-layer (LbL) technique, exhibit a responsive behavior and recovery times that are faster than those of most humidity sensors. The MXene/polyelectrolyte multilayer's thickness and sheet-to-sheet spacing fluctuate because of water molecules intercalating into (or deintercalation from) it in reaction to humidity variations. This changes the tunneling resistance between MXene sheets. Ellipsometry and quartz crystal microbalance are used to support this mechanism. By using a portable microcontroller for wireless sensing to continuously track human respiration, the rapid response was further illustrated (H. An et al. 2019).

Among the most popular tools for environmental monitoring applications are MXene humidity sensors because they are among the metrics that must be accurate to build habitats that are cozy, secure, energy-efficient, and with high precision.

10.2.3 Pressure Sensors

A pressure detector is a gadget for gauging liquid or gas pressure. Commonly expressed as force per unit area, pressure is the physical term used to characterize the force needed to stop a fluid from expanding. Usually acting as a transducer, a pressure sensor responds to applied pressure by generating a signal. In this chapter, a signal of such nature is considered as mechanical. Pressure sensors find extensive use in many daily applications including monitoring and control. Other indirectly monitored variables including fluid/gas flow, water level, speed, and height can also be tracked using pressure sensors. Along with these designations, pressure sensors are also referred to as pressure transducers, pressure indicators, pressure transmitters, piezometers, pressure senders, and manometers (H. Chen et al. 2022; Zheng et al. 2021).

Variations in technology, design, performance, appropriateness for some uses, and cost of pressure sensors can be somewhat significant.

A conservative estimate would put the number of technologies and pressure sensor manufacturers at above 50 and 300, respectively. A class of pressure sensors that are designed to measure dynamically and record extremely quick changes in pressure is also available. Two applications for this type of sensor are in an engine cylinder or a gas turbine measuring combustion pressure. Usually, these sensors are made from quartz and other piezoelectric minerals. Some pressure sensors turn on or off to act as pressure switches when a given pressure is attained. For example, a pressure switch can be used to turn on a water pump upon system discharge of water, lowering reservoir pressure (S. He et al. 2020; H. Shi et al. 2019).

S. Yue et al. (Su et al. 2022) stated their extraordinary capacity to conformally attach to the outermost layer of the skin and translate human actions into recordable electrical impulses have attracted a lot of interest for wearable and flexible pressure sensors (Figure 10.3). As a result, more and more research is being done to create

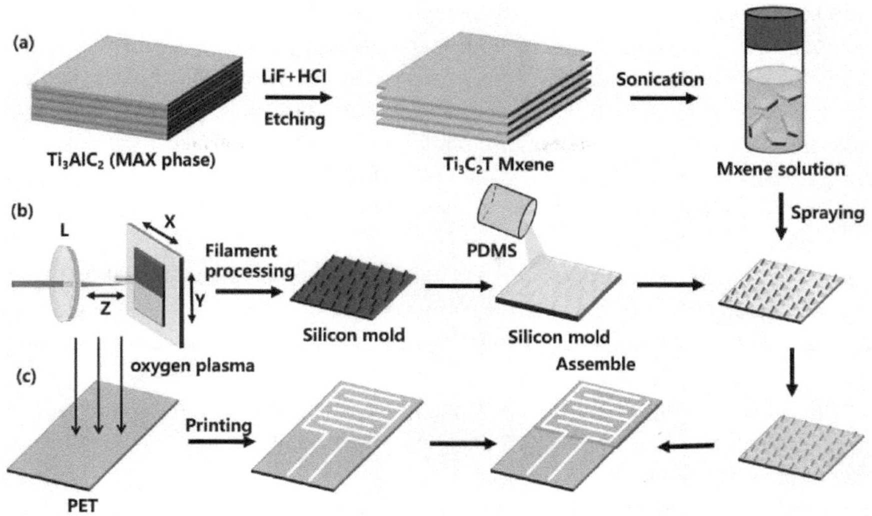

FIGURE 10.3 (a) The process of making $Ti_3C_2T_x$ MXene. (b) Creating a conductive film with a micro-pattern in MXene/PDMS. (c) The piezoresistive sensor based on MXene is put together.

Source: Adopted with permission from Su et al. (2022).

inexpensive, flexible, and highly sensitive detectors that can track a person's activity level. Here, two-dimensional transition metal carbides, nitrides, and carbonitrides, often referred to as MXenes, possess a hexagonal honeycomb-like structure. These materials, when produced by femtosecond filamentation pulses, result in high-performance flexible piezoresistive sensors. It is believed that the sensing mechanism is represented by the varying electrical routes over the upper interdigital terminals and the lower nanostructures sheets covered with MXene. The acquired sensor exhibits outstanding reliability and stability, a relatively quick reaction time of 0.61 kPa^{-1}., and great sensitivity. The sensor can successfully monitor minute physiological signals, identify finger gestures, and, thanks to the remarkable sensor performance described previously, find significant deformations during human movement. This shows the extensive spectrum of possible uses for the sensor in systems of physiological analysis, health monitoring devices, and human–machine interfaces (Gao et al. 2022).

Pressure sensors made of MXene are employed in countless common applications for monitoring and control. Additionally, they can be used to infer indirectly the values of other parameters including height, water level, fluid/gas movement, and speed.

10.3 MXENE-BASED HETEROSTRUCTURES FOR PHOTOVOLTAICS

10.3.1 Photodetectors

Electromagnetic radiation or light sensors are photodetectors, also called photosensors. Photodetectors come in a wide variety of forms, and they can be categorized on the basis of their modalities of detection and based on other performance metrics, such as spectrum response or other elements like photoelectric or photochemical reactions. Often present in semiconductor-based photodetectors is a p-n junction, which is a device that transforms light particles into electricity. Within the depleted area, the incident photons generate pairs of electrons and holes. Two examples of the numerous types of photodetectors are photodiodes and phototransistors. Solar cells turn a percentage of the light energy into electrical energy.

A fraction of the light energy is converted into electrical energy by solar cells (K. Liu, Sakurai, and Aono 2010).

It is possible to categorize objects based on how photodetectors detect them:

The photoelectric effect, also known as photoemission, occurs when the electrons in a material's conduction band transform into free electrons in a gas or vacuum or a gas.

Thermally, photons cause the electrons to transition into the mid-gap states and the subsequent decay of those states back into lower bands, which results in the creation of phonons and heat.

Photodetectors can be used in many ways. The total amount of light may be measured with a single sensor. It is feasible to measure and quantify the amount of light spread down a line by using a one-dimensional array of photodetectors, such as those included in spectrophotometers or line scanners. It is possible to utilize a (2D) array of photodetectors as an image sensor to record, recording images from the light pattern in front of it. Behind a lighting window that occasionally has an antireflective coating is frequently a photodetector or array. (M. Long et al. 2019; K. Liu, Sakurai, and Aono 2010).

According to researcher X. Zhang, the performance of the 3C-SiC/ZnO photodetector was greatly improved by tuning the separation between the movement and transfer of light-induced electric charges at the boundary through the piezo-phototropic effect (Figure 10.4) (X. Zhang et al. 2020). The researchers have demonstrated that the piezoelectric polarization charges produced at the interface between the materials may be effectively used to modify the energy band component and height of barriers of the 3C-SiC/ZnO heterostructure.

An ultrahigh photo-to-responsivity of 1.01×10^6 A/W and a specific detectivity of 1.41×10^{14} Jones were attained under a compressive stress of −0.381%, reflecting improvements of 117 and 84%, respectively. With a 75.9% improvement, a rapid response time of 24.1 ms was achieved, resulting in a significant improvement of 75.9%. The piezo-phototropic effect in semiconductor/semiconductor heterostructures refers to the phenomenon where the optical properties of these structures are influenced by mechanical strain, but it also offers a useful technique for enhancing the effect-based photoresponsive behavior in heterostructure photodetectors (Yan et al. 2020; X. Zhang et al. 2020).

Modern systems and applications need MXene-based photodetectors as a key component as they are the foundation of image sensors, biomedical imaging, spectrometry, ambient sensors in consumer electronics, and monitoring of food and manufacturing processes, to mention a few. MXene is particularly promising for future sensing and detection systems due to its broad variety of uses.

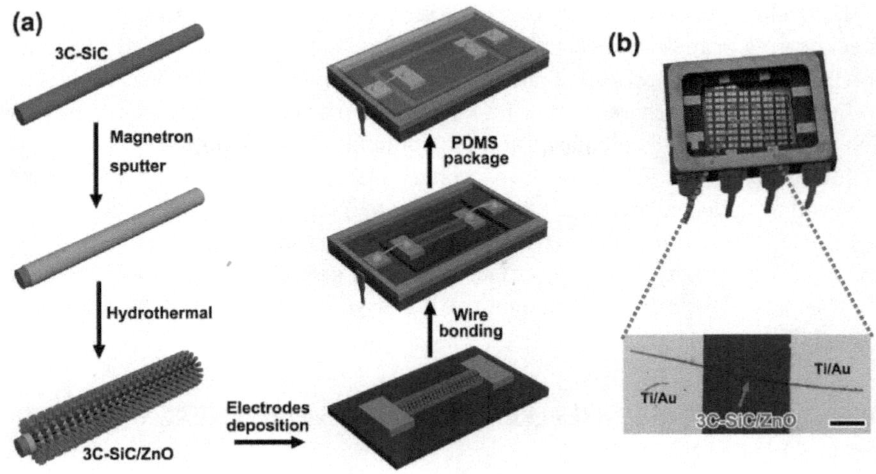

FIGURE 10.4 The development of a photodetector using a 3C-SiC/ZnO heterostructure: (a) Design for a photodetector's production and (b) A visual depiction of the finished device, mounted on a measuring cell, is shown.

Source: Adopted with permission from X. Zhang et al. (2020).

10.3.2 Solar Cell

A solar cell, commonly referred to as a photovoltaic cell, is a technical advancement that involves the direct conversion of light energy into electricity through the physical and chemical phenomenon known as the photovoltaic effect. An example of a device that changes its electrical characteristics in reaction to light is a photoelectric cell. Informally known as solar panels, photovoltaic modules are usually electrically built from separate solar cell components. The standard open-circuit voltage of a typical single-junction silicon solar power cell may generate an optimal open-circuit voltage ranging from 0.5 to 0.6 volts.

Photovoltaic cells are called solar cells, regardless of the fact that the light supply is natural or synthetic. They can produce energy, photo-detect visible light or electromagnetic radiation that falls within the visual spectrum, or measure range of wavelengths that are detectable by the human eye. They are commonly used to assess the intensity of light. Examples of such photodetectors are infrared (Grishina et al. 2017; Yun and Lee 2017).

For a photovoltaic (PV) cell to operate, it needs to possess the following three key qualities:

- the process of light absorption that yields plasmons and bound or unbound excitons (electron–hole pairs with holes in them).
- the categorization of charge carriers into different groups.
- One by one, the carriers are removed and added to an external circuit.

A photovoltaic absorber, in contrast, absorbs heat from the sun to produce either direct heating or indirect electricity. In contrast to a photovoltaic cell, a "photoelectrolytic cell" (also known as a photoelectrochemical cell) is a device that utilizes sunlight to directly separate water into hydrogen and oxygen. This technology was pioneered by Edmond Becquerel and is commonly used in conjunction with dye-sensitized solar cells. There are two primary methods for producing solar power: Photovoltaic cells and solar collectors (Garfi et al. 2022; Z. Wu et al. 2022).

MXenes have intriguing electrical, chemical, and structural characteristics. High electrical conductivity, mobility, surface functional groups, and adjustable chemistry are just a few of their exceptional qualities that have attracted a lot of interest in solar cell applications.

10.4 MXENE-BASED HETEROSTRUCTURES FOR ENERGY STORAGE DEVICES

Energy storage refers to the act of storing energy generated at a specific time for future usage in order to address the disparity between energy production and consumption. An accumulator or battery is a device that is designed to store energy. Kinetic, electrical, latent heat, chemical, gravitational potential, and radiation are among the several sources of energy. Energy storage refers to the process of transforming energy from complex or challenging-to-store forms into simpler or more convenient forms for storage.

Some systems only provide temporary energy storage, while others are far more reliable. At present, conventional and pumping hydraulic dams are the primary methods used for large-scale storage of energy. In the subject of bulk energy storage, conventional and pumped hydroelectric dams are currently dominant. System energy storage is the collective term for a number of methods for implementing storing energy on a wide scale inside an electrical power infrastructure (Sena Da Fonseca et al. 2014; Alami 2020).

Examples of energy storage commonly used include rechargeable batteries that hold chemical power that can be easily converted into electrical power for generating portable electronics. Ice storage containers are also used, which preserve ice that is formed using less expensive energy during the night to fulfill the high demand for cooling during the daylight hours. Hydroelectric power plants are another example, as they store energy in the form of gravity potential energy in storage tanks. Additionally, ice retention reservoirs are utilized for preserving ice formed using less expensive power during the nighttime to comply with the high cooling processes demand during the day. Rechargeable power sources are also employed for conserving chemical energy that can be readily converted into electricity for powering mobile equipment. (The ice is stored at extremely low energy levels and used for cooling during daylight hours.) Examples of fossil fuels include coal and gasoline, which contain energy that was conserved in the past; they are made up of energy previously derived from sunlight by organisms through the photosynthesis of creatures that subsequently perished, later died, were buried, and then through time were changed into these fuels. The similar approach that produces energies from fossils and organic source is also used to manufacture food, which is a sort of chemically stored energy (Pareek, Singh, and Rao 2018; W. Li et al. 2021).

There is an opportunity to reduce the usage of copper current collectors by exploring the use of heterogeneous sodium-ion preservation exhibits' potential for use in batteries and metal-ion capacitors that are available electrode materials which possess the ability to accommodate more sodium ions while maintaining a high charging rate and high volumetric capacity. This chapter explores the manufacturing procedure of permeable Ti_3C_2 MXene/CNT hybrid sheet membranes used in sodium-based devices for storing energy.

The heterogeneous structures emerged through an electrostatic interaction between positively charged one-dimensional carbon nanotubes (CNTs) and negatively charged two-dimensional MXene nanosheets ($Ti_3C_2T_x$). This technique successfully prevented the restacking of MXene nanosheets and established a distinct porosity structure. This resulted in the creation of the electrodes which are composed of MXene, an operational material, for storing ions of sodium, as it allowed ions to reach the electrode and facilitated electrolyte transport. The Ti_3C_2 MXene/CNTs porous films, designed specifically for sodium-ion storage, functioned as independent electrodes. These films exhibited a remarkable volumetric capacity of 421 mA

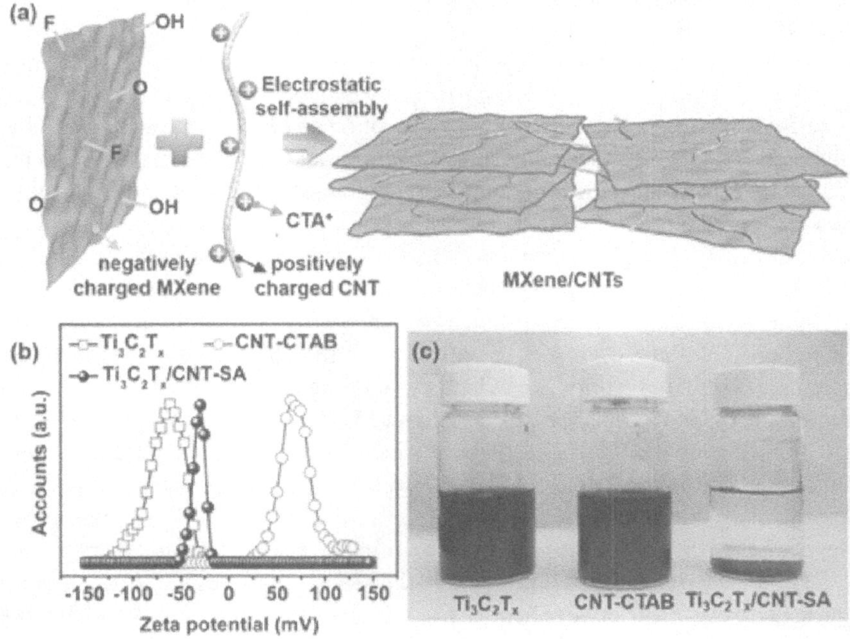

FIGURE 10.5 (a) The schematic and illustration depict the self-assembly mechanism of the porous MXene/CNT electrode. MXene nanosheets possess a negative charge as a result of their functional groups, while CTAB-grafted CNTs have a positive charge. (b) The zeta potential of $Ti_3C_2T_x$ nanosheets, carbon nanotubes (CNTs) with cetyltrimethylammonium bromide (CTAB) grafts, and self-assembled $Ti_3C_2T_x$/CNT-CTAB composites. (c) Digital pictures of $Ti_3C_2T_x$ composites, carbon nanotube (CNT-) coated with cetyltrimethylammonium bromide (CTAB) in aqueous solution, and $Ti_3C_2T_x$ suspension.

Source: Adopted with permission from X. Xie et al. (2016).

h cm^3 at a current density of 20 mAg^{-1}. Additionally, they revealed exceptional rate capabilities and excellent cycling stability (Figure 10.5).

In addition, sodium-ion cells were assembled utilizing the $Na_{0.44}MnO_2$ cathode and Ti_3C_2/CNTs anode in their initial condition. The creation of permeable MXene or CNT mixtures through straightforward self-assembly covers the way for the creation of materials consisting of electrodes with large sodium-ion storage volumetric capacities (Zong et al. 2020; Xiao et al. 2022).

The performance of supercapacitors has been enhanced by the implementation of a pseudocapacitive charge storage mechanism that exhibits electric double-layer behavior. In addition, the appropriate interlayer spacing and unique chemistry of MXene have facilitated batteries in achieving a large capacity, also allowing for rapid charge and discharge. Consequently, MXene shows significant potential for utilization in energy generation and uses for energy storage.

10.4.1 Li-Ion Batteries

A battery that uses lithium ion is a portable power source made up of electrodes, enables the transportation of the ions of lithium from the cathode to the anode via an electrical fluid throughout discharging, and reverses this procedure while recharging. Graphite is commonly employed as the cathode in lithium-ion batteries, while an intercalated lithium compound functions as the anode. Batteries made with lithium ion possess a notable density of energy, lack any memory operation, and have a minimal self-destruction percentage, with the exception of LFP batteries. It is feasible to design cells that place a focus on energy or power density. However, they also present a safety risk because they contain flammable electrolytes and can catch fire or explode if broken.

M. Stanley Whittingham developed the first rechargeable lithium-ion battery in the 1970s, which utilized a lithium-aluminum anode and a titanium disulfide cathode. John Goodenough made advancements in 1980 by using lithium cobalt oxide as a cathode John and improved upon this work in 1980. In 1985, Akira Yoshino created a prototype of a lithium-ion battery, building upon the earlier research conducted by John Goodenough, Rachid Yazami, M. Stanley Whittingham, and Koichi Mizushima during the 1970s and 1980s (Jin et al. 2016; T. Liu et al. 2022). In 1991, a team consisting of Sony and Asahi Kasei, led by Yoshio Nishi, successfully created a commercial lithium-ion battery (Jin et al. 2016; T. Liu et al. 2022). Lithium-ion batteries are being increasingly used in a burgeoning array of aeronautical and military application. Electric vehicles and portable devices regularly use them. Lithium-ion batteries come in a variety of chemistries, with varying performance, price points, and safety features. Handheld devices typically utilize lithium composite capacitors, which consist of a polymer gel serving as the electrolyte, the cathode element consists of lithium cobalt oxide ($LiCoO_2$), whereas the anode material is made of graphite. Lithium iron phosphate ($LiFePO_4$), lithium manganese oxide ($LiMn_2O_4$) spinel, lithium manganese-rich nickel cobalt oxide (LMR-NMC), and lithium nickel manganese cobalt oxide are prominent instances of lithium-based polymers. with longer lives and the possibility for greater rates ($LiNiMnCoO_2$). These batteries are frequently used in electric tools, medical devices, and other applications.

NMC and its related compounds are extensively utilized in the process of implementing electric power of transportation, which is a crucial technology, along with renewable energy, for mitigating automobiles' contribution to the release of greenhouse gases. Improperly recycled batteries can catch fire and produce dangerous waste, particularly when they include poisonous metals. Additionally, lithium mining presents substantial challenges due to its water-intensive nature in frequently dry places and the fact that other important minerals needed in batteries, like cobalt, are frequently conflict minerals. Researchers have been prompted by both environmental concerns to increase the effectiveness of minerals and find substitutes like iron-air batteries (Nitta et al. 2015; Y. Xie et al. 2014).

Lithium-ion battery research focuses on, among other things, cutting costs, enhancing safety, boosting energy density, extending lifetime, and quickening charging. Research for nonflammable electrolytes has been underway to improve safety because the organic solvents employed in the typical electrolyte are flammable and volatile. Aqueous lithium-ion batteries, polymer electrolytes, ionic liquids, ceramic solid electrolytes, and systems with high fluorine content are a few strategies (Chang et al. 2010; Karuppasamy et al. 2016).

The production of lithium-ion (Li-ion) batteries has experienced a substantial growth since their debut to the market in 1991 due to their exceptional performance, which may be attributed to their superior energy density, specific energy, efficiency, specific power, and extended lifespan (Horiba 2014). Originally, lithium-ion batteries found their first applications in consumer gadgets such as camcorders, mobile phones, and laptops. Subsequently, these batteries found utility in automotive applications, which have just emerged and are constantly growing, and ultimately in industrial applications such as energy storage. When considering energy storage applications, four cell chemistries are being regarded as promising options. The $LiFePO_4$/graphite cell chemistry has excellent safety characteristics; the $LiNi1XY_2CoXAlYO_4$/graphite cell chemistry offers a high specific energy, while the $LiMn_2O_4$/graphite cell chemistry utilizes naturally occurring, inexpensive, and abundant materials. Additionally, $LiNi1XY_2CoXAlYO_4$ is also used in this context. However, every cell chemistry has drawbacks, and the study of these technologies is still in its early stages. Thus, it is now premature to ascertain which cell chemistry will emerge as the leading contender for energy storage purposes. Prior to the determination of this course of action, a rigorous and sustained monitoring of scientific advancements, coupled with a meticulous evaluation of evolving political, economic, and social landscapes, is indispensable (Nitta et al. 2015; B. A. Lu et al. 2017).

Recent studies have demonstrated that MXenes are suitable for use as anode materials for Li-ion batteries (LIBs) because of their distinctive qualities, including good electrical conductivity, layered structure, and hydrophilicity. These make them interesting for applications involving energy production and storage.

10.4.2 Li-S Batteries

Lithium-sulfur (Li-S) electrodes are a variant of batteries that are rechargeable. Lithium-sulfur (Li-S) power sources belong to a distinct group of batteries that can be recharged. It stands out because of its high specific energy/. Li-S batteries are comparatively lightweight because lithium possesses an extremely low atomic weight, while sulfur exhibits an intermediate atomic weight (concerning water's, especially when compared to the density). In August 2008, Zephyr 6 utilized them in August 2008 to accomplish the lengthiest and highest altitude flight ever achieved by an unmanned solar-powered aircraft (at that point in time). Lithium-sulfur batteries may someday have the potential to replace lithium-ion cells due to their high demand in the future as they offer a greater energy density and lower cost. This is because sulfur is used in lithium-ion batteries instead of cobalt, which is more typical. Batteries made from lithium ion typically have particular strengths ranging from 150 to 260 Wh/kg. The energies of Li-S batteries are in the range starting from an effective 550 Wh/kg. The specific energies of Li-S batteries, ranging from 550 Wh/kg, surpass those of lithium-ion batteries, which typically fall between 150 and 260 Wh/kg. In 2017, Li-S batteries were successfully validated with up to a maximum of 1,500 charge and discharge cycles; however, there have not been any experiments testing the lifespan of commercial cycles utilizing electrolytes with low concentrations. None were available for sale in early 2021 (Noorpour and Tarighat 2021; Y. An et al. 2021).

One issue that has prevented acceptance is the polysulfide "shuttle" effect, which results in too few recharge cycles and a gradual leakage and slow release of material that is actively participating in a chemical reaction at the cathode. To fully make use of the capacity-enhancing impact of active mass, additional mass for a conducting agent is required due to the low conductivity of sulfur cathodes. Additional issues arise from the significant quantity of electrolyte needed and the increase of the sulfur cathode's volume when converting S to Li_2S.

A new variation on the rechargeable lithium flow batteries, the lithium-sulfur flow battery (LSFB), utilizes a cathode composed of sulfur or a sulfur compound in place of a lithium anode. In this study, G. Li et al. investigate the use of dimethyl trisulfide (DMTS), an organic sulfide, as a catholyte and multi-electron redox mediator in a solid sulfur cathode. The researchers demonstrate that DMTS can facilitate the discharge and charge processes without the need for conductive additives. The application of DMTS in a lithium-iron battery (LFB) is seen in Figure 10.6. The second configuration is alternatively referred to as a lithium-sulfur flow battery with redox targeting (RTLSFB). The initially available discharging capability of the LFB is 131.5 milliampere-hours per gram of dry matter of the active material (1.66 ampere-hours per liter), which decreases to 59 milliampere-hours per gram of dry matter of the active material after 40 cycles (0.75 ampere-hours per liter). The Real-Time Location System for Sulfur Batteries (RTLSFB) has significantly improved application performance. The preliminary discharging capability is demonstrated by 1225.3 mAh $gsulfur^{-1}$ (3.83 A h L^{-1}) and maintains a remaining capacity of 1030.9 mAh $gsulfur^{-1}$ (3.23 A h L1) after 40 cycles. The LFB's discharge and charge lifetime dramatically increased after sulfur was added, demonstrating that DMTS functions more efficiently in an RTLSFB as a mediator of redox reactions instead of as a catholyte in an LFB (D. Wang, Wu, and Gao 2021; S. Xu et al. 2018).

Lithium-sulfur (Li-S) devices with rechargeable capabilities have attracted considerable scientific attention due to their remarkable capacity and high density of energy. Their practical application in industry is currently limited due to various

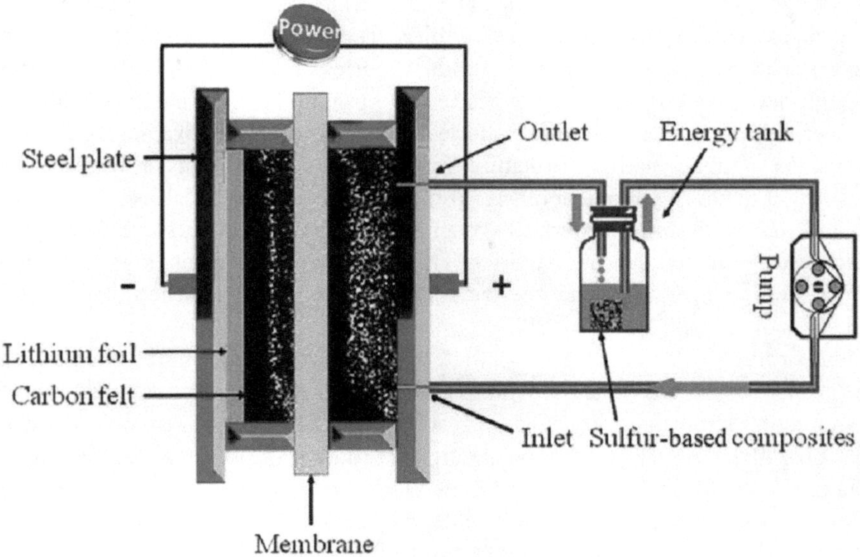

FIGURE 10.6 Schematic of the experimental RTLSFB.

Source: Adopted with permission from G. Li et al. (2018).

challenges, including the shuttle effect generated by soluble lithium sulfide species, the insulation capabilities of sulfur, and the quick decline in electrode capacity. Different approaches are being used to address these problems, including the identification of hosts with high conductivity and a large number of sites for polysulfide chemisorption, as well as modifying separators to impede physical as well as chemical transportation of polysulfides. MXenes, which are two-dimensional transition metal carbides, carbonitrides, and nitrides, possess strong conductivity, layered structure, and numerous surface terminations, making them excellent for containing the shuttling effects of polysulfides. As a result, extensive research has been conducted regarding the utilization of MXenes in Li-S power sources, specifically examining their application in separators, interfaces, and as conductive hosts for sulfur. These characteristics make MXene a very attractive material for energy production and storage applications.

10.4.3 Na-Ion Batteries

A sodium-ion battery (NIB or SIB) is a type of rechargeable battery that uses sodium ions (Na+) as the carriers of charge, in contrast to batteries made with lithium ions. Despite the substitution of lithium molecules, sodium compounds are utilized and the operating principle and cell design are quite like those of lithium-ion battery types that are readily available in the marketplace (L. Zhang et al. 2021; Zhou et al. 2021).

Sodium-ion batteries have garnered considerable academic and commercial attention as a potential supplementary technology to lithium-ion batteries during the

2010s and 2020s. This interest stems primarily from the notable environmental consequences, uneven geographical availability, and elevated expenses associated with several elements required for sodium-ion batteries. The most significant of them, though not necessary for many types of sodium-ion batteries, are cobalt, lithium, copper, and nickel. Sodium-ion batteries offer a significant advantage due to the abundant supply of sodium in nature. In comparison to lithium-ion batteries, this may result in cheaper manufacturing costs for sodium-ion batteries used in industry. As of 2020, sodium-ion batteries have a negligibly small market share. None of the electric cars employ salt ion batteries. The adoption of this technology is hindered by its low energy density and a restricted number of charge–discharge cycles. (Wu Zhang et al. 2021; C. He et al. 2021).

According to D. Hamani et al. (2011), the two NaxVO$_2$ polymorphs demonstrate the ability to reversibly interact with 0.5 sodium ions per unit formula. Additionally, it maintains more than 50% of its capacity centered about 1.6 to 1.7 volts while retaining its initial prismatic or octahedral sites during the entire process of sodium intake (Figure 10.7). Moreover, these electrodes exhibit exceptional capacity preservation despite exhibiting unpredictable voltage patterns. However, to completely understand and the examine the variation in voltage profile during cycling, it is necessary to conduct a comprehensive analysis of the physical properties of these NaxVO$_2$ phases, along with performing band structure calculations. These findings provide additional support for considering V or Ti-based compounds while searching for sodium insertion compounds with low voltage.

Due to the outstanding physicochemical properties of MXenes, one of the most auspicious or promising materials used as electrodes in devices that operate on sodium-ion technology is MXene-based material, sodium-ion batteries (SIBs),

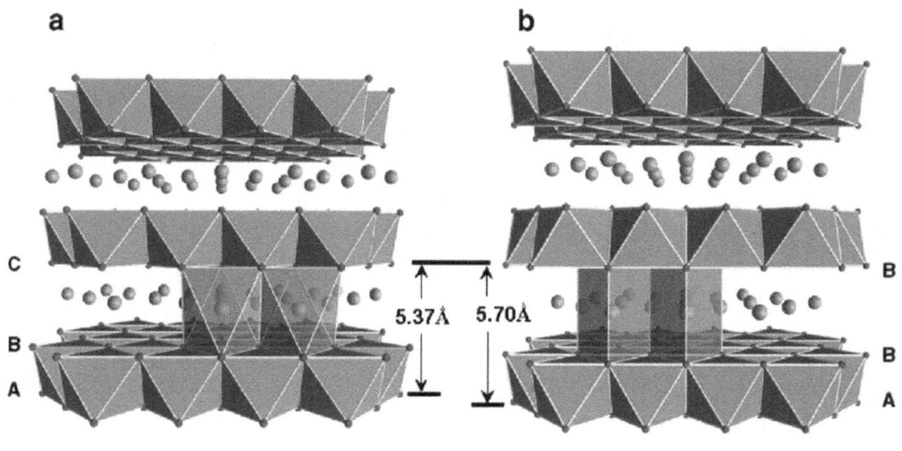

FIGURE 10.7 The structures of $NaVO_2(O_3)$ and $Na_{0.7}VO_2(P_2)$. The stacking sequences consist of layers with a negative charge (small grey scale dot on single layer), represented in grey. The surrounds of the Na ions are depicted in grey (spherical sphere between two layers).

Source: Adopted with permission from Hamani et al. (2011).

including sodium-sulfur batteries (SSBs) and sodium-ion capacitors (SICs). MXene is particularly promising for energy generating and storage applications because of these characteristics.

10.5 DRUG DELIVERY AND CANCER THERAPY (PHARMACEUTICAL TRANSPORTATION AND TREATMENT OF CANCER)

A rapidly and swiftly expanding field that now uses nanotube technology is drug delivery. Currently, drug delivery systems use polymers, dendrimers, and while liposomes are commonly used, carbon nanotubes provide the advantage of being effective structures with high drug-loading capacities and excellent cell penetration capabilities. The nanotubes can be swiftly absorbed by cells and have high aspect ratios for several functionalization attachments. They also offer a larger interior volume for use as a drug container. Carbon nanotubes can be fabricated with or without end caps due to their tubular nature, making the interior—where the medication is stored—more accessible. The half-life, clumping, and lack of solubility of carbon nanotube drug delivery devices are currently problems. In the meantime, solutions to all these issues are currently being sought to further the study the carbon nanotubes. The advantages of carbon nanotubes are still being used in the form of nano vectors for medication administration, where the effective demonstration of cell absorption of these structures resulted in audible impacts, indicating that the nanotubes can be less harmful as nano vehicles for pharmaceuticals. Encapsulating medications have also been shown to increase bioavailability, increase water dispersibility, and decrease toxicity. Encapsulation of molecules serves as a method of material storage in addition to providing security and controlled release for molecules that have been loaded. All of them result in a strong basis for drug delivery that can be further strengthened via study and comprehension. Greater water solubility, sustained half-life, reduced toxicity, and enhanced cellular penetration and uptake are currently unique but unproven concepts (X. Fan et al. 2020; Aissou, Braidy, and Veilleux 2022).

Cancer therapy is the term used to describe the use of one or more cytotoxic anti-neoplastic medications, sometimes known as chemotherapeutic agents, which are used as part of an approved treatment regimen for cancer. This phrase refers to a broad spectrum of the drugs, including those in broad categories like alkylating agents and antimetabolites. Most cancer cells can divide rapidly, which is how conventional chemotherapy agents function. It has been established that combination therapy, which administers several cytotoxic drugs, is superior to a single agent and is advantageous for the treatment of the illness as well as the statistics of survival and tumor response. Combination therapy is more effective at treating breast cancer that has spread, according to a Cochrane analysis. When toxicity and survival are considered, it is typically unknown if combination cancer therapy improves health outcomes (Z. L. Wang 2013; Zeng et al. 2022).

A type of cancer treatment known as targeted therapy focuses on certain biochemical distinctions between cancer and healthy cells. The development and progression of breast cancer were impeded by the initial targeted medications that inhibited the estrogen receptor molecule. An additional prominent example of

Bcr-Abl inhibitors is their application in the treatment of chronic myelogenous leukemia (CML). Targeted therapies are currently available for numerous highly prevalent cancer forms such as thyroid, bladder, breast, colorectal, kidney, leukemia, liver, lung, lymphoma, pancreatic cancer, and other types of cancer.

Cancer treatment effectiveness is influenced by the type and stage of the disease. When combined with surgery, chemotherapy for cancer has demonstrated favorable outcomes in the treatment of pancreatic cancer, breast cancer, osteogenic sarcoma, ovarian cancer, colorectal cancer, testicular cancer, and some different kinds of pulmonary malignancies. For some cancers, such as some leukemias, cancer therapy is curative; however, for most non-melanoma skin malignancies, it is either ineffective or unneeded.

The toxicity of cancer therapy to other body tissues frequently limits its efficacy. Even if cancer therapy does not offer a lasting cure, it may still be helpful to lessen symptoms like discomfort or shrink an untreatable tumor in the hopes that surgery will one day be doable (Vladimirova et al. 2020; Zhong et al. 2021; Anik et al. 2021).

According to B. Bahrami et al. (2017) the main antitumor therapeutic techniques include surgery, radiation, chemotherapy, and hormone therapy. However, a significant number of patients have found these techniques ineffectual because of the nonspecific targeting of cancer cells (Figure 10.8). The use of greater therapeutic doses to reach the tumor location is also essential due to the indiscriminate targeting of cancerous cells. Therefore, in order to efficiently access the site of the tumor, there are two primary challenges. There are two main approaches to treating cancer. The first is to prevent drugs from reaching healthy cells that are free of cancer. The second approach involves directly delivering medications to the tumor itself. The newly discovered method of delivering medications directly to cancer cells while minimizing drug diffusion into healthy cells utilizes nanoparticles (NPs). Utilizing NPs conjugated with ligands that selectively bind to cancer biomarkers linked with cancer is an effective therapeutic approach for treating cancer disorders. Studies have shown that the combination of nanocarriers with substances such as antibodies, nucleic aptamers, vitamins, peptides, and carbohydrates can effectively and accurately deliver medications to cancer cells, leading to a decrease in cancer. This research aims to assess the efficacy of various targeting strategies utilized by NPs in the transportation of drugs to malignant cells (Ren et al. 2022; Xin et al. 2016).

MXenes have been viewed as prospective options for cancer therapy and theragnostic treatment due to their abundance of desirable physicochemical properties, including strong electrical conductivity, particular large surface area, low toxicity, magnetism, luminescence, and excellent biocompatibility. These two-dimensional (2D) nanostructures possess the capacity to be highly efficient and noninvasive anticancer therapies because of their associated chemotherapeutic, synergistic, photothermal, and photodynamic properties. Their research has focused on investigating their targeted delivery of medication for treating cancer, as well as exploring the use of photothermal and chemo-photothermal therapy (PTT). MXenes have been utilized in biological sensing and bioimaging research due to their remarkable optical properties. Additionally, their high light-to-heat conversion efficiency makes them an ideal biocompatible and highly efficient nanoscale agent for photothermal therapy (PTT) devices. Several key challenges related to stability in physiological settings, controlled release of medicines, and biological degradation still require resolution.

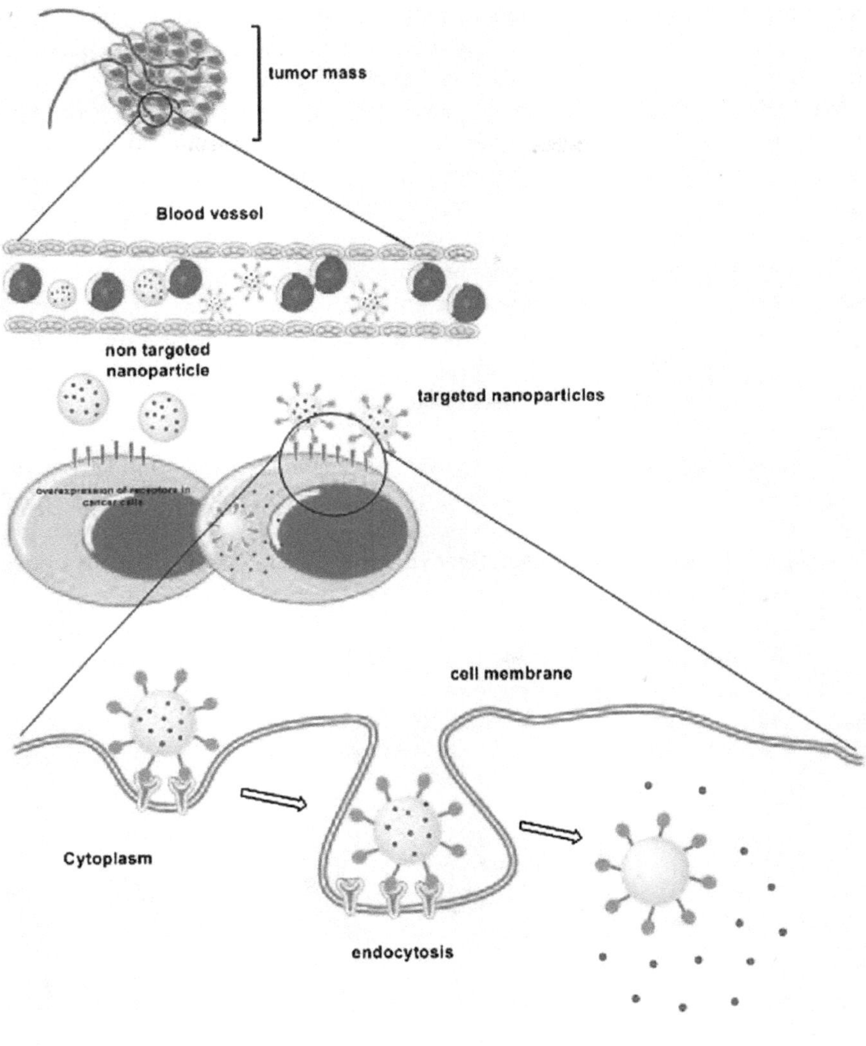

FIGURE 10.8 Active targeting enhances the concentration of nanoparticles at the tumor site. Active targeted nanoparticles, which are conjugated with ligands specific to tumor markers, are more efficient in reaching the tumor site compared to non-targeted nanoparticles. Through the endocytosis process, tumor cells take up active targeted NPs.

Source: Adopted with permission from Bahrami et al. (2017).

10.5.1 Tissue Engineering and Drug Delivery

Tissue engineering, a subdivision of biomedical engineering, utilizes engineering principles, cells, materials science, and relevant biochemical and physicochemical components to enhance, substitute, or mend many biological tissue categories. Tissue engineering encompasses more than just the use of cells and tissue scaffolds. It often entails the placement of cells on tissue scaffolds to generate fresh living tissue for therapeutic objectives. Initially regarded as a subset of biomaterials, this discipline has expanded in importance and breadth and is now recognized as a separate and independent domain (Cardoso et al. 2018; Tan and Thomas 2016).

While the term "tissue engineering" has a broad spectrum of uses, it is primarily used to refer to operations that involve the repair or replacement of entire tissues or specific portions of tissues such as bone, cartilage, bladder, muscle, blood vessels, and skin. In order for the tissues to function effectively, specific mechanical and structural characteristics are necessary. There have been efforts to carry out certain biochemical activities employing cells within a synthetic support system, which have also been referred to as a bioartificial liver or an artificial pancreas, for example. While the phrases "regenerative medicine" and "tissue engineering" are sometimes used interchangeably, practitioners of regeneration medicine prioritize the use of stem cells or progenitor cell lines over tissue transplantation (Kiyotake, Martin, and Detamore 2022).

Elastin-like peptides (ELPs) offer a versatile foundation for drug delivery and tissue engineering, according to B. Bulutoglu et al. (Bulutoglu et al. 2022a). Here, three therapeutic compounds to be employed in wound healing—stromal cell-derived growth factor 1 (SDF1), keratinocyte growth factor (KGF), and cathelicidin (LL37)—are genetically connected to micelle-forming ELP chains (Figure 10.9). A worldwide problem that is getting worse is chronic wounds. Combinatorial therapies are advantageous as they provide a regulated and prolonged release of active molecules, while also safeguarding these drugs from the adverse wound environment. The findings of this work indicate that the binding of the addition of growth factors KGF and SDF1 and the antibacterial protein LL37 to ELPs did not affect the

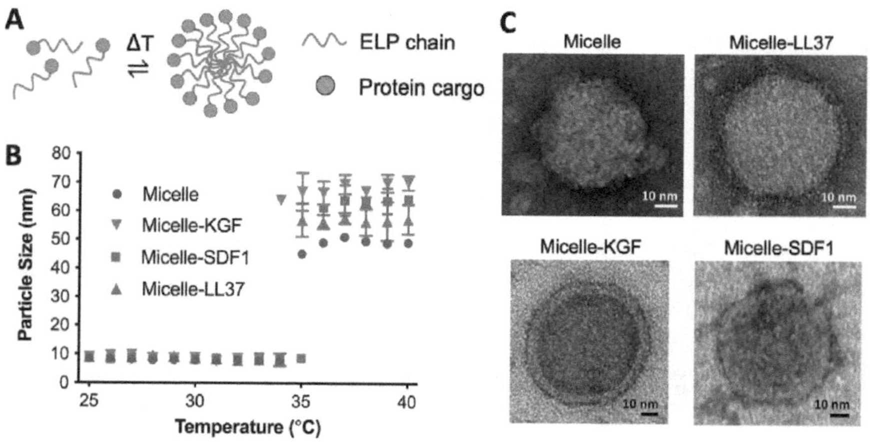

FIGURE 10.9 Design and characterization of particles. (a) The conceptual illustration of the production of ELP micelles in thermal response. (b) The temperature-dependent classification of particle size. (c) KGF, LL-37, and SDF-conjugated micelle nanoparticles as well as an empty micelle were photographed using transmission electron microscopy (10 nanometers is the scale bar).

Source: Adopted with permission from Bulutoglu et al. (2022b).

micelle composition. Furthermore, all three therapeutic components maintain their bioactivity when tested in laboratory conditions. Significantly, when applying the mixture of these micelle ELP nanomaterials to injuries in diabetic mice, an impressive 90% or more of the healing process is found. This is in contrast to when the therapies are administered in their unmodified states. This work is the first to investigate the combined targeting of multiple aspects of wound healing utilizing nanoparticles that were created (Gizaw et al. 2018; Bi et al. 2021).

Because of their exceptional physical and chemical qualities, unique flat shape, and usefulness in several areas, three-dimensional (3D) materials have caught the attention of interdisciplinary researchers in the realm of medicinal and therapeutic applications. Their vast surface area, low toxicity, biocompatibility, significant electrical conductivity, antibacterial activity, and hydrophilicity are only a few of the amazing characteristics of MXenes, 3D transition metal carbides, and nitrides. Although numerous studies have demonstrated the promising potential of MXenes for a wide range of biomedical purposes, including drug administration, tissue construction, regenerative medicine cancer therapy, bioimaging, and biosensing, there are still many challenging issues about their stability in physiological environments, sustained/controlled drug release, and biodegradability. The implementation of the skillfully created ultra-thin MXene nanosheets as potential biocompatible inorganic nanoplatforms for numerous biological purposes is made possible by the clinical translation of nanomedicine. MXenes are strong candidates for tissue engineering and regenerative medicine.

10.5.2 Photothermal and Photodynamic Therapy

Photothermal treatment (PTT) involves the use of electromagnetic radiation, specifically in the wavelengths which are infrared in nature, for the purpose of treating numerous medicinal illnesses such as cancer.

This technique and approach is a continuation of photodynamic treatment, in which the photosensitizer is stimulated by a particular wavelength of light. Upon activation, the sensitizer becomes stimulated and emits vibrational energy, which results in the destruction of the specific cells. Unlike photodynamic therapy, photothermal therapy does not rely on the presence of oxygen to interact with the target cells or tissues. Recent studies have demonstrated that photothermal therapy can utilize light with longer wavelengths, which possess lower energy levels and hence pose less harm to surrounding cells and tissues (P. Xu et al. 2018; Bai et al. 2022).

To induce cell death, a sort of phototherapy known as photodynamic therapy (PDT) uses light, a photosensitizing agent, and molecular oxygen (phototoxicity). Acne is frequently treated using PDT. It has demonstrated some efficacy in the treatment of viral infections, such as herpes, and is used clinically to treat a variety of medical problems, such as age-related macular degeneration, psoriasis, and atherosclerosis. Additionally, it treats malignant malignancies of the head and neck, bladder, lung, and some skin types. The technology has also been tested for the treatment of prostate cancer as well as has been used on canine models (S. Li et al. 2022; Song, Ma, and Zhan 2020).

It is acknowledged as a minimally intrusive, minimally harmful therapy approach. A photosensitizer is not necessary for other light- and laser-based treatments like intense pulsed light hair removal or laser wound healing and rejuvenation. To eliminate blood-borne viruses and microorganisms, photosensitizers have been used to sterilize water and blood, and they have also been studied for use in agriculture, including insecticides and herbicides. The benefits of photodynamic therapy are that it reduces the need for delicate surgery, provides prolonged recovery, limits creation of scar tissue, and causes minimum deformity. The resulting photosensitization of skin tissue is a side effect (Jing et al. 2022; B. Chen et al. 2017).

According to Y. Zhang et al. (Y. Zhang et al. 2022), the misuse of antibiotics has led to an increase in bacterial resistance to them. New treatments are therefore urgently required to take the place of conventional antibiotic therapy. In place of antibiotics, sterilizing techniques like photodynamic treatment and photothermal therapy are thought to be helpful against germs that are resistant to medications (Figure 10.10).

The study utilized curcumin as the photosensitizer and silica-coated gold nanorods having photothermal conversion effects as the photothermal carriers to create PTT/PDT dual-mode antibacterial nanocomposites (AuNRs@Cur). When comparing AuNRs@Cur to Cur and AuNRs, it was found that AuNRs@Cur had a greater yield of singlet oxygen and photothermal conversion efficiency. This significantly increased the utilization rate of the photosensitizer.

As a result, this markedly boosted the photoinactivation effect on bacteria. AuNRs@Cur suppressed S. aureus and E. coli at a higher rate under the dual irradiation of 405 and 808 nm, and the effects were compared to those of Cur or PDT alone. SEM and confocal fluorescence imaging were employed to illustrate the progression of the bacterial membrane structure. The bacterium can be killed by the

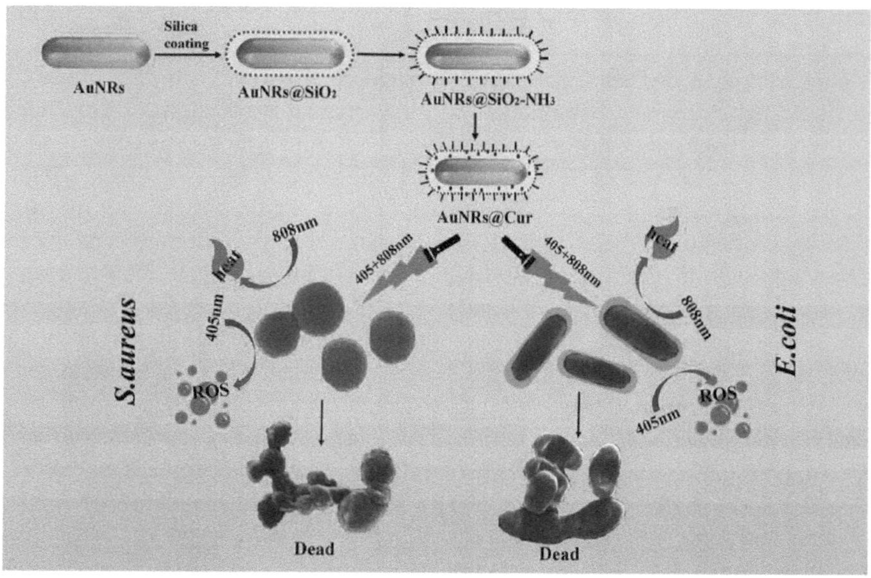

FIGURE 10.10 Diagram illustrating AuNRs@Cur's photodynamic inactivation of bacteria.

Source: Adopted with permission from Y. Zhang et al. (2022).

dual-wavelength (405 + 808 nm) laser irradiation, which generates reactive oxygen species and heat energy that damage the structure of the outer membrane. In addition, AuNRs@Cur showed minimal cytotoxicity and hemolytic activity. AuNRs@Cur nanocomposites demonstrate that the combined use of photodynamic and photothermal treatments can significantly enhance the bactericidal effect. As a result, the creation of AuNRs@Cur offers a novel viewpoint on pathogen infection caused by bacteria that are resistant to drugs and provides valuable nano-carrier expertise for next therapies (Kim et al. 2021; Morales et al. 2020; Q. Wu et al. 2021).

Nanomaterials are essential for the identification and treatment of illnesses. Two minimally invasive therapeutic approaches, photodynamic therapy and photothermal therapy (PTT), show promising results in the detection and treatment of cancer. Recent advances in photodynamic therapy and photothermal therapy (PTT) have made extensive use of both photothermal materials (such as noble metals, carbon materials, transition metal sulfur oxides, and up-conversion nanomaterials) and photodynamic materials (such as porphyrins, phthalein cyanogen, and other dye molecules) (PDT). Nanomaterials have also been used in cancer therapy as they are suitable for biocompatibility. This section summarizes the effects of diverse material types, material morphologies, synthesis procedures, and surface modifications on the outcomes of cancer therapy. The usage of nanoparticles in PDT and PTT was discussed, along with the advantages and disadvantages of each therapy in terms of preventing cancer.

10.6 SUMMARY

Titanium-based MXene is a novel category of two-dimensional (2D) transition metal carbides that has emerged, exhibiting a diverse array of uses in sensing, photovoltaics, and catalysis. MXene is a 2D material with a honeycomb-like pattern made up of transition metal atoms and carbon atoms. MXene, derived from titanium, demonstrates a substantial surface area and exceptional conductivity. MXene exhibits unique properties that make it very appropriate for use in sensing, photovoltaics, and catalysis. MXene has been employed in several sensing software, including as pressure sensors, gas sensors, and pH sensors. The immense array of surfaces of MXene makes it exceptionally well-suited for applications involving sensing. MXene's strong permeability makes it very suitable for sensing applications. MXene has also been employed in solar applications. The enormous surface area of MXene makes it exceptionally well-suited for solar power applications. MXene's strong conductivity makes it ideal for solar energy devices. MXene has also been employed in catalytic processes. The enormous surface area of MXene renders it a very suitable material for catalytic purposes as well. MXene's conductivity renders it an optimal substance for conversion.

10.7 COMPARISON TABLE

Materials compared are 2D titanium-based MXene, 2D titanium-based MXene/polymer, 2D titanium-based MXene/2D layered material, 3D titanium-based structure aerogel/hydrogel, along with other macroscopic materials. Conduct a thorough evaluation of various architectural designs and assess their effectiveness in the fields of monitoring, catalytic activity, solar energy systems, batteries for energy, and biomedicines.

Materials	Sensing	Catalysis	Photovoltaics	Energy Storage	Biomedicines
2D titanium-based MXene	Good	Good	Good	Good	Good
2D titanium-based polymer	Very good	Excellent	Excellent	Good	Good
2D-layered material (Graphine)	Excellent	Excellent	Excellent	Excellent	Excellent
2D-layered material (MXene)	Good	Excellent	Good	Good	Good
3D titanium-based aerogel	Excellent	Excellent	Excellent	Excellent	Excellent
3D titanium-based structure hydrogel	Good	Good	Excellent	Good	Good
Microscopic architectures	Good	Good	Good	Good	Good

REFERENCES

Aissou, Taki, Nadi Braidy, and Jocelyn Veilleux. 2022. "A New One-Step Deposition Approach of Graphene Nanoflakes Coating Using a Radio Frequency Plasma: Synthesis, Characterization and Tribological Behaviour." *Tribology International* 167 (December 2021): 107406. https://doi.org/10.1016/j.triboint.2021.107406.

Alami, Abdul Hai. 2020. "General Concepts BT—Mechanical Energy Storage for Renewable and Sustainable Energy Resources." *Advances in Science, Technology & Innovation*. http://www.springer.com/series/15883.

An, Hyosung, Touseef Habib, Smit Shah, Huili Gao, Anish Patel, Ian Echols, Xiaofei Zhao, Miladin Radovic, Micah J. Green, and Jodie L. Lutkenhaus. 2019. "Water Sorption in MXene/Polyelectrolyte Multilayers for Ultrafast Humidity Sensing." *ACS Applied Nano Materials* 2 (2): 948–55. https://doi.org/10.1021/acsanm.8b02265.

An, Yongkang, Shuangshuang Tan, Yu Liu, Kai Zhu, Lei Hu, Yaoguang Rong, and Qinyou An. 2021. "Designs and Applications of Multi-Functional Covalent Organic Frameworks in Rechargeable Batteries." *Energy Storage Materials* 41 (April): 354–79. https://doi.org/10.1016/j.ensm.2021.06.010.

Anik, Muzahidul I., M. Khalid Hossain, Imran Hossain, A. M. U. B. Mahfuz, M. Tayebur Rahman, and Isteaque Ahmed. 2021. "Recent Progress of Magnetic Nanoparticles in Biomedical Applications: A Review." *Nano Select* 2 (6): 1146–86. https://doi.org/10.1002/nano.202000162.

Bahrami, Behdokht, Mohammad Hojjat-Farsangi, Hamed Mohammadi, Enayat Anvari, Ghasem Ghalamfarsa, Mehdi Yousefi, and Farhad Jadidi-Niaragh. 2017. "Nanoparticles and Targeted Drug Delivery in Cancer Therapy." *Immunology Letters* 190 (July): 64–83. https://doi.org/10.1016/j.imlet.2017.07.015.

Bai, Qiang, Hongyang Luo, Shugao Shi, Shen Liu, Lina Wang, Fanglin Du, Zhugen Yang, Zhiling Zhu, and Ning Sui. 2022. "AuAg Nanocages/Graphdiyne for Rapid Elimination and Detection of Trace Pathogenic Bacteria." *Journal of Colloid and Interface Science* 613: 376–83. https://doi.org/10.1016/j.jcis.2022.01.046.

Bi, Yaqi, Di Feng, Shuangshuang Liu, Lingjie Jia, Yan Meng, Wei Xu, Jiayi Yang, Sida Liu, Baolong Wang, Yong Qin, and Xiuhan Li. 2021. "Solid Ion Channels Gel

Battery Driven by Triboelectric Effect and Its Integrated Self-Powered Foreign Matter Intrusion Detecting System." *Nano Energy* 83 (January): 105791. https://doi.org/10.1016/j.nanoen.2021.105791.

Bulutoglu, Beyza, Aylin Acun, Sarah L. Deng, Safak Mert, Elise Lupon, Alexandre G. Lellouch, Curtis L. Cetrulo, Basak E. Uygun, and Martin L. Yarmush. 2022a. "Combinatorial Use of Therapeutic ELP-Based Micelle Particles in Tissue Engineering." *Advanced Healthcare Materials* 11 (13): 1–10. https://doi.org/10.1002/adhm.202102795.

Bulutoglu, Beyza, Aylin Acun, Sarah L. Deng, Safak Mert, Elise Lupon, Alexandre G. Lellouch, Curtis L. Cetrulo, Basak E. Uygun, and Martin L. Yarmush. 2022b. "Combinatorial Use of Therapeutic ELP-Based Micelle Particles in Tissue Engineering." *Advanced Healthcare Materials* 2102795: 1–10. https://doi.org/10.1002/adhm.202102795.

Cao, Mao Sheng, Yong Zhu Cai, Peng He, Jin Cheng Shu, Wen Qiang Cao, and Jie Yuan. 2019. "2D MXenes: Electromagnetic Property for Microwave Absorption and Electromagnetic Interference Shielding." *Chemical Engineering Journal* 359 (August 2018): 1265–302. https://doi.org/10.1016/j.cej.2018.11.051.

Cardoso, Vanessa Fernandes, António Francesko, Clarisse Ribeiro, Manuel Bañobre-López, Pedro Martins, and Senentxu Lanceros-Mendez. 2018. "Advances in Magnetic Nanoparticles for Biomedical Applications." *Advanced Healthcare Materials* 7 (5): 1–35. https://doi.org/10.1002/adhm.201700845.

Chang, Chieh, Van H. Tran, Junbo Wang, Yiin Kuen Fuh, and Liwei Lin. 2010. "Direct-Write Piezoelectric Polymeric Nanogenerator with High Energy Conversion Efficiency." *Nano Letters* 10 (2): 726–31. https://doi.org/10.1021/nl9040719.

Chen, Biao, Enzuo Liu, Tingting Cao, Fang He, Chunsheng Shi, Chunnian He, Liying Ma, Qunying Li, Jiajun Li, and Naiqin Zhao. 2017. "Controllable Graphene Incorporation and Defect Engineering in MoS_2-TiO_2 Based Composites: Towards High-Performance Lithium-Ion Batteries Anode Materials." *Nano Energy* 33 (January): 247–56. https://doi.org/10.1016/j.nanoen.2017.01.034.

Chen, Huanyu, Rongzhi Chen, Shengjiong Yang, Dahu Ding, Xiaoping Li, Xinxin Long, Ting Zhao, Yuxuan Du, Meng Liu, Jihua Tan, and Yang Chen. 2022. "Sustainable Heterolytic Cleavage of Peroxymonosulfate by Promoting Fe(III)/Fe(II) Cycle: The Role of in-Situ Sulfur." *Chemical Engineering Journal* 446 (P3): 137257. https://doi.org/10.1016/j.cej.2022.137257.

Chen, Winston Yenyu, Xiaofan Jiang, Sz Nian Lai, Dimitrios Peroulis, and Lia Stanciu. 2020. "Nanohybrids of a MXene and Transition Metal Dichalcogenide for Selective Detection of Volatile Organic Compounds." *Nature Communications* 11 (1). https://doi.org/10.1038/s41467-020-15092-4.

Fan, Bingbing, Ning Li, Binzhou Dai, Siyang Shang, Li Guan, Biao Zhao, Xiaohui Wang, Zhongyi Bai, and Rui Zhang. 2020. "Investigation of Adjacent Spacing Dependent Microwave Absorption Properties of Lamellar Structural Ti_3C_2Tx MXenes." *Advanced Powder Technology* 31 (2): 808–15. https://doi.org/10.1016/j.apt.2019.11.035.

Fan, Xin, Cesar Torres-Luna, Marjan Azadi, Roman Domszy, Naiping Hu, Arthur Yang, and Allan E. David. 2020. "Evaluation of Commercial Soft Contact Lenses for Ocular Drug Delivery: A Review." *Acta Biomaterialia* 115: 60–74. https://doi.org/10.1016/j.actbio.2020.08.025.

Fu, Chiyu, Kai Wang, Wenyang Tang, Azadeh Nilghaz, Christopher Hurren, Xungai Wang, Weilin Xu, Bin Su, and Zhigang Xia. 2022. "Multi-Sensorized Pneumatic Artificial Muscle Yarns." *Chemical Engineering Journal* 446 (P3): 137241. https://doi.org/10.1016/j.cej.2022.137241.

Gao, Jiaojiao, Jinpeng Song, Yao Wang, Zhuo Wang, Jianping Song, and Xuedong Zhao. 2022. "Microstructures and Mechanical Properties of Functionally Graded TiCN–TaC

Ceramics Prepared by a Novel Layer Processing Strategy." *Ceramics International* 48 (12): 16990–96. https://doi.org/10.1016/j.ceramint.2022.02.254.

Garfi, Gaetano, Cédric M. John, Maja Rücker, Qingyang Lin, Catherine Spurin, Steffen Berg, and Samuel Krevor. 2022. "Determination of the Spatial Distribution of Wetting in the Pore Networks of Rocks." *Journal of Colloid and Interface Science* 613: 786–95. https://doi.org/10.1016/j.jcis.2021.12.183.

Gizaw, Mulugeta, Jeffrey Thompson, Addison Faglie, Shih Yu Lee, Pierre Neuenschwander, and Shih Feng Chou. 2018. "Electrospun Fibers as a Dressing Material for Drug and Biological Agent Delivery in Wound Healing Applications." *Bioengineering* 5 (1): 1–28. https://doi.org/10.3390/bioengineering5010009.

Grishina, Maria, Oleg Bol'shakov, Andrey Potemkin, and Vladimir Potemkin. 2017. "Benzo[1,2,5] Thiadiazole Dyes: Spectral and Electrochemical Properties and Their Relation to the Photovoltaic Characteristics of the Dye-Sensitized Solar Cells." *Dyes and Pigments* 144: 80–93. https://doi.org/10.1016/j.dyepig.2017.05.020.

Hamani, David, Mohamed Ati, Jean Marie Tarascon, and Patrick Rozier. 2011. "NaxVO2 as Possible Electrode for Na-Ion Batteries." *Electrochemistry Communications* 13 (9): 938–41. https://doi.org/10.1016/j.elecom.2011.06.005.

He, Chaohui, Yong Wang, Yang Chen, Xiaoqing Wang, Jiangfeng Yang, Libo Li, and Jinping Li. 2021. "Modification of the Pore Environment in UiO-Type Metal-Organic Framework Toward Boosting the Separation of Propane/Propylene." *Chemical Engineering Journal* 403 (July 2020): 126428. https://doi.org/10.1016/j.cej.2020.126428.

He, Shuangjiang, Yingqing Zhan, Jiaxin Hu, Guiyuan Zhang, Shumei Zhao, Qingying Feng, and Wei Yang. 2020. "Chemically Stable Two-Dimensional MXene@UIO-66-$(COOH)_2$ Composite Lamellar Membrane for Multi-Component Pollutant-Oil-Water Emulsion Separation." *Composites Part B: Engineering* 197 (February): 108188. https://doi.org/10.1016/j.compositesb.2020.108188.

Horiba, Tatsuo. 2014. "Lithium-Ion Battery Systems." *Proceedings of the IEEE* 102 (6): 939–950.

Jiang, Xiantao, Artem V. Kuklin, Alexander Baev, Yanqi Ge, Hans Ågren, Han Zhang, and Paras N. Prasad. 2020. "Two-Dimensional MXenes: From Morphological to Optical, Electric, and Magnetic Properties and Applications." *Physics Reports*. https://doi.org/10.1016/j.physrep.2019.12.006.

Jin, Seong, Jianlin Li, Claus Daniel, Debasish Mohanty, Shrikant Nagpure, and David L. Wood. 2016. "The State of Understanding of the Lithium-Ion-Battery Graphite Solid Electrolyte Interphase (SEI) and Its Relationship to Formation Cycling *." *Carbon* 105: 52–76. https://doi.org/10.1016/j.carbon.2016.04.008.

Jing, Ziwei, Qiuzheng Du, Xiaojian Zhang, and Yu Zhang. 2022. "Nanomedicines and Nanomaterials for Cancer Therapy: Progress, Challenge and Perspectives." *Chemical Engineering Journal* 446 (P3): 137147. https://doi.org/10.1016/j.cej.2022.137147.

Karuppasamy, K., P. Anil Reddy, G. Srinivas, Amit Tewari, Ramakant Sharma, X. Sahaya Shajan, and Dipti Gupta. 2016. "Electrochemical and Cycling Performances of Novel Nonafluorobutanesulfonate (Nonaflate) Ionic Liquid Based Ternary Gel Polymer Electrolyte Membranes for Rechargeable Lithium Ion Batteries." *Journal of Membrane Science* 514: 350–57. https://doi.org/10.1016/j.memsci.2016.05.010.

Kim, Ju Young, Joonam Park, Seok Hun Kang, Seungwon Jung, Dong L Shin, Myeong Ju Lee, Jimin Oh, et al. 2021. "Revisiting TiS_2 as a Diffusion-Dependent Cathode with Promising Energy Density for All-Solid-State Lithium Secondary Batteries." *Energy Storage Materials* 41 (March): 289–96. https://doi.org/10.1016/j.ensm.2021.06.005.

Kiyotake, Emi A., Michael D. Martin, and Michael S. Detamore. 2022. "Regenerative Rehabilitation with Conductive Biomaterials for Spinal Cord Injury." *Acta Biomaterialia* 139: 43–64. https://doi.org/10.1016/j.actbio.2020.12.021.

Li, Guochun, Liuqing Yang, Xi Jiang, Tianran Zhang, Haibin Lin, Qiaofeng Yao, and Jim Yang Lee. 2018. "A Multi-Electron Redox Mediator for Redox-Targeting Lithium-Sulfur Flow Batteries." *Journal of Power Sources* 378 (October 2017): 418–22. https://doi.org/10.1016/j.jpowsour.2017.11.066.

Li, Ran, Xiaoyuan Ma, Jianmin Li, Jun Cao, Hongze Gao, Tianshu Li, Xiaoyu Zhang, Lichao Wang, Qinghong Zhang, Gang Wang, Chengyi Hou, Yaogang Li, Tomás Palacios, Yuxuan Lin, Hongzhi Wang, and Xi Ling. 2021. "Flexible and High-Performance Electrochromic Devices Enabled by Self-Assembled 2D TiO_2/MXene Heterostructures." *Nature Communications* 12 (1): 1–11. https://doi.org/10.1038/s41467-021-21852-7.

Li, Shimei, Jian Zou, Longfei Tan, Zhongbing Huang, Ping Liang, and Xianwei Meng. 2022. "Covalent Organic Frameworks: From Linkages to Biomedical Applications." *Chemical Engineering Journal* 446 (P3): 137148. https://doi.org/10.1016/j.cej.2022.137148.

Li, Wanyang, Xuefeng Yang, Jupeng Xiao, and Qimin Hou. 2021. "Effect of WC Mass Fraction on the Microstructure and Friction Properties of WC/Ni60 Laser Cladding Layer of Brake Discs." *Ceramics International* 47 (20): 28754–63. https://doi.org/10.1016/j.ceramint.2021.07.035.

Liu, Kewei, Makoto Sakurai, and Masakazu Aono. 2010. "ZnO-Based Ultraviolet Photodetectors." *Sensors* 10 (9): 8604–34. https://doi.org/10.3390/s100908604.

Liu, Peijiang, Zhengjun Yao, Vincent Ming Hong Ng, Jintang Zhou, Ling Bing Kong, and Kan Yue. 2018. "Facile Synthesis of Ultrasmall Fe_3O_4 Nanoparticles on MXenes for High Microwave Absorption Performance." *Composites Part A: Applied Science and Manufacturing* 115 (October): 371–82. https://doi.org/10.1016/j.compositesa.2018.10.014.

Liu, Tongyu, Zijie Xu, Linlin Chen, Yali Zhang, Ming Wang, Yun Jia, and Yaoguo Huang. 2022. "Boosting Zinc Ion Storage Performance of Sandwich-like V_2O_5/Graphene Composite by Effectively Inhibiting Vanadium Dissolution." *Journal of Colloid and Interface Science* 613: 524–35. https://doi.org/10.1016/j.jcis.2022.01.057.

Long, Jiangyou, Qingfa Peng, Gaopan Chen, Yuliang Zhang, Xiaozhu Xie, Guoshun Pan, and Xiaofeng Wang. 2021. "Centimeter-Scale Low-Damage Micromachining on Single-Crystal 4H–SiC Substrates Using a Femtosecond Laser with Square-Shaped Flat-Top Focus Spots." *Ceramics International* 47 (16): 23134–43. https://doi.org/10.1016/j.ceramint.2021.05.027.

Long, Mingsheng, Peng Wang, Hehai Fang, and Weida Hu. 2019. "Progress, Challenges, and Opportunities for 2D Material Based Photodetectors." *Advanced Functional Materials* 29 (19): 1–28. https://doi.org/10.1002/adfm.201803807.

Lu, Bang An, Tian Sheng, Na Tian, Zong Cheng Zhang, Chi Xiao, Zhen Ming Cao, Hai Bin Ma, Zhi You Zhou, and Shi Gang Sun. 2017. "Octahedral PtCu Alloy Nanocrystals with High Performance for Oxygen Reduction Reaction and Their Enhanced Stability by Trace Au." *Nano Energy* 33 (January): 65–71. https://doi.org/10.1016/j.nanoen.2017.01.003.

Lu, Qixin, Hong Chen, Yuanming Zeng, Jiehui Xue, Xia Cao, Ning Wang, and Zhonglin Wang. 2022. "Intelligent Facemask Based on Triboelectric Nanogenerator for Respiratory Monitoring." *Nano Energy* 91 (September 2021): 106612. https://doi.org/10.1016/j.nanoen.2021.106612.

Morales, M. V., A. Guerrero-Ruiz, E. Castillejos, E. Asedegbega-Nieto, and I. Rodríguez-Ramos. 2020. "Taking Advantage of Sulfur Impurities Present in Commercial Carbon Nanofibers to Generate Selective Palladium Catalysts." *Carbon* 157: 120–29. https://doi.org/10.1016/j.carbon.2019.10.021.

Nitta, Naoki, Feixiang Wu, Jung Tae Lee, and Gleb Yushin. 2015. "Li-Ion Battery Materials : Present and Future." *Biochemical Pharmacology* 18 (5): 252–64. https://doi.org/10.1016/j.mattod.2014.10.040.

Noorpour, Mojtaba, and Amir Tarighat. 2021. "Molecular Dynamics Study of a Macromolecular Model of Hydrous Sodium Aluminosilicate Geopolymer with Charge Balancing by Extra-Structural Aluminum for Prediction of Its Properties." *Ceramics International* 47 (14): 19304–14. https://doi.org/10.1016/j.ceramint.2021.02.116.

Pareek, Avinash, Prashant Singh, and P. Narendra Rao. 2018. "Analysis and Comparison of Charging Time between Battery and Supercapacitor for 300W Stand-Alone PV System." *Proceedings of the 2018 International Conference on Current Trends towards Converging Technologies, ICCTCT 2018*, 1–6. https://doi.org/10.1109/ICCTCT.2018.8551164.

Qin, R., G. Shan, M. Hu, and W. Huang. 2021. "Two-Dimensional Transition Metal Carbides and/or Nitrides (MXenes) and Their Applications in Sensors." *Materials Today Physics*. https://doi.org/10.1016/j.mtphys.2021.100527.

Ren, Xiaoxiang, Xiao Chen, Zhen Geng, and Jiacan Su. 2022. "Bone-Targeted Biomaterials: Strategies and Applications." *Chemical Engineering Journal* 446 (P3): 137133. https://doi.org/10.1016/j.cej.2022.137133.

Selamneni, Venkatarao, Vivek Adepu, Harini Raghavan, and Parikshit Sahatiya. 2022. "Ultra-High Responsivity and Enhanced Trap Assisted Charge Transfer by Utilizing Ti_3C_2TX(MXene) as a Transport Layer for ReS_2 Based Flexible Broadband Photodetector: A Better Alternative to Graphene." *FlatChem* 33 (March): 100363. https://doi.org/10.1016/j.flatc.2022.100363.

Sena Da Fonseca, B., A. Vilão, C. Galhano, and J. A. R. Simão. 2014. "Reusing Coffee Waste in Manufacture of Ceramics for Construction." *Advances in Applied Ceramics* 113 (3): 159–66. https://doi.org/10.1179/1743676113Y.0000000131.

Shi, Hongyang, Mohammed Al-Rubaiai, Christopher M. Holbrook, Jinshui Miao, Thassyo Pinto, Chuan Wang, and Xiaobo Tan. 2019. "Screen-Printed Soft Capacitive Sensors for Spatial Mapping of Both Positive and Negative Pressures." *Advanced Functional Materials* 29 (23): 1–10. https://doi.org/10.1002/adfm.201809116.

Song, Youshuo, Yue Ma, and Ke Zhan. 2020. "Simulations of Deformation and Fracture of Graphene Reinforced Aluminium Matrix Nanolaminated Composites." *Mechanics of Materials* 142 (December 2019): 103283. https://doi.org/10.1016/j.mechmat.2019.103283.

Su, Yue, Kainan Ma, Fang Yuan, Jun Tang, Ming Liu, and Xu Zhang. 2022. "High-Performance Flexible Piezoresistive Sensor Based on Ti_3C_2Tx MXene with a Honeycomb-like Structure for Human Activity Monitoring." *Micromachines* 13 (6): 821. https://doi.org/10.3390/mi13060821.

Tan, B., and N. L. Thomas. 2016. "A Review of the Water Barrier Properties of Polymer/Clay and Polymer/Graphene Nanocomposites." *Journal of Membrane Science* 514: 595–612. https://doi.org/10.1016/j.memsci.2016.05.026.

Vladimirova, S. A., K. Ya Prikhodko, M. N. Rumyantseva, E. A. Konstantinova, A. S. Chizhov, N. O. Khmelevsky, and A. M. Gaskov. 2020. "Nanocrystalline Complex Oxides $NixCo_3$-XO_4: Cations Distribution Impact on Electrical and Gas Sensor Behaviour." *Journal of Alloys and Compounds* 828: 154420. https://doi.org/10.1016/j.jallcom.2020.154420.

Wang, Dongming, Xiaomei Wu, and Qiang Gao. 2021. "Novel Energy-Saving Window Coating Based on F Doped TiO_2 Nanocrystals with Enhanced NIR Shielding Performance." *Ceramics International* 47 (20): 28557–65. https://doi.org/10.1016/j.ceramint.2021.07.013.

Wang, Qi, Tao Ruan, Qingda Xu, Bin Yang, and Jingquan Liu. 2021. "Wearable Multifunctional Piezoelectric MEMS Device for Motion Monitoring, Health Warning, and Earphone." *Nano Energy* 89 (PA): 106324. https://doi.org/10.1016/j.nanoen.2021.106324.

Wang, Zhong Lin. 2013. "Triboelectric Nanogenerators as New Energy Technology for Self-Powered Systems and as Active Mechanical and Chemical Sensors." *ACS Nano* 7 (11): 9533–57. https://doi.org/10.1021/nn404614z.

Wei, Bei, Guancheng Xu, Jincheng Hei, Li Zhang, Tingting Huang, and Qian Wang. 2021. "CoFeP Hierarchical Nanoarrays Supported on Nitrogen-Doped Carbon Nanofiber as Efficient Electrocatalyst for Water Splitting." *Journal of Colloid and Interface Science* 602: 619–26. https://doi.org/10.1016/j.jcis.2021.06.045.

Wu, Qian, Lei Tan, Xiangmei Liu, Zhaoyang Li, Yu Zhang, Yufeng Zheng, Yanqin Liang, Zhenduo Cui, Shengli Zhu, and Shuilin Wu. 2021. "The Enhanced Near-Infrared Photocatalytic and Photothermal Effects of MXene-Based Heterojunction for Rapid Bacteria-Killing." *Applied Catalysis B: Environmental* 297 (April): 120500. https://doi.org/10.1016/j.apcatb.2021.120500.

Wu, Zhenhua, Shuai Zhang, Zekun Liu, Erzhen Mu, and Zhiyu Hu. 2022. "Thermoelectric Converter: Strategies from Materials to Device Application." *Nano Energy* 91 (November 2021): 106692. https://doi.org/10.1016/j.nanoen.2021.106692.

Xiao, Siyang, Kaixin Meng, Quan Xie, Linxin Zhai, Zhiping Xu, Hao Wang, and Chuanhua Duan. 2022. "Edge-Enhanced Ultrafast Water Evaporation from Graphene Nanopores." *Cell Reports Physical Science* 3 (6): 100900. https://doi.org/10.1016/j.xcrp.2022.100900.

Xie, Xiuqiang, Meng Qiang Zhao, Babak Anasori, Kathleen Maleski, Chang E. Ren, Jingwen Li, Bryan W. Byles, Ekaterina Pomerantseva, Guoxiu Wang, and Yury Gogotsi. 2016. "Porous Heterostructured MXene/Carbon Nanotube Composite Paper with High Volumetric Capacity for Sodium-Based Energy Storage Devices." *Nano Energy* 26: 513–23. https://doi.org/10.1016/j.nanoen.2016.06.005.

Xie, Yannan, Sihong Wang, Simiao Niu, Long Lin, Qingshen Jing, Yuanjie Su, Zhengyun Wu, and Zhong Lin. 2014. "Multi-Layered Disk Triboelectric Nanogenerator for Harvesting Hydropower." *Nano Energy* 6: 129–36. https://doi.org/10.1016/j.nanoen.2014.03.015.

Xin, Qingping, Yuan Zhang, Yue Shi, Hui Ye, Ligang Lin, Xiaoli Ding, Yuzhong Zhang, Hong Wu, and Zhongyi Jiang. 2016. "Tuning the Performance of CO_2 Separation Membranes by Incorporating Multifunctional Modified Silica Microspheres into Polymer Matrix." *Journal of Membrane Science* 514: 73–85. https://doi.org/10.1016/j.memsci.2016.04.046.

Xu, Pei, Weijia Fu, Yadong Hu, and Yunsheng Ding. 2018. "Effect of Annealing Treatment on Crystalline and Dielectric Properties of PVDF/PEG-Containing Ionic Liquid Composites." *Composites Science and Technology* 158: 1–8. https://doi.org/10.1016/j.compscitech.2018.01.039.

Xu, Song, Yuanyuan Cheng, Lan Zhang, Kaihang Zhang, Feng Huo, Xiangping Zhang, and Suojiang Zhang. 2018. "An Effective Polysulfides Bridgebuilder to Enable Long-Life Lithium-Sulfur Flow Batteries." *Nano Energy* 51 (June): 113–21. https://doi.org/10.1016/j.nanoen.2018.06.044.

Yan, Xiaohui, Gang Li, Zheyan Wang, Zhichao Yu, Kaiying Wang, and Yucheng Wu. 2020. "Recent Progress on Piezoelectric Materials for Renewable Energy Conversion." *Nano Energy* 77 (July): 105180. https://doi.org/10.1016/j.nanoen.2020.105180.

Yun, Ju Hui, and Jun Yeob Lee. 2017. "Benzoisoquinoline-1,3-Dione Acceptor Based Red Thermally Activated Delayed Fluorescent Emitters." *Dyes and Pigments* 144: 212–17. https://doi.org/10.1016/j.dyepig.2017.05.036.

Zeng, Wenfeng, Zimu Li, Hongzhong Chen, Xiaowei Zeng, and Lin Mei. 2022. "An Optimal Portfolio of Photothermal Combined Immunotherapy." *Cell Reports Physical Science* 3 (6): 100898. https://doi.org/10.1016/j.xcrp.2022.100898.

Zhang, Lupeng, Xinle Li, Mingrui Yang, and Weihua Chen. 2021. "High-Safety Separators for Lithium-Ion Batteries and Sodium-Ion Batteries: Advances and Perspective." *Energy Storage Materials* 41 (April): 522–45. https://doi.org/10.1016/j.ensm.2021.06.033.

Zhang, Wei, Hao Huang, and Roy Bernstein. 2022. "Zwitterionic Hydrogel Modified Reduced Graphene Oxide/ZnO Nanocomposite Blended Membrane with High Antifouling and Antibiofouling Performances." *Journal of Colloid and Interface Science* 613: 426–34. https://doi.org/10.1016/j.jcis.2021.12.194.

Zhang, Wu, Tiefeng Liu, Yao Wang, Yujing Liu, Jianwei Nai, Liang Zhang, Ouwei Sheng, and Xinyong Tao. 2021. "Strategies to Improve the Performance of Phosphide Anodes in Sodium-Ion Batteries." *Nano Energy* 90 (PA): 106475. https://doi.org/10.1016/j.nanoen.2021.106475.

Zhang, Xinglai, Hanwen Wang, Jian Zhang, Jiangxu Li, Zongyi Ma, Jing Li, Bing Leng, Pingjuan Niu, and Baodan Liu. 2020. "Enhanced Optoelectronic Performance of 3C–SiC/ZnO Heterostructure Photodetector Based on Piezo-Phototronic Effect." *Nano Energy* 77 (April): 105119. https://doi.org/10.1016/j.nanoen.2020.105119.

Zhang, Ying, Hongjun Yan, Jiawen Tang, Peiyuan Li, Rixiang Su, Haiyi Zhong, and Wei Su. 2022. "Dual-Mode Antibacterial Core-Shell Gold Nanorod@mesoporous-Silica/Curcumin Nanocomplexes for Efficient Photothermal and Photodynamic Therapy." *Journal of Photochemistry and Photobiology A: Chemistry* 425 (December 2021): 113722. https://doi.org/10.1016/j.jphotochem.2021.113722

Zhao, Peng, Bin Tang, Zixuan Fang, Feng Si, Chengtao Yang, and Shuren Zhang. 2021. "Improved Dielectric Breakdown Strength and Energy Storage Properties in Er_2O_3 Modified $Sr_{0.35}Bi_{0.35}K_{0.25}TiO_3$." *Chemical Engineering Journal* 403 (April 2020): 126290. https://doi.org/10.1016/j.cej.2020.126290.

Zheng, Xianhong, Qiaole Hu, Zongqian Wang, Wenqi Nie, Peng Wang, and Changlong Li. 2021. "Roll-to-Roll Layer-by-Layer Assembly Bark-Shaped Carbon Nanotube/Ti_3C_2Tx MXene Textiles for Wearable Electronics." *Journal of Colloid and Interface Science* 602: 680–88. https://doi.org/10.1016/j.jcis.2021.06.043.

Zhong, Mei'e, Jindiao Guan, Jingchun Sun, Xiaoqing Shu, Hao Ding, Liying Chen, Nan Zhou, and Zhubing Xiao. 2021. "A Cost- and Energy Density-Competitive Lithium-Sulfur Battery." *Energy Storage Materials* 41 (December 2020): 588–98. https://doi.org/10.1016/j.ensm.2021.06.037.

Zhou, Peng, Yaqiang Wu, Cong Wang, Huining Huang, Danning Xing, Yuanyuan Liu, Zeyan Wang, Peng Wang, Zhaoke Zheng, Hefeng Cheng, Ying Dai, and Baibiao Huang. 2021. "Tailoring the Composition and Structure of Ni_3S_2 by Introduction of Co Towards High Efficiency Energy Storage Device." *Chemical Engineering Journal* 403 (July 2020): 126285. https://doi.org/10.1016/j.cej.2020.126285.

Zong, Wei, Chao Yang, Lulu Mo, Yue Ouyang, Hele Guo, Lingfeng Ge, Yue E. Miao, Dewei Rao, Jiangwei Zhang, Feili Lai, and Tianxi Liu. 2020. "Elucidating Dual-Defect Mechanism in Rhenium Disulfide Nanosheets with Multi-Dimensional Ion Transport Channels for Ultrafast Sodium Storage." *Nano Energy* 77 (July): 105189. https://doi.org/10.1016/j.nanoen.2020.105189.

11 Summary and Future Directions

11.1 BRIEF SUMMARY

MXenes, a group of highly versatile 2D materials, demonstrate substantial potential across a wide array of applications. This analysis categorizes MXene precursors, with particular emphasis on carbide- and nitride-based materials derived from MAX and non-MAX phases, and explores various synthesis techniques, particularly exfoliation methods. $Ti_3C_2T_x$ MXene is specifically highlighted due to its straightforward production process, remarkable physical properties, and suitability for use in biological and environmental sensors. By assessing the costs, yields, and practical applicability of different synthesis methods, this overview underscores the increasing significance of MXenes in advanced material applications, emphasizing their versatility, stability, and promising future in diverse industrial sectors (Naguib, Barsoum, and Gogotsi 2021; Abbasi, Xiao, Peng, et al. 2021; Jayakumar et al. 2024; Wei et al. 2021; Rehman et al. 2024). Defect engineering plays a crucial role in enhancing the performance of MXenes, a class of two-dimensional transition metal carbides and nitrides, through the strategic introduction and manipulation of structural defects. MXenes are of considerable interest in contemporary materials research due to their unique mechanical, optical, electrical, magnetic, and stability properties. Utilizing defect engineering, it becomes possible to optimize and tailor these properties to meet the specific demands of various technological applications, thereby expanding the potential uses of MXenes in advanced technologies (Tang et al. 2022; Liu and Fan 2023; Wang et al. 2021; Hassan, Ghaffar, et al. 2024; Hassan, Li, et al. 2024).

MXenes, particularly $Ti_3C_2T_x$ and Ti_2CT_x, have exhibited significant potential as advanced materials owing to their exceptional mechanical, electrical, and surface properties. These attributes render them highly suitable for applications across a wide range of industries, including energy storage, photovoltaics, biomedicine, and catalysis. Ongoing research into their synthesis, characterization, and application further highlights the growing importance of MXenes in technology. As the field continues to evolve, future developments are expected to broaden their range of applications, offering promising opportunities across multiple sectors. MXenes, a class of two-dimensional transition metal carbides and nitrides, exhibit remarkable mechanical, electrical, and chemical properties, making them ideal candidates for making flexible and wearable electronics (Hantanasirisakul and Gogotsi 2023; Hassan, Li, et al. 2024). The integration of MXenes with polymers to form composite materials further enhances their mechanical, electrical, and thermal performance, surpassing the capabilities of pure polymers or MXenes alone. Furthermore, the straightforward solution-based fabrication method for these composites renders them highly feasible for a variety of practical applications. As research in this area

DOI: 10.1201/9781003319658-11

progresses, MXene-based composites are likely to contribute significantly to technological advancements across various industries (Hassan, Ghaffar, et al. 2024).

Recent developments in the synthesis of MXenes and van der Waals heterostructures, particularly those involving titanium-based MXenes such as $Ti_3C_2T_x$ and Ti_2CT_x, have demonstrated considerable potential. These materials exhibit outstanding metallic conductivity and hydrophilic surfaces, making them highly suitable for applications in energy storage, catalysis, electromagnetic shielding, and water splitting. However, challenges related to their structural stability and surface smoothness persist, necessitating the adoption of advanced techniques to improve their practical applicability. The fabrication of heterostructures combining MXenes with other two-dimensional materials aims to preserve the intrinsic properties of both materials, facilitated by weak intermolecular interactions. These heterostructures show significant promise for applications in catalysis, optoelectronics, and energy storage, thus offering promising avenues for future research and technological advancements (Wang et al. 2024; Abbasi, Xiao, Zhang, et al. 2021).

The successful commercialization of MXene-based derivatives requires addressing several key challenges, including surface roughness, deposition techniques, large-scale production, and long-term stability. The development of mixed-dimensional heterostructures that integrate 2D MXene derivatives with various low-dimensional materials (0D, 1D, 2D) is crucial to enhancing the functionality of MXenes (Ding et al. 2017). The unique surface properties of $Ti_3C_2T_x$ and Ti_2CT_x facilitate their interaction with other 2D materials in solvents, enabling versatile chemical interactions. This interaction allows for the application of various deposition techniques such as drop-casting, dip coating, and spin coating. These heterostructures aim to improve the stability and surface properties of MXenes, thereby expanding their potential applications in energy storage, particularly in supercapacitors, lithium-ion batteries, and sodium-ion batteries. This chapter provides a comprehensive examination of these deposition methods and their applications, offering valuable insights for researchers and professionals in the field (Ding et al. 2017; Kumar et al. 2023).

MXenes have significantly impacted materials science by providing exceptional properties that are being leveraged across a broad range of applications. Noteworthy advancements in MXene-based photodetectors, particularly $Ti_3C_2T_x$ when integrated into heterostructures with n-Si and InSe, highlight their potential to enhance sensing technologies. In the field of solar energy, the combination of MXenes with silicon in $Ti_3C_2T_x$-SiO2 heterostructures offers promising prospects for improving solar cell efficiency. Furthermore, the remarkable catalytic capabilities of MXenes are emphasized, particularly in electrocatalysis for water splitting and in photocatalysis for CO_2 conversion to methanol. These studies underscore the significant contribution of MXenes to advancing technology and facilitating sustainable solutions (Yu et al. 2019; Guo et al. 2018; Fu et al. 2019).

The study of 3D assemblies of $Ti_3C_2T_x$ and Ti_2CT_x, along with 2D assemblies of graphene and TMDCs, demonstrates their exceptional physical and chemical properties. These materials, synthesized through chemical vapor deposition for $Ti_3C_2T_x$ and Ti_2CT_x and mechanical exfoliation for graphene and TMDCs, show great promise for applications in electronics, energy storage, and advanced device fabrication. Advanced characterization techniques such as AFM, XPS, and Raman spectroscopy reveal their honeycomb lattice structures and elemental compositions. Both 3D and 2D assemblies exhibit unique characteristics that underscore their potential to drive advancements across multiple domains. The distinctive properties of MXenes, particularly in their 3D configurations, position them as promising materials for advanced applications in sensing, energy conversion, and catalysis. Their exceptional electrical conductivity, robust chemical stability, and large surface area make them highly suitable for various technological advancements (Sikdar et al. 2021; Zhao et al. 2021; Xiang et al. 2022; Luo et al. 2023; Bera et al. 2024). These characteristics enable MXenes to contribute to precise detection in sensing, improve energy conversion efficiency in photovoltaics, and facilitate effective catalytic processes. As a result, MXenes are expected to play a pivotal role in the development of cutting-edge technologies across diverse domains.

11.2 FUTURE DIRECTION

The future directions for MXenes in materials science and technology hold significant promise, with several key areas poised to drive continued innovation and expand their applications. One of the primary areas of focus will be the advancement of synthesis techniques and the optimization of defect engineering to improve the performance and stability of MXenes. This will be crucial in broadening their applicability across various industrial sectors. Efforts to enhance the structural stability and surface smoothness of MXenes are expected to facilitate their scalability and commercial viability.

Another promising avenue for future research lies in the integration of MXenes with other two-dimensional materials, such as graphene and transition metal dichalcogenides (TMDCs), to create heterostructures. This approach aims to leverage the combined properties of these materials, improving their performance for use in applications such as energy storage, catalysis, optoelectronics, and advanced sensing technologies. The refinement of deposition techniques, such as drop-casting, dip coating, and spin coating, will be essential in enabling the large-scale fabrication of these heterostructures while preserving their functional properties.

The potential for MXenes in flexible and wearable electronics represents another important research direction. When integrated with polymers to form composite materials, MXenes enhance the mechanical, electrical, and thermal properties of these materials, making them suitable for a variety of practical applications. As research on MXene-based composites continues, their role in advancing technology across industries, including healthcare, environmental monitoring, and consumer electronics, is expected to grow.

In the area of catalysis, MXenes' exceptional properties in electrocatalysis, particularly for water splitting, and photocatalysis, for CO_2 conversion, present exciting opportunities for sustainable energy production and environmental remediation. Future studies will focus on optimizing these catalytic processes and scaling them for industrial applications, contributing to the development of clean energy technologies.

For energy storage, MXenes are anticipated to play a pivotal role in the making of next generation of supercapacitors, lithium-ion batteries, and sodium-ion batteries. Research into mixed-dimensional heterostructures and the improvement of surface properties will continue to enhance their performance and long-term stability, making MXenes key materials for energy storage devices. These developments will enable MXenes to play a significant role in a wide range of applications, from energy storage and catalysis to flexible electronics and sustainable technologies, positioning them as crucial materials for future technological progress.

In conclusion, MXenes, particularly $Ti_3C_2T_x$ and Ti_2CT_x, exhibit exceptional mechanical, electrical, and chemical properties, making them highly promising for a variety of technological applications, including energy storage, catalysis, photodetectors, and flexible electronics. Despite their significant potential, challenges such as structural stability, surface smoothness, and scalability must be addressed for their widespread commercial use. The development of MXene-based heterostructures, incorporating other two-dimensional materials, is a critical area of research that can expand their applications in energy storage, catalysis, and optoelectronics. Furthermore, MXenes' ability to enhance the performance of composites for flexible and wearable electronics presents new opportunities for practical applications across industries.

Looking forward, future research will focus on refining synthesis methods, improving defect engineering, and advancing deposition techniques to support large-scale production. Additionally, significant efforts will be directed toward enhancing their catalytic properties for water splitting and CO_2 conversion, as well as optimizing their integration in energy storage systems. Ultimately, MXenes are poised to play a pivotal role in driving technological advancements and facilitating sustainable solutions across diverse industrial sectors.

11.3 SUMMARY

MXenes, a versatile class of 2D materials, exhibit significant potential across various applications due to their unique properties. Key findings highlight the importance of controlled synthesis and defect engineering to optimize their performance. Heterostructure formation with other 2D materials and polymer composites further enhances their capabilities, particularly for energy storage and flexible electronics. Challenges related to stability and scalability are being addressed through refined deposition techniques. Notably, MXenes show promise in sustainable energy solutions, especially in water splitting and CO_2 conversion. Future research is focused on overcoming these challenges to fully realize their potential in diverse industrial sectors.

REFERENCES

Abbasi, Nasir Mahmood, Yao Xiao, Ling Peng, Yanhong Duo, Lude Wang, Li Zhang, Bing Wang, and Han Zhang. 2021. "Recent advancement for the synthesis of MXene derivatives and their sensing protocol." *Advanced Materials Technologies* 6 (10):2001197.

Abbasi, Nasir Mahmood, Yao Xiao, Li Zhang, Ling Peng, Yanhong Duo, Lude Wang, Peng Yin, Yanqi Ge, Haiyong Zhu, and Bing Zhang. 2021. "Heterostructures of titanium-based MXenes in energy conversion and storage devices." *Journal of Materials Chemistry C* 9 (27):8395–8465.

Bera, Sumanta, Tapas Kumar Mondal, Yan-Kuin Su, and Shyamal Kumar Saha. 2024. "Sulfur-tuned MoS_2 quantum dot decorated Ti_3C_2Tx (MXene) electrode materials for high performance supercapacitor applications." *Journal of Alloys and Compounds* 985:174010.

Ding, Li, Yanying Wei, Yanjie Wang, Hongbin Chen, Jürgen Caro, and Haihui Wang. 2017. "A two-dimensional lamellar membrane: MXene nanosheet stacks." *Angewandte Chemie International Edition* 56 (7):1825–1829.

Fu, Hui-Chun, Vinoth Ramalingam, Hyunho Kim, Chun-Ho Lin, Xiaosheng Fang, Husam N. Alshareef, and Jr-Hau He. 2019. "MXene-contacted silicon solar cells with 11.5% efficiency." *Advanced Energy Materials* 9 (22):1900180.

Guo, Zhanglin, Liguo Gao, Zhenhua Xu, Siowhwa Teo, Chu Zhang, Yusuke Kamata, Shuzi Hayase, and Tingli Ma. 2018. "High electrical conductivity 2D MXene serves as additive of perovskite for efficient solar cells." *Small* 14 (47):1802738. doi: 10.1002/smll.201802738.

Hantanasirisakul, Kanit, and Yury Gogotsi. 2023. "Electronic and optical properties of 2D transition metal carbides and nitrides (MXenes)." *MXenes*:135–205.

Hassan, Muhammad, Abdul Ghaffar, Gang Lou, Ziang Miao, Zhengchun Peng, and Kemal Celebi. 2024. "Enhanced transport kinetics of electrochromic devices by $W_{18}O_{49}$ NW/Ti_3C_2Tx composite films." *Advanced Functional Materials*:2310535.

Hassan, Muhammad, Pingping Li, Ju Lin, Zihao Li, Muhammad S. Javed, Zhengchun Peng, and Kemal Celebi. 2024. "Smart energy storage: $W_{18}O_{49}$ NW/Ti_3C_2Tx composite-enabled all solid state flexible electrochromic supercapacitors." *Small*:2400278.

Jayakumar, Suresh, P. Chinnappan Santhosh, S. Ramakrishna, and A. V. Radhamani. 2024. "2D (Ti_3C_2Tx) MXene: A comprehensive review of advancements in synthesis protocols, applications in supercapacitors, sustainability targets and future prospects." *Journal of Energy Storage* 97:112741.

Kumar, Nilmani, Harpreet Singh, Madhu Khatri, and Neha Bhardwaj. 2023. "2D-transition metal carbides and nitrides: Materials for the next generation." In *Age of MXenes, Volume 1. Fundamentals and Artificial Intelligence: Machine Learning Interventions*, 1–25. ACS Publications.

Liu, Fu, and Zhanxi Fan. 2023. "Defect engineering of two-dimensional materials for advanced energy conversion and storage." *Chemical Society Reviews* 52 (5):1723–1772.

Luo, Shaohong, Tamador Alkhidir, Sharmarke Mohamed, Shoaib Anwer, Baosong Li, Jing Fu, Kin Liao, and Vincent Chan. 2023. "Investigation of interfacial interaction of graphene oxide and Ti_3C_2Tx (MXene) via atomic force microscopy." *Applied Surface Science* 609:155303.

Naguib, Michael, Michel W. Barsoum, and Yury Gogotsi. 2021. "Ten years of progress in the synthesis and development of MXenes." *Advanced Materials* 33 (39):2103393.

Rehman, Zia Ur, Khakemin Khan, Shanshan Yao, Mohsan Nawaz, Antonio Miotello, Mohammed A. Assiri, Tariq Bashir, Tensangmu Lama Tamang, and Muhammad Sufyan Javed. 2024. "Recent progress in MXene-based materials, synthesis, design, and application in lithium-sulfur batteries." *Materials Today Chemistry* 40:102200.

Sikdar, Anirban, Pronoy Dutta, Sujit Kumar Deb, Abhisek Majumdar, N. Padma, Subhradip Ghosh, and Uday Narayan Maiti. 2021. "Spontaneous three-dimensional self-assembly of MXene and graphene for impressive energy and rate performance pseudocapacitors." *Electrochimica Acta* 391:138959.

Tang, Yi, Chenhui Yang, Xingtao Xu, Yunqing Kang, Joel Henzie, Wenxiu Que, and Yusuke Yamauchi. 2022. "MXene nanoarchitectonics: Defect-engineered 2D MXenes towards enhanced electrochemical water splitting." *Advanced Energy Materials* 12 (12):2103867.

Wang, Jiaqi, Ganceng Yang, Yanqing Jiao, Haijing Yan, and Honggang Fu. 2024. "Subtle 2D/2D MXene-based heterostructures for high-performance electrocatalytic water splitting." *Small Methods*:2301602.

Wang, Xiaotong, Jingjing Chen, Dajian Wang, and Zhiyong Mao. 2021. "Defect engineering to boost the lithium-ion storage performance of $Ti_3C_2T \times$ MXene induced by plasma-assisted mechanochemistry." *ACS Applied Energy Materials* 4 (9):10280–10289.

Wei, Yi, Peng Zhang, Razium A. Soomro, Qizhen Zhu, and Bin Xu. 2021. "Advances in the synthesis of 2D MXenes." *Advanced Materials* 33 (39):2103148.

Xiang, Zhen, Xiao Wang, Xiang Zhang, Yuyang Shi, Lei Cai, Xiaojie Zhu, Yanyan Dong, and Wei Lu. 2022. "Self-assembly of nano/microstructured 2D Ti_3CNTx MXene-based composites for electromagnetic pollution elimination and Joule energy conversion application." *Carbon* 189:305–318.

Yu, LePing, Abdulaziz S. R. Bati, Tom S. L. Grace, Munkhbayar Batmunkh, and Joseph G. Shapter. 2019. "Ti_3C_2Tx (MXene)-silicon heterojunction for efficient photovoltaic cells." *Advanced Energy Materials* 9 (31):1901063.

Zhao, Zifang, Shuai Wang, Fang Wan, Zhiwei Tie, and Zhiqiang Niu. 2021. "Scalable 3D self-assembly of MXene films for flexible sandwich and microsized supercapacitors." *Advanced Functional Materials* 31 (23):2101302.

Index

N

O

P

Q

R

S

T

U

V

W

X

Y

Z

FSC
www.fsc.org
MIX
Papier | Fördert gute Waldnutzung
FSC® C083411

Kolibri 360 GmbH
Ferdinand-Jühlke-Straße 7
99095 Erfurt, Deutschland
produktsicherheit@kolibri360.de